International Conference on Applied Photonics and Electronics (InCAPE 2017)

EPJ Web of Conferences Volume 162 (2017)

Port Dickson, Malaysia
9 - 10 August 2017

Editor:

Mohamad Halim Abd. Wahid

ISBN: 978-1-5108-5254-9

Printed from e-media with permission by:

Curran Associates, Inc.
57 Morehouse Lane
Red Hook, NY 12571

CURRAN ASSOCIATES INC. proceedings .com

Some format issues inherent in the e-media version may also appear in this print version.

Printed by Curran Associates, Inc. (2018)

For additional information, please contact EDP Sciences – Web of Conferences at the address below.

EDP Sciences – Web of Conferences
17, Avenue du Hoggar
Parc d'Activité de Courtabœuf
BP 112
F-91944 Les Ulis Cedex A
France

Phone: +33 (0) 1 69 18 75 75
Fax: +33 (0) 1 69 28 84 91

contact-edps@webofconferences.org

Additional copies of this publication are available from:

Curran Associates, Inc.
57 Morehouse Lane
Red Hook, NY 12571 USA
Phone: 845-758-0400
Fax: 845-758-2633
Email: curran@proceedings.com
Web: www.proceedings.com

TABLE OF CONTENTS

EPJ Web of Conferences 162, 00001 (2017)
InCAPE2017 DOI: 10.1051/epjconf/201716200001

The International Conference on Applied Photonics and Electronics 2017 (InCAPE2017) has been successfully organized on 9th August 2017 at Avillion Port Dickson by School of Microelectronic Engineering and Advanced Communication Engineering (Centre of Excellence), Universiti Malaysia Perlis and supported by IEEE Photonics Society and IEEE UniMAP Student Branch.

In line with the development on the electronic-photonic integration, InCAPE is dedicated to cutting edge research that addresses scientific needs of academic researchers and industrial professionals to explore new horizons of knowledge on various electronics and photonics related fundamentals and application.

InCAPE 2017 attracted about 150 participants with 7 invited speakers and 2 keynote speakers. 87 technical papers were classified into 10 tracks: i) Nanophotonics, Plasmonics & Nonlinear Optics, ii) Photodetectors, Sensors, System and Imaging & Optical Fiber Technology, iii) Optical Fiber Technology, iv) Optical Communications and Systems, v) Biophotonics, MEMS and NEMS, Electronic / Microelectronic Fabrication, vi) Electronic Devices & Microelectronic Fabrication, vii) Electronic Devices, Solar Cell & High Power Source, viii) Integrated Circuits & System-on-Chip, ix) System on Chip, Signal Processing, x) Electromagnetic Field Theory & Telecommunication.

The successful organization of InCAPE 2017 is the results of the commitment and dedication of all committee members and strong support from a wide spectrum of very generous sponsors.

Statement of Peer review

In submitting conference proceedings to EPJ Web of Conferences, I certify to the Publisher that I adhere to the **Policy on Publishing Integrity** of the journal in order to safeguard good scientific practice in publishing.

1. All articles have been subjected to peer review administered by the proceedings editors.
2. Reviews have been conducted by expert referees, who have been requested to provide unbiased and constructive comments aimed, whenever possible, at improving the work.
3. Proceedings editors have taken all reasonable steps to ensure the quality of the materials they publish and their decision to accept or reject a paper for publication has been based only on the merits of the work and the relevance to the journal.

Title, date and place of the conference

Title: International Conference on Applied Photonics and Electronics 2017 (InCAPE2017)
Date: 9-10 August 2017
Place: Avilion, Port Dickson, N. Sembilan, Malaysia.

Proceedings editor(s):

DR. MOHAMAD HALIM ABD. WAHID

Date and editor's signature

31 OKTOBER 2017

 DOI: 10.1051/epjconf/201716201001

Optimization of SPR signals: Monitoring the physical structures and refractive indices of prisms

Wan Maisarah Mukhtar[1,*], *Razman* Mohd Halim[2], *Hazirah* Hassan[1]

[1]Faculty of Science and Technology, Universiti Sains Islam Malaysia (USIM), Bandar Baru Nilai, 71800 Nilai, Negeri Sembilan, Malaysia
[2]National Metrology Institute of Malaysia, Lot PT 4803, Bandar Baru Salak Tinggi, 43900 Sepang, Selangor

Abstract. Surface plasmon resonance (SPR) can only be achieved if sufficient energy is provided at the boundary between metal and dielectric. An employment of prism as a light coupler by using Kretschmann configuration is one of the alternative for the production of adequate energy to be generated as surface plasmon polaritons (SPP). This work is carried out to investigate the effect of physical structure of the prism and its refractive index to the excitation of SPPs. A 50nm gold thin metal film with dielectric constant of ε=-12.45i+1.3 was deposited on the hypotenuse surface of the prisms. The physical structures of the prisms were varied such as triangular, conical, hemispherical and half cylindrical. These prisms were classified into two types of refractive indices (RI), namely n=1.51(type BK7) and n=1.77(type SF11). Based on SPR curve analyses, we discovered that strong SPR signals which consist of 82.98% photons were excited as SPPs can be obtained by using type-BK7 prism with physical structures of hemispherical or half cylindrical. From the view of selectivity ability as sensors, the usage of type-SF11 prisms (half cylindrical and hemispherical) able to enhance this impressive feature in which sharp SPR curves with small FWHM values were obtained. In conclusion, apart from properties of thin film materials, the physical structure of prisms and their RI values play crucial roles to obtain optimum SPR signal. High sensitivity SPR sensor can be established with the appointment of type-BK7 prisms (hemispherical or half cylindrical shape) as light couplers.

1 Introduction

An evolution of plasmonic technology is growth rapidly nowadays due to its ability in light manipulation which make it a suitable candidate in the development of optoelectronic device technology [1-5]. Plasmonic wave properties that react at the interface of the optical component which used nanophotonic structures made it as one of the favourable sensing techniques, have been widely used as surface plasmon resonance (SPR) sensor [6-8].

To enhance the sensitivity of SPR sensor, researchers introduced nanomaterials such as chitosan [9,10], zinc oxide [11,12], gold nanoparticle [13,14], graphene [15,16] and graphene oxide [17,18]. Obviously, the employment of those materials results positive responds where the sensitivity of sensors is successfully increased. The synthesis processes of these materials involve complicated procedure and costly materials such as seed-mediated growth technique, plasma enhance chemical vapor deposition and Hummer's method [14, 17,18].

Apart of hiring nanomaterials, an optimization of SPR setup able to generate maximum excitation of surface plasmon polaritons. Basic SPR setup consists of laser, lens, polarizer, prism, thin film and photodetector [19]. Each component contributes its own significant role. Laser light must be oriented in p-polarized mode so that the electrons' interaction between metal and dielectric will create SPPs oscillations due to the applied electromagnetic field [20]. Note that suitable types of metal thin film with appropriate thicknesses, usually 50nm also need to be considered [21].

In free space optics, light energy is collected from open beams and surfaces and sent on to the photodetector. SPR involved a free space optics application in which an appropriate light coupling method need to be employ so that sufficient energy for the excitation of surface plasmon polaritons can be produced. One of the common alternatives is the usage of prism as light coupler. The prism is designed to couple a substantial fraction of the power contained in a laser beam into a metal thin film [22]. Ignoring the presence of prism will result unsuccessful SPR phenomena. To date, there are two types of prism coupling configuration for SPR application which are familiarly used; namely Kretschmann configuration [23] and Otto configuration [24]. Kretschmann configuration is preferable due to its simplicity.

In this work, the important control parameters that influence the strength of SPR signal are studied for the improvement of SPP excitations. We investigate the

* Corresponding author: wmaisarah@usim.edu.my

effect of physical structure and refractive index (RI) of prisms on SPR. We believe that the shape of prism and its RI value play significant roles to enhance the SPR signal, in which leads to the development of a low cost and high sensitivity optical sensor.

2 Theory

Surface plasmon is the localized charge density oscillations that propagate along the boundary between a thin metal film and a dielectric medium. The excitation of surface plasmon polaritons (SPP) can be established as the conservation of both energy and momentum occurs. To achieve this, the relation between energy (in terms of the angular frequency ω) and the momentum (in terms of the wavevector in the propagation direction k_x) need to be analysed. An important characteristics of surface plasmons is that for a given energy $\hbar\omega$, the wavevector k_x is always larger than the wavevector of light in free space as illustrated in Figure 1. The light line ω/c is plotted as dashed lines. The strong coupling between light and surface charges increased the momentum of SPP; where the electrons along the metal surface has been dragged by the light field. This phenomenon describes the condition where SPP on a plane interface cannot be excited by light of any frequency that propagates in free space. Excitation of SPP by light is only possible if a wavevector component of the exciting light can be increased over its free-space value. One of the alternative solution is by created the evanescent field at the interface between a medium with refractive index > 1, where the light will be tilted by a factor of n since $\omega=ck/n$. This situation is shown in Figure 1, which portrays the SPP dispersion with the free-space light line and the tilted light line in glass with various refractive indices such as n_1, n_2 and n_3 where $n_1 < n_2 < n_3$. As the light lines cross the SPP curve, the energy and momentum are conserved which resulting the occurrence of SPR phenomenon [22].

The relationship of the surface plasmon wave vector (k_{spp}) between the metal and dielectric medium is given as

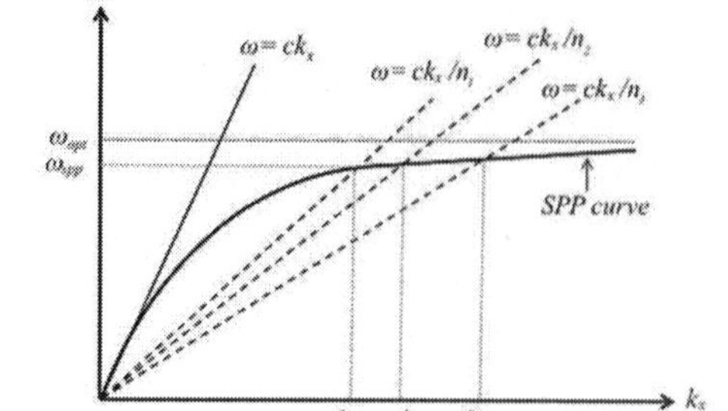

Fig. 1. SPP dispersion with the free-space light line and the tilted light line in glass with various refractive indices such as n_1, n_2 and n_3 where $n_1 < n_2 < n_3$.

$$k_{spp} = k_o \sqrt{\frac{\varepsilon_{metal}\varepsilon_{dielectric}}{\varepsilon_{metal} + \varepsilon_{dielectric}}} \quad (1)$$

where k_o is the free space wave vector of the optical wave, ε_{metal} the dielectric functions of metal and $\varepsilon_{dielectric}$ is the dielectric functions of metal and the dielectric medium [25]. During SPR, the energy of the incident light can be strongly coupled from photons onto the surface plasmon wave which decays via heat dissipation along the propagation path. The relationship between k_x and angle of incident light can be expressed as follows [26]:

$$k_x = k_o n_{glass} \sin\theta_{inc} \quad (2)$$

where n_{glass} is the refractive index of the glass prism and θ_{inc} is the angle of incidence [17]. The conditions for the surface plasmon excitation at the interface between the metal and dielectric layers in a glass-metal-dielectric are given by

$$k_x = k_{spp} \quad (3)$$

Resonance is achieved by matching the projection of the wave vector of the incident light in the direction of the interface and wave vector of the surface plasmon oscillations as expressed in Eq. (3). Using a prism with dielectric constant value larger than 1 with the incident angle is larger than the critical angle, value of k_x can be determined. As the incident light satisfies the resonance condition, a strong absorption dip in the reflectance can be observed.

SPP wavelength, λ_{SPP} is a wavelength during surface plasmon resonance phenomena which represents by a sharp dip curve. The SPP wavelength can be calculated from the derivation of Eq. 1, which expressed in Eq. 4 and 5 as follow [26]:

$$\frac{2\pi}{\lambda_{spp}} = \frac{2\pi}{\lambda_0}\sqrt{\frac{\varepsilon_{metal}\varepsilon_{dielectric}}{\varepsilon_{metal} + \varepsilon_{dielectric}}} \quad (4)$$

$$\lambda_{spp} = \lambda_0 \sqrt{\frac{\varepsilon_{metal} + \varepsilon_{dielectric}}{\varepsilon_{metal}\varepsilon_{dielectric}}} \quad (5)$$

3 Methodology

A 50nm gold thin film with dielectric constant of ε=-12.45i+1.3 was deposited on the hypotenuse surface of the prisms. Two types of commercialized prisms were used, namely a BK-7 prism with refractive index of 1.51 and SF11 prism with refractive index of 1.77. A He-Ne laser source with 632.8nm excitation wavelength was strike on the prism. To observe the role of prisms' shape as light coupler; various designs of prism were employed namely triangular, conical, hemispherical and half cylindrical. Fig.2 shows an illustration of SPR setup, where a half cylindrical prism was employed as light

 DOI: 10.1051/epjconf/201716201001

coupler. The resonant excitation of surface plasmons was characterized by a dip in the reflectance curves which refers to the energy of the incident light that is partially transferred to plasmons [22]. In this work, the characterization of SPR characteristics was conducted by calculating the depth of SPR curves. A good SPR sensor with high sensitivity can be evaluated by studying the SPR curve depth. The deeper the depth of SPR curve, the more the SPPs were excited [2,6,14]. A full-width-half-maximum (FWHM) analysis is carried out to investigate the selectivity capability of SPR sensor. Sharper SPR dip will produce smaller value of FWHM, which leads to the excellent selectivity capability due to the improvement of sensor resolution [27].

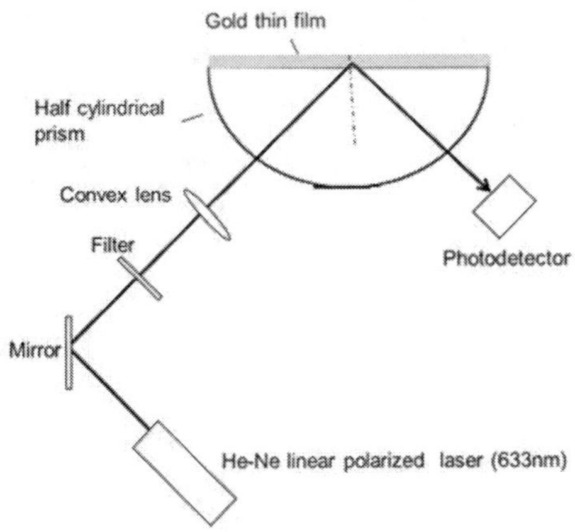

Fig. 2. Illustration of SPR setup using half cylindrical prism as light coupler.

4 Results and discussions

Fig.3 depicts the SPR curves for various structures of SF11 prisms consist of conical, hemispherical, half cylindrical and triangular structures; with refractive index of 1.77 (type SF11). Strong SPR signals were obtained for all prisms where two groups of SPR curves were obviously seen. The cone and triangular prisms displayed same SPR curves criteria, as well as hemisphere and half cylindrical prisms. To facilitate the analyses, these groups were classified as follows: Group 1 consists of conical and triangular prisms; Group 2 comprises of hemispherical and half cylindrical prisms. These groups exhibited significant differences of SPR curves. For Group 1, the values of minimum reflectance, R and SPR curve depth, ΔR were obtained as R=0.0363 and ΔR=0.7741 respectively where the resonant angle is obtained at θ_R=28.6935° and FWHM=1.44. Meanwhile for Group 2, the values obtained were R=0.0233, ΔR=0.7684, θ_R=35.9900° and FWHM=0.83.

The usage of SF11 prisms with conical and triangular shapes able to generate strong SPR signal where 77.41% of input beam was excited as SPP. Group 1 capable to produce stronger SPR signal which is 1.201% better than Group 2. The capability of SPR sensor is also determined by its selectivity [29]. We found that Group 2 (hemisphere and half cylindrical prisms) exhibited smaller values of FWHM in which indicated an excellent selectivity capability as compare to Group 1. The contradiction between these parameters demonstrates that the performance of SPR sensor can be categorized into two parts, namely sensitivity (due to the strong excitation of SPPs) and selectivity (due to the sharpness of the SPR curves).

Subtle differences were obtained as SF11 prisms (n=1.77) were replaced with BK7 prisms (n=1.51) where the resonant angles were red-shifted about 35% appeared as depicted in Fig.4. Obviously, the hemispherical and half cylindrical prisms exhibited identical SPR properties as well as triangular and cone. By applying the same approach of prism grouping (classification of Group 1 and Group 2 based on prism's structure), the performance of BK7 prisms as light coupler were investigated. For Group 1, the value of minimum reflectance, R and SPR curve depth, ΔR were obtained as R=0.0404 and ΔR=0.8212 respectively with the resonant angle at θ_R=43.1658° and FWHM=1.72. Meanwhile for Group 2, the values obtained were R=0.0310, ΔR=0.0.8298, θ_R=43.7688° and FWHM=1.16.

Contrary with SF11 prisms, strong SPR signals were obtained as type BK-7 prisms consists of half cylindrical and hemispherical structures were employed where about 82.98% of photons were generated as SPPs. Group 2 results better SPR signal which is 1.036% of reflected light than Group 1. FWHM analyses explained that excellent selectivity of SPR sensing properties can be obtained by using hemispherical and half cylindrical prisms in which the values of FWHM were smaller than conical and triangular prisms. By employing hemispherical or half cylindrical type-BK7 prisms, it can be concluded that high sensitivity and excellent selectivity SPR sensor can be developed. Table 1 lists SPR parameters which were obtained using two types' prisms with different RI.

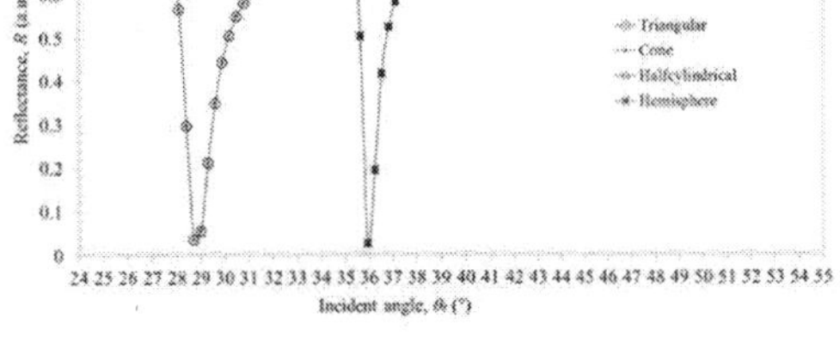

Fig. 3. SPR curves for various structures of SF11 prisms (n=1.77) consist of conical, hemispherical, half cylindrical and triangular.

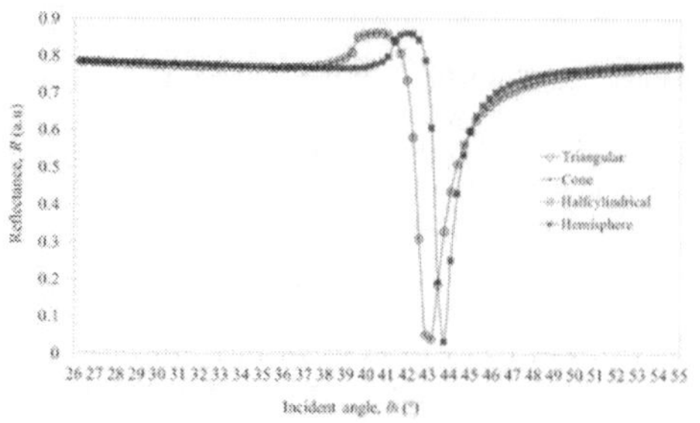

Fig.4. SPR curves for various structures of BK7 prisms (n=1.51) consist of conical, hemispherical, half cylindrical and triangular

Table 1. Important parameters of SPR signals

Type	Shape	Minimum reflectance, R (a.u)	Resonant angle, θ (°)	SPR curve depth, ΔR (a.u)	FWHM, $\Delta\theta$ (°)
SF11 Prism (n=1.77)	Triangular	0.0363	28.6935	0.7741	1.6200
	Cone	0.0363	28.6935	0.7741	1.6200
	Halfcylindrical	0.0261	35.9900	0.7684	0.8800
	Hemisphere	0.0261	35.9900	0.7684	0.8800
BK7 Prism (n=1.51)	Triangular	0.0404	43.1658	0.8212	1.7200
	Cone	0.0404	43.1658	0.8212	1.7200
	Halfcylindrical	0.0310	43.7688	0.8298	1.1600
	Hemisphere	0.0310	43.7688	0.8298	1.1600

Fig.5 displays the properties of SPR as the RI of prisms were manipulated. Different SPR angles were substantially observed as different RI of prisms employed. The usage of Group 1 prisms with SF11 type results SPR angles at 28.6935° (Fig. 5(a)). By replacing the prism with BK7-type, the SPR angles were shifted about 50.43% to 43.1658°. The employment of BK7 prisms able to produce high sensitivity SPR sensor due to their capability to excite more SPPs, which is 6.08% than SF11. The appointment of Group 2 prisms results the shifting of SPR angle about 21.61% from 35.9900° to 43.7688°. This outcome proves that apart of

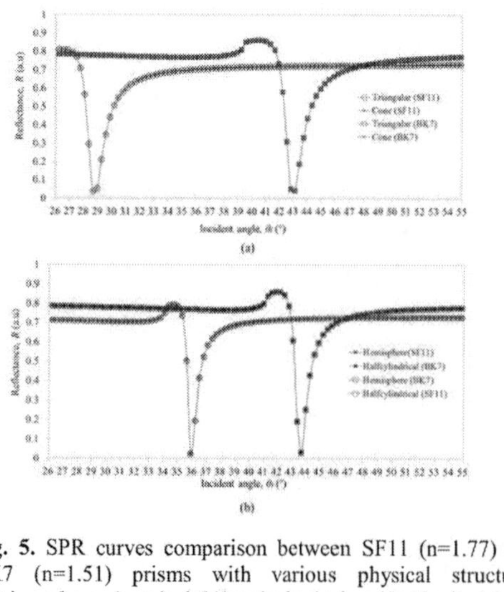

Fig. 5. SPR curves comparison between SF11 (n=1.77) and BK7 (n=1.51) prisms with various physical structures (a)triangular and conical (b)hemispherical and half cylindrical.

thin film thicknesses, the physical structure and the value of RI of the prisms also play major roles for the generation of maximum amount of SPPs to optimize the SPR signal. The location of SPR angles are mainly influenced by RI's value of the light coupler. High sensitivity sensor due to strong SPR signal can be obtained by using Group 2 type-BK7 prism where 82.98% of photons were excited as SPPs (Fig. 6). From the view of excellent selectivity function as sensors [28], the usage of Group 2 type-SF11 prisms able to enhance this impressive feature in which sharp SPR curves with small FWHM values were obtained as summarized in Fig. 7.

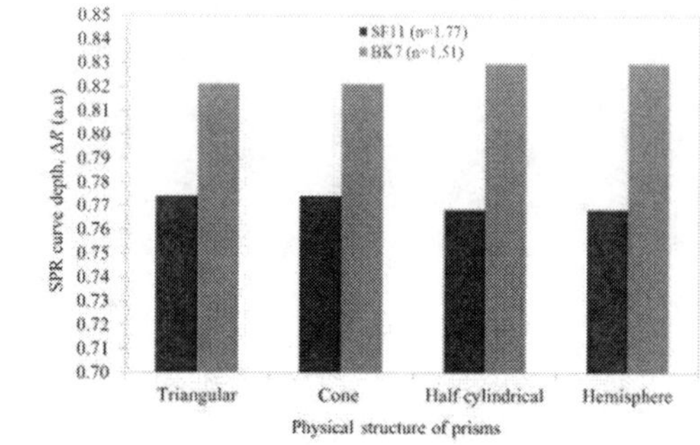

Fig. 6. Analysis of SPR curve depth versus various types of physical structure of prisms.

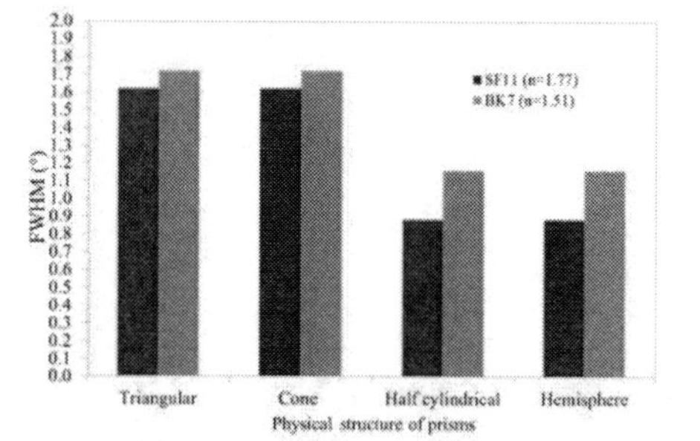

Fig. 7. FWHM analysis of SPR curves versus various types of physical structure of prisms

5 Conclusions

In this study, we successfully demonstrated the significant roles of physical structures of the prisms and their RI value which contribute to the remarkable SPR signals. The employment of half cylindrical or hemispherical prisms with low RI using BK7-type (n=1.51) enhance the excitation of SPPs up to 8% in comparison with SF11-type (n=1.77). This feature is one of the crucial criteria for the development of optical sensor. We believe that the output of this research will benefits to the rapid evolution of SPR optical sensing field.

The authors would like to acknowledge the support of the Universiti Sains Islam Malaysia and Malaysian Ministry of

DOI: 10.1051/epjconf/201716201001

Education for funding this work under grant FRGS/2/2014/SG02/USIM/03/1. Knoll Group from Max Planck Institute for Polymer Research, German is also acknowledged for the Winspall 3.02 simulation software.

References

1. R. P. Dwivedi, H. S. Lee, J. H. Song, S. An, S, E. H. Lee, Opt Commun, **284** 1418 (2011)
2. W. M. Mukhtar, S. Shaari, A. A Ehsan, P. S. Menon, Opt Mat Exp, **4** 424 (2014).
3. V.E.Babicheva, A. V. Lavrinenko, Optics Communications, **285** 5500 (2012).
4. R. Thomas, Z. Ikonic, R. W. Kelsall, Photonic Nanostruct, **10** 183 (2012).
5. W. M. Mukhtar, S. Shaari, P. S. Menon, H. A. Razak, 2015 IEEE Regional Symposium, 2015, p.1.
6. W. M. Mukhtar, S. Shaari, P. S. Menon, Optoelectron Adv. Mat, 7 9 (2013).
7. J. Homola, S. S Yee, G. Gauglitz, Sensor Actuat B-Chem, **54** 3 (1999).
8. A. G. Brolo, R. Gordon, B. Leathem, K. L. Kavanagh, Langmuir, **20** 4813 (2004).
9. L. G. Zamfir, I. Geana, S. Bourigua, L. Rotariu, C. Bala, A. Errachid, N. Jaffrezic-Renault, Sensor Actuat B-Chem, **159** 178 (2011).
10. Y. W. Fen, W. M. M. Yunus, N. A. Yusof, Sensor Actuat B-Chem, **171** 287 (2012).
11. C. C. Chang, N. F. Chiu, D. S. Lin, Y. Chu-Su, Y. H. Liang, C. W. Lin, Anal Chem, **82** 1207 (2010).
12. E. Sachet, M. D. Losego, J. Guske, S. Franzen, J. P. Maria, Appl Phys Lett, **102** 05111 (2013).
13. S. Ko, T. J. Park, H. S. Kim, J. H. Kim, Y. J. Cho Biosens Bioelectron, **24** 2592 (2009).
14. W. M. Mukhtar, S. Shaari, P. S. Menon, Adv Sci Lett, **19** 1412 (2013).
15. S. Szunerits, N. Maalouli, E. Wijaya, J. P. Vilcot, R. Boukherroub, Anal Bioanal Chem, **405** 1435 (2013).
16. L. Wang, C. Zhu, L. Han, L. Jin, M. Zhou, S. Dong, Chem Commun, **47** 7794 (2011).
17. T. Wu, S. Liu, Y. Luo, W. Lu, L. Wang, X. Sun, Nanoscale, **3** 2142 (2011).
18. S. K. Mishra, S. N. Tripathi, V. Choudhary, B. D. Gupta, Sensors Actuat B- Chem, **199** 190 (2014).
19. W. M. Mukhtar, P. S. Menon, S. Shaari, M. Z. Malek, A. M. Abdullah, J Phys Conf Ser, **431** 012028 (2013).
20. W. B. Lin, N. Jaffrezic-Renault, A. Gagnaire, H. Gagnaire, Sensors Actuat A-Phys, **84** 198 (2000).
21. S. Zeng, K. V. Sreekanth, J. Shang, T. Yu, C. K. Chen, F. Yin, K. T. Yong, Adv Mat, **27** 6163 (2015).
22. L. Novotny, B.Hecht, Principles of Nano-Optics. Cambridge University Press. United Kingdom. (2006).
23. S. Liu, M. Li, C. Gong, H. Gao, Laser Phys, **26** 065006 (2016).
24. S. Maegawa, J. Yamaguchi, F. Itoigawa, T. Nakamura, Tribol Lett, **62** 1(2016).
25. W. M. Mukhtar, S. Shaari, P. S. Menon, Adv Sci Lett, **19** 66 (2013).
26. M. S. Islam, A. Z. Kouzani, X. J. Dai, W. P. Michalski, Biomed Signal Proces, **6** 147(2011).
27. A. Paliwal, A. Sharma, M. Tomar, V. Gupta, Sensors Actuat B-Chem, **216** 497 (2015).
28. N. Cennamo, G. D'Agostino, M. Pesavento, L. Zeni, Sensors Actuat B-Chem, **191** 529 (2014).

 DOI: 10.1051/epjconf/201716201002

Synthesis methods of gold nanoparticles for Localized Surface Plasmon Resonance (LSPR) sensor applications

Nurul Diyanah Samsuri, *Wan Maisarah* Mukhtar*, *Affa Rozana* Abdul Rashid, *Karsono* Ahmad Dasuki, and *Awangku Abdul Rahman* Hj. Awangku Yussuf

Faculty of Science and Technology, Universiti Sains Islam Malaysia, 71800 Malaysia

Abstract. Gold nanoparticles (GNPs) have been known as an excellent characteristic for Local Surface Plasmon Resonance (LSPR) sensors due to their sensitive spectral response to the local environment of the nanoparticle surface and ease of monitoring the light signal due to their strong scattering or absorption. Prior the technologies, GNPs based LSPR has been commercialized and have become a central tool for characterizing and quantifying in various field. In this review, we presented a brief introduction on the history of surface plasmon, the theory behind the surface plasmon resonance (SPR) and the principles of LSPR. We also reported on the synthetization as well of the properties of the GNPs and the applications in current LSPR sensors.

1 Introduction

Surface plasmon resonance (SPR) is a charge-density oscillation that may exist at the interface of two media with dielectric constants of opposite signs, for instance, a metal and a dielectric. The charge density wave is associated with an electromagnetic wave, the field vectors of which reach their maxima at the interface and decay evanescently into both media. This surface plasma wave (SPW) is a TM-polarized wave (magnetic vector is perpendicular to the direction of propagation of the SPW and parallel to the plane of interface) [1]. Due to the multidisciplinary nature of SPR, major areas of applications of SPR has been reported such as for measurement of physical quantities (displacement [2] and angular position [3]), chemical sensing (monitoring distillation processes [4]), and biosensing (first demonstrated in 1983 [5]).

Historically, in 1902, an experiment conducted by Wood which he had illuminated a metallic diffraction grating with polychromatic light was the first documented observation of surface plasmon. In 1958, an increased drop in reflectivity was observed by Thurbadar [6] as a thin metal film was being illuminated on a substrate without related this observation to SPR phenomena. In 1968, Otto described Thurbadar's observation by showing the excitation of surface plasmon resulted the drop in the reflectivity in the attenuated total reflection method [7]. This explanation also being supported by Kretschmann and Raether, which explained the same excitation of surface plasmon in the same year but with different configurations [8]. Thus, these observations of Otto, Kretschmann and Raether had entrenched and introduced the surface plasmon into modern optics [9]. The advantages of surface plasmon started being known in early 1980s when the method was used for detecting gases [10]. Ever since, various SPR sensors have been investigated.

In recent years, sensors based localized surface plasmon resonance (LSPR) had drawn more interest [11]. LSPR is also a SPR phenomenon but instead of existed in thin film metal, they are in the metallic nanoparticles (MNPs) form [12]. The physical properties of LSPR is reported to be more sensitive than the traditional bulk metal thin film based SPR sensors [13] due to the nanoscale size of the particles [14]. These improvements had attracted more significant development of the high sensitivity LSPR sensors. Among the MNPs, gold nanoparticles (GNPs) have been demonstrated as a great characteristic for a LSPR sensor due to its solitary optical properties [15].

2 Basic Theory

Surface plasmon is the free electrons at the interface of two materials located at a metal surface where free conduction electrons are abundant. This condition obeys the metal dielectric interface by Maxwell's equations [16] which can mathematically be described by an electric field $\boldsymbol{E}$:

$$E = E_0 e^{-k_{y_2} y} \exp(j\omega t - jk_x x) \qquad (1)$$

where E_0 is the amplitude of the electric field, ω is the angular frequency, $\boldsymbol{k}$ is the wavevector, $(\boldsymbol{x},\boldsymbol{y})$ is the position vector and $j = \sqrt{-1}$.

* Corresponding author: wmaisarah@usim.edu.my

DOI: 10.1051/epjconf/201716201002

The propagating electron that occurring at the interface between metal and dielectric experienced an evanescence wave. Evanescent waves happened when incident wave is reflected at an angle that greater than the critical angle thus attenuated total reflection (ATR) occurred as shown in Figure 1.

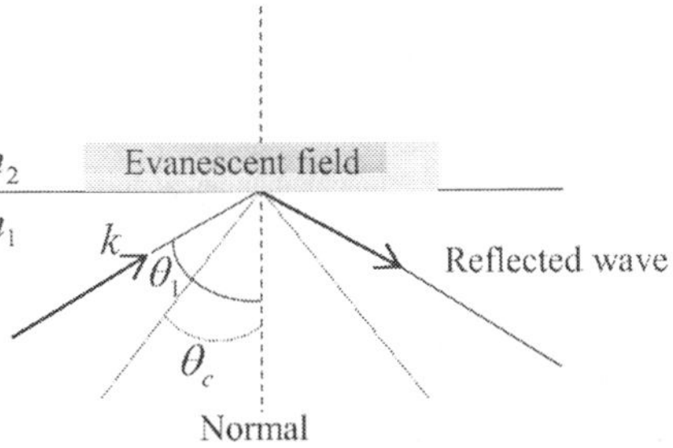

Fig. 1. Shown an interface consisting of two media with refractive indices $n_2 < n_1$. An evanescent field generated when the angle of incidence is larger than the critical angle [17].

3 Principles of LSPR

LSPR is a phenomenon when the electromagnetic field of the light induces a collective coherent oscillation of the surface conduction electron in resonance with the frequency of light when incident light interacts with MNPs [18]. The excitation of LSPR occurred when a charge separation happened between the free electron and ionic metal core due to the interaction between the electric field of light and free electron in the nanoparticles. Thus, the restoring force (Coulomb repulsion among the free electron) forced the free electron to shift in different direction [19].

The phenomenon of LSPR shows a great absorption of light. Since each MNPs shows varies absorption properties hence the properties of LSPR is effected by the size, shape and the dielectric properties of the each MNPs [23]. As LSPR is exceptionally susceptible to the refractive index change in the local dielectric medium, the alteration of the peak wavelength or peak absorbance in the absorbance spectrum of MNPs in sensor system is exerted as a measure of the LSPR sensor feedback [35].

The Mie solution to Maxwell's equations can be applied to defined the light absorption of gold nanosphericals (GNSs) affected by LSPR [38]. Based on the Mie theory, the extinction cross-section, C_{ext} for separated spherical nanoparticles with a radius R (that is smaller than the wavelength of light λ ($R/\lambda < 0.1$)), can be indicated as:

$$C_{ext} = \frac{24\pi^2 R^3 N \varepsilon_m^{\frac{3}{2}}}{\lambda} \frac{\varepsilon_i}{(\varepsilon_r + 2\varepsilon_m)^2 + \varepsilon_i^2} \quad (2)$$

where ε_m is the dielectric constant of the surrounding medium, ε_r is the real part of the dielectric function of the MNPs, ε_i is the imaginary part of the dielectric function of the MNPs and N is the number of spheres per unit volume.

According to Equation (2), the plasmon absorption band appears when $\varepsilon_r = -2\varepsilon_m$. Figure 2 shows the excitation of LSPR for the GNSs.

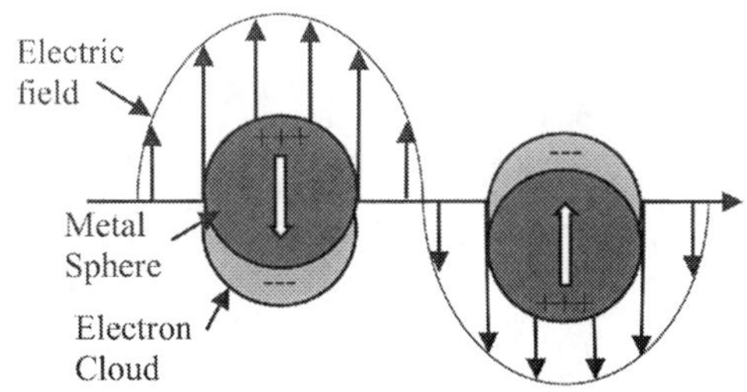

Fig. 2. Excitation Schematic illustration of LSPR excitation for GNSs [22].

4 Synthesis of Gold Nanoparticles

In prior years, citrate reduction method is being used for yielding GNPs by adding the GNPs into the boiling gold salt ($HAuCl_4$) solution where the citrate solution is known, thus, allowing the size of the GNPs to be controlled by varies the ratio between the gold salt and citrate solution [23]. In this section, a brief review of several typical synthesis methods is discussed.

4.1. Chemical Method

The simplest method for synthesize a GNPs is through a chemical method. Usually, the synthesized monodisperse spherical GNPs is range between 10 to 20 nm in diameter [24]. The GNPs is produced in the presence of reducing agents such as amino acid, ascorbic acid or citrate [25], which then reacted with a small quantity of hot gold salt. The citrate ions which act as reducing and capping agents helps the formation of colloidal gold. To produce a larger particle, the amount of sodium citrates need to be decreased so that the remaining amount will not be sufficient in reducing the gold. As the quantity of sodium citrate decreased, the quantity of the citrate ions needed for stabilizing the particles in the solution decreased. Hence, the small particles will start to accumulate into bigger particles [26].

In early 1990s, Brust and Schiffrin contrive the Brust method which used in synthesizing GNPs in an organic solution which are not mixable with water such as toluene. Tetra-octylammonium bromide (TOAB) solution is being reduced with the gold salt in toluene together with sodium borohydride ($NaBH_4$) which act as both an anti-coagulant and reducing agent. The GNPs produced usually in 2 to 6 nm in diameter. Here, the

 DOI: 10.1051/epjconf/201716201002

TOAB and $NaBH_4$ both also acted as the stabilizing agent and phase transfer catalyst. In 1994, the Burst method is chosen as the popular synthesize technique as it gives the control in both particle diameter and grain size distribution also functionalization of particle surface with alkanethiols which helps in stabilized the GNPs [27]. This resulted in the color change from orange to brown in the reaction [28]. Purification of GNPs stabilized with dodecanethiol from TOAB was recorded by Schriffin [29].

4.2 Seed-mediated Growth Method

Due to its simplicity of the experimental procedure [30], seed-mediated growth has been the most favored and extensively used among the reported GNPs synthesis method. Plus, this method enable researchers to produce a high number of nanoparticles [31], the access in controlling the size of the particle [32] and the ease in structural alterations [33]. Basically, seed-mediated growth involved two important steps. Firstly, by producing a seed particles (reduced in gold salts with strong reducing agent) which then added to a metal salt solution in presence of weak reducing agent. Plus, to quicken the growth and avoid further nucleation of the GNPs, a structure directing agent also been added to the solution. Lastly, by varies the concentration of the seeds, reducing and structure directing agents; the geometry of the GNPs can be modified [34].

4.3 Biological Method

In recent years, the advancement of 'green chemistry' technology in GNPs synthesis is widely been progressed to study the biological applications [35]. By using different types of microorganisms, various morphology and chemical composition of GNPs have been synthesized and been studied in several medical technology areas [36] due to the safe, non-toxic and eco-friendly procedures [37]. These microorganisms being categorized into two group which are intracellular and extracellular depends on the location of the GNPs are growth [38]. Besides, plant extracts also being studied and used to synthesize a uniform size and shape of GNPs [39].

5 Properties of Gold Nanoparticles

The quantum size effect in GNPs resulted in individual electron transition energy levels. An external ligands, magnetic fields and electrolyte ions can be used to modulate the quantized capacitance charging behaviour of GNPs. This gives likely applications in electronic devices and electrochemical labels [40]. As for the redox activity, GNPs promoted SPR phenomena [41]. This phenomenon happened when a collective oscillation of conduction electrons across the nanoparticles occurred due to the excitation of the resonant from the incoming photon [42].

The colours from the GNPs is resulted from the resonance condition which meet at visible wavelength [43]. Besides size, SPR also is affected by the solvent, ligand and temperature [42].

A spectral shift created by solvent refractive index changes that are parallel with Mie theory have been reported by Murray et al. [44]. These core changes are crucial in affecting the shifting in the surface plasmon band (SPB) energy to a lower energy with electron deficiency and higher energy with surplus electronic charges [45]. Electronic dephasing can also affect the SPR [46] and has been suggested rather electron–phonon, only electron–electron interactions are implicated [47].

Nevertheless, the phenomena of both procedure in the individual GNPs has been validated from the femtosecond light scattering by experimental data [48]. A significant red-shifting (from ~520 to ~650 nm) and the solution changing from red to blue colour because of the interparticle plasmon coupling are the results from the aggregation of nanoparticles since SPR frequency is responsive to the vicinity of other nanoparticles [49]. This occurrence made GNPs a great particle for colorimetric sensors.

A theoretical calculation of the extinction coefficient of GNPs have been recorded by El Sayed et al. [41]. From the UV-Vis/extinction spectra, the size and concentration of GNPs have been certified both theoretically and experimentally [50]. Also, various stabilizing ligands of the extinction coefficients of the GNPs have been reported [51].

6 Applications in LSPR Sensors

Analyte adsorption which caused the local dielectric charges increased to the wavelength shift in the LSPR spectrum is the basic principle of LSPR based sensors. The conduction of LSPR assays happened in both solution phase [52] and with surfaces filmed with nanoparticle monolayers [53]. When an interaction occurred between monoclonal antibodies with GNPs functionalized and analytes, the absorption maxima of LSPR was redshifted [54]. Plus, the number of ligands is proportional to the wavelength shift [54]. Nevertheless, most GNPs based SPR sensors have been made to be immobilized nanoparticles onto surface [55]. To enhance the sensitivity of SPR, GNPs have been introduced onto the sensing surface due to its metal film on the surface, high dielectric constants and the electromagnetic coupling between GNPs [56]. For instance, by immobilizing an MIP gel with implanted GNPs, the level of dopamine in nanomolar concentration can be detected by using a gold film-coated chip [57]. Plus, as supports for assisted GNPs in identifying analytes such as human serum albumin, numerous substrates such as sol-gel matrix, optical fibers, quartz and ITO glass have been utilized [58]. Lastly, in discovering insulin, an immobilized hydroxyl/thiol-functionalized fourth generation PAMAM dendrimer capsulated GNPs were attached onto maleimide terminated SAMs [59]. An enhance sensitivity with detection limit of 0.5 pM and increased stability were resulted from the GNPs-modified dendrimers. By studying human serum samples with good correlation to standard methods by comparing normal and diabetic patients, the sensor was further certified [60]. An alteration happened due to an increased in surface absorption band to the visible colour change caused by the GNPs aggregation. Mirkin et al. had utilized the above principles by using an oligonucletide functionalized GNPs both in surface and dispersions to develop a calorimetric sensor for DNA hybridization assay [59]. Meanwhile, other reported SPR based sensor used the GNPs aggregation to detect proteins [61] and lectin [62].

6 Conclusion

Based on the review, as compared to the traditional SPR sensors, LSPR GNPs based have been determined to be more sensitive and sensible in sensors applications due to its enhance optical properties. Hence, many researches had shown interest in studying on the development and improvement of GNPs based LSPR sensors in enhancing its sensor ability.

The authors would like to acknowledge the support of Malaysian Ministry of Higher Education (MOHE) through Universiti Sains Islam Malaysia (USIM) under grant USIM/FRGS/FST/32/51514.

References

1. S. S. Yee, Surface plasmon resonance sensors: review, **54**, 3–15 (1999)
2. G. Margheri, A. Mannoni, F. Quercioli, *Proc. SPIE* **2783**, 211–220 (1996)
3. J.K. Schaller, R. Czepluch, C.G. Stojanoff, *Proc. SPIE* **3098**, 476–486 (1997)
4. E.G. Ruiz, I. Garcez, C. Aldea, M.A. Lopez, J. Mateo, J. Alonso-Chamarro, S. Alegret, Sensors and Actuators A **37–38,** 221–225 (1993)
5. B. Liedberg, C. Nylander, I. Lundstro¨m, Sensors and Actuators **4** 299–304 (1983)
6. T Turbadar, *Proc Phys Soc* **73**, 40 (1959)
7. A Otto, Zeits Phys **216**, 398 (1968)
8. E Kretschmann, H Raether, Z Naturforsch **2135–2136**, 2135–2136 (1968)
9. H Raether, Springer Tracts Mod Phys **111,** 1 (1988)
10. C Nylander, B Liedberg, T Lind, Sensor Actuator **3**, 79 (1982)
11. M. Potara, A.M. Gabudean, S. Astilean, Journal of Materials Chemistry **21**, 3625–3633 (2011)
12. A.D. McFarland, R.P. Van Duyne, *Nano Letters* **3**,1057–1062 (2003)
13. A.V. Kabashin, P. Evans, S. Pastkovsky, W. Hendren, G.A. Wurtz, R. Atkinson,R. Pollard, V.A. Podolskiy, A.V. Zayats, Nature Materials **8** ,867–871 (2009)
14. S. Eustis, M.A. El-Sayed, Chemical Society Reviews **35**, 209–217 (2006)
15. X. Huang, S. Neretina, M.A. El-Sayed, Advanced Materials **21**,4880–4910 (2009)
16. R. O. B. P. H. Kooyman, Physics of Surface Plasmon Resonance, 15–34.
17. C. Soc, M. Schnippering, S. R. T. Neil, R. Mackenzie, P. R. Unwin, Chem Soc Rev 207–220 (2011)
18. J.N. Anker, W.P. Hall, O. Lyandres, N.C. Shah, J. Zhao, R.P. Van Duyne, Nature Materials **7**, 442–453 (2008)
19. J. Cao, T. Sun, K. T. V. Grattan, Sensors and Actuators B : Chemical, **195**, 332–351 (2014)
20. K.M. Mayer, S. Lee, H. Liao, B.C. Rostro, A. Fuentes, P.T. Scully, C.L. Nehl, J.H.Hafner, ACS Nano **2**, 687–692 (2008)
21. K.M. Mayer, J.H. Hafner, Chemical Reviews **111**, 3828–3857 (2011)
22. K.A. Willets, R.P. Van Duyne, Annual Review of Physical Chemistry, Annual Reviews, Palo Alto **58**, 267–297 (2007)
23. V.V.R. Sai, T. Kundu, S. Mukherji, Biosensors & Bioelectronics **24**, 2804–2809 (2009)
24. N. Sharma, G. Bhatt, P. Kothiyal, IJPBR **3(2)**, 13–27 (2015)
25. X. Ji, X. Song, J. Li, Y. Bai, W. Yang, X. Peng, Journal of the American Chemical Society **129**, 13939-13948 (2007)
26. R.K. Gangwar, V.A. Dhumale, D. Kumari, U.T. Nakate, S.W. Gosavi, R.B. Sharma, S.N. Kale, S. Datar, Materials Science and Engineering : C **32**, 2659-2663 (2012)
27. M. Giersig, P. Mulvaney, The ACS journal of surfaces and colloids **9**, 3408-3413 (1993)
28. M. Faraday, Philosophical Transactions of the Royal Society of London **147**, 145-181 (1857)
29. C.A. Waters, A.J. Mills, K.A. Johnson, D.J. Schiffrin, Chemical Communications, 540-541 (2003)
30. X. Huang, S. Neretina, M.A. El-Sayed, Advanced Materials **21**, 4880–4910 (2009)
31. TK. Sau, CJ. Murphy, J Am Chem Soc **126**, 8648–8649 (2004)
32. C. Murphy, A. Gole, S. Hunyadi, C. Orendorff, Inorg Chem **45**, 7544–7554 (2006)
33. M El-Sayed, Acc Chem Res **37**, 326–333 (2004)
34. R. C. Price, R. L. Whetten, J. Am. Chem. Soc. **127**, 13750 (2005)
35. S.K. Boruah, P.K. Boruah, P. Sarma, C. Medhi, O.K. Medhi, Advanced Materials Letters **3** ;6, 481-486 (2012)
36. O.V. Salata, Journal of Nanobiotechnology **2** ;1, 3 (2004)

37. O. A. Sadik, F. Osonga, I. Yazgan, V. Kariuki, D. Luther, A. Jimenez, P. Le, RSC Adv. (2015)
38. S.S. Shankar, A. Ahmad, R. Pasricha, M. Sastry, Journal of Materials Chemistry **13** ;7, 1822-1826 (2003)
39. P. Elia, R. Zach, S. Hazan, S. Kolusheva, Z. Porat, Yehuda, International Journal of Nanomedicine **9**, 4007-4021 (2014)
40. C. Subramaniam, T. Pradeep, J. Chakrabarti, Phys. Rev. Lett. **95**, 164501 (2005)
41. P. K. Jain, K. S. Lee, I. H. El-Sayed, M. A. El-Sayed, J. Phys.Chem. B **110**, 7238 (2006)
42. G. Mie, Ann. Phys **25**, 377 (1908)
43. P. K. Jain, X. Huang, I. H. El-Sayed, M. A. El-Sayed, Acc.Chem. Res. **41**, 1578 (2008)
44. S. Link, M. B. Mohamed, M. A. El-Sayed, J. Phys. Chem. B **103**, 3073 (1999)
45. W. Rechberger, A. Hohenau, A. Leitner, J. R. Krenn, B. Lamprecht, F. R. Aussenegg, Opt. Commun. **220**, 137 (2003)
46. S. Link, M. A. El-Sayed, J. Phys. Chem. B **103**, 4212 (1999)
47. T. Itoh, T. Asahi, H. Masuhara, Appl. Phys. Lett. **79**,1667 (2001)
48. K. H. Su, Q. H. Wei, X. Zhang, J. J. Mock, D. R. Smith, S. Schultz, Nano Lett. **3**, 1087 (2003)
49. S. Srivastava, B. L. Frankamp, V. M. Rotello, Chem. Mater. **17**, 487 (2005)
50. N. G. Khlebtsov, Anal. Chem. **80**, 6620 (2008)
51. X. O. Liu, M. Atwater, J. H. Wang, Q. Huo, Colloids Surf. B **58**, 3 (2007)
52. K. J. Fahnestock, M. Manesse, H. A. McIlwee, C. L. Schauer, R. Boukherroub, S. Szunerits, Analyst **134**, 881 (2009)
53. J. J. Deng, Y. Song, Y. A. Wang, J. W. Di, Biosens. Bioelectron. **26**, 615 (2010)
54. P. Englebienne, A. Van Hoonacker, M. Verhas, *Analyst* **126**, 1645 (2001)
55. A. Prabhakar, S. Mukherji, *Lab Chip* **10**, 3422 (2010)
56. J. Wang, *Small* **1**, 1036 (2005)
57. J. Matsui, K. Akamatsu, N. Hara, D. Miyoshi, H. Nawafune, K. Tamaki, N. Sugimoto, Anal. Chem. **77**, 4282 (2005)
58. F. Frederix, J. M. Friedt, K. H. Choi, W. Laureyn, A. Campitelli, D. Mondelaers, G. Maes, G. Borghs, Anal. Chem. **75**, 6894 (2003)
59. R. A. Reynolds, C. A. Mirkin, R. L. Letsinger, J. Am. Chem. Soc. **122**, 3795 (2000)
60. M. Frasconi, C. Tortolini, F. Botre, F. Mazzei, Anal. Chem. **82**, 7335 (2010)
61. D. C. Hone, A. H. Haines, D. A. Russell, Langmuir **19**,7141 (2003)
62. S. Watanabe, K. Yoshida, K. Shinkawa, D. Kumagawa, H. Seguchi, Colloids Surf. B **81**, 570 (2010)

 DOI: 10.1051/epjconf/201716201003

Optimization of reactive-ion etching (RIE) parameters for fabrication of tantalum pentoxide (Ta_2O_5) waveguide using Taguchi method

M. Firdaus A. Muttalib[1, 2,*], *Ruiqi Y.* Chen[1], *S. J.* Pearce[1], and *Martin D. B.* Charlton[1]

[1]School of Electronics and Computer Science, University of Southampton, Southampton, SO17 1BJ, United Kingdom.
[2]School of Bioprocess Engineering, University Malaysia Perlis, Arau 02600, Malaysia.

Abstract. In this paper, we demonstrate the optimization of reactive-ion etching (RIE) parameters for the fabrication of tantalum pentoxide (Ta_2O_5) waveguide with chromium (Cr) hard mask in a commercial OIPT Plasmalab 80 RIE etcher. A design of experiment (DOE) using Taguchi method was implemented to find optimum RF power, mixture of CHF_3 and Ar gas ratio, and chamber pressure for a high etch rate, good selectivity, and smooth waveguide sidewall. It was found that the optimized etch condition obtained in this work were RF power = 200 W, gas ratio = 80 %, and chamber pressure = 30 mTorr with an etch rate of 21.6 nm/min, Ta_2O_5/Cr selectivity ratio of 28, and smooth waveguide sidewall.

1 Introduction

Recently there has been an increasing interest in fabrication and characterization of waveguide layer and photonics crystal (PC) slab structure for investigating various optical nonlinear effects such as optical parametric amplification process [1], supercontinuum generation [2] and harmonic generations [3]. The planar waveguide and the PC slab structure are a desirable structure to realize a compact optical components device (e.g., tunable laser source). Among the available materials, the tantalum pentoxide (Ta_2O_5) has received much attention due to its high refractive index ($n > 2$), low absorption over a wide wavelength region 0.3 to 10 μm [4], high nonlinear refractive index $n_2 = (7.23\pm0.36)\times10^{-19}$ m^2/W [5], large value of chi 3 [6] and high optical damage threshold [7]. Recently, researchers have shown optical non-linear behaviour and demonstrated supercontinuum generation [8] and OPO [9] in Ta_2O_5 waveguides, suggesting a good stability and high damage threshold limits.

In order to make waveguide geometry and PC slab structure, Ta_2O_5 material has to be etched. However, it has been found that Ta_2O_5 material is quite challenging to be etched thus smooth vertical waveguide sidewall is difficult to be realized. The smooth vertical waveguide sidewall of waveguide is essential for low propagation loss transmission. A dry etching technique using a reactive-ion etching (RIE) etcher can be used to etch Ta_2O_5 as it is able to produce finer features with higher anisotropic etch profile. The RIE etcher has several variables factors (i.e., RF power, gas ratio and process pressure) that correspond to output parameters (i.e., etch rate, selectivity, surface roughness and sidewall angle). It requires many experiments run to find optimum setting thus increases development time and cost. Thus design of experiment (DOE) of Taguchi method can be employed for the optimization process. This method is more systematic and efficient approach for determining near optimum settings, with small number of experiments configurations due to the use of orthogonal array.

In this work, design of experiment (DOE) of Taguchi method was employed to optimize the RIE process parameters for fabrication of Ta_2O_5 waveguide for a high etch rate, good selectivity, and smooth waveguide sidewall using RIE etcher in Southampton Nanofabrication Centre. A mixture of trifluoromethane (CHF_3) and argon (Ar) gas ratio was used with chromium (Cr) hard mask in this experiment. Optimum conditions setting of Ta_2O_5 etching recipe have been determined for smooth vertical sidewall, high etch rate and selectivity.

2 Experiment

2.1. Samples preparation

6-inches silicon wafers with <100> orientation, coated with 2 μm silicon dioxide (SiO_2) was used as an initial wafer. A first layer of 400 nm Ta_2O_5 and second layer of 1 μm positive resist (Shipley S1813) were deposited on the wafer. Waveguide pattern from a mask was exposed by an EVG 620T mask aligner. The exposed resist was then dissolved by a developer (Shipley MF319), and hard baked at 120 °C for 30 minutes. Wet etching (Cr etchant) was used to transfer the pattern into Cr metal layer and the residual resist was then removed by Acetone. The sample was diced into 2 x 2 cm small chip

* Corresponding author: firdausmuttalib@unimap.edu.my

 DOI: 10.1051/epjconf/201716201003

size. Samples were then inspected by optical microscope to reject defective chips prior to the etching experiments.

2.2. RIE system configuration

The etching of Ta_2O_5 experiments were carried out in OIPT Plasmalab 80 RIE etcher. This system has equipped with fluorine based reactive gas. It consists of parallel plate electrode powered with 13.56 MHz RF system. The substrate is placed on top of lower electrode, connected with RF generator. The opposite electrode is located on the roof of the chamber connected to ground. Plasma contains ions and neutral species is generated by exposing a neutral gas (Ar) to the strong RF potential power at low pressure. Ions generated from plasma are accelerated in the electric field, bombarding with surface of the material. The neutral species reacts chemically with samples and the material can be etched away. The RIE takes advantage of both the physical and the chemical etching mechanism and etch rates that are faster with an anisotropic etching profile are achieved. This RIE machine has a range limitation of the RF power and the chamber pressure which its optimum operating power is at 200 W and minimum pressure is at 15 mTorr respectively. A quartz cover plate was used in this experiment.

2.3. Experimental parameters

In our experiment, three factors such as RF power (W), gas ratio (%) and chamber pressure (mTorr) and levels were selected to optimize high etch rate, good selectivity and the smooth vertical waveguide sidewalls. Based on preliminary experiment data and machine operational limitation, range and levels of the variable were selected as shown in Table 1.

Table 1. Range and level of parameters selected in this experiment.

Factor	RF power (W)	Gas ratio (%) $CHF_3/(CHF_3+Ar)$	Chamber pressure (mTorr)
Level 1	100	20	15
Level 2	150	50	30
Level 3	200	80	45

A combination of CHF_3 and Ar gas ratio was used to etch Ta_2O_5 material. Gas ratio of 20, 50, and 80 % were a mixture of CHF_3 = 10 sccm and Ar = 40 sccm, CHF_3 = 25 sccm and Ar = 25 sccm, and CHF_3 = 40 sccm and Ar = 10 sccm respectively. In this case L9 orthogonal array was selected; nine different runs were executed as shown in Table 2. The temperature of substrate plate and the etching time were kept constant at 20 °C and 10 minutes respectively.

Table 2. A L9 orthogonal array

Trial No.	RF power (W)	Gas ratio (%) $CHF_3/(CHF_3+Ar)$	Chamber pressure (mTorr)
1	100	20	15
2	100	50	30
3	100	80	45
4	150	20	15
5	150	50	30
6	150	80	45
7	200	20	15
8	200	50	30
9	200	80	45

2.3. Experimental procedure

The experiment was executed according to experimental configuration as listed in Table 2. In this experiment, initial thickness of Ta_2O_5 and Cr layer of the pre etching sample is measured by ellipsometer and Dektak profiler respectively. The process chamber is cleaned by striking oxygen plasma for 5 minutes to remove any residual polymers deposited on a wall inside chamber process from previous etching run. The process chamber cleanliness is essential to obtain a repeatability of etching process. After venting procedure, a chamber lid is opened and the sample is then placed at a centre of the quartz cover plate. The chamber lid is closed and pumped down. Once a 0 mTorr pressure is achieved, a selected recipe is then executed. During the etching process, a reflected power and a DC bias value at process window is closely monitored to check if any unstable process occurs. When the etching process is finished, the process chamber is vented and the sample is unloaded. The thickness of Ta_2O_5 and the etch depth of the post etching sample is then measured by the same method that is mentioned above. An etching rate and a selectivity of the each experiment runs are calculated as described by Figure 1 and equations 1, 2, and 3. Figure 2 shows the SEM image of pre etching process of the waveguide samples. The commercial software Minitab® has been used to analyse experimental results based on Taguchi method to determine optimum level of RF power, gas ratio and chamber pressure for smooth vertical waveguide sidewall and high etch rate and selectivity. Finally, a separate verification test was performed to fabricate ridge Ta_2O_5 waveguide with the obtained optimum parameters in this work. The sample etch profile and waveguide sidewall roughness was observed with a scanning electron microscope (SEM).

 DOI: 10.1051/epjconf/201716201003

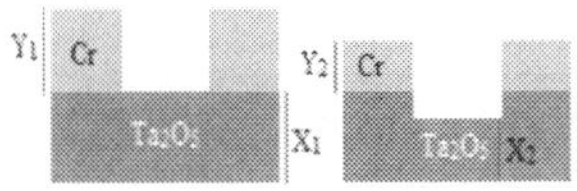

Fig. 1. Schematic of pre and post etching of waveguide sample for etch and selectivity calculation.

$$Ta_2O_5\ etch\ rate = (X_1 - X_2)\ /\ Etching\ time \quad (1)$$

$$Cr\ etch\ rate = Y_1\text{-}(Y_2\text{-}(X_1 - X_2))\ /\ Etching\ time \quad (2)$$

$$Selectivity = Ta_2O_5\ etch\ rate\ /\ Cr\ etch\ rate \quad (3)$$

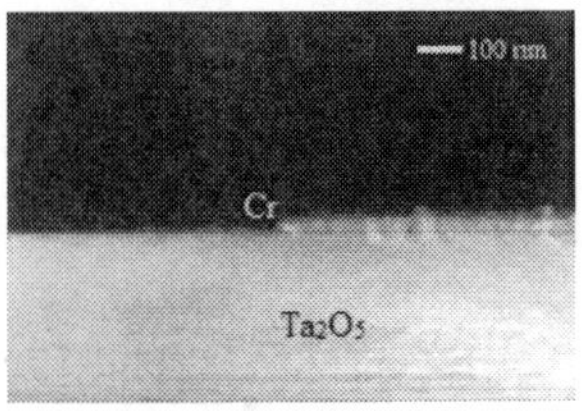

Fig. 2. SEM image of pre-etching process waveguide sample with 50 nm Cr hard mask and 400 nm Ta_2O_5 layer.

3 Results and Discussion

The experimental results of etching rate and selectivity of different experiment runs as listed in Table 2 are summarized in Table 3.

Table 3. Experimental results

Trial No.	Ta_2O_5 etch rate (nm / min)	Cr etch rate (nm / min)	Selectivity (Ta_2O_5/Cr)
1	7.5	0.7	11.1
2	8.8	1.2	7.4
3	6.9	1.2	5.8
4	13.2	0.8	17.6
5	13.4	1.6	8.5
6	15.3	1.4	11.2
7	17.1	2.1	8.2
8	15.4	0.5	30.4
9	21.6	0.8	28

An advantage of Taguchi method is its design experiment is orthogonal which mean of each level can easily be calculated thus influential rank of three factors can be estimated as shown in Table 4.

Table 4. Mean of each level of three factors

Parameter	Level	RF power (W)	Gas ratio (%)	Chamber pressure (mTorr)
Ta_2O_5 etch rate	1	7.7	12.6	12.7
	2	14.0	12.5	14.5
	3	18.0	14.6	12.5
	Delta	10.3	2.1	2.0
	Rank	1	2	3
Cr etch rate	1	1.0	1.2	0.8
	2	1.2	1.1	0.9
	3	1.1	1.1	1.6
	Delta	0.2	0.1	0.8
	Rank	2	3	1
Ta_2O_5 / Cr selectivity	1	8.1	12.3	17.6
	2	12.4	15.4	17.7
	3	22.2	15.0	7.5
	Delta	14.1	3.2	10.2
	Rank	1	3	2

In Table 4, delta is the value of maximum mean minus minimum mean. This delta value indicates an influential factor for each parameter response. For Ta_2O_5 etch rate the rank of three factors is RF power > Gas ratio > Pressure. For Cr etch rate case the rank is Pressure > RF power > Gas ratio and for the selectivity the rank is RF power > Pressure > Gas ratio.

Ta_2O_5 etch rate and selectivity were chosen for detailed study. As can be seen, the rank of each influential factor for each parameter response was different. The aim of this work is to obtain vertical sidewall of Ta_2O_5, thus etch rate and selectivity were focused for detailed study. It can be observed from Figure 3(a) that RF power has significant influence on Ta_2O_5 etch rate. It is found that increasing RF power increases Ta_2O_5 etch rate. This trend is as expected which higher RF power provides large voltage difference (higher DC bias) that accelerates positive ions and neutral species to bombard the material to be etched with high energy. Besides, higher concentrations of reactive species will be formed with increasing RF power.

The Gas ratio and chamber pressure factors are less significant compared to RF power on Ta_2O_5 etch rate as indicated in Figure 3(a). As can be observed from the graph, etch rate of Ta_2O_5 increased with increasing gas ratio. High gas ratio means that high flow rate of CHF_3 reactive gas is injected into chamber compared to Ar gas. It could be explained that the increase of CHF_3 flow rate

DOI: 10.1051/epjconf/201716201003

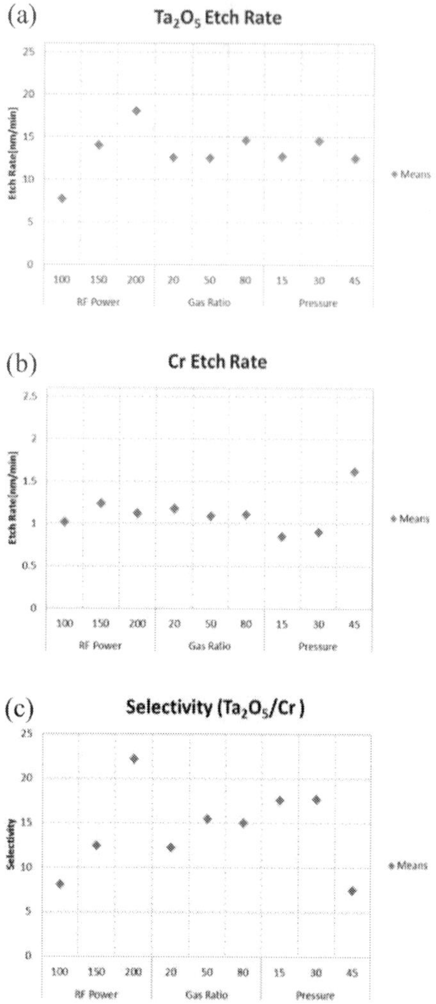

Fig. 3. Trend charts of levels of factors versus parameters. (a) Ta_2O_5 etch rate (b) Cr etch rate and (c) Selectivity (Ta_2O_5 / Cr).

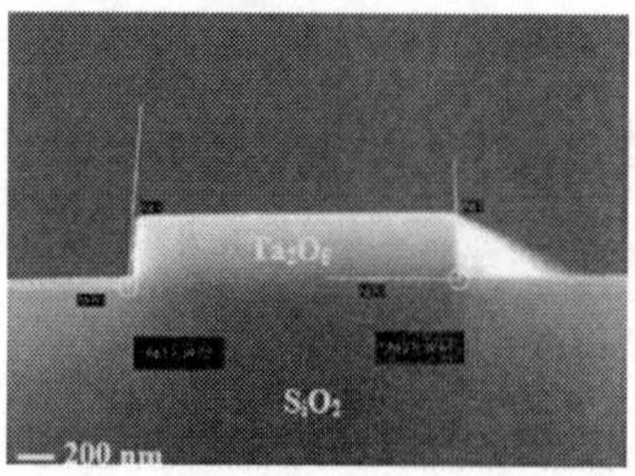

Fig. 4. SEM image of post-etching sample of optimized parameters (Trial No. 9).

increase the density of F atoms, hence increase the Ta_2O_5 etch rate. From the graph we found that Ta_2O_5 etch rate increases as the chamber pressure increases. This phenomenon could be due to the increase in gas density (positive ions and neutral species) thus enhance the etch rate. However, the etch rate was eventually dropped at chamber pressure of 45 mTorr. Beyond a certain limit, gas density in the process chamber result in more collisions of ions and species which can reduce mean free path of bombarding ions, thus reduce etch rate. Furthermore, more collisions of ions may affect the number of ions attacking vertically towards the sample (loss of direction) thus anisotropic profile could be degraded. Based on the experimental results, optimum values for three variable factors for high etch rate and selectivity are RF power of 200 W, gas ratio of 80 % (CHF_3 = 40 sccm, Ar = 10 sccm) and chamber pressure of 30 mTorr. As can be observed in Table 2, the combination of optimum values were considered in L9 orthogonal array (Trial No. 9). The etching profile of this sample was observed by SEM as displayed in Figure 4. Based on the figure, about 87~89 ° of smooth vertical sidewall is obtained. The etch rate of these parameters is 21.6 nm/min. We carried out the verification experiment using the same process parameters with etching time of 20 minutes with the aim to fabricate Ta_2O_5 ridge waveguide (100 % of 400 nm Ta_2O_5 layer to be fully etched).

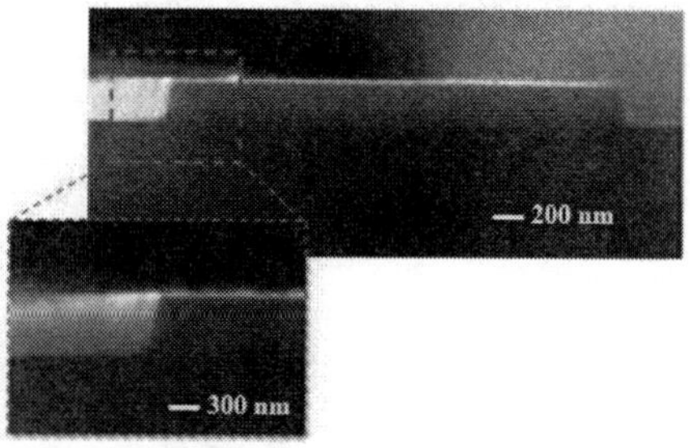

Fig. 5. SEM image of verification run samples with 20 minutes of etching time.

Figure 5 shows a smooth sidewall of the etching profile for this verification run. However, it should be noted that the obtained sidewall angle is about 83~85 ° less vertical from the previous run with identical process parameters and shorter etching time. The less vertical sidewall must have been due to longer etching time. These optimized parameters can be used for fabrication of Ta_2O_5 waveguide in this work, but are not suitable for PC slab structure formation. This is because the PC slab requires highly smooth vertical sidewall of PC structure to minimize low loss transmission [10]. A superior inductively coupled plasma (ICP) etching can be utilized to obtain higher etch rate and smooth vertical sidewall angle [11].

4 Conclusions

In this paper, DOE of Taguchi method was employed to investigate the optimum etching recipe for fabrication of Ta_2O_5 by using OIPT Plasmalab 80 RIE etcher. The variable factors studied were RF power, gas ratio and chamber pressure. Based on the experimental data, it was found that the optimum conditions for RF power, gas ratio and process pressure were 200 W, 80 % and 30 mTorr respectively. With the obtained optimum setting, smooth sidewall angle of 87~89 ° with 21.6 nm/min etch

DOI: 10.1051/epjconf/201716201003

rate and selectivity of 28 were demonstrated. However, waveguide sidewall angle was slightly decreased to 83~85 ° when the etching time was increased to 20 minutes maybe due to loading effect.

References

1. X. Liu, R. Osgood, and Y. Vlasov, Nat. Photonics, **4** (2010).
2. I.-W. Hsieh et al., Opt. Express, **15** (2007).
3. B. Corcoran, C. Monat, C. Grillet, and D. Moss, Nat. Photonics, **3** (2009).
4. P. S. Dobal, R. S. Katiyar, Y. Jiang, R. Guo, and a S. Bhalla, J. Raman Spectrosc., **31** (2000).
5. C.-Y. Tai et al., Opt. Express., **12** (2004).
6. R. Y. Chen, M. D. B. Charlton, and P. G. Lagoudakis, Opt. Lett., **34** (2009).
7. J. Jasapara, A. Nampoothiri, W. Rudolph, D. Ristau, and K. Starke, Phys. Rev. B, **63** (2001).
8. C. M. Netti, *Proceeding of. SPIE*, **5714**, (2005).
9. R. Y. Chen, M. D. B. Charlton, and P. G. Lagoudakis, *Proceeding of SPIE*, **7781** (2010).
10. A. L. Pyayt, B. Wiley, Y. Xia, A. Chen, and L. Dalton, Nat. Nanotechnol., **3**, (2008).
11. Muttalib, M. F. A., Chen, R. Y., Pearce, S. J., & Charlton, M. D., J. Vac. Sci. Technol. A, **32**(2014).

Enhancement of Lithium Niobate nanophotonic structures via spin-coating technique for optical waveguides application

Makram A. Fakhri[1*], *M.Halim* A. Wahid[2], *Ban A.* Badr[3], *Suad* M. Kadhim[2], *Evan* T. Salim[3], *Uda* Hashim[4] and *Zaid* T. Salim[3]

[1]Laser and Optoelectronic Department, University of Technology, 10001 Baghdad, Iraq
[2]Semiconductor Photonics & Integrated Lightwave Systems (SPILS), School of Microelectronic Engineering, University Malaysia Perlis, Pauh Putra Campus, Arau 02600 Malaysia
[3]Laser Science Branch, University of Technology, 10001 Baghdad, Iraq
[4]Institute of Nano Electronic Engineering, University Malaysia Perlis, 01000 Kangar, Perlis, Malaysia

Abstract. This work is dedicated to investigation of temperature effects in Lithium Niobate (LiNbO3) nanostructures. The LiNbO3 nanostructures were deposited on glass substrate by spin-coating technique. LiNbO3 was set down at 3000 rpm for 30 sec and annealed from 100 to 600 °C. The structures were characterized and analyzed by scanning electron microscopy (SEM) and ultra-violet visible (UV-vis) spectrophotometer. The measured results have showed that by increasing annealing temperatures, the structures start to be more crystallized and be more homogenized until the optimum arrangement was achieved. Once this was accomplished, it's applicable for optical waveguides development. Eventually, it starts to be less crystallization and non-homogeneous. Energy gap was recorded to be at average value of 3.9 eV.

1 Introduction

Lithium Niobate (LN) is a very important optical material which is widely used by the photonics industry mainly due to its excellent electro/acousto-optical properties [1-3]. Lithium niobate (LiNbO3) is an important ferroelectric material because of its excellent piezoelectrical, electrooptical, pyroelectrical and photo-refractive properties [4, 5]. It's widely used as a polar material for photonic applications [6, 7]. In addition it's employed in nonlinear optics for frequency conversion and in telecommunication for electro-optic modulation [8, 9]. Furthermore, it is an attractive material for the fabrication of optical wave-guide devices [10, 11]. Moreover, direct and indirect energy band gap of LiNbO3 is reported to be in the range of 3.5 - 4.7 eV which depends on LiNbO3 concentrations. These changes are attributed to several parameters like grain size, composition and defects [12-14]. There are different known approaches to synthesize undoped LiNbO3 nanocrystals using soft-chemistry [15], pulsed laser deposition [16, 17], RF sputtering [18] and hydrothermal methods [19, 20]. LiNbO3 is also studied in its heterojunction structure, photonic crystal and single phase thin film [21-25], and surface acoustic wave device (SAW) [26]. LiNbO3 waveguides are employed broadly in many functional electro-optic and acousto-optic devices [27]. The waveguide structures are essential in many integrated-optic devices especially at high frequency region.

This work reports the preparation of LiNbO3 nanostructures by exploiting the spin-coating technique. The characterization and analysis have been elaborated as a function of LN's annealing temperatures. The refractive index analysis is the main work on LiNbO3 optical waveguides because of refraction coefficients between the base and deposit samples will ensure access to total internal reflection that gives better realization of optical waveguides. The refractive index is measured and calculated in order to fit the best application of optical waveguides. Thus, the focus in the study is to prepare the LN thin film and to enhance towards a high quality film with optical properties closely like a single crystal LiNbO3 wafers.

2 Experimental details

LiNbO3 films were prepared mainly by adding Nb2O5 (ultra-pure, 99.99%) and citric acid (CA) without further purification. The solution was then mixed with Li2CO3

* Corresponding author: mokaram_76@yahoo.com

 DOI: 10.1051/epjconf/201716201004

and Ethylene Glycol. The molar ratio between Li2Co3 and Nb2O5 was 1:1 in order to maximize the formation of LiNbO3 stoichiometry phase. Firstly, the Li2Co3, Nb2O5, and citric acid were dissolved in Ethylene Glycol with simultaneous heating and stirring at 90 oC for 48 hours. Obtaining homogeneous and crack-free films of LiNbO3 requires the precursor to be deposited by spin coating technique on quartz substrates at a spinning speed of 3000 RPM for 30 Sec and 0.5 Ml/L molarity concentration. In total, seven layers were prepared.

The film was further dried at 120 °C for 5 min and calcined at 250 °C for 30 min in static air and oxygen atmosphere with the intention of removal the organics. Finally, the film was annealed from a constant step variation of 100 to 600 °C. The structural evolution of the as-prepared thin films was examined using a high-resolution X-ray diffraction (HR-XRD) system equipped with Cu-Kα-radiation of wavelength λ = 0.15418 nm at 40 kV and 30 mA. The thickness of the annealed samples was measured using scanning optical reflectometery. The scanning electron microscopy (SEM) was employed to investigate the surface morphology of LiNbO3 and the atomic force microscopy (AFM) was utilized to observe the surface roughness of the LiNbO3 films. The optical properties were inspected by means of double-beam Ultra-Violet (UV-vis) spectrophotometer.

3 Results and discussion

3.1 Morphological studies

Figure 1 shows SEM images of $LiNbO_3$ nanostructures deposited on quartz substrates at different annealed temperatures. The density of nucleation for the $LiNbO_3$ thin films was not uniform on the flat substrate with dissimilar temperatures. At least molarity concentration notifies the emergence of a high proportion of the pores and voids. These pores were observed as a result of impurities impacts like Nb_2O_5. The $LiNbO_3$ thin films grew smoothly at 500 °C that's led to good distribution better than other annealing temperatures as shown in Fig. 1. On the other hand, the structure is more homogenous at higher molar concentration [14]. This suggests that 500 °C leads to increase the regular distribution of $LiNbO_3$ nano and micro structures. A closer examination of these samples indicates that our synthesized film like ice layers shape of morphology.

Fig. 1. Surface morphology of $LiNbO_3$ nanostructures with different annealing temperatures. (a) 400 °C , (b) 500 °C, and (c) 600 °C

3.2 Optical properties

The transmission spectra of $LiNbO_3$ nanostructures at different annealing temperatures are shown in Fig. 2. It was measured that the transmission increased with the annealing temperature. This may be due to the increasing of grain size and decreasing thickness. Similar result was also recorded in other work [28]. The values of transmission were about 70-80% with the annealing temperature range of 100 to 600 °C. The deposited samples were white to brown in color and showed high transmittance. The high value of transmittance is attributed to the excessive ($LiNbO_3$) ions at interstitial site that increased its transparency level.

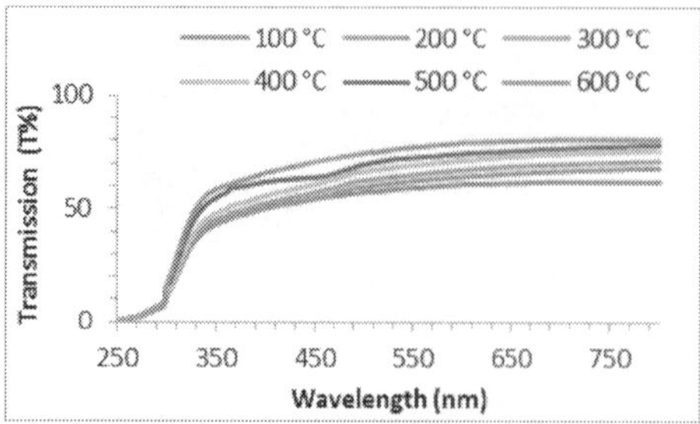

Fig. 2. Transmission spectra of $LiNbO_3$ nanostructures at different annealing temperatures

Figure 3 shows the reflectance of $LiNbO_3$ nanostructures in the wavelength range of 250-800 nm at room temperature. The reflection spectra were found to decrease with annealing temperature due to the increment in the surface roughness that related to crystals grain size.

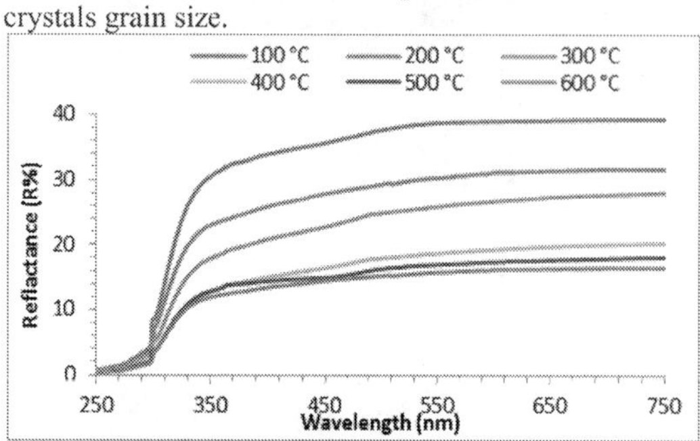

Fig. 3. Reflection spectra of $LiNbO_3$ nanostructures at different annealing temperatures

The energy band gap (E_g) was established as a function of photon energy in Fig. 4. All curves were singularized at a specific point around 3.25 eV. However, the E_g was extrapolated and recorded at different values. The value was noted to increase with annealing temperatures due to shrinkage in the grain size.

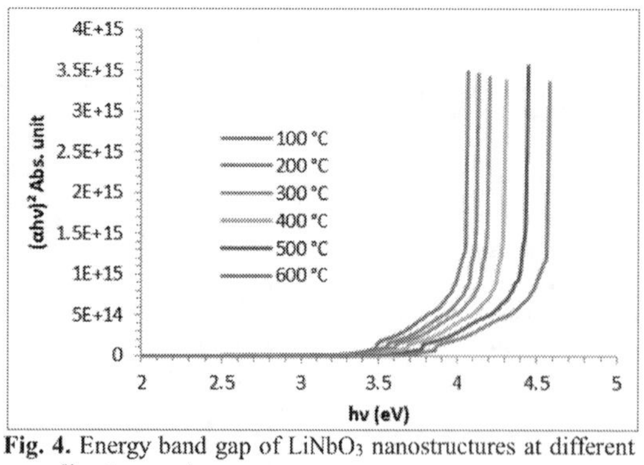

Fig. 4. Energy band gap of $LiNbO_3$ nanostructures at different annealing temperatures

The refractive index (n) was determined from a transmittance spectrum as a function of the wavelength at different annealing temperatures. The spectrum showed that there was a decrease in the refractive index with incident wavelength from 2.32 to 2.56 which related to the drop in the reflectance as depicted in Fig. 5.

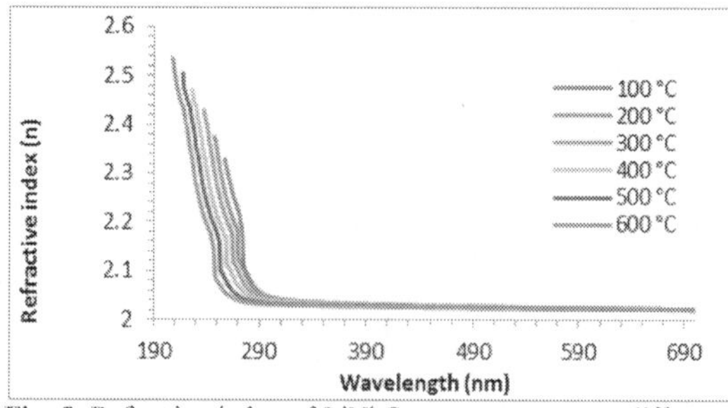

Fig. 5. Refractive index of $LiNbO_3$ nanostructures at different annealing temperatures

Beside that an increment in the refractive index could be realized in the UV region with the annealing temperature as shown in Fig. 6. This behavior may be incorporated to growth in packing density as a result of the oxidation process during heat treatment. All the higher values of refractive index are suitable for optical waveguide fabrication.

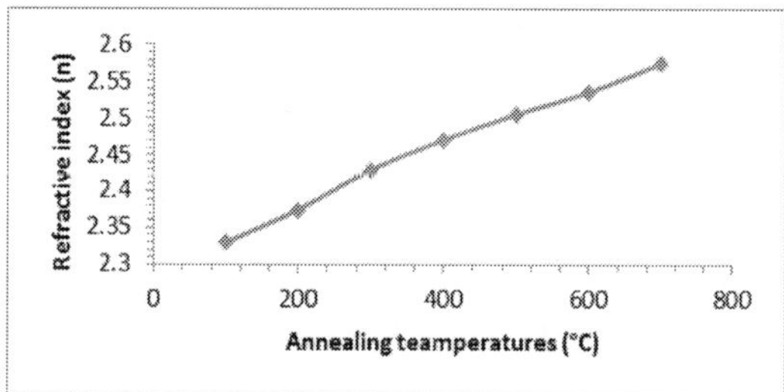

Fig. 6. Refractive index with concentration of $LiNbO_3$ nanostructures at different annealing temperatures

4 Conclusion

The LiNbO3 nanostructures have been chemically prepared by spin-coating technique. SEM shows diameter of grain size from 64 to 167 nm. Optical properties give high values of transmission which is about 89 - 96% and the measured energy band gaps are 3.6, 3.85, 4 and 4.2 eV. We found an approximate match between the energy band gap in the values calculated using the transmission spectra. The refractive index is determined from the transmission spectrum and established its appropriateness for application in optical waveguides that needs high value.

References

1. Makram A. Fakhri, Y. Al-Douri, U. Hashim and E. T. Salim, Australian Journal of Basic and Applied Sciences **9**, 128-133 (2015).
2. Makram A Fakhri, Y Al-Douri, Uda Hashim and Evan T Salim, Advanced Materials Research **1133**, 457-461 (2016).

 DOI: 10.1051/epjconf/201716201004

3. Evan T. Salem, Surface Review and Letters **20**, 1350046-1-1350046-6 (2013).
4. J. Zhang and X. Zhang, Optics Communications **320**, 56–59 (2014).
5. L. Cao, A. Aboketaf,Z. and Wang,S. Preble, Optics Communications **330**, 40–44 (2014).
6. H. Chen, , T. Lv, A. Zheng and Y. Han, Optics Communications **294**, 202–207 (2013).
7. P. Ganguly, Optics Communications **285**, 4347–4352 (2012).
8. B. Knabe, D. S. Tze, T. Jungk , M. Svete, W. ssenmacher, W. Mader and K. Busem, Status Solidi (a) **208**, 857–862 (2011).
9. S. Mamoun , A.E. Merad and L. Guilbert, Computational Materials Science **79**, 125-131 (2013).
10. C. Thierfelder , S. Sanna, A. Schindlmayr, and W. G. Schmidt, Phys. Status Solid (C) **7**, 362–365 (2010).
11. A. Tumuluri, K .L. Naidu and K.C.JamesRaju, International Journal of Chem Tech Research CODEN (USA): IJCRGG **6**, 3353-3356 (2014).
12. O. P. Nautiyal1 and S. C. Bhatt, American Journal of Materials Science **1**, 1-4 (2011).
13. Makram A. Fakhri, Y. Al-Douri, U. Hashim, Evan T. Salim, Deo Prakash and K. D. Verma, Applied Physics B **121**, 107–116 (2015).
14. Makram A. Fakhri, U. Hashim, Evan T. Salim and Zaid T. Salim, J Mater Sci: Mater Electron **27**, 13105–13112 (2016).
15. M.Liu, D Xue and K Li, Journal of Alloys and Compounds **449**, 28-31 (2013).
16. X. Wang ,Y.Liang, S. Tian, W. Man and J. Jia, Journal of Crystal Growth **375**, 73–77 (2013).
17. Makram A Fakhri, Y Al-Douri, U Hashim and Evan T Salim, Solar Energy **120**, 381-388 (2015).
18. V. Iyevlev and A. Kostyuchenko, Journal of Materials Science: Materials in Electronics **22**, 1258-1263 (2011).
19. Evan T. Salim, International Journal of Nanoelectronics and Materials **5**, 95-100 (2012).
20. M. Aufray, S. Menuel, Y. Fort, J. Eschbach, D. Rouxel and B. Vincent, Journal of Nanoscience and Nanotechnology **9**, 4780–4785 (2009).
21. Evan T. Salim, Indian Journal of Physics **87**, 349–353 (2013).
22. Y. Guo , Ch. Divin A. Myc, F. L. Terry, J.R. Baker, T. B. Norris, and J. Y. Ye, Opt. Express **16**, 11741–11749 (2008).
23. H. Lu, B. Sadani, G. Ulliac, N. Courjal, C. Guyot, J. -M. Merolla, M. Collet, F. I. Baida,and M.-P. Bernal, OPTICS EXPRESS, **20**, 20884-20893, (2012).
24. J. Son, S. S. Orlov, B. Phillips, L.Hesselink, Journal of Electroceramics, **17**, 591-595, (2006).
25. AbdulhadiKadhim, Evan T. Salim, Saeed M. Fayadh, Ahmed A. Al-Amiery, Abdul Amir H. Kadhum, and Abu BakarMohamad, The Scientific World Journal, Article ID 490951, 6 page (2014).
26. K.S. Kao, C.C. Cheng, Y.C. Chen, Y.H. Lee, Applied Physics A, **76**, 7, 1125-1127 , (2003).
27. G-h. Shao, X-s.Song, F. Xu, and Y-q. Lu1, OPTICS EXPRESS , **20**, 19343-19348, (2012).
28. Ibrahim R. Agool, Evan T. Salem, Marwa A. Muhsien, International Journal of Modern Physics B, **25**, 1081-1089, (2011).
29. Z. T Salim, U Hashim, MKM Arshad, Makram A.Fakhri, ET Salim, Materials Research Bulletin **86**, 215-219, (2017).
30. Evan T Salim, Marwa S Al Wazny, Makram A Fakhry, Modern Physics Letters B, **27**, 1350122-7-1350122-1, (2013).
31. David E. Zelmon and D. L. Small, J. Opt. Soc. Am. B, **14**, 3319 – 3322, (1997).
32. T.Ghosh, B.Samanta ,P.C. Jana, and P. Ganguly, Fiber and Integrated Optics, **31**, 1-10, (2012).
33. D.L. Zhang , Q. Zhang, C.X. Qiu, W.H. Wong, E. Y. B. Pun, Optical Materials Express, **4**, 2215-2220, (2014).
34. S. Nakamura, Mukai, T., Senoh, M., Jpn. J. Appl. Phys. 30 L1705, (1991).

EPJ Web of Conferences 162, 01005 (2017)
InCAPE2017
DOI: 10.1051/epjconf/201716201005

The structure and optical properties of Lithium Niobate grown on quartz for photonics application

Makram A. Fakhri[1*], *M.Halim* A. Wahid[2], *Suad* M. Kadhim[1], *Ban* A. Badr[3], *Evan* T. Salim[4], *Uda* Hashim[5] *and Zaid* T. Salim[5]

[1]Laser and Optoelectronic Department, University of Technology, 10001 Baghdad, Iraq
[2]Semiconductor Photonics & Integrated Lightwave Systems (SPILS), School of Microelectronic Engineering, University Malaysia Perlis, Pauh Putra Campus, Arau 02600 Malaysia
[3]Laser Science Branch, University of Technology, 10001 Baghdad, Iraq
[4]Institute of Nano Electronic Engineering, University Malaysia Perlis, 01000 Kangar, Perlis, Malaysia

Abstract. This paper focuses on the structure and optical properties of Lithium niobate (LN) deposited on quartz substrate by sol-gel method. The solution was prepared at different molarity concentration and annealed at 500 °C. The LNs were analyzed by UV-visible, X-ray diffraction (XRD) and Scanning Electron Microscope (SEM). The results show that as the molar concentration increases, the films crystallization becomes more atomically distributed and the energy band gap approaches 4 eV. Transmission spectra were taken in the wavelength range of 400-1100 nm. Hexagonal structure was observed with good atoms distribution and homogeneous crystallization. There were two phases of LNs recorded in this work.

1 Introduction

Lithium Niobate (LiNbO3) has recently attracted photonics industries owing to its optical material properties such as electro optic and acoustic optic. These properties have been investigated and showed uniqueness thus became attractive candidate in the nanophotonics area [1, 2]. LiNbO3 single crystal is an excellent material for non-linear optics and electro-optics applications. It is one of the key materials for optical based technologies because of its large second-order nonlinearities [3]. Because of its remarkable linear and nonlinear optical properties, chemical and mechanical stability makes lithium niobate (LN) an attractive host material for application in photonic crystal devices [4]. Moreover LN has high electro-optical coefficient and low optical losses thus offers suitability in optical communication systems, which are widely used for applications in microwave telecommunications, memory units, electro optics, modulators, optical switches, waveguides, beam deflectors, second harmonic generation, surface acoustic waves (SAW) devices, parametric optical converters and data transmission [5-10].

LiNbO3 has hexagonal structure with lattice parameters a = 0.5147 nm and c = 1.3862 nm [11]. However, LiNbO3s have several potential advantages from technological or commercial view point of integrated optics [12]. Optical waveguides comprised of hetero-materials efficiently confine propagating light because of large difference in refractive indices between the film and the substrate [13]. Therefore, LiNbO3s have been grown by several techniques including liquid-phase epitaxy [14], Ion Beam Sputtering [15], RF magnetron sputtering [16], molecular beam epitaxy [17], metal–organic chemical vapor deposition [18], pulsed-laser deposition (PLD) [19], thermal plasma spray chemical vapor deposition [20], electron-cyclotron-resonance (ECR) plasma sputtering [21], electron beam irradiation [22] and pulsed laser ablation (PLA) [23].

In the present work, the effect of molarity concentration on the growth of LiNbO3 on quartz substrate was investigated. Generally the structure and optical behaviors was measured and analyzed in correspondent to variable molarity. Sol-gel method was utilized due to its simplicity, economic and adequately precise. Quartz was chosen as a substrate because of its compatibility with a wide variety of processing techniques and other optical properties like band gap, refractive index, extinction coefficient, dielectric constant and Urbach energy.

* Corresponding author: mokaram_76@yahoo.com

2 Methodology

The preparation procedure of $LiNbO_3$ was performed by mixing Nb_2O_5 (ultra-purity, 99.99%) and citric acid (CA) without further purification. The solution was added into Li_2CO_3 and Ethylene Glycol. The molar ratio between Li_2CO_3 and Nb_2O_5 was the same as to maximize the formation of $LiNbO_3$ stoichiometry phase. Firstly, the $LiCO_3$, Nb_2O_5, and citric acid was dissolved in Ethylene Glycol with heating and stirring at 90 °C for 48 hr. The obtained solution was directly deposited on quartz substrate of 1×1 cm^2 at 3000 rpm for 30 sec. After each coating, the film was dried in air at 120 °C for 10 min and was calcined at 250 °C for 1 h to remove the organics. The process was performed several times until the desired film thickness was achieved. Finally, the films were annealed at 500 °C for 2 hr. The prepared film was clear, transparent, white to brown color and has a good adhesive properties and smooth surface. After that it was calcined at 500 °C for 2 hr in static air and oxygen atmosphere to remove the organics. The optical properties were measured via transmittance (T) spectrum in the range of 300-1100 nm double-beam Ultr-Violet (UV-vis) spectrophotometer (Shimadzu UV-Vis 1800, Japan). The incident photon energy was calculated as a function of wavelength (λ) by using [24]

$$E_g(eV)=1240/\lambda(nm) \quad (1)$$

where λ is wavelength. The energy dependence of absorption coefficient (α) near the band edge for band to band and excition transition could be described by Tauc formulas [25]

$$(\alpha h\nu)=B(h\nu-E_g)r \quad (2)$$

where B is a constant inversely proportional to amorphous, r is constant equals 2, 3, 1/2, 3/2 depending on the material and the type of the optical transition $\alpha h\nu$. If the straight portion of the plot of $(\alpha h\nu)^{1/r}$ against ($h\nu$) is extrapolated to $(\alpha h\nu)^{1/r}=0$, the intercept gives the energy gap value. The absorption coefficient (α) for each wavelength was calculated by [26]

$$\alpha=2.303(A/t) \quad (3)$$

where A is absorbance and t is transmittance.

The structural evolution of the as-prepared nanostructures was examined using high-resolution X-ray diffraction (HR-XRD), equipped with Cu-Kα-radiation of wavelength λ = 0.15418 nm at 40 kV and 30 mA. The thickness of the annealed samples was studied using scanning optical reflectometer while the surface structure was monitored by the scanning electron microscopy (SEM).

3 Results and discussion

3.1 Optical properties

The transmission spectra of $LiNbO_3$ nanophotonics films at different molarities concentrations are shown in Fig. 1. It was found that the transmission decreased as molarities concentration increased due to increment in deposition rate and structural thickness. The values of transmission were ranged between 48-96% with molarities concentration of 1.00 to 0.1 mol/L, similar results were found in other work [27]. The lower value of transmittance is attributed to excessive $LiNbO_3$ ions existing at interstitial sites that probably absorb the light. The moderately high transmittance throughout UV-Vis regions makes a good material for optoelectronics application.

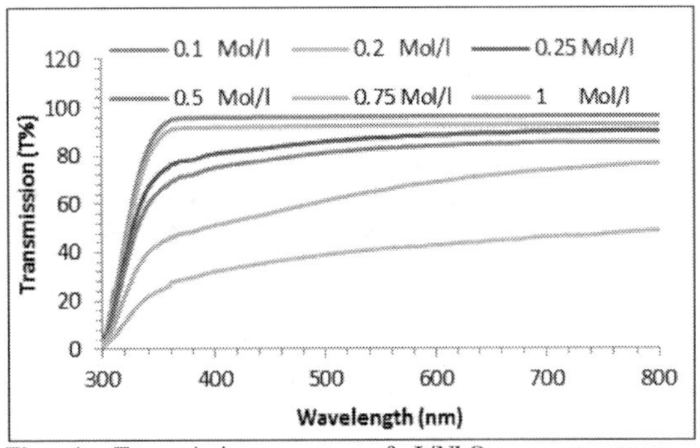

Fig. 1. Transmission spectra of $LiNbO_3$ nanostructures prepared at different molarities concentration

The optical reflectance (R %) of $LiNbO_3$ nanophotonics was measured using double-beam UV-vis instrument. Fig. 2 shows the reflectance of $LiNbO_3$ nanophotonics films within the wavelength range of 300-700 nm at room temperature. It's clearly observed that optical reflection increased with molar concentration. This is due to the fact that when a beam of light illuminates on a dielectric thin film the light ray is partially reflected and transmitted, depending on energy of incident photon and film thickness since the prepared material consists of charged particles, bound and loose electrons, ionic elements, etc.

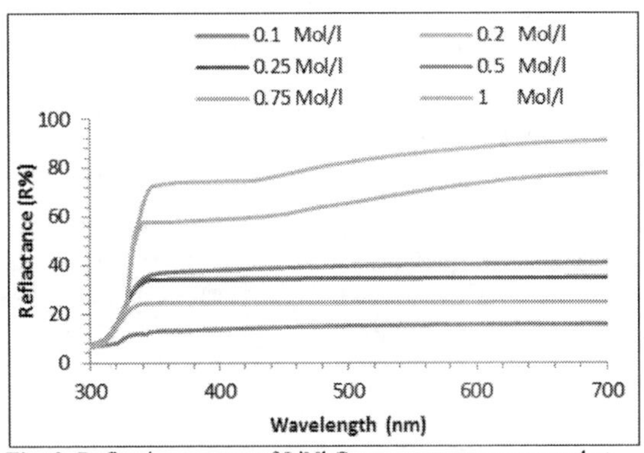

Fig. 2. Reflection spectra of $LiNbO_3$nanostructures prepared at different molarities concentration

At infrared and visible light frequencies, the contribution comes only from the displacement of the electron. In dielectric solids, there is a specific energy separating the valence band and the conduction band, known as the energy band-gap. It values were estimated and found by a plot as shown in Fig. 3. It could be shown that the band gap increases with the molarities concentration due to decreases in material grain size. In dielectrics materials the main absorption process resulting from the interaction of optical radiation and the electrons. If the energy of the incident light beam equals to the energy required in exciting an electron to the higher energy level, the photon may be absorbed taking into account that the electron may be free electron or core electron in the materials. In case of $LiNbO_3$, in the visible region no excitation can accrue, since the electrons are tightly restricted that the energy in the ultraviolet region is required. The obtained results are synchronized with experimental value in other work [24].

In the field of optical thin films deposition, the light wavelength is mostly larger than inter atomic distance. Therefore the light wave interaction with the materials are middling over a lot of unit cells. As a result, the optical properties within thin films could be depicted by using optical parameters or optical constants.

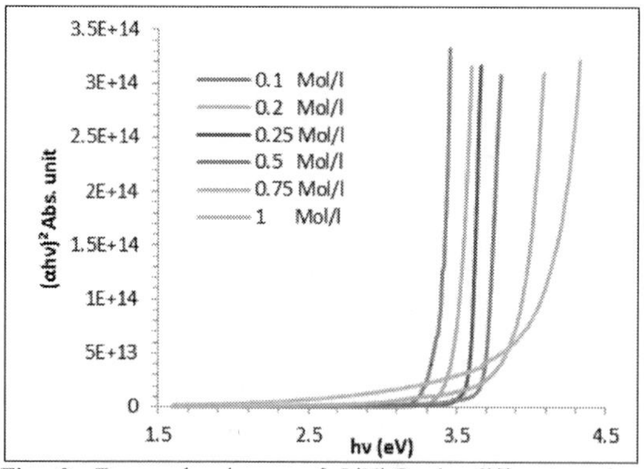

Fig. 3. Energy band gap of $LiNbO_3$ in different molar concentrations

3.2 Structural properties

The XRD result of $LiNbO_3$ nanostructures deposited on quartz substrates grown by sol–gel method is shown in Fig. 4. The XRD pattern was obtained from $LiNbO_3$ thin films deposited on quartz substrates at optimum condition (1Ml/L).The crystal structure of $LiNbO_3$ nanostructures is found to have hexagonal structure. It is observed that the peaks at 2θ = 23.634, 32.637, 34.674, 48.355, 53.106, and 55.879 correspond to (012), (104), (110), (024), (116) and (122) planes. Hence, the crystal structure will be more crystallized with higher molarity concentration. All the reflection peaks could be indexed to the hexagonal structure with lattice parameters a = 5.145, c= 13.858, which were very close to the reported data in the A.Z. Simoes et al [28]. The thin film remained polycrystalline structure after annealing at 500 ^{0}C for 2 h in static air. Two phases of Lithium niobate could be recognized which is Δ and δ phases. It is noted that a preferred Δ phase exists in the film and performed orientation of (012). The XRD clearly indicates the presence of a small amount of secondary Li deficient phase ($LiNb_3O_8$) at all molar concentrations. This phase is originated from an interface reaction between the Oxygen and $LiNbO_3$. The phase could be detected by XRD owing to its high crystallization temperature and observed at peaks 2θ = 24.407, 30.262 and 35.981 correspond to (400), (410) and (212) planes. This is attributed to annealing effect in static air and oxygen atmosphere. There was an impurity site of Nb_2O_5 at peaks 2θ = 24.433 and 31.623 which correspond to (-105) and (014) planes. The planes of Nb_2O_5 were not considered as a major plane that may change our observation significantly.

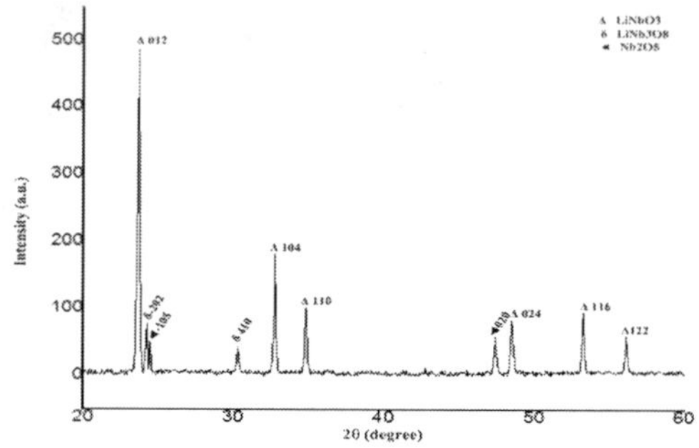

Fig. 4. XRD spectra of $LiNbO_3$ in optimum condition of higher molarity concentration (1 Mol/L)

3.3 Morphological studies

It is an interesting parameter for integrated-optic and optoelectronic applications. Figure 5 shows SEM images (5x5 μm) of $LiNbO_3$ nanostructures deposited on quartz substrates at different molarity concentrations. The density of nucleation for the $LiNbO_3$ thin films was not uniform on the flat substrate. Higher molarity concentration (optimum condition) depicts more homogenousity and leads to a better distribution than the lower molar concentration level. As discussed earlier, this suggests that higher molarity concentration directs the $LiNbO_3$ nanostructures toward regular spread.

An overview image at low magnification illustrates that the obtained samples are quite uniform in shape. A closer examination of these samples indicates that our synthesized film like ice layers shape of morphology.

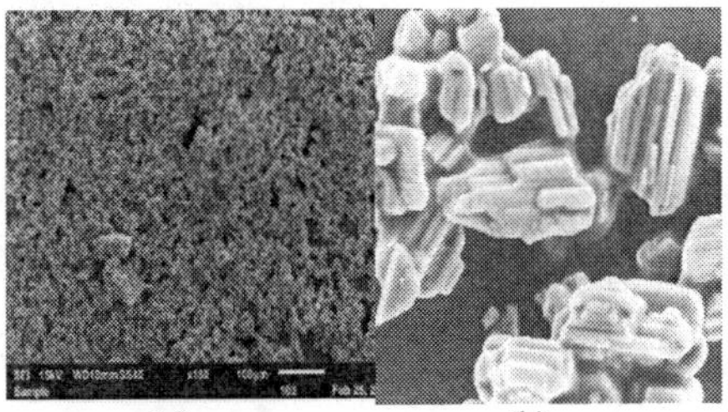

(a) (b)

Fig. 5. Surface morphology of $LiNbO_3$ nanostructures at optimum condition of molarity concentration (a) 1 Mol/L at high magnification, (b) 1 Mol/L at low magnification

4 Conclusion

The $LiNbO_3$ nanostructures have been chemically prepared by spin-coating technique. SEM images elucidate that the nanostructures will be more homogenous as the molarity concentration increases. Based on XRD results, the $LiNbO_3$ nanostructures have polycrystalline in nature. The highest intensity is shown at orientation (012) at $2\theta = 23.634$, a significant increment at the spin coating speed of 3000 rpm. As expected, the structure is more crystalline with higher molar concentration. Optical properties analysis provides energy band gap values between 2.88 to 3.975 eV for different molarity concentrations.

References

1. D. Janner, D. Tulli, M. Jofre, D. Yudistira, S. Balsamo, M. Belmonte, and V. Pruneri, IEEE Journal of Selected Topics in Quantum Electronics,**19**, 34006-34016, (2013).
2. Makram A Fakhri, Y Al-Douri, U Hashim, Evan T Salim, Solar Energy, **120**: 381-388, (2015).
3. P. Kumar, S.M. Baru, S Perero, R.M.L.Sai, I Bhamik, S. Ganesamoorthy, A.K. Karnal, J. Phys. **75**, 1035-1040, (2010).
4. M.R.R. Gesualdi, C. Jacinto, T. Catunda, M. Muramatsu, V. Pilla, Appl. Phys. B, **93** , 879–883 (2008).
5. Marwa Abdul Muhsien,Evan T. Salem, Ibrahim R. Agool, Haidar Hamed Hamdan, Applied Nanoscience **4**: 719–732 (2014).
6. M.-P. Bernal, N. Courjal, J. Amet, M. Roussey, C.H. Hou, Optics Communications **265**, 180–186 (2006).
7. V. Ievlev, M. Sumets, A. Kostyuchenko, N. Bezryadin, J Mater Sci: Mater Electron **24** , 1651–1657 (2013).
8. H-S. Jung, Journal of the Optical Society of Korea **16**, 47-52 (2012).
9. Makram A. Fakhri, Y. Al-Douri, U. Hashim, E.T. Salim, D. Prakash, K. D. Verma, Applied Physics B **121**: 107–116 (2015).
10. D. Janner1, D.Tulli , M. Garc´ıa-Granda, M. Belmonte, and V.Pruneri, Laser & Photon. Rev **3**, 301–313 (2009).
11. V. Ievlev , V. Shur , M. Sumets and A. Kostyuchenko, Acta Metallurgica Sinica (English Letters) **26**, 630-634, (2013).
12. N. E. Stankovaa, S. H. Tonchevb, E. Gyorgyc, G. Socolc, I. Mihailescuc, Journal of Optoelectronics and Advanced Materials **6**, 1345 – 1348, (2004).
13. Ch. Wang,M. J. Burek, Z. Lin, H. A. Atikian, V. Venkataraman, I. Huang, P. Stark and M. Lončar, OPTICS EXPRESS **22**, 30924-30933, (2014).
14. C. Dubs , J-P. Ruske , J. Krulich , A. Tünnermann, Optical Materials **31**, 1650–1657, (2009).
15. Makram. A. Fakhri, Y. Al-Douri, Uda. Hashim, Evan. T. Salim, Australian Journal of Basic and Applied Sciences **9**, 128-133, (2015).
16. V. Ievlev, M. Sumets, A. Kostyuchenko, N. Bezryadin, Journal of Materials Science: Materials in Electronics **24**, 1651-1657, (2013).
17. Abdulhadi Kadhim, Evan T. Salim, Saeed M. Fayadh, Ahmed A. Al-Amiery,Abdul Amir H. Kadhum, and Abu Bakar Mohamad, *The Scientific World Journal*, Article ID 490951, 6 paes, (2014).
18. Y. Akiyama, K. Shitanaka, H. Murakami, Y-S. Shin, M, Yoshida, N.Imaishi, Thin Solid Films **515**, 4975–4979, (2007).
19. Ji-Won Son, Sergei S. Orlov, Bill Phillips, Lambertus Hesselink, Journal of Electroceramics **17**, 591-595, (2006).
20. Makram A. Fakhri, U. Hashim, Evan T. Salim, Zaid T. Salim, J Mater Sci: Mater Electron **27**, 13105–13112, (2016).
21. H. Akazawa, M. Shimada, Vacuum **80**, 704–707, (2006).
22. M.R. Beghoul , A. Boudrioua , R. Kremer , M.D. Fontana , B. Fougere , C. Darraud , J.C. Vareille , P. Moretti, Optical Materials **31**, 136–142, (2008).
23. F. Meriche, A. Boudrioua, R. Kremer, E. Dogheche, E. Neiss-Clauss, R. Mouras, A. Fischer M-R. Beghoul, E. Fogarassy, N. Boutaoui, Optical Materials **32**,1427–1434, (2010).
24. Evan T. Salem, Ibrahim R. Agool, Marwa Abdul Muhsien, International Journal of Modern Physics B **25**, 3863-3869, (2011).
25. N. Ghobadi, Ghobadi International Nano Letters **3**, 2-4, (2013).
26. M. Caglar , Y. Caglar, S. Ilican, Journal of optoelectronics and advanced materials **8**, 1410 – 1413, (2006).
27. S. Ilican , Y. Caglar , M. Caglar , F. Yakuphanoglu, Applied Surface Science **255**, 2353-2359, (2008).
28. A.Z. Simoes, M.A. Zaghetea, B.D. Stojanovica, C.S. Riccardi, A. Ries, A.H. Gonzaleza, J.A. Varelaa, Materials Letters **57** , 2333 – 2339, (2003).

 DOI: 10.1051/epjconf/201716201006

Multiple lasing Brillouin Fiber Laser with the implementation of reflective Fiber Bragg Grating in a ring cavity configuration

A. Zakiah Malek[*,1,2], *N. A. M.* Ahmad Hambali[**,1,2], *M. H. A.* Wahid[1,2], *M. M.* Shahimin[3], *A.* Harun[1] and *N.* Ali[1,2]

[1]School of Microelectronics Engineering, Universiti Malaysia Perlis (UniMAP), 02600 Arau, Perlis, Malaysia
[2]Semiconductor Photonics & Integrated Lightwave Systems (SPILS), School of Microelectronic Engineering, Universiti Malaysia Perlis, Pauh Putra Main Campus, 02600 Arau, Perlis, Malaysia
[3]Department of Electrical and Electronic Engineering, Faculty of Engineering, National Defence University of Malaysia (UPNM), Kem Sungai Besi, 57000 Kuala Lumpur.

Abstract. A multi-wavelength Brillouin fiber laser with the implementation of reflective fiber Bragg grating is developed. In this ring cavity architecture, the fiber laser system is performed by the stimulated Brillouin scattering effect and four-wave mixing process. An improvement in the number Brillouin Stokes signal and average of the optical signal to noise ratio were generated through the reflection process of fiber Bragg grating. Due to the optimization of output coupling ratio from 10% to 60%, the highest Brillouin Stokes signal of 17 and average optical signal to noise ratio of 14.7 dB were recorded at 60% of the output coupling ratio. The amplified Brillouin pump power of 16.5 dBm at 1550 nm of Brillouin pump wavelength was injected into the laser system. A constant line spacing of 0.08 nm at each of the Brillouin Stokes signals was also recorded.

1 Introduction

Multi-wavelength fiber lasers can be defined as laser sources that emitting light more than one signal or multiple Brillouin Stokes (BS) signals from a main single wavelength light source [1]. The multi-wavelength generation is commonly capable producing an accurate narrow linewidth that made them offered several high potentials in an optical communication system especially in dense wavelength division multiplexing (DWDM) application [2]. As mentioned in a previous paper [2], the operation of DWDM system was manipulated from the extreme growth of multiple signals in order to attain an efficient operation and consequently to make the DWDM system as one of the attractive technology among the researchers. Other notable technologies that applied the multi-wavelength lasers are mostly found in microwave signal processing [3], optical sensors, optical spectroscopy, current monitors, short optical pulses [4] and high capacity optical fiber communication networks [3].

Recently, the requirement for multiwavelength Brillouin fiber laser (MWBFL) is highly desirable for the optical communication system due to its ability to counter the limitations in the multiwavelength Brillouin-Erbium fiber laser (MWBEFL) and multiwavelength Brillouin-Raman fiber laser (MWBRFL). In such a situation, the major drawback of MWBEFL is related to the intrinsic generation of self-lasing cavity modes at the peak of the Erbium-doped fiber (EDF) gain [5] meanwhile a spectrum broadening effect in the MWBRFL cavity is another issue need to be considered [6]. Consequently, these limitations have a major impact on the formation of multi-wavelength generation. In an effort to increase the performance of the laser system, the hybrid MWBFL can counter these problems by allowing a higher coherent light source with narrow downshifted signal through stimulated Brillouin scattering (SBS) effect [7] and thus offer a solution to produce a larger number BS signal.

The direction of MWBFL studied was demonstrated in both linear and ring cavity configurations. Various ring cavity configurations of MWBFL system have been previously reported in [1, 8-9]. In this reported work [1], the ring cavity was successfully produced up to 8 BS signal with 0.08 nm line spacing by a combination of forward and backward output signals. Meanwhile, 0.16 nm of line spacing in between two consecutive odd or even BS signals was attained when the forward and backward direction were discriminated from the laser system. Another invention also proposed by M. R. Nurdik and his team [8]. This ring cavity BFL was mainly constructed by using the same type and length (25 km of SMF-28) of the Brillouin gain medium as previously reported in [1]. The performance of laser system was achieved with 10 BS signals and 7 anti-Stokes signals at a line spacing of 0.088 nm by launching 11.29 dBm BP power at 1510 nm of BP wavelength. Besides, at a BP wavelength of 1550 nm, the limitation of multi-wavelength generation occurred

* Corresponding author: *amirazakiahmalek@gmail.com and **azuramalini@unimap.edu.my

 DOI: 10.1051/epjconf/201716201006

when only 5 BS signals were extracted at 50% output coupling ratio.

A linear cavity of BFL with multiple wavelength outputs was also reported in [10]. The optimization of output coupling ratios was varied from 70% to 99% in order to control the amount of light propagation. The effect of 95:5 coupling ratio produced up to 11 BS signals and anti-Stokes signals simultaneously with a line spacing of 0.08 nm. A maximum BP power of 11.7 dBm with 1550 nm BP wavelength was chosen to inject into the laser cavity. In this paper, as compared to our previous configurations [1, 8-10], we successfully reported the increment number of BS signals with an acceptable average optical signal to noise ratio (OSNR) value. The employment of a fiber Bragg grating's (FBG) reflectivity offers a direct impact to produce better lasing performance in a ring cavity. The reflectivity and selective wavelengths of FBG enhance the SBS effect in the Brillouin gain medium to generate a higher nonlinear effect with 17 BS signals and 14.7 dB of average OSNR value. A constant line spacing of 0.08 nm was also observed.

2 Experimental Setup

The architecture of ring cavity MWBFL with the implementation of a reflective FBG is illustrated in Fig. 1. It consists of an external tunable laser source (TLS), a unit of erbium-doped fiber amplifier (EDFA), 8 km of single mode fiber (SMF) that acts as a Brillouin gain medium, 3-port of optical circulator, 2x2 optical coupler and 3 dB bandwidth of 5 nm FBG with 70% reflectivity is used as another fiber loop mirror. Based on the FBG's properties, it only reflected the 5 nm wavelengths back at the centre wavelength of 1550 nm, while others wavelengths would be filtered and eliminated from the laser system.

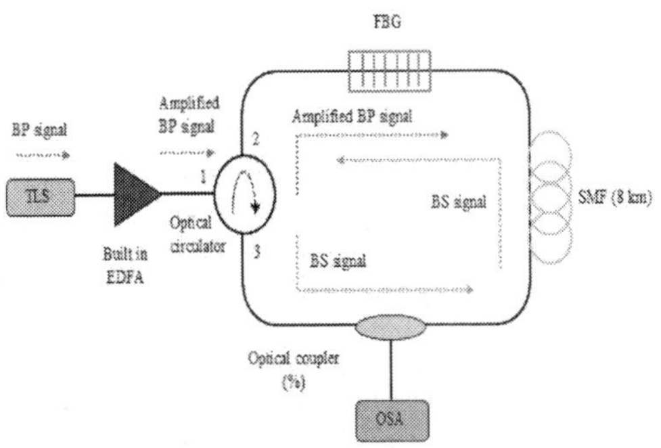

Fig. 1. The proposed ring cavity MWBFL system with the implementation of a reflective FBG.

In this experiment, the main nonlinear phenomenon called as SBS effect initiated in the SMF once the intensity of Brillouin pump (BP) signal is sufficient enough to excite the atoms. Without sufficient high BP power from the external TLS, the BP signals would be absorbed by the EDFA instead of being amplified by it. After being amplified, the high BP signals oscillated into the port-2 via port-1 of the optical circulator. From port-2 of the optical circulator, the amplified BP signals propagated directly into the FBG in order to remove the unwanted wavelengths according to the FBG's properties. About 30% of amplified BP signals continued to transmit into the SMF while another 70% of amplified BP signals reflected in the opposite direction.

In the SMF, the SBS effect only generated when the Brillouin threshold condition is achieved. The first downshifted BS signal would propagate in the opposite direction of amplified BP signals and enter the FBG for the reflection process. Later, the first BS signal oscillated to the optical circulator and circulated back into the FBG and SMF to complete clockwise and anti-clockwise light propagation. This propagation continues until the power of subsequent BS signals is high enough to overcome the Brillouin threshold power. In other words, the first BS signal acts as a BP power to generate the second BS signal, and this propagation continues as long as the Brillouin gain is equal or greater than the cavity loss. This MWBFL system has a line spacing of approximately 0.08 nm which depends on the type of optical fiber (SMF). The output BS signal is extracted from the optical coupler form 10% to 60% and is monitored by using an optical spectrum analyzer (OSA) with 0.02 nm resolution bandwidth.

3 Results and Discussion

The corresponding optical spectrum of FBG's reflection profile was investigated as shown in Fig. 2. The 3 dB bandwidth of 5 nm is utilized in order to enhance the signal quality by having 70% of reflectivity. The improvement of light intensity generated within 1547 nm to 1553 nm. There is a stop band of wavelengths in which filtered certain wavelengths from propagated into the Bragg grating. Thus, only desired wavelengths that fulfilled the Bragg's condition are strongly back-reflected.

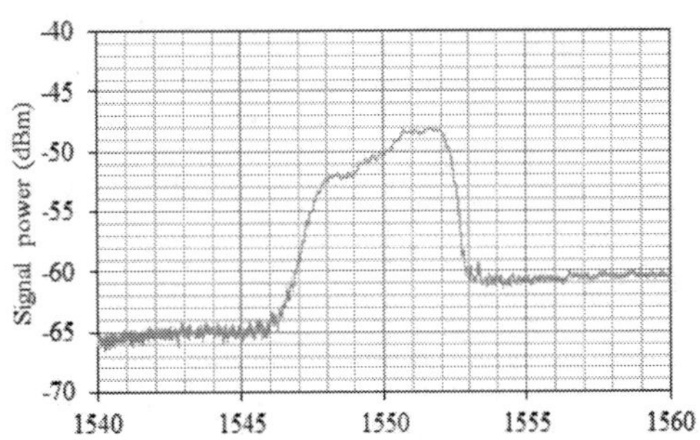

Fig. 2. The optical spectrum of FBG's reflection profile.

As the input light of 15.5 dBm with a broad spectrum was propagated along the core of the optical fiber, there was only 1547 nm to 1553 nm that back-reflected light at maximum reflectance at Bragg wavelength, λ_B.

DOI: 10.1051/epjconf/201716201006

Meanwhile, the transmitted light would eliminate from the spectrum due to Bragg's condition. It should be noted that the light propagation in the FBG was not only occurred in the back-reflected direction but also occurred in the forward direction.

Next, the effect of amplified BP power on the generation of multi-wavelength was investigated as depicted in Fig. 3. In this investigation, the amplified BP power is considered as the input signal after being amplified by the EDFA. The amplified BP power was varied from 12.0 dBm to 16.5 dBm whereas the BP wavelength was fixed at 1550 nm and 60% of the output coupling ratio was used to extract the output signal. Referring to Fig. 3, the number of BS signals increases as the amplified BP power gradually increases. This condition can be explained as the EDFA provides high Brillouin gain for initiating the maximum SBS effect. Due to the increment of the amplified BP power, the SBS effect generates efficiently to produce the downshifted BS signal in the opposite direction and thus more multi-wavelength generation was observed.

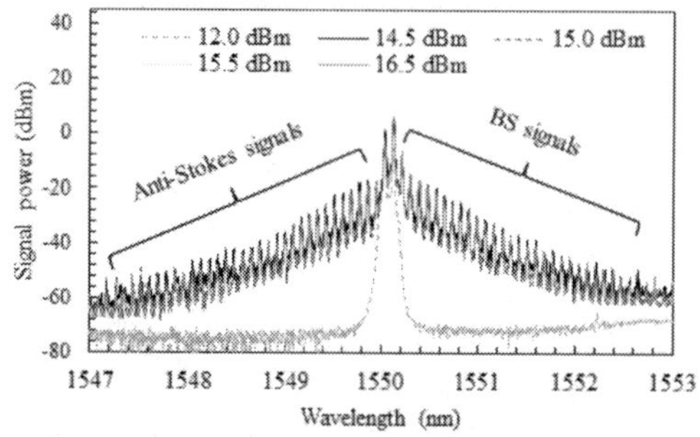

Fig. 3. Output spectra for different amplified BP powers at 1550 nm of BP wavelength by utilized 60% of the output coupling ratio.

At 12.0 dBm of amplified BP power, the optical spectrum indicates only 1 BS signal because of low injected input power. The BS signal generated once the Brillouin threshold power is satisfied [11]. In this case, at 12.0 dBm of amplified BP power, the laser system is considered as around Brillouin threshold condition. When the amplified BP power was increased further, it was found that the generation of multi-wavelength was started to generate with 8 BS signals at 14.5 dBm. It was noticed that the amount of power at each of the BS signals were dropped due to low Brillouin gain. By increasing the amplified BP powers from 15.0 dBm to 16.5 dBm, more multi-wavelength were generated. The most generated BS signals of 17 were found at maximum amplified BP power of 16.5 dBm. Besides that, it was also noticed that the anti-Stokes signals were generated at the left side of the optical spectra for 14.5 dBm to 16.5 dBm. The formation of anti-Stokes signal happened based on the interaction of multiple four-wave mixing (FWM) process. The unidirectional propagation in the laser system produced the anti-Stokes signal where the BS signal interacts with a pump beam to create other new signals (anti-Stokes signal). The number of anti-Stokes signals is lower compared to BS signals because the FWM process depends on the BS signals. The anti-Stokes signals increases as the BS signals increases.

The relationship of number BS signal with different amplified BP powers by utilizing 8 km of SMF is illustrated in Fig. 4. In this study, the BS signal experienced a flat trend before having a dramatic growth for all proposed output coupling ratios. At amplified BP powers of 12.0 dBm to 13.5 dBm, only 1 BS signal was recorded for all case output coupling ratios. At amplified BP power of 14.0 dBm to 16.5 dBm, the number BS signals were dramatically increased. This observation has resulted because of the higher Brillouin gain property accumulated higher SBS effect. This effective Brillouin gain medium provides maximum light intensities to lase the MWBFL system. Thus, this oscillated light resulted in the increment of the generated BS signals. The highest BS signal of 17 was recorded at 16.5 dBm of amplified BP power. In terms of output coupling ratio factor, the BS signals improved as the coupling ratios were increased. This also corresponds to the higher output coupling ratio influenced the increment of cavity loss. Therefore, the oscillating energy experienced inside the ring cavity becomes lesser. As a result, the lowest number BS signal of 4 was found at 10% output coupling ratio with 16.5 dBm of amplified BP power. Meanwhile, the highest number BS signal of 17 was attained at 60% optimum coupling ratio by launching 16.5 dBm of amplified BP power.

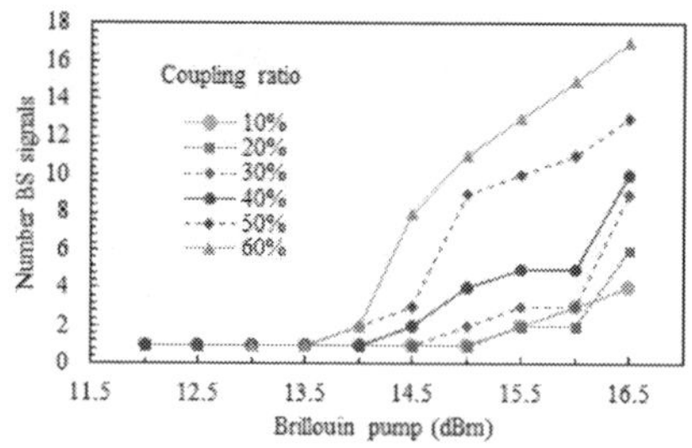

Fig. 4. Number BS signals as a function of amplified BP powers at 1550 nm of BP wavelength.

Fig. 5 depicts the average of OSNR with different amplified BP powers. Due to the effect of additional FBG's reflection, the average of OSNR gradually increases as the increment of amplified BP powers and output coupling ratios. The sufficient amount of amplified BP power that shared at each of the BS signals was the main factor that affected this abrupt change. At high amplified BP power, the laser system provided a high Brillouin gain which results in increasing of the OSNR. Therefore, high amplified BP power was required to supply the input power to reduce the noise level of the MWBFL system in which led to increases of OSNR. Moreover, the higher average OSNR is also owing to the implementation of additional FBG in which enhances the signal quality by having a reflection process. In this experiment, the average of OSNR was measured by comparing the peak power at each of the BS signals to the highest noise floor level of the signal [12]. The deduction of both signals is defined as the OSNR. Since the MWBFL system produced a multi-wavelength generation, the OSNR for each of the BS signals was calculated and then divided into the total number BS signals in order to calculate the average OSNR. Overall, the lowest average of OSNR was obtained around 11.1 dB for 12.0 dBm of amplified BP power at 10% of the output coupling ratio. Meanwhile, a higher average of OSNR about 14.7 dB was achieved for 16.5 dBm of amplified BP power and 60% of the output coupling ratio.

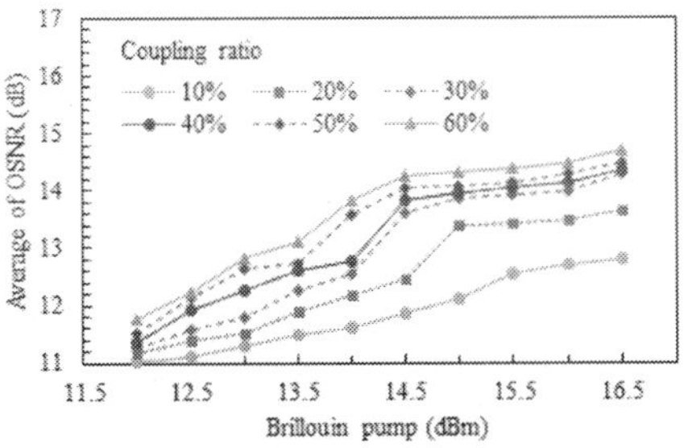

Fig. 5. Average of OSNR with different amplified BP powers for all cases output coupling ratios.

4 Summary

We have successfully demonstrated a ring cavity MWBFL system with the implementation of a reflective FBG that can be used at 1550 nm in the optical communication window. At amplified BP power of 16.5 dBm with a corresponding 60% of the output coupling ratio, 17 BS signals and 14.7 dB average of OSNR were generated by the combination process of multiple SBS and FWM effect. The output BS signals were constantly separated by 0.08 nm. The implementation of FBG inside the ring cavity can effectively be used in the laser system especially in the Brillouin fiber laser to increase the number BS signals. Moreover, due to the reflection process of FBG, an acceptable average of OSNR was recorded which could be used for several potential applications in the optical communication. For future work, a high reflection of FBG can be utilized to improve the average OSNR in order to meet the requirements of high capacity optical communication networks.

This work was fully supported by the Ministry of Higher Education, Malaysia under research grant # FRGS/9003-00532#. The authors would like to thank the School of Microelectronics Engineering, Universiti Malaysia Perlis (UniMAP) especially SPILS for their support in this work.

References

1. M. R. Shirazi, M. Biglary, S. W. Harun, K. Thambiratnam and H. Ahmad, J. of Optic A: Pure and Applied Optics **10**, 5 (2008).
2. A. Aloisio, F. Cevenini and V. Izzo, IEEE Trans. Nucl. Sci. 51, 526-531 (2004).
3. Y. Shen, X. Zhang and P. A. Robert, J. Lightw. Tech., **23**, 1860-1865 (2005).
4. Y. Shen, X. Zhang and K. Chen, J. of Lightw. Techn. **23**, 1860 (2005).
5. S. J. Tan, S. W. Harun, N. M. Ali, A. Ismail and H. Ahmad, IEEE J. of Quantum Electro., **49**, 595-598 (2013).
6. M. Ali. Toor, N. A. M. Ahmad. Hambali, A. Mansoor, M. Ajiya and Z. Yusoff, J. of Optics **17**, 025502 (2015).
7. M. P. Fok and C. Shu, Optics Exp., **14**, 2618-2624 (2006).
8. M. R. Nurdik, N. S. Abdul Rahim, and M. K. Abd Rahman, Brillouin fiber laser with 200-nm tuning range, *IEEE 2nd Internatinal Conference on Photonics (ICP)*, p1-4 (2011).
9. B. A. Ahmad, A. W. Al-Alimi, A. F. Abbas, M. Mokhtar, S. W. Harun and M. A. Mahdi, IEEE Photonics J., 1087-1094 (2012).
10. M. R. Shirazi, N. S. Shahabuddin, S. N. Aziz, K. Thambiratnam, S. W. Harun and H. Ahmad, Laser Phys. Let., **5**, 361-363 (2008).
11. B. A. Ahmad, Gh. Mamdoohi, A. F. Abas, M. Mokhtar and M. A. Mahdi, *International Conference on Photonics (ICP)*, 195-198 (2012).
12. A. W, Al-Alimi, M. H. Yaacob, A. F. Abas, M. A. Mahdi, M. Mokhtar and M. H. Al-Mansoori, IEEE Photonics J., **5**, (2013).

DOI: 10.1051/epjconf/201716201007

Non-Halal biomarkers identification based on Fourier Transform Infrared Spectroscopy (FTIR) and Gas Chromatography-Time of Flight Mass Spectroscopy (GC-TOF MS) techniques

Gunawan Witjaksono[1,*], *Irwan* Saputra[2], *Marsad* Latief[3], *Irwandi* Jaswir[2], *Rini* Akmeliawati[3], and *Almur* Abdelkreem Saeed Rabih[1]

[1]Department of Electrical and Electronic Engineering, Universiti Teknologi PETRONAS, Seri Iskandar, 32610, Perak Darul Ridzuan, Malaysia
[2]International Institute for Halal Research & Training (INHART), International Islamic University Malaysia (IIUM)
[3]Department of Mechatronics Engineering, International Islamic University Malaysia (IIUM)
P.O Box 10, 50728 Kuala Lumpur, Malaysia

Abstract. Consumption of meat from halal (lawful) sources is essential for Muslims. The identification of non-halal meat is one of the main issues that face consumers in meat markets, especially in non-Islamic countries. Pig is one of the non-halal sources of meat, and hence pig meat and its derivatives are forbidden for Muslims to consume. Although several studies have been conducted to identify the biomarkers for non-halal meats like pig meat, these studies are still in their infancy stages, and as a result there is no universal biomarker which could be used for clear cut identification. The purpose of this paper is to use Fourier Transform Infrared Spectroscopy (FTIR) and Gas Chromatography- Time of Flight Mass Spectroscopy (GC-TOF MS) techniques to study fat of pig, cow, lamb and chicken to find possible biomarkers for pig fat (lard) identification. FTIR results showed that lard and chicken fat have unique peaks at wavenumbers 1159.6 cm^{-1}, 1743.4 cm^{-1}, 2853.1 cm^{-1} and 2922.5 cm^{-1} compared to lamb and beef fats which did not show peaks at these wavenumbers. On the other hand, GC/MS-TOF results showed that the concentration of 1,2,3-trimethyl- Benzene, Indane, and Undecane in lard are 250, 14.5 and 1.28 times higher than their concentrations in chicken fat, respectively, and 91.4, 2.3 and 1.24 times higher than their concentrations in cow fat, respectively. These initial results clearly indicate that there is a possibility to find biomarkers for non-halal identification.

1 Introduction

Authenticity of food has become a paramount concern for both food industries and consumers [1]. The food industries have legal compliance toward selling the right food for the right price, maintaining well-defined quality and using safe ingredients [2]. On the other hands, for consumers transparency in food specification is an ever increasing demand and essential due to religion credence, hard-earned fortunes [3] and some people's allergy to some types of foods [4]. For raw meat markets, the common practice is that consumers rely on their visual inspection to differentiate between meat species. However, in most cases meat types differ subtly in the visual appearance, which make it more difficult to differentiate [5]. Furthermore, in case of mixed and processed meat the differentiation based on appearance will be quite impossible [6]. Regulatory agencies as a third party also concern about the food authenticity due to that food adulteration could bring revenue losses to the state. Research scientists have been conducting continuous studies to support the regulatory and commercial initiatives to develop analytical techniques to detect the fraudulent foods and the ingredients that breach the labelled marks [7]. From the religious perspective, most religions in the world urge their followers to comply with certain dietary laws. For instance, Islam encourages Muslims to consume only Halal (lawful) foods, which exclude some types of food such as pig meat and its derivatives like lard, ham, bacon, pork and sausage, except in very extreme conditions. Halal food is expected to reach 20% of the world trade as the Muslims are expected to make 30% of the world populations by the year 2025 [8]. Judaism has a defined dietary law (kashrut) for the Jewish to follow, which also prohibits pig meat and its derivatives [9], and in Hindu religion consumption of beef and its products is not allowed [10]. In many countries, food manufacturers use lard to adulterate oil ingredients due to the cheap price of lard [11]. In 2013, there was a big scandal that hit Europe, where beef had been replaced by horsemeat, and the later had been sold with beef labelling. In addition to Europe, Asia also witnessed scandals in Malaysia where chocolate and pharmaceutical preparations were found to contain pork traces in 2013, and 2014, respectively [12]. In some other Asian

* Corresponding author: gunawan.witjaksono@utp.edu.my

countries like India it is common to find buffalo meat mixed with other meat species due to its lower cost and availability, which will be very hard to identify especially if it is mixed with beef or if it is the processed form [13]. Beside meat markets, some manufacturers of dairy adulterate butter with some cheap fats like chicken fats, lard and mutton fats to gain more profits [14]. Due to aforementioned issues, food authentication has become a paramount concern for everyone. Various methods have been developed for food species identification. Some of these methods include protein based methods like electrophoretic techniques, chromatographic, immune assays and DNA as a molecular detection method [15]. In addition, electronic nose, polymerase chain reaction and spectroscopic techniques such as ultraviolet (UV), near infrared (NIR), mid infrared (MIR), visible light, and Raman spectroscopy are widely used [16]. IR based techniques like Fourier Transform Infrared Spectroscopy (FTIR) is reliable for meat sample identification due to its capability of using as fingerprint technique when it is combined with powerful chemometric techniques [17]. This paper reports a study on Pig fat (lard), chicken fat, cow fat and lamb fat using FTIR and Gas Chromatography-Time of Flight Mass Spectroscopy (GC-TOFMS) to find non-halal biomarkers for halal identification.

2 Methods and Samples Preparation

Two methods were used to study the non-halal biomarkers. Firstly FTIR was used, and then the results were confirmed by using GC-TOFMS.

2.1. Fourier Transform Infrared Spectroscopy (FTIR)

Infrared (IR) spectroscopy is a rapid, non-destructive, sensitive and easy sample preparation technique that is being used for many years in food sector for analysis and classification of foods. It is used either in near infrared (NIR) or in mid infrared bands (MIR) [1]. The advantage of using Infrared (IR) techniques for meat samples identification is that they provide much information, with a unique absorption bands for the functional groups that unique to the meat sample studied [3]. Fourier Transform Infrared Spectroscopy (FTIR) is a modern technique derived from the conventional IR spectroscopy, and it works in the MIR region (400 – 4000 cm^{-1}). It has a powerful capability for food analysis and investigating its individual components due to the high energy throughput, excellent wavenumber reproducibility and accuracy [18].

2.1.1 FTIR meat samples preparation and fats extraction

Total sixteen meat samples from pig, chicken, lamb and cow were collected from a local slaughterhouse at Gombak Market in Malaysia. Each meat type contains five samples. The preparation started firstly by washing the samples by distilled water to remove any contamination that might stuck on the surface of the meat samples. Then, the meat samples were cut into small element (1 cm x 1 cm) and kept at -20°C until they are used for the fat extraction process to obtain fats of the different samples. Pig fat (Lard) and chicken fat, lamb fat, and cow fat were extracted by rendering the adipose tissues according to the method described by Che Man et al [19]. All chemicals used in this experiment were of analytical grade. The pure extracted fats were then analyzed by means of FTIR spectroscopy.

2.1.2 GC-TOF MS meat samples preparation and fats extraction

For this study total nine meat samples of pig, chicken and cow meats were collected, by having three samples for each meat type. Preparation of the meat samples and also the extraction of the fats have been conducted using the same method mentioned above for the FTIR measurements. After extracting the pure fats, each fat sample (50 mg) were dissolved in 0.8 mL hexane. The mixture was then mixed for 1 min using a vortex mixer, and then stored in the dark at -18 °C prior to GC–TOF MS analysis.

3 Results and discussion

3.1 FTIR Results

Nicolet iS50 FTIR Spectrometer was used to acquire the full spectrum in the mid infrared region (400 – 4000 cm^{-1}). The number of scans was fixed to 32 with a resolution of 4 cm^{-1}. The measurement was calibrated against the background air. The whole FTIR spectrum which corresponds to the stretching of the functional groups that present in the fat of the species [17] studied is shown in Fig.1. The graph shows the average spectrum of three samples for each of lard, chicken fat, cow fat and lab fat.

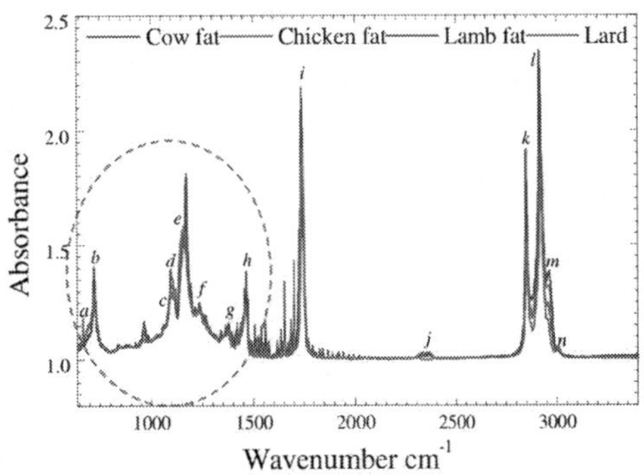

Fig. 1. FTIR Spectra of lard, chicken fat, cow fat and lmb fat

From the spectra it was found that lard is different from the other fats in some peaks. The spectra shown in **Fig.1** could be divided into two regions: the higher

DOI: 10.1051/epjconf/201716201007

wavenumbers of more than 1500 cm^{-1} and the finger print region which is the region below 1500 cm^{-1}.

3.1.1 FTIR peaks in Region one: >1500 cm^{-1}

At the wavenumbers higher than 1500 cm^{-1}, lard is different from the other fats in term of the absorbance at the peaks *n*, *m* and *j* as shown in **Fig. 1**. **Fig. 2** shows the closer look at the points *n* and *m*. in the point m (shown by the red-dotted circle in **Fig. 2**) lard has shown a small shoulder, while chicken has a broader shoulder.

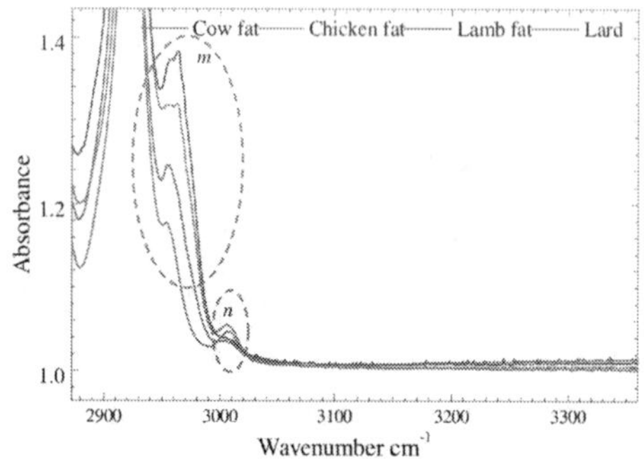

Fig. 2. FTIR Spectra of lard, chicken fat, cow fat and lamb fat at *m* and *n* regions

On the other hands, cow fat exhibited single peak and lamb fat has demonstrated double peaks. At the region referred by *n* only lard and chicken fats have shown peaks, while cow and lamb fats have no peaks. Lard is again different from the other fats in the region marked by *j* in **Fig. 1**. At his region lard has shown an inverted peak as explained further in **Fig. 3**.

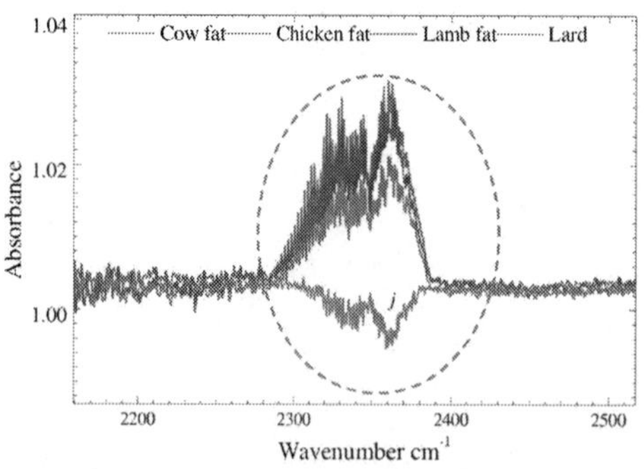

Fig. 3. An inverted peak for lard at wavenumber 2361.05 cm^{-1}

3.1.2 FTIR peaks in Region two: <1500 cm^{-1}

For the wavenumbers less than 1500 cm^{-1}, lard is different from the other fats at the peak *a*, *c*, *d*, *e*, and *h* as shown in Figure 4. However chicken also exhibited same peaks except at the peak denoted by *a*, which is unique for lard. Hence at the fingerprint region *a* could be the most suitable peak number to be used for lard identification.

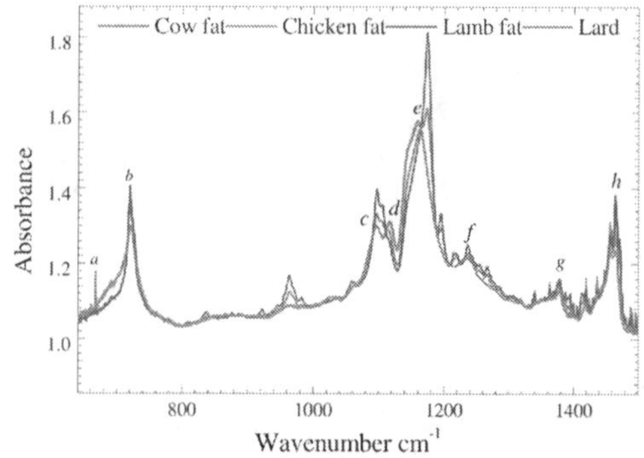

Fig. 4. Fingerprint region (wavenumber <1500 cm^{-1}) for lard identification

Fig. 5 shows representing the absorbance of the lard, cow, lamb and chicken fats using different reference levels. For clarity of differentiation in the graph only cow fat has the same original reference level, while chicken reference level is raised one level above the cow fat reference by adding 1 to the original absorbance values of the chicken fat, lamb fat is two levels above the cow fat reference by adding 2 to the original absorbance values of the lamb fat, and finally lard is three levels above the original reference line by adding 3 to the original absorbance values of lard.

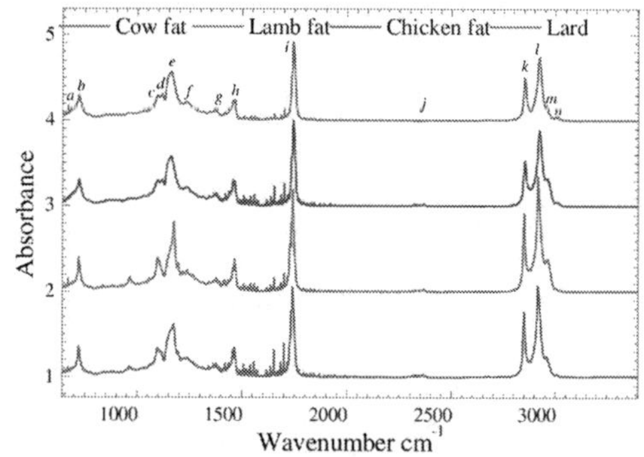

Fig. 5. FTIR Spectra of lard, chicken fat, cow fat and lamb fat

3.2 Gas Chromatography-Time of Flight Mass Spectroscopy (GC-TOF MS) Results

Techniques based on chromatography offer rapid and reliable sample separation for compounds analysis. GC-TOF MS is known as a powerful tool for the analytical analysis of complex samples [1]. For this research Agilent 7693 B GC coupled with TOF MS was used with hp-5ms column. The study was conducted on total nine samples (three samples each) of lard, cow, lamb and chicken fats to analyze the aromatic hydrocarbons. Each fat type contains three samples. Table 1 shows the obtained average (for three samples) area covered by each hydrocarbon which represents the composition weightage for the different fat types.

Table 1. Aromatic hydrocarbons compositions for lard, chicken and cow fats

Hydrocarbon compound	Area %		
	Lard	Chicken fat	Cow fat
2,4-Imidazolidinedione, 5-[3,4-bis[(trimethylsilyl)oxy]phenyl]-3-methyl-5-phenyl-1-(trimethylsilyl)-	0.08299	0.04853	0.141035
Benzene, 1,2,3-trimethyl-	21.33433	0.085155	0.233378
Benzene, 1-methyl-3-(1-methylethyl)-	0.597023	0	0
Benzene, 1-methyl-4-(1-methylpropyl)-	0.013952	0	0.018403
Benzene, 1-methyl-4-propyl-	0.787343	0	0
Benzene, 2-ethyl-1,4-dimethyl-	0.374043	0.432713	0.068181
Decane	21.33433	0	21.167
Decane, 4-methyl-	1.286363	0	0.95659
Hexanedioic acid, bis(2-ethylhexyl) ester	0.583767	0	0
Indane	0.125046	0.008597	0.054052
Naphthalene, 1,2,3,4-tetrahydro-2-methyl-	0.055843	0	0
Nonane, 2-methyl-	0.098341	0	0
Octane, 2,3,7-trimethyl-	0.037965	0	0
p-Cymene	0.447551	0.116017	0
Tridecane	0.22617	0	0
Undecane	10.3596	8.062533	8.382467
Benzene, (2-methyloctyl)-	0	2.472467	2.657267
Benzene, 1-ethyl-4-methyl-	0	0.035777	0.102822
Benzene, 1-methyl-3-propyl-	0	0.680724	0
Cyclohexane, butyl-	0	0.529553	0
Naphthalene, 1,2,3,4-tetrahydro-	0	0.06464	0.138087
o-Cymene	0	0.121686	0
1-Dodecanol, 3,7,11-trimethyl-	0	0	0.222381
Benzene, 1-ethyl-2,4-dimethyl-	0	0	0.062808
Benzene, 1-ethyl-3-methyl-	0	0	0.110763
Heptacosane	0	0	0.056515
Nonane, 2-methyl-	0	0	0.062937
Squalene	0	0	0.342603
Benzene, 4-ethyl-1,2-dimethyl-	0	0.541307	0

From Table 1 it can be seen that lard can be differentiated from the other fats in many hydrocarbon compositions. Following are some of the hydrocarbons that have higher percentage area in lard as compared to the other fats: Benzene, 1, 2, 3-trimethyl-; Benzene, 1-methyl-3-(1-methylethyl)-; Benzene, 1-methyl-4-propyl-; Hexanedioic acid, bis (2-ethylhexyl) ester; p-Cymene; Tridecane; Undecane. By using chemometric analysis techniques such as PCA or PLS these data could be analyzed to cluster and differentiate the lard from the other fats.

4 Conclusion

This paper demonstrated identification of lard discrimination from cow, chicken and lamb fats. Two methods based on spectroscopy (FTIR) and chromatography (GC-TOF MS) techniques were used for the measurement and analysis of these fats discrimination. FTIR results have shown that lard is different from the other fats by having a peak at a wavenumber 668.1cm^{-1}, where the rest of the fats do not have peak at this point. In addition, in the range of wavenumber 2950 – 2990 cm^{-1} lard exhibited shoulder peak, while cow and lamb fats have shown sharp peak, and chicken has shown double peaks. An inverted peak was observed for the lard at the wavenumber 2361.05 cm^{-1}in contrary to the normal peaks seen for the rest of the fats. GC-TOF MS results have confirmed that lard is different for the other fats in terms of the volatile hydrocarbon compounds, where compounds such as Benzene, 1-methyl-3-(1-methylethyl)-; Hexanedioic acid, bis(2-ethylhexyl) ester and p-Cymene have shown very high compositional percentage in lard compared to other fats.

The authors of this paper acknowledge the financial support by Universiti Teknologi PETRONAS (UTP) under STIRF-UTP, fund/ project code: 0153AA-F71
Ministry of Higher Education, Malaysia and International Islamic University Malaysia (IIUM) under the scheme of MyRA INCENTIVE RESEARCH GRANT SCHEME (MIRGS), Project code: MIRGS13-01-002-0002.

References

1. A. Rohman and Y. B. C. Man Food Reviews International, **18**, pp. 97-112, (2013).
2. W. Kamm, F. Dionisi, C. Hischenhuber, and K.-H. Engel, Food Reviews International, **17**, pp. 249-290, (2001).
3. E. Kurniawati, A. Rohman, and K. Triyana, Meat Science, **96**, pp. 94-98, (2014).
4. B. Mamikoglu, Otolaryngology - Head and Neck Surgery, **133**, pp. 534-537, (2005).
5. N. Z. Ballin, F. K. Vogensen, and A. H. Karlsson, Meat Science, . **83,** pp. 165-174, (2009).
6. S. A. Sarah, W. N. Faradalila, M. S. Salwani, I. Amin, S. A. Karsani, and A. Q. Sazili, Food Chemistry, **199**, pp. 157-164, (2016).
7. G. Downey, Advances in Food Authenticity Testing, 1st Edition ed.: Woodhead Publishing, (2016).
8. H. I. Abdullah Amqizal, H. A. Al-Kahtani, E. A. Ismail, K. Hayat, and I. Jaswir, Food Control, **78**, pp. 297-303, (2017).
9. Schröder and M. J.A., Food Quality and Consumer Value: Delivering Food that Satisfies, (2003).
10. K. Bonne and W. Verbeke, Agriculture and Human Values, **25**, pp. 35-47, (2008).
11. A. A. Aida, Y. B. Che Man, C. M. V. L. Wong, A. R. Raha, and R. Son, Meat Science, **69**, pp. 47-52, (2005).
12. C. Flaudrops, N. Armstrong, D. Raoult, and E. Chabrière, Journal of Food Composition and Analysis, **41**, pp. 104-112,(2015).
13. B. G. Mane, S. K. Mendiratta, A. K. Tiwari, and K. N. Bhilegaokar, Food Analytical Methods, **5**, pp. 296-300,(2012).
14. A. F. Nurrulhidayah, S. R. Arieff, A. Rohman, I. Amin, M. Shuhaimi, and A. Khatib, International Food Research Journal, **22**, pp. 832-839, (2015).
15. W. Xiang, Y. Shang, Q. Wang, Y. Xu, P. Zhu, K. Huang, et al., Food Chemistry, **234**, pp. 472-478, (2017).
16. L. M. Reid, C. P. O'Donnell, and G. Downey, Trends in Food Science & Technology, **17**, pp. 344-353, (2006).
17. H. Rahmania, Sudjadi, and A. Rohman Meat Science, **100**, pp. 301-305,(2015).
18. A. Bendini, L. Cerretani, F. D. Virgilio, P. Belloni, M. Bonoli-Carbognin, and G. Lercker, Journal of Food Quality, **30**, pp. 424 - 437, (2007).
19. Y. B. Che Man, Z. A. Syahariza, M. E. S. Mirghani, S. Jinap, and J. Bakar, Food Chemistry, **90**, pp. 815-819,(2005).

 DOI: 10.1051/epjconf/201716201008

Influence of electromagnetic (EM) waves polarization modes on surface plasmon resonance

Noor Faezah Murat[1], *Wan Maisarah* Mukhtar[1*], *P Susthitha* Menon[2], *Affa Rozana* Abdul Rashid[1], *Karsono* Ahmad Dasuki[1] and *Awangku Abdul Rahman* Awangku Yussuf[1]

[1]Faculty of Science and Technology, Universiti Sains Islam Malaysia, 71800 Nilai Malaysia
[2]Institute of Microengineering and Nanoelectronics (IMEN), Universiti Kebangsaan Malaysia, 43600 Bangi Malaysia

Abstract. Optical sensor based on surface plasmon resonance are widely used for various applications. In this paper, an optimization of EM waves polarization modes for the occurrence of SPR by using a modified optical waveguiding assembly has been discussed. A 633 nm of linearly polarized He-Ne laser was appointed as a light source. The polarization modes were controlled by rotating the position of He-Ne laser. It was found that the optimum SPR signal with reflectance, R = 0.21 a.u and SPR angle, $\theta_{SPR} = 48°$ can be obtained as the position of laser was rotated at 90°, with the presence of polarizer. The location of the critical angle was observed at $\theta_C = 46°$. For the validation purpose, a simulation based on Fresnel equation was conducted where the percentage difference between the experimental and theoretical results were analyzed. The percentage difference between experimental and theoretical results for both θ_{SPR} and θ_C were 9.09% and 15.00% respectively. In conclusion, an excellent agreement between the theoretical and experimental findings proved the stability of this SPR setup, where the EM waves polarization can easily be controlled by manipulating the position of laser.

1 Introduction

Surface plasmon resonance (SPR) technique has been used as one of the optical detection due to its powerful device for monitoring interactions in biological system [1]. The application also had been developed widely in the past two decades in areas such as environmental protection, biotechnology, chemical analytes, food safety and security [2, 3]. Surface plasma wave (SPW) is an electromagnetic excitation existing on the boundary of a dielectric and a metal [4].

SPR is an optical method for measuring the refractive index of very thin layers of material adsorbed on a metal [5]. This can be concluded that SPR is a phenomenon that involving the interaction of matter and polarized light while surface plasmon polariton (SPP) is an excitation of photon in SPR. It can detect the small change of the refractive index on the metal surface by showing a sharp dip of curved which represent the minimum reflectance in SPR phenomenon [6]. The sharp dip of curve is called the SPR curve. It happens when p-polarized light strikes a thin-film metal under the condition of total internal reflection (TIR) [7].

The main objective of this research is to study the influence of EM waves polarization modes on SPR by controlling the position of linearly polarized He-Ne laser. SPR experimental assembly consists of laser with wavelength of 633 nm, a polarizer, a convex lens, a prism, and a rotational stage [8].

2 SPR Phenomenon

SPR phenomenon is a coupled electromagnetic field charge density oscillation that occurs on the boundary of metal-dielectric which have different sign (positive sign: dielectric, negative sign: metal) of dielectric constant as shown in Fig. 1 [9]. Among the numerous possible metals that can be used are silver and gold. Silver has the sharpest SPR resonance peak while gold shows excellent surface stability are commonly employed [10].

Other than that, the SPR phenomenon also take place when the wave vector of the incidence beam and the surface plasmon are equal and the reflected light intensity becomes completely attenuated. The promising potential of SPR sensors is that they are very high sensitivity to a change in the refractive index of the sensing medium [11].

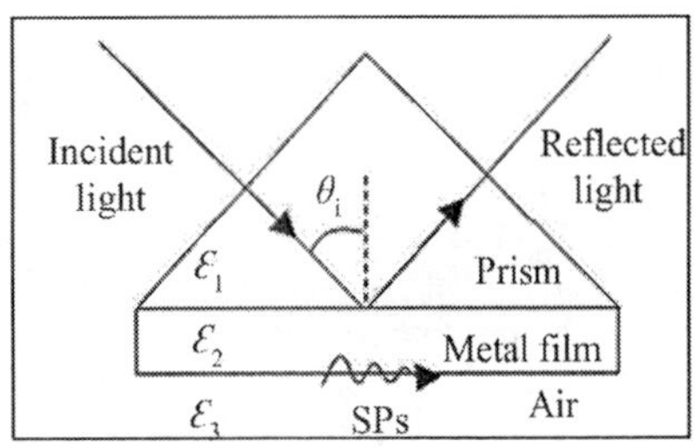

Fig. 1. SPR phenomenon [9].

* Corresponding author: wmaisarah@usim.edu.my

2.1 Maxwell Equation's

Maxwell Equation's describes light interaction with matter. It's also used to show that in order for a point charge to produce electromagnetic waves, charge must accelerate. Maxwell's equations in a medium are given by [12]:

$$\nabla.\vec{D} = \rho \quad (1)$$

$$\nabla.\vec{B} = 0 \quad (2)$$

$$\nabla \times \vec{E} = -\partial\vec{B} / \partial t \quad (3)$$

$$\nabla \times \vec{H} = J + \partial\vec{D} / \partial t \quad (4)$$

where is $\vec{D}$ the dielectric displacement, ρ is the density of charges, $\vec{B}$ is the magnetic field, $\vec{E}$ is the electric field, $\vec{H}$ is the magnetic field, and J is the current density. All the equations above represent Ampere's law, Faraday's law of induction, Gauss's law for electric and magnetic field respectively.

2.2 Evanescent Wave

Evanescent wave plays an important role in nano-optics field [13]. The evanescent wave does not transmit energy across the interface unless an absorbing or scattering species is presented in the lower index medium close to the boundary. This phenomenon happens when the wave strikes the medium at an angle greater than the critical angle, θ_c shown in Fig. 2 below [14, 15].

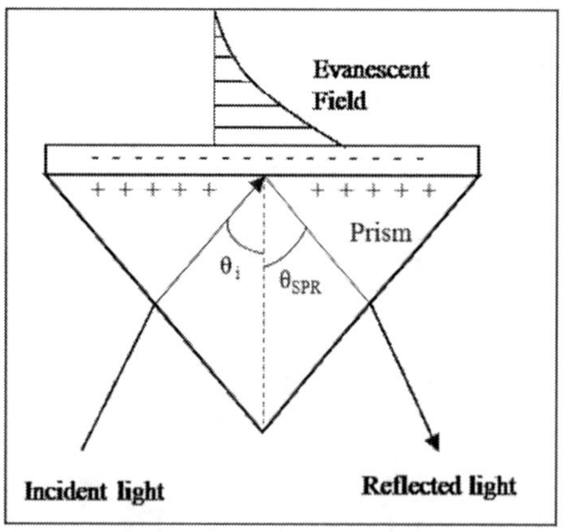

Fig. 2. The presence of evanescent field after light undergo total internal reflection [16].

2.3 EM Polarization Mode

Linear polarization can be classified into Transverse Magnetic mode (TM) or p-polarized and Transverse Electric mode (TE) or s-polarized with respect to the normal direction of the surface that is contained by the plane and the ray direction which is the propagation vector [17, 18]. To recognize the polarization modes either s-polarized (horizontal polarization) or p-polarized (vertical polarization) value of reflectivity is observed. For s-polarized light, at first, the reflectivity increases gradually with the increment of incident angles. As the angles approaching 90°, the value of reflectivity increases more rapidly. Meanwhile, for the p-polarized light the reflectivity increases after it declines to zero [19].

It is crucial to understand the characteristic of light polarization in order to create SPR phenomena. EM waves polarization modes is one of the important requirements for SPP propagation to establish and influence the SPR phenomena. Since SPP can only be excited in p-polarization mode, the presence of Brewster angle, θ_B able to indicate type of polarization mode [20]:

$$\theta_B = \tan^{-1}(n_2 / n_1) \quad (5)$$

where n_1 and n_2 are refractive indexes of two different media.

At $\theta_I = 0°$, p-polarization reflection is positive and has magnitude as pictured in Fig. 3. As θ_I increases, the magnitude decreases until it vanishes at the θ_B. As θ_I increases beyond θ_B, the reflection becomes negative and its magnitude increases until it reaches unity at the θ_C. The wave undergoes TIR when θ_I is greater than θ_C [20].

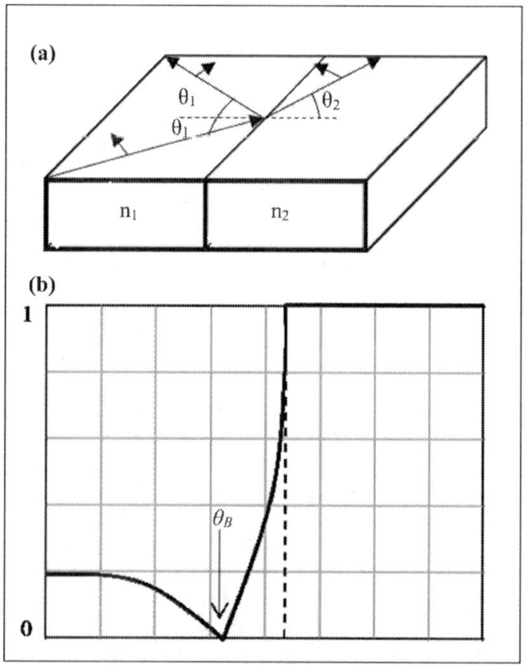

Fig. 3. (a) Magnitude of p-polarized waves for internal reflection **(b)** The magnitude vanished at the Brewster angle [20].

The power reflectance, $\Re$ and power transmittance, T are defined as the ratio of power flow (along a direction normal to the boundary) of the reflected and transmitted waves to that incident wave. Since reflected and incident waves propagate in the same medium and make the same angle with normal to the surface, it follows that $\Re = |r|^2$. The power reflectance at normal incidence for p-polarization at internal reflection is therefore [20]:

DOI: 10.1051/epjconf/201716201008

$$\Re = \left(\frac{n_1 - n_2}{n_1 + n_2} \right)^2 \quad (6)$$

3 Methodology

We started the experiment by building the SPR experimental setup. The setup consisted of few optical components as illustrated in Fig. 4. The linearly polarized He-Ne laser with wavelength of 633 nm (Brand: Newport) was used for the light excitation. A polarizer (Brand: SIGMA KOKI) was placed between the laser and mirror with the distance of 18 cm and 7 cm respectively. The triangular prism (refractive index of 1.51) was positioned on top of mini labjack (Brand: Newport) which act as prism holder with distance of 26.5 cm from the mirror. The photodetector (Brand: Newport) was used to detect the reflected power in constant distance of 13 cm from the prism. The reading of the reflected power was measured by the power meter (Brand: Newport) which connected to the silicon photodetector.

Two types of prisms were employed in this experiment which were 50 nm gold coated prism and bare prism (without gold layer on top of the surface) as shown in Fig. 5 [21, 22]. The function of bare prism is to determine the location of θ_C and θ_B. The presence of θ_B become an indicator of the propagation of p-polarized mode of the EM waves. Once the polarization modes were confirmed as p-polarized, the gold was deposited on the prism for the investigation of SPR signal.

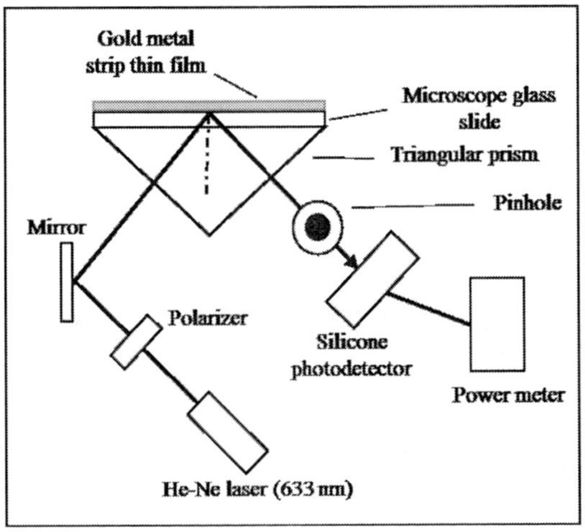

Fig. 4. Modified optical waveguiding assembly.

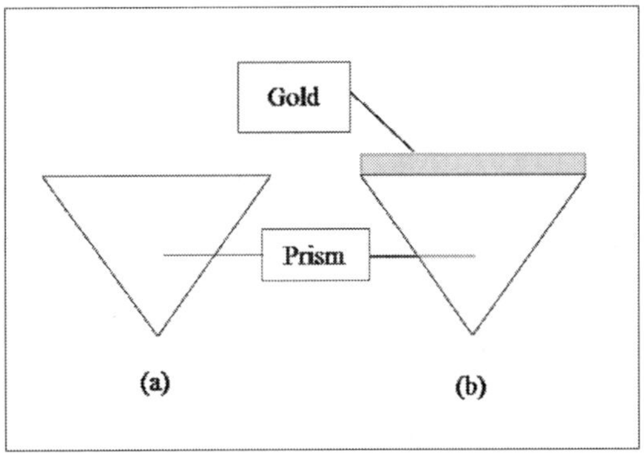

Fig. 5(a). Bare prism **(b)** gold-coated prism in Kretschmann configuration.

It is compulsory to excite p-polarization light as optical excitation of SPP since there are no solutions that exist for the case of s-polarization light [13]. In order to confirm that it was p-polarization light, the linearly polarized He-Ne laser source was rotated clockwise at three different angles such as at 0°/180°, 90°/270° and 45° as illustrated in Fig. 6.

The next procedure was to perform an angular interrogation technique by rotating the prism in counter clockwise direction with an increment of 2° for each reading. The range for the incident angles were between 20° < θ < 60° as illustrated in Fig. 7. The reading of power reflected was taken from power meter when photodetector was used to detect the reflected laser from the prism. The value of reflectance was calculated using the equation below:

$$R(a.u) = \frac{P_{reflected}(\mu W)}{P_{incident}(\mu W)} \quad (7)$$

where R is the reflectance, $P_{reflected}$ is the reflected power and $P_{incident}$ is the incident power [8].

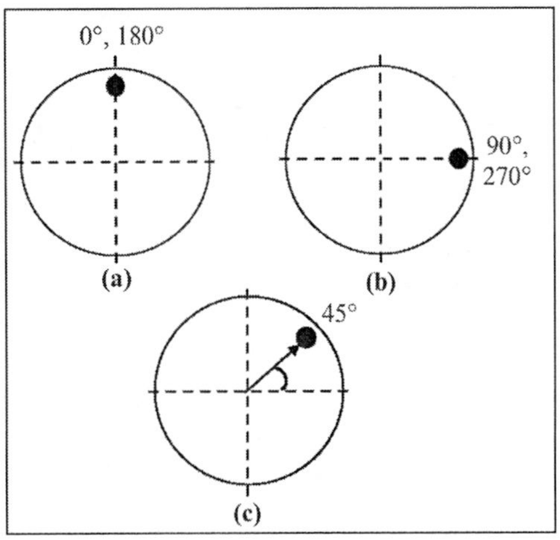

Fig. 6. Rotational angle of He-Ne laser.

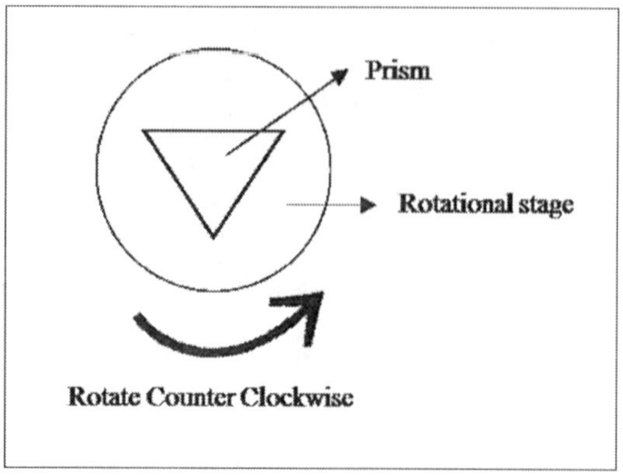

Fig. 7. Rotating the prism in counter clockwise direction.

The quality of the SPR signal was examined by calculating the FWHM of the SPR curve, as follow [22, 23]:

$$\Delta\theta = \theta_2 - \theta_1 \qquad (8)$$

where $\Delta\theta$ is FWHM, θ_1 and θ_2 are incident angle of SPR curve depth. As mentioned, FWHM represents the width of a reflection curve.

We performed a simulation by using WINSPALL 3.02 simulation developed by Knoll Group, Max-Plank Institute that based on a normalized Fresnel equation to compare the experimental and the calculated result. The results of the comparison represent the competency and accuracy of our SPR setup [22].

4 Results and Discussions

The occurrence of SPR can be observed as the position of laser was set at angle of 90°/270° (Fig. 6 (b)). As the position of laser were varied into 0°/180° (Fig. 6 (a)) and 45° (Fig. 6 (c)), SPR signals were not observed. These results show that the p-polarized light can be obtained as the position of laser was at 90°/270°.

Fig. 8 shows the comparison between Brewster angle and critical angle for both experimental and simulation approaches. Due to the absence of a thin film layer, SPR cannot be observed thus results in no penetration of the surface plasmon waves evanescently [2]. The Brewster angle for simulation and experimental were 26° and 30° respectively with percentage difference of 15.38%. The critical angle for simulation was 40° and red shifted for experimental which was 46° with percentage difference of 15%. We expected that the location of SPR dips should be beyond these angles since the excitations of SPP must occur above the critical angles [24]. Table 1 summarizes the theoretical and experimental results for Brewster angle and critical angle of simulation and experimental results.

Table 1. Values of Brewster angle and critical angle obtained by simulation and experimental approaches

Angle	Simulation	Experimental	Percentage Difference (%)
Brewster	26°	30°	15.38
Critical	40°	46°	15.00

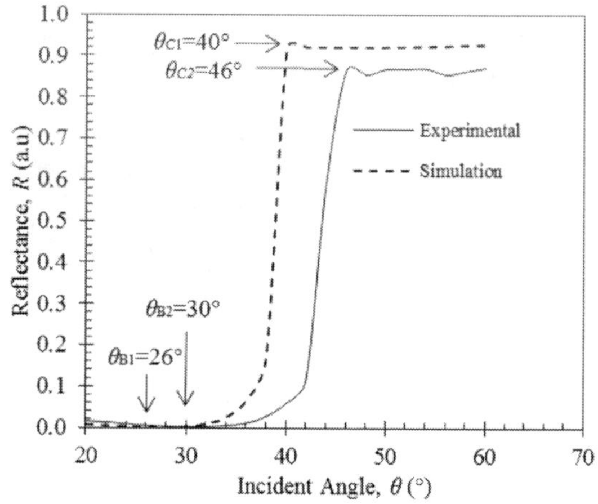

Fig. 8. Location of Brewster angle and critical angle for experimental and simulation results.

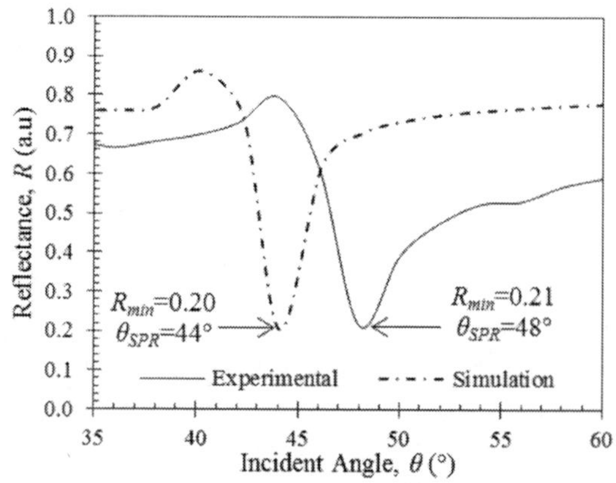

Fig. 9. Comparison of SPR curves of experimental and simulation results with the presence of gold thin layer deposited on the hypotenuse side of prism with thickness of 50nm.

Table 2. Summarization of SPR curve depth and FWHM

Type	SPR Curve Depth, ΔR (a.u)	FWHM (°)
Simulation	0.66	2.1
Experimental	0.59	2.0

The SPR curves were analyzed by studying the graph of reflectance against incident angle as portrayed in Fig. 9. By comparing the experimental and simulation of SPR results, the optimum thickness of gold was obtained at t = 50 nm. It shows that the curve dip of experimental result was red shifted for about 4° compared to simulation which is from 44° to 48°. According to the simulation results, the SPR curve depth was obtained as ΔR = 0.66 a.u while for experimental result ΔR = 0.59 a.u with percentage difference of 10.60% was acquired. A good concurrent between simulation and experimental results had been proved based on FWHM analysis where the percentage difference between both of them were very small, about 4.7% as depicted in Table 2.

5 Conclusion

In this study, the influence of EM waves polarization modes on SPR has been investigated theoretically and experimentally. We introduced a simple approach for the observation of SPR by controlling the position of He-Ne laser. An excellent agreement between experimental and theoretical results indicate the validity and stability of our SPR setup.

The authors would like to acknowledge the support of Malaysian Ministry of Higher Education (MOHE) through Universiti Sains Islam Malaysia (USIM) under grant USIM/FRGS/FST/32/51514.

References

1. W. M. Mukhtar, S. Shaari, P. S. Menon, Adv. Sci Lett, **19**, 66-64 (2013)
2. J. Homola, Anal. Bioanal. Chem., **377**, 528–539, (2003)
3. Y. Chen, H. Ming, Photonic Sensors, **2**, 37–49 (2012)
4. A. V. Zayats, I. I. Smolyaninov, A. A. Maradudin, Phys. Rep., **408**, 131–314 (2005)
5. P. Pattnaik, Appl. Biochem. Biotechnol., **126**, 79–92 (2005)
6. K. M. Byun, N.-H. Kim, J. W. Leem, J. S. Yu, Appl. Phys. B Lasers Opt., **107**, 1–6 (2012)
7. A. Kausaite, M. Van Dijk, J. Castrop, A. Ramanaviciene, J. P. Baltrus, J. Acaite, A. Ramanavicius, Biochem. Mol. Biol. Educ., **35**, 57–63 (2007)
8. W. M. Mukhtar, S. Shaari, P. S. Menon, Opto & Adv. Mat., **7**, 9–13 (2013)
9. Z. Yang, D. Gu, Y. Gao, Opt. Commun., **329**, 180–183 (2014)
10. B. Lee, S. Roh, J. Park, Opt. Fib. Tech., **15**, 209-221 (2009)
11. S. Szunerits, N. Maalouli, E. Wijaya, J. P. Vilcot, R. Boukherroub, Anal. Bioanal. Chem., **405**, 1435–1443 (2013)
12. H. D. Young, R. A. Freedman, *University Physics with Modern Physics*, **13**, 1052 (2012)
13. L. Novotny, B. Hecht, *Principles of Nano-Optics*, Cambridge University Press, United Kingdom (2006)
14. T. Thio, Am. Sci. **94**, 40–47 (2006)
15. P. L. Marston, Acoust. Soc. Am., **111**, 2378 (2002)
16. Y. Tang, X. Zeng, J. Liang, Chem. Educ., **87**, 7 (2010)
17. D. Halliday, R. Resnick, J. Walker, Fundamentals of Physics, **9**, 1230-1233 (2010)
18. D. Sarid, W. Challener, Modern Introduction to Surface Plasmons: Theory, Mathematica Modelling and Applications, **1**, 18-19 (2010)
19. W. A. A. Luhaybi, IUP, 12 (2014)
20. B. E. A. Saleh, M. C. Teich, Fundamentals of Photonics, **2**, 212-214 (2007)
21. W. M. Mukhtar, S. Shaari, A. A. Ehsan, P. S. Menon, Opt Mat Exp, **4**, 424 (2014)
22. N. F. Murat, W. M. Mukhtar, A. R. A. Rashid, K. A. Dasuki, A. A. R. A. Yussuf, *IEEE-ICSE2016 Proc. 2016*, **2**, 244–247 (2016)
23. S. Y. Wu, H. P. Ho, *Proc. 2002 IEEE Hong Kong Electron Devices Meet*, 63–68 (2002)
24. W. M. Mukhtar, P. S. Menon, S. Shaari, M. Z. A. Malek, A. M. Abdullah, J. Phys. Conf. Ser., **431**, 12028 (2013)

Free-space optics mode-wavelength division multiplexing system using LG modes based on decision feedback equalization

Angela Amphawan[1,2*], *Alaan* Ghazi[1,3] and *Aras* Al-dawoodi[3]

[1]Optical Computing and Research Laboratory, School of Computing, Universiti Utara Malaysia, 06010 Sintok, Kedah, Malaysia
[2]Research Laboratory of Electronics, Massachusetts Institute of Technology, 02139 Cambridge, Massachusetts, USA
[3]Computer Science Department, College of Science, Kirkuk University, 00964 Kirkuk, Iraq

Abstract. A free-space optics mode-wavelength division multiplexing (MWDM) system using Laguerre-Gaussian (LG) modes is designed using decision feedback equalization for controlling mode coupling and combating inter symbol interference so as to increase channel diversity. In this paper, a data rate of 24 Gbps is achieved for a FSO MWDM channel of 2.6 km in length using feedback equalization. Simulation results show significant improvement in eye diagrams and bit-error rates before and after decision feedback equalization.

1 Landscape

Free Space Optics (FSO), a license-free technology which offers higher bandwidth than conventional radio communications. It can be installed quickly at a lower cost compared to laying fiber optic cables [1]. Based on these advantages, FSO has emerged as a promising solution for the last mile bottleneck considering the scarcity of radio bandwidth despite various approaches for optimization of the radio spectrum [2, 3]. FSO also offers an alternative transmission medium to fiber optics for rural and geographically-challenged locations where cabling may be an issue [4].

With the tremendous growth in the Internet technology, the demand for higher bandwidth to support more applications and a larger number of subscribers is expected to continue escalating in the near future [5, 6]. Therefore, enhancing the capacity of FSO is necessary for high-speed transfer of a large amount of data. Currently, numerous techniques such as wavelength division multiplexing (WDM) [7], time division multiplexing (TDM) [8], and orthogonal frequency division multiplexing (OFDM) [9] have been employed in enhancing data transmission capacity in FSO systems.

Nevertheless, as a result of the exponential growth in data bandwidth, existing multiplexing strategies may not be sufficient to accommodate future traffic growth. An emerging multiplexing scheme, known as *mode division multiplexing* (MDM), has been used to transport independent channels simultaneously using eigenmodes [10]. In MDM, eigenmodes are used in MDM to drive the propagation of a number of channels on different modes generated by various mechanisms such as photonic crystal fibers [11], optical signal processing [12, 13], few mode fiber [9, 14], modal decomposition methods [15, 16], laser designs [17, 18] and spatial light modulators [19, 20]. MDM has been used for switching [21] and for multiplexing [10] in FSO systems.

Although FSO is an advantageous technology, there are some impediments to the viable bandwidth. Obviously, strict alignment requirement and adverse atmospheric weather conditions are major barriers [1]. Rudimentarily, the performance of FSO could be supported by modulation techniques [22] and estimation of transmitting power levels and wavelengths [23]. Apart from that, various multiplexing schemes may be combined to improve channel diversity as well as to reduce outage probability. In [24], WDM and MDM over RoFSO have employed simultaneously for improving the capacity.

* Corresponding author: alaanghazi@gmail.com

 DOI: 10.1051/epjconf/201716201009

Furthermore, in order to make FSO more feasible in respect to high capacity performance, photonic and electronic equalization schemes need to be integrated and explored, which will scale FSO better than through being duplicated using simple hardware [25]. In this paper, we explicate an initiative that uses decision feedback equalization (DFE) for MWDM using LG modes from a vertical cavity surface-emitting laser (VCSEL) as data carriers over FSO. The analyses of the MWDM system performance before and after the DFE will be analyzed.

In the remaining part of this paper, Section 2 reports on the MWDM modelling using LG modes over FSO. Section 3 discusses the simulation results of the MWDM system and the discussion is concluded in Section 4.

2 Simulation of FSO MWDM System

A FSO MWDM system of seven Laguerre-Gaussian (LG) modes based on DFE has been simulated in OptSim [26] and exhibited illustratively in Fig. 1. The FSO MWDM system consists of three parts; transmitter, FSO channel, and receiver. The input signal is generated using a pseudo-random binary sequence (PRBS) from the data generator at 3 Gbps. which is modulated to non-return-to-zero (NRZ) sequence, and is connected to a seven-segmented spatial vertical cavity surface emitting laser array with each segment represented as SpVCSELs. The SpVCSELs operate on two wavelengths; 1550.12 nm and 1551.92 nm and seven LG modes; LG 0 0, LG 0 1, LG 0 2, LG 1 1, LG 1 2, LG 2 0 and LG 2 1. Each SpVCSELs has an array emitting one LG mode on two wavelengths. Given 3 Gbps is transmitted by each mode and each wavelength. Thus, with 7 modes and 2 wavelengths per mode, the total data rate is 7 x 2 x 3Gbps = 42 Gbps. The FSO channel is 2.6 km long.

The FSO transmitter and receiver aperture is set to 20 cm, beam divergence is 1 μrad, and laser power is 0 dBm and based on clear weather conditions. After FSO, the seven receivers were used to retrieve the signals from FSO.

The main objective of decision feedback equalization is to compensate the ISI. The DFE comprises of feed-forward filter and feedback filter as seen in Fig. 2. The first part of DFE is a linear filter (feed-forward filter). A Linear filter was set as the parameter using tap number 9. The other part of DFE is a nonlinear filter (feedback filter) using tap 12 with MMSE-optimization. The optimization of MMSE is operated between the actual data and the target data, based on the error of MMSE will update the weight of tap filter to get a better result. Moreover, the performance of eye diagram and BER were analyzed for the LG modes before and after feedback filter. The main objective of decision feedback equalization is to enhance the signal received from the transmitted signals and reduce ISI. The results will be discussed in the next section.

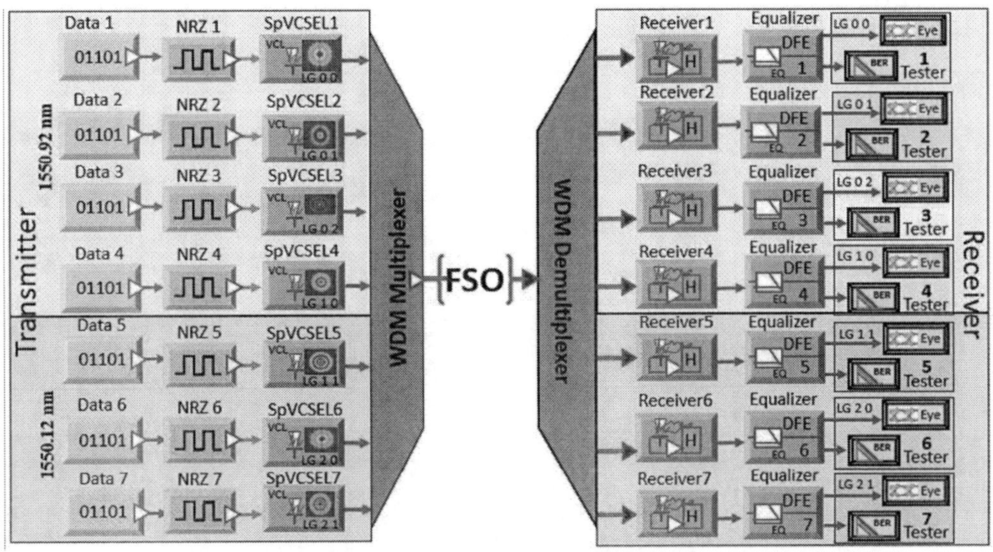

Fig. 1. FSO MWDM system in conjunction with decision feedback equalization (DFE).

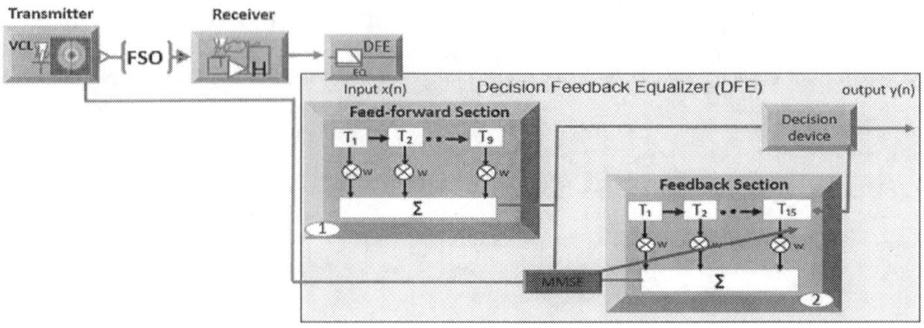

Fig. 2. Feed forward and feedback taps in DFE at the receiver

3 Simulation Results

Significant results were discovered from the BER and eye diagram upon application of DFE to improve the bandwidth of the FSO system.

Fig. 3 represent the eye diagrams of the first four channels using LG 0 0, LG 0 1, LG 0 2 and LG 1 0 modes before and after DFE. Fig. 4 represent the eye diagrams of the first four channels using LG 1 1, LG 2 0 and LG 2 1 modes before and after DFE. Fig. 3 and Fig. 4 illustrate that the eye openings for all channels on the seven LG modes have been widened and jitter is reduced.

Another performance metric is the BER. The average values for BER prior to DFE and with DFE for distances between 1 km and 3 km are represented in Fig, 5(a, b). Meanwhile, the BER values for LG mode (0 0) at 1 km is 1.71 x 10^{-2} before DFE (as seen in Figure 5(a) and Figure 5(b)). After DFE, it reaches up to 9.28 x 10^{-11}. Meanwhile, at 2 km, the BER is 1.63 x 10^{-2} before DFE and after DFE, the BER is reduced to 1.10 x 10^{-12}. Further, LG (0 0) at 2.6 km is 1.84 x 10^{-2} before DFE and after DFE it yields to 3.92 x 10^{-10} Also, LG (2 0) mode in Figure 5(a, b) show the BER values at 1 km is 1.71 x 10^{-2} before DFE and after DFE, the BER improves to 2.87 x 10^{-103}. At 2 2 km, the BER is x 10^{-2} before DFE and after DFE, it decreases to 9.03 x 10^{-25}. Meanwhile, the BER at 2.6 km is 2.31 x 10^{-2} without DFE and with DFE, it is enhanced to 5.43 x 10^{-12}.

Additionally, Fig. 5(a, b) shows the BER for LG 2 1. At 1 km, it is 1.79 x 10^{-2} before DFE and after DFE, the BER improves significantly to 8.44 x 10^{-135}. At 2.6 km, it is 2.04 x 10^{-2} before DFE and after DFE it becomes 3.84 x 10^{-17}. For LG (2 1), at 2.6 km it is 4.62 x 10^{-12} before feedback equalization and after feedback equalization it becomes 3.92 x 10^{-10}. The results of the BER values have been enhanced after using DFE and after increasing the distance, the BER increases as seen in Figure (a, b). The maximum distance attainable with DFE is 2.6 km with acceptable BER.

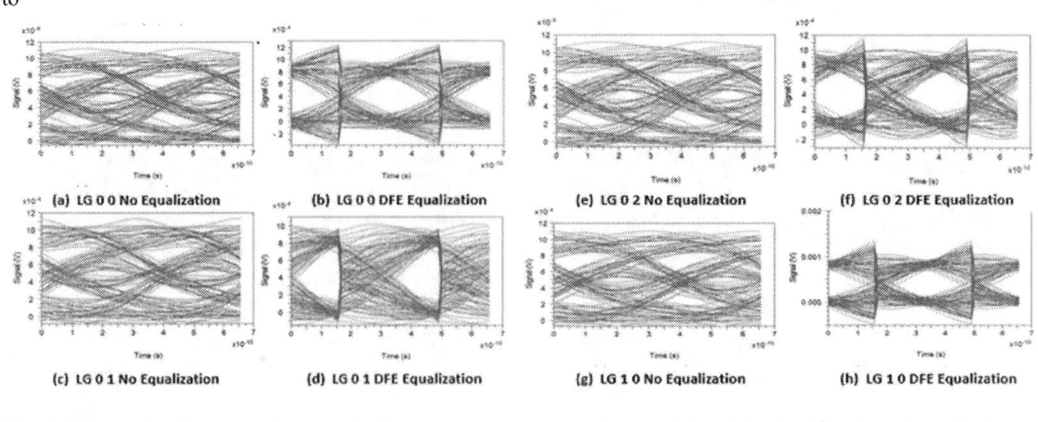

Fig. 3. Effect of DFE on eye diagrams for first four channels on LG 0 0, LG 0 1, LG 0 2, LG 1 0 modes over 2.6 km FSO: (a) LG 0 0 without DFE (b) LG 0 0 with DFE (c) LG 0 1 without DFE (d) LG 0 1 with DFE (e) LG 0 2 without DFE (f) LG 0 2 with DFE (g) LG 1 0 without DFE (h) LG 1 0 with DFE

DOI: 10.1051/epjconf/201716201009

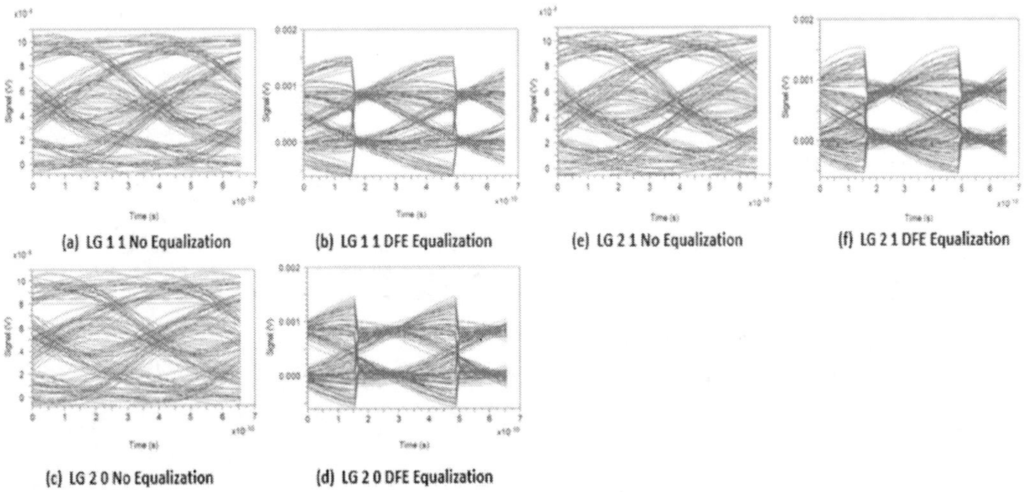

Fig. 4. Effect of DFE on eye diagrams for last three channels on LG 1 1, LG 2 0 and LG 2 1 modes over 2.6 km FSO: (a) LG 1 1 without DFE (b) LG 1 1 with DFE (c) LG 2 0 without DFE (d) LG 2 0 with DFE (e) LG 2 1 without DFE (f) LG 2 1 with DFE

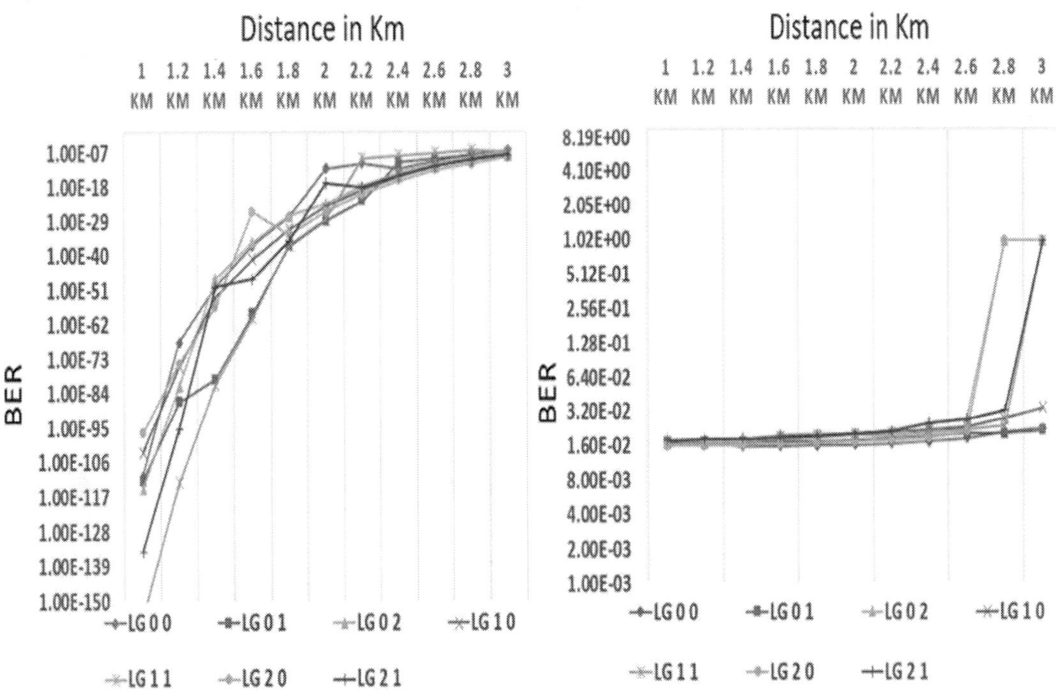

(a) (b)

Fig. 5. BER of seven LG modes (a) without DFE (b) after DFE

4 Conclusion

Simulation results demonstrate that with employment of DFE, eye diagrams and BERs for a FSO MWDM system over varying distances using LG modes have been improved. LG modes are successfully recovered at the receiver and 24 Gbps data transmission over 2.6 km has been attained.

References

1 Zabidi, S.A., et al., *Int. Conf. on Comp. and Comm. Eng. (ICCCE)*, Kuala Lumpur, Malaysia, pp. 1-5, (2010).
2 Liu, Z.Li, C.: Wireless Networks, , **23**, 2, pp. 453-466. (2017).
3 Amphawan, A., Omar, M.N. & Din, R. J Wireless Com Network 2012, **356**, (2012)
.4 Enayati, S.Saeedi, H., IEEE Comm. Let., **20**, (9), pp. 1824-1827.(2016).
5 Cisco, Cisco Visual Networking Index: Forecast and Methodology, 2016-2021 (2017).
6 Nisar, K., et al.: Journal of Network and Computer Applications, **36**, 2, pp. 933-948, (2013).
7 Shahidinejad, A., et al.: Reviews in Theoretical Science, **2**, 3, pp. 201-210, (2014).
8 Wang, D., et al., Int. Soc. for Optics and Photonics, 2016, edn.), pp. 100201F-100201F-100209, (2016).
9 Amphawan, A., et al., J. of the Eur. Opt. Soc. Europe, 9, (2014).
10 Amphawan, A., et al., IEEE Int. Coll. on Signal Proc. and its Applications (CSPA), pp. 145 - 149,(2015).
11 Amphawan, A., et al.: J. Mod. Opt., 2013, **60**, 20, pp. 1675-1683, (2013).
12 Arik, S.O., et al., IEEE Signal Processing Magazine, 2014, **31**, 2, pp. 25-34, (2014).
13 Amphawan, A., et al., Journal of Modern Optics, 2012, **59**, 20, pp. 1745-1752, (2012).
14 Tsekrekos, C.P.Syvridis, D., IEEE Photonics Technology Letters, , **24**, 18, pp. 1638-1641, (2012).
15 Martin, L., et al., **7**, pp. 44995, (2017).
16 Amphawan, A.O'Brien, D., Proceed. IEEE International Conf. on Photonics (ICP2010), pp. 1-5. (2010).
17 Amphawan, A.Fazea, Y.: J. of the Eur. Opt. Soc.-Rapid Publ., **12**, 1, pp. 12, (2016).
18 Xueyuan, D., et al., Laser Physics, **25**, 9, pp. 095102, (2015).
19 Carpenter, J.Wilkinson, T.D., J. of Lightw. Tech., **30**, 12, pp. 1978-1984, (2012).
20 Amphawan, A., Optics express, , **19**, 23, pp. 23085-23096, (2011).
21 Amphawan, A.Chaudhary, S. *Int. Conf. on Optical and Photonic Eng. (icOPEN 2015), Proc. of SPIE*, **9524**, (2015).
22 Wang, Z., et al., IEEE Photonics J., **1**, 6, pp. 277-285, (2009).
23 Borah, D.K.Voelz, D.G., J of Lightw. Tech., **27**, 18, pp. 3965-3973, (2009).
24 Chaudhary, S., et al.: Int. J. of Comp. App. , **93**, 1, (2014).
25 Fazea, Y.Amphawan, J. of Opt. Comm, **37** ,4, pp.331-413, (2016).
26 'Synopsis OptSim 5.2 ', (2010).

Respiration rate detection based on intensity modulation using plastic optical fiber

Zawawi Mohd Anwar[1,*], *Nor Ziran* Nurul Sufia[1], and *Manap* Hadi[2]

[1]Faculty of Electrical & Electronics Engineering, Universiti Malaysia Pahang, 26600 Pekan, Malaysia
[2]Faculty of Engineering Technology, Universiti Malaysia Pahang, 26600 Pekan, Malaysia

Abstract. This paper presents the implementation of respiration rate measurement via a simple intensity-based optical fiber sensor using optical fiber technology. The breathing rate is measured based on the light intensity variation due to the longitudinal gap changes between two separated fibers. In order to monitor the breathing rate continuously, the output from the photodetector conditioning circuit is connected to a low-cost Arduino kit. At the sensing point, two optical fiber cables are positioned in series with a small gap and fitted inside a transparent plastic tube. To ensure smooth movement of the fiber during inhale and exhale processes as well as to maintain the gap of the fiber during idle condition, the fiber is attached firmly to a stretchable bandage. This study shows that this simple fiber arrangement can be applied to detect respiration activity which might be critical for patient monitoring.

1 Introduction

This paper presents the development of a simple and low cost respiration sensor based on intensity modulation technique using fiber displacement loss between two separated plastic optical fibers. Two plastic fibers are applied; the one connected to the light source is the emitting fiber and the other fiber attached to the photodetector is the receiving fiber. The objective of this study is to determine the potential sensor arrangement using plastic optical fiber that can be applied to detect the respiration rate from the chest movement.

A brief review on currently available techniques for measuring respiration process using optical and non-optical devices is discussed in brief. This is followed by a description of the sensor configuration presented in this paper. Related theoretical information is also presented to describe the fiber displacement effect on the light intensity.

2 Review on current optical fiber sensor for respiration measurement

The development of various sensor approaches for respiration measurement are available in many research articles. Some popular devices for this particular application are capacitive sensor [1], low-power radar system [2], vision system [3] and optical fiber technique using temperature sensor at nostril [4] and macro-bend sensor embedded into a stretchable textile [5].

For the first type of respiration detection using capacitive sensor [1], two capacitive sensors were designed and fabricated for detecting chest or abdominal circumference. The sensor is placed on a respiration belt and can be applied to detect chest variation if maximum 60 mm. In theory, capacitive sensors work by measuring changes in an electrical property called capacitance, commonly with two conductive objects with a space between them respond to a voltage difference applied to them. As the respiration process takes place, the changes in capacitance value are monitored, which in this application representing the changes in chest circumference.

For the second device type [2], the respiration is detected alongside with the heartbeat using a 5 GHz radar system. This technique applied the concept of Doppler's effect to detect the chest movement during breathing without any physical contact with the human body. As a result, subject under study can be allowed to move freely inside a controlled room during the monitoring process without changing their normal respiration activity. However, this technique suffers from the harmonics effect of the respiration signal which may interfere with the detection of heartbeat rate and thus reduce its accuracy.

Another interesting research is based on the video image system [3]. In this study, the image and signal processing techniques were used to extract the chest and abdominal movement's information from a series of recorded video using a single camera. The data was stored for further offline analysis to determine the respiration rate based on the time sequence of the recorded video. As applied for other detection application using camera system, this technique is rather expensive for normal health monitoring needs. Moreover, frequent calibration process is important to guarantee the accuracy of the

* Corresponding author: mohdanwar@ump.edu.my

DOI: 10.1051/epjconf/201716201010

obtained measurement results, which could be time-consuming.

As far as optical fiber technique is concerned, the respiration monitoring can be implemented via the heat detection at the nostril and/or the chest movement. The first technique is realized using an optical temperature sensor placed at the nostril [4]. A sensing material called thermo-chromic microcapsule pigment was selected as the detection surface because this material changes its colour from red to white upon the absorption of heat (during exhale cycle). The intensity of the reflected light was monitored at the output which was corresponding to the changes in the pigment film colour. Second optical sensor technique is based on the macro-bend technique [5] using optical fiber cable. In this case, the optical fiber was stitched on a flexible garment taking the "U-shape" form. The elasticity of the garment allows the attached fiber to experience changes in the curve radius, thus changes the intensity of the detected light. As the breathing process take place, the light intensity also varies according to the chest expansion.

In this paper, a rather similar fiber optic material is used but the light intensity variation is altered due to fiber displacement gap changes that represent the chest movement during respiration process. In order to identify the suitable initial gap for the fiber installation on the stretchable bandage, several experiments have been conducted. Finally, the assembled optical sensor is connected to suitable LED driver and signal conditioning circuits before the output of the circuit is displayed on the monitor screen via Arduino kit interface.

3 Theory on fiber displacement gap effect on light intensity

Many previous studies have presented the relationship between the light intensity of a plastic fiber due to different fiber movement; namely longitudinal gap, lateral offset, tilting angle offset and reflective gap between fiber end tip and a reflective mirror [6]. For the explanation on the optical sensor implemented in this study, only the effect of longitudinal gap between the emitting and receiving fibers is further discussed. Particularly for this fiber movement, the light intensity at the receiving fiber is dependent on multiple parameters such as refractive index of the medium at the fibre gap (nm), refractive index of the applied fiber core (n_c) and cladding (n_{cl}), radius of the fibre core (a) as well as the separation distance the sensor gap (z).

Figure 1 illustrates this relationship which shows the main parameters involved in the theoretical estimation. The equation is obtained from preliminary study by Tsuchiya et. al. [7] :

$$\frac{P_o}{P_i} = \frac{16K^2}{(1+K)^4} \cdot \left(1 - \frac{z}{4a} \cdot K \cdot (2 \cdot \Delta)^{0.5}\right) \qquad (1)$$

Where:

$$K = \frac{n_c}{n_m}$$

Po - the output power at the receiving fibre due to displacement.

Pi - the output power at the receiving fibre at zero gap.

n_c - the refractive index of the fibre core.

n_{cl} - the refractive index of the fibre cladding.

n_m - the refractive index of the medium between the fibres

z - the longitudinal gap between input and output fibre.

a - the fibre core radius.

Δ - the refractive index difference between core and cladding (n_c - n_{cl})

For the applied fiber type in this study (Mitsubishi Rayon simplex POF), the numerical aperture, NA = 0.5, fiber core radius, a = 0.49 mm, core refractive index, n_c = 1.492, cladding refractive index, n_{cl} = 1.402, and the refractive index difference, Δ = 0.09. These results in the calculated value of K = 1.0642. As the emitting and receiving fibers are connected together in air, which refractive index, nm = 1.000293 ≅ 1.00, the normalized output intensity estimation versus longitudinal displacement using Eq. 1 above is illustrated in Figure 2.

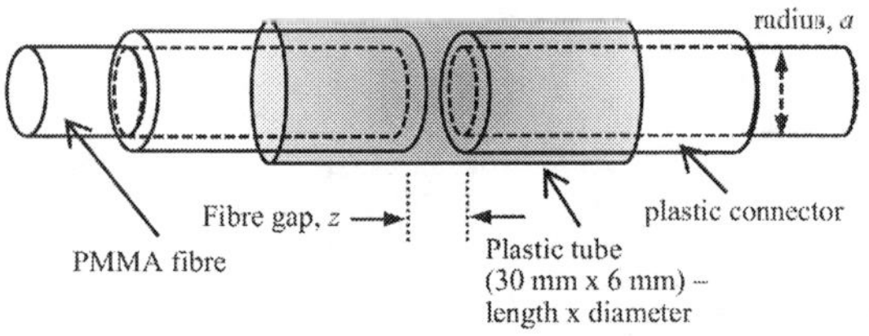

Fig. 1. Side-polish optical fiber bending sensor.

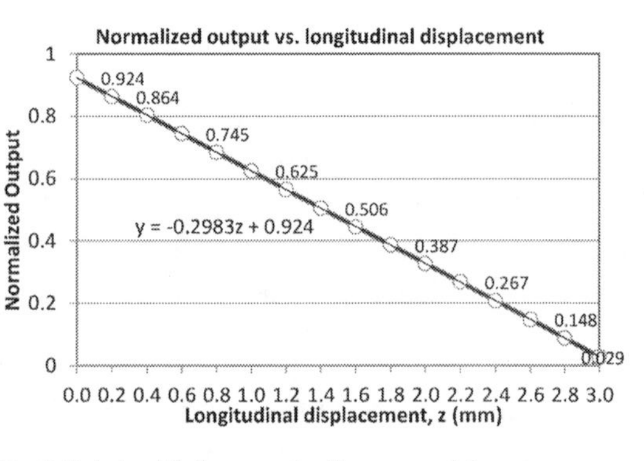

Fig. 2. Relationship between the fiber gap and the output intensity.

From the relationship between the normalized output and the longitudinal displacement of the emitting and receiving fiber end tips as shown in Figure 2, it was observed that a linear relationship is obtained for a displacement gap between 0.0 mm to 3.0 mm. Additional fiber gap will result in poor light intensity at the receiving fiber, thus should be avoided in the sensor development.

4 Sensor configuration of this investigation

As previously mentioned, fiber gap displacement technique based on intensity-modulation between the emitting and receiving fibers has been implemented as a respiration sensor aimed to monitor the chest expansion during inhalation and exhalation processes. In terms of the sensor arrangement, it is important to have a consistent output readings for a certain period of time (for continuous monitoring purpose), so that the light intensity variation from the optical sensor is only due to the chest diameter variation. Other associated factors, such as variation in light intensity brightness, photodetector sensitivity and supply voltage fluctuation could lead to sensor measurement inaccuracy. In this study, we only focused on the sensor assembly on a stretchable material (i.e. standard clinical bandage) to ensure a firm sensor attachment on its surface and basic laboratory tests to verify the sensor arrangement suitability for the aimed application.

In order to have a sufficient light reception at the receiving fiber before being attached on the bandage, the fiber gap (end tip distance between emitting and receiving fibers) of less than 3 mm is necessary for the sensor design. From our initial study of the fiber displacement effect using a single axis translational stage (Thorlabs), a poor light reception is produced for a fiber gap of more than 3 mm, which is corresponding to the light intensity estimation in Figure 2. The emitting fiber was fixed inside the plastic tube using a heat gun glue, while the receiving fiber cable was purposely allowed to move freely inside the tube. This arrangement is important to ensure the produced light attenuation is mainly from the fiber movement in longitudinal direction, but not due to light loss from fiber offset in lateral and angular directions. For the experimental work presented in this paper, the applied fiber was Mitsubishi Rayon GHV4001. This is a multimode plastic optical fibre (PMMA) which has a core/cladding diameter of 1 mm, numerical aperture at 0.5 and with step index profile.

The overall sensor schematic is shown in Figure 3.

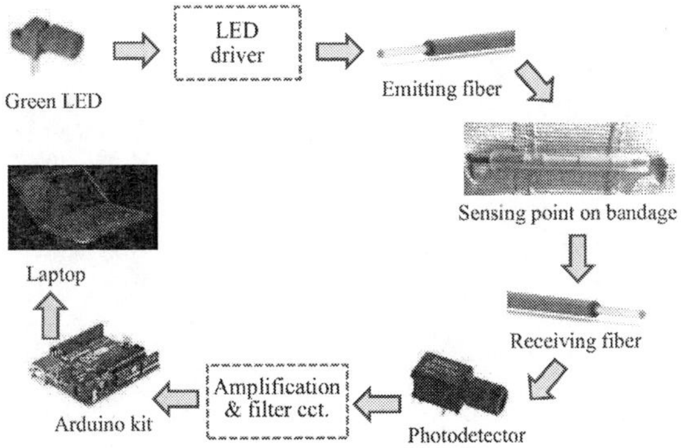

Fig. 3. Fiber optic respiration sensor schematic.

The respiration sensor proposed in this paper is based on the theoretical study and parameter calculation of light intensity variation with respect to different fiber gap as illustrated in Figure 2. In terms of the sensor preparation, the fiber cable was cut using a hot knife at a pre-determined length. Then, the jacket of the fiber was removed with a length of 1 cm at each fiber end tip. The fiber end tips were then polished using a sand paper with different grids before being inserted into a plastic optical fiber (POF)-type v-pin connector to ensure good light acceptance into the fiber. The connector has 1.0 mm and 3.2 mm inner and outer diameter, respectively. Both emitting and receiving fibers were held together using a plastic tube to allow the fiber to displace in longitudinal direction with minimum lateral offset.

The bandage (with the optical sensor on it), was then placed encircled a large plastic container as shown in Figure 4 to represent the respiration process in laboratory scale test rig.

Fig. 4. Respiration sensor placement on a circle object.

5 Results and analysis

In the experimental study, the output of the optical sensor is the light intensity detected by a photodetector (with embedded lens), which has a spectral range between 400 and 1100 nm. The optical sensor is supplied by a high brightness green LED with a peak wavelength of 530 nm. The maximum detected voltage from the designed circuit is only at 50 mV, which is too small to be displayed on a standard desktop monitor. In order to amplify the signal, a signal conditioning circuit with an amplification gain of 100 is fabricated and applied before the amplified signal is inserted into the Arduino kit analogue pin.

The output responses from this study are shown in Figure 5. To study the capability of the sensor arrangement to maintain its initial fiber gap, the movement of the fiber gap is tested at two different time responses (e.g. fast and slow movement). After multiple trials, the results show that, the respiration sensor is able to return its initial position after being stretched for several times. Moreover, the bandage can also be attached to different body sizes without changing the initial fiber gap because the receiving fiber position on the bandage can be adjusted by altering related segment of the bandage.

In the current stage of this study, the sensor has yet to be tested on any human volunteers. Nevertheless, it will be necessary in the future to carry out such experiments to prove the performance and reliability of the proposed optical sensor technique. Another concern is that, although the use of intensity modulation technique, i.e. fiber longitudinal displacement is not considered as advanced sensing methods such as fiber grating and fiber interferometry counterparts, the implementation of this method is considered new for this particular health monitoring application.

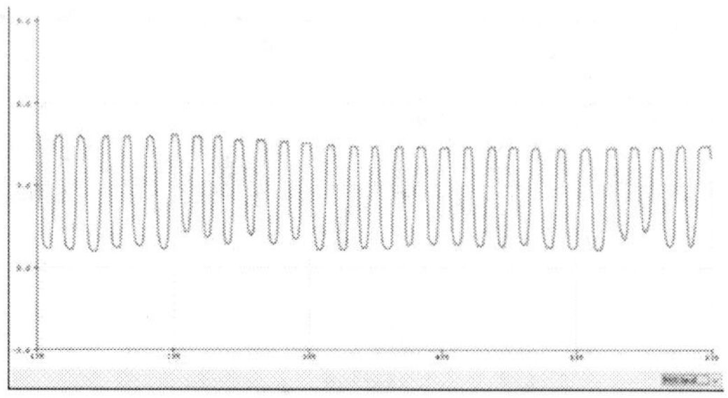

(a) fast response (3 sec. per cycle)

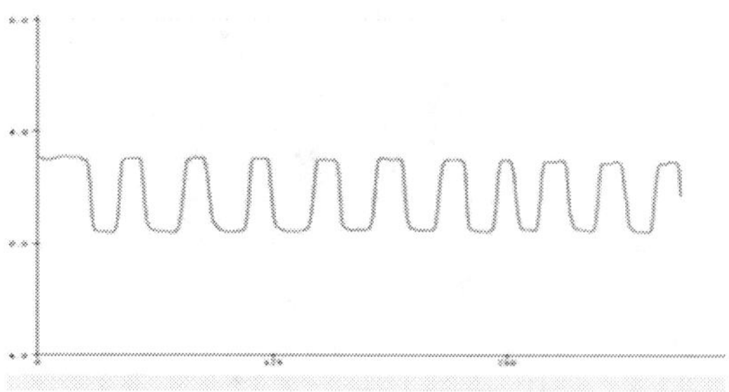

(b) slow response (8 sec. per cycle)

Fig. 5. Output response of the optical sensor during the natural and stretch conditions to represent the chest movement

As a conclusion, the development of a respiration monitoring sensor using plastic optical fiber based on intensity modulation technique has been presented. The inhalation and exhalation process is detected from the longitudinal fiber displacement, due to chest movement where the sensor is located via a stretchable bandage. The fiber gap for sensor assembly was obtained from a theoretical estimation involving different parameters of the applied fiber. The experimental study of the sensor showed a consistent output level during idle and stretched condition.

The authors wish to thank Universiti Malaysia Pahang for the financial support on the research activities via internal research grant RDU150364.

References

1. C.R. Merritt, H. Troy Nagle, E. Grant, IEEE Sensors Journal. **9**, 1, pp. 71-78 (2004)
2. X. Yanming, L. Changzhi, L. Jenshan, IEEE Sensors Journal. **7**, 7, pp. 1042-1043 (2007)
3. K.S. Tan, R. Saatchi, H. Elphick, D. Burke, *Proc. of the Int Symp. on Communication Systems Networks and Digital Signal Processing*, **7**, pp. 770-774 (2010)
4. W.J. Yoo, K.W. Jang, J.K. Seo, J.Y. Heo, J.S. Moon, J.H. Jun, J.Y. Park, B. Lee, Optical Review. **18**, 1, pp. 132-138 (2011)
5. A. Grillet, D. Kinet, J. Witt, M. Schukar, K. Krebber, F. Pirotte, A. Depre, IEEE Sensors Journal. **8**, 7, pp. 1215-1222 (2008)
6. O. Ziemann, J. Krauser, P. E. Zamzow, W. Daum, *Optical Short Range Transmission Systems.* **2**nd ed., Springer, Berlin, Germany, pp. 259-268 (2008)
7. H. Tsuchiya, H. Nakagome, N. Shimizu, S. Ohara, Applied Optics. **16**, 5, pp. 1323-1331 (1977)

 DOI: 10.1051/epjconf/201716201011

Q-switched erbium doped fiber laser using antimony telluride-polyvinyl alcohol (Sb_2Te_3-PVA) as saturable absorber

Muhammad Quisar Lokman[1*], *Fauzan* Ahmad[1], and *Sulaiman Wadi* Harun[2]

[1]Malaysia-Japan Institute of Technology, Universiti Teknologi Malaysia, Jalan Sultan Yahya Petra, 54100 Kuala Lumpur, Malaysia
[2]Department of Electrical Engineering, Faculty of Engineering, University of Malaya, 50603 Kuala Lumpur, Malaysia

Abstract. Q-switched erbium doped fiber laser was demonstrated using antimony telluride (Sb_2Te_3) as saturable absorber (SA). The SA was fabricated by adding Sb_3Te_2 powder into PVA suspension and left dry in room temperature for two days. Then, the SA was sandwiched in between two FC/PC fiber ferrules, which can provide easy integration and flexibility into the laser cavity. Stable and self-started Q-switched laser operates at 1531 nm center wavelength. The laser repetition rate increased from 54.5 kHz to 88.4 kHz and pulse duration decreased from 6.84 μs to 4.58 μs as the pump power increased. A signal to noise ratio value of 55 dB was achieved at pump power 130 mW. At the maximum pump power, the average output power and pulse energy are 0.26 mW and 2.78 nJ.

1 Introduction

Topological insulator materials has captured many research interest due to the outstanding mechanical, electronic and optical properties that are highly promising in a variety of potential applications, for example electronics [1], fiber laser [2-3], solar cell [4] and automotive [4]. Antimony telluride (Sb_2Te_3) is one of the significant topological insulator due to narrow bandgap which is approximately 0.2 eV [5], a valuable characteristic, especially in fiber laser application for pulse generation. Besides, Sb_2Te_3 has an outstanding electronics properties [6] and thermoelectric properties for [7] that are highly beneficial for electrical applications.

Fiber laser is one of essential tool that can fulfill a wide range application from scientific, medical and industrial application. In fiber laser, there are two ways to generate fiber laser, either by active or passive technique. In active techniques, a modulator is used to modulate the loss in the cavity. This technique can promise more stable pulse, higher repetition rate, and shorter pulse width but are costly to execute [8]. For passive technique, saturable absorber (SA) are incorporated to modulate the loss in the fiber laser cavity. This simple and low cost techniques also can generate a stable pulse depending the on pump power to tune the pulse repetition rate [9].

On the basis of limited study on Sb_2Te_3 saturable absorber (SA) fabrication. There are several fabrication techniques to fabricate SA including mechanical exfoliation, drop cast and thin film or polymer composite, but only mechanical exfoliation of Sb_2Te_3 has been reported [10]. In mechanical exfoliation technique, most of the techniques are using bulk Sb_2Te_3 instead of powder form. The main drawbacks of this technique is its' difficulty to control the desired Sb_2Te_3 layer and thickness that need to be transferred on the FC/PC fiber ferrule [11]. To embed into PVA polymer, using powder is a suitable ways in order to ensure the Sb_2Te_3 strongly bind with the polymer. Using this method, concentration and thickness of the SA can be controlled. There are several kinds of polymer material, for example polymethylmethacrylate (PMMA), polyvinyl alcohol (PVA), polyethylene oxide (PEO) and polyimide that can be used as polymer host [12].

In this paper, we successfully demonstrated generation of Q-switched EDFL using Sb_2Te_3 based SA. The Sb_2Te_3 SA embedded in PVA thin film was sandwiched in between FC/PC fiber ferrule to generate stable pulsed.

2 Experiment Details

2.1. Fabrication of Sb_2Te_3-PVA Saturable Absorber (SA)

SA was prepared by dissolving 25 mg Sb_2Te_3 powder [Sigma Aldrich, -325 Mesh, 99.6% trace metal basis] into the PVA suspension and stirred for an hour, followed by sonication for 15 minutes until the powder completely dissolves. Then, the mixture was poured into the petri dish and left dry in ambient temperature. After 2 days, thin film was slowly peeled from the petri dish and cut into 2 mm^2 then attached onto the FC/PC fiber

* Corresponding author: muhammadquisar@gmail.com

ferrule. Fig 1 shows the Sb_2Te_3–PVA SA attached on the FC/PC fiber ferrule.

Fig. 1. Sb_3Te_2 – PVA saturable absorber.

2.2 Experimental setup

Fig 2 shows the configuration of the Q-switched erbium doped fiber laser (EDFL) based on Sb_2Te_3–PVA SA, which consists of a 980 nm laser diode (LD), 980/1550 nm wavelength division multiplexing (WDM), 2.4 m Erbium-doped fiber (EDF), isolator and 90/10 output coupler.

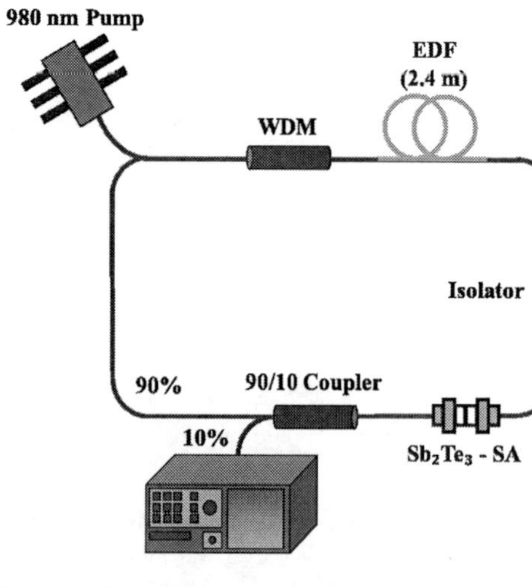

Fig. 2. Configuration setup of the Q-switched EDFL.

The SA was cut into a small pieces (2 mm^2) and sandwiched in between two FC/PC fiber ferrule. The EDF has a core and cladding diameters of 4 µm and 125 µm, with a numerical aperture of 0.16 and erbium ion absorption 23 dB/m at 980 nm. The laser is pumped by a 980 nm LD via the WDM. An isolator is incorporated in the cavity to ensure unidirectional propagation of the laser. The light is extracted from the cavity by 90/10 coupler, 90% of the lights are propagates in the cavity and 10 % are used for output. An optical spectrum analyzer (YOKOGAWA: AQ6370B), oscilloscope (QW Instek, GDS-3352) and radio frequency spectrum analyzer (Anritsu, MS2638A) coupled with 1.2 GHz InGaAs photodetector (Thorlabs, DET01CFC) were used to measure optical spectrum, pulse train and signal to noise ratio of the pulse. An optical power meter is used to measure the output power. The total cavity length of the laser cavity is measured to be approximately 13 m.

3 Results and Discussion

A self-starting Q-switched pulse were achieved at pump power 39.2 mW. Q-switched performance was observed when the pump power is tuned from 39.2 mW until 65.5 mW. Beyond 65.5 mW, the pulse generated started to become unstable and diminished.

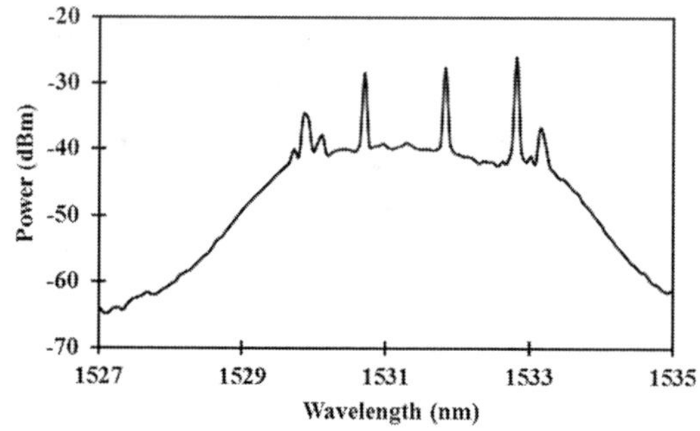

(a)

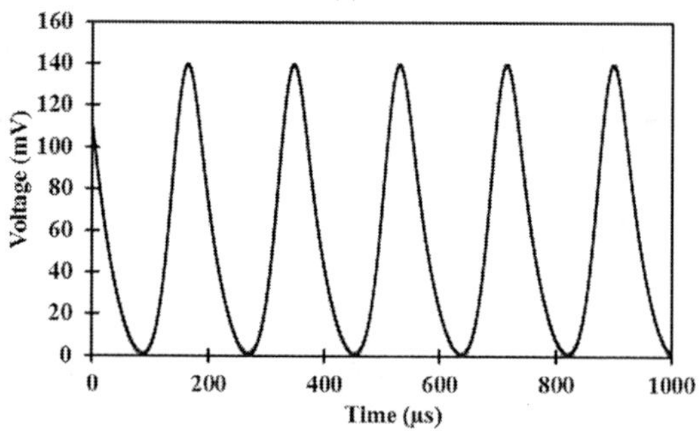

(b)

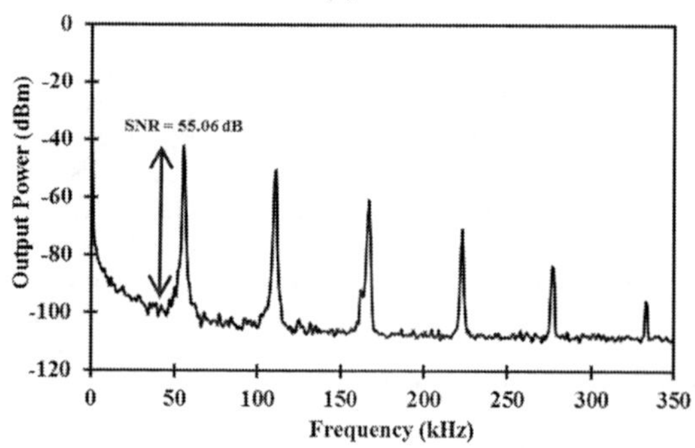

(c)

Fig. 3. (a) Output spectra (b) Pulse train and (c) RF spectrum at 39.2 mW pump power.

Fig 3 (a) displays the optical spectrum at pump power of 39.2 mW. The laser operates at center

DOI: 10.1051/epjconf/201716201011

wavelength 1531 nm with a 3 dB spectral bandwidth of 4.68 nm. To verify the Q-switching pulses, the SA was removed from the fiber ferrule and no pulses was observed when the pump power was tuned over a wide range. Fig 3(b) shows the typical oscilloscope traces of the Q-switched pulse trains at pump power of 39.2 mW. The pulses shows a uniform and no distinct amplitude with peak to peak duration of 186 µs. In Fig 3(c) shows the output RF spectrum at 39.2 mW matching with the 54.48 kHz repetition rate. The signal to noise ratio (SNR) of the fundamental peak to the pedestal extension is estimated 55.06 dB.

Fig 4 (a) represent the pulse repetition rate and pulse duration as the function of pump power. It is observed that repetition rate is increased from 54.5 mW to 69.0 mW as the pump power increased 39.2 mW until 65.5 mW. The pulse width are reduced from 6.84 µs to 5.11 µs when the pump power increased. Fig 4 (b) shows the average output power and calculated pulse energy at 39.2 mW to 65.5 mW. The average output is raised from 0.17 mW to 0.24 mW as increased the pump power. The pulse energy are increased from 3.08 nJ until 3.43 nJ at 39.2 mW to 65.5 mW.

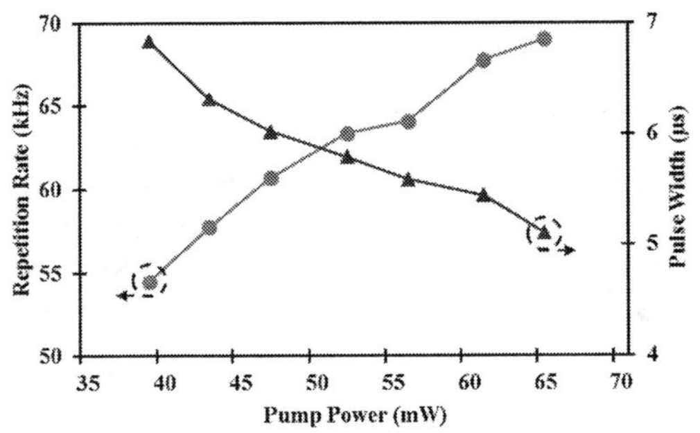

(a)

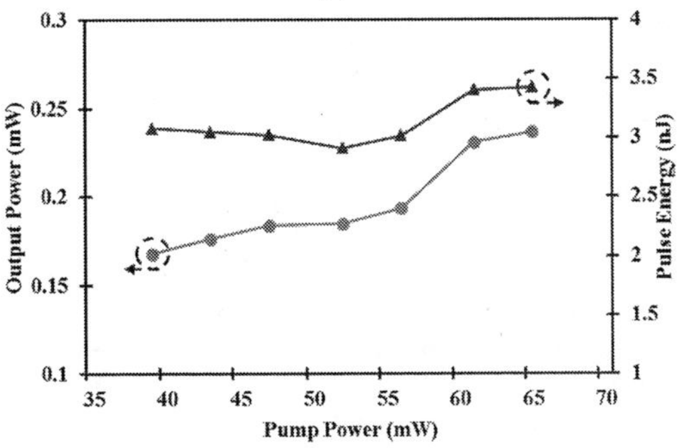

(b)

Fig. 4. (a) Repetition rate and pulse width, and (b) Output power and pulse energy at pump power 39.2 mW to 65.5 mW.

3 Conclusion

In summary, Q-switched EDFL using Sb_2Te_3 are demonstrated in this paper. The Sb_2Te_3 was prepared by mixing with PVA suspension and SA was peeled after dried in ambient for 2 days. Then, the SA is cut into small piece and sandwiched between the fiber ferrule to generates pulse laser. The Q-switched operates at the center wavelength of 1531 nm and a stable pulse is achieved at the pump power 39.2 mW to 65.5 mW. At the maximum pump power, the output power of 0.24 mW and pulse energy of 3.43 nJ are produced.

References

1. Chen, Y. P. In SPIE Defense, Security, and Sensing. *International Society for Optics and Photonics.* (2012)
2. Li, K., Song, Y., Yu, Z., Xu, R., Dou, Z., & Tian, J. Laser Physics Letter. **12** (10), 105103 (2015)
3. Yan, P., Lin, R., Ruan, S., Liu, A., Chen, H., Zheng, Y., ... & Hu, J. Scientific reports. **5**, 8690 (2015)
4. Peng, H., Dang, W., Cao, J., Chen, Y., Wu, D., Zheng, W., ... & Liu, Z. Nature chemistry, **4** (4), 281-286 (2012)
5. Park, N. W., Lee, W. Y., Hong, J. E., Park, T. H., Yoon, S. G., Im, H., ... & Lee, S. K. Nanoscale research letters, **10** (1), 20 (2015)
6. Zhu, M., Wu, L., Rao, F., Song, Z., Ren, K., Ji, X., ... & Feng, S. Applied Physics Letters, **104** (5), 053119 (2014)
7. Goncalves, L. M., Alpuim, P., Rolo, A. G., & Correia, J. H. Thin Solid Films, **519** (13), 4152-4157 (2011)
8. Lü, X., Han, Q., Liu, T., Chen, Y., & Ren, K. Laser Physics, **24** (11), 115102 (2014)
9. Wang, Y., & Xu, C. Q. Progress in Quantum Electronics, **31** (3), 131-216 (2007)
10. Sotor, J., Sobon, G., Grodecki, K., & Abramski, K. M. Applied Physics Letters, **104** (25), 251112 (2014)
11. Chandra Paul. Passive Q-switched and Mode-locked Fiber Lasers Using Carbon-based Saturable Absorbers, *Fiber Laser* (2016)
12. Bonaccorso, F., & Sun, Z. Optical Materials Express, **4** (1), 63-78 (2014)

Effect of decision feedback equalizer taps on 3x6- channel mode-wavelength division multiplexing system performance in multimode fiber

Tendai Masunda[1,*], *Angela* Amphawan[1,2], and *Aras* Al-dawoodi[3]

[1] School of Computing, Universiti Utara Malaysia Sintok, Kedah, Malaysia
[2] Research Laboratory of Electronics, Massachusetts Institute of Technology,Cambridge, Massachusetts, USA.
[3] Computer Science Department, College of Science, Kirkuk University, 00964 Kirkuk, Iraq

Abstract. In this paper, the effect of tap configurations in a decision feedback equalizer is investigated for controlling modal dispersion in a 60Gb/s 18-channel mode division multiplexing - wavelength division multiplexing system in a 2.5km multimode fiber on three Laguerre-Gaussian modes and six wavelengths centered at 1550nm. Spectrum analyser diagrams and bit error rate measurements show satisfactory performance for feed forward – feedback tap combinations (10,5), (14,0), (15,6) and (15,10) by effectively suppressing higher order modes.

1 Introduction

Today's optical networks are the backbone of modern communications networks, moving data traffic across distances of just a few kilometers to stretch across continents. There has been an exponential growth in Internet users and the demand for higher bandwidth as users interact more spontaneously on the Internet using applications that require large data transfers and high interactivity such as for video conferencing, video streaming, voice-over-Internet-Protocol (VoIP), television streaming, social media, online gaming and others [1-3]. Digital content is growing and network subscribers require high data transfer and interactivity.

Various multiplexing schemes have been investigated for increasing the bandwidth of current optical networks, such as wavelength division multiplexing (WDM) [4], orthogonal frequency division multiplexing (OFDM) [5], optical code division multiple-access system (OCDMA) [6] and polarization division multiplexing (PDM) [7]. There has also been a compelling vision to accommodate traffic growth by devising ways that can allow us to transmit large amounts of data in a single fiber using modes of an optical fiber. The future solution lies in mode division multiplexing (MDM) which allows independent data streams to be transmitted on multiple modes of an optical fiber, which are then retrieved by multiple receivers [8].

* Corresponding author: tendaimasunda@gmail.com

2 Related Work

MDM complements WDM, OFDM and other multiplexing schemes which have been the workhorse of major telecommunication companies in order to increase the aggregate capacity. MDM has been realized through through holographic wavefront generation [9-12], multi-core fibers [13-15], mode-selective fiber couplers [16-18], laser cavity design [19-21] or via optical signal processing [17, 22, 23].

The frontier of MDM lies in its prowess to unleash the physical capacity boundaries of multimode fiber (MMF) by confining modal coupling which instigates modal dispersion and inter-symbol interference, thus restricting the achievable bandwidth [24]. To suppress mode coupling, several adaptive equalization schemes have been introduced in MDM. Adaptive equalizers work efficiently in an unknown environment. These filters extract useful information from noisy data and employ recursive algorithms for continuous adjustment within a changing environment for example when signal statistics are not defined priori, and also when there is a spectral overlap between the signal of interest and noisy signal. Equalizers that have been used thus far in MDM include least-mean-square algorithm which reduce the mean square error between the received and target channel impulse response [25], recursive least-squares which recursively finds the coefficients that minimize a weighted linear least squares cost function relating to the received and target channel impulse response [26],

 DOI: 10.1051/epjconf/201716201012

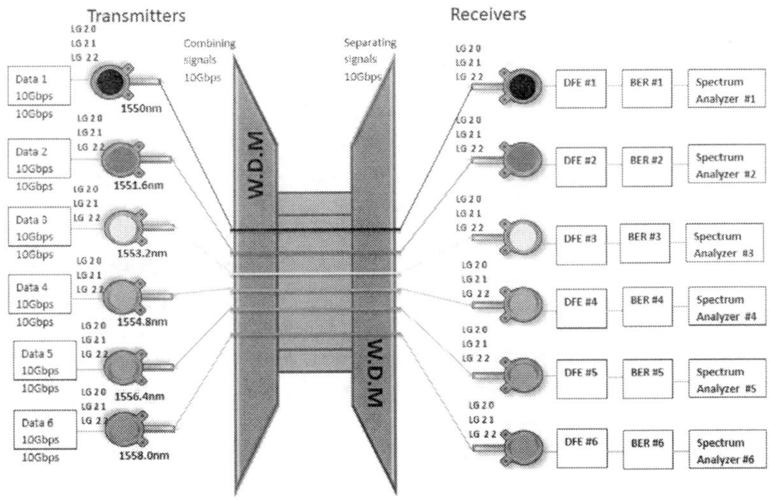

Fig. 1. Schematic diagram for Decision Feedback Equalization in DQPSK MWDM system using MMF.

space-time block coding exploiting space and delay diversity [27], constant modulus algorithm [28] and multi-modulus blind equalization [29]. RLS is preferable for next-generation MDM systems with faster convergence time compared to LMS.

3 Contribution

In contrast to the aforementioned equalization schemes, in this paper, electronic dispersion compensation (EDC) using decision feedback equalization (DFE) is numerically investigated for mode-wavelength division multiplexing (MWDM) in MMF. The objective of this paper is to analyze the relationship between different configuration of taps in the DFE and the resulting BER in order to establish the ideal combination of taps to restore the original signal after distortion due to mode coupling in a 60Gbps DQPSK WDM-MDM system through a 2.5km MMF. In DFE, fixed tap length is conventionally used in noise control. To cancel the undesired noise from a primary noise source, the equalizer compensates for the uncertain effects after the receiver. Moving the DFE to the receiver side has several advantages. First, the Transmitter design is simplified, enabling a low-power design. Second, the equalizer coefficients can directly be adjusted at the receiver end, where they can be measured, which does not require a back-channel to the transmitter.

4 MWDM System Design with DFE in MMF

Fig. 1 shows the proposed design using DFE for a MWDM system propagating independent data on three modes per wavelength for a total of six wavelengths modulated using differential quadrature phase-shift keying (DQPSK). The proposed equalizer for the system is modeled using Synopsis OptSim optical communications suite [30]. Each channel is connected to a DFE which parameters include feed forward taps number and feedback taps number. Various combination of taps produce a corresponding bit error rate (BER) and this study shall establish the four best tap combinations that will produce satisfactory BER in the 2.5 km 60Gbps DQPSK MWDM optical system.

The transmitter consists of six spatial vertical cavity surface-emitting laser (VCSEL) arrays which are incorporated in the DQPSK transmitter and are directly modulated by an electrical signal. The VCSEL is driven by pseudo-random binary sequence (PRBS) electrical signals and modulated to non-return-to-zero pulses at its input and a fixed bias current. The six VCSELS are centered on wavelengths 1550 nm, 1551.6 nm, 1553.2 nm, 1554.8 nm, 1556.4 and 1558.0 nm. Each VCSELs generate a laser beam composed of three different Laguerre-Gaussian modes LG 2 0, LG 2 1 and LG 2 2. The output of all VCSELs are multiplexed and propagated through a parabolic graded MMF with a radius of 25 μm spanning 2.5 km in length.

Table 1. DFE feed forward and feedback tap combinations under investigation

Feed Forward Taps Number	Feedback Tap Number
10	5
14	0
15	6
15	10
1	0
2	2
5	2
11	11

Coupling between modes is modelled by a coupling matrix [31] :

$$C = \begin{bmatrix} c_{i,k} & c_{i,k+1} & \cdots & c_{i,n} \\ c_{i+1,k} & c_{i+1,k+1} & \cdots & c_{i+1,n} \\ \cdots & \cdots & \cdots & \cdots \\ c_{m,k} & c_{m,k+1} & \cdots & c_{m,n} \end{bmatrix} \quad (1)$$

where i describes the source mode, k describes the destination mode, m describes the highest source mode and n describes the highest destination mode. At the receiver side, a demultiplexer splits the power from the signals to six DQPSK spatial photodetectors which are each connected to a DFE. The DFE structure is based on transversal finite-impulse response (FIR) filter. The signal transformation is described by the following equation:

$$y(t) = \sum_{d=1}^{N} C_d x\left(t - i\Delta T\right) \quad (2)$$

where $x(t)$ is incoming signal, $y(t)$ is the output signal, d is the tap number, N is the total number of taps, C_d is the weight coefficients for the d-th tap and ΔT is the tap delay.

5 Simulation Results

The DFE corrects the received modal distribution in association with a priori modal distribution. The feedforward and feedback tap combinations under investigation are shown in Table 1. The output from each DFE is analyzed to determine the improvement in the BER.

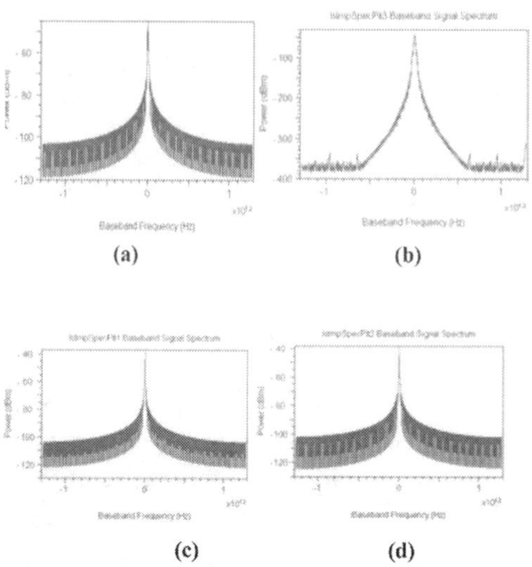

Fig. 2. Spectrum Analyzer diagrams for various tap combinations: (a) (10,5), (b) (14,0), (c) (15,6) and (d) (15,10)

The analysis on several combinations of feed forward tap number and feedback tap number in the decision feedback equalizer revealed that when the set of (10,5), (14,0), (15,6) and (15,10) feed forward-feedback taps were employed, satisfactory BER was achieved as compared to many other combinations and in effect, significantly mitigated mode coupling. This led to the best DMG confinement and lowest BER. The spectrum analyzer diagrams for these combinations are shown in Fig. 2. The lowest noise levels were found to be -120dBm and the highest peaked at -40dBm making the system highly efficient in terms of signal loss. Comparing the spectrum analyser diagrams in Fig.2 it is evident that the decision feedback equalizer with the set of (10,5) feed forward-feedback taps achieved the least signal power loss with -45.342dBm compared to the corresponding other 3 each exhibiting signal power losses of -38.63dBm, -40.137dBm and -37.8dBm. The smaller the signal power loss the better the performance of the mode transmission system.

The BER based on the feed forward and feedback tap numbers are shown in Table 2. 60Gbps data transmission was successfully achieved at a BER of 1.1519×10^{-29} for feed forward - feedback tap combination (10,5), 1.2769×10^{-26} for feed forward - feedback tap combination (14,0), 2.33×10^{-23} for feed forward - feedback tap combination (15,6) and 1.16×10^{-35} for feed forward - feedback tap combination (15,10) for a six-channel WDM-MDM system at a central wavelength of 1550nm and 10Gbps per channel. However, feed forward – feedback tap combinations (1,0) (2,2) (5,2) and (11,11) yielded the worst BERs ranging from

5.1974x10^{-3} to 5.2624 x '10^{-11} which in effect are not acceptable BERs.

The results demonstrate satisfactory equalization of power in MMF in order to alleviate mode coupling and modal dispersion.

Table 2. Results on comparison of number of DFE taps and BER

Feed forward tap number	Feedback tap number	BER
10	5	1.15 x 10^{-29}
14	0	1.28 x 10^{-26}
15	6	2.33 x 10^{-23}
15	10	1.16 x 10^{-35}

6 Conclusion

DFE for a DQPSK MWDM architecture using three modes LG 2 0 ,LG 2 1 and LG 2 2 and 6 wavelengths achieved the lowest BER at 2.33x 10^{-23} using 15 feed forward taps and 6 feedback taps. The DFE optical equalizer compensates for the effects of mode coupling and achieved 60Gbps through a 2.5km-long MMF a central wavelength of 1550nm. The DFE suppress higher order modes and effectively excites only four DMG per wavelength.

References

1. "Cisco White Papers," San Jose, California (2015)
2. Deloitte, Digital Democracy Survey (2016)
3. K. Nisar *et al.*, Journal of Network and Computer Applications, **36**, no. 2, pp. 933-948 (2013)
4. A. Amphawan *et al.*, Network and Complex Systems, **2**, no. 3, pp. 1 – 9 (2012)
5. S. Chaudhary *et al.*, Optik - Int. Journal for Light and Electron Optics, **125**, no. 18, pp. 5196–5198 (2014)
6. H. Al-Khafaji *et al.*, IEICE Electronics Express, **10**, no. 5, p. 20130044 (2013)
7. S. Aneesh *et al.*, Optical and Quantum Electronics, **48**, no. 5, pp. 1-10 (2016)
8. A. Amphawan, Opt. Eng., **50**, p. 102001 (2011)
9. N. Bozinovic *et al.*, Science, **340**, no. 6140, pp. 1545-1548 (2013)
10. J. Carpenter *et al.*, J. Lightw. Technol., **30**, no. 12, pp. 1978 - 1984 (2012)
11. A. Amphawan, Opt. Express, **19**, no. 10, pp. 9056-9065 (2011)
12. A. Amphawan, J. Mod. Opt., **59**, no. 5, pp. 460-469 (2012)
13. Y. Wu *et al.*, Opt. Express, **23**, no. 16, pp. 20867-20875 (2015)
14. T. Sakamoto *et al.*, J. Lightw. Technol., **33**, no. 6, pp. 1175-1181 (2015)
15. B. J. Puttnam *et al.*, *2016 21st European Conference on Networks and Optical Communications (NOC)*, pp. 29-34 (2016)
16. K. J. Park *et al.*, Opt. Express, **24**, no. 4, pp. 3543-3549 (2016)
17. D. Lee *et al.*, J. Lightw. Technol., **34**, no. 8, pp. 1754-1761 (2016)
18. A. Amphawan *et al.*, J. Mod. Opt., **60**, no. 20, pp. 1675-1683 (2013)
19. A. Amphawan *et al.*, *Int. Conf. on Opt. and Photon. Engineering (icOPEN)*, **9524**: SPIE (2015)
20. Y. Fazea *et al.*, J. Optical Commun. vol. **36**, pp. 327–333 (2015)
21. A. Amphawan *et al.*, Journal of the European Optical Society-Rapid Publications, **12**, no. 1, p. 12 (2016)
22. S. Ö. Arık *et al.*, J. Lightw. Technol., **34**, no. 11, pp. 2867-2880 (2016)
23. A. Amphawan *et al.*, J. Mod. Opt., **50**, no. 20, pp. 1745-1752 (2012)
24. P. J. Winzer, *39th European Conference and Exhibition on Optical Communication (ECOC 2013)*, pp. 1-4 (2013)
25. K. Choutagunta *et al.*, Journal of Lightwave Technology, **35**, no. 12, pp. 2451-2463 (2017)
26. Y. Weng *et al.*, Optical Fiber Technology, **36**, pp. 155-180 (2017)
27. Y. Weng *et al.*, Optical Fiber Technology, **36**, pp. 231-236 (2017)
28. X. Li *et al.*, *Optical Fiber Communication Conference*, Los Angeles, California, p. W2A.45: Optical Society of America (2017)
29. Y. Ren *et al.*, IEEE Transactions on Wireless Communications, **16**, no. 5, pp. 3151-3161 (2017)
30. RSoft, "OptSim 5.2," ed (2010)
31. G. Li *et al.*, Advances in Optics and Photonics, **6**, no. 4, pp. 413-487 (2014)

Simulation realization of 2-D wavelength/time system utilizing MDW code for OCDMA system

M. S. A. Azura, *C. B. M.* Rashidi, *S. A.* Aljunid, *R.* Endut, and *N.* Ali

Advanced Communication Engineering, Centre of Excellence, School of Computer and Communication Engineering, Universiti Malaysia Perlis, Kangar, 01000, Perlis, Malaysia.

Abstract. This paper presents a realization of Wavelength/Time (W/T) Two-Dimensional Modified Double Weight (2-D MDW) code for Optical Code Division Multiple Access (OCDMA) system based on Spectral Amplitude Coding (SAC) approach. The MDW code has the capability to suppress Phase-Induce Intensity Noise (PIIN) and minimizing the Multiple Access Interference (MAI) noises. At the permissible BER 10^{-9}, the 2-D MDW (APD) had shown minimum effective received power (*Psr*) = -71 dBm that can be obtained at the receiver side as compared to 2-D MDW (PIN) only received -61 dBm. The results show that 2-D MDW (APD) has better performance in achieving same BER with longer optical fiber length and with less received power (P*sr*). Also, the BER from the result shows that MDW code has the capability to suppress PIIN ad MAI.

1 Introduction

Optical Code Division Multiple Access (OCDMA) technology is one of the talented technologies to execute all-optical networks. It is a group of multiplexing and internetworking technologies that encodes/decodes signal through employing straightforward and cost-effective inactive optical component such as the signal multiplexing, routing and switching can be implement easily [1]. OCDMA can be realized in one-dimension (1-D) and two-dimension (2-D) encoding through spatial/spectral domain. The execution assessment of OCDMA framework in view of wavelength/time (W/T) code has been investigated by measuring the estimations of bit error rates for various number of active users [2]. The network capacity of the OCDMA is in high cardinality, compatible code with various bandwidth demands, throughput and acceptable at 10^{-9}-bit error rate (BER) [2, 3, 4]. The OCDMA systems undergo from assured noise such as phase induce-intensity noise (PIIN) shot noise and thermal noise [5]. Besides, multiple access interference (MAI) is the major performance degradation mainly when a large number of users are occupying in the OCDMA systems. Then, the most significant consideration is the code design for decreasing contribution at the MAI to the optical power receives [6]. In OCDMA systems, there are several techniques can be implemented to eliminate MAI. Spectral Amplitude Coding (SAC) is the one that can be used in this system. SAC schemes activate at bit rate, and as a result the condition for receiver bandwidth is tranquil [7, 8]. Here, the performance of 2-D MDW codes for OCDMA systems to achieve higher performance possible through suppressing PIIN and decreasing the MAI was analyzed [9, 10].

* Corresponding author: amyazura92@gmail.com

2 Development of 2-D Modified Double Weight (MDW) code

2-D MDW codes are also signified by using the $M \times N$ matrix form. The 2-D MDW codes with M as a number of wavelengths, N as a temporary codes length, W as a weight, λ_a and λ_c are auto-correlation and cross-correlation values respectively. There are denoted by $(M\ x\ N, W, \lambda_a, \lambda_c)$. The j_{th} user's 2-D codes $C^j_{M,N}$, in equation (1), is a matrix of M row vectors $d^j_{k,N}$ related to the temporary spreading; $d^j_{1,N} = c^j_{k,1}, c^j_{k,2}, \dots, c^j_{k,N-1}, c^j_{k,N}$ where $c^j_{k,i} \in \{0,1\}$ and k is the emitted wavelength of $k \in \{1,\dots,M\}$ [5, 7, 9].

$$C^j_{M,N} = \begin{bmatrix} d^j_{1,N} \\ d^j_{2,N} \\ \cdot \\ \cdot \\ \cdot \\ d^j_{M-1,N} \\ d^j_{M,N} \end{bmatrix} \quad (1)$$

The signals $r_{k,N}(t)$ are sum of the temporary spreading data of Fu user carried on the wavelength λ_k and expressed as $r_{k,N}(t) = \sum_{j=1}^{Fu} b^j_i(t) d^j_{k,N}$, where $b^j_i(t)$ is i_{th} user data bit of j_{th} and Fu is the number of users. The

M signals $r^j_{k,N}(t)$ are multiplexed and the total signal $R_{M,N}(t)$ transmitted on optical fiber is expressed as $M \times N$ [5, 7, 9].

$$R_{M,N}(t) = \begin{bmatrix} r_{1,N(t)} \\ r_{2,N(t)} \\ \cdot \\ \cdot \\ \cdot \\ r_{M-1,N(t)} \\ r_{M,N(t)} \end{bmatrix} \quad (2)$$

2-D MDW optical CDMA network are made of M (transmitter) and N (receiver). $A_{g,h}$ correspond to the code where $g \in (1,2,3,\ldots,M-1)$ and $h \in (1,2,3,\ldots,N-1)$. X_g and Y_k are spectral and spatial encoding respectively. Table 1 shows the 2-D MDW code projections [5, 7, 9].

Table 1. 2-D MDW Code Projections $k_1 = 4$ and $k_2 = 2$.

$X_{g,h}/Y_g$	[000011011]	[011000110]	[110110000]
$\begin{bmatrix}0\\1\\1\end{bmatrix}$	$\begin{bmatrix}000000000\\000011011\\000011011\end{bmatrix}$	$\begin{bmatrix}000000000\\011000110\\011000110\end{bmatrix}$	$\begin{bmatrix}000000000\\110110000\\110110000\end{bmatrix}$
$\begin{bmatrix}1\\1\\0\end{bmatrix}$	$\begin{bmatrix}000011011\\000011011\\000000000\end{bmatrix}$	$\begin{bmatrix}011000110\\011000110\\000000000\end{bmatrix}$	$\begin{bmatrix}110110000\\110110000\\000000000\end{bmatrix}$

$X = [x_0, x_1, x_2, \ldots, x_{M-1}]$ and $Y = [y_0, y_1, y_2, \ldots, y_{N-1}]$ are the code sequence of 1-D MDW codes. The 2-D MDW cross-correlation can be derived through four characteristic matrices of $A^{(d)}$, where $d \in (0,1,2,3)$ are defined as:

$$A^0 = Y^T X \quad (3)$$

$$A^1 = Y^T \bar{X} \quad (4)$$

$$A^2 = \bar{Y}^T X \quad (5)$$

$$A^3 = \bar{Y}^T \bar{X} \quad (6)$$

The cross-correlation between $A^{(d)}$ and $A_{g,h}$ can be defined as:

$$R^{(d)}(g,h) = \sum_{i=0}^{M-1} \sum_{j=0}^{N-1} a_{ij}^{(d)}\, a_{(i+g)(j+h)} \quad (7)$$

Where $a_{ij}^{(d)}$ is $(i,j)_{th}$ of $A^{(d)}$ and $a_{(i+g)(j+h)}$ is $(i,j)_{th}$ of $A_{g,h}$; $g \in (1,2,3,\ldots,M-1)$ and $h \in (1,2,3,\ldots,N-1)$.

Table 2 shows the cross-correlation of the 2-D MDW code generated from $R^{(d)}(g,h)$. From the table, $R^{(3)}(g,h)$ has nonzero value when $g \neq 0 \cap h \neq 0$. Furthermore, when $g \neq 0$ and $h \neq 0$, the values of $R^{(0)}(g,h)$, $R^{(1)}(g,h)$, $R^{(2)}(g,h)$ and $R^{(3)}(g,h)$ indicates the specific relationships. Cross-correlation function is defined by using $R^{(3)}(g,h)$ to eliminate influence due to $A_{g,h}$ from $R^{(0)}(g,h)$, $R^{(1)}(g,h)$, and $R^{(2)}(g,h)$ when $\neq 0 \cap h \neq 0$. New derived 2-D MDW cross-correlation function is expressed as;

Table 2. 2-D MDW Cross-Correlation.

$X_{g,h}$	$R^{(0)}(g,h)$	$R^{(1)}(g,h)$	$R^{(2)}(g,h)$	$R^{(3)}(g,h)$
$g = 0$, $h = 0$	$k_1 k_2$	0	0	0
$g = 0$, $h \neq 0$	k_1	k_1	0	0
$g \neq 0$, $h = 0$	k_2	0	$k_2(k_1$-1)	0
$g \neq 0$, $h \neq 0$	1	1	k_1-1	k_1-1

$$R^{(0)}(g,h) - R^{(1)}(g,h) - \frac{R^{(2)}(g,h)}{(k_1-1)} + \frac{R^{(3)}(g,h)}{(k_1-1)} = \begin{cases} k_1 k_2, & \text{for } g = 0 \text{ and } h = 1 \\ 0, & \text{otherwise} \end{cases} \quad (8)$$

3 2-D Wavelength/Time (W/T) Signal-to-Noise Analysis

Phase-induced intensity noise (PIIN), shot noise and thermal noise are three types of noises that have to be considered in the performance analysis. To identify the thermal lights produces, the photodiode can be expressed as follows [9]:

$$< i^2 > 2eIB + I^2 B\tau_c + \frac{4K_b T_n B}{R_L} \quad (9)$$

The noise power of I^2_{PIIN} can be expressed as below.

$$\langle I^2_{PIIN} \rangle = \frac{B_r \Re^2 P^2_{Sr}}{M\Delta f k_2^2 (MN-1)^2} \begin{bmatrix} k_1\, k_2\, (MN-1)^2 + \\ [k_2\,(W-1)(M-1)]^2 \end{bmatrix} \quad (10)$$

The short noise current can be written as follow:

$$\langle I^2_{short} \rangle = 2eB_r \left\{ \frac{\Re^2 P^2_{Sr}}{M k_2 (MN-1)} [k_1\, k_2 (MN-1) + 2k_1(W-1)(N-1) + 2k_2(W-1)(M-1) + 4(W-1)(M-1)(N-1)] \right\} \quad (11)$$

The thermal noise power is:

$$\langle I^2_{thermal} \rangle = \frac{4K_b\, T_n\, B_r}{R_L} \quad (12)$$

Total photocurrent output from receiver I_r:

$$\langle I_r^2 \rangle = \left[\frac{\Re\, P_{ST}\, k_1}{M}\right]^2 \quad (13)$$

The signal-to-noise ratio (SNR) equation at receiver is developed as follows:

$$SNR = \frac{I_r^2}{\langle I^2_{PIIN} \rangle + \langle I^2_{short} \rangle + \langle I^2_{thermal} \rangle} \quad (14)$$

Therefore, bit error rate (BER) can be expressed in terms of the signal-to-noise ratio (SNR) [9]:

$$BER(M) = \frac{1}{2} erfc\left(\sqrt{\frac{SNR}{8}}\right) \quad (15)$$

where $erfc$ is:

$$erfc(x) = \frac{2}{\sqrt{x}} \int_x^{\infty} e^{(-z^2)} dz \quad (16)$$

4 Simulation Results and Discussions

Table 3. Table of link parameters.

Parameters Used in Numerical Calculation	
PD quantum efficiency	η=0.6
Spectral width of broadband light source	$\Delta\lambda$=30 nm(Δ=3.75 THz)
Operating wavelength	λ_0=1.55μm
Electrical bandwidth	B=311 MHz
Data transmission rate	R_b=622 Mbps
Receiver noise temperature	T_n=300 K
Receiver load resistor	R_L=1030 Ω
Boltzmann's constant	K_b=1.38x10^{-23} W/K/Hz
Electron charge	e=1.6021764x10^{-19} coulombs
Light velocity	C=3x10^8 m/s

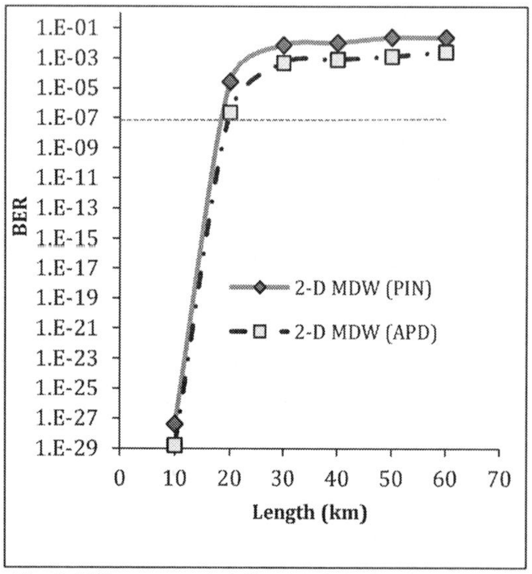

Fig. 1. BER versus optical fiber length for 2-D MDW codes.

Table 3 is the system parameters used to find the numerical results, while Figure 1 demonstrates the graph of BER versus optical fiber length in kilometer (km) for the 2-D MDW codes. In this simulation, 622 Mbps of bit rate is used. The number of length used is from 10 km until 60 km. According to the simulation and the result obtained, the graph increase for both bit rate due to the increasing of the optical fiber length. The system performances of BER 10^{-9} 2-D MDW (APD) at 19 km of fiber length have better performance compared to 2-D MDW (PIN) at fiber length of 18 km. This is because of the distance described by the number of dispersion and the signal loss.

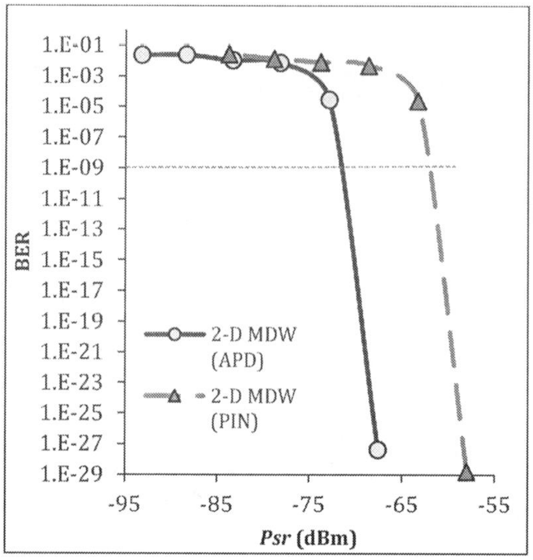

Fig. 2. BER versus effective received power (*Psr*) for 2-D MDW codes.

Figure 2 illustrates the curve of BER versus effective received power (*Psr*) for 2-D MDW codes. The values of effective received power (*Psr*) are varied from -95 dBm until -40 dBm. The number of bit rate used is 622Mbps. Then the length of optical fiber used is 10 km to 60 km. It can be seen that, 2-D MDW (APD) code requires -71 dBm more effective source power to achieve normal optical transmission requirements compared to 2-D MDW (PIN), which requires only -61 dBm. This situation occurs when the losses of fiber optic cable caused the *Psr* to decrease while the BER is increased.

5 Conclusion

From the numerical results, the *Psr* value for 2-D MDW (APD) codes is – 71 dBm while 2-D MDW (PIN) codes is -61 dBm. Therefore, photodiode APD are more effective compared to photodiode PIN because of photodiode APD are increase according their gain. In addition, the theoretical calculation demonstrates the number of BER 10^{-9} so, it's easily to suppress PIIN and eliminate MAI.

The author would like to acknowledge this work was supported by the High Ministry of Education Malaysia under the Fundamental Research Grant Scheme (FRGS) under grant #9003-00456.

References

1. C. B. M. Rashidi, S. A. Aljunid, F. Ghani, H. A. Fadhil, M. S. Anuar, A. R. Arief, Optik-International Journal for Light and Electron Optics, **125**, 17, pp. 4889-4894 (2014)
2. V. Jyoti, R. S. Kaler, Optik, pp. 851-857 (2011)
3. A. R. Arief Jamil Abdullah, S. A. Aljunid, A. M. Safar, J. M. Nordin, R. B. Ahmad, Fiber Optics and Optical Communication, **51**, 6 (2012)

4. C.C. Yang, Optical Fiber Technology, **14**, pp. 134-142 (2007)
5. A. R. Arief, S. A. Aljunid, M. S. Anuar, M. N. Junita, R. B. Ahmad, F. Ghani, *3rd International Conference on Photonics,* (2012)
6. C. B. M. Rashidi, S. A. Aljunid, M. S. Anuar, Hilal A. Fadhil, F. Ghani, Journal of Theoritical and Applied Information Technology, **61** (2014)
7. A. R. Arief, S. A. Aljunid, M. S. Anuar, M. N. Junita, R. B. Ahmad, F. Ghani, *IEEE International Conference in Control System, Computing and Engineering (ICCSCE)*, pp. 488-494 (2011)
8. C. C. Yang, IEEE Photonics Technology Letters, **17**, 6, pp. 1343-1345, (2005)
9. A. R. Arief, S. A. Aljunid, M. S. Anuar, M. N. Junita, R. B. Ahmad, Optik - International Journal for Light and Electron Optics, pp. 3786-3793 (2013)
10. N. Din Keraf, S. A. Aljunid, A. R. Arief, M. S. Anuar, C. B. M. Rashidi, P. Ehkan, M. N. Nurol, *International Conference on Electronic Design (ICED),* (2014)

DOI: 10.1051/epjconf/201716201014

Passively Q-switched of EDFL employing multi-walled carbon nanotubes with diameter less than 8 nm as saturable absorber

Siti Nur Fatin Zuikafly[1,*], *Fauzan* Ahmad[1], *Mohd Haniff* Ibrahim[2] and *Sulaiman Wadi* Harun[3]

[1]Malaysia Japan International Institute of Technology (MJIIT), Universiti Teknologi Malaysia, 54100 Kuala Lumpur, Malaysia
[2]Faculty of Electrical Engineering, Universiti Teknologi Malaysia, 81310 Skudai, Johor, Malaysia
[3]Photonic Research Centre, University of Malaya, 50603 Kuala Lumpur, Malaysia

Abstract. The paper demonstrates passively Q-switched erbium-doped fiber laser implementing multiwalled carbon nanotubes (MWCNTs) based saturable absorber. The paper is the first to report the use of the MWCNTs with diameter less than 8 nm as typically, the diameter used is 10 to 20 nm. The MWCNTs is incorporated with water soluble host polymer, polyvinyl alcohol (PVA) to produce a MWCNTs polymer composite thin film which is then sandwiched between two fiber connectors. The fabricated SA is employed in the laser experimental setup in ring cavity. The Q-switching regime started at threshold pump power of 103 mW and increasable to 215 mW. The stable pulse train from 41.6 kHz to 76.92 kHz with maximum average output power and pulse energy of 0.17 mW and 3.39 nJ are produced. The shortest pulse width of 1.9 μs is obtained in the proposed experimental work, making it the lowest pulse width ever reported using MWCNTs-based saturable absorber.

1 Introduction

As an alternative to solid state laser, ultrafast fiber laser has been the centre of attention these past few years. Q-switching techniques in generating pulsed laser enable shorter pulse duration as well as higher peak power as compared to mode-locked regime [1]. Passive modulation in Q-switching operation is preferred over the active one due to its simplicity and compactness [2]. Saturable absorber (SA) is used to modulate the intracavity losses without the aid of any external modulator, hence the simplicity implied. Multiwalled carbon nanotubes (MWCNTs) have its own advantages over other materials for the use as SA for its intrinsic properties. These properties include good thermal stability and mechanical strength [3] besides its versatility in fibre laser configurations [4]. Saturable absorbers based on carbon nanotubes have been proven to yields fast response time in sub-picosecond [5] as well as wide working range in broadband operation [6] due to its' broad diameter distribution [4]. Carbon nanotubes (CNTs) are comprised of covalently bonded carbon atoms in cylindrical forms. Multi-walled carbon nanotubes are CNTs with more than one graphene cylinders in its structure, making it stronger mechanically, thermally and spectroscopic-wise [7]. Furthermore, the fabrication of MWCNTs is comparably simpler than other reported SAs. Its multi-walled structure makes it far less sensitive towards environment changes [8].

Following the increased interest in carbon nanotubes, many works have been reported on MWCNT-based passive saturable absorber demonstrating different laser performance in Q-switched regime. The broadband properties of MWCNTs for pulsed laser generation utilized MWCNTs with diameter of 10 to 20 nm had successfully demonstrated at 1 μm [9], 1.5 μm [3, 10] and 2 μm [10] region. A paper reported by Kasim et al. [9] utilizes MWCNTs-polyethylene oxide (PEO) composite as SA in 1 micron region, yielding a maximum of 24 kHz repetition rate and 143.5 nJ of pulse energy operating at central wavelength 1060.2 nm. The shortest pulse width is at 12.18 μs at maximum input pump power of 65.72 mW. Ahmad et al. [3] resides the same MWCNTs-PEO based SA in Erbium-Doped Fiber laser (EDFL), producing a stable pulse train with repetition rates ranging from 4.5 kHz to 20.0 kHz as well as maximum pulse energy of 15.3 nJ and shortest pulse width of 8.8 μs. Meanwhile Ahmad et al. [10] applied MWCNTs-PVA in EDFL to generate maximum repetition rate of 29.9 kHz, maximum average output power of 1.49 mW, pulse energy of 49.8 nJ and shortest pulse width of 3.49 μs. In 2 micron region, polyvinyl alcohol (PVA) is used as the host polymer in a MWCNTs-PVA composite film with the ratio of 1:4. The lasing threshold of the laser is very high at 1591.3 mW, tunable to a maximum of 2100 mW [11].

In this paper, a 1.9 μs of pulse width is produced, which is the shortest ever reported using MWCNTs based passive saturable absorber. The results may be caused by the diameter of the MWCNTs used as the diameter of the carbon nanotubes play an important role in the SA's absorption characteristics during laser generation [12-13].

* Corresponding author: fauzan.kl@utm.my

DOI: 10.1051/epjconf/201716201014

2 Material Preparation

The mixture of MWCNTs (diameter of < 8 nm and length distribution ~1 to 2 μm) with a 1% sodium dodecyl sulfate (SDS) in deionized water was ultrasonicated for 1 hour at 50 W to form a homogeneous solution. The solution was then centrifuged at 1000 rpm to remove large particles of undispersed MWCNTs. To prepare the host polymer, 1 g of polyvinyl Alcohol (PVA) (Sigma Aldrich) was dissolved in 120 ml De-ionized (DI) water with the aid of magnetic stirrer at room temperature. A MWCNTs/PVA composite was prepared with the solution ratio of 1:1 and ultrasonicated for several hours. The MWCNTs/PVA solution was poured into a petri dish and left to dry at room temperature for about 48 hours to produce film with thickness of around 50 μm as shown in Fig. 1.

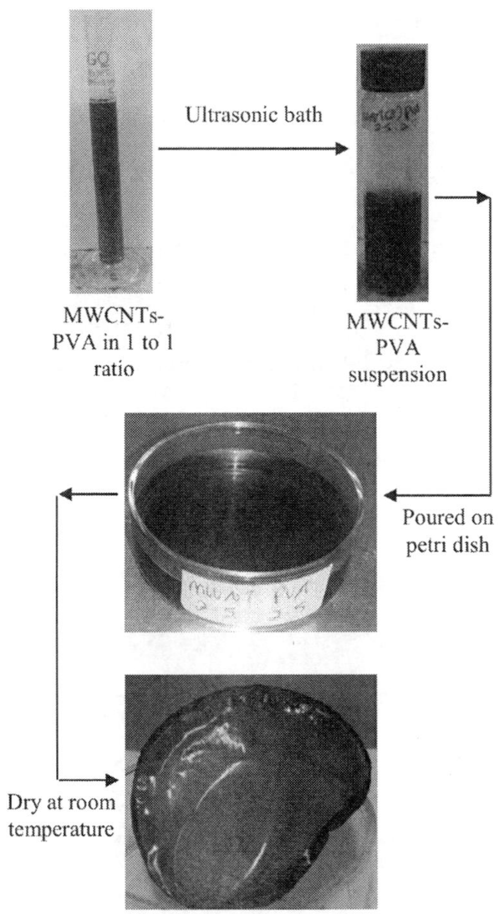

Fig. 1. Preparation of MWCNTs-PVA SA.

3 Experimental Set Up

The experimental setup for the Q-switched EDFL consists of a 1 m long erbium-doped fiber (EDF), a 1480/1550 wavelength division multiplexer (WDM), an isolator, the fabricated SA, and a 95/5 output coupler. The isolator is placed to ensure uni-directional propagation of the light oscillating the cavity. The instrument is arranged in ring configuration as shown in Fig. 2, pumped with a 1480 nm laser diode via the WDM. The output of the Q-switched fiber laser is observed through an oscilloscope (OSC), optical power meter (OPM) and optical spectrum analyser (OSA). The output is fed through a 3 dB coupler from the 5% tapped laser output while the other 95% remains in the cavity.

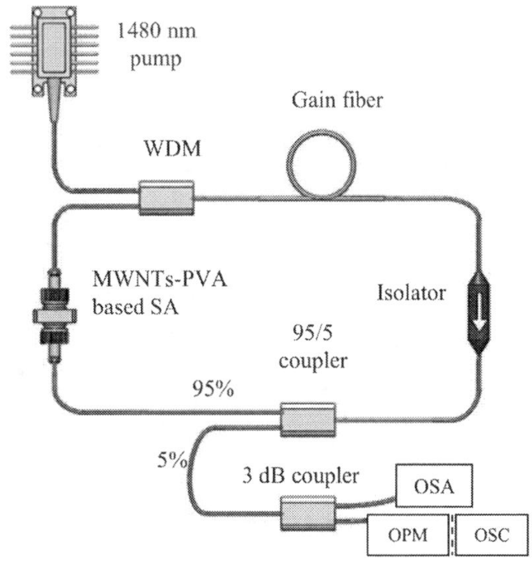

Fig. 2. Schematic configuration of the Q-switched EDFL.

4 Results and Discussions

The EDFL started lasing in Q-switched regime at around 103 mW pump power. The threshold input pump power is considered high compared to other reported works [3, 10], this is due to opaque properties of PVA which when combined with carbon black material, require higher pump power to fully saturate the SA. Beyond the maximum power, the pulse laser is no longer observed. However, reducing the pump power to 215 mW and less enable Q-switching operation to be accomplished again. Fig. 3 shows the central operating wavelength of the laser output of 1558.2 nm at maximum input pump power with 3-dB bandwidth of 0.4 nm. Fig. 4(a), 3(b), 3(c), and 3(d) shows the oscilloscope traces of the pulse generated at four different pump powers which is 131 mW, 187 mW, and 215 mW, with the respective single pulse envelope, respectively. The recorded pulse train shows an acceptable intensity fluctuation when the light propagates in the laser cavity for every cavity round trip. The pulse repetition rate at 215 mW with 76.92 kHz

 DOI: 10.1051/epjconf/201716201014

generated the shortest pulse width of 1.9 μs. The proposed works produced the shortest pulse width using MWCNTs based SA in EDFL compared to the reported works in [3,10] and other SAs. For example, the paper reported by Luo et al. [14] using molybdenum disulphide (MoS_2) as SA produced shortest pulse duration of 5.4 μs. Topological insulators as passive saturable absorber was demonstrated using bismuth selenide (Bi_2Se_3) and bismuth telluride (Bi_2Te_3), yielding pulse width of 1.9 μs and 13 μs respectively [15,16].

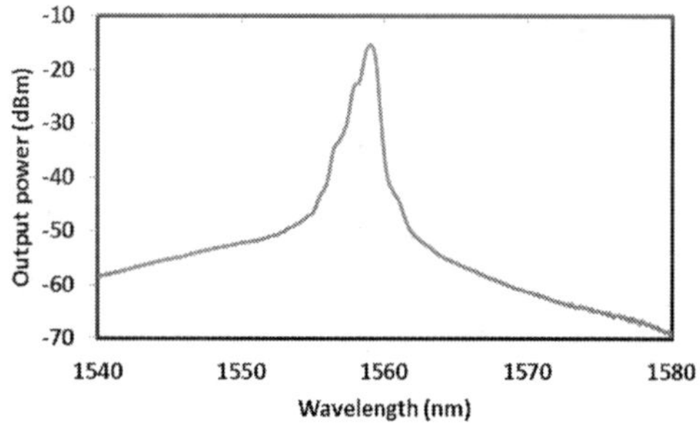

Fig. 3. OSA trace at 215 mW.

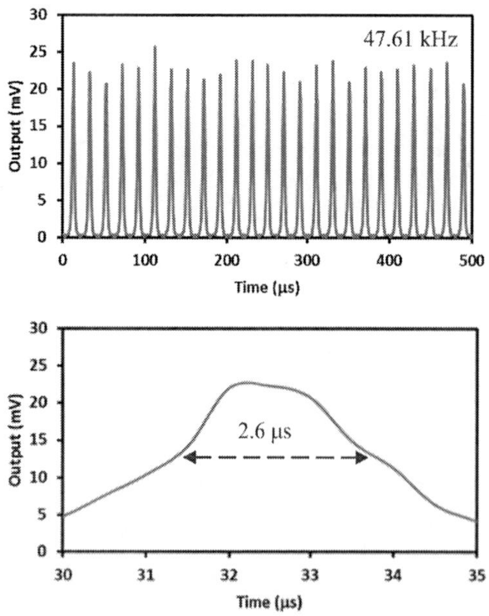

(a)

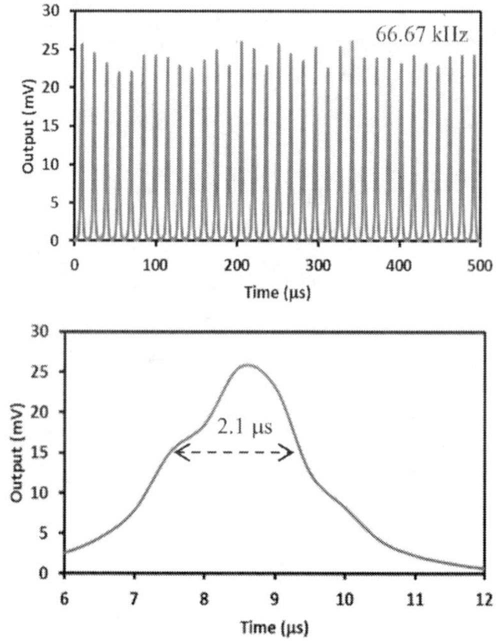

(b)

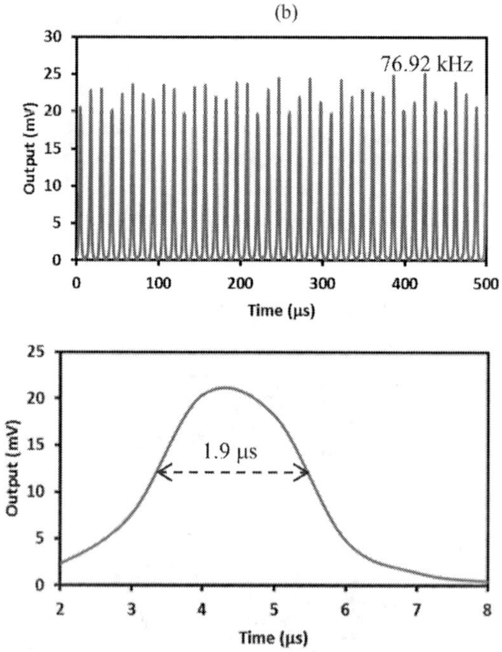

(c)

Fig. 4. Pulse train at (a) 131 mW, (b) 187 mW, and (d) 215 mW with respective single pulse envelope.

The repetition rate and the average output power of the pulsed laser increases with the increasing pump power. Higher pump power supplied to the cavity means shorter time is taken for the inversion number of the gain medium to reach the threshold, which then resulted in higher calculated peak power [2]. As the pump power is tuned from 103 mW to 215 mW, the repetition rate increases almost steadily from 41.6 kHz to 76.92 kHz. The repetition rate at incident pump power of 215 mW, is highly comparable to other previously reported works on MWCNTs-based SAs [4, 9, 11] [17-19] The average output power of 0.17 mW is obtained at maximum pump power. The relationship between the repetition rate and average output power with respect to the pump power is clearly shown in Fig. 5.

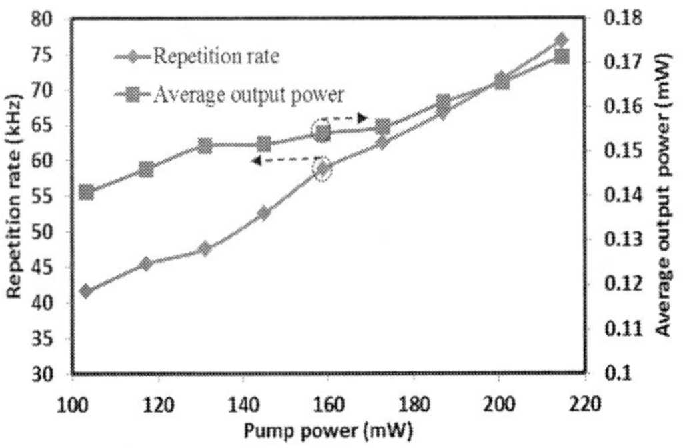

Fig. 5. Pulse repetition rate and average output power versus pump power MWNTs/PVA polymer composite (<8nm).

Fig. 6 shows the pulse width and pulse energy as a function of the pump power. Increasing the pump power makes the pulse width shorter while the pulse energy becomes larger. As the pump power is increased, the energy required for the transition of carriers from the valence band to the conduction band is higher, thus shortening the time taken for the light round trip in the cavity. This in turn contributes to the increasing calculated pulse energy. The pulse width decreases from 3.3 μs to 1.9 μs, while the pulse energy is increased to a maximum of 3.39 nJ with the increment of the pump power from threshold to 215 mW. The characteristics of the proposed Q-switched laser are in agreement with other reported Q-switched lasers [17-20].

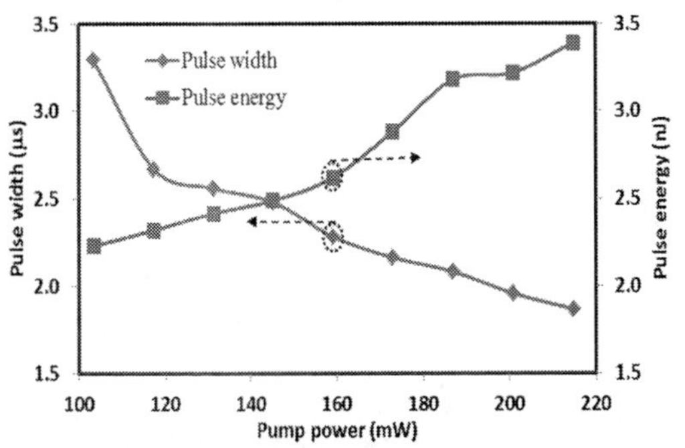

Fig. 6. Pulse width and pulse energy versus pump power MWNTs/PVA polymer composite (<8nm)

4 Conclusion

A stable passively Q-switched EDFL incorporating MWCNT composite polymer thin film with diameter of less than 8 nm as saturable absorber is demonstrated. To the best of our knowledge, the laser performance generated a 1.9 μs pulse width, the shortest pulse duration reported using MWCNTs based SA. The maximum repetition rate of 76.92 kHz is obtained with a maximum of 3.39 pulse energy.

References

1. U. Keller. *Laser Physics and Application* **11** (New York, Springer-Verlag 2007)
2. L. Liu, Z. Zheng, X. Zhao, S. Sun, Y. Bian, J. Liu, J. Zhu. Optics Comm. **294**, 267 (2013)
3. F. Ahmad, H. Haris, R. M. Nor, N. R. Zulkepely, H. Ahmad, S. W. Harun. Chin. Phys. Lett **31**, 034204 (2014)
4. A. Martinez, Z. Sun. Nature Photonics 7, 842 (2013)
5. M. Jung, J. Koo, Y. M. Chang, P. Debnath, Y. W. Song, J. H. Lee. Laser Phys. Let. 9, 669 (2012)
6. M. Chernysheva, A. Rozhin, Y. Fedotov, C. Mou, R. Arif, S. M. Kobtsev, E. M., Dianov, S. K. Turitsyn. Nanophotonics 6, 1 (2017)
7. R. J. Young, L. Deng, L. Gong, I. A. Kinloch. Nature Photonics 7, 842 (2013)
8. Y. G. Wang, L. Zhang, X. C. Lin. Optik 124, 4465 (2013)
9. N. Kasim, A. H. H. Al-Masoodi, F. Ahmad, Y. Munajat, H. Ahmad, S. W. Harun. Chin Phys. Let. 12, 031403 (2014)
10. H. Ahmad, M. F. Ismail, S. N. M. Hasan, F. Ahmad, M. Z. Zulkifli, S. W. Harun. Applied Optics 53 (2014).
11. I. M. Babar, M. B. S. Sabran, Z. Jusoh, H. Ahmad, S. W. Harun, A. Halder, M. C. Paul, S. Das, S. K. Bhadra. Ukr. J. Phys. Opt. 15, 173 (2014)
12. S. Kivisto, T. Hakulinen, A. Kaskela, B. Aitchison, D. P. Brown, A. G. Nasibulin, E. I. Kauppinen, A. Harkonen, o. G. Okhotnikov. Optics Express 17, 2358 (2009)
13. J. W. Nicholson, R. S. Windeler, D. J. DiGiovanni. Optics Express 15, 9176 (2007)
14. Z. Luo, Y. Huang, M. Zhong, Y. Li, J. Wu, B. Xu, H. Xu, Z. Cai, j. Peng, J. Weng. J. of Lightwave Tech. 32, 4077 (2014)
15. Z. Yu, Y. Song, J. Tian, Z. Dou, H. Guoyu, K. Li, H. Li, X. Zhang. Optics Express 22, 11508 (2014)

16. Y. Chen, C. Zhao, S. Chen, J. Du, P. Tang, G, Jiang, H. Zhang, S. Wen, D. Tang. IEEE J. of Selected Topics in Quantum Electron. 20, 0900508 (2013)

17. S. W. Harun, M. A. Ismail, F. Ahmad, M. F. Ismail, R. M. Nr, N. R. Zulkepely, H. Ahmad. Chin. Phys. Let. 29 (2012)

18. M. Z. Razak, Z. S. Saleh, F. Ahmad, C. L. Anyi, S. W. Harun, H. Arof. Optical Eng. 55, 106112 (2016)

19. A. Al-Masoodi, M. H. M. Ahmad, h. Arof, A. M. Banabila, S. W. Harun. Optoelectron. and Advanced Materials 9, 1104 (2015)

20. A. Al-Masoodi, M. F. Ismail, , F. Ahmad, N. Kasim, Y. Munajat, H. Ahmad, S. W. Harun. Microwave and Optical Tech Lett. 26, 1770 (2014)

Chemical tapering of polymer optical fiber

Affa Rozana Abdul Rashid[1,*], *Amna Afiqah* Nasution[1], *Aisyah Hanim* Suranin[1], *Nur Athirah* Taib[1], *Wan Maisarah* Mukhtar[1], *Karsono* Ahmad Dasuki[1], and *Abang Annuar* Ehsan[2]

[1] Faculty of Science and Technology, Universiti Sains Islam Malaysia, 71800, Nilai, Negeri Sembilan, Malaysia
[2] Institute of Microengineering and Nanoelectronics, Universiti Kebangsaan Malaysia, 43600 UKM Bangi, Selangor, Malaysia

Abstract. Polymer optical fibers (POFs) have significant advantages over numerous sensing applications. The key element in developing sensor is by removing the cladding of the fiber. The use of organic solvent is one of the methods to create tapered POF in order to expose the core region. In this study, the etching chemicals involved is acetone, methyl isobutyl ketone (MIBK), and acetone-methanol mixture. The POF is immersed in 100%, 80%, and 50% of acetone and MIBK dilution. In addition, the mixture of acetone and methanol is also used for POF etching by the ratio 2:1 of the volume. Acetone has shown to be the most reactive solvent towards POF due to its fastest etching rate compared to MIBK and acetone-methanol mixture. The POF is immersed and lifted from the solution for a specific time, depending on the power loss properties for the purpose of producing unclad POF. In comparison to silica fiber optic, the advantages of POF in terms of its simple technique and easy handling enable it to produce unclad POF without damaging the core region. The surface roughness of the POF is investigated under the microscope after being immersed into different solvent. This method of chemical tapering of POF can be used as the fundamental technique for sensor development. Next, the unclad fiber is immersed into ethanol solutions in order to determine the reaction of unclad POF towards its surrounding. The findings show that this particular sensor is sensitive towards concentration changes ranging between 10 wt% to 50 wt%.

1 Introduction

Polymer optical fibres (POFs) also known as plastic optical fibres possess many advantages over conventional silica or glass optical fibre, especially in sensing application. The wide application of POFs in sensor design tends to originate from the development of polymer technology. Their advantages which include low cost or low cost competitive system over the glass or silica optical fibre have initiated a great interest among the scientific community to further explore POF. POF is commonly found in the multi-mode fibres with large diameter, thus allowing the propagation of multiple modes [1, 2]. The diameter of POF must be reduced by removing the cladding of the fiber to develop sensor based polymer optical fiber. The etched region of POF becomes more sensitive to the surrounding due to the evanescent wave in the cladding [3]. Etching refers to the reduction process of localised diameter, which came within any profile from linear and gradual changes to abrupt steps [4]. There are several ways to remove the cladding of POF, in which one of them is described as hot drawing method but this method does not efficiently remove the cladding because the fiber will break before any elongation occurs [4]. Hence, the best way to remove the cladding of POF is by immersing chemical etching in the organic solvent. The penetration depth of the evanescent field and the power proportion were observed to increase within the tapered region as the fiber is immersed in the testing solution. Evanescent wave in fiber optic occurs when a part is light up, whereby the incident between different refractive index of core and cladding is transmitted into the cladding rather than being reflected back which is based on the principle of total internal reflection [5]. In this particular kind of evanescent wave absorbance sensing phenomena, the measurement is calculated based on the loss of propagating light at the interface of core and cladding medium [6]. The method of etching tapers is performed using acetone, MIBK, and acetone-methanol mixture for the purpose of developing step-index POF in this project.

2 Methodology

The POF model adopted in this experiment is known as step-index Polyethylene Jacketed Optical. The diameter of cladding and core must be set at 1000 μm and 980 μm, respectively. Three types of solution have been used in the etching of polymer optical fiber, namely acetone, MIBK and acetone-methanol mixture. 4cm of the jacket of the fiber is removed and placed inside a testing solution. In the first step, the POF is immersed in pure solution of acetone and MIBK. Next, the fiber is neutralized with de-ionize water. The experiment is repeated using dilute solution of 80% and 50% acetone and MIBK, respectively. On top of that, the mixture of

* Corresponding author: affarozana@usim.edu.my

DOI: 10.1051/epjconf/201716201015

acetone and methanol is also used in the ratio with the volume (2:1) [7] in order to study the etching effect of mixture towards the unjacketed POF. The preparation of different solutions which include ethanol should be properly done in order to study the effect of absorbing liquid on the unclad POF. In this process, the dilution of 10wt%, 20wt%, 30wt%, 40wt%, and 50wt% of each ethanol concentration are prepared.

3 Result and Discussions

3.1 Effect of Acetone on Etching Rate

Acetone is capable of removing the cladding of polymer which is composed by fluoropolymer with PMMA as its core. Fig. 1 shows the etching rate of POF that is immersed in acetone without dilution. As can be seen in the figure, it takes around 10 s to reach the core of the POF fiber. Next, the power loss increases abruptly as the acetone reaches and etches the core part after 10 s. Hence, it can be concluded that it only a short time is needed to reach the core but the diameter of the fiber becomes unpredictable and fragile as a result of pure acetone. Fig. 2 and Fig. 3 show the etching rate of the fiber in acetone with concentration percentage of 80% and 50%. The surrounding environment needs to be controlled which include humidity, temperature, pressure, and others. The fiber breaks and damages at a very short time because the power loss increases rapidly. The time taken to lift the fiber decreases as the concentration increases.

For 80% concentration of acetone, it takes around 1 min before the loss of the power output to abruptly increase. Meanwhile, the fiber does not show any abrupt change even though the fiber is immersed in the solution for 2 hours for 50% concentration of acetone as shown in Fig. 3. The graph of the power loss slightly increases from 12.13 mW to 12.27 mW with the time recorded at 85 s to 90 s, respectively. However, the power output still fail to show any changes even with 2 hours of immersion.

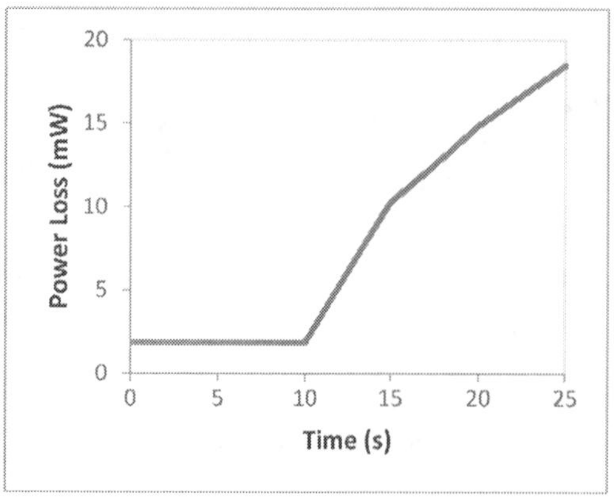

Fig. 1. Graph of power loss vs etching time of POF in 100% Acetone

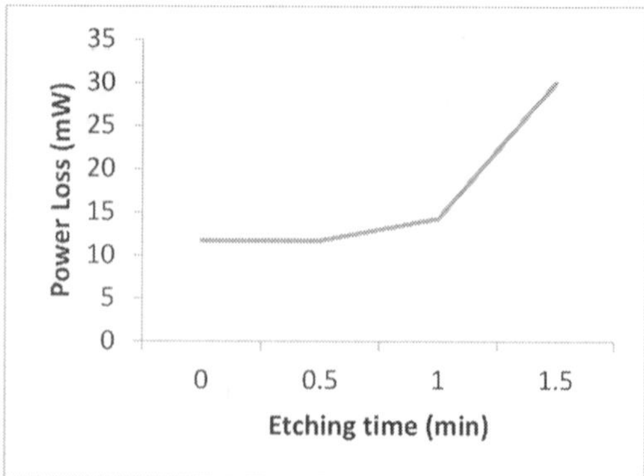

Fig. 2. Graph of power loss vs etching timeof POF in 80% Acetone

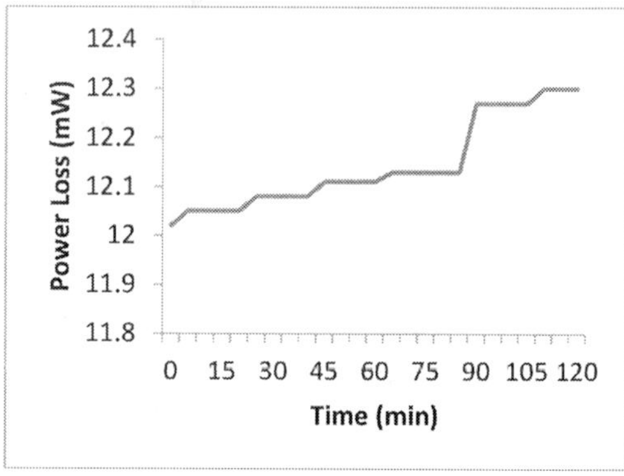

Fig. 3. Graph of power loss vs etching time of POF in 50% Acetone

A more concentrated acetone is more likely to dissolve or affect the surface fiber, which will result in tapered POF. Fig. 4 shows the macroscopic view of POF with 80% of acetone immersion, and the diameter of the fiber is observed to decrease to 990 μm. In Fig. 5, the immersion of 50% acetone gives a smoother surface compared to higher concentration of acetone. Moreover, the collision molecules of the acetone with the surface of POF become less reactive when the dilution of the acetone increases. This factor seems to result in a smooth and clear surface after it is lifted from the solution.

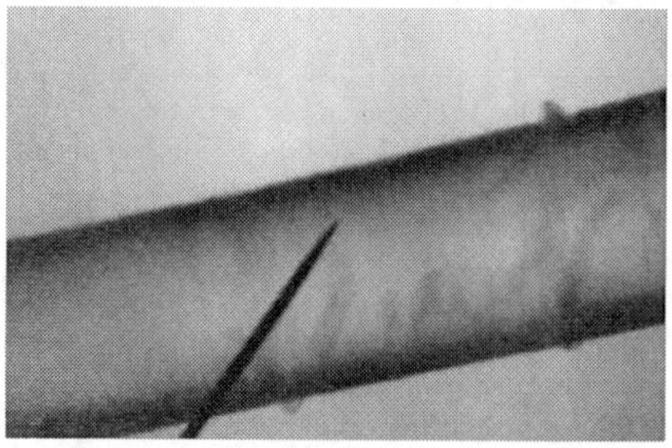

Fig. 4. Microscopic image of POF with immersion of 80% Acetone

DOI: 10.1051/epjconf/201716201015

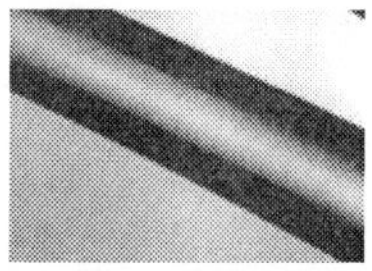

Fig. 5. Microscopic image of POF with immersion of 50% Acetone

3.2 Effect of MIBK on Etching Rate

MIBK is also an organic solvent that is very efficient in removing the cladding of polymer optical fiber such as acetone. Fig. 6 shows that around 210 s is needed to reach the fiber core when the solution for immersion is pure MIBK. Acetone appears to react faster compared to MIBK. These organic solvents can be used in the etching process of POF because it becomes susceptible to brittle stress fracture for a short while. In 80% concentration, the power loss increases after 3 min and it reaches the fibers core around 4.5 min. Next, the power loss increases abruptly as it reaches and etches the core as shown in Fig. 7. For 50% of MIBK, the time taken to etch POF is longer as the concentration of the solution decreases. The power loss becomes much faster when the fiber is immersed in 80% compared to 50% of MIBK, in which it takes around 10.07 min to reach the fibers core as shown in Fig. 8.

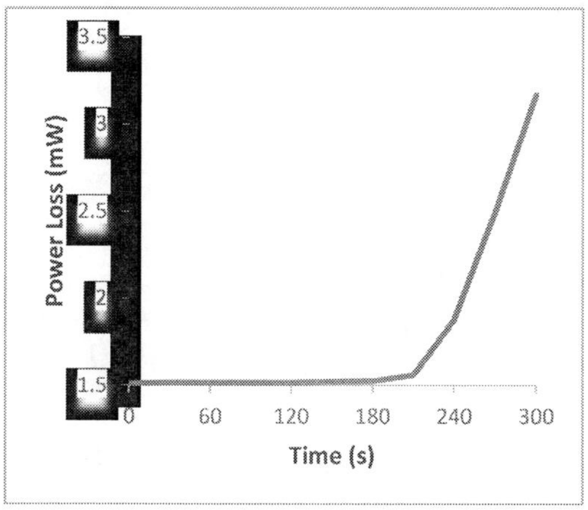

Fig. 6. Graph of power loss vs etching time of POF in 100% MIBK

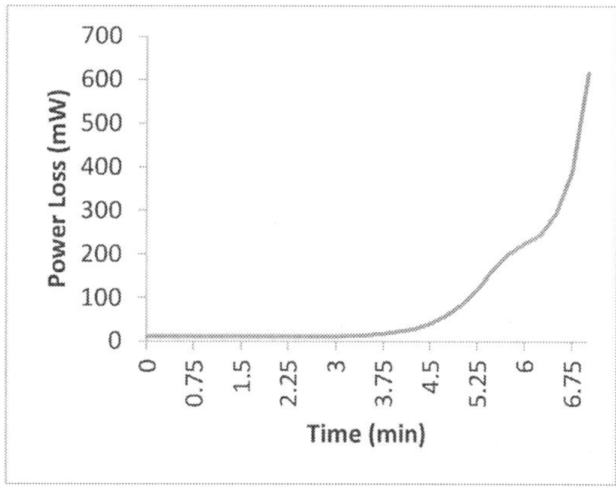

Fig. 7. Graph of power loss vs etching time of POF in 80% MIBK

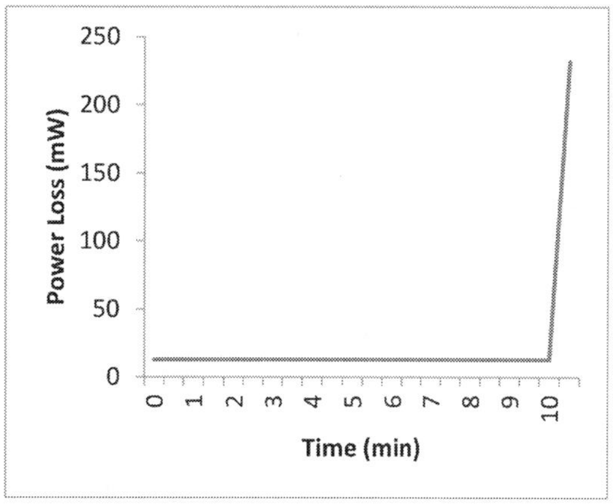

Fig. 8. Graph of power loss vs etching time of POF in 50% MIBK

Fig. 9 shows the microscope image of unclad POF when it is immersed in 50% of MIBK solution. As can be seen, it has a rougher surface compared to when it is immersed in 50% of acetone. The surface roughness in the immersion of 50% and 80% of MIBK is quite similar. The POF respectively reduce the diameter to 970 µm and 980 µm for 50% and 80% of MIBK.

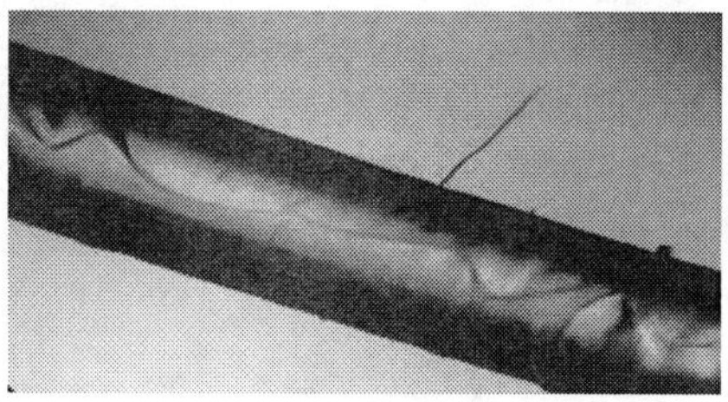

Fig. 9 Microscopic image of POF with immersion of 80% MIBK

 DOI: 10.1051/epjconf/201716201015

3.3 The Effect of Acetone-Methanol Mixture on Etching Rate

Fig. 10 shows POFs that are etched through the immersion in a mixture of acetone and methanol by ratio (2:1). The power output loss stops to increase slowly after 50 minutes of immersion, and no changes can be observed until the two hours of immersion ends at 503.5 mW. Acetone plays an important role in etching due to its extremely reactive nature towards the polymer. For this particular solution, the mixture of the solution managed to produce tapered POF but with longer time compared to acetone and MIBK. Generally, acetone which is ketone can be mixed with any solvent including methanol. The function of methanol in the solution is to enhance the reactivity of the solvent towards the fiber surface.

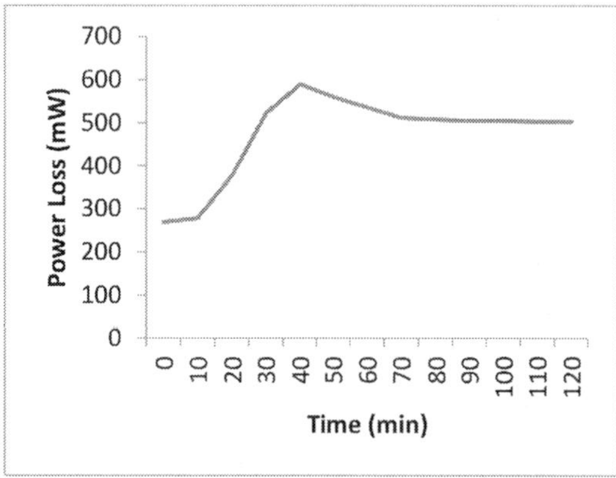

Fig. 10 Graph of power loss vs etching time of POF in the mixture of Acetone and Methanol (40:20)

For the surface structure of acetone-methanol mixture, the fiber fails to etch smoothly based on the presence of residue on the surface of POF after being immersed into the solution after 2 hours as shown in Fig. 11. This is in contrast with the microscopic image of acetone shown in Fig. 4, which produces smooth and clean surface after being immersed in 50% of acetone with the same 2 hours duration as applied in acetone-methanol mixture.

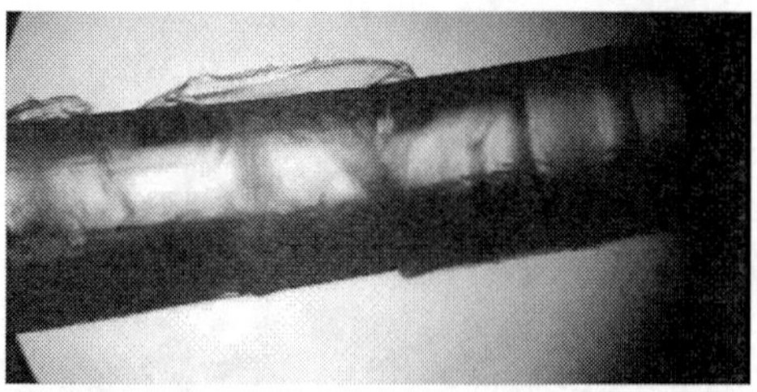

Fig. 11. Microscopic image of POF with immersion of mixture of acetone and methanol

3.4 Measurements of Ethanol Solution on Unclad POF

The initial loss of this sample is fixed at 11.16 dBm when immersed in ethanol concentration. POF sample unclad with 100% of acetone solution is immersed in different concentrations of ethanol solution. Fig. 12 shows the graph of power output reading versus the concentration of ethanol, whereby higher concentration tends to produce lower power output. Moreover, the evanescent absorption coefficient increases when the concentration increases. The evanescent field penetrates into both solution and evanescent wave absorbance occurs when the unclad POF is immersed in ethanol solution, in which it will increase as the concentration of solution increases [8]. The refractive index of the ethanol also increases when the concentration percentage is increased [9]. According to Khijwania and Gupta [10], the increase of penetration depth can result in the increase of absorbance of evanescent field by the absorbing medium, which can lead to the increase of sensitivity of the sensor.

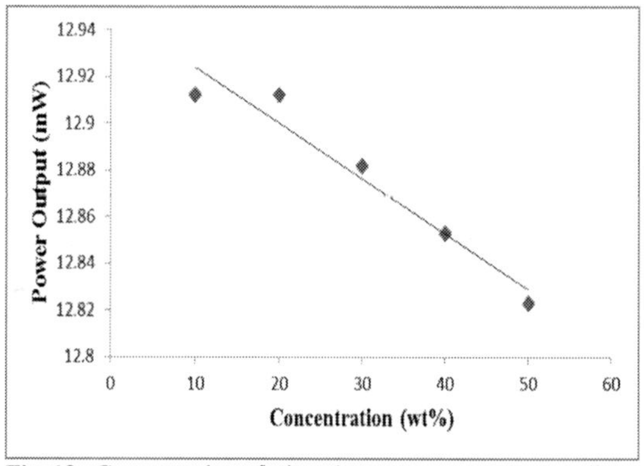

Fig. 12. Concentration of ethanol versus power output reading

4 Conclusion

In the evanescent wave fiber optic sensor, the cladding of the optical fiber is removed and made to be in direct contact with the evanescent wave. Moreover, any materials which are capable of absorbing the particular wavelength of the incoming light inside the optical fiber will take the energy out of the evanescent wave. Hence, the higher the concentration of the acetone and MIBK, the faster is the time taken for the fiber to break because the fiber becomes very brittle easily. Acetone which acts as the etching solution can give a smoother surface compared to MIBK and mixture acetone-methanol. Therefore, it is better for the etching solution to dilute with distilled water even though the etching rate will be longer. Finally, the surface quality will be good for sensing application because the rough surface which is formed on POF will affect the waveguiding process.

This project was sponsored by the Kementerian Pendidikan Malaysia (KPM) through grant scheme (USIM/RAGS/FST/36/50215) and Universiti Sains Islam Malaysia (USIM) through grant schemes (PPP/USG-0115/FST/30/12315).

References

1. Kara Peters, Smart Material and Structures, 20 013002, pp. 17 (2011)

2. Kuzyk, Mark G., *Polymer Fiber Optics: Materials, Physics, and Applications (1st Ed)* pp. 7-10 (Florida: CRC Press, 2007)

3. Batumalay, M; Lokman, A; Ahmad, F; Arof, H; Ahmad, H; Sulaiman, W.H., Sensor Journal, 13(12), pp 4702-4705 (2013)

4. D.F. Merchant, P.J. Scully, and N.F. Schmitt., Sensor and Actuators, 76: 365-371 (1999)

5. SM John, Evanescent wave fibre optic sensors: theory – Shodhganga (2011)

6. N. Punjabi, J. Satija, and S. Mukherji, Smart Sensors, Measurement and Instrumentation, pp. 25-45 (2015)

7. X. Hu, C. F. J. Pun, H. Y. Tam et al., Optics Express. **22** (15) : 18807-18817 (2014)

8. P Suresh Kumar, CPG Vallabhan, VPN Nampoori, VN Sivasankara Pillai and Radhakrishnan, Journal of Optics A: Pure and Applied Optics, 4 pp. 247-250 (2002)

9. J. Nowakowska, *The refractive indices of Ethyl Alcohol and Water Mixtures,* Loyola University Chicago (1939)

10. S.K Khijwania, B. D. Gupta., Optic Communication, 152(4-6), pp. 259-262 (1998)

 DOI: 10.1051/epjconf/201716201016

32 channel DQPSK DWDM-PON for local area network using dispersion compensation fiber

Yousef Fazea[1,2]* *and Angela* Amphawan[1]

[1]Optical Computing and Research Laboratory, School of Computing, Universiti Utara Malaysia, 06010 Sintok, Kedah, Malaysia
[2]Internetworks research Laboratory, School of Computing, Universiti Utara Malaysia, 06010 Sintok, Kedah, Malaysia

Abstract. In the contest to improving optical channel rates in dense wavelength multiplexing (DWDM) service to realize 40Gb/s, different adjustment and coherent detection plans are being investigated. In this paper, we present differential quaternary phase-shift keying (DQPSK) balance for the arrangement of a 32-channel DWDM in a Local Area Network (LAN).

1 Introduction

The media communications industry is hustling to support optical channel rates in dense wavelength multiplexing (DWDM) product and services to 40Gbps and more [1-4]. Legacy infrastructure in Local Area Networks (LANs) has faced genuine difficulties in pleasing to popular cloud infrastructures and the expansion in data traffic due to diverse services such as, triple play services applications which involves video, audio, data and many more bandwidth-intensive applications [4-7]. Capacity-limited routers in LAN will benefit essentially from the expansion to 100Gbps. Nonetheless, 40Gbit/s DWDM deployment in LANs is not a panacea. There are a few difficulties, such as, multivendor interconnectivity, compatibility issues with the current layout, scattering resilience, and cost viability for widespread deployment [8, 9]. The 40Gbit/s DWDM is going to be deployed firstly in core and long-haul networks, which makes it importantly required in optical networking transport [10, 11]. In order to address this, numerous modulation format and coherent detection schemes are being investigated. For long-haul 100Gbps DWDM, dual polarization quadrature phase shift keying (DP-QPSK) is envisaged to be the standard modulation format.

In this paper, we introduce differential quaternary phase-shift keying (DQPSK) modulation for DWDM deployment in a LAN. This paper proceed as follows; Section 2 describes the DWDM model and parameters followed by Section 3 where the simulation results are analyzed. Conclusion of this paper is presented in Section 4.

Table 1. Fiber parametrs

Fiber	Attenuation. (dB/km)	Dispersion. (ps/km-nm)	Dispersion slope (ps/km-nm^2)	Effective core area (μm^2)
SMF	0.2	17	0.075	70
DCF	0.5	-85	-0.3	22

Table 2. Parameters Configuration

Data date	40Gbit/s
Seq. length	64
Sample per Bit	256
Wavelength Spacing	200GHz
Length	50Km
Power	7.5dBm

2 Simulation and Model Description

* Corresponding author: yosiffz@internetworks.my

DOI: 10.1051/epjconf/201716201016

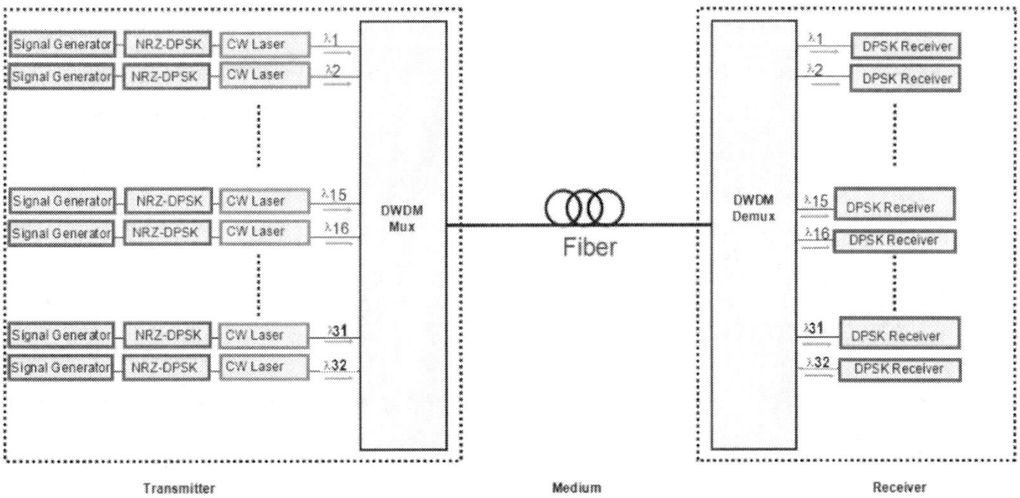

Fig.1. Simulation of 32 channels DWDM-PON

Fig. 1 presents a DQPSK DWDM model in a LAN, simulated using Synopsis OptSim 5.2 [12]. The network maybe divided into three parts, the transmitter, the channel and the receiver. The central frequency is 190THz with a spacing of 200GHz and transmitter power of 7.5 dBm. The parameters for the optical fiber are given in Table 1 whereas the transmission parameters are presented in Table 2.

3 Result Analysis and Discussions

The DWDM model has been evaluated using several metrics measurement as the eye diagram, and optical signal-to-noise ratio (OSNR). The eye opening is proportional to the transmitted time delay, therefore as can be observed in Figure 2, narrow eye-opening result to low propagation delay whereas the OSNR is shown in Figure 3.

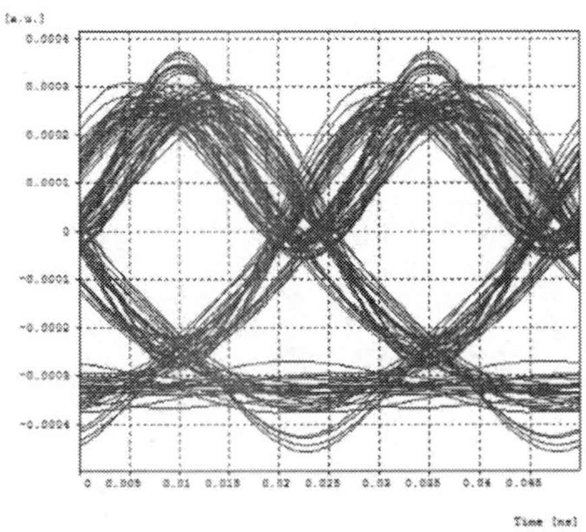

Fig. 2. (a) eye diagram of channel 1

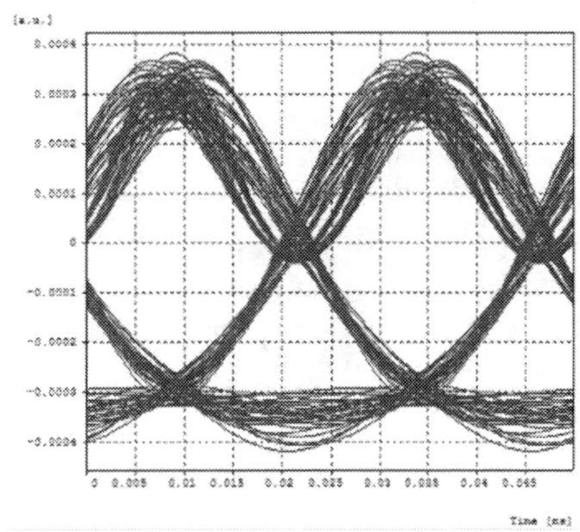

Fig. 2. (b) eye diagram of channel 16

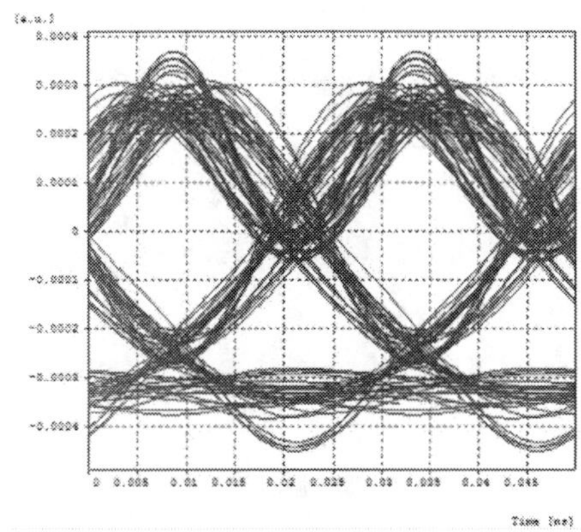

Fig. 2. (c) eye diagram of channel 32

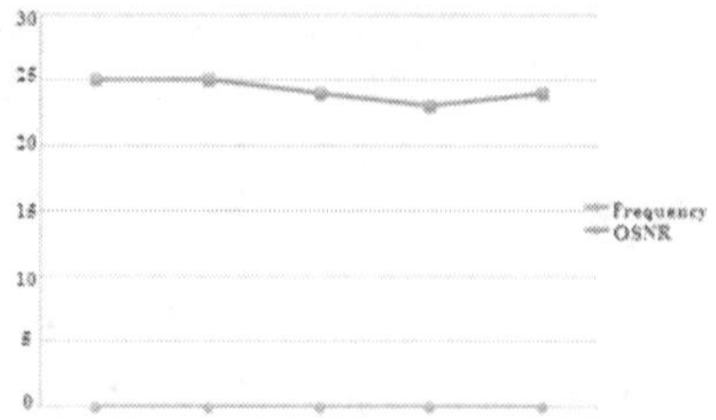

Fig. 3. Frequency and OSNR

4 Conclusion

A 40Gbps 32-channel DWDM has been achieved using DQPSK modulation. The DQPSK DWDM system has potential for increasing current capacity-constrained routers in LAN.

References

1. F. Cavaliere, L. Giorgi, and R. Sabella, Ericsson Review, (2013)
2. Y. Fazea and A. Amphawan, Journal of Optical Communications, vol. **36**, pp. 327-333 (2015)
3. Y. Fazea, A. Amphawan, and H. Abualrejal, Advanced Science Letters (2016)
4. A. Amphawan and Y. Fazea, Journal of the European Optical Society-Rapid publications, vol. **11** (2016)
5. S.-J. Park, C.-H. Lee, K.-T. Jeong, H.-J. Park, J.-G. Ahn, and K.-H. Song, *Journal of Lightwave Technology*, **22**, p. 2582 (2004)
6. J.-P. Elbers, *Optical Fiber Communication (OFC), collocated National Fiber Optic Engineers Conference, 2010 Conference on (OFC/NFOEC)*, 2010, pp. 1-3 (2010)
7. F. Council and F. Penetration, *Status North America*, (2011)
8. D. Fritzsche, D. Breuer, A. Ehrhardt, J. Yeh, Z. Zhang, and W. I. Way, *National Fiber Optic Engineers Conference*, 2009, p. NWD4 (2009)
9. J. Choi, M. Yoo, and B. Mukherjee, *Journal of Optical Communications and Networking*, **2**, pp. 38-50 (2010)
10. G.-K. Chang, A. Chowdhury, Z. Jia, H.-C. Chien, M.-F. Huang, J. Yu, *et al.*, *Journal of Optical Communications and Networking*, **1**, pp. C35-C50, (2009)
11. Y. Ma, Y. Qian, G. Peng, X. Zhou, X. Wang, J. Yu, *et al.*, *Optical Fiber Communication Conference*, 2012, p. PDP5D. 7 (2012)
12. I. Rsoft Design Group, *OptSim User Guide* (2010)

Optimizing bandwidth of fiber optical parametric amplifier with different combinations of higher-order dispersion coefficient signs

N. Othman[1], *N. S. M.* Shah[1,*], *K. G.* Tay[1], and *N. A.* Cholan[1]

[1]Faculty of Electrical and Electronic Engineering, Universiti Tun Hussein Onn Malaysia, 86400 Batu Pahat, Malaysia

Abstract. This paper investigates the influence of different combinations of second-order (β_2), fourth-order (β_4) and sixth-order (β_6) dispersion coefficient signs (negativity/positivity) to the fiber optical parametric amplifier gain performance. The numerical simulation has exploited the fourth-order Runge-Kutta method to solve the coupled amplitude equations which basically represent the parametric process of four-wave mixing. In the normal regime at which β_2 is positive, the fiber optical parametric amplifier exhibits a poor gain spectrum and it is well performed once the pump is presumed to be in anomalous regime i.e. β_2 is negative. The effect of β_4 is significant when the signal wavelength is further from the pump wavelength and a wide amplification bandwidth is produced when the fiber have positive β_4 in the anomalous regime. As for β_6, unfortunately in this simulation its influences are hardly can be seen probably because of its small value regardless its sign, but, from phase-mismatch equation it is shown that its impact is prominent when the signal is positioned way far from the pump. All in all, this shown that the signs combination of the higher-order dispersion coefficients are significant in order to optimize the FOPA amplification bandwidth.

1 Introduction

The nonlinear effect of FWM has been utilized in many applications such as the parametric oscillators [1], FWM-assisted lasers [2], FWM-cascaded [3], optical generators [4] and filter [5]. The ability of FWM effect to amplify a signal wave together with the generation of an idler wave made the effect convenient for those mentioned applications. One of the applications which also exploited the nonlinear effect of FWM is the fiber optical parametric amplifier (FOPA). The FOPA researches in the communication systems have greatly developed since its capability in providing adjustable center frequency and adjustable gain spectra have surpassed the conventional amplifiers i.e. Raman amplifier and Erbium-doped fiber amplifier [6].

There are several ways to optimize the FOPA gain spectrum for instance by increasing the pump power to obtain high parametric gain [7] and place the pump close to zero dispersion frequency to attain wide amplification bandwidth [8]. Either way, one can also include the higher-order dispersion coefficient of a fiber in order to optimize the FOPA gain spectrum. The higher-order dispersion coefficients have been proven to be important in obtaining the desire amplification bandwidth [9]. The involvement of dispersion coefficients up to sixth-order shows significant improvement of the amplification bandwidth [10]. However, the signs of these coefficients are needed to be well combined or else it will result in substandard gain spectrum. Hence, in this paper the effects of the combination of signs of the second-order, fourth-order and sixth-order dispersion coefficients to the FOPA gain spectrum were investigated. During the numerical simulation, the FOPA is assumed to be experience losses and the pump is depleted.

2 Mathematical model

The FOPA amplifies the signal by utilizing the effect of FWM which caused by the third-order nonlinear susceptibility. The coupled amplitude equations are given by [11]

$$\frac{dA_p}{dz} = i\gamma\left[A_p\left(\left|A_p\right|^2 + 2(|A_s|^2 + |A_i|^2)\right)\right] + 2i\gamma A_s A_i A_p^* \exp(i\Delta\beta z), \quad (1)$$

$$\frac{dA_s}{dz} = i\gamma\left[A_s\left(|A_s|^2 + 2\left(\left|A_p\right|^2 + |A_i|^2\right)\right)\right] + i\gamma A_i^* A_p^2 \exp(-i\Delta\beta z), \quad (2)$$

$$\frac{dA_i}{dz} = i\gamma\left[A_i\left(|A_i|^2 + 2\left(\left|A_p\right|^2 + |A_s|^2\right)\right)\right] + i\gamma A_s^* A_p^2 \exp(-i\Delta\beta z), \quad (3)$$

are basically represented the process of FWM, where γ is the fiber nonlinearity and A_j for $j \in \{p, s, i\}$ are the amplitudes of the pump, signal and idler, meanwhile A_j^* are the corresponding complex conjugate. Noteworthy, the fiber attenuation can be taken into account by adding term $-(\alpha/2)A_j$ to the right hand side of each equations.

* Corresponding author: shahida@uthm.edu.my

The phase-mismatch, $\Delta\beta$ along the fiber length z is generally expressed as [11]

$$\Delta\beta = 2\sum_{m=1}^{\infty}\frac{\beta_{2m}}{(2m)!}(\Delta\omega)^{2m}, \quad (4)$$

where $\Delta\omega = \omega_p - \omega_s$ is the frequency detuning and $\omega_k = 2\pi c/\lambda_k$ for $k \in \{p, s\}$ are the frequency of pump and signal in terms of wavelength, λ_k. The odd order of dispersion coefficients do not contribute to the gain spectrum, that is why only even order are taken into account [11]. The quantity of $\beta_{2m} = d^{2m}\beta(\omega)/d\omega^{2m}$ with β as mode-propagation constant expanded by Taylor series, can be related to the higher-order dispersion coefficient calculated at zero dispersion frequency, ω_0 such as [9]

$$\beta_{2m} = \sum_{n=2m}^{\infty}\frac{\beta_{n0}}{(n-2m)!}(\omega_p - \omega_0)^{n-2m}. \quad (5)$$

Since this paper will investigate the linear phase-mismatch curve with the involvement of second-order, fourth-order and sixth-order dispersion coefficient denoted by β_2, β_4 and β_6, respectively, thus by combining (4) and (5) for $m = 1$ to 3, the $\Delta\beta$ can be written as

$$\Delta\beta = \beta_2(\Delta\omega)^2 + \frac{\beta_4}{3}\left[2(\omega_p - \omega_0)^2 + \frac{1}{4}(\Delta\omega)^2\right](\Delta\omega)^2 + \frac{\beta_6}{24}\left[(\omega_p - \omega_0)^4 + \frac{1}{15}(\Delta\omega)^4\right](\Delta\omega)^2. \quad (6)$$

The β_2 sign is depends on the position of pump i.e. negative if the pump wavelength is in anomalous regime and positive if the pump wavelength is in normal regime. In practice, the positivity or negativity of the β_4 and β_6 at the respective pump wavelength are depend on the fiber dispersion profile. However, in this paper their signs are assumed to be negative or positive depending on the combination.

In order to compute the FOPA parametric gain, firstly, the Eqs. (1)-(3) were solved by using built-in function of Matlab 2014, *ODE45* which basically adopting the numerical method named fourth-order Runge-Kutta. Then, the parametric gain for the respective wavelength was obtained by dividing the output with the input signal power such as

$$G(\mathrm{dB}) = 10\log\left(\frac{P_{s,\mathrm{out}}}{P_{s,\mathrm{in}}}\right), \quad (7)$$

and generally P_s is the square of A_s i.e $P_s = |A_s|^2$ for both input and output.

3 Results and Discussion

This section simulates the impacts of the different combinations of β_2, β_4 and β_6 signs on FOPA gain spectrum. The pump power, P_p of 1 W was launched into the 500 m long highly-nonlinear dispersion-shifted fiber (HNL-DSF) to amplify the signal with the power of -40 dBm. Note that, the HNL-DSF has the zero dispersion wavelength (ZDW) at 1556.5 nm. The fiber attenuation and nonlinearity are 0.82 dB/km and 11.5 $\mathrm{W^{-1}km^{-1}}$ correspondingly. Given that $\beta_2 = 1.970 \times 10^{-2}$ $\mathrm{ps^2/km}$, $\beta_4 = 6.231 \times 10^{-5}$ $\mathrm{ps^4/km}$ and $\beta_6 = 1.178 \times 10^{-8}$ $\mathrm{ps^6/km}$, the sign of the respective higher-order coefficients were manipulated depending on the combinations. There are eight combinations of β_2, β_4 and β_6 signs [9]. These combinations are basically can be divided into two categories, firstly, when the pump is placed in the normal regime of HNL-DSF dispersion profile which results in positive value of β_2 and secondly, when the pump is lauched in the anomalous regime at which the β_2 value is negative. In order to effectively investigate the effects of the higher-order dispersion coefficients sign to the FOPA gain spectrum, the position of pump is fixed at $\lambda_p = 1558$ nm and it is assumed to be in the normal or anomalous regimes, depending on the case of study.

Fig. 1 pictures the FOPA gain spectra when the pump is assumed to be in the normal regime. It is worth noting that since the total gain spectrum of FOPA is symmetric with respect to λ_p, here the half gain spectra were plotted. It can be seen that FOPA experiences poor gain due to the domination of the positive β_2 value in (6) thus results in large phase-mismatch especially when the signal wavelength offsets further from the pump. Note that, the real parametric gain occurs when $\Delta\beta$ is lies in between $-4\gamma P_p$ to 0. In Fig. 2, the corresponding $\Delta\beta$ of the gain spectra in Fig. 1 were plotted against the $\lambda_s - \lambda_p$. Evidently, the reason of poor gains are because of the $\Delta\beta$ curve lies outside the proper area, especially the curves in Fig. 2(a) and Fig. 2(b). In addition, it is worth to mention that the positive β_4 is also contributing to this large phase-mismatch. The positive values of both β_2 and β_4 have monopolized the $\Delta\beta$ and even the negative value of β_6 is unable to encounter it, which explain why the almost similar spectrum is obtained in Fig. 1(a) and Fig. 1(b). On the other hand, Fig. 2(c) and Fig. 2(d) show that with the negative value of β_4, some parts of $\Delta\beta$ curve lie within $-4\gamma P_p$ and 0 notably when the λ_s is far from the λ_p which produces the parametric gain at the respective wavelength; see Fig. 1(c) and Fig. 1(d). Unfortunately, in these figures, the influence of β_6 is not clear which evidenced by the no changes of the resulted gain spectrum regardless the sign of β_6. This is probably because the

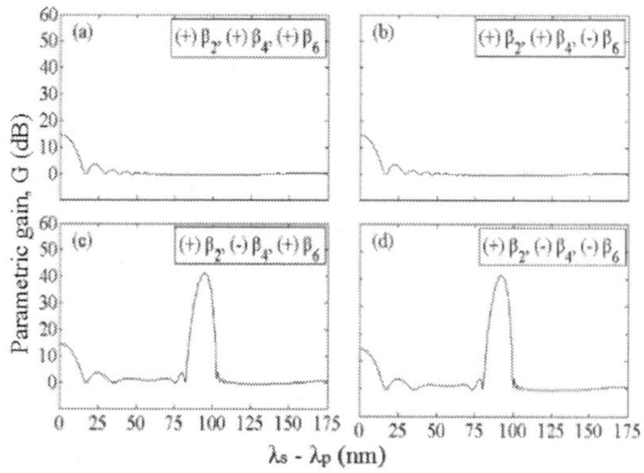

Fig. 1. Half gain spectrum of the FOPA when λ_p is assumed to be in normal regime.

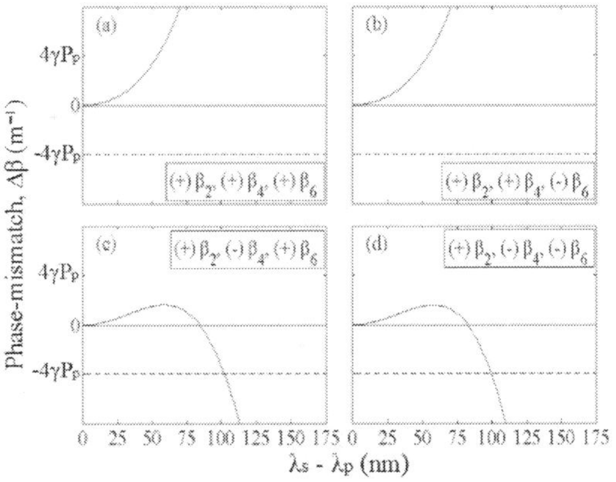

Fig. 2. The phase-mismatch of the FOPA gain spectrum corresponding to Fig. 1.

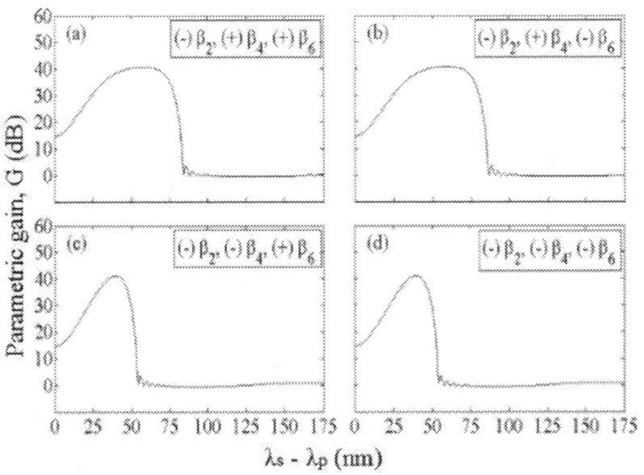

Fig. 3. Half gain spectrum of the FOPA when λ_p is assumed to be in anomalous regime.

value of β_6 is very small and consequently gives no effect to (6).

Next, in Fig. 3 the pump is presumed to be positioned in the anomalous regime in which the value of β_2 is negative. It exhibits that the FOPA can perform efficiently in this regime. The obtained amplification bandwidth implies that the $\Delta\beta$ definitely lies in the range of $-4\gamma P_p$ to 0 and Fig. 4 has proven it. As seen in Fig. 3(a) and Fig. 3(b), the positive value of β_4 causes the FOPA to amplify the signal at the larger λ_s then results in the high parametric gain and subsequently the FOPA bandwidth. This also portrays that the value of β_4 is significant when the distance of pump and signal wavelength is large enough. Similar with previous resulted gain spectra, the changes of β_6 sign from positive to negative does not affect the $\Delta\beta$ curve. Additionally, in Fig. 4(c) and Fig. 4(d), the effect of β_6 regardless its sign has been suppressed by the domination of the negative values of both β_2 and β_4, which then results in almost similar spectrum in Fig. 3(c) and Fig. 3(d). Although the effects of β_6 to the FOPA gain spectra cannot be clearly seen in this simulation results, it is believed that if the value of β_6 is carefully tailored, the large amplification bandwidth is possibly obtained [10].

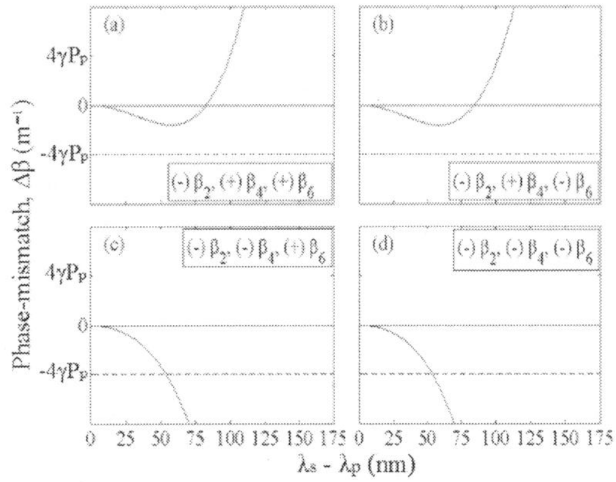

Fig. 4. The phase-mismatch of the FOPA gain spectrum corresponding to Fig. 3.

4 Conclusion

This paper has numerically investigated the effects of the different combination of β_2, β_4 and β_6 signs to the FOPA gain spectrum. In order to realize that, the well-known coupled amplitude equations have been solved by using fourth-order Runge-Kutta method in Matlab 2014. When the pump is placed in normal regime, the positive value of β_2 causes the large phase-mismatch and the consequence is worst if it is combined with the positive β_4. Nonetheless, the FOPA performance is practically works well in the anomalous regime in which β_2 is negative. This is because the phase-mismatch in this regime is small particularly when FOPA is operated close to ZDW. Noteworthy, the influence of β_4 and β_6 is more significant when signal is launched at the wavelength far from the pump. Yet, the effect of β_6 sign is hardly can be observed because of β_6 small value and the effect is almost disappeared if it is combined with both positive or both negative values of β_2 and β_4, since their impact are dominant in (6). To summarize, the combination of $\beta_2 > 0$, $\beta_4 > 0$ and $\beta_6 > 0$ is resulted in a poor gain spectrum as the phase-mismatch is very large for this combination. Generally, the FOPA amplification bandwidth can be optimized if the values of β_2, β_4 and β_6 are carefully tailored and their signs combination are well chosen.

This work was supported in part by Centre for Graduate Studies Universiti Tun Hussein Onn Malaysia (UTHM) and Research Supporting Grant Scheme (RSGS) Vot U102 granted by UTHM.

References

1. B. Sun, K. Hu, D. Chen, Y. Wei, S. Gao, S. He, J. Light. Technol. **30**, 1937 (2012)
2. N. A. Cholan, M. H. Al-Mansoori, A. S. M. Noor, A. Ismail, M. A. Mahdi, Appl. Phys. B Lasers Opt. **115**, 251 (2014)
3. N. A. Cholan, M. H. Al-Mansoori, A. S. M. Noor, A. Ismail, M. A. Mahdi, Opt. Express **21**, 6131 (2013)

 DOI: 10.1051/epjconf/201716201017

4. N. S. M. Shah, M. Matsumoto, Opt. Commun. **284**, 4687 (2011)
5. D. Xu, Y. Cao, Z. R. Tong, J. P. Yang, Optoelectron. Lett. **12**, 417 (2016)
6. M. E. Marhic, P. A. Andrekson, P. Petropoulos, S. Radic, C. Peucheret, M. Jazayerifar, Laser Photon. Rev. **9**, 50 (2015)
7. H. Pakarzadeh, M. Taghizadeh, M. Hatami, J. Nonlinear Opt. Phys. Mater. **25,** 1650023 (2016)
8. P. S. Maji, P. R. Chaudhuri, IEEE Photonics J. **7**, 1 (2015)
9. S. K. Chatterjee, S. N. Khan, P. R. Chaudhuri, Opt. Commun. **332**, 244 (2014)
10. P. Dainese, G. S. Wiederhecker, A. A. Rieznik, H. L. Fragnito, H. E. Hernández-Figueroa, *SBMO/IEEE MTT-S International Microwave and Optoelectronics Conference Proceedings* 2005, 92 (2005)
11. M. E. Marhic, *Fiber Optical Parametric Amplifiers, Oscillators and Related Devices* (Cambridge University Press, UK, 2008)

 DOI: 10.1051/epjconf/201716201018

Polymer optical fiber tapering using chemical solvent and polishing

L. S. Supian[1*], *Mohd* Syuhaimi Ab-Rahman[1], and *Norhana* Arsad[1,2]

[1]Department of Electrical and Electronic Engineering, Faculty of Engineering and Built Environment
[2]Institute of Microelectronic and Nanotechnology, Faculty of Engineering and Built Environment
Universiti Kebangsaan Malaysia, 43000, Bangi, Selangor, Malaysia

Abstract. A method for developing polymer optical fiber (POF) directional coupler is introduced where the initial procedure includes using chemical solvent to remove the cladding, and bare out the core in order to align the unclad center of the fiber with other similar fiber to develop a coupler. The process is safe, simple, inexpensive and require low operation skill. The etched fiber offers improvement to the performance of various POF devices, i.e, couplers and sensors. Instead of relying only on silica or glass fiber, POF now can be used as an alternative to improve the network performance in short distance communication system. The measurement parameters laid out offer great outcomes. However, the couplers intended to be developed is yet to be realized, where deeper research and various experiments are needed in order to develop a simple but optimum performance coupler that can be used for various applications.

1 Introduction

Couplers are common passive optical device that functions to combine two or more wavelengths or signals into one single optical fiber. There have been several methods on tapering polymer optical fiber for short-haul communication device. The existing method to develop the coupler is fused coupler [1], one of the most popular techniques of manufacturing a coupler. The process is simple where at least two fibers are brought close in proximity at a certain length and fused at the particular area. The results will produce a coupler where the middle part of the fiber has diameter of around ~1.0 mm – 1.2 mm. The narrower cores increase the spread of the evanescent field in the cladding and this assist the coupling process. The drawback emerges through this method where it is quite impossible to control the spacing of the two cores that held close in proximity and the length of the coupling region accurately [2,3].

In this particular study, chemical etching technique is used to fabricate a directional optical coupler. A quite similar but different technique [4] is already developed where the fibers are embedded into a solid material such as plastic and the flat surface is then polished until the cladding of the fibers is removed at 4 microns thick. A D-shape fiber results and the plastic is removed. The cladding is thinned out by etching using hydrofluoric acid. Two D-shape fibers are joint longitudinally along the flat areas using index-matching epoxy-resin. This technique shows better precision in terms of spacing between the two fibers and the coupling length than fused-taper method [5]. A technique of removing the cladding of polymethylmethacrylate (PMMA) polymer optical fiber using chemical solvent is used to create etched tapers at certain length in the middle part of the fiber. In consequence of tapering of the multimode optical fiber cladding, higher modes of the fiber are removed while some modes are redistributed. This technique is conventionally applicable to glass fiber. However, POF is becoming widely used in optical communication system and this would be a jump start for fabricating highly-potential coupler using this technique.

2 Fabrication method/experiment

Polymer optical fiber used in this work has diameter core of ø = 980 μm and diameter cladding of ø = 20 μm thick. Chemical solvent i.e. acetone is used in this studies in order to remove cladding layer. Due to the effectiveness of the solvent to impair and remove the cladding layer as reported by the research done by Merchant, Scully and Schmitt [2], pure 100% acetone is used in the experiment without any additional liquid or solvent involved or change the concentration of the solvent.

The first set of experiment involves a flat-base plate of 80 mm based (x_1 mm) × 25 mm height (y_1 mm) as in Fig. 3(a). Pure solvent of 10 ml are used and polymer fiber strand of about 30 mm are held stationary at both ends which the center part of the polymer fiber is dipped thoroughly inside the acetone. Time varies for each of the polymer fiber dipped inside the solvent. Trials of more than 20 secs failed as the fiber at the center section becomes susceptible to brittle stress fracture at certain time i.e. threshold time. Break will precede the brittleness once the fiber held is moved rigorously or tension is applied onto the polymer fiber in etching. One factor that contributes to the breaking of the fiber is the bending of the strand of fiber influenced by the height of the side of the plate. Here, the temperature is taken as default and not taken into consideration. After each of the etching process is done, the affected region is polished evenly in order to remove the cladding layer effectively. Fig. 1. shows fiber after etching process and the fiber is polished along the affected region. Cloudy surface is observed around the affected region.

* Corresponding author: cawa711@gmail.com

 DOI: 10.1051/epjconf/201716201018

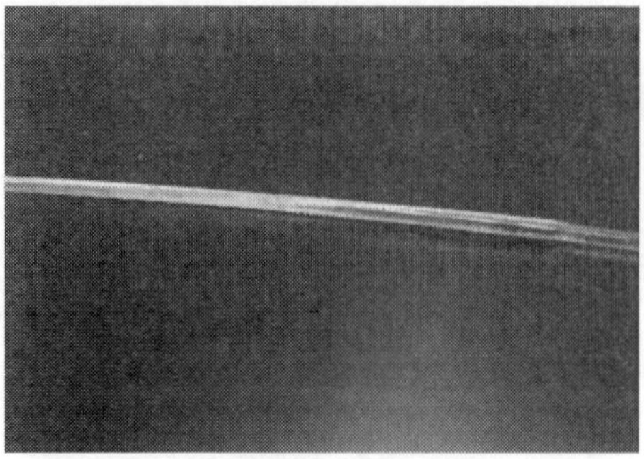

Fig. 1. Etched and polished fiber outer structure

The second set of the experiment involves different flat-base plate of 60 mm based (x_2 mm)× 10 mm height (y_2 mm)as in Fig. 3(b). Similar amount of pure solvent are used i.e. 10 ml. Due to the lower bending of the fiber strand as a factor of the lower height of the side plate, longer time are taken from initial time of 5 secs up till 35 secs for each of the polymer fiber dipped into the pure solvent. Again, the temperature is taken as default and not taken into consideration. For the second and third set of experiment, polishing is excluded. Fig. 2. shows fiber after etching process and no polishing is applied on the etched region. Less cloudy surface is observed compared to the previous polished fiber.

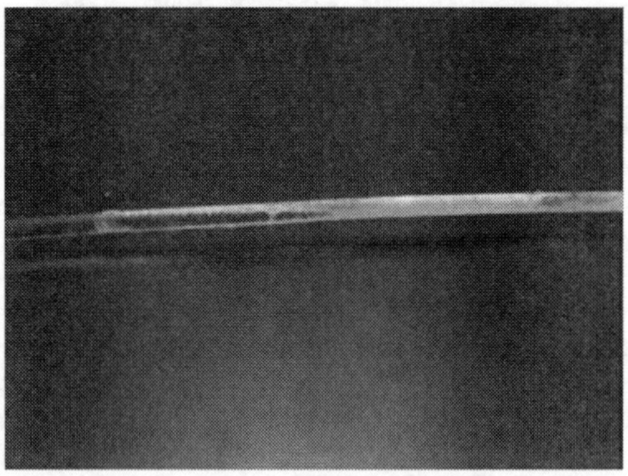

Fig. 2. Etched and unpolished fiber outer structure

After each of the experiment is completed, 1 ml of isopropyl alcohol and 1 ml of distilled water are used to neutralize and wash the waste of the outer cladding that have been etched before any measurement is taken.

3 Results and analysis

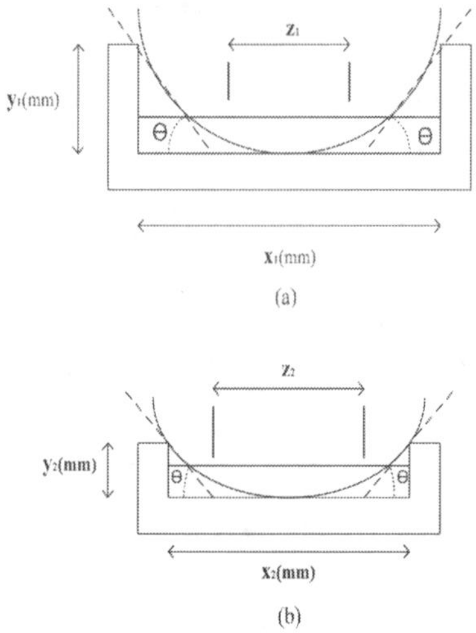

Fig. 3 (a). Diagram of side-etching using flat-base plate 80 mm based (x_1 mm) × 25 mm height (y_1 mm), **Fig.3 (b).** Diagram of side-etching using flat-base plate 60 mm based (x_2 mm) × 10 mm height (y_2 mm).

Fig. 3. (a). shows the diagram of the etching process using the flat base plate of size of 80 mm based × 25 mm height. The bending of the fiber influences the etching process since the fiber has to be held stress-free. If the fiber is under bending stress, once the fiber gets brittle, the fiber is easier to break into two as in Fig. 4. The angle Θ_1 in Fig. 3. (a) is considerably big compared to Θ_2 in Fig. x. (b) and the angle influences the bending applied onto the fiber. Fig. 3. (b) shows the diagram of the etching process using flat base plate of size 60 mm based × 10 mm height.

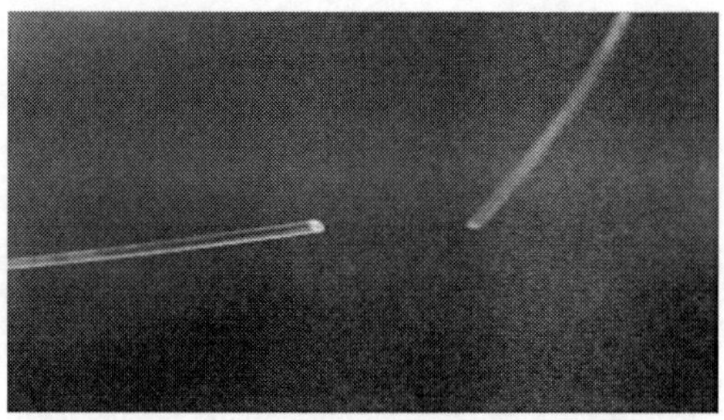

Fig. 4. Broken fiber due to brittleness and bending stress

Less bending stress is applied on the fiber and the fiber is held with small tension applied while the fiber is dipped in chemical solvent. Fig. 4. shows the fiber broke into two due to the factor of brittleness and bending stress applied onto the fiber in process of etching.

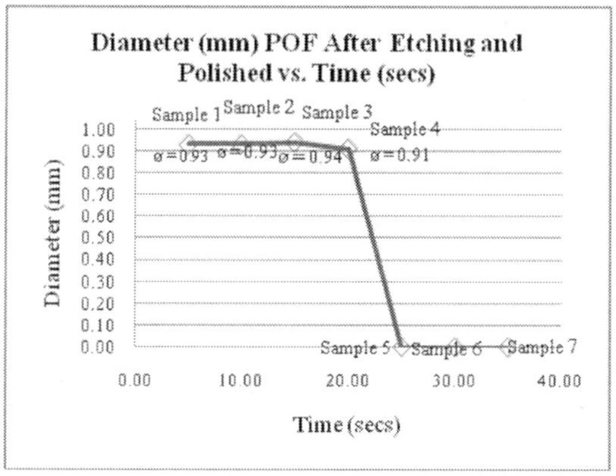

Fig. 5. Time (secs) vs Diameter (mm) for etched and polished POF samples using base-plate 80 mm × 25 mm.

Fig. 5. above is based on experiment done using flat-base plate of 80 mm based × 25 mm height. In addition to etching, for this particular experiment, polishing is done onto the surface of the affected region in order to ensure that the cladding layer is taken off entirely. However, it is crucial to notice that after 20 secs of dipping the fiber strand in the solvent, the samples broke. Analysis shows that it is due to the stress-fracture bending when the fiber region that exposed to the solvent gets brittle and resulted in breaking the fiber strand into two.

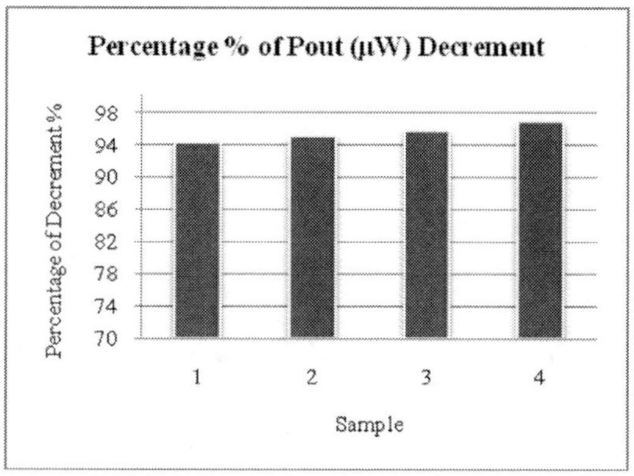

Fig. 6. Decrement in term of percentage of Output power (µW) after samples are etched and polished.

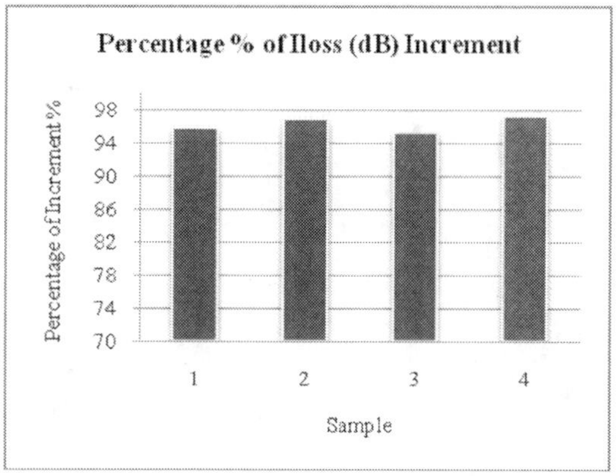

Fig. 7. Increment in term of percentage of Insertion loss (dB) after samples are etched and polished.

Fig. 6. and Fig. 7. show the after etching loss measurement for the four samples that didn't break. As the etching process gets longer i.e more than 5 secs, the output power decreased. When the light signal is transmitted through the fiber, the light spread out at the affected region and this might contribute to the decrement of the output power. Due to this behavior, the insertion loss increases.

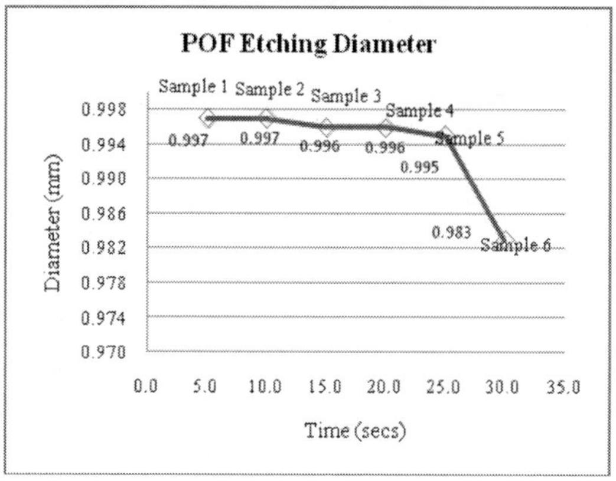

Fig. 8. Time (secs) vs Diameter (mm) for etched and unpolished region of POF samples using base-plate 60 mm × 10 mm.

For the etching process done using flat-base plate of 60 mm based × 10 mm height, six samples are successfully etched with longer time taken that is 30 secs as in Fig. 8. This experiment excludes the polishing of the etched area and let alone only the etching process to dissolve the cladding layer. The physical attribute of the plate contributed to the less bending of the fiber strand and less stress-fracture effect on the fiber as it gets brittle when dipped in the solvent.

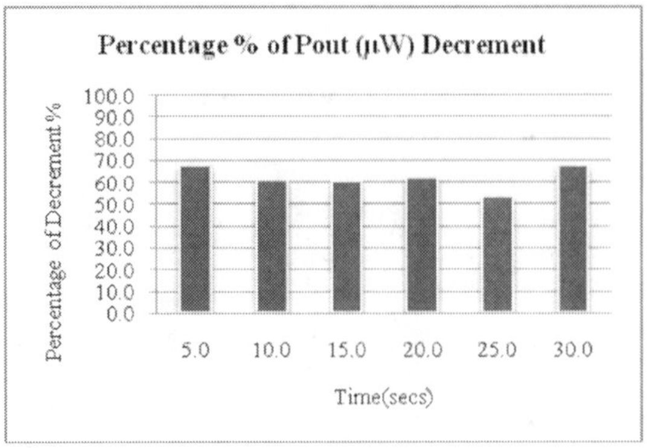

Fig. 9. Decrement in term of percentage of Output power (μW) after samples are etched.

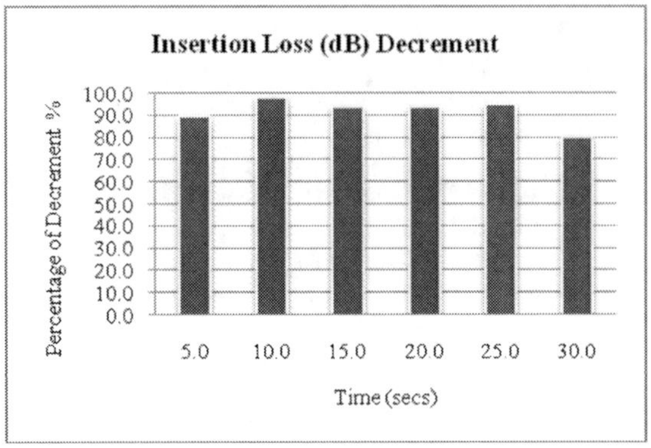

Fig. 10. Increment in term of percentage of Insertion loss (dB) after samples are etched.

Fig. 9. and Fig. 10. show the output power decrement and increment of the insertion loss measurement taken after the etching process. It can be seen that the output power decrement is not much differ from each other. Same behavior goes for insertion loss measurement taken. Since no polishing is applied onto the surface of the etched region, the behavior of the light transmission differs from the previous experiment. However, cloudy surface can be observed surrounded the etched region. This is due to the cladding stripping or etching is in progress, thus, unclear surface or cloudy surface is detected around the affected region. Longer time have to be applied during etching process to ensure that the cladding layer is fully taken off appropriately. This is proved by the diameter of the fiber after etching.

4 Discussions

The affected region that has been tapered using the solvent left only the pure core material section and the properties of the PMMA remains as flexible. The etching process produces through this method have the same optical properties like other multimode fiber tapers [4]. In large core multimode optical fiber such as POF, the number of guided modes can be approximated as explained in Senior [5],

$$M = (kr_cN)^2 = (kr_c(n_o^2 - n_l^2)^{1/2})^2 \quad (1)$$

where k is the fiber propagation constant, r_c is the radius of the core and N is the numerical aperture, n_0 and n_l are the refractive indices of the core and cladding respectively. Losses of a coupler consist of the desired splitting loss and the undesired excess loss that exists beyond the splitting loss. Typical excess losses are on the order of 1dB [6]. The etching technique produces a tapered fiber that requires no alignment and has a constant attenuation of low-order modes. Modal redistribution length of the fiber is about few hundred meters thus the effect of tapers is local length [2]. If bare core is surrounded by an absorbing medium such as a liquid, the effective absorption coefficient of the evanescent field is inversely proportional to the radius [7].Thus, in this research, a coupler can be developed by aligning two bare core in close proximity to each other and to lessen the loss by the evanescent field effect, liquid such as glue having appropriate refractive index will be used surrounding the two fibers.

According to Huang [8], the simplest model for the coupled waveguide system is a directional coupler which consists of two uniform, parallel waveguides placed in close proximity. However, exact analytical solutions for practical directional coupler are difficult to find. Coupled mode theory gives simple analytical solutions that provide insight into the mode coupling process in directional coupler. When two guided modes are brought into close proximity, the modes will couple to each other as the results of evanescent fields.

5 Conclusion

The research presented explains the initial step in developing directional coupler using etching method. Apart from previous parameters analyzed, other parameters that are considered for future development are cost effective and environmental friendly fabrication, uniformity of the device, and excess loss. The applications of the fabricated device include vehicle network, sensor system, wavelength division multiplexing, smart housing network and video surveillance network for home security. The etching process done in realizing the coupler device is simple, low cost and requires only several equipment and it is a safe procedure since the solvent used doesn't involve harmful concentrated acids. Future research includes producing waist diameters less than ø = 0.98 to ensure optimum mode coupling. The measured parameters of the tapered POF offer significant advantages especially for performance of coupler that will be developed in the future.

DOI: 10.1051/epjconf/201716201018

References

1. J. Crisp., *Introduction to fiber optics* (1996).
2. D.F.Merchant, P.J. Scully, N.F.Schmitt, *Chemical tapering of polymer optical fibre*, (Elsevier, 1998).
3. *Understanding optical communication*, (IBM Corporation, International Technical Support Organization, 1998).
4. A.W. Snyder, J.D.Love, *Optical waveguide theory*, (Chapman and Hall, 1983).
5. J.M.Senior, *Optical fibre communication principles and practice*, (Prentice-Hall, London, 1985).
6. J. Powers, *An introduction to fiber optic systems, 2nd ed.*, (MacGraw-Hill, 1999).
7. V.Ruddy, Fiber and Integrated Optics, **9(2)**, 143-151 (1990).
8. W.P.Huang, Optical Society of America, **11 (3)** (1994).

High performance mixed optical CDMA system using ZCC code and multiband OFDM

Nawawi N. M. [1,*], Anuar M. S.[1], Junita M. N [1] and Rashidi C. B.M.[1]

[1]Optical Research Group, Advanced Communication Engineering Centre of Excellence (ACE-COE), School of Computer & Communication Engineering, Universiti Malaysia Perlis (UNIMAP), 02600 Perlis, Malaysia.

Abstract. In this paper, we have proposed a high performance network design, which is based on mixed optical Code Division Multiple Access (CDMA) system using Zero Cross Correlation (ZCC) code and multiband Orthogonal Frequency Division Multiplexing (OFDM) called catenated OFDM. In addition, we also investigate the related changing parameters such as; effective power, number of user, number of band, code length and code weight. Then we theoretically analyzed the system performance comprehensively while considering up to five OFDM bands. The feasibility of the proposed system architecture is verified via the numerical analysis. The research results demonstrated that our developed modulation solution can significantly enhanced the total number of user; improving up to 80% for five catenated bands compared to traditional optical CDMA system, with the code length equals to 80, transmitted at 622 Mbps. It is also demonstrated that the BER performance strongly depends on number of weight, especially with less number of users. As the number of weight increases, the BER performance is better.

1 Introduction

In the past two decades, optical CDMA spaces have been studied extensively. An OCDMA attracts considerable attention due to asynchronous access capability [1] and provides security in transmission network as its distinguished each channel by a specific optical code [2]. Many research on finding the most suitable form of optical CDMA coding system has been widely investigated for coherent and incoherent optical multiple access techniques [3][4][5]. Recently, there are a number of published studies describing the hybrid technology by combining OCDMA system with different technologies to take benefits of both technologies. Such systems are wavelength division multiplexing (WDM) [6][7][8], subcarrier multiplexing (SCM) [9] and OFDM[10]. The application of the advanced modulation technique in the higher speed optical CDMA transmission system has become a new research hotspot.

OFDM modulation technique has aroused extensive research interests due to its resistance to inter symbol interference (ISI) caused by a dispersive channel [11]. It is a well-known developing technology for high data rates system. This tool spreads the data to be transmitted in the medium over huge number of subcarriers, therefore can efficiently use the allocated bandwidth. Compared with the conventional modulation mode on-off keying, OFDM has the following advantages: high spectral efficiency, dynamic bandwidth allocation and adaptive bit rate functionalities, and high tolerance to multi-path interference, channel dispersion and frequency-selective fading [1][12][13].

Motivated by the advance contribution in modulation technologies, this project aims to explore and embed a new modulation scheme into optical domain based on multiband QAM-OFDM. This technique is then combined with OCDMA system to achieved large bandwidth efficiencies while maintaining the high data rates system. Moreover, the maximum channel capacity can be enhanced by sending multiple OFDM bands and/or increasing OCDMA codewords. Comparison between conventional OCDMA with multiband/catenated-OFDM-OCDMA is provided.

The rest of this paper is structured as follows. Section 2 describes the system architecture of an OCDMA transmission network based on ZCC to transmit catenated

* Corresponding author: norizan@unimap.edu.my

band of OFDM signals. End of this section discuss the theoretical model for performance analysis. Results and discussion is provided in Section 3. Finally, some conclusions are drawn in Section 4.

2 Operation principle

Fig. 1 schematically depicts our proposed design system that exploits OCDMA-ZCC and catenated OFDM technologies. The developed system employs *n*-bands OFDM signals with ZCC codeword simultaneously over optical channel. As can be seen in the figure, each input data streams are passed through OFDM modulators to generate corresponding OFDM signals. Each subcarrier of OFDM signal encoded with 16-QAM symbol and generated at a center frequency f_1, ..., f_n, respectively. OFDM modulator which then modulates a digital signal into multiple orthogonal subcarriers.

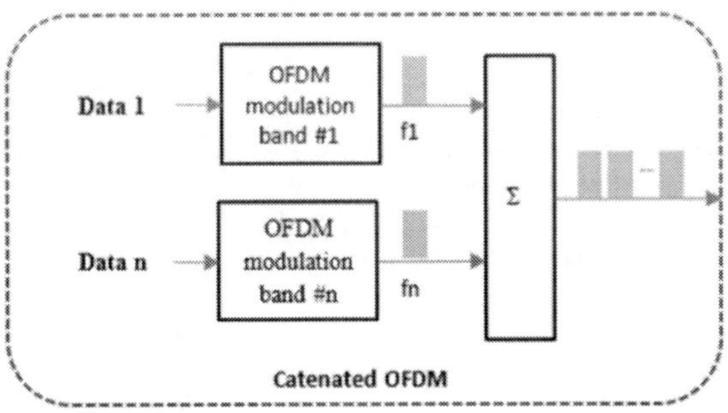

Fig. 1 Multiband catenated OFDM

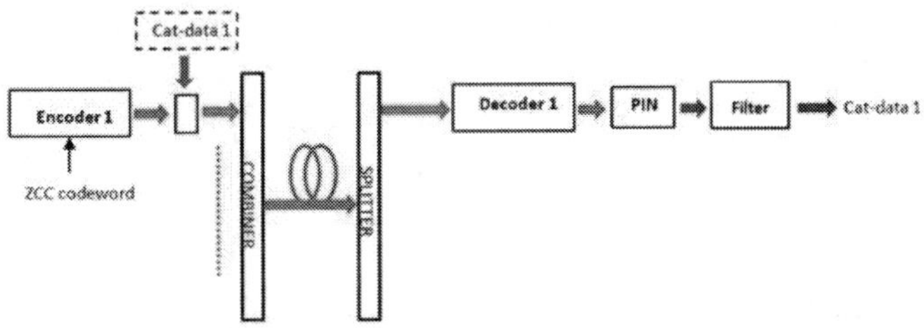

Fig. 2 Catenated-OFDM Optical CDMA based on ZCC architecture

DOI: 10.1051/epjconf/201716201019

Next, the sum of the OFDM signal bands then optically modulated onto sequence through external Mach Zehnder modulator (MZM), each code sequence is assigned with different ZCC code structure as shown in Fig. 2. The *n*th OFDM band at *k* subcarriers according to [14] is expressed as

$$s(t)_{cat} = \sum_{k=1}^{N-1} X_n[k].e^{(j2\pi f_k t)} \sum_{n=0}^{N_{sb}} m_n.e^{j2\pi n f_m t} \quad (1)$$

where f_k and f_m are the frequencies of *k*th subcarrier and *m*th band frequencies respectively, $X_n[k]$ denotes the OFDM information symbol.

Then the modulated code sequences are combined together via an optical combiner and transmitted through the optical fiber. At the receiver, an optical splitter is used to split the modulated code sequences to various paths according to the number of receivers. Only the matched code sequences are decoded while the unmatched code sequences will be filtered out. Another advantage using the proposed catenated OFDM scheme is the overall spectrum can be readily partitioned with electrical filter at the receiver to separate each band. Each channel that brings mutltiband OFDM signals is assigned a particular code sequence.

The signal to noise ratio (SNR) and BER for combining catenated OFDM signals with ZCC OCDMA structure is given by

$$SNR = \frac{(R.\frac{P_r w}{L}.\sum_{k=0}^{N_{sc}-1} x_k e^{j2\pi f_k t}.\sum_{n=1}^{N_{sb}} m_n e^{j2\pi n f_m t})^2}{2e(R.\frac{P_r w}{L}.\sum_{k=0}^{N_{sc}-1} x_k e^{j2\pi f_k t}.\sum_{n=1}^{N_{sb}} m_n e^{j2\pi n f_m t})B + \frac{4kTB}{R_L}} \quad (2)$$

where L and w are ZCC code length and weight respectively. ZCC code properties were calculated based on equation in [15].

$$BER = \frac{\sqrt{M}-1}{\sqrt{M}\log_2\sqrt{M}} erfc\sqrt{\frac{3\log_2 M}{2(M-1)}SNR} \quad (3)$$

3 Result and Discussion

In this section, we evaluate the performance, in terms of bit error rate, of the proposed modulation scheme using multiband OFDM and OCDMA-ZCC. The numerical results are obtained based on the expressions derived in Section 2. The BER is evaluated with different number of parameters consisting of the effective power, number of users and code length. The transmission of catenated-OFDM signal, whose parameters according to Table 1 is analyzed.

Table 1. Parameters for numerical analysis

PARAMETER	VALUE
Number of subcarrier	256
Number of band	1 to 5 bands
Modulation Mapping, M	16-QAM
Data rate	622 Mbps
Number of bands	1 - 5
Number of samples	65536
Responsivity, R	0.7
Bandwidth	311 MHz
Load Resistor, R_L	1030
Electron charge, e	$1.6\text{x}10^{-19}$ C
Boltzmann's constant, k	$1.38\text{x}10^{-23}$ J/K
Absolute temperature, T	300 K

Firstly, in Fig. 3, the BER of the proposed scheme and conventional OCDMA based on ZCC code are demonstrated versus the number of permissible users with different value of effective power. It is seen that the use of new scheme helps to significantly increase the number of users compared to original OCDMA-ZCC system. More specifically, for BER less than $10E^{-9}$, the catenated-OFDM/OCDMA-ZCC system with 5 bands achieved more than 200 number of users while only 110 number of user is allowed for case of OCDMA-ZCC with P_r = -10dBm. BER performance degrades with respects to lower effective power. Hence, it is worth noting that, for proposed scheme at larger OFDM bands (5 bands), the BER performance is far better than traditional OCDMA-ZCC at same effective power level. By transmitting 5 bands simultaneously, the available electrical bandwidth is fully utilized and increase the spectral efficiency of the system.

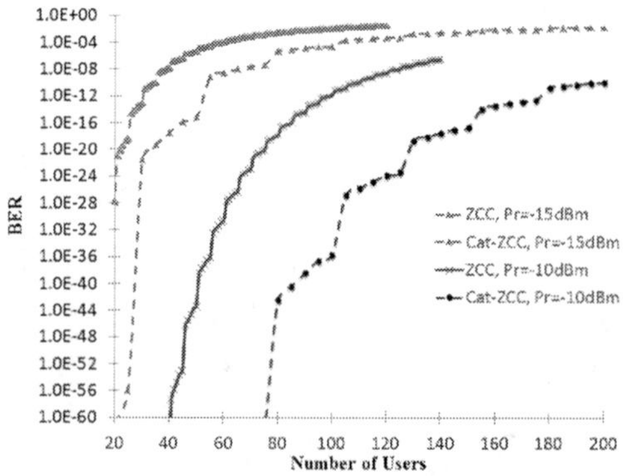

Fig. 3. Comparison of number of users for catenated-OFDM/OCDMA-ZCC with conventional OCDMA-ZCC

Secondly, to scrutinize the effect of code length on number of user performance, the model has been run for different code length in range of 10 to 80 and compared with traditional OCDMA-ZCC system. As shown in Fig. 4, the number of user dramatically increases with the

 DOI: 10.1051/epjconf/201716201019

increased code length for catenated-OFDM at 5 bands. The number of user improves approximately 80% between catenated-OFDM and conventional ZCC for a given code length of 80. Therefore, the catenated-ZCC technique shows better performance for larger OFDM bands at same bandwidth allocation.

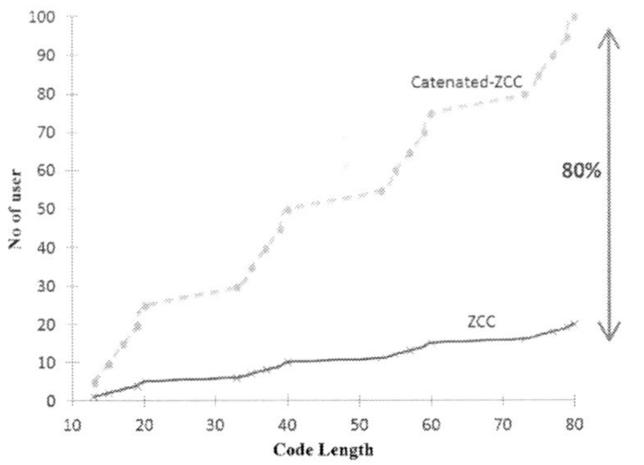

Fig. 4. Comparison of code length for catenated-OFDM/OCDMA-ZCC with conventional OCDMA-ZCC

Finally, Fig. 5 quantifies the effect number of users for different number of weight. Previous results are based on number of weight, w=4. Here, we consider four value of weight; w=2, w=4, w=6 and w=8. It is demonstrated that the BER performance strongly depends on number of weight, especially with less number of users. As the number of weight increases, the BER performance is better. This is due to more power density will be added to the overall spectrum while increasing the number of weight. However, when number of weight larger, bigger bandwidth is required and hence, increase the cost of the system.

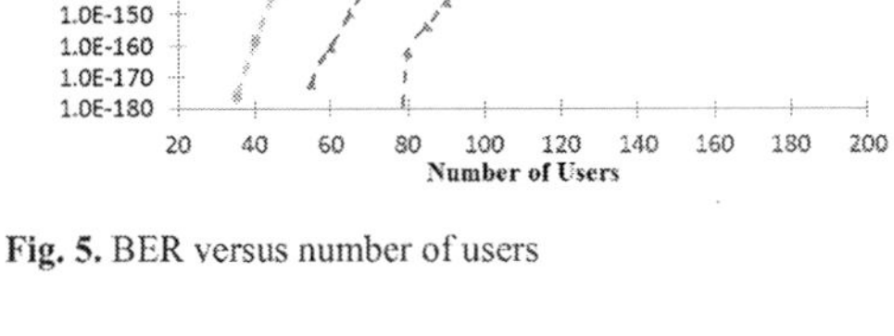

Fig. 5. BER versus number of users

4 Conclusion

In this paper, we have proposed high spectral efficiency modulation technique employing multiband catenated-OFDM technique and ZCC OCDMA system. The developed system is able to provide better BER performance at their respective parameters, higher permitted number of users and high spectral efficiency. To the best of our knowledge, it is the first time the performance of OCDMA system is numerically analyzed using multiband OFDM signals. We have also comprehensively investigated the impacts of number of OFDM bands, on the performance. The feasibility of our proposed scheme is verified by numerical evaluation and proves that overall system capacity enhancement approximately 80% with code length of 80. The numerical results also show that inter band interferences is main factor that effect on performance of the proposed systems in sending more bands simultaneously.

References

1. X. Wang, *7th IEEE Int. Conf. Optoelectron. Fiber Opt. Photonic, 2004*, pp. 1–6 (2004)
2. M. S. Anuar, S. A. Aljunid, A. R. Arief, M. N. Junita, and N. M. Saad, *Opt. - Int. J. Light Electron Opt.*, **124**, no. 4, pp. 371–375 (2013)
3. T. Hussein, S. A. Aljunid, H. Adnan, R. A. Ahmad, and N. M. Saad, *Opt. Fiber Technol.*, **17**, no. 4, pp. 273–280 (2011)
4. M. S. Anuar, S. A. Aljunid, N. M. Saad, and S. M. Hamzah, *Opt. Commun.*, **282**, no. 14, pp. 2659–2664 (2009)
5. N. D. Keraf, S. A. Aljunid, A. R. Arief, and P. Ehkan, Advanced Computer and Communication Engineering Technology, pp. 129-140 (2014)
6. I. A. M. Ashour, S. Shaari, and P . Susthitha Menon and Hesham A . Bakarman, Photonic Technology Laboratory, **8**, no. 10, pp. 1718–1729 (2012)
7. F. Hou and M. Yang, *Opt. - Int. J. Light Electron Opt.*, **125**, no. 15, pp. 4150–4153 (2014)
8. N. Ahmed, M. A. Rashid, M. Islam, *26th International Telecommunication Networks and Applications Conference (ITNAC)*, (2016)
9. M. N. Junita, S. A. Aljunid, Thanaa Hussein Abd, M. S. Anuar, A. R. Arief, R. A. Rahim, *2011 2nd International Conference on Photonics* (2011)
10. A. O. Aldhaibani, S. A. Aljunid, M. S. Anuar, and A. R. Arief, J. Theor. Appl. Inf. Technol., **66**, no. 2, pp. 634–637 (2014)
11. M. Singh and K. Kaur, *Ijircce*, **2**, no. 12, pp. 4793–4797 (2013)
12. I. B. Djordjevic and B. Vasic, *Opt. Express*, **14**,

no. 9, pp. 3767–3775 (2006)

13. Y. Yin, W. Lan, T. Lin, C. Wang, and M. Feng, **35**, no. 2, pp. 258–264 (2017)
14. A. I. Technology, N. M. Nawawi, M. N. Junita, A. K. Rahman, S. A. Aljunid, and C. Engineering, **91**, no. 1 (2016)
15. M. S. Anuar, S. A Aljunid, A. R. Arief, and N. M. Saad, *7th Int. Symp. High-Capacity Opt. Networks Enabling Technol. HONET 2010*, pp. 128–132 (2010)

Optimization of AMI-MDM-RoFSO under atmospheric turbulence

Sushank Chaudhary[1] and *Angela* Amphawan[1,2]

[1]Optical Computing and Research Laboratory, School of Computing, Universiti Utara Malaysia, 06010 Sintok, Kedah, Malaysia
[2]Research Laboratory of Electronics, Massachusetts Institute of Technology, 02139 Cambridge, Massachusetts, USA

Abstract. Radio over Free Space (Ro-FSO) is promising candidate for providing ubiquitous digital services especially in rural areas. This work investigates the performance of MDM of two 5Gbps-10GHz data channels over FSO link using LP 01 and LP 02 modes under the effect of atmospheric turbulences. The signal to noise ratio (SNR), total received power, modal decomposition at receiver at the receiver is also reported. The reported result shows the successful transmission of two channels with acceptable SNR over FSO link under atmospheric turbulences.

1 Introduction

The plethora of wireless network platforms includes numerous operators with independent radio base stations and networks. The diversity of wireless services leads to excessive investment on infrastructure and redundancy of equipment, which may affect cellular architecture efficiency. According to reports by the International Telecommunication Union (ITU), cellular services witnessed a remarkable growth of 7.5 billion subscribers in 2013 [1]. Despite this, radio frequency is spectrum is still a scarce commodity.

The limitation of radio frequency amidst the growth in mobile subscribers has motivated the development of Ro-FSO, a technology that transmits multiple RF signals through a high-speed optical carrier without investing excessively on optical fiber cabling [2, 3]. As Ro-FSO utilizes optical carriers to exploit various electromagnetic spectrum for mobile networks, it efficiently eliminates congestion issues and licensing of RF signals prevalent in the current wireless networks. Ro-FSO is a multi-faceted technology that combines radio-over-fiber (RoF) and free-space optical (FSO) technologies. RoF is used for centralizing processes including RF handoff, up-down conversion, coding, switching, and multiplexing among the base stations [4]. FSO is used for transporting data signals at a higher bandwidth through atmosphere than radio signals. Thus, Ro-FSO is efficient for integrating radio and optical networks [5].

However, the performance of Ro-FSO can be affected by atmospheric turbulences arising from fog, rain and snow [6]. In order to achieve high capacity in Ro-FSO while optimizing network resources, researchers have highlighted the use of advanced modulation formats [7]. Alternate mark inversion (AMI) has been used for improving the capacity of fiber transmission. AMI is a bipolar encoding system which has binary 0 as the neutral voltage and binary 1 as the alternating positive and negative voltages. However, AMI has not been exploited in free-space optical network systems. Another bandwidth-enhancing strategy, known as *mode division multiplexing* (MDM), is used for enhancing the capacity of free-space optical networks [8-12] by leveraging the abundance of fiber eigenmodes for transmitting various channels by generating specific modes by means of optical signal processing [13-15] , spatial light encoding [16-18], few mode fibers [19].

The current study aims to utilize both aforementioned strategies, specifically a new combination of AMI and MDM in Ro-FSO systems for increasing the capacity and transmission distance, which thus far has not been covered in any previous studies. In order to achieve this objective, a simulation of an MDM system on Linearly Polarized (LP) 01 and 02 modes using AMI-encoded signals will be performed over a 3 km FSO link to optimize the parameters of the system under various atmospheric conditions.

2 System Description

MDM-RoFSO modelled in OptiSystem^{TM}. AMI is used for 5 Gbps data encoding while its line codes are

* Corresponding author: sushankchaudhary@gmail.com

DOI: 10.1051/epjconf/201716201020

generated using delay-and-add and delay-and-subtract operations [20, 21]. Data is transmitted through pulse amplitudes of on-off-keying (OOK) while impressing its phase modulation using mathematical operations of binary input signal x(n) = [22] and output signal y_{AMI} (n) as:

$$y_{AMI}(n) = x(n) - x(n-1) \quad (1)$$

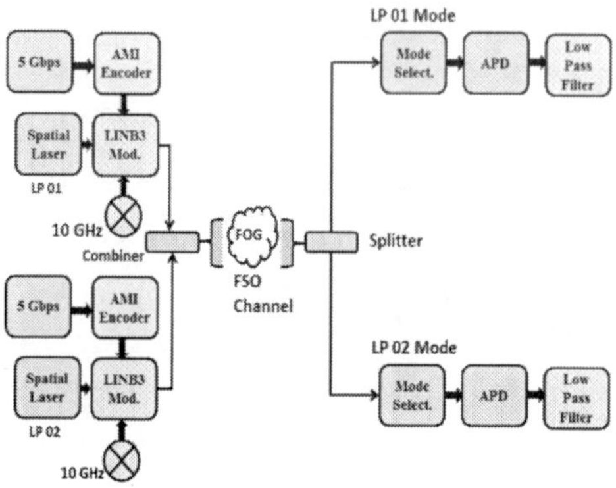

Fig. 1. Proposed Ro-FSO system

The two channels are launched from LP 01 mode and LP 02 mode, where y_{AMI} is the AMI sequence, modulated using a Lithium Niobate dual-port optical modulator and 10GHz radio signal through a 850 nm spatial laser respectively after propagation through a multimode fiber.

The two modes are transmitted over a 3 km FSO along with the pre and post compensation using optical amplifiers. The FSO link equation is given as [24]:

$$P_{Received} = P_{Transmitted} \frac{d^2_R}{(d_T + \theta_R)^2} 10^{\alpha R/10} \quad (1)$$

where d_R is the receiver aperture diameter, d_T is the transmitter aperture diameter, θ is the beam divergence, R the range, and α the atmospheric attenuation.

The performance of the proposed Ro-FSO system under specified atmospheric conditions is investigated by adjusting the atmospheric attenuation value for different atmospheric conditions, i.e, 0.14 dB for clear weather conditions, 4 dB for hazy weather, 9 dB/km for thin fog, 13 dB/km for light fog, 16 dB/km for thick fog, and 22 dB/km for heavy fog [25, 26].

An Avalanche photodetector with low pass filter is used for recovering the original baseband transmitted signal. The mean squared error minimization of the estimate of sum of electric fields and the output field is performed [27] to retrieve the modal composition at the receiver [27].

3 Results and Discussion

As shown in Fig. 2, the signal to noise ratio (SNR) at the receiver side for Channel 1 and Channel 2 is measured with respect to FSO range under various atmospheric turbulences. The following SNR values are calculated for channel 1 under different atmospheric conditions: 71.11 dB, 54.32 dB, and 22.31 dB for hazy weather; 68.21 dB, 38.32 dB, and 0 dB for light fog; 67.34 dB, 27.41 dB, and 0 dB for moderate fog; and 66.23 dB, 18.43 dB, and 0 dB for heavy fog at the FSO range of 500 m, 1625 m, and 5000 m respectively.

Fig. 3 shows the following values of the total power calculated at the receiver side with respect to FSO range for channel 1 under various atmospheric conditions: -17.35dBm, -45.10 dBm, and -91.67 dBm for hazy weather; -22.11 dBm, -61.78 dBm, and -100 dBm for light fog; -25.10 dBm, -71.28 dBm, and -100 dBm for moderate fog; and -28.05dbm, -81.36 dBm, and -100 dBm for heavy fog at the FSO range of 500 m, 1625 m, and 5000 m respectively.

From Fig. 2 and Fig. 3, it is evident that the FSO link reach is extended under various atmospheric conditions: 3900 m in hazy weather, 2700 m in light fog, 2100 m in moderate fog, and 1600 m in heavy fog with acceptable SNR values for channel 1. The largest amount of power for channel 1 is coupled into LP 01 mode, as shown in Fig. 4.

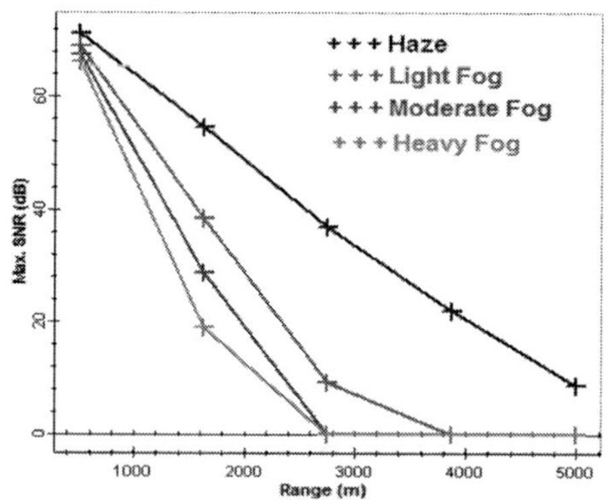

Fig. 2. SNR versus range under different weather conditions (Channel 1)

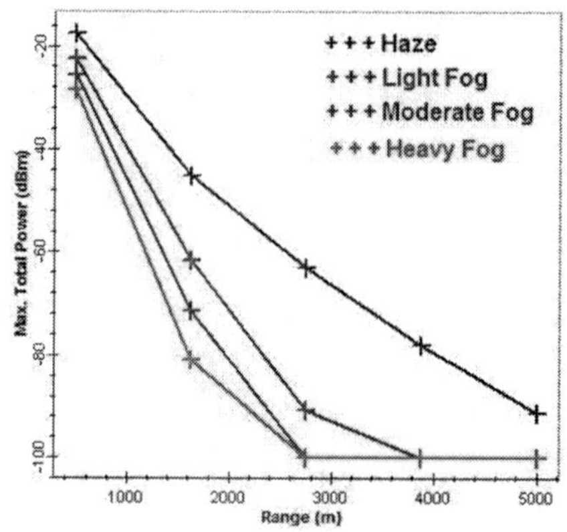

Fig. 3. Total received power versus range (Channel 1)

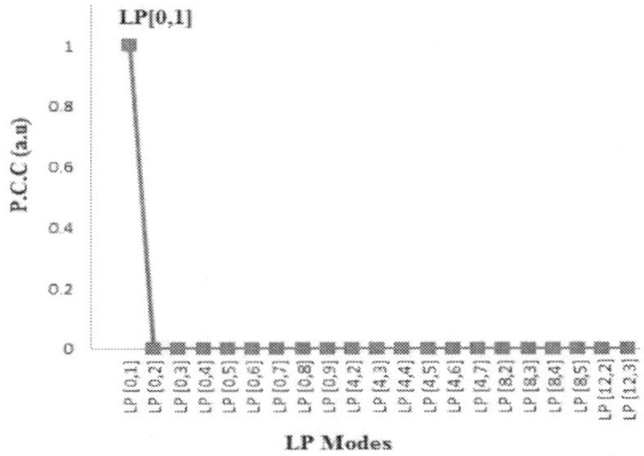

Fig. 4 Power coupling coefficients in terms of LP Modes for Channel 1

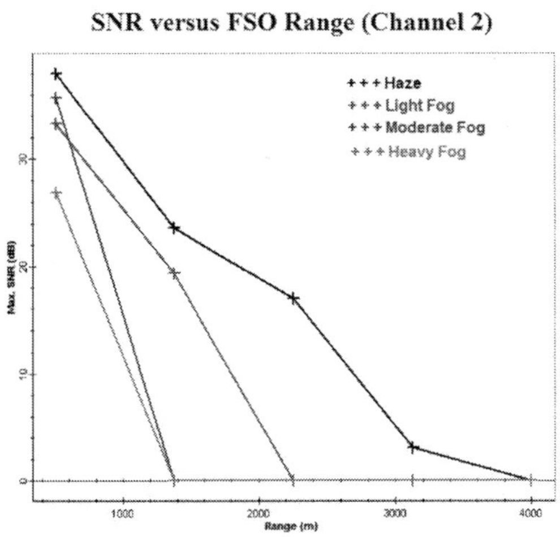

Fig. 5 SNR versus range under different weather conditions (Channel 1)

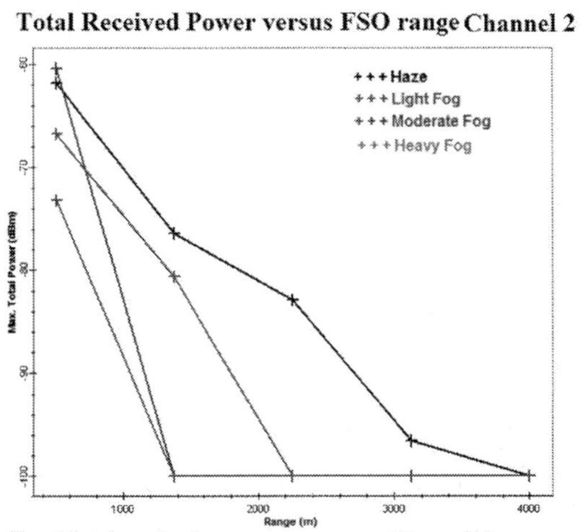

Fig. 6 Total received power versus range (Channel 2)

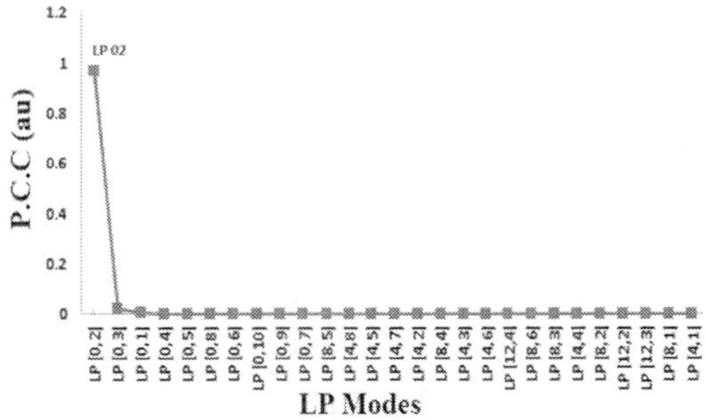

Fig 7 Power coupling coefficients in terms of LP Modes for Channel 2

Similarly, the following SNR values are calculated for channel 2 under different atmospheric conditions: 37.91 dB, 17.23 dB, and 0 dB for hazy weather; 33.87 dB, 0 dB, and 0 dB for light fog; 35.68 dB, 0 dB, and 0 dB for moderate fog; and 26.81 dB, 0 dB, and 0 dB for heavy fog at the FSO range of 500 m, 2250 m, and 4000 m respectively as shown in Fig. 5. Consequently, Fig. 7 shows the following values of the total power calculated at the receiver side with respect to FSO range for channel 1 under various atmospheric conditions: -61.23 dBm, -82.45 dBm, and -100 dBm for hazy weather; -66.45 dBm, -100 dBm, and -100 dBm for light fog; -60.23 dBm, -100 dBm, and -100 dBm for moderate fog; and –73.10 dbm, -100 dBm, and -100 dBm for heavy fog at the FSO range of 500 m, 2250 m, and 4000 m respectively.

From Fig. 5 and Fig. 6, it is evident that the FSO link reach for channel 2 is extended under various atmospheric conditions: 2200 m in hazy weather, 1375 m in light fog, 1050 m in moderate fog, and 900 m in heavy fog with acceptable SNR values. The largest amount of power for channel 1 is coupled into LP 02 mode, as shown in Fig. 7.

4 Conclusion

Numerical simulations demonstrate that MDM system on LP 01 mode and LP 02 mode using AMI-encoded signals are viable for extending the range of the FSO link with acceptable SNR. The results show that the for channel 1, FSO links prolong to 3.9 km in hazy weather whereas the links prolong to 2.7 km in case of light fog. In case of moderate fog, the FSO links prolong to 2.1 km and to 1.6 km in heavy fog. For channel 2, FSO link prolongs to 2200 m in hazy weather whereas 1375 m in light fog. In case of moderate fog, FSO link prolongs to 1050 m and to 900 m in heavy fog.

References

1. I. Data, "Statistics Division Telecommunication Development Bureau ITU,"The world in 2013: Ict facts and figures,"" *ITU Int. Telecom. Union* (2013)

2. A. Amphawan, S. Chaudhary, and V. Chan, *Journal of the European Optical Society-Rapid publications,* **9** (2014)
3. K. Kazaura, K. Wakamori, M. Matsumoto, T. Higashino, K. Tsukamoto, and S. Komaki, *Communications Magazine, IEEE,* **48**, no. 2, pp. 130-137 (2010)
4. V. Thomas, M. El-Hajjar, and L. Hanzo, *Performance improvement and cost reduction techniques for radio over fiber communications* (2015)
5. Angela Amphawan, S. Chaudhary, R. Din, and M. N. Omar, *IEEE International Colloquium on Signal Processing and its Applications (CSPA), 6-9th March 2015*, Kuala Lumpur (2015)
6. S. Chaudhary and A. Amphawan, " *Journal of Optical Communications,* **35**, no. 4, pp. 327-334 (2014)
7. K. Mishina *et al.*, *Optics Express,* vol. 15, no. 12, pp. 7774-7785 (2007)
8. A. Amphawan and S. Chaudhary, *International Conference on Optical and Photonic Engineering (icOPEN2015)*, pp. 95242H-95242H-6: (2015)
9. I. M. Fazal *et al.*, *Optics letters,* **37**, no. 22, pp. 4753-4755 (2012)
10. H. Huang *et al.*, *Optical Fiber Communication Conference*, p. OTh4G. 5: Optical Society of America (2013)
11. H. Huang *et al.*, *Optics letters,* **39**, no. 2, pp. 197-200 (2014)
12. Y. Ren *et al.*, *Optics letters,* **38**, no. 20, pp. 4062-4065 (2013)
13. A. Amphawan, *Optics express,* **19**, no. 23, pp. 23085-23096 (2011)
14. A. Amphawan, V. Mishra, K. Nisar, and B. Nedniyom, *Journal of Modern Optics,* **59**, no. 20, pp. 1745-1752 (2012)
15. S. O. Arik, J. M. Kahn, and K.-P. Ho, *Signal Processing Magazine, IEEE,* **31**, no. 2, pp. 25-34 (2014)
16. A. Amphawan, *Optics express,* **19**, no. 10, pp. 9056-9065 (2011)
17. A. Amphawan, *Journal of Modern Optics,* **59**, no. 5, pp. 460-469 (2012)
18. J. Carpenter and T. D. Wilkinson, *Journal of Lightwave Technology,* **30**, no. 12, pp. 1978-1984 (2012)
19. R. Ryf *et al.*, *Optical Fiber Communication Conference*, p. PDPB10: Optical Society of America (2011)
20. O. Leclerc *et al.*, *Journal of Lightwave Technology,* **21**, no. 11, p. 2779 (2003)
21. P. J. Winzer and R.-J. Essiambre, *Proceedings of the IEEE,* **94**, no. 5, pp. 952-985 (2006)
22. "Tools of the Trade," **20**, Thorlabs, Ed., ed (2010)
23. A. Ghatak and K. Thyagarajan, *An introduction to fiber optics* (1998)
24. E. J. McCartney, "Optics of the atmosphere: scattering by molecules and particles," *421 p.,* **1** (1976)
25. I. I. Kim, B. McArthur, and E. Korevaar, *Proc. SPIE,* **4214**, pp. 26-37 (2001)
26. A. K. Majumdar, *Free-Space Laser Communications*: pp. 57-108 (2008)
27. A. Amphawan and D. O'Brien, *IEEE Int. Conf. Photon. (ICP)*, Langkawi, pp. 1-5: (2010)

EPJ Web of Conferences **162**, 01021 (2017)
InCAPE2017
DOI: 10.1051/epjconf/201716201021

Fiber-wireless for smart grid: A survey

Radzi NAM[1,*], *Ridwan* MA[1], *Din* NM[1], *Abdullah* F[1], *Mustafa* IS[1], *l-Mansoori* MH[2]

[1] Centre for Photonics Technologies, Universiti Tenaga Nasional, lalan Ikram-Uniten, 43000 Kajang, Malaysia

[2] Faculty of Engineering, Sohar University, Oman

Abstract. Smart grid allows two-way communication between power utility companies and their customers while having the ability to sense along the transmission lines. However, the downside is such, when the smart devices are transmitting data simultaneously, it results in network congestion. Fiber wireless (FiWi) network is one of the best congestion solutions for smart grid up to date. In this paper, a survey of current literature on FiWi for smart grid will be reviewed and a testbed to test the protocols and algorithms for FiWi in smart grid will be proposed. The results of number of packets received and delay vs packet transmitted obtained via the testbed are compared with the results obtained via simulation and they show that they are in line with each other, validating the accuracy of the testbed.

1 Introduction

An interconnected network that is used to deliver electricity from suppliers to consumers is what is called as "electrical grid". It works in a way where electrical power is produced at the generating stations to be carried to demand centers via high-voltage transmission line and further to individual customers via distribution lines.

Despite of being an engineering marvel, to date electrical grid is being stretched to its capacity. This is due to the population growth and more modern electrical appliances are being added in every household, such as high-definition televisions, laptops and wireless telephones. These modern appliances are more sensitive to variations in electric voltage causing the entire electric grid to be overused and fragile.

Therefore, an improved electricity supply chain that is called as "Smart Grid" is being introduced to maximize the throughput of the system while reducing the energy consumption. Figure 1 shows the differences between conventional electrical grid and Smart Grid where additional features are introduced in Smart Grid such as monitoring, analysis, control, and communication capabilities. Devices located along the power lines and on premises are able to interact with each other, allowing for two-way communication between utility and customers. Hence, Smart Grid is able to respond digitally to the ever-changing electric demand.

The term used to denote an automated two-way communication between a smart meter and a utility data center is called as Advanced Metering Infrastructure (AMI). AMI uses bidirectional communication to provide energy management data such as consumption data and outage reports as well as control information data such as alerts and equipment settings.

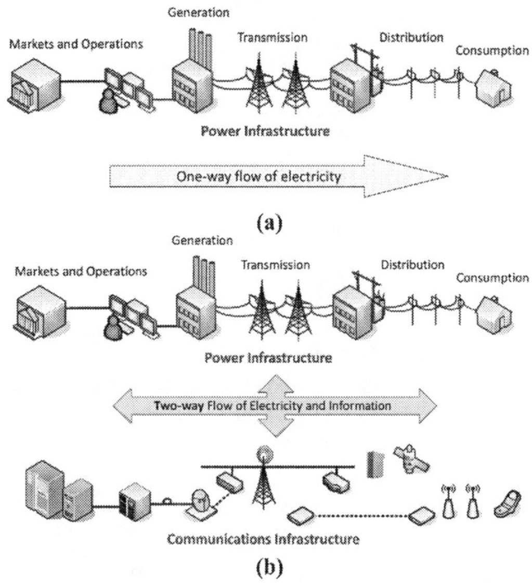

Fig. 1. a) Conventional electric grid b) Smart Grid

AMI comprises of four tiers; home network, smart meter, concentration point and utility data centre. In home network, Home Area Network connects smart appliances to a smart meter for data collection to measure real time energy consumption. The collected data from smart meter then traverse to a concentration point such as substation or communication tower as part of smart grid. Afterwards, data flows from concentration point to the metering data management systems (MDMS) located in utility data centre via private

* Corresponding author: asyikin@uniten.edu.my

network. The MDMS will then process and manage data on energy consumption and fault detection.

However, the downside of Smart Grid is such, when the smart devices are transmitting data simultaneously, it results in network congestion [1]. Challenges occur in ensuring the reliability and timeliness of the data transmitted over these networks. Therefore, advance techniques such as cognitive radio [2] and fiber wirelesss (FiWi) are developed to fully utilize the capability of smart grid wireless networks.

Cognitive radio networks allow unlicensed devices to transmit in unused "spectrum holes" in licensed bands without causing harmful interference to authorized users. It configures the radio for different combinations of protocol, operating frequency, and waveform. However, cognitive radio needs to deal with two primary issues which are hidden primary users [3] and spread spectrum primary users [4], both of which lead a cognitive radio to incorrectly decide that a spectrum block is empty. Hence, it has higher probability to cause the signals to interfere with the licensed primary user.

This does not happen with FiWi as the optical side is able to provide reliable transmission for smart meter and intelligent sensor data, while the wireless side allows flexible access to remote locations and broad coverage. Therefore, this paper discusses in details on the integration of FiWi and smart grid. The remainder of this paper is structured as follows. Section 2 introduces the fundamental of FiWi networks. The state of the art of FiWi networks in smart grid is then reviewed and we highlight selected FiWi testbeds in the literature. We introduce our proposed testbed in the next section and the final section concludes the paper.

2 Fundamental FiWi Networks

FiWi is an integration between fiber and wireless network that is also known as the endgame for the broadband access network. Fiber and wireless are combined together to achieve high bandwidth as well as mobility in the network.

Typical architecture of FiWi is depicted in Figure 2, where the wired side consists of a basic configuration of Passive Optical Network (PON) as it is the dominant broadband access network emerging up to date. FiWi consists of Optical Line Terminal (OLT) at the central office that is connected to multiple Optical Network Units (ONUs) at the customer's sides via optical fiber and passive optical splitter. The ONUs are further connected to end users wirelessly.

PON for FiWi at the wired side can generally be divided into two types; Wavelength Division Multiplexing (WDM) and Time Division Multiplexing (TDM) PON. WDM PON allows each ONU to operate at a different wavelength to avoid collision. In order to receive the data transmitted in multiple channels, a tunable receiver or a receiver array is required at the OLT. It also requires each ONU to use a fixed transmitter operating at a different wavelength, which would result in an inventory problem. Although the inventory problem can be solved by using tunable transmitters, these devices are costly, making the solution cost-ineffective.

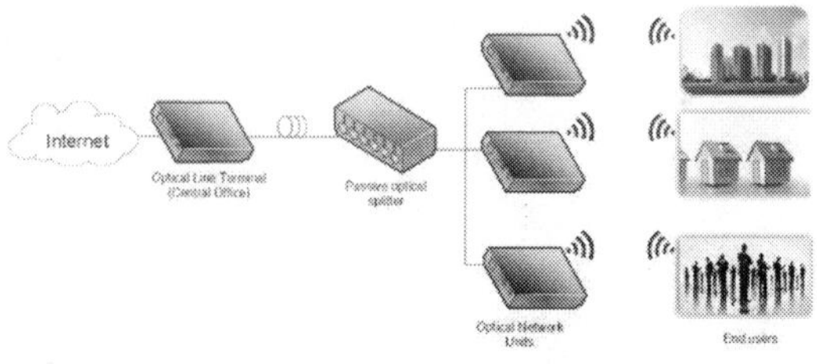

Fig. 2. FiWi architecture

On the other hand with TDM, OLT allocates a time slot or a transmission window for data transmission in ONU. Upon the arrival of its time slot, the ONU will send out its buffered packets at the full transmission rate of the upstream channel. If there are no frames in the buffer to fill the entire time slot, idles are transmitted. TDM is considered as cost effective approach but the arbitration mechanism is more complex compared to WDM.

The two most popular wireless technology used in FiWi are WiFi and WiMax. WiFi that is using IEEE Standard 802.11 offers low bandwidth; 54Mbps for IEEE 802.11a whereas 11Mbps for 802.11b and 54Mbps for 802.11g. The range is also limited, typically up until 100m, which is why it is mainly used for wireless local area networks. In WiFi, central authority known as access point is required to manage the network. It is gaining its popularity due to its flexibility to multihop.

On the other hand for WiMax, IEEE Standard 802.16 is used. WiMax provides high bandwidth up to 75Mbps in a range of 3 to 5km. However, as the distance gets longer, the bandwidth reduces to 20-30Mbps due to the fact that WiMax does not work proficiently for non-line-of-sight communications. WiMax is typically used for metropolitan area networks with base station is required to manage the network. The downside is such, WiMax is more suitable for single hop.

Data is transmitted in FiWi networks via these two techniques; Radio over Fiber (RoF) or Radio and Fiber (R&F). RoF is a technique where radio signals are transmitted over optical fiber to provide communication service. It is an analog communication scheme where the signals are simply converted from electrical to optical and vice versa. One of the advantages of RoF is that only minimal modifications are required at the base stations or access points since Radio Frequency (RF) signals are transmitted to remote antenna as it is.

DOI: 10.1051/epjconf/201716201021

On the other hand for R&F, different media access control (MAC) protocols are used in fiber and wireless link. Although it requires major modifications at base stations or access points, since it uses two different protocols for two different links, it solves the problems caused by insertion of the optical distribution system in wireless networks.

For FiWi in smart grid, different combination of techniques are used at the fiber link, wireless link and for data transmission. The combinations are discussed in the following section.

3 State of Art

3.1 FiWi for Smart Grid

To date, there are still very limited literature on utilizing FiWi for smart grid. Among the earliest paper on FiWi for smart grid found in the literature is Uber-FiWi published 2011 by Maier et al. [5]. Uber-FiWi combines a big fiber ring network at the wired side to interconnect the distribution management system to either EPON network (for urban area) or WiMax base station (for suburban or rural area). Total power consumption and total cost of various scheduling algorithms are studied using this hybrid network.

In 2013, Zaker et al. [6] has proposed a Fiber-Wireless Sensor Networks (Fi-WSN) gateway design for smart grid. In the design proposed, TDM Ethernet PON (EPON) serves as back-end of the communication network whereas the WSN forms the front-end. Due to the fact that data in smart grid needs to be treated differently as according to their urgency, an algorithm has been proposed in the paper to allow differentiation between high priority and low priority packets.

In 2014, Ghassemi et al. [7] has proposed the usage of RoF networks for smart grid. RF signals are distributed from a control unit called headend (or OLT) to remote antenna units (RAU) (or ONU) so that the complex signal processing functions (for modulation, synchronization, multiplexing, coding, etc.) are centralized. This simplifies RAU and has greatly reduced the system installation and operational cost. This system can be used for both TDM and WDM PON as well as for both WiFi and WiMax technology.

As can be observed, there are still limited researches done in studying FiWi for smart grid as FiWi is still considered as a new technology. Due to this reason, a reconfigurable FiWi testbed is needed to study the best protocols, algorithms and topology for smart grid.

3.2 FiWi Testbed

Numerous FiWi testbed are developed for various reason ranging from evaluations, comparison, review and enhancement. Digitized RoF FiWi testbed has been developed in [8] where the equipment used are modulated vertical cavity laser, wireless signal generator, digitized sampling oscilloscope, photodetector, digital to analog converter, arbitrary waveform generator and vector signal analyser.

Pang et. al [9] proposed a FiWi testbed, which transmits on W-Band, between 75-110 GHz frequency band. The testbed proposed uses hardware such as 16 QAM optical baseband transmitter, 100GHz photodetector, Erbium-doped fiber amplifier, W-Band horn antenna, low-noise amplifier, W-band balance mixer, local oscillator, Rohde and Schwarz signal synthesizer, analog to digital converter and digital signal processing-based receiver.

Both digitized FiWi testbed and W-band testbed are purely hardware-based testbed. With purely hardware-based testbed, it is not easy to reconfigure the testbed to study on the protocols and algorithm, making it less suitable to be used in FiWi for smart grid study.

There are also implementations of network virtualization in the FiWi. Network virtualization is a combination of hardware and software. Dai et. al [10] applies network virtualization to hide the differences between fiber network to the wireless network. In terms of software, the testbed contains virtual resources that provides bandwidth, computing capacity, storage and virtual networks.

Meng et. al [11] proposed a Modified Weighted Round Robin (MWRR) algorithm based on the model of FiWi network virtualization and this testbed uses MATLAB as their programming platform.

However, since network virtualization is partially software-based testbed, it is not able to capture some of the non-linearization effects.

Lim et. al [12] proposed a testbed incorporating liquid-crystal-on-silicon (LCoS)-based programmable optical processor (POP) in the remote node to study the performance of WDM-based 60 GHz millimetre FiWi link. This testbed uses programmable tools where the POP is flexible, robust and enables simple future system upgrade. POP used in this testbed is to de-multiplex the interleaved channel before distributing to base station with error free transmission. Although POP is a programmable module, however its application only limited at the remote node and not the overall FiWi network architecture.

A fully fast reprogrammable FiWi testbed is needed to study the most suitable protocol and algorithm in smart grid network. The next section discusses on a proposed FiWi tesbed that uses software defined radio (SDR) to make it fast reconfigurable, simple and scalable.

DOI: 10.1051/epjconf/201716201021

4 Fast Reconfigurable FiWi Testbed

Figure 3 shows the fast reconfigurable FiWi testbed proposed where the major components used are SDRs for OLT, ONU and end users. We chose SDR as the main component of our testbed as it has the ability to support multiple functionality simply by modifying the software without the need of changing the hardware, making the study of protocols and algorithms to be simpler, faster and cost savvy.

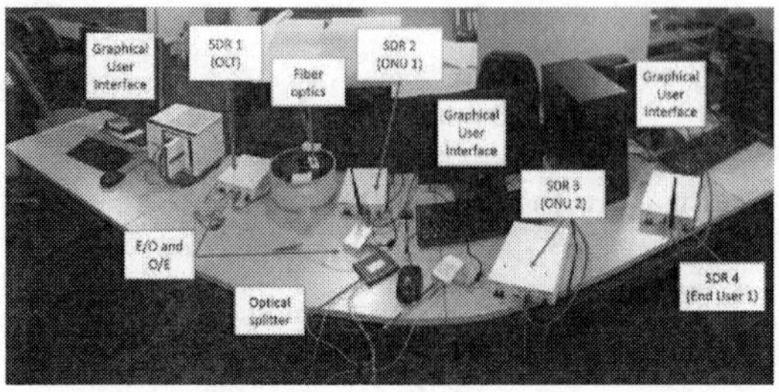

Fig. 3. FiWi Testbed Proposed

The SDR chosen for our testbed is Universal Software Radio Peripheral (USRP) by National Instrument (NI). A wide range of USRPs is available in the market, but we have chosen USRP-2922 particularly as it meets our testbed requirements which are to transmit at 2.5GHz frequency band and to have simultaneous transmit/receive feature.

The USRP that acts as OLT is first connected to electrical/optical converter before it goes to the 20km single mode fiber spool, then to a passive optical splitter to split the data into multiple ONUs. From the splitter, the data needs to go through an optical/electrical converter before it is connected to USRP that acts as the ONUs. From ONU's USRP to end users' USRP, antenna are needed as the transmission is done wirelessly.

Every SDR is further connected to a computer equipped with a Graphical User Interface (GUI) as shown in Figure 4 to monitor traffics coming in and out. The study of protocols and algorithms require the program to be modified via Labview at affected USRP without having to modify the hardware and the results can then be observed at the GUIs. Among the parameters that can be obtained are throughput, delay and jitter.

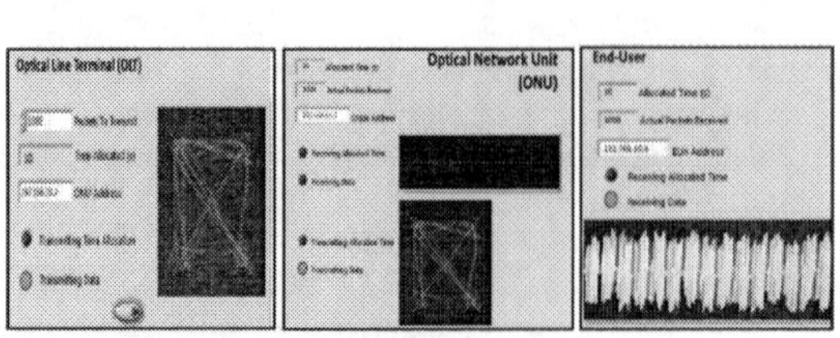

Fig. 4. Graphical User Interface

Figure 5 shows number of packets transmitted versus number of packets received using static bandwidth allocation algorithm implemented in this FiWi testbed. It shows that up until 11,000 packets are received for both methods via simulation and via testbed, validating the results achieved using our FiWi testbed.

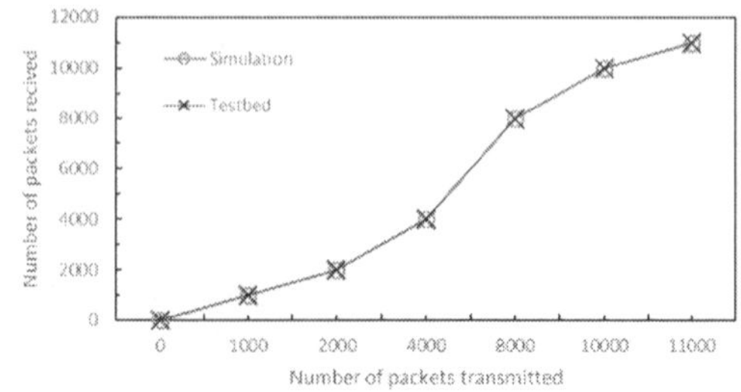

Fig. 5. Number of packets received vs number of packets transmitted via simulation and testbed

Figure 6 shows number of packets transmitted versus delay using static bandwidth allocation algorithm implemented in this FiWi testbed. It shows that with both method, delay increases as the number of packets transmitted increases and the results via simulation and testbed are in line with each other.

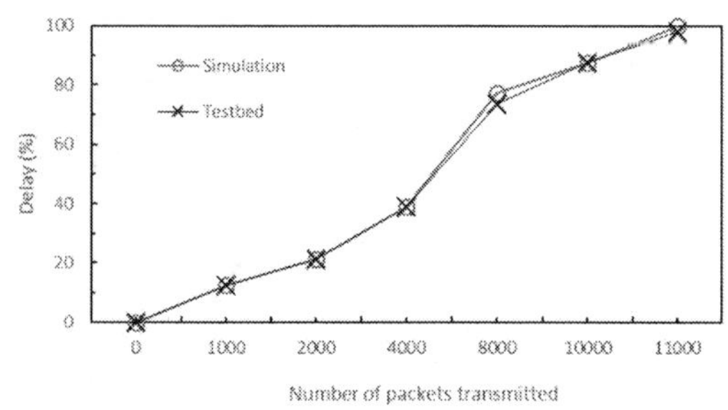

Fig. 6. Number of packets transmitted vs delay for simulation and testbed

5 Conclusion

In this paper, we have introduced the fundamental of smart grid and FiWi networks and we have reviewed the state of art of FiWi networks for smart grid as well as selected FiWi testbed. We have also introduced our proposed fast reconfigurable FiWi testbed with the number of packets received and delay results, which show that the testbed is validated as the results obtained via simulation and testbed are in line with each other.

References

1. The Independent Business Voice for the Environment, (2014)
2. Hiew, Y. K., Aripin, N. M., Jayavalan, S., & Din, N. M., *2013 IEEE Student Conference on Research and Development (SCOReD)*, pp. 26-30 (2013)
3. Attar, A. L. Haoming, Q. Pang, and V. C. M. Leung, IEEE Commun. Mag., **50**, no. 6, pp. 107–113 (2012)
4. Salvo P, Cuomo F, Abbagnale A. *2010 IEEE Global Telecommunications Conference (GLOBECOM 2010)* Dec 6, pp. 1-5 (2010)
5. Ding L, Gao K, Melodia T, Batalama SN, Pados DA, Matyjas JD. IEEE Transactions on Wireless Communications,12(11):5394-405 (2013)
6. Maier, M., *13th International Conference on Transparent Optical Networks (ICTON)*, pp. 1-4 (2011)
7. Zaker, N., Kantarci, B., Erol-Kantarci, M. and Mouftah, H.T., *2013 IEEE International Conference on Communications Workshops (ICC)*, pp. 863-867 (2013)
8. Ghassemi, A., Gulliver, T.A., Cioffi, J.M. and Karagiannidis, G.K., *IEEE Global Communications Conference (GLOBECOM)*, pp. 2605-2611 (2014)
9. C. Lim, Y. Yang, A. Nirmalathas, *IEEE Photonics Global Conference*, pp. 1-5 (2012)
10. X. Pang, A. Caballero, A. Dogadaev, V. Arlunno, L. Deng, R. Borkowski, J.S. Pedersen, D. Zibar, X. Yu, I.T. Monroy IEEE Photonics Society, **4**, Number 3 (2012)
11. Q. Dai, G. Shou, Y. Hu, Z. Guo, *IEEE International Conference on Communication 2013*, pp. 858- 862, (2013)
12. Meng, G. Shou, Y. Hu, Z. Guo, IEEE Information and Communications Technology, pp 1-6 (2014)
13. C. Lim, C. Pulikkaseril, K.L. Lee, Journal of Lightwave Technology, **30**, No.19, pp.3110-3117 (2012)

 DOI: 10.1051/epjconf/201716201022

Cardinality enhancement utilizing Sequential Algorithm (SeQ) code in OCDMA system

C. A. S. Fazlina[1,*], *C. B. M.* Rashidi[1], *A. K.* Rahman[2] and *S. A.* Aljunid[1]

[1]Centre of Excellence, [1]Advanced Communication Engineering, Centre of Excellent School of Computer & Communication Eng, Universiti Malaysia Perlis, (UniMAP) Kangar, 01000, Kangar, Perlis
[2]Department of Electrical and Electronic Eng, Universiti Malaysia Sarawak, (UNIMAS), Sarawak

Abstract. Optical Code Division Multiple Access (OCDMA) has been important with increasing demand for high capacity and speed for communication in optical networks because of OCDMA technique high efficiency that can be achieved, hence fibre bandwidth is fully used. In this paper we will focus on Sequential Algorithm (SeQ) code with AND detection technique using Optisystem design tool. The result revealed SeQ code capable to eliminate Multiple Access Interference (MAI) and improve Bit Error Rate (BER), Phase Induced Intensity Noise (PIIN) and orthogonally between users in the system. From the results, SeQ shows good performance of BER and capable to accommodate 190 numbers of simultaneous users contrast with existing code. Thus, SeQ code have enhanced the system about 36% and 111% of FCC and DCS code. In addition, SeQ have good BER performance 10^{-25} at 155 Mbps in comparison with 622 Mbps, 1 Gbps and 2 Gbps bit rate. From the plot graph, 155 Mbps bit rate is suitable enough speed for FTTH and LAN networks. Resolution can be made based on the superior performance of SeQ code. Thus, these codes will give an opportunity in OCDMA system for better quality of service in an optical access network for future generation's usage

1 Introduction

There have three types of multiple accesses; there are TDMA (Time Division Multiple Access), FDMA (Frequency Division Multiple Access) and CDMA (Code Division Multiple Access) [1]. In communication, the demand for traffic information is leading to increase of capacity and functionality. The characteristics such as bandwidth, transmitted information and data bit rate in security, are critical to the evolution of the OCDMA system [2]. Access and multiplexing techniques have allowed the simultaneous existence of multiple users on the network sharing with the same optical domain. Optical Code Division Multiple Access (OCDMA) techniques can provide moderate security communication. OCDMA also allowed to have multiple number of users to access optical network by sharing the bandwidth mechanism [3]. The main advantages of OCDMA compared with others techniques are the high capacity for higher connectivity, have more flexible bandwidth usage, higher granularity and scalability within optical networks, capable improved crosstalk performance, asynchronous access, and have potential for improved system security as number of users on the network has specific address code [4]. In OCDMA system, a bit of "1" is to be transmitted as an assigned code and a bit of "0" does not transmit anything [5]. In OCDMA, the most important consideration is code design. The code which is designed improperly and higher number of simultaneous users can be degraded the system performance due to presence Multi Access Interference (MAI). MAI is the interference from other users transmitting at the same time, which will limit the effective error probability with the presence of noise in the overall system [6]. In the overall system, MAI is the interference between user's transmissions at the same time, which will limit the effective error probability with the presence of noise [7]. In OCDMA, Phase Induced Intensity Noise (PIIN) is deeply related the MAI. PIIN will increase between code sequences because of inappropriate cross correlation among the address sequences. PIIN depends on the number of interfering users and cannot be improved by increasing the transmitted power [8]. Most codes have been proposed for OCDMA system such as Modified Double Weight (MDW), Modified Frequency Hopping (MFH), Flexible Cross Correlation (FCC), Dynamic Cyclic Shift (DCS) and Hadamard code, respectively. In this paper, a new algorithm called Sequential Algorithm (SeQ) code has been developed. This proposed code was developed based on Tridiagonal matrix code with parity of even and odd were involved in order to enhance the limitations occurs in the existing codes.

Corresponding author: eng_fazz@yahoo.com

1.1 Development of Sequential Algorithm (SeQ) Code

The new class of SeQ code design is concerned about mathematical preliminaries to design codes for any given number of users and weights to have minimum code length. Since, optical codes may be represented as vectors; linear algebra has been used for the design of these codes. The codes are treated as rows of a matrix of finite dimension; hence, the methods of design and analysis of matrix algebra is best suited in these cases. Since, the codes are unique, the code matrix is required to have a finite value of its rank for a given set of users. The in phase auto-correlation functions and the cross-correlation functions of the codes can easily be defined in term by the dot and cross product of the rows of the matrix. The auto-correlation of each codeword $A_i = (a_1, a_2, a_3, ..a_N)$ and the cross-correlation between any two distinct codeword $A_i = (a_1, a_2, a_3 ..a_N)$ and $B_j = (b_1, b_2, b_3 .. b_N)$, respectively. Optical code are family of K (for K users) binary [0, 1] sequences of length N, code weight W (the number of "1" in each codeword) and the maximum cross-correlation, λ_{xym}.

$$\lambda_a(\tau) = \sum_{i=1}^{N} a_i a_{i+\tau} = W \; for \; \tau = 0 \quad (1)$$

$$\lambda_{ab}(\tau) = \sum_{i=1}^{N} a_i b_{i+\tau} \leq 1 \; for \;\; \tau = 0 \quad (2)$$

Hence, major bottleneck in the successful implementation of all optical networks is basically when all the users try to transmit the data simultaneously. It is noteworthy when designing coding sequences such that, it must cause least overlapping between data chips. Now, let $A = \{a_n\}$ and $B = \{b_n\}$ be the sequences of length N such that the vector form can be written as [8];

$$\left.\begin{aligned} \{a_n\} &= '0' \, or \, '1', i = 0,..N-1 \\ \{b_n\} &= '0' \, or \, '1', i = 0,..N-1 \end{aligned}\right\} \quad (3)$$

Since x_n is a {0, 1} binary sequence, the maximum value of $\lambda_a(\tau)$ in Eq (1) is for $\tau = 0$ and is equal to W, the code weight of the sequence can be expressed as;

$$\lambda_a(0) = W \quad (4)$$

The codes described by Eq (3) can also be represented in vector form as;

$$\left.\begin{aligned} A &= \{a_i\} = for \; i = 0, 1, \dots N-1 \\ B &= \{b_i\} = for \; i = 0, 1 \dots N-1 \end{aligned}\right\} \quad (5)$$

where, X and Y are vectors of length N with elements as defined by Eq (5). In terms of the vectors A and B, Eqs (1) and (2) can be written as;

$$\begin{aligned} \lambda_a(0) &= aa^T = W \\ \lambda_{ab}(0) &= ab^T \end{aligned} \quad (6)$$

where, A^T and B^T denote the transpose of vectors A and B, respectively.

The code weight of the codes is given by the inner (dot) product of the vector representing the codes. The length of the codes can also be controlled by controlling the number of columns of the code matrix with the present of bit parity of even and odd elements. The first A_i^{th} row for the first K user of the code matrix can be written as;

$$A_i = \overbrace{1\,1..1}^{W} \;\; \overbrace{0...0}^{x_i} \quad (7)$$

where, the first row of A_i is define as;

$$W + x_i \qquad for \; 1, 2, \dots K \quad (8)$$

Assume that, $r = (W-\lambda_{ab})$ and x_i can be derived as;

$$\begin{aligned} x_i &= Kr + \lambda_{ab} - r(i-1) - W \\ &= r(K-i) \end{aligned} \quad (9)$$

In summary, all the parameters of the codes can be defined in term of the matrix representing the codes set. When a set of code parameters is chosen, a code can be constructed with good auto and cross-correlation so that the interference called MAI and PIIN among users can be suppressed effectively by decoding the signals. This requires that the address codes must satisfy, where all address code word maybe easily distinguished from a possibly shifted version of every other code word.

From the basis of the above discussions, the construction of an optical codes having value of K, w, λ_a consists of the following steps:

1). For a given number of users K, and weight w, form a set of Non-Zero In Phase Cross Correlation code with minimum length as given by Eq (7).

2). The K rows of the code matrix that give the K optical CDMA codes having Non- Zero In-Phase Cross-Correlation (IPCC), weight w and minimum length.

This procedure will now be explained with the help of an example:

Assume that, it is desired to generate a set of minimum length of SeQ code for users = 4, weight = 3 and λ_a equal = 1.

Table 1. Codeword of SeQ for W=3, K=4 and $\lambda_a \leq 1$

Basic	Parity
(1 2 4)	1 1 0 1 0 0 0
(2 3 5)	0 1 1 0 1 0 0
(3 4 6)	0 0 1 1 0 1 0
(4 5 7)	0 0 0 1 1 0 1

Notice that the code weight of each row more than two, and the relation between N and K for SeQ code can be express as:

$$\mathrm{N} = \mathrm{K} + 2W - 3 \quad (10)$$

1.2 Signal to Noise Ration Performance Analysis

The Signal Noise Ratio (SNR) is defined as the average signal to noise power, SNR = $\left[\frac{I^2}{\sigma^2}\right]$, where σ^2 is the average power of noise which is given by,

$$\sigma^2 = \langle i_{shot}^2 \rangle + \langle i_{PIIN}^2 \rangle + \langle i_{thermal}^2 \rangle \quad (11)$$

Eq. (10) can be expressed as,

$$\sigma^2 = 2eBI + I^2 B\tau_C + \frac{4K_b T_n B}{R_L} \quad (12)$$

Where, e is the electron charge, I is the average photocurrent, I^2 is the power spectral density for I, B is the noise equivalent of electrical bandwidth, K_b is the Boltzmann constant, T_n is the absolute receiver noise temperature and R_L is the receiver load resistor. Power Spectrum Density (PSD) will be calculated and the photodiode current I and can be written as follows;

$$I = \mathcal{R} \int_0^{\infty} G(v)\, dv \qquad (13)$$

$\mathcal{R}$ represents as the responsively of the photo-detectors. Hence the photo current I can be expressed as;

$$I = \Re \left[\frac{P_{sr}[W]}{N} \right] \qquad (14)$$

The power of *Shot* noise can be written as;

$$\langle i_{Shot}^2 \rangle = 2eB\mathcal{R} \left[\frac{P_{sr}}{N} \right] [3W + 1] \qquad (15)$$

The Phase Induced Intensity Noise (PIIN) noise will dominate the broadband sources. Hence, with PSD from each user is the same; the calculation for receiver PIIN noise directly from the total PSD of each photodiode. By using an Eq. (15) the PIIN noise at the receiver output is given by;

$$\langle i_{PIIN}^2 \rangle = B(I_1^2 \tau_{c1} + I_1^2 \tau_{c2}) = I^2 * \tau_C * B \qquad (16)$$

Where, I_1 and I_2 are the average photodiode currents, τ_{c1} and τ_{c2} are the coherence times of light incident on each photodiode. The variance of the PIIN noise at the receiver can be expressed as;

$$\langle i_{PIIN}^2 \rangle = \frac{B\mathcal{R}^2 P_{sr}^2 KW}{N^2 \Delta v} [3W + 1] \qquad (17)$$

Since all equation shown the properties of SeQ code are unique and independent of each other, Eq. (17) is also independent of the active users' data, consequently, proposed coding system does not depend on the timing of transitions in the data and it applied to the asynchronous system. *Thermal* noise is given as;

$$\langle i_{Thermal}^2 \rangle = \frac{4K_b T_n B}{R_L} \qquad (18)$$

From Eqs. (14), (15), and (17), the SNR for the proposed SeQ code in the ODCMA coding system is defined by;

$$SNR = \frac{\left[\frac{\Re P_{sr} W}{N}\right]^2}{\left[\frac{2eB\Re P_{sr}}{N}\right][3W+1] + B\Re\left[\frac{P_{sr}^2 KW}{N^2 \Delta V}\right][3W+1] + \frac{4K_b T_n B}{R_L}} \qquad (19)$$

Since, there no pulses are sent for the data bit *'0'* an assuming that the noise distribution is Gaussian, thus, the corresponding BER can be obtained as follows;

$$BER = \frac{1}{2} efrc \left(\sqrt{\frac{SNR}{8}} \right) \qquad (20)$$

Finally, Eqs. (19) and (20) will be used for the numerical calculation for an evaluation of the proposed OCDMA coding system utilizing SeQ code. The performance of the SeQ code has been compared numerically with the existing OCDMA codes such as FCC and DCS codes. The numerical parameters used in calculations are shown in Table 2.

Table 2: Typical parameters for theoretical calculations.

Parameter	Value
Electron's charge	$e = 1.60217646 \times 10^{-19}$ coulombs
PD quantum	η = 0.75
Electrical bandwidth	B = 80 MHz
Boltzmann constant	$K_b = 1.38 \times 10^{-23}$ W/K/Hz
Receiver noise	$T_n = 300K$
Receiver load resistor	$R_L = 1030\ \Omega$
Data transmission rate	$R_b = 155$ Mbps
Broadband line width	Δλ = 3.75 THz

2 Results and Discussion

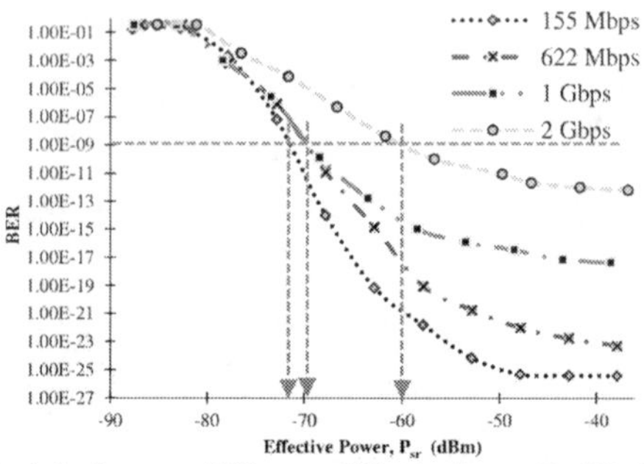

Fig. 1. Performance BER versus Effective Power, P_{sr} (dBm) of 155 Mbps, 622 Mbps, 1 Gbps and 2 Gbps for SeQ (W=4) code.

Fig. 1 illustrates the plots of performance of BER reflect to the effective power, P_{sr} (dBm) of varied 155 Mbps, 622 Mbps, 1 Gbps and 2Gbps for SeQ code. These result was analyzed by utilizing Optisystem, respectively. The result shows, the effective power P_{sr} obtained from -90 dBm until - 30 dBm. From the results, the performance of BER for 155 Mbps shows the excellent performance in comparison with 622 Mbps, 1 Gbps and 2Gbps. When the effective power, P_{sr} larger (P_{sr} > -72 dBm) the performance of BER are increasingly decrease in comparison with when effective power P_{sr} lower (P_{sr} < -72 dBm) the BER performance are worse. At communication error floor BER = 10^{-09}, the effective received power P_{sr} are -72 dBm, -60 dBm for 155 Mbps and 1 Gbps data bit rate. Whereas, the effective received power P_{sr} = -69 dBm for data bit rate 622 Mbps and 1 Gbps. From the plots graph, 155 Mbps bit rate are suitable enough speed for FTTH and LAN networks.

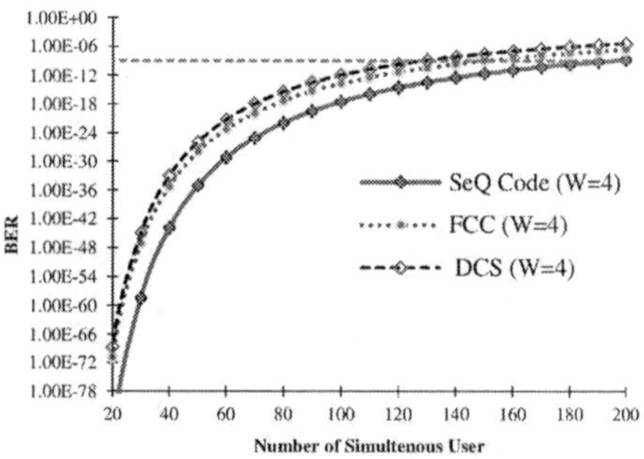

Fig. 2. Performance BER versus Number of Simultaneous Users, K for various OCDMA codes.

The plots graph demonstrates performance of BER versus number of simultaneous users, K for SeQ (W=4),

FCC (W=4) and DCS (W=4) OCDMA codes considered of 200 number of users at effective power, P_{sr} -10 dBm and 155 Mbps bit rate, respectively. At typical error rate of optical communication system BER 10^{-09}, the system performance of BER become increasingly degrade as the number of simultaneous users increased. This performance analyzed from Eq. (20). Fig. 2 shows performance of SeQ was superior in comparison with FCC (W=4) and DCS (W=4). Whereas, SeQ Code (W=4) can accommodate 190 number of users compared contrast to FCC (W=4) and DCS (W=4) can support 140 and 90 number of users, respectively. Since each code sequence is assigned to specific BER = 10^{-09} error floor, the SeQ (W=4) code has cardinality enhancement about 36% and 111% of FCC (W=4) and DCS (W=4). Thus, the SeQ code presents excellent performance due to the arrangement of the code algorithm and phase cross-correlation. This performance of this system is revealed the improvement significantly, because the total loss is reduced as the AND detection technique requires less number of filters in the decoder. Hence, this results as an evidence that SeQ code can enhance the performance of the OCDMA system better than others existing codes.

3 Conclusions

The performance of OCDMA system by utilizing SeQ code have been performe. The system degradation due to PIIN can be suppressed by using SeQ code results in enhancing BER performance. The proposed SeQ code have excellent of BER performance contrast with effective received power, P_{sr} based on the data bit rate where 155 Mbps, 622 Mbps, 1 Gbps and 2 Gbps. The performance of the proposed SeQ code achieves high cardinality number of simultaneous user's in comparison with FCC (W=4) and DCS (W=4) code. Moreover, the SeQ code arrangement with less cross correlation value able to mitigates the PIIN and MAI, which capable to enhance the overall system performance.

This project was supported by the Government of Malaysia under the Fundamental Research Grant Scheme at Universiti Malaysia Perlis, Malaysia under the research grant # 9003-00456.

References

1. L. Verma and G. Singh, Int. J. of Sci., Eng. And Tech. Res.**3**,6 (2014).
2. G. D. F. Guimarães, J. Electromagn. Anal. Appl., **5**, 2, pp. 49–57, (2013).
3. T. H. Abd, S. A. Aljunid, H. A. Fadhil, R. A. Ahmad, and N. M. Saad, Opt. Fiber Technol., **17**, 4, pp. 273–280, (2011).
4. A. O. Aldhaibani, S. A. Aljunid, M. S. Anuar, A. R. Arief, and C. B. M. Rashidi, Opt. Fiber Technol., **22**, pp. 7–12, (2015).
5. S. A. Aljunid, F. Ghani, H. A. Fadhil, and M. S. Anuar, " J. of Applied Sci. Res., **8**,12, 5614-5626, (2012).
6. T. H. Abd, S. A. Aljunid, H. A. Fadhil, I. F. Radhi, and N. M. Saad, J. Opt. Comm., **33**, 4, pp. 279–283, (2012).
7. C. B. M. Rashidi, S. A. Aljunid, F. Ghani, H. A. Fadhil, M. S. Anuar, and A. R. Arief, Optik, **125**, 17, pp. 4889–4894, (2014).
8. N. D. Keraf and M. N. Nurol, Proceeding of Malaysia University Conference Engineering Technology 2014, pp 463-467,(2015).

DOI: 10.1051/epjconf/201716201023

Channel impulse response equalization scheme based on particle swarm optimization algorithm in mode division multiplexing

Shaymah Yasear[1,*] and *Angela* Amphawan[1,2]

[1]School of Computing, Universiti Utara Malaysia, Sintok, Kedah, Malaysia
[2]Research Laboratory of Electronics, Massachusetts Institute of Technology, Cambridge, Massachusetts, USA

Abstract. Mode division multiplexing (MDM) technique has been introduced as a promising solution to the rapid increase of data traffic. However, although MDM has the potential to increase transmission capacity and significantly reduce the cost and complexity of parallel systems, it also has its challenges. Along the optical fibre link, the deficient characteristics always exist. These characteristics, damage the orthogonality of the modes and lead to mode coupling, causing Inter-symbol interference (SI) which limit the capacity of MDM system. In order to mitigate the effects of mode coupling, an adaptive equalization scheme based on particle swarm optimization (PSO) algorithm has been proposed. Compared to other traditional algorithms that have been used in the equalization process on the MDM system such as least mean square (LMS) and recursive least squares (RLS) algorithms, simulation results demonstrate that the PSO algorithm has flexibility and higher convergence speed for mitigating the effects of nonlinear mode coupling.

1 Introduction

The unstoppable and increasing demand for data traffic requires providing higher bandwidth to end users[1]. To cope with the rapid increase of data traffic; for many years the researchers focused on the development and improve the efficiency of optical fibre transmission system [2].

The MDM transmission system is considered as a promising solution. However, in the optical MDM system, when the light ray propagates through the MMF; it spreads into multiple paths (modes). Some of these modes (called low-order modes) travel at low angles to the fibre axis. These modes have shorter paths compared to the other modes (high-order modes) which travel at larger angles to the core axis and they have longer paths. Therefore, these modes travel through the fibre with different group velocities (i.e., differential mode group delays (DMGD)), and arrive to the receiver at different times[3]. During propagation of these modes, and due to the fibre manufacturing imperfections (e.g., bending, stresses and twisting), the orthogonality of the modes will damage and the modes will couple with each other, causing the so-called mode coupling [4]. Each time the mode coupling occurs, the power is leaked from one data symbol launched into a particular mode to the adjacent symbol so-called cross-talk. This causes the symbols to spread over time. This phenomenon is known as modal dispersion (MD). The modal dispersion will lead to overlapping the neighboring symbols of signal while it transmits to the receiver. The signal is then no longer usable. This phenomenon is known as inter-symbol interference (ISI) [5]. In the optical communication, each pulse represents a symbol and each symbol or several symbols representing digital information (bit). Due to the mode coupling effects, the receiver cannot restore the transmitted symbols and recover their information. The mode coupling is responsible for the bandwidth limitation of the transmitted signal and may lead to significant increase in bit error rate (BER) in long transmission distances [6].

To mitigate the effects of mode coupling in the MDM system, several techniques have been proposed [7, 8]. However, the hardware-based equalizer cannot provide an acceptable performance with time-varying channels, due to the implementation complexity. Moreover, a simple linear equalizer is difficult to meet the basic requirements of the system. In this case, an equalizer based on an adaptive algorithm must be used.

The adaptive equalizer is implemented using algorithms that have conditions where it is periodically check and compensate or equalize the transmission channel characteristics. In addition to, restore the transmitted symbols and recover their information. However, the traditional adaptive equalization algorithms, that have been used with MDM system; such as least mean square (LMS), recursive least squares (RLS) have drawbacks such as slow convergence, speed, poor stability, high computational complexity and inaccurate equalization result [4, 9]. Furthermore, the performance of linear equalization algorithm is degraded in channels having

*Corresponding author: shayma1985akram@gmail.com

large eigenvalue spread [10]. For these reasons, the use of these algorithms has given rise to the limited performance of the optical MDM system. Therefore, it is worth investigating an alternative algorithm which is able to cope with a large eigenvalue spread of the channel. One possible solution that has been proposed to overcome these issues is using the particle swarm optimization (PSO) algorithm. The remainder of this paper is as follows. Section 2 presents MDM system, Section 3 reviews the other conventional algorithms, while Section 4 and Section 5 discusses the result of equalization schemes and the conclusion respectively.

2 Related work

To date, most studies realized on using an equalizer based on LMS and RLS algorithms to mitigate the effect of mode coupling in MDM system [11].

2.1. Equalization schemes in MDM system

In[12], LMS and constant modulus algorithm (CMA) have been proposed to be used with MDM system. Although the LMS algorithm has a low computational complexity, its training sequence adds an overhead and thus reduces the performance of the system. Moreover, the CMA requires a suitable learning step; the equalizer converges slowly with small steps and cannot converge with large steps for accurate estimation [8].In [13], Koebele et al., discuss the use of CMA to update the equalizer coefficients, although, CMA provides low computational complexity, it has a poor performance. This is due to the instability and a slower convergence rate compared with LMS algorithm[14, 15]. In [16], proposed Single-Carrier frequency domain equalization (SC-FDE), has low computational complexity; compared to time domain equalization (TDE). However, FDE it still suffers from high hardware complexity. This is due to large fast Fourier transform (FFT) size determined by the channel impulse response spread[17].In [4], the performance of adaptive FDE based on LMS and RLS algorithms has been reviewed. LMS algorithm suffers from low cyclic prefix (CP) efficiency and higher SERs compared with RLS. In the other hand, RLS, suffers from high computational complexity and high cost and power consumption. In [17], a single-stage FDE based on RLS algorithm (FD-RLS) has been proposed. According to the authors, the purposed single-stage adaptive FD-RLS algorithm provides low hardware complexity compared with the two-stage method. However, the RLS algorithm still suffers from instability and high computational complexity.

2.2. Current PSO-based equalization schemes in wireless radio system

The PSO algorithm has been proposed by Kennedy and Eberhart in 1995 [25], aims to seek the optimal solution, in a search space. Recently, several studies [17, 26-28] have investigated the use of PSO algorithm in adaptive equalization processes. These studies demonstrated that the PSO algorithm outperforms other traditional algorithms that have been used in the adaptive equalization of the MDM system. The PSO algorithm provides a far more efficient performance in a channel with a large eigenvalue spread compared to traditional algorithms. The convergence speed of PSO algorithm is faster than LMS algorithm. Furthermore, the PSO algorithm is able to reduce the BER and provides a high convergence rate compare to LMS and RLS algorithms. In nonlinear channel; the PSO-based algorithms outperform the LMS algorithm, especially in heavily-distorting channels. For this reason, this study will focus on investigating the possibility of improving the performance of MDM system by minimizing the effect of mode coupling by using adaptive equalization based on PSO algorithm. In[18], the PSO algorithm was used to train the functional link artificial neural networks (FLANNs) equalization, in order to enhance its classification capability. According to the authors, the PSO algorithm provides a better result compare to LMS and BP algorithm. In [19], the performance of a different versions of PSO algorithm was studied and compared with the LMS-based algorithm. The study demonstrated that, PSO algorithms converge faster than LMS algorithm. In [20], a new resource management technique for the orthogonal frequency division multiplexing (OFDM) technique was proposed based on PSO algorithm, to solve the problem of sub-channel resource allocation in downlink of OFDMA system. In their study, the authors assumed the channel is constant during allocation. With this assumption, the PSO-based technique provided better performance compared to other methods. the performance of adaptive equalizer based on LMS, RLS and PSO algorithms in wireless mobile communication was compared. The algorithms were tested with the help of three parameters, MSE, BER and time of convergence (TOC). The results of the study demonstrated that, compare to LMS and RLS algorithms, PSO algorithm is more efficient in reducing the MSE and BER. In [19], an adaptive frequency-domain (FD) equalizer for the single-carrier frequency division multiple access (SC-FDMA) system was designed based on modified PSO algorithm. In[10], the PSO algorithm integrated with the LMS algorithm to improve the decision feedback equalization (DFE). According to the authors, the proposed technique reduces the complexity and provide better performance compared to LMS and RLS algorithms.

The optical MDM system shares some similarities with the wireless radio MIMO system. In optical MDM system the manufacturing imperfections in the MMF lead to mode coupling, giving rise to chromatic dispersion (CD) and modal dispersion (MD). This induces multipath propagation similar to that in wireless radio MIMO systems caused by scattering and Doppler effects. The success of PSO-based equalization algorithms in wireless radio systems and the similarity in the multipath characteristics of both optical MDM and wireless radio MIMO systems motivates the adaptation of the PSO-based equalization algorithm for optical MDM systems.

3 Simulation model

3.1. Modelling MDM system

In order to evaluate the performance of PSO equalization scheme, a MDM system is simulated using Optisystem simulation software. Figure 1 shows the configuration of the simulated link.

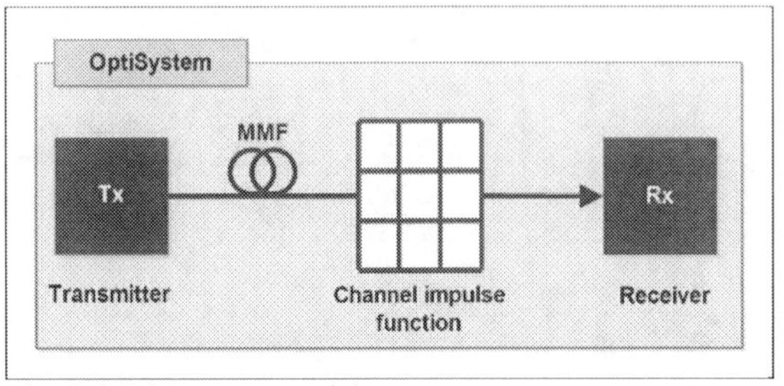

Fig. 1. MDM system model

MDM model consists of spatial optical transmitter (Tx) which includes optical source or transmitter and a transverse mode generator. The type of mode that have been used in this experiment is Laguerre Gaussian (LG). LG modes are extensions of the simple Gaussian mode shape to higher order modes in a cylindrical coordinate system, and are often used to describe the transverse mode shapes at the output of a VCSEL. The analytical representation of the mode LG_{ml} is given by equation (1)[21].

$$\psi_{ml}(r,\phi)=\alpha.(\frac{2r^2}{w_0^2})^{\frac{L}{2}}.L_m^l(\frac{2r^2}{w_0^2}).\exp(-\frac{r^2}{w_0^2})$$

$$\exp(j\frac{\pi r^2}{\lambda R_o}).\begin{Bmatrix}\cos(L\phi), l>=0\\ \sin(L\phi), l<0\end{Bmatrix} \quad (1)$$

where α is a normalization constant, $L = |l|$, λ is the field wavelength, and is a generalized Laguerre polynomial. At the beam waist, the inverse of R_o is zero, indicating a flat phase front. At any distance to the left or right of the waist the beam begins to diverge and R_o becomes finite.

The resulted optical signal is transmitted through MMF component to the receiver (Rx). In this study, a MMF with a core diameter of 62.5μm has been used to guide 2 spatial LG modes, namely, LG01, LG11 at 1550nm. The modulation type used in this experiment is NRZ. Table 1 summarize the parameters of MDM system model.

Table 1. Parameters of MDM system model

Parameter	value
MMF length	2km
Wavelength	1550nm
Bit rate	40GB/s
modulation type	NRZ
Mode type	LG
Core	62.5

In the receiver, the photodetector is used to convert the optical signal into the electrical current. The output of MDM system model, which represents a signal with mode coupling effects, is used as an input signal to be recovered by using PSO equalization scheme. Next section will discuss in details the PSO equalization scheme.

3.2. Modelling PSO equalization scheme

In this study, the PSO equalization scheme has been proposed to improve the performance of MDM system by improving the mitigation of mode coupling and increase a bandwidth. The proposed scheme has been implemented at the receiver side for post-compensation, using Matlab platform. Figure 2 illustrates the PSO equalization scheme.

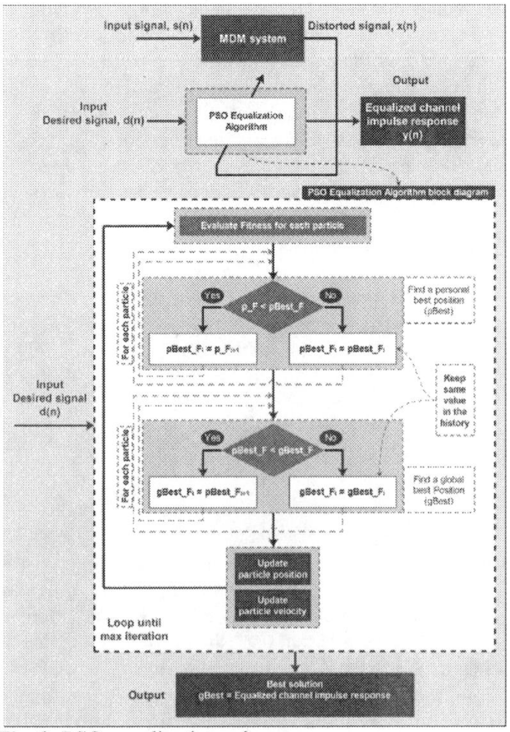

Fig. 2. PSO equalization scheme

The input of MDM system at time n, is the channel impulse function, s(n). The output of MDM system model, x(n) which represents a signal with mode coupling effects, termed as the channel impulse response in the time domain. The desired signal, d(n) is assumed to be a Gaussian pulse, which is a standard ideal pulse shape[22]. The Gaussian pulse is shaped with mean = 5.5, and variance = 2.0, d(n) and x(n) are the inputs to the PSO equalization scheme.

The block diagram of PSO equalization scheme consists of: the evaluation fitness block to evaluate all particles (potential solutions). In order to find the particle with the best fitness value (pBest_F) in each iteration, which is called personal best position (pBest), the current fitness value (p_Fi+1) will be compared with the (pBest_Fi) in the history. If p_Fi+1< pBest_Fi then pBest_Fi = p_Fi+1. Furthermore, to find the particles with the best fitness value between all the particles (gBest_F), known as global best position (gBest), the current fitness value (pBest_Fi) value, will compared with the (gBest_Fi) in the history. If pBest_Fi< gBest_Fi+1 then gBest_Fi = pBest_Fi+1. The particle velocity and position will be updated by using (2) and (3) respectively.

$$v_i(t+1) = \omega . v_i(t) + c_1 . r_1 (pbest_i(t) - x_i(t)) + c_2 . r_2 (gbest(t) - x_i(t)) \quad (2)$$

$$x_i(t+1) = x_i(t) + v_i(t+1) \quad (3)$$

where c1 is cognitive parameter and c2 is a social parameter that propel solutions toward personal best and global best respectively; r1 and r2 are random numbers with uniform distribution, ω is an inertia weight (w) used to control the velocity. The output of PSO equalization algorithm (gBest) is the equalized channel impulse response, becomes as close as possible to a desired signal d(n). Table 2 summarize the parameters' values of PSO algorithm.

Table 2. Parameters' values of PSO algorithm

Parameter	Value
c_1	2.0
c_2	1.5
w_0	1.0
w	0.98
Max iteration	50
Swarm size	20
Terminal value	<= 0.00001

4 Result

The performance of PSO equalization scheme was compared with equalization based on LMS and RLS algorithms. Two channels were used as a distorted signal, similarly, two Gaussian signals were used as a desired signal. The mean square error (MSE) has been used to measure the differences between desired signal and distorted signal, before equalization and between desired signal and equalized impulse response after equalization. Figure 3 shows the MSE before and after equalization using LMS, RLS and PSO algorithm. From the results it can be clearly seen that the PSO algorithm outperforms LMS and RLS algorithms with less MSE values for both channel 1 and channel 2. Figure 4 shows the differences between impulse responses of channel 1 and channel2, before and after performing PSO equalization scheme. Before the equalization, the symbols of channel 1 and channel 2 have been overlapped with each other, while after PSO equalization the impulse responses of channel 1 and channel 2 have been recovered as shown in Figure 5.

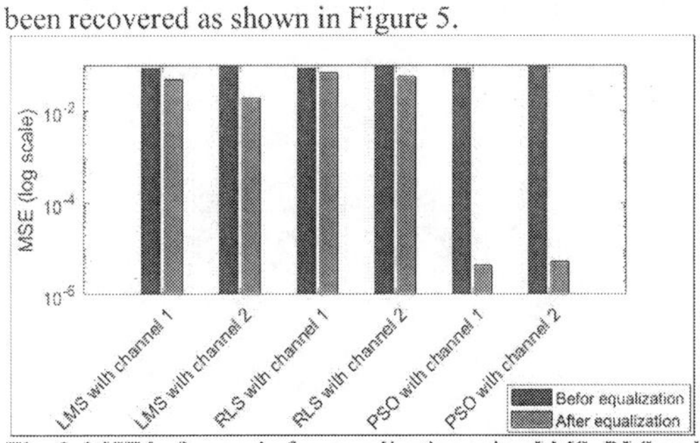

Fig. 3. MSE before and after equalization using LMS, RLS and PSO algorithms

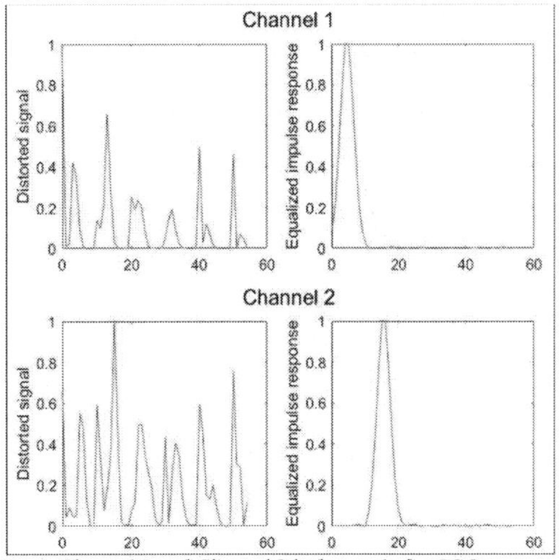

Fig. 4. Channel 1 and Channel 2 before and after PSO equalization scheme

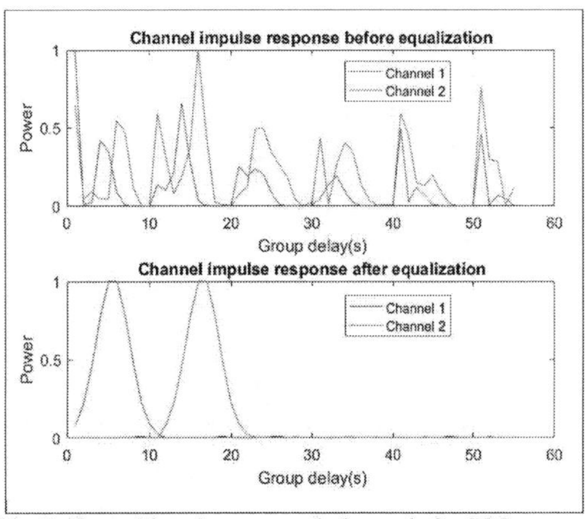

Fig. 5. Channel impulse response before and after PSO equalization scheme

5 Conclusion

An PSO equalization scheme is proposed and simulated for MDM system with MMF of length 2km. It is shown that the proposed PSO equalization scheme significantly reduces the mode coupling effects compared to LMS and RLS algorithms.

References

1. Cisco, *CISCO White paper*, (2015)
2. B. Batagelj, V. Janyani, and S. Tomažič, *Informacije MIDEM,* **44**, no. 3, pp. 177-184 (2015)
3. B. Woodward and E. B. Husson, *Fiber optics installer and technician guide*. (John Wiley & Sons, 2006)
4. S. O. Arik, J. M. Kahn, and K. P. Ho, IEEE Signal Processing Magazine, **31**, no. 2, pp. 25-34 (2014)
5. D. Richardson, J. Fini, and L. Nelson, Nature Photonics, **7**, no. 5, pp. 354-362 (2013)
6. N. Bai and G. Li, Optics express, **22**, no. 4, pp. 4247-4255 (2014)
7. A. Amphawan and N. M. A. Al Samman, *SPIE Optical Engineering & Applications*, **8847**, pp. 88470Y-88470Y-6 (2013)
8. T. Masunda and A. Amphawan, *Journal of Optical Communications* (2016)
9. C. Yue, X. Kewen, and L. Jianfei, Advances in Information Sciences and Service Sciences, **5**, no. 9, p. 114 (2013)
10. N. Iqbal, A. Zerguine, and N. Al-Dhahir, Signal Processing, **108**, pp. 1-12 (2015)
11. X. Xiang et al., *2015 14th International Conference on Optical Communications and Networks (ICOCN),* pp. 1-3 (2015)
12. R. Ryf et al., *Optical Fiber Communication Conference and Exposition (OFC/NFOEC), 2011 and the National Fiber Optic Engineers Conference* pp. 1-3 (2011)
13. C. Koebele et al., *European Conference and Exposition on Optical Communications* (2011)
14. S. A. A. Khan and S. A. Sheikh, *The Second International Conference on Electrical, Electronics, Computer Engineering and their Applications (EECEA2015)* p. 163 (2015)
15. L. Yan and G. Hu, IEEE Photonics Journal, **8**, no. 2, pp. 1-11 (2016)
16. N. Bai and G. Li, IEEE Photonics Technology Letters, **24**, no. 21, pp. 1918-1921 (2012)
17. Y. Weng, X. He, and Z. Pan, *SPIE OPTO*, pp. 97740B-97740B-12 (2016)
18. S. Yogi, K. Subhashini, J. Satapathy, and S. Kumar, *2010 IEEE International Conference on Communication Control and Computing Technologies (ICCCCT),* pp. 725-730 (2010)
19. N. Sharma, A. K. Tarcar, V. A. Thomas, and K. Anupama, Information Sciences, **182**, no. 1, pp. 115-124 (2012)
20. A. Amphawan, Y. Fazea, and M. Elshaikh, *Advanced Computer and Communication Engineering Technology: Proceedings of ICOCOE* (2015)
21. H. A. Sulaiman, M. A. Othman, M. F. I. Othman, Y. A. Rahim, and N. C. Pee, Eds. Cham: Springer International Publishing pp. 355-363 (2016)
22. C. E. Webb and J. D. Jones, *Handbook of Laser Technology and Applications: Laser design and laser systems*. (CRC Press, 2004)

EPJ Web of Conferences **162**, 01024 (2017) DOI: 10.1051/epjconf/201716201024
InCAPE2017

Analysis of the effect of BER and Q-factor on free space optical communication system using diverse wavelength technique

A. A.Anis[1,*], *C .B. M. Rashidi*[1], *A. K. Rahman*[2] , *S. A. Aljunid*[1], *and N. Ali*[1,3]

[1]Advanced Communication Engineering, Centre of Excellence - School of Computer & Communication Eng, Universiti Malaysia Perlis, (UniMAP) Kangar, 01000, Kangar, Perlis
[2]Department of Electrical and Electronic Eng, Universiti Malaysia Sarawak, (UNIMAS), Sarawak
[3]Semiconductor Photonics & Integrated Lightwave Systems (SPILS), School of Microelectronic Engineering, Universiti Malaysia Perlis, Pauh Putra, 02600 Arau, Perlis, Malaysia

Abstract. This paper presents the analysis of the effect of BER and Q-factor on free space optical communication system using the diverse wavelength technique. The visibility is gradually decreased due to attenuation effects such as weather conditions as the bit error rate increases. Despite this disadvantage, FSO performance could be further improved by using a wide range of wavelengths of the laser beam. The results show good performance for 1550 nm wavelength, where the wavelength can accommodate visibility for 6 km at link range of 1 km, and Q-factor 25 at the system performance BER = 10^{-124} could improve the system by 67% compared to 850 nm and 785 nm wavelength ,which could enhance the system by only 2%. This clearly show that the superior capabilities of 1550 nm wavelength in handling attenuation problems and capability to reduce the power loss compared to 850 nm and 785 nm wavelength.

1 Introduction

Free Space Optics (FSO) has ultimately become a major leading topic in communication systems nowadays. It is a new technology that utilizes a light beam propagating through the atmosphere. The signal transmission is from the transmitter to transfer data received on the other side of the two point communication system. The FSO communication is also called as Optical Wireless Communication, Fibreless, or Laser Communication [1].

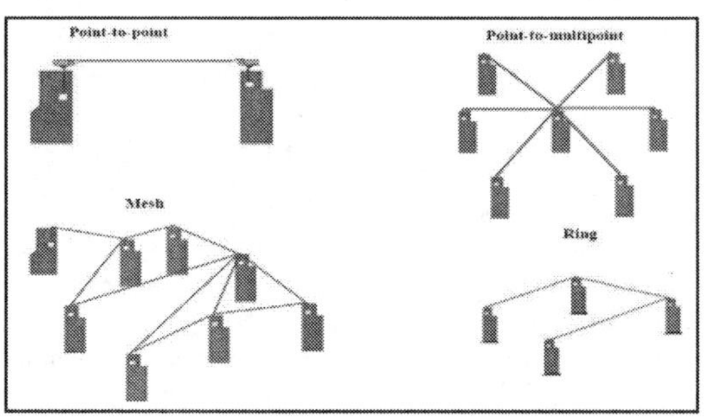

Fig. 1. Terrestrial FSO links [2]

Fig. 1 shows the types of conventional terrestrial FSO links. The high carrier frequency of FSO in the range of 20 THz to 375 THz enables it to transfer data at really high rates. As FSO is offering a wide broadband service, it needs Line of Sight (LOS) communication between the transmitter and receiver [3]. It is due to the principle of light which travels in straight line. It can be implemented as the other choice for an Optical Fiber replacement. Modulation from a laser beam is used to transmit optical binary information. Modulation bandwidth of FSO is better than the optical fiber communication as the FSO system is capable to transfer more data at a time with faster rate. The output modulated optical signal transmits through the atmosphere from a transmitter to a receiver [4]. Fig. 2 demonstrates the transmitter and receiver of a FSO system. Both the transmitter and the receiver is connected by the atmosphere as the transmission medium of the optical signals.

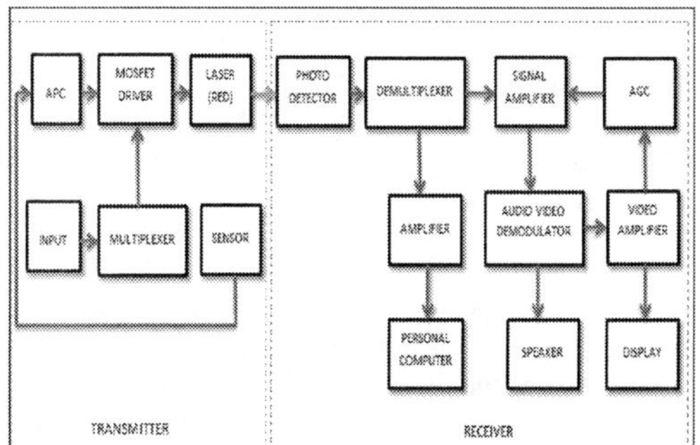

Fig. 2. The primary parts of a FSO system. FSO must consist of a transmitter and a receiver.

* Corresponding author: anis_afina@ymail.com

There are various FSO employed situations lately under research. The examples are ground-to-ground, satellite uplink/downlink, inter-satellite or deep space probes to ground, ground-to-air e.g. UAV, HAP / air-to-ground terminal [5]. The increased in the application number of wireless communication have many disadvantages. Such drawbacks are limitation of power, bandwidth regulations, high data rates etc. Nonetheless, FSO has various benefits to be shown. They are no licensing requirements for link utilization, absence of radio frequency radiation hazards, no need of road digging as in the case of optical fiber, wider bandwidths, and low power consumption [6-8]. FSO is affected by attenuation of the atmosphere caused by the unsteady weather conditions such as haze, rain, and fog. These natural phenomena cause attenuation in the transmitted signal. This leads to high bit error rate or signal loss at the receiver end [2]. When a laser beam transmits via the atmosphere, it experiences power loss caused by multiple known factors. One of them is absorption and scattering loss as mentioned in Beer's Law [9].

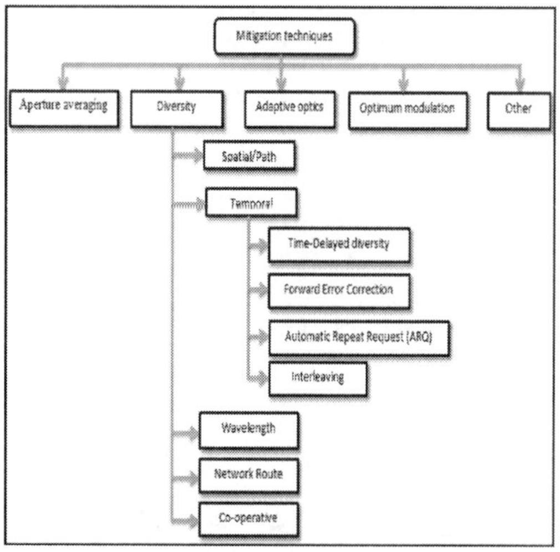

Fig. 3. Tree diagrams showing the various atmospheric mitigation techniques.

The photon of the beam bouncing with haze particles in the mid-air caused a condition called as the absorption. It caused the link margin, distance and the availability of the link to reduce greatly [1]. Scattering is the scattering of a beam of transmission in a range of directions as an impact of physical interactions. When a particle interrupts an electromagnetic wave, some of the energy of the wave is scattered by the particle and re-radiated into a solid angle centered on it. The scattered light is polarized. It is also on the same wavelength as the incident wavelength. Thus there is no loss of energy to the particle [10]. All of these atmospheric conditions act to attenuate the light. They also could block the light path in the atmosphere. The performance of FSO links can be analyzed by prediction of the attenuation factor in terms of visibility. Therefore, employment of FSO requests a thorough and detailed study of local weather patterns. Fig. 3 above shows the tree diagram of various atmospheric mitigation techniques to overcome the limitation of FSO. The study is focused more on the effects of haze on the FSO systems. Weather conditions are to be studied in detail before installing FSO systems. This is to reduce the effects of the atmosphere, and also to ensure that the transmitted power is sufficient and minimal losses during bad weather. The performance of FSO system can be evaluated in many ways, such as by analyzing the BER and Q-factor [11]. These two parameters are closely related and they will be explained more in the budgets link section.

1.1 Optical Wavelength Approach

1.1.1 780-850 nm Wavelength

These wavelengths are well suited for FSO operations. Approximately 850 nm, reliable, inexpensive, high-performance transmitter and detector components are readily available. The wavelength is normally used in network and transmission equipment. Highly sensitive silicon (Si) avalanche photodiode (APD) detector technology and advanced vertical-cavity surface-emitting laser (VCSEL) technology can be employed to operate in this wavelength.

1.1.2 1520-1600 Wavelength

These wavelengths are suitable for free-space transmission. Nevertheless, the components are generally costly. The detectors are less sensitive and have a smaller receive surface area when compared with Si APD detectors that work in the 850 nm wavelength. Lastly, 50–65 times as much power can be transferred in 1520–1600 nm than can be transferred at 780 –850 nm for the same eye safety classification.

1.2 Essential of Free Space Optic Link Budget

The haze on FSO systems can be discovered by determining the scattering coefficient. The scattering coefficient is directly affected by the visibility conditions. The low visibility means the concentration and size of the particles are more compared to average visibility. Thus, scattering and attenuation may be triggered easier in low visibility conditions. The atmospheric attenuation can be calculated based on the scattering coefficient. The attenuation coefficient has contributions from the absorption and scattering of laser photons by different aerosols and gaseous molecule in the atmosphere. Since the laserbit wavelengths used (typically 785 nm, 850 nm and 1550 nm) were chosen, and they fall inside the transmission windows within the atmospheric absorption spectra, the contributions of absorption to the total attenuation coefficient were found to be very small [12]. Thus, the effects of scattering dominate the total attenuation coefficient. The type of scattering is determined by the size of particular atmospheric particle with respect to the transmission laser wavelength. The scattering coefficient can be expressed as a function of the visibility and wavelength.

Theoretically, high scattering coefficient caused less visibility due to higher concentration and bigger particle sizes of haze along the path of laser beam. Equation (1) is used to determine the scattering coefficient in hazy days.

$$\beta = \frac{3.91}{V}\left(\frac{\lambda}{550nm}\right)^{-q} \quad (1)$$

where :
V= visibility in kilometers,
λ = wavelength in nanometers
q = the size distribution of the scattering particles

$$q = \begin{cases} 1.6 & \text{for } V>50 \text{ Km} \\ 1.3 & \text{for } 6 \text{ Km}<V<50 \text{ Km} \\ 0.585V^{1/3} & \text{for } V<6 \text{ Km} \end{cases}$$

The visibility depends on several parameters. They are the degree of coherence of the source, on the difference in length between the paths as well as on the detector location with respect to the source. The effectiveness and availability of FSO systems is reduced by low visibility. It can occur during a specific time period within a year or at specific times of the day. Low visibility is termed when the concentration and size of the particles are higher compared to average visibility. Hence, the more scattering and attenuation may cause the low visibility conditions [13]. The following Beer's Law equation (2) is derived to describe the atmospheric attenuation (5) :

$$\tau(R) = -\frac{P(R)}{P(O)} = e^{-(\beta)R} \quad (2)$$

where
τ(R)= transmittance at range R,
P(R) = laser power at R,
P(0) = laser power at the source, and
β = attenuation (per unit length)
The conversion of Equation (2) can be done as it is in linear scale, in logarithms scale :

$$\tau(R) = -10\log\frac{P(R)}{P(O)} \quad (3)$$

$$= 10\log e^{\beta R}$$

$$= -10\log e^{-\beta R} \quad (4)$$

The equations to calculate the atmospheric attenuation on hazy days can be obtained by substituting Equation (1) into (4).

$$\tau(R) = \log e^{[\left(\frac{3.91}{V}\right)*\left(\frac{\lambda}{550nm}\right)^{-q}*R]} dB \quad (5)$$

The total of received power is calculated using Equation (6):

$$P_{received} = P_{transmit}\frac{A_{receiver}}{(\theta R)^2}e^{-\alpha R} \quad (6)$$

where:

$P_{transmit}$ = transmitted power (Watt)
$A_{receiver}$ = area of the receiver (m^2)
θ = beam divergence in radian (rad)
R = distance between the transceiver (m)
α = average atmospheric attenuation coefficient (dB/km)

The total of received power could be compared to the total of power transmitted and the area of the collection aperture. It is inversely proportional to the square of the beam divergence and the square of the link range.

Table 1 shows the specifications of FSO system used in this paper. Data calculations and simulations are also made using the specifications set.

Table 1. FSO Evaluation parameters.

Parameter	Value
Transmission power	70 mW
Beam Divergence	1 mrad
Receiver Dimension	20 cm
Receiver sensitivity	2000 nW
Wavelength	785 nm, 850 nm, 1550 nm

2 Results and Discussion

Fig. 4 demonstrates the relationship between BER and FSO visibility with various wavelengths. The possible minimum visibility that system can be achieved wills a better performance in an FSO link system. It is obviously evident that 1550 nm wavelength has the lowest BER at 10^{-16} and 2.6 km visibility. Meanwhile, at wavelength 850 nm and 785 nm, the lowest BER are 10^{-6} and 10^{-4} at 4 km and 4.2 km visibility respectively. From the analysis that has been done, the 1550 nm wavelength shows the best performance regarding the lowest BER. Less signal power is loss along the transmission link when the BER is lower which directly improves the performance of the system. This means that higher wavelength system has better performance.

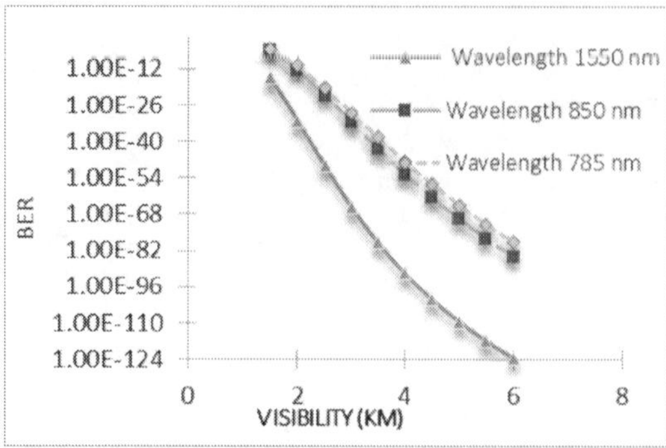

Fig. 4. Graph of visibility against bit error rate (BER) with adverse wavelength of 1550 nm, 850 nm, and 785 nm.

Fig. 5 demonstrates that maximum Q-factor keeps increasing as the visibility increases. The range of visibility studied in this paper is from 1.5 km to 6 km and the link range used is 1 km. At visibility of 6 km, the wavelength 1550 nm achieves the Q-factor of 24 compared to the wavelength of 785 nm and 850 nm as they are able to achieve the Q-factor of just 19 and 20 respectively. The wavelength of 1550 nm is chosen as the best wavelength to be used in the system when maximum Q-factor and visibility becomes higher. Thus the hybrid FSO system with wavelength 1550 nm could be improved by 21% compared to 5% improvement by wavelength 850 nm and 785 nm.

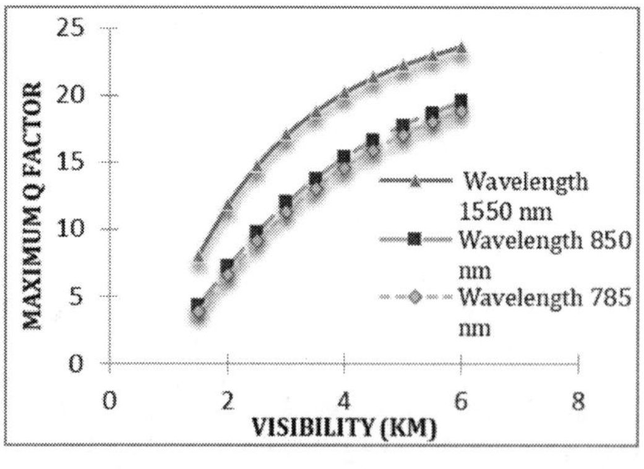

Fig. 5. Graph of visibility against maximum Q-factor with adverse wavelength of 1550 nm, 850 nm, and 785 nm.

The eye diagram is a useful tool for the qualitative analysis of signal used in digital transmission. Using wavelength 1550 nm in Fig. 6, the system has a large eye opening between the top and bottom level compared with wavelength 785 nm, where the eye diagram shown in Fig. 7 is closer and smaller. The height of the eye-opening at the specified time interval shows the noise margin or noise immunity of the system. Based on these eye diagrams, it can be shown that 1550 nm wavelength produced a better performance compared to shorter wavelength because more error is produced due to lower value of Q-factor. As the value of Q factor increases, less signal power is loss. Thus promoting the link performance of FSO system. Another reason of using 1550 nm is because of it is compatibility with current technology and device.

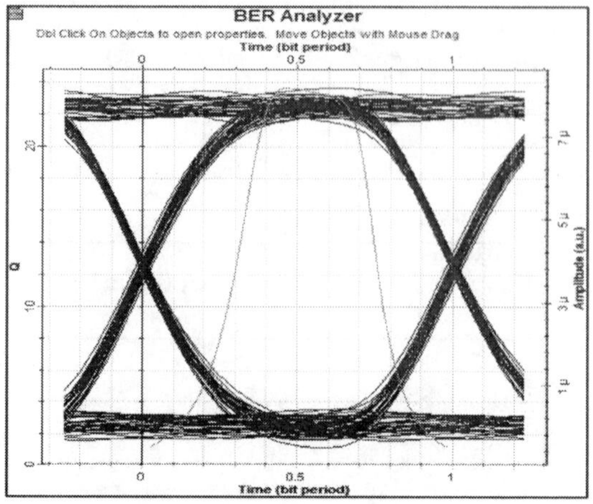

Fig. 6. Eye diagram of Q-factor for 1550 nm signal

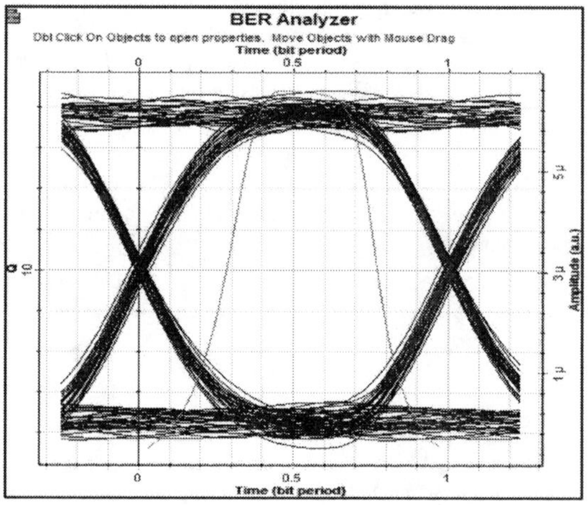

Fig. 7. Eye diagram of bit error rate for 1550 nm signal

3 Conclusions

The FSO systems reveal that communication system with wavelength 1550 nm at BER 10^{-124} is the best suited compared to the 850 nm and 785 nm wavelength. The result had shown that 1550 nm wavelength with low bit rate can produced optimum Q-factor of 25. It can be concluded that, optical signal with higher bit rate is a superior choice to be used for higher output and less bit error rate. Moreover, the attenuation effect could also be reduced greatly.

This work was supported by the High Ministry of Education Malaysia under the Fundamental Research Grant Scheme (FRGS) # 9003-000556

DOI: 10.1051/epjconf/201716201024

References

1. A. Alkholidi, K.S. Altowij, Optical Communications Systems, 41-75 (2012)
2. H. Kaushal, G. Kaddoum, IEEE Communications Surveys & Tutorials, **19**, Issue 1 57-96 (2015)
3. Rashidi, C.B.M., Aljunid, S.A., Anuar, M.S., Fadhil, H.A., Ghani, F., Journal of Theoretical and Applied Information Technology. **61**, Issue 1, Pages 155-159 (2014)
4. A.L. Puryear, *Proc. SPIE*, **7464,** 74640J (2009)
5. P. Vishwakarma, J.P. *IJAIEM*, **2**, 379-382 (2013)
6. Rashidi, C.B.M., Aljunid, S.A., Ghani, F., Fadhil, H.A., Anuar, M.S. Journal of Applied Sciences Research. **8**, Issue 12, pp 5614-5626 (2012)
7. A.K. Rahman, S.A. Aljunid, M.S. Anuar, H.A. Fadhil, Key Engineering Materials, **594**, 1027-1031 (2014)
8. X. Zhu, J. Kahn, IEEE Transactions on Communication, **50,** 1293-1300 (2002)
9. A. Alkholidi, K.S. Altowij, Free Space Optical Communications – Theory and Practices, 159-212 (2014)
10. A.H. Hashim, *Modelling Intersatellite Optical Wireless Communication System, Universiti Teknologi Malaysia*, 1-65 (2009)
11. A.K. Rahman, S.A. Aljunid, M.S. Anuar, H.A. Fadhil, Journal of Theoretical and Applied Information Technology, **60**, 1, 362-367 (2014)
12. K. Dev, R. Nebuloni, C. Capsoni, *10th EuCAP on IEEE*, 1-5 (2016)
13. Rashidi, C.B.M., Aljunid, S.A., Fadhil, H.A., Anuar, M.S. *4th International Conference on Photonics*, Pages 166-168 (2013)

Noise suppression system of OCDMA with spectral/spatial 2D hybrid code

Rima, Matem[1], *S. A.* Aljunid[1], *M. N* Junita[1], *C. B.M* Rashidi[1]., *Israa* Shihab Aqrab[1]

[1]Advanced Communication Engineering Centre of Excellence
School of Computer and Communication Engineering (ACE CoE -SCCE),
Universiti Malaysia Perlis (UniMAP),Perlis, Malaysia,

Abstract. In this paper, we propose a novel 2D spectral/spatial hybrid code based on 1D ZCC and 1D MD where the both present a zero cross correlation property analyzed and the influence of the noise of optical as Phase Induced Intensity Noise (PIIN), shot and thermal noise. This new code is shown effectively to mitigate the PIIN and suppresses MAI. Using 2D ZCC/MD code the performance of the system can be improved in term of as well as to support more simultaneous users compared of the 2D FCC/MDW and 2D DPDC codes.

1 Introduction

OCDMA has attracted huge research interest according to its advantages features like huge and effective the maximum amount of utilization, higher secure , with high of access of network and dynamic bandwidth assignment. The focus on the development of OCDMA systems is to improve the cardinality, mitigate PIIN (Phase Induced Intensity Noise), as well as suppress MAI (Multiplied Access Inference). Many schemes has been proposed in accordance in the OCDMA system detection such as, frequency hopping [1, 2], the time spreading [3–4], spectral amplitude coding (SAC) [5–6] and code of spatial [7]. To have a good OCDMA detection of a code sequence should have the minimum cross-correlation and maximum auto-correlation properties for maintenance.

Several approaches of OCDMA which have been proposed to overcome Phase Induced Intensity Noise [PIIN] reduction and improvement of system of performance [8-9] particularly, in Spectral Amplitude Coding (SAC) OCDMA. For instance, one dimensional (1D) proposed set of code sequences as in [10-11]. The performance is enhanced but the main drawback of the one dimensional code still where the code length become too long as simultaneous users' number increases. To overcome the drawbacks of 1D code, various schemes of 2D (time/spectral, spectral/spatial, etc.) codes have been proposed [12–13]. In comparison with one dimensional sequence for code (1D), the system performance has greatly been improved as well as accommodating more users number within two dimensional codes for (2D) [14].

It became as one of causes to augment enormous research on two dimensional of sequences for (2D) code as it has been proposed in [12][15], to enhance the system of performance and to minimize the Phase Induced Intensity Noise [PIIN]. In the present paper, we do analyze new 2D hybrid code OCDMA based on 1D ZCC code and 1D MD code to examine the effect of Phase Induced Intensity Noise [PIIN] in the system performance.

2 2D Hybrid ZCC/MD Code

The 2D-MD code can be constructed based on the combination of 1D ZCC code [15] and 1D MD code [16].Let$X\ \{x_0, x_1, x_2, \ldots, x_{M-1}\}$,$Y\ \{y_0, y_1, y_2, \ldots, y_{N-1}\}$ with code weight k_1, k_2 represent 1D-ZCC code and 1D-MD code respectively. Noted that $w_1, w_2$are the code size of 1D-ZCC code and 1D-MD code respectively, so the code length of X, Y are$M = k_1 w_1, N = k_2 w_2$. The code size of 2D-ZCC/MD is$W = w_1 w_2$. Table 1 exhibits some example for sequences of 2D hybrid ZCC/MD codes as follows $k_1 = 3, k_2 = 2$ where 2D hybrid ZCC/MD codes can be produced by $g = \{1,2,3, \ldots, M\}$ and$h = \{1,2,3, \ldots, N\}$.Y_g stands for spreading patterns of spectral while X_h is the patterns for space.

$$Zcc = \begin{bmatrix} 101000100000 \\ 010010001000 \\ 000000010101 \end{bmatrix} \quad MD = \begin{bmatrix} 100001 \\ 010010 \\ 001100 \end{bmatrix} \qquad (1)$$

* Corresponding author: rymann90@hotmail.com

 DOI: 10.1051/epjconf/201716201025

Table 1. 2D Hybrid ZCC/MD Code for $k_1 = 3$ and $k_2 = 2$ Sequences

$A_{g,h}$	[101000100000]	[010010001000]	[000000010101]
$\begin{bmatrix}1\\0\\0\\0\\0\\1\end{bmatrix}$	$\begin{bmatrix}101000100000\\000000000000\\000000000000\\000000000000\\000000000000\\101000100000\end{bmatrix}$	$\begin{bmatrix}010010001000\\000000000000\\000000000000\\000000000000\\000000000000\\010010001000\end{bmatrix}$	$\begin{bmatrix}000000010101\\000000000000\\000000000000\\000000000000\\000000000000\\000000010101\end{bmatrix}$

2D ZCC/MD code cross-correlation is derived from 4 matrices of characteristic (d), $d \in (0,1,2,3)$ has been produced in [8] to achieve the same assumption for the cross correlation property as $A\,(d)$ for 2D Hybrid ZCC/MD codes can be described as follows :

$$A^0 = Y^T X \quad (2)$$
$$A^1 = Y^T \bar{X} \quad (3)$$
$$A^2 = \overline{Y^T} X \quad (4)$$
$$A^3 = \overline{Y^T} \bar{X} \quad (5)$$

Whereas parameters $\bar{X}$ as well as $\bar{Y}$ are representing the whole enchantment of X & Y appropriately.

The cross correlation of hybrid 2D ZCC/MD code $A^{(d)}$ and $A_{g,h}$ is shown as in the table below:-

$$R^{(d)}(g,h) =) \textstyle\sum_{M=1}^{M-1} \sum_{N=1}^{N-1} a_{ij}^{(d)} a_{(i+g)(j+h)} \quad (6)$$

Whereas $a_{ij}^{(d)}$ is the $(i,j)_{th}$ of $A^{(d)}$ and $a_{(i+g)(j+h)}$ is the $(i,j)_{th}$ of $A_{g,h}$. Table 2 represents the cross correlation is within two different codes $A^{(d)}$and $A_{g,h}$ for 2D Hybrid ZCC/MD code obtained the equation above.

Table 2. The correlation of 2D code for ZCC/MD

$A_{g,h}$	$R^{(0)}$	$R^{(1)}$	$R^{(2)}$	$R^{(3)}$
$g = 0, h = 0$	k_1k_2	0	0	0
$g = 0, h \neq 0$	0	k_1k_2	0	0
$g \neq 0, h = 0$	0	0	k_1k_2	0
$g \neq 0, h \neq 0$	0	0	0	k_1k_2

The cross correlation of $A_{0,0}^{0}$ and $A_{g,h}$ can be expressed as follows:

$$R^{(0)}(g,h) = \sum_{i=1}^{M-1} \sum_{j=1}^{N-1} a_{i,j}^{(0)} a_{i,j}(g,h) = \begin{cases} k_1k_2 & for\; g = 0, h = 0 \\ 0 & otherwise \end{cases} \quad (7)$$

PIIN (Phase Induced Intensity Noise), Shot Noise and Thermal Noise are considered as three types of noise which one considered in analysis of performance.

BER is basically calculated on Gaussian approximation. The photodiode is employed to detect thermal lights and to obtain the photocurrent as in [18, 19]:

$$\langle i_{noise}^2 \rangle = \langle i_{PIIN}^2 \rangle + \langle i_{shot}^2 \rangle + \langle i_{thermal}^2 \rangle = BI^2\tau + 2eBI \quad (8)$$

Whereas I stands for the average photocurrent, e is an abbreviation for the electron charge, B means the electrical bandwidth, K_b is constant of Boltzmann, T_n is the unlimited temperature, $R_L$is the load resistance and τ is the time for coherence of the light which expressed as [9]:

$$\tau = \frac{\int_0^\infty S(v)^2 dv}{\left[\int_0^\infty S(v)\right]^2} \quad (9)$$

Where $S(v)$ indicates the density for single sideband power spectral. Since the system of 2D ZCC/MD code has the zero cross property of correlation, there is no overlapping in a spectra of divergent users, thus the effect of MAI have been ignored. Accordingly, by applying the identical method as in [10] where each user probability equally sending bit 1 is equal. We will obtain the output currents of PD at receiver (0,0) is depend on the cross correlation between $A_{0,0}^{(0)}$and $A_{g,h}$as follows:

$$I = R\int_0^\infty G(v)dv = \int_0^\infty \frac{P_{sr}}{k_2\Delta v} \textstyle\sum_1^W d(w) R^0(i,j) U(v,i) dv \quad (10)$$

$$= \frac{P_{sr}\,\Delta v}{k_2 \Delta v M}\left[k_1k_2 + \textstyle\sum_1^W d(w)R^0(i,j)\right] = \frac{RP_{sr}k_1}{M} \quad (11)$$

So the equations of PIIN, shot noise and thermal noise could be written as follow as:

$$\langle i_{PIIN}^2 \rangle = B\frac{R^2P_{sr}^2 w_2 k_1}{\Delta v W} \quad (12)$$

$$\langle i_{shot}^2 \rangle = 2eB\frac{RP_{sr}w_2}{W} \quad (13)$$

$$\langle i_{thermal}^2 \rangle = \frac{4K_bT_nB}{R_L} \quad (14)$$

So:

$$\langle i_{noise}^2 \rangle = \frac{R^2P_{sr}^2 w_2 k_1}{\Delta v W} + 2eB\frac{RP_{sr}w_2}{W} + \frac{4K_bT_nB}{R_L} \quad (15)$$

where R is the being responsive of the given photo-diode given by $R = \eta e/hv_0$, η is the efficient. quantum of the photo-diode, and h is the constant of Plank . P_{sr} is the effective source power at the receiver, k_1 and k_2 are the weights of code for spectral and spatial accordingly, W represents the active users number, N and M are the code lengths for spatial and spectral code sequences as well .

Note: that the sending probability bit "1" at different time for every user is being ½ so:

$$\langle i_{noise}^2 \rangle = \frac{R^2 P_{sr}^2 w_2 k_1}{2\Delta v W} + \frac{eBRP_{sr} w_2}{W} + \frac{4K_b T_n B}{R_L} \quad (16)$$

At the receiver, the SNR can be obtained as:

$$SNR = \frac{I^2}{\langle i_{noise}^2 \rangle} = \frac{\left[\frac{RP_{sr} w_2}{W}\right]^2}{\frac{R^2 P_{sr}^2 w_2 k_1}{2\Delta v W} + \frac{eBRP_{sr} w_2}{W} + \frac{4K_b T_n B}{R_L}} \quad (17)$$

The BER can then be calculated from SNR as follows [2]:

$$BER = \frac{erfc\left(\sqrt{SNR/8}\right)}{2} \quad (18)$$

Where

$$erfc = \frac{2}{\sqrt{\pi}} \int_0^{\infty} \exp(-y^2) dy \quad (19)$$

Table 3. Used parameters in analytical calculations.

PD quantum efficiency	$R = 0.75$
Spectral width of broadband light source	$\Delta\lambda = 30nm$ ($\Delta\lambda = 3.75\ THZ$)
Operating wavelength	$\lambda_0 = 1.55\mu m$
Electrical bandwidth	$B = 320MHz$
Data transmission rate	$R_b = 622Mbps$
Receiver noise temperature	$T_n = 300K$
Receiver load resistor	$R_L = 1030\Omega$
Boltzmann's constant	$K_b = 1.38 \times 10^{-23}\ W/K/Hz$
Electron charge	$e = 1.60217646 \times 10^{-19}\ coulombs$
Light velocity	$C = 3 \times 10^8\ m/s$

3 Result and Discussions

Table 3 represents the analysis of the newly proposed parameters system performance for 2-D ZCC/MD hybrid OCDMA. The results are numerically shown in the following figures.

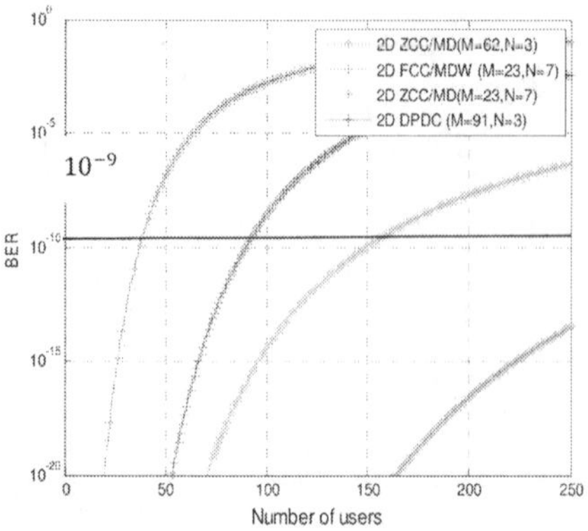

Fig.1. Simultaneous users number which includes only the PIIN noise versus BER

Fig.1 indicates the BER variation versus simultaneous users number for 2D ZCC/MD (M=63, N=3) compared to 2D FCC/MDW (M=23, N=7) and 2D DPDC (M=91, N=3) taking into account effective power Psr=-10dBm and data rate Rb=2.5Gbps. By neglect shot noise and thermal noise, and considering only PIIN, the result clearly indicates that 2D ZCC/MD code represents the best performance, where for the BER equal to 10^{-9} the proposed code can be support more than 154 of simultaneous users while 2D DPDC can support 98 users and 2D FCC/MDW 43 user. The figure indicates also that the variation of code length show that for 2D ZCC (M=63, N=3) can accommodate 154 users where for 2D ZCC (M=23, N=7) can accommodate until 400 users, so the variation of code length has an effect to the system performance where the proposed code represents always the best cardinality comparing to other codes. The ideal BER is approximately considered beneath 10^{-9} error floor with high cardinality. The increase of number of users causes deterioration of the system performance. PIIN noise represents the main effect on the system performance because of increase of code sequences the through higher interference [17] .

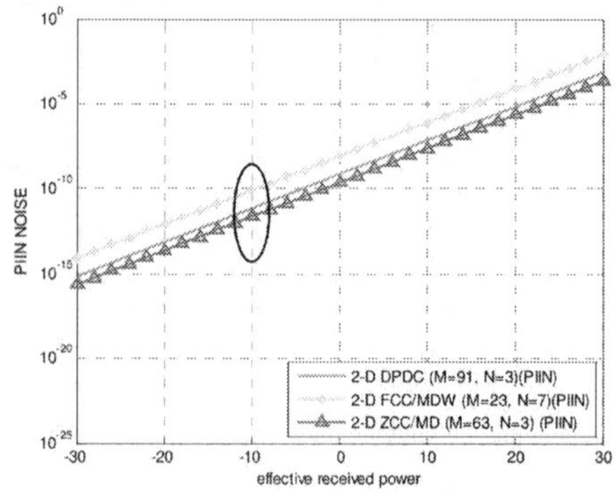

Fig.2. PIIN noise versus effective power when number of user is 80 and data rate s 622Mbps.

DOI: 10.1051/epjconf/201716201025

Fig. 2 represents dominant phase-induced intensity noise (PIIN) versus effective power that spreads from -30dBm to 30dBm at 622Mbps for 2D ZCC/MD (M=63, N=3), 2D FCC/MDW (M=23, N=7) and 2D DPDC (M=91, N=3). It can be obviously shown the effective increase of the power which leads to increase of noise (PIIN) . Comparing 2D ZCC/MD (M=63, N=3), 2D FCC/MDW (M=23, N=7) and 2D DPDC (M=91, N=3) codes, PIIN is optimally suppressed by 2D ZCC/MD code. From the figure, below -10dBm PIIN can be maximally mitigated. The betterment of OCDMA system performance could be obtained when Psr has low value as possible [17]. The 2D ZCC/MD (M=63, N=3), 2D FCC/MDW (M=23, N=7) reaches optimum PIIN suppression is at -10dBm.

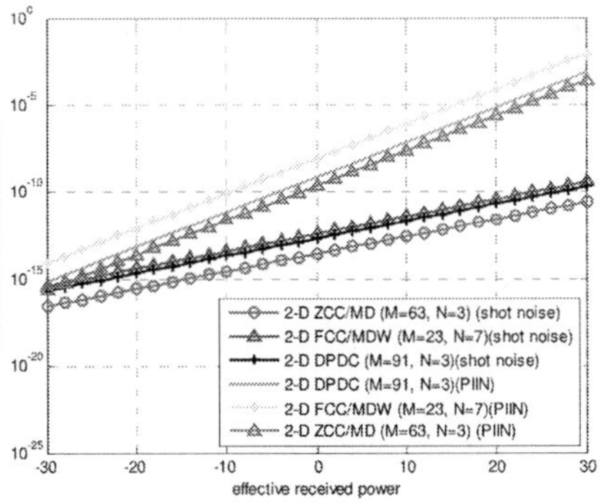

Fig. 3. Phase-Induced Intensity Noise and shot noise versus Psr of effective power within identical codes.

Fig. 3 represents the link between the Phase-Induced Intensity Noise [PIIN] and shot noise versus the source of effective power (Psr). The simultaneous users number is set at 250 and Psr value is various between -30 dBm and 30 dBm range for 2D ZCC/MD (M=63, N=3), 2D FCC/MDW (M=23, N=7) and 2D DPDC (M=91, N=3). From the figure it can be observed that the increase of the effective power Psr produces an increase of PIIN and Shot noise where it attaints the higher values for all codes. For Psr above -10dBm Ipiin is dominant and Ishot become negligible. For Psr=-10dBm, the BER float below 10^{-9} , Ipiin and Ishot equal to $7.31x10^{-12}, 2.77x10^{-15}$ respectively for 2D ZCC/MD code. 2D ZCC/MD is better than 2D FCC/MDW and 2D DPDC.

4 Conclusions

In the present paper, we study a current/advanced code of 2D hybrid ZCC/MD in term of noise suppression. We analyze the effect of phase-induced intensity [PIIN] and shot noise on the newly developed of 2D hybrid ZCC/MD OCDMA systems compared to 2D hybrid FCC/MDW and 2D DPDC where 2D hybrid ZCC/MD represents a better performance with effective power fixed at -10dBm and it can provide a higher users number when BER equal to 10^{-9} . PIIN is the main factor in 2D ZCC/MD and others performance degradation. 2D ZCC/MD represents a better performance and effectively mitigates the PIIN.

References

1. H. Fathallah, Rusch, L. A., & S. LaRochelle, J. Lightwave Tech., **17**(3), p 397, (1999).
2. E. Inaty, Shalaby, H. M., & Fortier, P., IEEE Trans. Comm., **53** (2), p 323-334, (2005).
3. F. R. Chugh, J.A. Salehi, & V. K. Wei., IEEE Trans. Inf. Thery, **35**(3), (1989).
4. C.S. Weng, & J. Wu., J. Lightwave Tech., **19**(2), p.186. (2001).
5. E.D. Smith, R. J. Blaikie, & D. P. Taylor, IEEE Trans. Comm., **46**(9),p 1176-1185. (1998).
6. A.T. Hussein, S. A. Aljunid, S. Junid, H. Adnan, & R. Badlishah, Aust. J. Basic & Appl. Sci., **6** (3), p112 (2012).
7. K.I. Kitayama, IEEE J. on Sel. Areas in Comm., **12**(4), p762-772. (1994).
8. M. S. Anuar, S. A. Aljunid, N. M. Saad & S. M. Hamzah, Optics Comm., **282**(14), p.2659-2664. (2009).
9. Z. Wei, H. M. H. Shalaby, & H., Ghafouri-Shiraz, J. Lightwave Tech., **19**(9), p 1274-1281. (2001).
10. S. A. Aljunid, M. D. A. Samad, M. Othman, M. H. Hisham, A. H. Kasiman, & M. K Abdullah, *IEEE 7th Malaysia International Conference on Communication, 2005,* **1**, pp. 5 (2005).
11. J. F. Huang, & Y.T Chang., *Communication Networks and Services Research Conference, 2006. CNSR 2006. Proceedings of the 4th Annual*, pp. 8 (2006).
12. S. Wan, X. He & J .Sun, Optics Comm., **282**(24), p4776-4778. (2009).
13. B. C. Yeh, C. H.Lin., C.L. Yang, & J. Wu., *J. Lightwave Tech.*, **27**(13), 2420-2432. (2009).
14. N. D. Keraf, S. A. Aljunid, A. R. Arief, M. N. Nurol, M. S. Anuar, C. B. M. Rashidi, & , P.Ehkan, *2014 IEEE 5th International Conference on Photonics*, pp. 166-168. (2014).
15. M.S. Anuar, S.A. Aljunid, N.M. Saad. IJCSNS Int. J. Comp. Sci. and Netw. Sec., **6** (12), pp. 180–184. (2006).
16. R.A. Kadhim, H. A. Fadhil, S. A. Aljunid & M. S Razalli. Optics Comm., **329**, pp. 28-33. (2014).
17. A. Arief, S. Aljunid, M. Anuar, M.N. Junita, R. Ahmad, and F. Ghani, *IEEE 3rd International Conference on Photonics*. pp. 275–279. (2012).
18. B.C. Yeh, C.H. Lin & J. Wu. J. Opt. Comm. Netw. **2**(9), pp. 653-661. (2010).
19. Z. Wei and H. Ghafouri-Shiraz, J. Lightwave Tech., **20**, pp. 1284-1289. (2002).

EPJ Web of Conferences **162**, 01026 (2017)
InCAPE2017 DOI: 10.1051/epjconf/201716201026

High range free space optic transmission using new dual diffuser modulation technique

A.K Rahman[1], *N* Julai[1], *M* Jusoh[2], *C.B.M.*Rashidi[2], , *S.A* Aljunid[2], *Anuar* M.S[2], *M.F* Talib[2] , *Nurdiani* Zamhari[1], *S.K* Sahari[1], *K.F.* Tamrin[1], *Rudiyanto* P. Jong[1], *D.N.A* Zaidel[1], *N.A.A* Mohtadzar[1], *M.R.M* Sharip[1], *Y.S* Samat[1]

1 Faculty of Engineering, Department of electrical electronic, University Malaysia Sarawak, Malaysia
2 Advance Communication Engineering, Centre of Excellence, SCCE, University Malaysia Perlis, Malaysia

Abstract. Free space optical communication fsoc is vulnerable with fluctuating atmospheric. This paper focus analyzes the finding of new technique dual diffuser modulation (ddm) to mitigate the atmospheric turbulence effect. The performance of fsoc under the presence of atmospheric turbulence will cause the laser beam keens to (a) beam wander, (b) beam spreading and (c) scintillation. The most deteriorate the fsoc is scintillation where it affected the wavefront cause to fluctuating signal and ultimately receiver can turn into saturate or loss signal. Ddm approach enhances the detecting bit '1' and bit '0' and improves the power received to combat with turbulence effect. The performance focus on signal-to-noise (snr) and bit error rate (ber) where the numerical result shows that the ddm technique able to improves the range where estimated approximately 40% improvement under weak turbulence and 80% under strong turbulence.

1 Introduction

Free space optical communication is an attractive alternative over fiber optical communication where provides high bandwidth, fast-installation and high security [1].

However FSOC is suffering with the atmospheric turbulence which can lead the laser beam (a) beam wanders, (b) beam spreading and (c) scintillation. The beam wanders is refer to the condition beam deflected randomly through the changing refractive index oddities. Meanwhile beam spreading refers to conditions where the beam spread more than diffraction estimated predict. Lastly, for scintillation it affected the phase front of the beam can vary and resulting fluctuation irradiance or well known as intensity signal. The combination of all these effect will cause both the spatial and temporal experience random fluctuations in refractive index of the variability of element factor such as temperature, pressure and wind variations along the optical propagation path through the channel [2-9].

In this paper we make the comparison performance dual-detection approach with the conventional approach in term of turbulence strength. In section 2 we outline the model dual-detection approach with assume using OOK modulation. In section 3 we derive the theoretical analysis for SNR and BER. In section 4 results and discussion and section 5 presents the conclusion

2 Dual Diffuser Modulation Technique

The system employs two transmitters and On-Off Shift Keying (OOK) modulation as reference for conventional system. When the first transmitter sends binary '1', the second transmitter which is set in compliment condition will send the binary '0' in simultaneously and vice versa.

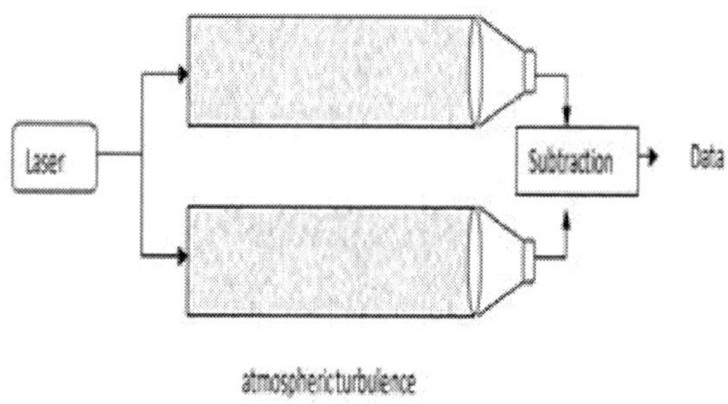

Fig. 1. Dual Diffuser Modulation technique setup in free space optical communication

Meanwhile at the receiver part, the signal will go through the subtractor for the differential detection process. Here we assume the ideal subtractor condition where no losses signal occurs during subtraction. Therefore, the signal output will become bit '1' for sending binary '1' and bit '-1' for sending binary '0'.

This condition leads to modification on conventional OOK modulation particularly in improving signal threshold detection and reduce power loss.

3 Theory Of Analysis

Based on the Rytov theory, the field of the plane wave in random media is given [10]:

$$E_o(\vec{r}) = A_o(\vec{r}) \exp\left(i\varphi_o(\vec{r})\right) \quad (1)$$

Where $A_o(\vec{r})$ is amplitude of the laser beam in the atmosphere without the presence of turbulence. φ_o and $E_o(\vec{r})$are the phase and laser beam profile, respectively.

 DOI: 10.1051/epjconf/201716201026

However in the presence of turbulence, the scintillation effect is more dominant which change the refractive index and as a consequence lead to intensity fluctuations. Therefore this condition directly change the laser beam profile [10,11]:

$$E_o(\vec{r}) = A_o(\vec{r})\exp(i\varphi_o(\vec{r})) = E_o(\vec{r})\exp(\Phi) \quad (2)$$

Where $A(r)$ is amplitude of the laser in the turbulence atmospheric, ϕ is the exponential factor due to turbulence given as:

$$\Phi = Ln\left(\frac{A(\vec{r})}{A_o(\vec{r})}\right) + i[\phi(\vec{r}) - \phi_o(\vec{r})]$$

$$= \chi + iS \quad (3)$$

Where X represents the fluctuation of the log-amplitude of the field and S is the phase fluctuations. This equation predicts a fluctuation in intensity and phase of receiving signal due to turbulence is weak and assuming that the refractive index structure coefficient C_n^2 is symmetrical along the beam path, Rytov suggested a variance for the log-irradiance given as [6]:

$$\sigma_{ln\,IR}^2 = \langle(ln\,I - \langle ln\,I\rangle)^2\rangle = 1.23C_n^2k^{7/6}L^{11/6} \quad (4)$$

Where k is wave number, $k=2\pi/\lambda$, L is propagation distance and I is intensity signal. However in practical FSO for plane wave in weak condition and for the symmetrical turbulence of the log-amplitude fluctuation is given by [9, 10]:

$$\langle\chi^2\rangle = 0.31C_n^2k^{7/6}L^{11/6} \quad (5)$$

By neglecting other source noises, we assume the only dominant noise source is atmospheric turbulence which gives the log-amplitude as:

$$\chi = Ln\left(\frac{A(\vec{r})}{A_o(\vec{r})}\right) = ln\left[\frac{A_o(\vec{r})+A_n(\vec{r})}{A_o(\vec{r})}\right] = ln(1+\varepsilon) \quad (6)$$

Where $A_n(\vec{r})$ is the amplitude of noise and $\varepsilon = A_n(\vec{r})/A_o(\vec{r})$ is the ratio of the amplitude of noise to signal. Both SNR and BER are used to evaluate the quality of the communication systems. The BER performance depends on the average received power, the scintillation strength, receiver noise and threshold bits (1 or 0) detection. Here the dual-detection approach is a modification from the conventional OOK detection that using direct detection where improves power received and eliminated the need of complex adaptive threshold. The SNR for case only dominant turbulence noise in term of mean signal and noise intensity I_o and $<I_n>$ given as:

$$SNR = \frac{I_o}{\langle I_n\rangle} = \frac{\langle A_o^2(\vec{r})\rangle}{\langle A_n^2(\vec{r})\rangle} = \frac{1}{\langle\varepsilon^2\rangle} \quad (7)$$

So, the BER for OOK modulation can be written as [10]:

$$BER = \frac{1}{2}erfc\left(\frac{Q}{2}\right) = \frac{e^{(-\frac{Q^2}{2})}}{Q\sqrt{2\pi}} \quad (8)$$

where $erfc()$ the complementary error function and Q is defined as:

$$Q = \frac{A_1 - A_0}{\sigma_1 + \sigma_0} \quad (9)$$

Where A_1, σ_1 are the mean signal and noise at the receiver for a bit '1' respectively, and for a bit '0' the mean signal and noise are denoted as A_0, σ_0.

For conventional OOK bit detection,

$$A_1 = < A_o(\vec{r}) >, A_o = 0 \; and \; \sigma_1 + \sigma_0 = < A_o(\vec{r}) >$$

$$SNR = \frac{<A_0^2(\vec{r})>}{<A_n^2(\vec{r})>} = Q^2 \quad (10)$$

However for case dual-detection approach,

$$A_1 = < A_o(\vec{r}) >, A_o = < -A_o(\vec{r}) > \; and \; \sigma_1 + \sigma_0 = < A_o(\vec{r}) >$$

$$SNR = \frac{<2A_0^2(\vec{r})>}{<A_n^2(\vec{r})>} \quad (11)$$

For weak turbulence model, ε is very small thus (6) is given as [10]:

$$\chi = \ln(1+\varepsilon) \approx \varepsilon \quad (12)$$

When turbulence is strong, the SNR and χ can be relating using Tailor series for function $f(x) = (e^x - 1)^2$ and can be approximately to:

$$SNR = \frac{1}{\langle\chi^2 + \chi^3 + \cdots\rangle} \cong \frac{1}{\alpha\langle\chi^2\rangle} \quad (13)$$

4 Results and Discussion

In this numerical result section, we use parameters wavelength 1550nm, Cn^2 = 1 x10-15 for weak turbulence and Cn2 = 1 x10-14 for strong turbulence and range propagation 5km for all calculations.

In Figure 2 shows the SNR in dB response to turbulence strength. From the graph pattern, we can know that as the distance increase, the SNR also decrease. The SNR is influenced by turbulence strength. Therefore by using the dual-detection approach the SNR can improve approximately 3 dB or 17% compare to the conventional approach.

In Figure 3 shows the BER versus range propagation in response to weak turbulence. By using the standard communication for acceptable BER at 10-9, the conventional approach only reaches the range propagation 2.5km but the dual-detection approach can reach further to 3.5km which approximately 40% improvement.

 DOI: 10.1051/epjconf/201716201026

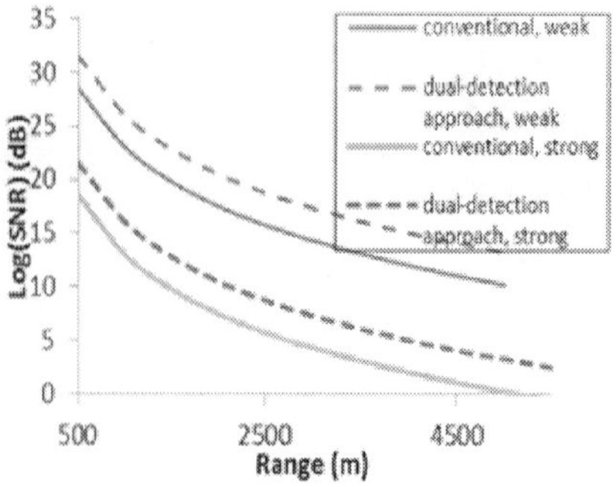

Fig. 2: SNR (dB) versus range propagation

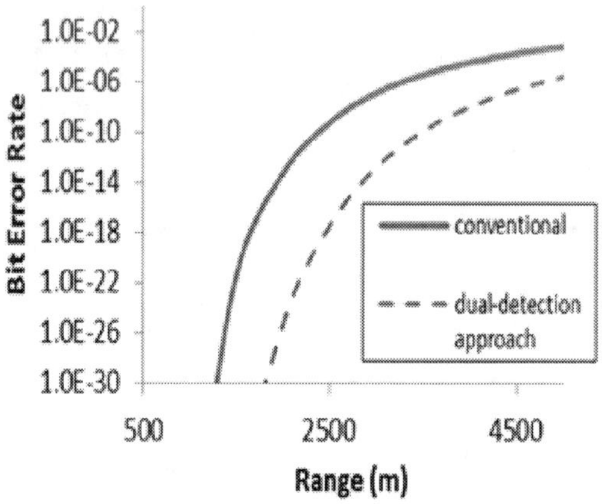

Fig. 3: BER versus range propagation for weak turbulence condition

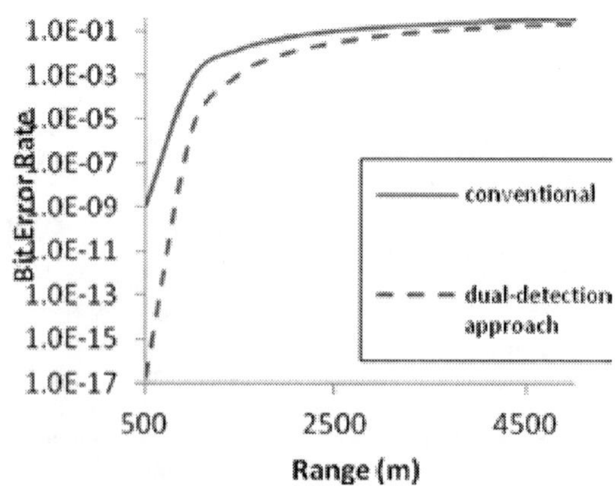

Fig. 4: BER versus range propagation for strong turbulence condition

In Figure 4 shows the BER versus range propagation for strong turbulence condition. Here we assume the value of α is 2. In range of good quality communication (BER=10-9), the conventional approach only reaches the range propagation about 0.5km but the dual-detection approach still manage to reach 0.9km where about 80% improvement.

5 Conclusions

This paper focuses on the turbulence effect that deteriorates the performance of FSOC using new technique that is DDM. The comparison performance measured in SNR and BER between two approaches that are conventional (OOK-direct detection) and DDM. Analyses that carried out shows that the proposed DDM technique improve the margin SNR and BER. The range of propagation can be predicted improved 40% in weak turbulence and 80% in strong turbulence.

This research is supported under UNIMAS MyRA - Special Short Term Grant (SpSTG) scheme.

References

1. Kim, Isaac, Opt. Eng. SPIE, **37**, 12, pp. 3143–3155 (1998).
2. N. Namazi, R. Burris, and G. C.Gilbreath, Proc. of SPIE, **5892**, pp. (58920T) 1-15, (2005).
3. S. Trisno, I. Smolyaninov, S. D. Milnerb, and C. C. Davis, Proc. of SPIE, **5892**, pp. (589215) 1-10, (2005).
4. M. C. R. Cordeiro, C. P. Colvero, and J. P. von der Weid, Int. Conf. on Microwave and Optoelec., pp. 393-395, (2005).
5. M. Uysal, J. Li, and M. Yu, IEEE Transaction on Wireless Communications, **5**, 6, pp. 1229-1233, (2006).
6. I. B. Djordjevic, B. Vasic, and M. A. Neifeld, IEEE Photon. Techn. Let. **18**, 14, pp. 1491-1493, (2006).
7. M. Tariq, Fiber-Optics and Photonics Technology Conference, IEEE, pp. 21-21, (2007).
8. X.Zhu, and J.M Kahn, IEEE Transactions on Communication, **50**, 8, pp.1293-1300, (2002).
9. J. Li, and M. Uysal, Global Telecommunications Conference, **5**, pp. 2654- 2658,(2003).
10. X. Guoliang, Z. Xuping, W. Junwei, and F. Xiaoyong, Proc. SPIE, Optical Transmission, Switching, and Subsystems, **5281**, pp. 816-823, (2004).
11. L. C. Andrews, R. L. Phillips, and C. Y. Hopen, Laser Beam Scintillation with Applications, SPIE Optical Engineering Press,Bellingham, Washington, (2001).

The effect of total noise on two-dimension OCDMA codes

Layth A. Khalil Al Dulaimi [1], *R.Badlishah* Ahmed [2], *Naimah* Yaakob[3], *Syed* A. Aljunid [4], *Rima* Matem[5]

[1, 3] Embedded Networking and Advance Computing Cluster, School of Computer & Communication Engineering, Universiti Malaysia Perlis (UniMAP) Tingkat 1, Kampus Tetap Pauh, Putra 02600 Arau, Perlis.
[2] Faculty of Informatics and Computing, Universiti Sultan Zainal Abidin, 21300 Kuala Terengganu, Terengganu, Malaysia.
[4,5] Center of Excellent Advanced Communication Engineering, School of Computer & Communication Engineering, Universiti Malaysia Perlis (UniMAP), 01000 kangar, Perlis.

Abstract. In this research, we evaluate the performance of total noise effect on two dimension (2-D) optical code-division multiple access (OCDMA) performance systems using 2-D Modified Double Weight MDW under various link parameters. The impact of the multi-access interference (MAI) and other noise effect on the system performance. The 2-D MDW is compared mathematically with other codes which use similar techniques. We analyzed and optimized the data rate and effective receive power. The performance and optimization of MDW code in OCDMA system are reported, the bit error rate (BER) can be significantly improved when the 2-D MDW code desired parameters are selected especially the cross correlation properties. It reduces the MAI in the system compensate BER and phase-induced intensity noise (PIIN) in incoherent OCDMA The analysis permits a thorough understanding of PIIN, shot and thermal noises impact on 2-D MDW OCDMA system performance. PIIN is the main noise factor in the OCDMA network.

1 Introduction

The major technical challenge for today's communication network systems is to take high bandwidth since the volume of information produced increases rapidly. Optical code division multiplexing access (OCDMA) is one technique of such promising technology that takes advantage of excess bandwidth in single mode optical fiber with low information rate into high rate optical sequences followed by optical transmitter to obtain random, asynchronous access without centralized control among users [1].

The important to an effective CDMA system is the choice of efficient statement code sequences with high autocorrelation and zero or fix cross-correlation properties for encoding and decoding of the source bits [2].

The autocorrelation amplitude must be very high and the cross correlation and cross correlation amplitude between two users must remain as low as possible. High autocorrelation amplitude enables the detection of the orginal signal and low cross-correlation amplitude supports in reducing interference from users [3,4]. Wide research has been made to discover the impact optical or codes for non-coherent optical CDMA systems.

The 2-D coding allows large cardinality by reducing MAI. Fixed in-phase spectral cross-correlation equal to one spectral coding is reduce the effect of PIIN [4]. 2-D diluted perfect difference code has good effect on the system performance significantly by minimizing cross-correlation and transceiver structure with MAI cancellation property [5]. The 2-D hybrid codes has properties of cancelling MAI and suppressing PIIN. 2-D perfect difference code (PDC) one of the code that can suppress PIIN and eliminate MAI, 2-D MQC with fixed cross-correlation results in complete elimination of MAI by differential detection and PIIN suppression and 2-D m-sequence code spreads in wavelength and time domain using optical delay lines [5,6]. Even though these codes achieve fixed in-phase and optimized spectral cross-correlation, PIIN still contributes to the degradation of system performance.

* Corresponding author: layth.a.khalil@gmail.com

2 Performance Analysis of 2-D MDW Code

2.1 Signal to Noise Ratio Equation

The 2-D W/T MDW code is W/T OCDMA code which was developed in 2013 by Arief et al [7]. This code has a good cross-correlation property, optimizing data transmission rate, capable of suppressing PIIN, eliminating MAI and enhancing system performance of the OCDMA system architecture. The code also has the advantage of high cardinality and low effective transmitted power. The 2-D W/T MDW OCDMA code is derived from the 1-D MDW OCDMA [8,9]. The serious requirement in minimizing interference which is non-exclusive to all code groupings is the low cross-correlation and high auto-correlation with low side flaps. These are important parameters which will be changed so the effect on the framework execution can be understood; the synchronous number of clients, the number of Hamming weight, the bit rate, the effective transmitted power and the noise power [10].

In this section, signal to noise ratio (SNR) and bit error rate (BER) performance for incoherent SAC optical CDMA has been studied for 2-D MDW code design. MAI is mainly due to thermal noise, shot noise and phase-induced intensity noise. The effect of dark current for photodiodes has been neglected here. As mention before there are three types of noise that are taken into consideration in the performance analysis. The effect of the dark current is neglected [4] because it is so small. The BER calculation is based on the Gaussian approximation. The photo-diode, which is used to detect thermal lights, produces the photodiode noise that can be expressed as follows [9]:

$$i = 2eIB + I^2 B\tau_c + \frac{4K_b T_n B}{R_L} \tag{1}$$

The Power Spectral Density (PSD) of the received optical signal is written as [9]:

$$r(f) = \frac{P_{sr}}{\Delta f k_2} \sum_{w=1}^{W} d(w) \sum_{i=0}^{M-1} \sum_{j=0}^{N-1} a_{ij}(w) \times$$

$$\left\{ u\left[f - f_0 - \frac{\Delta f}{2M}(-M+2i) \right] - u\left[f - f_0 - \frac{\Delta f}{2M}(-M+2i-2) \right] \right\} \tag{2}$$

where P_{sr} is the effective power of a broadband source at the receiver, Δf is the bandwidth of the source, k_2 is the code weight of the time-chip code sequence. w is the number of simultaneously active users, d(w) is the data bit of the w_{th} user which can either be ''1'' or ''0''. M is the code length of a spectra code sequence, N is the code length of the spatial code or time sequence, a_{ij} is an element of the w_{th} user's code word [7].

The modulated 2-D W/T MDW code carries data to the decoder. The *u (f)* represents the unit step function that can be expressed as [9]:

$$u(f) = \begin{cases} 1, f \geq o \\ 0, f\langle 0 \end{cases} \tag{3}$$

PSD that exists at the PD0, PD1, PD2 and PD3 of the receiver by using the cross-correlation between the code-word $A_{00}^{d}, A_{g,h}$ is calculated. The PSDs for a single bit period of the optical signals at PD0, PD1, PD2 and PD3 of the receiver are represented by Arief et al.[9]:

$$G_0(f) = \frac{P_{sr}}{\Delta f k_2} \sum_{k=1}^{W} d(w) \sum_{i=0}^{M-1} \sum_{j=0}^{N-1} a_{ij}^{(0)} a_{ij}(w) \left\{ u\left[f - f_0 - \frac{\Delta f}{2M}(-M+2i) \right] - u\left[f - f_0 - \frac{\Delta f}{2M}(-M+2i-2) \right] \right\} \tag{4}$$

$$G_1(f) = \frac{P_{sr}}{\Delta f k_2} \sum_{k=1}^{W} d(w) \sum_{i=0}^{M-1} \sum_{j=0}^{N-1} a_{ij}^{(1)} a_{ij}(w) \left\{ u\left[f - f_0 - \frac{\Delta f}{2M}(-M+2i) \right] - u\left[f - f_0 - \frac{\Delta f}{2M}(-M+2i-2) \right] \right\} \tag{5}$$

$$G_2(f) = \frac{P_{sr}}{(k_1 - 1)\Delta f k_2} \sum_{k=1}^{W} d(w) \sum_{i=0}^{M-1} \sum_{j=0}^{N-1} a_{ij}^{(2)} a_{ij}(w) \left\{ u\left[f - f_0 - \frac{\Delta f}{2M}(-M+2i) \right] - u\left[f - f_0 - \frac{\Delta f}{2M}(-M+2i-2) \right] \right\} \tag{6}$$

$$G_3(f) = \frac{P_{sr}}{(k_1 - 1)\Delta f k_2} \sum_{k=1}^{W} d(w) \sum_{i=0}^{M-1} \sum_{j=0}^{N-1} a_{ij}^{(3)} a_{ij}(w) \left\{ u\left[f - f_0 - \frac{\Delta f}{2M}(-M+2i) \right] - u\left[f - f_0 - \frac{\Delta f}{2M}(-M+2i-2) \right] \right\} \tag{7}$$

k_1 is the code weight of a spectra code sequence. For a worst scenario analysis, we set $d(w) = 1$. The $G_0(f)$, $G_1(f)$, $G_2(f)$ and $G_3(f)$ are used to obtain output current from each arm of the detector. Average output currents PD0, PD1, PD2 and PD3 from the receiver are written as [8,9]:

$$I_0(f) = \Re \int_0^{\infty} G_0(f)df =$$

$$\frac{\Re P_{sr}}{M k_2} \left\{ k_1 k_2 + k_1 \frac{(W-1)(N-1)}{(MN-1)} + k_2 \frac{(W-1)(M-1)}{(MN-1)} + \frac{(W-1)(M-1)(N-1)}{(MN-1)} \right\} \tag{8}$$

$$I_1(f) = \Re \int_0^{\infty} G_1(f)df = \frac{\Re P_{sr}}{M k_2} \left\{ k_1 \frac{(W-1)(N-1)}{(MN-1)} + \frac{(W-1)(M-1)(N-1)}{(MN-1)} \right\}$$

(9)

$$I_2(f) = \Re \int_0^{\infty} G_2(f)df$$

$$= \frac{\Re P_{sr}}{Mk_2}\left\{k_2 \frac{(W-1)(M-1)}{(MN-1)} + \frac{(W-1)(M-1)(N-1)}{(MN-1)}\right\} \quad (10)$$

$$I_3(f) = \Re \int_0^{\infty} G_3(f)df = \frac{\Re P_{sr}}{Mk_2}\left\{\frac{(W-1)(M-1)(N-1)}{(MN-1)}\right\} \quad (11)$$

Where $\Re$ is the response of the photodiode, $\Re = \eta e / hf_0$, η is the quantum efficiency of the PD, h is the Plank's constant and f_0 is the central frequency of the incident light. The average photocurrent output from the receiver is defined as the average output current of (PD0-PD1) minus (PD2-PD3) of the receiver and was expressed as [8,9]:

$$I_r = (I_0 - I_1) - (I_2 - I_3) = \frac{\Re P_{sr} k_1}{M} \quad (12)$$

Mixing an incoherent light field's incident which causes intensity noise at PD output, PIIN is the result of phase different between users. The power of PIIN that exists in a photocurrent of the receiver can be written as:

$$I^2_{PIIN} = B_r I_r^2 \tau_r = \frac{B_r \Re^2 P_{sr}^2 \left[k_1 k_2 (MN-1)^2 + k_2 (W-1)(M-1)\right]}{2M\Delta f k_2^2 (MN-1)^2} \quad (13)$$

The shot noises are coming from the receiver, the average output currents of the receiver are independent of each other. The power of shot noise current can be represented by:

$$I^2_{shot} = 2eB_r(I_0 + I_1 + I_2 + I_3)$$

$$= 2eB_r \frac{\Re P_{sr}}{Mk_2(MN-1)}\left\{\begin{matrix} k_1 k_2 (MN-1) + 2k_1 (W-1)(N-1) + \\ 2k_2 (W-1)(M-1) + 4(W-1)(M-1)(N-1)\end{matrix}\right\}$$

(14)

Noise power thermal noise; the variance of thermal noise power is defined as:

$$I^2_{thermal} = \frac{4K_b T_n B_r}{R_L} \quad (15)$$

From equation 12 the total photocurrent output from the receiver I_r is given by:

$$I_r^2 = \left[\frac{\Re P_{sr} k_1}{M}\right]^2 \quad (16)$$

2-D W/T MDW code SNR function was developed from (13),(14),(15) and (16) as:

$$SNR = \frac{I_r^2}{I^2_{PIIN} + I^2_{shot} + I^2_{thermal}} \quad (17)$$

BER can be expressed in terms of the signal-to-noise ratio (SNR) as:

$$BER(M) = \frac{1}{2} erfc(\sqrt{\frac{SNR}{8}} \quad (18)$$

The photodiode is used to detect thermal lights produces the photocurrent noise. .PIIN, shot noise and thermal noise are considered in performance analysis. BER calculation is based on Gaussian approximation.

2.2 Results and Discussion

In the theoretical calculation, the link parameter of the Equation 18 that has been used is shown in Table 4.1. These parameters are the same as used in [8].

The 2-D W/T MDW OCDMA code sequence uses the notation of $(M, N, \lambda_a, \lambda_c)$ to denote M as the number of wavelengths, N is temporal code length, λ_a and λ_c are auto-correlation and cross-correlation values.

Table 1. Parameters Used in Numerical Calculation [8].

Parameters Used in Numerical Calculation	
PD quantum efficiency	η=0.75
broadband light source Spectral width	Δλ=30nm(Δλ=3.75THz)
Operational wavelength	λ_c=1.55μm
Electrical bandwidth	B=311 MHz
Data transmission rate	R_b =622 Mbps, 1 Gbps, 1.5 Gbps, 2 Gbps
Receiver noise temperature	T_n =300K
Receiver load resistor	R_L=1030Ω
Boltzmann's constant	K_b=1.38×10^-23 W/K/Hz
Electron charge	e=1.60217646×10^-19 coulomb
Light velocity	$C = 3\times10^8$ m/s

The total noise consists of PIIN, Shot noise which is always low and negligible above -30dBm and thermal noise Fig. 1 details the total noise of 2-D MDW (M=63, N=3), 2-D PDC (M=57, N=3) versus P_{sr} below -30 dBm the constant curves are observed. 2-D MDW shows lowest total noises power curve possible compare to others. Low total noise results in high SNR and low BER.

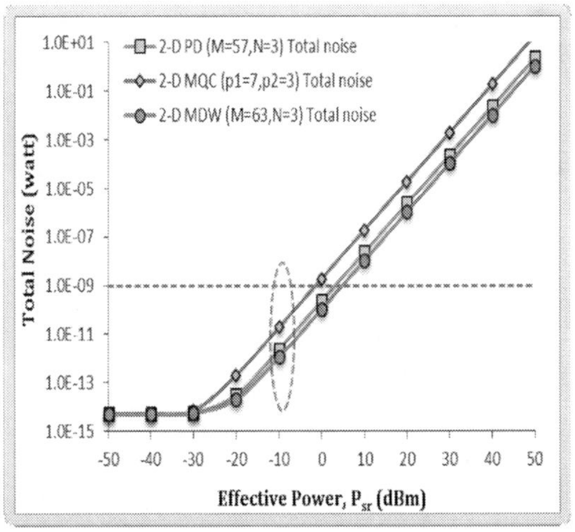

Fig. 1. Total noise BER versus effective power (P_{sr}) for 2-D MDW *(M=63,N=3)*, 2-D PDC*(M=57,N=3)*, 2-D MQC$(p_1=7,p_2=3)$.

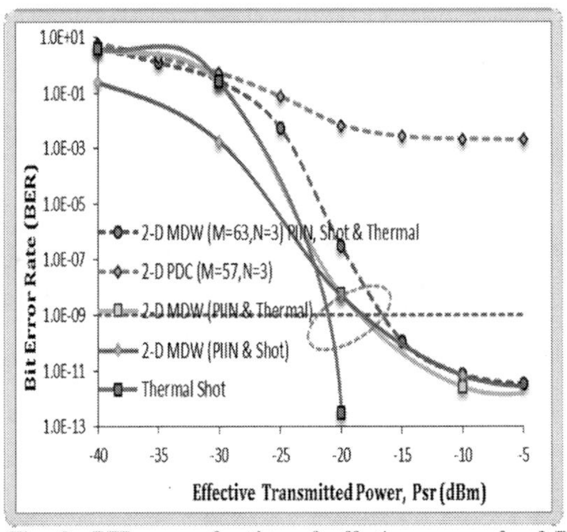

Fig. 2. BER as a function of effective power for 2-D PDC(M=57, N=3) and 2-D MDW(M=63, N=3).

Fig. 2 shows the effect of 2-D MDW (M=63, N=3) BER curves of PIIN combine with shot noise, PIIN + thermal noise and finally combine of the three noises for the P_{sr} starts from -40 to -5 dBm on the effective power slop.

Table 2. 2-D MDW OCDMA PIIN, Shot and Thermal noise versus effective power.

Code	Psr(dBm)	Lost(dBm)
2-DMDW (M=63,N=3)PIIN, Shot & Thermal noise	-17.5	-
2-DMDW (M=63,N=3)PIIN & Shot noise	-18.5	1.0
2-DMDW (M=63,N=3)PIIN & Thermal noise	-18.3	0.8
2-DMDW (M=63,N=3) Shot & Thermal noise	-20.3	2.8
2-DPDC (M=57,N=3)PIIN, Shot &Thermal noise	-	-

From Table 2, PIIN noise alone causes -2.8 dBm power penalty lost; 16% lost. Whereas shot and thermal noise contribute -1 dBm and -0.8 dBm towards power penalty lost 5.7% and 4.6% respectively. The 2- D PDC (M=57, N=3) code sequence does not fulfil the 10^{-9} BER error floor. OCDMA Asynchronous nature causes overlapped pulses results in PIIN increases. Performance is sensitive to PIIN in comparison to Shot and Thermal noise. PIIN is the main contributor towards system degradation. 2-D MDW OCDMA is designed with the property of PIIN suppression .

3 Conclusion

PIIN is main limitation factor to the system performance as compare to the shot noise and thermal noise. When P_{sr} is low, thermal noise source becomes the main factor that limits the system performance. Shot noise always low and negligible above -30dBm. The lower total noises power results in high SNR and low BER. It is also been observed that thermal noise is much more influential than shot noise. 2-D MDW code achieves low optical transmission requirement (P_{sr}) at -17.5 dBm in comparison to 2-D PDC and 2-D MQC code.

References

1. R. Kadhim, H. Fadhil , S. Aljunid, M. Razalli. ICCSCE. pp. 90-94 (2014).
2. N. Keraf , S. Aljunid, P. Ehkan, A. Safar. Journal of Theoretical & Applied Information Technology. **15.** 94(1) (2016).
3. G. Kaur, & S. Singh. Review on Optical Code Division Multiple Access Systems, (2016).
4. I. Ahmed, S. Aljunid., C. Rashidi, & L. Al Dulaimi. ECIT'16, **3**(1). 73-76 (2016).
5. N. Ahmed. Doctoral dissertation, Universiti Malaysia Perlis (UniMAP)) (2012).

DOI: 10.1051/epjconf/201716201027

6. Briley, B. E. An introduction to fiber optics system design. Elsevier. ISBN 9781483293868, 1-352, (2016).
7. I. Ahmed, S. Aljunid, C. Rashidi, L. Al dulaimi. Australian Journal of Basic and Applied Sciences. **11** (1), 83-90 (2017).
8. A. Arief, S. Aljunid, M. Anuar, R. Ahmad, M. Junita, M. Rashidi. In Key Engineering Materials Trans Tech Publications. **594**. 1052-1058 (2014).
9. A. Arief, S. AlJunid , M. Anuar , Junita M. , R. Ahmad, F. Ghani. ICP. **1**. 275-279 (2012).
10. Z. Cao, C. Yang, Z. Zhou. Journal of the Optical Society of Korea. **20**(6). 694-7 (2017).

 DOI: 10.1051/epjconf/201716201028

Analysis of hybrid subcarrier multiplexing of OCDMA based on single photodiode detection

N. A. A Ahmad[1,*], *M. N* Junita[1], *S.A.* Aljunid[1], *C.B.M.* Rashidi[1], and *R.* Endut[1]

[1]Advanced Communication Engineering, Centre of Excellence, School of Computer and Communication Engineering, Universiti Malaysia Perlis, Arau, 01000, Perlis, Malaysia.

Abstract. This paper analyzes the performance of subcarrier multiplexing (SCM) of spectral amplitude coding optical code multiple access (SAC-OCDMA) by applying Recursive Combinatorial (RC) code based on single photodiode detection (SPD). SPD is used in the receiver part to reduce the effect of multiple access interference (MAI) which contributes as a dominant noise in incoherent SAC-OCDMA systems. Results indicate that the SCM OCDMA network performance could be improved by using lower data rates and higher number of weight. Total number of users can also be enhanced by adding lower data rates and higher number of subcarriers.

1 Introduction

Recently, researchers have shown an increased interest in optical code division multiple access (OCDMA). OCDMA is an optical-based multiple-access method which can transmit the data simultaneously from several users over the network and can be distinguished by its signature of the code based on specific detection. In addition, OCDMA scheme is good in combining the unlimited bandwidth of fiber with the flexibility of CDMA technique to accomplish high-quality transmission without overwhelming the traffic management or system synchronization [1]. Besides that, OCDMA also have numerous advantages. For instance, it provides secure transmission, ability to support asynchronous access network, does not need network protocol and affordable cost [2]. However, OCDMA also have some disadvantages. The major drawback of OCDMA is, the performance of the system degrades owing to the effects of multiple access interference (MAI) [3].

Thereby, Spectral amplitude coding (SAC) approaches have been proposed. It is a flexible implementation of optical access networks for long-haul transmission. It gives a better quality transmission with capability to reduce the effects of multiple access interference (MAI). Moreover, for synchronization, no electronic are needed. Then, it offers a low cost, all optical processing, simplified network control, and bandwidth efficiency. An effective SAC-OCDMA system can be achieved by choosing efficient codes for encoding the source bits. In order to select a family code with effectively, it needs to distinguished based on their features such as code length (N), cross-correlation, code weight (w) and number of users (K) [4]. In this work, we utilize Recursive Combinatorial (RC) code because it has minimum code lengths, minimum cross correlation which is equal to one and its flexibility in choosing code parameter [5].

Subcarrier multiplexing (SCM) technique has become increasing demand in OCDMA system since it gives the independency of different channels [6]. This allows for great flexibility in the choice of modulation schemes. There are increasing demands for high data rates with low propagation loss, so the best solution to fulfil the request is by combine subcarrier multiplexed with optical code division multiple accesses to the fiber optic communication system. SCM method becomes high attention because of their advantages. For instance, it able to operate asynchronously, increase in privacy, and increase its capacity when in bursty network[3]. In addition, it is flexible in the allocation of channels.

However, to reduce the effects of MAI, single photodiode detection (SPD) is applied in the SAC-OCDMA system. SPD detection is proposed because it offers cost-effective and less system complexity compared with other types of SAC detection techniques. The rest of the paper is organized as follows: In Section 2, we introduce system design based on SPD detection technique. In Section 3, we analyse the results for hybrid SCM OCDMA system based SPD technique. Finally, the conclusion is given in Section 4.

* Corresponding author: n.amirahamniahmad@gmail.com

 DOI: 10.1051/epjconf/201716201028

2 Single Photodiode Detection (SPD)

In OCDMA systems, the detection schemes affect the design of transmitter and receivers. The types of detection are divided by two which are coherent system and incoherent system. Incoherent system is unipolar sequence for the code sequence whereas coherent system is bipolar sequence for the code words. In our system, we applied incoherent system since the coherent system is phase sensitive, which would be practically difficult to implement in this system.

Fig. 1 illustrates the implementation of SCM OCDMA structure at the transmitter part. Here, data from users are modulated into electrical signal within different specific frequencies. These RF frequencies are then combined using electrical adder before being modulated into optical signal codes via a Mach Zender Modulator. The optical RC codes are executed by splitting the broadband optical sources into smaller wavelengths, and encoded it back in the form of codes using an encoder. Optical combiner is then used to combine the optical codes before being transmitted via a single mode fiber.

Fig. 2 shows the implementation of SPD detection at the receiver. The structure is simple compared to other detections, such as AND detection. It only needs a couple of decoder and photo detector rather than AND subtraction detection where it requires two pairs of decoder and photo detector. After optical subtraction, the output is either code weight power unit for active user or zero power units for interferers. This implies that the interference signals are suppressed in the optical domain before the conversion of the signals to the electrical domain, as a result, the proposed SPD scheme alleviates both PIIN and MAI in the optical domain. Moreover, the two interference signals at the optical subtractor are assumed to be equal and cancel each other out. However, practically, the interference signals differ slightly at the optical subtractor and results in a small amount of optical power to reach the photodiode.

At the photodetector, the photo diode only detects and filters the clean chips or the non-overlapping chips. Then, the effects of MAI are successfully reduced in the optical domain because there are only the required signal spectra are filtered. Then, there is no subtraction process in SPD technique. It is because; the photo detector only detects the clean optical pulses and converts it into electrical domain. The electrically modulated signal is then passed through a Band Pass Filter (BPF) to remove the unwanted frequency. In order to remove the higher frequency, a low pass filter (LPF) is assigned. The original data signal is detected at the output of LPF using a BER analyser.

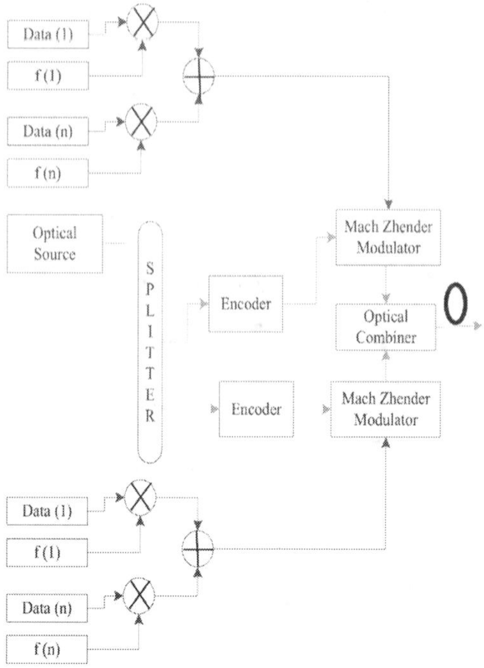

Fig. 1. Transmitter part of hybrid SCM SAC- OCDMA system

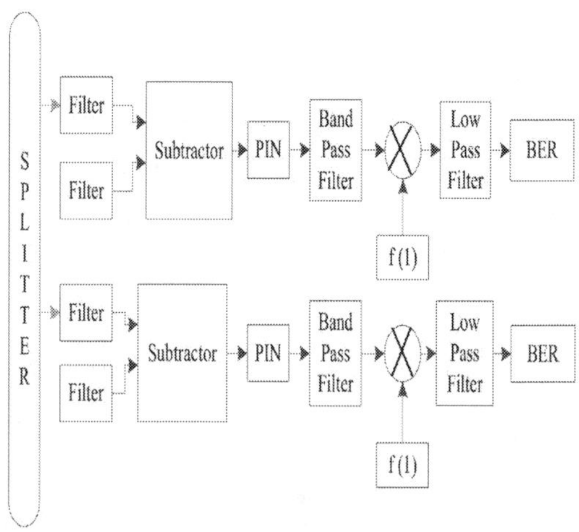

Fig. 2. The receiver part hybrid SCM SAC-OCDMA system based on SPD detection technique

3 Theoretical Development of Hybrid SCM OCDMA System Based on SPD Technique

The performance of RC code in hybrid system can be analysed by describing the properties of RC code which are the cross correlation and auto correlation function of the code take into account. In equation 1, Cd(*i*) represents the *i*th element of *A*th RC code sequence. So that, the properties of RC code based on SPD method can be illustrated as:

$$\sum_{i=1}^{L} C_d(i)C_e(i) = \begin{cases} W ; for\ d=e \\ 1 ; for\ d \neq e \end{cases} \quad (1)$$

Where C_d and C_e depict the RC code sequences, meanwhile L depicts the code length and W depicts the number of the weight. At the receiver part, the photo detector only detects and decodes the non-overlapping chips. Therefore, the total incident power during one period at the input of the photo detector *PIN 1* for the SPD detection technique can be expressed by:

$$\int_0^{\infty} G_{dd}(v)dv = \int_0^{\infty} \frac{P_{sr}}{\Delta v} \sum_{k=1}^{K} d_k \sum_{i=1}^{L} C_d(i)\, C_e(i) u\left[\frac{\Delta v}{L}\right] dv$$

$$= \frac{P_{sr}}{\Delta v}\left[\frac{\Delta v}{L}\right] \sum_{k=1}^{K} d_k \sum_{i=1}^{l} C_d(i)C_e(i) \quad (2)$$

The photodiode current, I indicate the signal from the desired user and can be displays as in eq (3):

$$I = I_{dd} = \Re \int_0^{\infty} G_{dd}(v)dv \quad (3)$$

Then, substitute the equation (4) in (5), the equation becomes:

$$I_{dd} = \Re \frac{P_{sr}}{L}(w-1) \quad (4)$$

Based on equation (6), (W-1) represents the power spectral density of the overlapping chips or cross correlation are subtracted at the received power signal. Thereby, the photodiode only detects and filters the PSD of the non-overlapping chips.

For the noise variances, there are three types of considered noises which are shot noise, thermal noise and intermodulation noise while PIIN noise is neglected. The PIIN noise is neglected because it only appears when the optical signal is incident more than one on the photo detector surface. The shot noise can be expressed as:

$$I_{shot}^2 = 2eB\, I_{dd}$$

$$I_{shot}^2 = 2eB\, \Re \frac{P_{sr}(w-1)}{L} \quad (5)$$

The thermal noise [3] can be expressed as:

$$I_{thermal}^2 = \frac{4K_b T_N B}{R_L} \quad (6)$$

The inter-modulation noise [8] can be expressed as:

$$I_{IMD}^2 = P_{sr}^2\, \Re^2 m_{n,k}^6 \left[\frac{D_{111}}{32} + \frac{D_{21}}{64}\right] \quad (7)$$

The Signal Noise Ratio or SNR equation of the hybrid SCM OCDMA using RC code using SPD method can be expressed as:

$$SNR = \frac{\left(\frac{\Re P_{sr}(W-1)}{L}\right)^2 m_{n,k}^2}{eB\Re\frac{P_{sr}(W-1)}{L} + \frac{4K_b T_n B}{R_L} + P_{sr}^2 \Re^2 m_{n,k}^6 \left[\frac{D_{11}}{32} + \frac{D_{21}}{64}\right]} \quad (8)$$

Based on equation (10), the formula to calculate the bit error rate (BER) can be written as:

$$BER = \frac{1}{2} erfc\left(\sqrt{\frac{SNR}{8}}\right) \quad (9)$$

$$\text{Here, } erfc = \frac{2}{\sqrt{\pi}} \int_x^{\infty} \exp(-z^2)dz \quad (10)$$

Table 1. List of parameter used in the analysis

Symbol	Parameter	Value
η	The efficiency of photodetector quantum	0.6
Δv	The source of Linewidth broadband	3.75THz
λ_0	The wavelength of operating	1550 nm
B	Electrical bandwidth	80/331 MHz
R_b	Data Bit Rate	155/622 Mbps
T_n	The temperature of Receiver noise	300K
R_L	Receiver load resistor	1030 Ω
e	Electron charge	1.6×10^{-19} C
k_b	Boltzmann's constant	1.38×10^{-23} J/K

4 Analysis of Results and Discussion

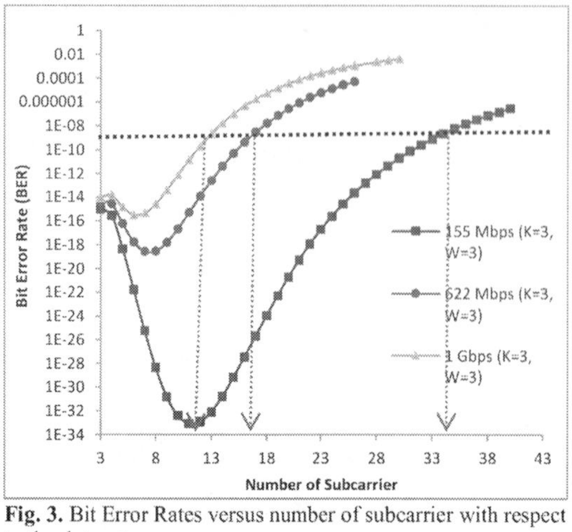

Fig. 3. Bit Error Rates versus number of subcarrier with respect to the data rate

Fig. 3 depicts the relationship between the bit error rate and number of subcarriers with respect to various numbers of data rates such as 155 Mbps, 622 Mbps and 1 Gbps. As can be seen in Figure 3, when lower data rates are assigned in the system, it can occupy large number of subcarriers. For instance, at the threshold BER value of 10^{-9}, a system with 155Mbps per sub channel can accommodate about 35 subcarriers. As data rate increases to 622 Mbps, only 17 subcarriers can be hold in one system meanwhile at data rate 1 Gbps, only 12 subcarriers can be included in the same system. The system capacity will increase if the numbers of subcarriers increase and the value of bit rate decrease rather than having less number of subcarriers but with higher bit rates. Thereby, the performance of hybrid SCM system can be maintained above threshold BER value 10^{-9}.

Fig. 4 presents the performance of hybrid SCM system by using different bit rates per sub channel with varying effective transmitted power. As can be seen in Figure 4, the performance of BER for bit rate 155 Mbps is better compared to bit rate 622 Mbps. At the BER value 10^{-9}, the effective power for bit rate 155 Mbps is about -18 dBm while for the bit rate 622 Mbps, the effective power is about -14 dBm for the same system. Increasing in data rates will increase the electrical bandwidth, then it will degrade the signal to noise ratio. Thus, the performance of the hybrid SCM system will degrade as the data rate increase. As a conclusion lower data rates will increase the system performance.

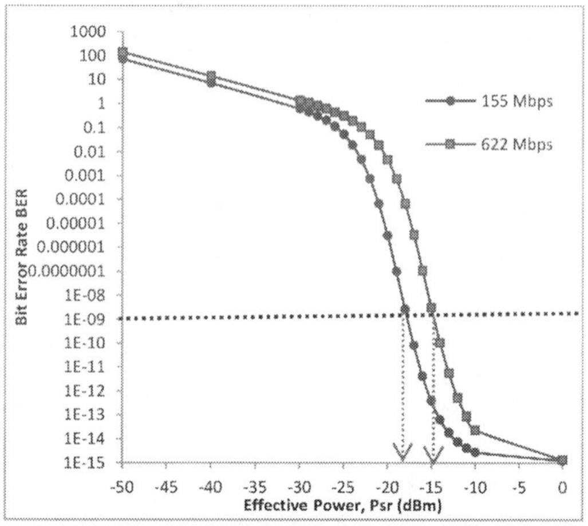

Fig. 4. Bit Error Rates versus Effective Power with respect to bit rates

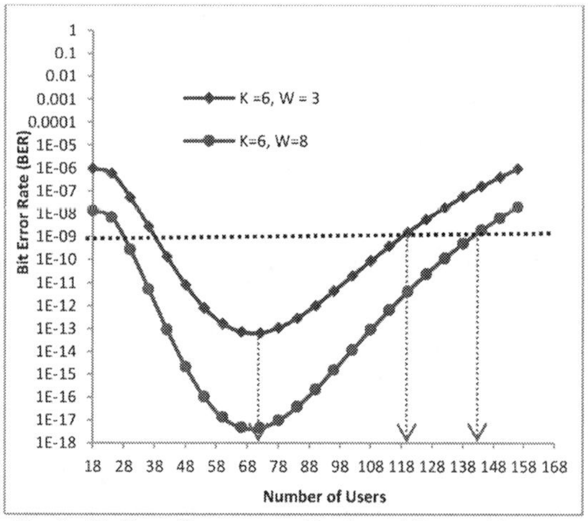

Fig. 5. Bit Error Rates versus Number of Users with respect to number of weight

Fig. 5 depicts the performance of hybrid SCM system with various numbers of weights. The number of optical codes is fixed to six, meanwhile the number of subcarrier is varied. By multiplying the number of code sequence with the number of subcarriers channels, we will get the total number of users. The definition number of weight in OCDMA system is the number of "ones" in code sequences will represent the number of weights. As shown in Figure 5, at BER 10^{-9}, for the weight three, the total number of users is 120 whereas for the weight eight, the total number of users is 143. A part from that, the optimum BER for width three and width eight are 70 users. Thereby, increasing the number of weight will increase the information-bearing signal power and will enhance the performance of hybrid system.

DOI: 10.1051/epjconf/201716201028

5 CONCLUSION

This paper presents the hybrid SCM system by implementing SPD technique. The SPD technique have been proposed and analysed to evaluate the performance of the hybrid SCM system by utilized Recursive Combinatorial (RC) code. Based on the results, it can be concluded that, the performance of hybrid system can be increased by reducing the bit rates and increasing the number of weight. As the bit rate decreases, the bandwidth of the system decreases, then the performance will increase. SPD detection technique gives better performance for hybrid system as it reduces the effect of MAI at the optical domain. Thus, it can support higher number of users by increasing the number of subcarrier channel without affecting the number of code lengths and optical codes. In addition, its bandwidth becomes more efficient.

The authors would like to thank the Ministry of Higher Education (MOHE) Malaysia for the funding under the Fundamental Research Grant Scheme (FRGS).

REFERENCES

1. S. G. Abdulqader, H. A. Fadhil, and S. A. Aljunid. J. Theor. Appl. Inf. Technol., **67**(2) , pp. 368–372, (2014).
2. H. M. R. Al-Khafaji, S. A. Aljunid, A. Amphawan, H. A. Fadhil, and A. M. Safar. *J. Eur. Opt. Soc., vol. **8**, (2013).*
3. Somaya A. Abd E, Mottaleb, Heba A. Fayed, Ahmed Abd El Aziz, and Moustafa H.Aly. IOSR J. Electron. Commun. Eng., vol. **9**(2), pp. 55–60, (2014).
4. M. Z. Norazimah, H. M. R. Al-Khafaji, S. A. Aljunid, and H. A. Fadhil. ICP 2012 - 3rd Int. Conf. Photonics 2012, Proc., no. October, pp. 199–203,(2012).
5. J. M. Nordin, S. A. Aljunid, A. M. Safar, A. R. Arief, R. A. Rahim, and R. B. Ahmad. Aust. J. Basic Appl. Sci., vol. 8, no. **1**, pp. 343–350, (2014).
6. R.K.Z. Sahbudin, M.K. Abdullah, M.D.A. Samad, M.A. Mahdi and M. Ismail. J. of Applied. Sci,vol **7**, pp.1942-1947, (2007)
7. S. G. Abdulqader, S. A. Aljunid, H. M. R. Al-khafaji, and H. A. Fadhil. J. Commun., vol. **9(2)**, pp. 99–106,(2014).
8. T. H. Abd, S. A. Aljunid, H. A. Fadhil, I. F. Radhi, R. B. Ahmad, and M. A. Rashid. Sci. Res. Essays, vol. 7(11), pp. 1262–1272, (2012).

DOI: 10.1051/epjconf/201716201029

Numerical modelling on stimulated Brillouin scattering characterization for Graphene-clad tapered silica fiber

Hui Jing Lee[1,*], *Fairuz* Abdullah[1], and *Aiman* Ismail[1]

[1]Center for Photonics Technologies, College of Engineering, Universiti Tenaga Nasional, Malaysia

Abstract. This paper presents finite numerical modelling on the cross-sectional region of tapered single mode fiber and graphene-clad tapered fiber. Surface acoustic wave propagation across the tapered surface region on tapered single mode fiber has a high threshold power at 61.87 W which is challenging to overcome by the incident pump wave. Surface acoustic wave propagation of fiber surface however made tapered wave plausible in the optical sensor application. This research introduces graphene as the cladding layer on tapered fiber, acoustic confinement occurs due to the graphene cladding which lowers the threshold power from 61.87 W to 2.17 W.

1 Introduction

Stimulated Brillouin Scattering (SBS) is an optical scattering process caused by the non-linearity (χ^3) of a medium. The incident wave which is also known as the pump wave has to be higher than threshold power to generate acoustic wave [1]. The process of sound vibration fluctuates the refractive index of optical wave and scatters the injected pump wave. This effect creates a Stoke wave propagating in the opposite direction [2]. The scattering effect will transfer power from pump wave to Stoke wave therefore loss in energy is expected due to the fiber vibration. Because of this, acoustic frequency shift happens therefore lowers the frequency of Stoke wave from the original pump wave frequency. The Stoke wave will generate a strong optical Brillouin gain which can be used to amplify the weak injected wave [3].

Threshold power has both its functionality and limitation. In the communication field, SBS limitation could be signal loss, power saturation and backward propagation [4]. Hence, downside of Stoke wave is that it might saturates the amplifier and creates noise in the transmitter. Suppression of SBS can be done with several methods: one option would be to ensure the pump power is lower than the threshold power, the other option would be to increase the threshold power without dispersion penalty [5]. On the contrary, SBS with low threshold power has its functionality in lasers and amplifiers [6].

SBS has enormous potential in the application of sensing and optical communication. Low threshold power and short fiber length are two major indicators for a good sensor. In previous research, extensive experiments have been done to measure acoustic frequency shift across fibers. The focus is now on the threshold power therefore feasibility of different fibers in SBS applications are investigated [7]. Chalcogenide (ChG) fiber which secured acoustic confinement in the fiber core shows a low threshold power at 52 mW [8]. On the contrary, recent finding in SBS across tapered Silica fiber presented a high threshold power (60 W) which is undesired in SBS application [9]. This happens because Surface Acoustic Wave (SAW) at tapered waist region has minimum interaction with optical wave. Hence, gold and silver are two types of cladding materials proposed to overcome the high threshold power of tapered fiber.

A tapered single mode fiber (TSMF) is a heated and stretched standard single mode fiber which comprises of the non-tapered region, transition region and tapered waist region. An ideal TSMF diameter of tapered waist region around 0.8 ~ 3.0 μm has smooth fiber surface and constant diameter [10-12]. TSMF has great plausibility in the sensor application. Light propagation across both adiabatic and non-adiabatic tapered fibers were both reported [13]. TSMF is also confirmed to have high sensitivity and is made possible to detect the change in protein concentration [14]. In addition, TSMF also has its functionality as a magnetic field sensor such that the magnetic fluid functions as the cladding layer for tapered fiber [15]. Also, functionality of TSMF was found in the refractive index based sensors and surface plasmon resonance (SPR) transducers [16, 17].

Studies shown that there are significant differences in the characteristics of tapered fiber clad with varied materials. One common similarity is that they are feasible in sensor application. For example, functionality of TSMF clad with Indium Tin Oxide (TIO) as a humidity sensor was shown [18]. Recent studies focus on graphene material which received great attention due to its uniqueness in optical, electrical, chemical and mechanical properties. The main focus would be on its optical properties with 97.7 % of light transmittance rate [19]. Yao presented that graphene-clad microfiber enhances the evanescent field of fiber, it possesses ultrahigh nonlinearity therefore successfully generates Four Wave

* Corresponding author: LHjing@uniten.edu.my

Mixing (FWM) which microfiber fails in such case [20]. Optical light was reported to confine at the graphene clad region, it is therefore applicable in the fiber optic non-linear devices. Also, finding revealed uric acid detection using graphene-clad tapered fiber (GCTF) [21]. Functionality of GCTF in glucose detection was brought into discussion. Consequently, the functionality of graphene oxide clad fiber in aqueous ethanol sensing was proved [22]. Recent finding proved the functionality of graphene oxide clad tapered fiber bio-sensor for DNA detection [23]. Furthermore, tapered fiber partially covered with graphene layer is proved to enhance its optical conductivity [24].

Study regarding TSMF is scanty, realizing the gap in the extant literature, more research work is needed to investigate the potential of TSMF in SBS. However, so far none of the numerical solution tool has been used to model the exact SBS behaviour across TSMF or fiber with the variation of cladding materials [25]. The potential functionality of graphene-clad tapered fiber in SBS in yet to be studied. Hence, this paper performs numerical modelling using Finite Element Method (FEM) to study the stimulated Brillouin characterization on a graphene-clad tapered silica fiber.

FEM is a promising tool in computational Science and Engineering [26]. FEM has two approachable methods namely the Variational Method and Galerkin's Method. Generally, Galerkin's Method is preferable as it is more accessible for cases where Variational Method fails to handle [27]. Thorough literature reveals that FEM is a mature numerical modelling tool for fiber modelling such as for hollow fiber, inextensible fibers and ceramic matrix continuous [28-30].

2 Stimulated Brillouin Scattering Formulation

Fundamental of SBS characterization is based on the optical and acoustic wave interaction. When the incident pump wave exceeds the threshold power, acoustic wave will be generated. Fluctuation in the refractive index of optical wave will scatter the injected pump wave hence a Stoke wave propagates in the opposite direction occurs. Hence, due to the scattering effect, transfer power is transferred from the pump wave to Stoke wave. In SBS characterization, firstly the optical waveguide is solved then the acoustic waveguide is generated under a mode-matching condition. Based on the optical-acoustic interaction, the Brillouin gain and threshold power are calculated.

The fundamental optical mode property is solved as:

$$\Delta_t^2 + \left(\frac{2\pi}{\lambda}\right)^2 (n^2 - n_{\text{eff}}^2)E = 0 \qquad (1)$$

where λ is the wavelength of incidence light, n and n_{eff} are the refractive index and effective refractive index of fundamental optical wave, respectively. E represents the optical spacial distribution whereas Δ_t signifies the transverse Laplacian operator in the reference of x and y plane [31].

Using n_{eff} and λ from Equation (1), the optical wave propagation constant is solved [32]:

$$\beta_{optic} = \frac{2\pi n_{\text{eff}}}{\lambda} \qquad (2)$$

SBS characterization is performed under a mode-matching condition. The β_{optic} in Equation 2 is then used to solve for the acoustic wave propagation by applying the Braggs Grating mathematical derivation to define the diffractive indices which refract and reflect at the interfaces:

$$k_{acoustic} = 2\beta_{optic} \qquad (3)$$

This explains that the acoustic field is based on the n_{eff} solved from Equation 1, the backward scattered wave explains the formation of Stoke wave [33]. Therefore, the acoustic velocity, V_{ac} is hence defined as:

$$V_{ac} = \frac{\omega}{k_{acoustic}} \qquad (4)$$

where ω is the angular energy. Hence, using the acoustic velocity, V_{ac} solved from Equation (4), the Brillouin shift for acoustic wave at fundamental mode is solved [34]:

$$v_B = \frac{2n_{\text{eff}}V_{ac}}{\lambda} \qquad (5)$$

The acoustic velocity, V_{ac} is also used to solve for the acoustic frequency shift v_{ac}:

$$V_{ac} = v_{ac}\sqrt{\frac{1}{v_1^2} - \frac{1}{v_2^2}} \qquad (6)$$

where v_1 is the core velocity and v_2 is the clad velocity.

Formulation of Brillouin gain spectrum (BGS) is expressed as:

$$g^m(v) = g_p \frac{\frac{\Delta v_{ac}{}^{(m)2}}{2}}{[\,v_s - \left(v_p - v_{ac}^{(m)}\right)]^2 + \frac{\Delta v_{ac}{}^{(m)2}}{2}} \Gamma_{ij} \qquad (7)$$

where $v_{ac}^{(m)}$ is the m-th order acoustic frequency shift, $\Delta v_{ac}{}^{(m)}$ represents the Full Width Half Maximum (FWHM) of the m-th order acoustic mode, v_p is the pump frequency and v_s is the stoke frequency. Γ_{ij} signifies the overlapping factor between optical waveguide and acoustic waveguide:

$$\Gamma_{ij} = \frac{\int [|H_i^2(x,y)||U_j(x,y)|^2]\, dA}{\int |H_i(x,y)|^4\, dA \int |U_j(x,y)|^2 dA} \qquad (8)$$

where $H_i(x, y)$ is the optical wave at fundamental mode and $U_j(x, y)$ is the acoustic wave displacement.

The fundamental mode of v_{ac} is equivalent to v_B. The Brillouin peak gain is achieved under condition where the difference between pump and stoke frequency is equivalent to the Brillouin shift, v_B [35, 36].

$$g_B(v) = g_p = \frac{4\pi^3 n_{\text{eff}}^8 p_{12}^2}{c\lambda_p^3 \rho_o v_B\, \Delta v_B} \qquad (9)$$

where ρ_{12} signifies the elasto-optic coefficient, ρ_o is the fiber core density, c represents speed of light and Δv_B is the FWHM of acoustic wave.

Substituting the Brillouin gain solve for the SBS threshold power:

$$P_{th} = \frac{\kappa A_{\text{eff}}}{g_p L_{\text{eff}} K} \tag{9}$$

where κ is the numerical factor manipulated by fiber length and is approximated to a constant value ≈ 21. A_{eff} and L_{eff} are the effective area and effective length of fiber, respectively. K is the polarization factor range between 1-2 [37].

3 Results and Discussion

This section firstly presents the fundamental optical and acoustic waveguides of tapered fiber (TF) and graphene-clad tapered fiber (GCTF) using FEM programmed in Fortran. The numerical output of the optic and acoustic are then used to calculate the Brillouin gain and threshold power parameters using Matlab. Design parameters for silica fiber and graphene are listed in Table 1 [36, 38].

Table 1. Design parameters for silica fiber and graphene

Properties	Silica Fiber	Graphene
$V_{longitudinal}$ (m/s)	5933	19, 700
V_{shear} (m/s)	3764	12, 700
ρ (kg/m^3)	2202	1750

3.1 Optical wave

FEM is used to solve for the optical wave propagation for TF and GCTF using Equation 1 and 2. The outer region for TF and GCTF are air and graphene, respectively. Table 2 lists the output parameters for the fundamental optic mode of TF and GCTF.

Table 2. Optical properties of TF and GCTF

Properties	TF	GCTF
n_{core}	1.4447	1.4447
n_{clad}	1	2
n_{eff}	1.1748	1.9327
β_{optic} (µm^{-1})	4.8621	8.2080

Figure 1 (a), (b), and (c) show the fundamental Hy, Hx and Hz optical mode for TF with fiber radius = 0.5 µm. Total internal reflection is observed in Figure 1 (a) with the fundamental optic mode dominating in the y direction. The optic confinement is shown in Figure 1 (d) with peak intensity at 2.7.

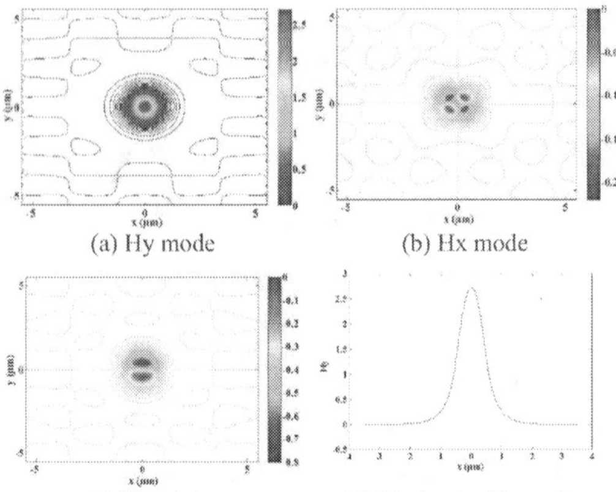

(a) Hy mode (b) Hx mode

(c) Hz mode (d) 2d plot for Hy mode

Fig 1. Fundamental optical mode for TF with R = 0.5 µm

Figure 2 (a), (b), and (c) show the fundamental Hy, Hx and Hz optical mode for GCTF with fiber radius equivalent to 0.5 µm and clad with 0.1 µm of graphene layer. Optical confinement is observed at the fiber surface, optic wave is trapped at the graphene layer due to it's higher refractive index [20]. Figure 2 (d) shows 2d plot of Hy mode at -3.52 peak intensity. The strong confinement is contributed by the graphene layer.

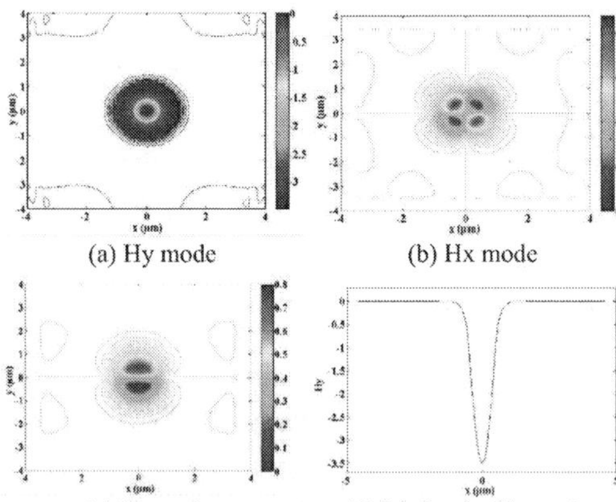

(a) Hy mode (b) Hx mode

(c) Hz mode (d) 2d plot for Hy mode

Fig 2. Fundamental optical mode for GCTF with R = 0.5 µm.

3.2 Acoustic wave

Table 3 lists the acoustic wave output parameters. Based on Equation 3, $k_{acoustic}$ is solved by referring to the β_{optic} result from Table 2. GCTF has a higher acoustic velocity and higher acoustic frequency shift as compared to TF due to the high graphene velocity. Acoustic velocity of SAW for TF is recorded as 3306 m/s. Such result is in agreement with the SAW theory in which the velocity of propagation has to be less than the shear velocity (3764 m/s) and longitudinal velocity (5933 m/s).

Table 3. Acoustic properties of TF and GCTF

Properties	TF	GCTF
$k_{acoustic}$ (µm^{-1})	9.632	16.416
v_B (GHz)	5.237	10.198
V_{ac} (GHz)	3306	3822

Significant difference can be observed in the acoustic result of TF and SMTF. SAW propagates on the fiber surface for TF, whereas acoustic confinement in the core of GCTF. Figure 3 (a), (b), and (c) show the Uz, Ux and Uy acoustic mode for TF. Figure 3 (d) verifies the result where two sharp peaks occur at 0.5 and -0.5 which indicate the fiber surface.

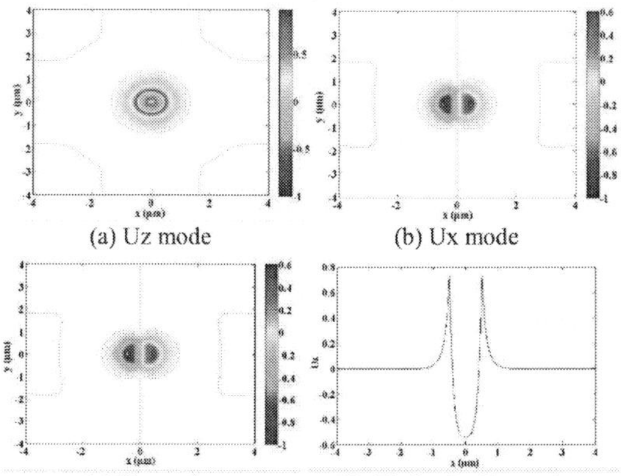

(a) Uz mode (b) Ux mode
(c) Uy mode (d) 2d plot for Uz mode

Fig 3. Fundamental acoustic mode for TF with R = 0.5 µm

0.1 µm of graphene is then clad around TF, the acoustic wave confines from surface acoustic wave to hybrid acoustic wave (HAW). This can be explained in Equation 6 in where clad velocity has to be higher than core velocity to secure propagation of HAW. Figure 4 (d) confirms the mode propagation of HAW for tapered fiber clad with graphene layer.

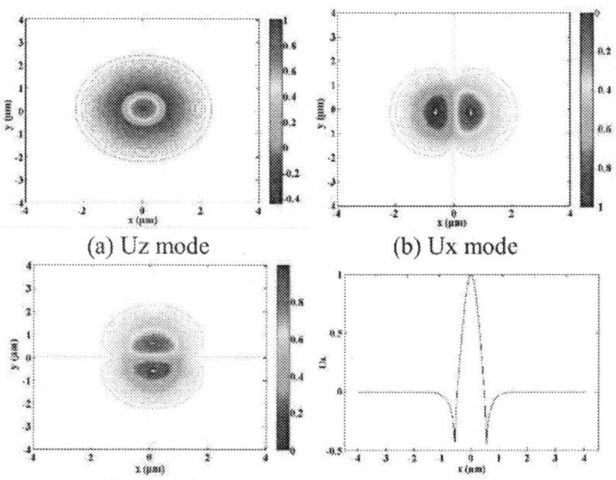

(a) Uz mode (b) Ux mode
(c) Uy mode (d) 2d plot for Uz mode

Fig 4. Fundamental acoustic mode for GCTF with R = 0.5 µm

3.3 Brillouin gain and threshold power

In this subsection, the Brillouin gain and threshold power of TF and GCTF are calculated and analyzed. Table 4 lists the Brillouin gain and threshold power of TF and GCTF, calculation is performed based on Equation 9 and 10. Graphene cladding confines the acoustic mode propagation hence reduces the threshold power from 61.97 W down to 2.17 W.

Table 4. Brillouin gain and threshold power of TF and GCTF

Properties	TF	GCTF
v_B (GHz)	5.237	10.198
Δv_B (GHz)	20	19
g_B (m/W)	1.419 x 10^{-12}	3.909 x 10^{-11}
A_{eff} (m^2)	1.788 x 10^{-13}	1.732 x 10^{-13}
P_{th} (W)	61.87	2.17

Graphene is a unique element with outstanding technical and mechanical properties. However, graphene properties used in this research are pre-defined to be in an ideal situation. In real time application, graphene properties might vary based on the preparatory method. Assuming an ideal case, graphene contribution in Stimulated Brillouin Scattering characterization is highlighted. Results demonstrate that graphene enhances the Brillouin gain therefore reduces the threshold power for tapered fiber. Such contribution would be interesting to be further explored in the real time experiment.

4 Conclusion

This research work was undertaken to perform Stimulated Brillouin Scattering characterization on the cross-sectional region of tapered fiber and the change in result when graphene is clad on the fiber surface. One significant finding to emerge from this research work is that graphene layer lower the threshold power of a tapered fiber from 61.87 W down to 2.17 W. Such result highlights that tapered fiber and graphene-clad tapered fiber with the gap in threshold power are applicable in the optical sensing and laser/amplifiers, respectively. It is recommended that experimental work can be done to identify the Brillouin scattering across graphene-clad tapered fiber to have a better understanding of graphene properties in SBS.

We thank Professor Azizur Rahman from City University London, United Kingdom, Dr and Dr. Siamak Emami from Shahid Baheshti University, Iran for the advice. Also, we thanks Universiti Tenaga Nasional internal grant RJO10289176 for the support.

References

1. Brillouin, Léon. Ann. Phys.(Paris) **17**, no. 88-122,21 (1922)
2. Keiser, Gerd. *Optical fiber communications*. John Wiley & Sons, Inc. (2003)
3. Agrawal, Govind P. *Nonlinear fiber optics*. Academic press (2007)
4. Hicks, John W. U.S. Patent 4,778,238, (October 18, 1988)
5. Monro, Tanya M., Yvonne D. West, Daniel W. Hewak, N. G. R. Broderick, and D. J. Richardson. Elec. Lett. **36**, no. 24: 1998-2000 (2000)
6. Woodward, R. I., E. J. R. Kelleher, S. V. Popov, and J. R. Taylor. Opt. Lett. **39**, no. 8: 2330-2333 (2014)
7. Cherif, Rim, A. B. Salem, T. S. Saini, A. Kumar, Ravindra K. Sinha, and M. Zghal. Opt. Eng. **54**, no. 7 (2015)
8. Hu, Kai, Irina V. Kabakova, Thomas FS Büttner, Simon Lefrancois, Darren D. Hudson, Sailing He, and Benjamin J. Eggleton. Opt. Lett. **39**, no. 16: 4651-4654 (2014)
9. Beugnot, J.-Charles, S. Lebrun, G. Pauliat, H. Maillotte, V. Laude, and T. Sylvestre. Nat. com. **5** (2014)
10. M-Hernández, Carlos, D. Monzón-Hernández, I. Hernández-Romano, and J. Villatoro. *Opt. exp.* **23**, no. 17 x: 22141-22148 (2014)
11. Kerttula, Juho, V. Filippov, V. Ustimchik, Y. Chamorovskiy, and Oleg G. Okhotnikov. *Opt. exp.* **20**, no. 23: 25461-25470 (2012)
12. Harun, S. W., K. S. Lim, C. K. Tio, K. Dimyati, and H. Ahmad. Opt-Int J. for L. E. Opt. **124**, no. 6: 538-543 (2013)
13. Hartung, A., F. Wirth, and H. Bartelt. In *Proceedings to Progress in Electromagnetics Research Symposium*, pp. 255-258 (2011)
14. Yadav, T. K., Narayanaswamy, R., Bakar, M. A., Kamil, Y. M., & Mahdi, M. A. Opt. exp., *22*(19), 22802-22807 (2014)
15. Layeghi, A., H. Latifi, and O. Frazão. *23rd International Conference on Optical Fiber Sensors*, pp. 91570Z-91570Z (2014)
16. González-Cano, A., Navarrete, M.C., Esteban, Ó. and Díaz-Herrera, N., Sen. J., **14**(3), pp.4791-4805 (2014)
17. P. Wang, G. Brambilla, M. Ding, Y. Semenova, Q. Wu, G. Farrell,J. Opt. Soc. Am. B **28** (5) 1180–1186 (2011)
18. Ascorbe, J., Corres, J.M., Arregui, F.J. and Matias, I.R., Sen. J., pp. 1916-1919 (2014)
19. E. W. Hill, A. Vijayaragahvan, K. Novoselov, Sen. J., IEEE **11** (12) 3161–3170 (2011)
20. B. Yao, Y. Wu, Q. Feng, Z. Wang, Y. Rao, Y. Chen, K. S. Chiang, Opt. Soc. A. pp. 4–6 (2015)
21. M. Batumalay, S. W. Harun, F. Ahmad, R. Md Nor, N. R. Zulkepely, H. Ahmad, Sens. J., IEEE 14 (5) 1704–1709 (2014)
22. H. Qiu, S. Xu, S. Jiang, Z. Li, P. Chen, S. Gao, C. Zhang, D. Feng, App. Surf. Sc. **329** 390–395 (2015)
23. A. Shabaneh, S. Girei, P. Arasu, W. Rahman, A. Bakar, A. Sadek, H. Lim, N. Huang, M. Yaacob, Opt. Com. **331** 320–324 (2014)
24. B. Yu, Y. Huang, M. Ding, Y. Ran, B. O. Guan, *Fifth Asia Pacific Optical Sensors Conference, International Society for Optics and Photonics*, pp. 1–4 (2015)
25. A. Gorbach, A. Marini, D. Skryabin, Opt. let. **38** (24), 5244–5247 (2013)
26. E. Madenci, I. Guven, Springer, (2015)
27. Y. Zhang, C. Yu, X. Fu, D. Li, W. Jia, W. Bi, Measurement **51** 310–314 (2014)
28. J. A. Dehkordi, S. S. Hosseini, P. K. Kundu, N. R. Tan, Chem. Prod. & Proc. Mod. **11** (1) 11–15 (2016)
29. P. Wriggers, J. Schröder, F. Auricchio, Mod. & Sim. in Eng. Sc. **3** (1) 25 (2016)
30. M. Braginsky, C. Przybyla, Comp. Str. **136** 538–545 (2016)
31. Y. S. Mamdem, X. Pheron, F. Taillade, Y. Jaoüen, R. Gabet, V. Lanticq, G. Moreau, A. Boukenter, Y. Ouerdane, S. Lesoille, et al.,*Comsol conference* (2010)
32. Z. Zhu, D. J. Gauthier, R. W. Boyd, Sc. (N.Y.) **318** (5857) 1748–1750 (2007)
33. M. R. Mokhtar, Ph.D. thesis, Univ. of Southampton (2005)
34. K. Virta, K. Mattsson, J. Sc.Com. **61** (1) 90–118 (2014)
35. C. G. Poulton, R. Pant, B. J. Eggleton, JOSA B **30** (10) 2657–2664 (2013)
36. S. Sriratanavaree, Ph.D. thesis, City University London (2014)
37. C. Lee, S. Chi, IEEE P. Tech. Lett. **12** (6) 672–674 (2000)
38. Baimova, Y. A., Dmitriev, S. V., Savin. A. V., Kivshar, Y.S., Phy. of the Sol. State. 54 (4) 866-874 (2012)

Influence of fiber parameters to the erbium-doped fiber amplifier (EDFA) gain: A theoretical modeling

N. A. B. Ahmad[1,2,*], S. H. Dahlan[1,2], and N. A. Cholan[1,2]

[1]Research Centre for Applied Electromagnetics, Universiti Tun Hussein Onn Malaysia (UTHM), Batu Pahat, Johor, Malaysia.
[2]Faculty of Electrical and Electronic Engineering, Universiti Tun Hussein Onn Malaysia (UTHM), Batu Pahat, Johor, Malaysia.

Abstract. This paper studies the theoretical analysis of erbium-doped fiber (EDF) amplification. The EDF amplification is investigated in order to understand its behavior and identify the relevant parameters that are used for formulating the amplification process related to spontaneous and stimulated emission. Results show that under certain condition, the selected fiber parameters such as fiber length, EDF pump power, signal power and pump direction would significantly contribute to the EDFA gain enhancement.

1 Introduction

Multiwavelength fiber lasers have attracted great interest due to their applications in optical sensors, microwave optics, spectroscopy and optical communication systems to name a few. One technology where these fiber lasers can be realized is by a combination of two optical gains namely erbium-doped fiber (EDF) and Brillouin amplification or commonly known in literature as multiwavelength Brillouin-erbium fiber lasers (BEFLs). Compared with other technologies, multiwavelength BEFLs have advantages such as low threshold power and ultra-narrow linewidth [1]. Literature suggests that performances of multiwavelength BEFLs can be improved when the residual waves are recycled back into the laser cavity [2]. Characteristics of residual waves are studied thoroughly in [3-4] to gain higher number of channels. This paper intents to optimize the EDF amplification in the multiwavelength BEFLs to improve the performance further. Therefore in this work, erbium-doped fiber amplifier (EDFA) parameters for different pumping powers and different EDF length are investigated to realize a high performance EDFAs.

2 Mathematical Modelling of EDFA

The basic configuration of a typical EDFA is illustrated in Fig. 1. An EDFA consists of a wavelength division multiplexer (WDM) coupler to combine the pump with the input signal light. The combined pump and signal light is injected into a 20 m EDF and an optical isolator is placed after the EDF, ensuring unidirectional operation of the EDFA. The pump source is of a 980 nm laser diode.

Fig. 2 illustrates the energy levels and transitions of Er^{3+} ions in the EDF when it is pumped with the 980 nm laser diode. It is based on a two-level assumption. At first, the ions absorb the pump power, enabling them to transit to the high level. The ions stayed at the level for a period of τ_3 before jumping to the meta-stable state, resulting in emission of photons [5]. Due to the small value of τ_3, it can be neglected and hence simplifies the three-level system to the two-level [6]. A radiative transition towards the ground energy level then takes place. As a result, photons at 1550 nm are emitted. The rate equations, eqs. (1) – (3) describe the effects of absorption, stimulated and spontaneous emission on the ions populations of the ground state N_1 and metastable states N_2 [7-8]

$$\frac{dN_1}{dt} = -[W_{sa}(z)+W_P(z)]N_1 + \left[\frac{1}{\tau_{21}}+W_{se}(z)\right]N_2, \quad (1)$$

$$\frac{dN_2}{dt} = [W_{sa}(z)+W_P(z)]N_1 - \left[\frac{1}{\tau_{21}}+W_{se}(z)\right]N_2, \quad (2)$$

$$N_1(z)+N_2(z)=N_T \quad (3)$$

where τ_{21} is the spontaneous emission lifetime, N_T is the dopant density of the EDF, W_{sa}, W_{se}, and W_P are the stimulated absorption, stimulated emission and pumping rates respectively. Further elaborations can be found in [9] and summarizes as

$$W_{sa}(z) = \sum_{v} \frac{\xi_{sa}(v)}{A_{eff}hv}[P_S + P^+_{ASE}(z,v)+P^-_{ASE}(z,v)]\eta(v)dv, \quad (4)$$

$$W_{se}(z) = \sum_{v} \frac{\xi_{se}(v)}{A_{eff}hv}[P_S + P^+_{ASE}(z,v)+P^-_{ASE}(z,v)]\eta(v)dv, \quad (5)$$

$$W_P(z) = \frac{1}{A_{eff}}\frac{\xi_{pa}(v_P)}{hv_P}[P^+_P(z)+P^-_P(z)]\eta(v_P), \quad (6)$$

where P_S is the signal power, P^+_P and P^+_{ASE} represent the forward propagating of EDF pump powers and amplified spontaneous emission (ASE) powers at frequency v, respectively, while P^-_P and P^-_{ASE} are the backward propagating of EDF pump powers and ASE powers at frequency v, correspondingly, ξ_{sa}, ξ_{pa}, ξ_{se}, and ξ_{pe} are

*Corresponding author: nurulati89@gmail.com

the absorption and emission cross sections at the pump and signal wavelength. A_{eff} is the effective area of the EDF and h is the Plank's constant. Meanwhile, $\eta(v)$ is the overlapping factor between each radiation and fiber fundamental modes.

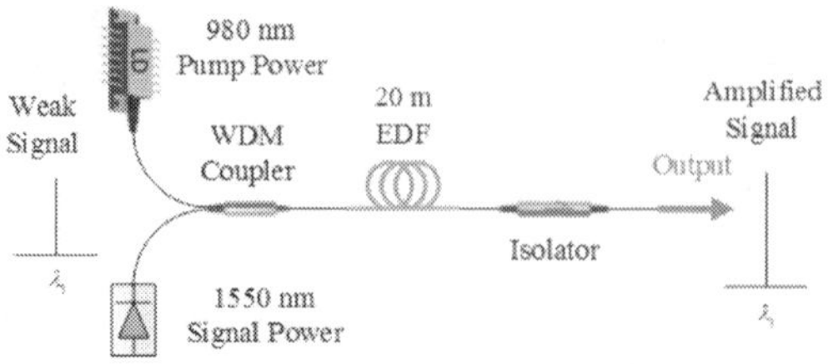

Fig. 1. Basic configuration of the EDFA.

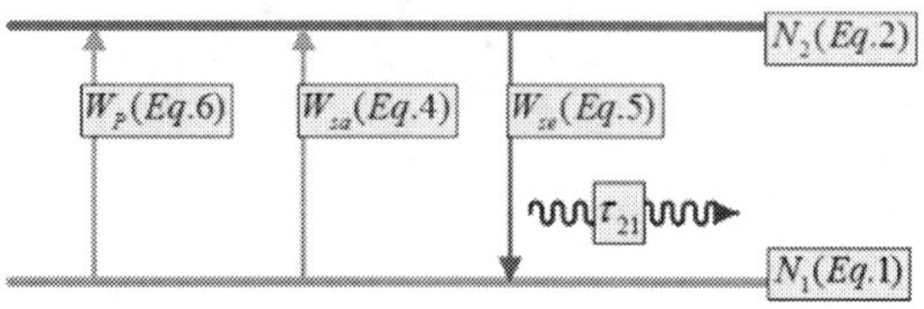

Fig. 2. Energy level diagram of Er^{3+} doped in silica.

Propagation of light through the EDF is shown in Fig. 3(a). The light in the EDF is assumed to propagate as a number of optical beams of frequency bandwidth Δv. The wave propagation of signal P_S, pump $P_P^{\pm}$ and noise power $P_{ASE}^{\pm}$ can be expressed as [10]

$$\frac{dP_S(z)}{dz} = (N_2\xi_{se} - N_1\xi_{sa})\eta(v_S)P_S(z), \quad (7)$$

$$\frac{dP_P^{\pm}(z)}{dz} = \pm(N_2\xi_{pe} - N_1\xi_{pa})\eta(v_P)P_P^{\pm}(z), \quad (8)$$

$$\frac{dP_{ASE\_n}^{\pm}(z)}{dz} = \pm[N_2\xi_{se}(v) - N_1\xi_{sa}(v)]\eta(v)P_{ASE\_n}^{\pm}(z) \pm d, \quad (9)$$

where $d = 2hv\Delta vN_2\eta(v)\xi_{se}(v)$ is the ASE power caused by the spontaneous emission.

In order to solve for power from the eqs. (1) – (6), the EDF with length z is divided into segments along the forward and backward propagation as illustrated in Fig. 3(b). Each segment contains its own population densities. The pump, signal and ASE powers which propagate in the first segment (segment 0) are solved by using the initial condition as shown in Table 1. Then, the power of the earlier segment is used as the input for the next consecutive segments. The eqs. (7) - (9) are numerically solved through a trial and error technique known as relaxation method. This method makes iterative adjustment to the solution where an initial set of boundary is chosen for the first integration. Utilizing this method, accuracy of 0.01% is achieved for P_S, $P_P^{\pm}$, and $P_{ASE}^{\pm}$ [11-12].

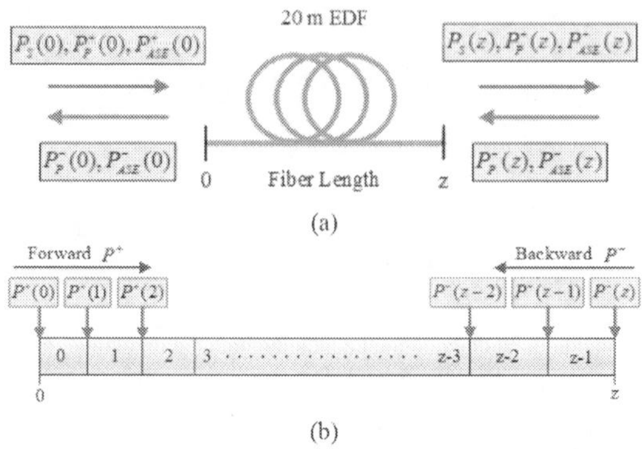

Fig. 3. (a) Lightwave propagation along the EDF.
(b) Schematic of the numerical fiber model.

Table 1. Initial condition.

Initial condition	Operating wavelength (nm)
$P_{P1}(0) = P_P$	$\lambda = 980$
$P_S(0) = P_S$	$\lambda = 1550$
$P_{ASE}^+(0,v) = P_{ASE}^-(z,v) = 0$	$1450 \leq \lambda \leq 1600$
$P_{ASE}^+(z,v) = P_{ASE}^-(0,v) \neq 0$	$1450 < \lambda < 1600$

3 Results and Discussions

A series of numerical simulations are performed for the EDFA modelling using parameters of an EDF from OFS Model MP980. These parameters are $N_T = 3 \times 10^{23}$, $N_{2\_initial} = 10^{-10}$, $L_{EDF} = 20$ m, $\Delta v = 5 \times 10^{10}$ Hz, $\tau_{21} = 10^{-2}$ s, $\xi_{pa} = 0$, $\xi_{pe} = 3.28 \times 10^{-25}$ m^2, $A_{eff} = 50$ μm^2, $c = 3 \times 10^8$ m/s, and $h = 6.63 \times 10^{-34}$ Js.

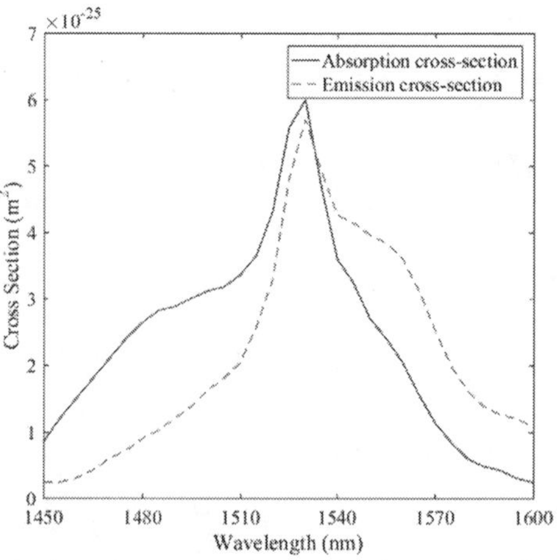

Fig. 4. Absorption and emission cross sections spectra of ASE signals.

DOI: 10.1051/epjconf/201716201030

The ξ_{sa} and ξ_{se} for the EDF of different wavelengths are obtained by using curve fitting method as shown in Fig. 4. Notice that each wavelength has its own absorption and emission coefficients since photons can be absorbed to generate carriers via a stimulated absorption process while at the same time, new photons can be generated owing to the stimulated emission process. Fig. 5 depicted the ASE power of both backward and forward ASE along the EDF for forward pumping configurations. During the simulation, we found that $P_{ASE}^{-}(z)$ must be lower than 0.01 mW.

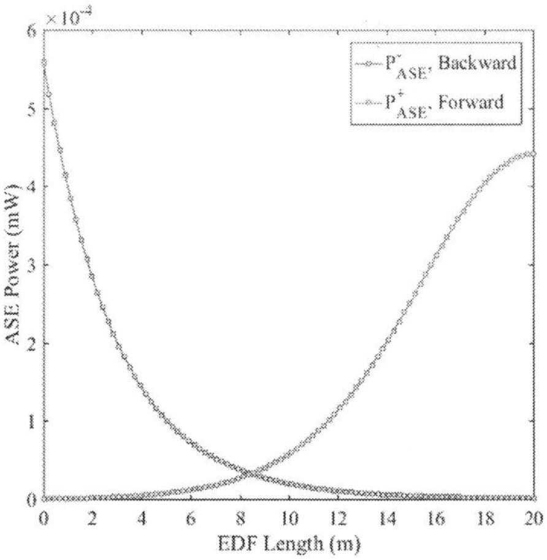

Fig. 5. Simulation output of forward and backward ASE.

The variation of gain, $G = P_S(z)/P_S(0)$ for different EDFs length and pump power is then investigated as shown in Fig. 6. With the EDF pump power increased from 30 mW to 100 mW for 20 m EDF, the gain in Fig. 6(a) was observed to increase along with the increasing EDF length. However, the gain started to decrease after reaching a maximum value. The reason is that the pump propagates through the EDF gets absorbed and hence the power reduces. After propagating a certain distance, the power becomes too small to create population inversion. As a result, the EDF would start to absorb the signal rather than amplify the signal. Therefore, there is an optimum EDF length to achieve maximum gain with variation of EDF pump power. Fig. 6 (b) depicted the variation of gain for different EDF length with the EDF pump power of 100 mW. For EDF length from 10 m to 20 m, the gain is expected to increase as the EDF pump power increase. When the EDF pump power increase, the population inversions are created gradually. No more Er^{3+} ions are available when all the Er^{3+} ions are excited, and thus the gain would saturate.

Further we study the EDFA gain by plotting EDFA gain versus wavelength with varying signal power from -40 dBm to -10 dBm as depicted in Fig. 7. The results were obtained using a 20 m OFS MP980 of EDF and a 100 mW pump power at 980 nm wavelength. Initially, the gain spectrum of an EDFA is not uniform across the bandwidth. A narrow gain peak around 1530 nm and a relatively flat region between 1540 nm and 1560 nm can be observed in Fig. 7. When the input signal power increased from -40 dBm to -10 dBm, the optical gain at 1550 nm decreases from 37 dB to 15 dB, resulting by gain saturation. Apart from that, the spectral shape of the optical gain also changes with the input signal power. When the input signal power increase, the gain tends to tilt toward the longer wavelength side while the short wavelength peak around 1530 nm is suppressed. This situation has to be considered in designing gain-flattening techniques.

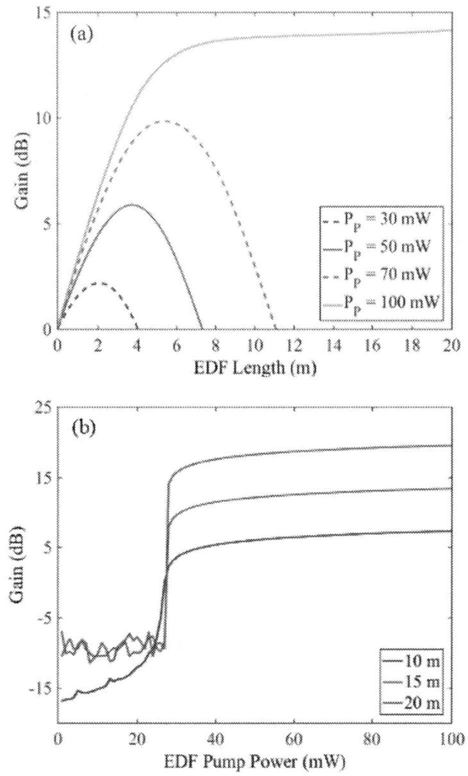

Fig. 6. The variation of gain with (a) EDF length and (b) Pump power

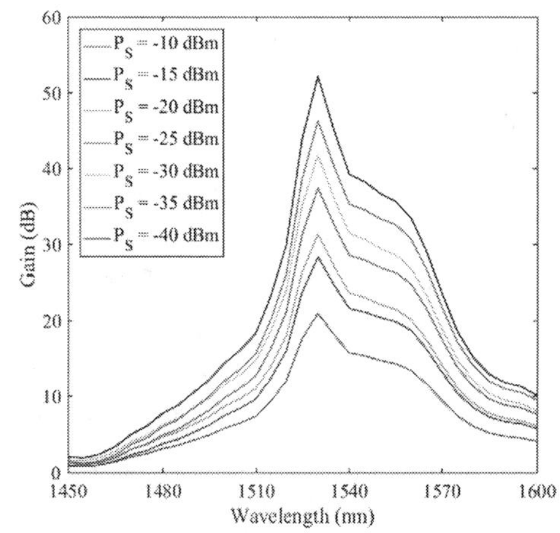

Fig. 7. EDFA gain at different input signal power with fixed EDF length of 20 m and pump power of 100 mW

DOI: 10.1051/epjconf/201716201030

4 Conclusion

We have theoretically studied the effect of EDF parameters to the gain performance of C-band EDFAs. Theoretical analysis and simulation are carried out using the coupled rate and propagation equations, taking into accounts both the ground and excited states. The results show that the gain of EDFA is strongly dependent on the EDF length, pump power and input signal. When the pump power is sufficient enough, the EDFA shows that it could be operated in saturation regimes, leading to the maximum gain.

This work was supported in part by Fundamental Research Grant Scheme (FRGS) Vot 1622 and Research Acculturation Collaborative Effort (RACE) Vot 1509. Both grants are granted by the Ministry of Higher Education (MOHE) Malaysia. This work as well supported in part by Centre for Graduate Studies Universiti Tun Hussein Onn Malaysia (UTHM).

References

1. M. Chen, Z. Meng, Y. Zhang, J. Wang, W. Chen, IEEE Photonics J., **7**, 7029071 (2015)
2. J. Tang, J. Sun, L. Zhao, T. Chen, T. Huang, Y. Zhou, Opt. Express, **19**, 14682–14689 (2011)
3. N.A. Cholan, M.H. Al-Mansoori, A.S.M. Noor, A. Ismail, M.A. Mahdi, Opt. Comm., **329**, 163–167 (2014)
4. N.A.B. Ahmad, S.H. Dahlan, N.A. Cholan, *2016 IEEE 6th International Conference on Photonics (ICP)*, 7510021 (2016)
5. O. De Varona, M. Steinke, D. Kracht, J. Neumann, P. Wessels, Opt. Express, **24**, 24883-24895 (2016)
6. K.S. Thyagarajan, A. Ghatak, *John Wiley & Sons*, **10** (2007)
7. M. Li, K. Liu, W. Jing, G.-D. Peng, J. Opt. Soc. Korea, **14**, 14-21 (2010)
8. Y. Yuan, Y. Yao, J.J. Xiao, Y. Yang, J. Tian, C. Liu, J. Appl. Phys.,**115**, 043102 (2014)
9. X. Dong, P. Shum, N.Q. Ngo, H.-Y. Tam, X. Dong, J. Lightwave Technol., **23**, 1334-1341 (2005)
10. S.D. Emami, H.A. Abdul-Rashid, H. Ahmad, A. Ahmadi, S.W. Harun, Ukr. J. Phys. Opt., **13**, 74-81 (2012)
11. R. Hui, M. O'Sullivan, *Academic Press,* (2009)
12. T. Kasamatsu, Y. Yano, T. Ono, J. Lightwave Technol., **20,** 1826–1838 (2002)

Two dimension MDW OCDMA code cross-correlation for reduction of phase induced intensity noise

Israa Sh. Ahmed[1], *Syed* A. Aljunid [2], *Junita* M. Nordin[1], *Layth A. Khalil Al Dulaimi*[2], *Rima* Matem[1]

[1] Center of Excellent Advanced Communication Engineering, School of Computer & Communication Engineering, Universiti Malaysia Perlis (UniMAP), 01000 Kangar, Perlis.

[2] Embedded Networking and Advance Computing Cluster, School of Computer & Communication Engineering, Universiti Malaysia Perlis (UniMAP) Tingkat 1, Kampus Tetap Pauh, Putra 02600 Arau, Perlis.

Abstract. In this paper, we first review 2-D MDW code cross correlation equations and table to be improved significantly by using code correlation properties. These codes can be used in the synchronous optical CDMA systems for multi access interference cancellation and maximum suppress the phase induced intensity noise. Low P_{sr} is due to the reduction of interference noise that is induced by the 2-D MDW code PIIN suppression. High data rate causes increases in BER, requires high effective power and severely deteriorates the system performance. The 2-D W/T MDW code has an excellent system performance where the value of PIIN is suppressed as low as possible at the optimum P_{sr} with high data bit rate. The 2-D MDW code shows better tolerance to PIIN in comparison to others with enhanced system performance. We prove by numerical analysis that the PIIN maximally suppressed by MDW code through the minimizing property of cross correlation in comparison to 2-D PDC and 2-D MQC OCDMA code.scheme systems.

1. Introduction

Codes with low cross correlation are required in spectral-amplitude-coding Optical Code Division Multiple Access (CDMA) systems since these codes eliminate multi access interference and also suppress the effect of phase-induced intensity noise. However, little research work has been done on such codes although codes with ideal low cross correlation have been studied for many years [1]. Nevertheless, in spectral-amplitude-coding systems, the inherent phase-induced intensity noise (PIIN) and many others affect the whole system performance. PIIN is due to the spontaneous emission of the broadband source, and its effect is proportional to the power of the created photocurrent [2]. Thus, the summation of different users' signals at the receiver's end fluctuates in intensity, which increases the variance of the received signal [3].

To suppress it, the value of should be kept as minor as possible. Therefore, the codes with low cross correlation become attractive. However, the performance is still limited by PIIN [4]. Earlier studies have used various methods to reduce PIIN in SAC-OCDMA systems. Some researchers have reduced the number of MAI to minimize the in-phase cross-correlation. Others have attempted different approaches in solving the problem, such as using saturated semiconductor optical amplifiers (SOA) in the noise-cleaning process [6]. Newly, the authors have proposed the modified-AND subtraction detection technique to overcome the impact of PIIN and MAI in incoherent SAC-OCDMA systems by dividing the spectrum of the utilized code sequences at the decoder branch [6].

Optical code-division numerous entry (OCDMA) is a guaranteeing method to bursty movement network, What's more, it gives asynchronous and security transmissions the middle of different clients [7]. "around various sorts of OCDMA schemes, incoherent frameworks need to be pulled in more consideration over sound ones, because of simpler acknowledgment about encoders Furthermore decoders. However, since those embraced 1-D optical code sequences would unipolar over incoherent systems, those code length ought to make made large will hold preferred auto- What's more cross-correlations[8].

Therefore, 2-D codes need aid recommended should get great correlation aspects for short code length by using other dimensions, for example, such that wavelength, space, thus. Some writers produce 2-D optical codes by creating specific calculations [9] alternately bringing focal point of the maturely produced 1-D optical codes clinched alongside a joined time-spreading also frequency-hopping plan. These codes bring huge cardinality, yet the lengths from claiming fiber delay lines utilized within the coding units would very much longer, Also multiple-access impedance (MAI).

* Corresponding author: israashihab83@gmail.com

2. Review of 2-D MDW Code Cross Correlation

The 2-D W/T MDW OCDMA synchronous number of clients is dictated by the duplication of wavelengths M and time—chips N. All in all, an expanded number of clients results in an extension in code length, effective power, reducing the bit rate, the increment in impedance acquainted with the framework which causes framework decay. The number of Hamming weight or code weight (w) depends on the size of the chosen codes. High Hamming weight of "1", in the code sequence, indicates an increase in short optical pulses scattered along the spectrum. The size of the code is indicated by the code length indicator. Usually, high code size will result in an increase in code length [8]. A good code sequence should have high code sequences, reasonable code length, low code weight and BER below 10^{-9} error floor. Bit rate or data rate is a measure of a number of bits transmitted per second over a fiber channel. Bit rate reflects the information carrying capacity of a fiber communications link. High data rate will increase the BER and deteriorate the system performance. Noise power is the noise interference that exists in the system other than the desired signal. he wavelength-time code cross-correlation is generated differently compared to the 1-D OCDMA code[9].

Let;

$X=[x_0,x_1,x_2,...,x_{M-1}]$

$Y=[y_0,y_1,y_2,...,y_{N-1}]$, be the code sequence of the 2-D MDW OCDMA code.

The cross-correlation of 2-D W/T MDW can be developed by proffering the four characteristic matrices of $A^{(d)}$, where d is the subset of 0,1,2 or 3, $d \in \{1,2,3\}$, are defined as:

$A^{(0)} = Y^T X$, $A^{(1)} = Y^T \overline{X}$, $A^{(2)} = \overline{Y} X$ and $A^{(3)} = \overline{Y^T}\overline{X}$.

The $\overline{X}$ and $\overline{Y}$ are the complement of $X=[x_0,x_1,x_2,...,x_{M-1}]$ and $Y=[y_0,y_1,y_2,...,y_{N-1}]$ respectively. Example, if X is equal **[000011011]** the complement of X will be the complement of all elements in X which is $\overline{X}$ = **[111100100]**.

The cross-correlation of 2-D W/T MDW code $A^{(d)}$ and $A_{g,h}$ can be represented by [10]:

$$R^{(d)}(g,h) = \sum_{i=0}^{M-1}\sum_{j=0}^{N-1} a_{ij}^{(d)} a_{(i+g)(j+h)} \quad (1)$$

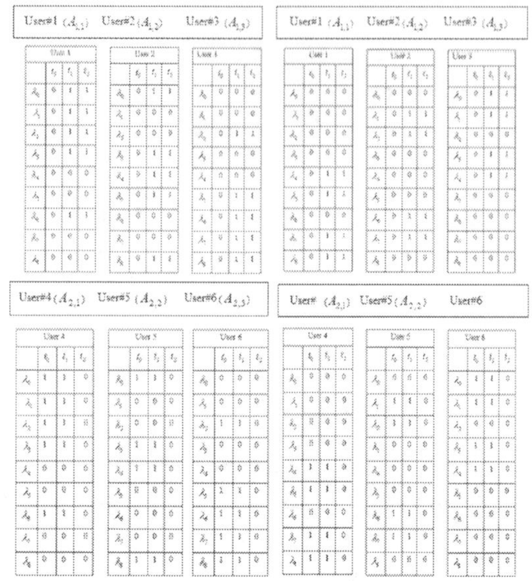

Fig.1. Generate Cross-Correlation of 2-D W/T MDW OCDMA code for $d=2$.

Where $a_{ij}^{(d)}$ is $(i,j)_{th}$ of $A^{(d)}$ and a(i+g)(j+h) is the element $(i,j)_{th}$ of $A_{g,h}$, g ϵ (1,2,3,.., $M-1$) and h ϵ (1,2,3,......, $N-1$).

From Equation 1, the cross-correlation Table 1 of the code was developed. In Fig. 1 the cross-correlation which was developed for d is equal to 0; $A^{(2)} = \overline{Y} X$. If $k_1 = 4$ and $k_2 = 2$. The complete calculation can be done by setting the following element, d=0, 1,2,3, for the example that uses d = 2. From the Equation (1) and Figure 1, $A^{(2)} = \overline{Y} X$ where $k_1 = 4$ and $k_2 = 2$ it is necessary to multiply the right and the left side of the above of n Figure 1, then Table 1 of cross correlation is the result as shown.

The multiplication configuration can be simplified as follow:

$$\left.\begin{matrix} A_{1,1}\times A_{1,1} \\ A_{1,2}\times A_{1,2} \\ A_{1,3}\times A_{1,3} \\ A_{2,1}\times A_{2,1} \\ A_{2,2}\times A_{2,2} \\ A_{2,3}\times A_{2,3} \end{matrix}\right\} 0 \qquad \left.\begin{matrix} A_{1,1}\times A_{1,2} \\ A_{1,1}\times A_{1,3} \\ A_{2,1}\times A_{2,3} \\ A_{2,1}\times A_{2,2} \\ A_{2,2}\times A_{2,1} \\ A_{2,3}\times A_{2,3} \end{matrix}\right\} 0$$

$$\left.\begin{matrix} A_{1,1}\times A_{2,1} \\ A_{1,2}\times A_{2,2} \\ A_{1,3}\times A_{2,3} \\ A_{2,1}\times A_{1,1} \\ A_{2,2}\times A_{1,2} \\ A_{2,3}\times A_{1,3} \end{matrix}\right\}4 \qquad \left.\begin{matrix} A_{1,1}\times A_{2,2} \\ A_{2,1}\times A_{1,2} \\ A_{1,2}\times A_{1,3} \\ A_{1,3}\times A_{2,1} \\ A_{2,3}\times A_{1,2} \\ A_{2,2}\times A_{1,1} \end{matrix}\right\}1$$

Table 1. The calculation of 2-D W/T MDW code cross-correlation.

$X_{g,h}$	$R^{(2)}_{(g,h)}$	$R^{(2)}_{(g,h)}$
g = 0 ,h = 0	0	0
g= 0 ,h ≠ 0	0	0
g ≠ 0 ,h = 0	$k_2(k_1\text{-}1)$	4
g ≠ 0 ,h ≠ 0	$k_1\text{-}1$	1

Table 2 shows the derivation of the equation (1), this table is the calculation of 2-D W/T MDW code cross-correlation. A unique 2-D code cross-correlation function was generated from this table. From Table 2, the tabulated cross-correlation of 2-D W/T MDW OCDMA codes were generated from $R^{(d)}(g,h)$ of different d values, where $R^{(3)}(g,h)$ had a non-zero value only when $g\neq 0\cap h\neq 0$ and $R^{(3)}(g,h)$ were developed. In the end, 2-D W/T MDW OCDMA cross-correlation function was derived by using $R^{(3)}(g,h)$ to eliminate the influence due to $A_{g,h}$ from $R^{(0)}(g,h)$, $R^{(1)}(g,h)$ and $R^{(2)}(g,h)$, at $g\neq 0\cap h\neq 0$

Table 2. Cross-Correlation of 2-D W/T MDW OCDMA code.

$X_{g,h}$	$R^{(0)}_{(g,h)}$	$R^{(1)}_{(g,h)}$	$R^{(2)}_{(g,h)}$	$R^{(3)}_{(g,h)}$
g = 0 ,h = 0	$k_1\,k_2$	0	0	0
g= 0 ,h ≠ 0	k_1	k_1	0	0
g ≠ 0 ,h = 0	k_2	0	$k_2(k_1-1)$	0
g ≠ 0 ,h ≠ 0	1	1	k_1-1	k_1-1

The $R^{(1)}_{(g,h)}$ and $R^{(2)}_{(g,h)}$ consist of the non-zero value when g =0 ∩h ≠0, g=0∩h=0 respectively.

Hence, this can be used to eliminate the influence caused by $A_{g,h}$ on $g=0\cap h\neq 0$, $g\neq 0\cap h=0$ from $R^{(0)}(g,h)$ and subsequently the following equation is developed [11]:

$$R_1^{(0)}(g,h)-R_1^{(1)}(g,h)-\frac{R_1^{(2)}(g,h)}{(k_1-1)}=\begin{cases}k_1k_2, for, g=0 and, h=0\\ 0, otherwise\end{cases} \tag{2}$$

the new 2-D W/T MDW OCDMA cross-correlation equation is developed as:

$$R^{(0)}(g,h)-R^{(1)}(g,h)-\frac{R^{(2)}(g,h)}{(k_1-1)}+\frac{R^{(3)}(g,h)}{(k_1-1)}=\begin{cases}k_1k_2, for, g=0 and, h=0\\ 0, otherwise\end{cases} \tag{3}$$

This equation is used for MAI mitigation and PIIN suppression in the overall 2-D W/T MDW OCDMA code.

PIIN is the result of mixing an incoherent light field's incident which causes intensity noise at PD output []. The power of PIIN that exists in a photocurrent of the receiver can be written as [11, 12]:

$$I^2_{PIIN}=B_rI_r^2\tau_r=\frac{B_r\Re^2P_{sr}^2\left\lfloor k_1k_2(MN-1)^2+k_2(W-1)(M-1)\right\rfloor}{2M\Delta fk_2^2(MN-1)^2} \tag{4}$$

$$I^2_{thermal}=\frac{4K_bT_nB_r}{R_L} \tag{5}$$

$$I^2_{shot}=2eB_r(I_0+I_1+I_2+I_3)$$

$$=2eB_r\frac{\Re P_{sr}}{Mk_2(MN-1)}\left\{k_1k_2(MN-1)+2k_1(W-1)(N-1)+2k_2(W-1)(M-1)+4(W-1)(M-1)(N-1)\right\} \tag{6}$$

$$SNR=\frac{I_r^2}{I^2_{PIIN}+I^2_{shot}+I^2_{thermal}} \tag{7}$$

$$BER(M)=\frac{1}{2}erfc(\sqrt{\frac{SNR}{8}} \tag{8}$$

here $\Re$ is the response of the photodiode, $\Re=\eta e/hf_0$, η is the quantum efficiency of the PD, h is the Plank's constant and f_0 is the central frequency of the incident light.

3. Results and Discussion

The In the theoretical calculation, the link parameter of the Equation (10) that has been used is shown in Table 3 These parameters are the same as used in Arief & Abdullah (2013).

Table 3. Parameters Used in Numerical Calculation [11].

Parameters Used in Numerical Calculation	
PD quantum efficiency	η=0.75
Spectral width of broadband light source	$\Delta\lambda$=30nm($\Delta\lambda$=3.75THz)
Operating wavelength	λ_o=1.55μm
Electrical bandwidth	B=311 MHz
Data transmission rate	R_b =0.622 Gbps, 1.1 Gbps, 2.5 Gbps and 10 Gbps.
Receiver noise temperature	T_n =300K
Receiver load resistor	R_L=1030Ω
Boltzmann's constant	K_b=1.38×10^-23 W/K/Hz
Electron charge	e=1.60217646×10^-19 coulomb
Light velocity	C=3×10^8m/s

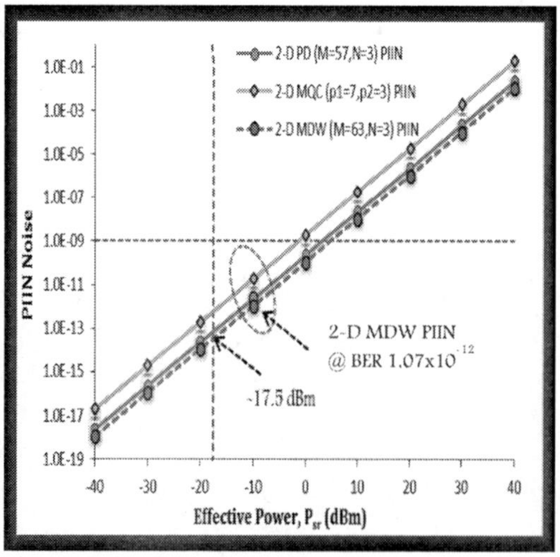

Fig. 2. BER PIIN noise versus effective power (P_{sr}) for 2-D MDW*(M=63,N=3)*, 2-D PDC*(M=57,N=3)*, 2-D MQC*(p_1=7,p_2=3)*.

In figure 2, it observed that PIIN curves are linearly increasing at constant slope as P_{sr} expanded. The BER increases by almost two order of magnitude system performance as P_{sr} is increased by a factor of 10 dBm. From the figure which that shows 2-D MDW improves 2.15 and 17.94 folds compare to 2-D PDC and 2-D MQC and details in Table 2. 2-D MDW code successfully suppresses the PIIN results in BER at the lowest curve in comparison to 2-D PDC and 2-D MQC code. At-10 dBm below 10^{-9} error floor, 2-D MDW, 2-D PDC and 2-D MQC PIIN values are 1.08 $\times 10^{-1}$, 2.5 $\times 10^{-13}$ and 2×10^{-11} respectively. Code with lowest PIIN suppuration will improve the BER and overall system performance. Because of the property of cross-correlation so can achieve the optimum PIIN suppuration when the P_{sr} is at -17.5 dBm .

Table 4. PIIN Improvement

Code	PIIN at -10.0 dBm	Improvement
2-D MDW*(63x3)*	1.07×10^{-12}	2.15 and 17.94 fold smaller compare to 2-D PDC and 2-D MQC.
2-D PDC*(57x3)*	2.30×10^{-12}	
2-D MQC*(7x3)*	1.92×10^{-11}	

Figure 3 shows the effect of bit rate on PIIN for 2-D MDW (*M=3, N=108*) at different data bit rates, 0.622 Gbps, 1.1 Gbps, 2.5 Gbps and 10 Gbps. The linear curves show an increase in PIIN noise as the data bit rate increase. Maximum PIIN suppression for low bit rate. The magnitude of PIIN system performance Suppressed by the factor of 1.3, 3.2 and 6.9 as bit rate decreases; 10.0, 5.0, 1.1 and 0.662 Gbps. High bit rate reduces total photocurrent and increases total noise power.

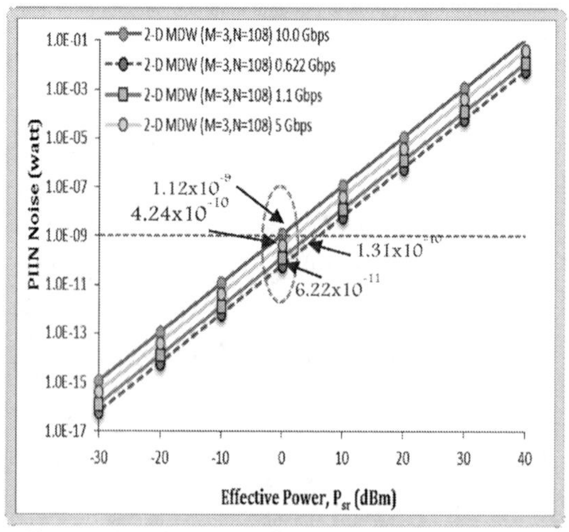

Fig. 3. PIIN noise versus effective power (P_{sr}) for 2-D MDW (M=3, N=108) at a different data rate; 0.622 Gbps, 1.1 Gbps, 2.5 Gbps and 10 Gbps.

4. Conclusion

Analyzing the impact of PIIN suppression on 2-D MDW OCDMA system compare to others of 2-D OCDMA

DOI: 10.1051/epjconf/201716201031

systems is done in this paper. 2-D MDW code properties improved the ability to mitigation MAI and maximally suppression of PIIN at the low optimum effective transmitted power. The property of minimum cross-correlation caused optimum PIIN suppression. PIIN is the dominant factor in 2-D MDW and others performance deterioration. However, 2-D MDW scheme has better acceptance applied to suppression PIIN. The 2-D MDW code property restrains the system degradation. Successfully suppressed PIIN by magnitude of 2.15 and 17.94 with comparison to 2-D PDC and 2-D MQC code

References

1. J. Huang, C. Yang. IEEE Transactions on Comm. , **54** ,1, pp 150-8 (2006).
2. H. Al-Khafaji, S. Aljunid, A. Amphawan, H. F adhil, A. Safar. Int. J. of Eng. and Tech. (IJET). **5**, 1, pp 95-100 (2013).
3. N. Keraf, S. Aljunid, A. Arief, M. Nurol, M. Anuar, C. Rashidi, P. Ehkan.. *IEEE 5th International Conference in Photonics,* p166-168 (2014).
4. H. Fathallah,., A. Bentrcia,., & H. Seleem. IET Communications, **8**(13), 2393-2403. (2014).
5. S. Panda, U. Bhanja. Int. J. of Electro. and Comm. Eng., **9**, 11, 2015
6. H, Al-Khafaji. Novel detection techniques for SAC-OCDMA system. Doctoral dissertation, Universiti Malaysia Perlis (UniMAP) (2014).
7. S. Singh. Optical Fiber Tech.. **31**;32:123-8 (2016).
8. I. Ahmed, S. Aljunid, C. Rashidi, L. Al Dulaimi. 3rd *International Conference Electronic Design (ICED),* 2016 ,159-163 (2016).
9. I. Ahmed, S. Aljunid., C. Rashidi, & L. Al Dulaimi. SSRN Electronic Journal (IJCCIE), **3**,1, p73-76 (2016).
10. I. Ahmed, S. Aljunid, C. Rashidi, L. Al dulaimi.. Australian J. of Basic and Appl. Sci. **11**,1,pp 83-90 (2017).
11. A. Arief, S. AlJunid , M. Anuar , Junita M. , R. Ahmad , F. .Ghani 2012 IEEE 3rd International Conference on Electronics Design, pp.275-279 (2012).
12. A. Arief, S. AlJunid , M. Anuar , Junita M. , R. Ahmad . J International for Light and Electron Optics, **24**, 19, (2012).

DOI: 10.1051/epjconf/201716201032

Performance study on the effect of filter curve in CWDM System for the access network

N. Ali[1,2,3*], *N. A* Rahman[1], *N. A. M.* Ahmad Hambali[1,2], and *C. B. M.* Rashidi[3]

[1]School of Microelectronics Engineering, Universiti Malaysia Perlis (UniMAP), 02600 Arau, Perlis, Malaysia
[2]Semiconductor Photonics & Integrated Lightwave Systems (SPILS), School of Microelectronic Engineering, Universiti Malaysia Perlis, Pauh Putra Main Campus, 02600 Arau, Perlis, Malaysia
[3]Advanced Communication Engineering, Centre of Excellence-School of Computer and Communication Engineering, Universiti Malaysia Perlis

Abstract. This paper presents the study on the effect of filter variation on the coarse wavelength division multiplexing (CWDM) system. The filter curve will affect the performance of the CWDM system due to changes of received power lever and isolation of the signal. The significant impact on the received power level and isolation can be found when the required signal is isolated from unwanted signal by the steep curve of filter. As a result, BER of $1.0\text{x}\ 10^{-12}$ was obtained corresponding to receive power level of -24.27 dBm with isolation of 23.22 dB. When the wavelength spacing is reduced to 1nm, the isolation is only 11.30 dB and BER increased to $5.49\text{x}10^{-7}$ with a received power of -15.39 dBm.

1 Introduction

Internet growth and expansion has becoming dense in access network [1, 7]. The migration of copper to optical network is already an obsolete solution. The access network is not for long haul transmission system, which loss in terms of power penalty will not contribute an impact to data quality during transmission. The main consideration is the cost of the system because the optical network is expensive [2, 8]. Utilizing previous installed network can save cost in optical system. The CWDM system gives opportunity for the existing network installed to be fully utilized. It is considered as a cost effective option which more network can be installed in one transmission system and the cost of the equipments i.e., multiplexer and filter is economical compare to other WDM system[3]. The filter used in the CWDM system is normally wide bandwidth where the price is less. The filter is not required to be narrow due to the wider wavelength spacing between channels. The wide filter shape also prevent lost due to the wavelength shifting occur during transmission. The isolation of the signal is determined by the filter shape. Filter with steeper shape produces higher isolation. The spacing of the channel also can reduce the isolation level if the spacing between the channels is small.

For CWDM, spacing is larger and so is the filter bandwidth. Hence, inter channel crosstalk is unlikely to happen in this system. Isolate the adjacent signal will also reduce the received power and the transmission distance as well. Therefore, it is important to study the filter effect on performance of the access network for CWDM system. In this paper, the received power level and isolation of the signal is investigated where the variation of these parameters contribute by the filter curve. The shape of the filter in the system setup (refer to Figure 1) is the main contributor to the isolation and the received power level of the system. These parameters will contribute on the performance of the system which BER of the system is investigated due to changes of isolation and received power level.

In this paper, testing was done using standard SDH data rate at 2.5 Gbps. The parameters under study were received power and isolation with respect to filter's curve

2 Experiment Setup

The experiment setup to analyze the performance of the new CWDM system is shown in Figure 1.

* Corresponding author: norshamsuri@unimap.edu.my

 DOI: 10.1051/epjconf/201716201032

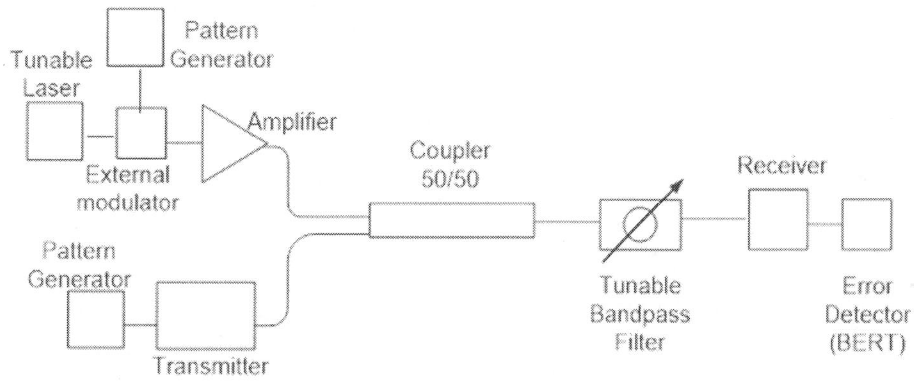

Fig. 1. Experiment setup

This setup is a point to point transmission where the distance of the transmission is based on power budget calculation. The main blocks of the experiment are the transmitter, coupler, tuneable bandpass filter and the receiver. Coupler acts as the multiplexer while the tuneable bandpass filter acts as the de-multiplexer. The main signal is modulated by the external modulator, while the adjacent signal is modulated from the 2.5Gbps transmitter. The main signal has to be amplified since the external modulator attenuates the transmitted power by 8.3dB. Both signals are then, multiplexed by coupler before being filtered by the tuneable bandpass filter to recover back the main signal. The performance of this setup is determined by the BER value detected by BER tester at the receiving side. The BER calculation is based on the 95% confidence level at 1e-12 error rates with 2 minutes requirement for every data collected.

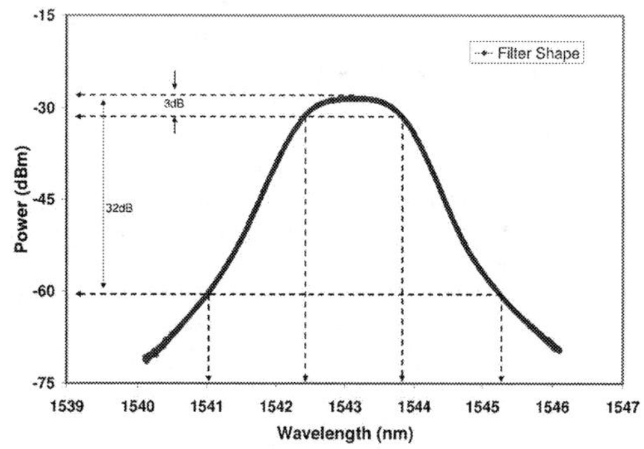

Fig. 2. Broad filter curve

3 Result and Discussion

Figure 2 below shows the spectrum of the filter where the 3 dB bandwidth of the filter is approximately 1.5 nm. The curve of the filter is shown to be steeper at the upper side of spectrum and less steep at the leg of the spectrum. From the graph, the insertion loss of the filter can be determined by investigating the spectrum or the shape of filter. The changes of the receive power level is depending on the form of the filter shape. In order to isolate the adjacent signal, the filter will be adjusted to the minimum power level of adjacent signal

To this extend, the character of the filter slope contribute most to the changes in the received power level. The isolation of signal is more effective when the isolation is done between the 3 dB points of the curve (at -35 dBm as shown in Figure 2) and at 31 dB point (at -60 dBm as shown in Figure 1) of the curve. This is the range where the curve of the filter is steeper and the isolation of adjacent signal will not reduce the power level of received signal, hence higher isolation value can be achieved. The system is tested using 2.5 Gbps data rate to check the system stability which introduces the broad filter. Normally the access networks support data rates up to 622 Mbps.

From the study, the new design provides up to 1.6 nm wavelength spacing as compared to normal CWDM system with minimum spacing of 20 nm[4]. The effect of filter's curve to the system is shown in Figure 3. The graph shows that when the wavelength spacing is reduced, the performance of the system is also reduced as shown by the increase of BER.

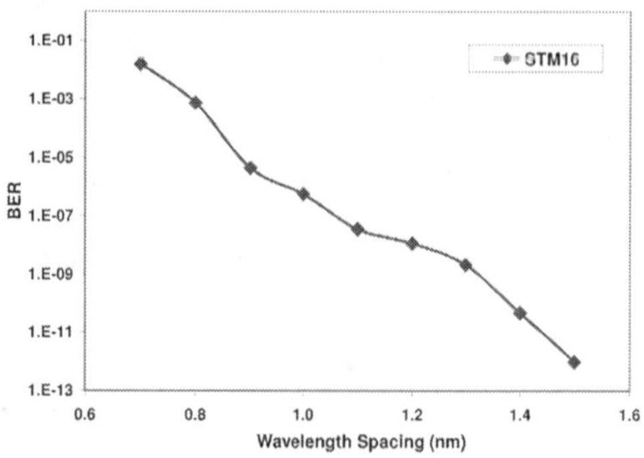

Fig. 3. Wavelength spacing effect on the performance of BER

The receive power level and the isolation of the signal is considered as the main design parameters because these are the main performance indicators for any optical receiver. Figure 4 presents the performance of the CWDM system when the isolation of the signal is reduced. From the graph, the minimum isolation that can be detected by the receiver with the least acceptable BER (≈10-11) [5] any errors is 17.24 dB. However, this data should be verified with the total received power since both of these parameters are interrelated. The isolation of the signal can be smaller than 17.24 dB but the received power level should be increased in order to maintain the data quality. The reason is that, the receiver has minimum sensitivity level, if the received power is lower than the sensitivity value, the receiver cannot distinguish between the actual signal or noises, even if the isolation of the signal is still acceptable.

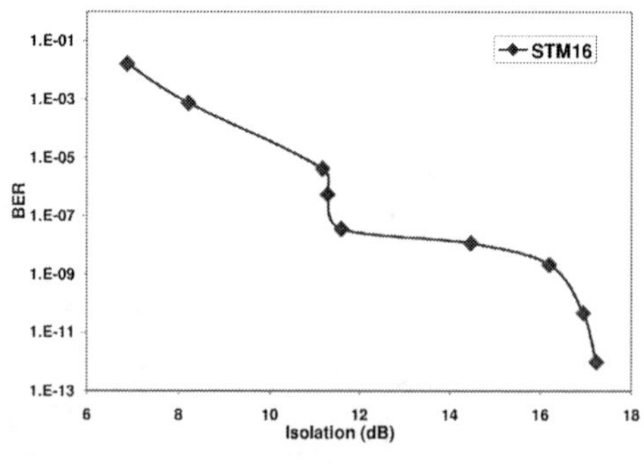

Fig. 4. The performance of isolation adjacent signal with respect to BER

On the other hand, the received power level can also be as small as -10.41 dBm as shown in Figure 5. In order to maintain the signal quality, the isolation of the signal should be bigger to ensure the receiver detects the signal correctly. These values are affected by the filter shape, where the steeper the shape of the filter curve, the better the quality of signal received [6]. Therefore, different type and characteristic of filter will give different value of isolation and receive power level and, it hence produced a different quality of performance.

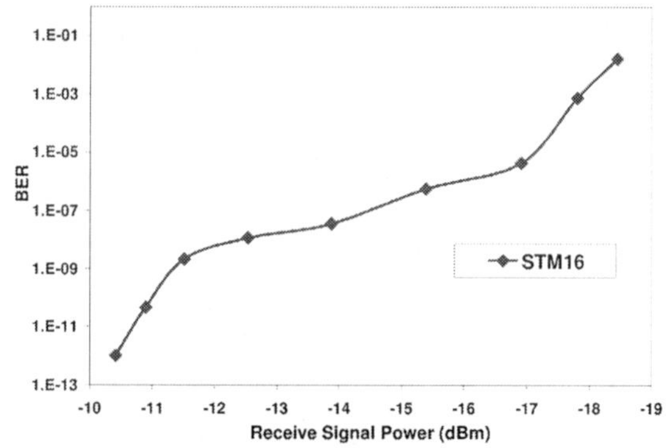

Fig. 5. The BER performance affected by the total received power.

Figure 6 shows the distance that can be establish by this new design CWDM system. The new design can reach up to 40 km with acceptable BER performance of the system compared to the normal CWDM system which only support up to10 km; this is improvement of 400% distance covered. This is done based on power transmitted around -7 dBm with power margin of 13 dB. The other advantage of this new design it utilizes wavelength spacing of 1.6 nm as compared to the normal design at distance of 40 km. By implementing this new system, the cost for the each link can be reduced and the load of the link can be increased in order to support the high capacity access network.

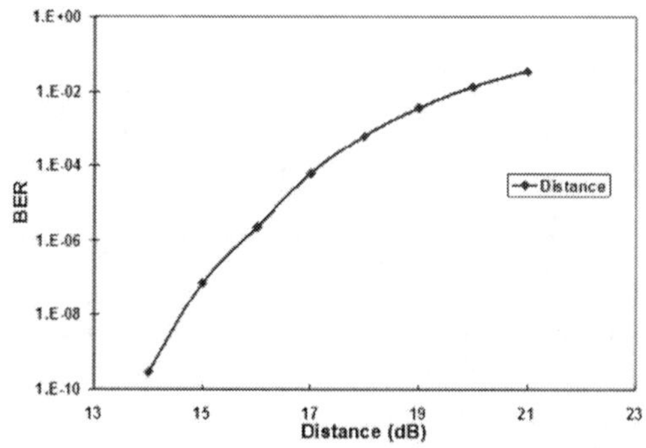

Fig. 6. Distance of transmission system due to BER performance

Conclusion

The study demonstrates the effect of filter curve on the performance of the new CWDM system on the access network. The new system can achieved a distance of 40 km at 1.6 nm spacing with data rate of 2.5 Gbps. Furthermore, the new design optimizes the channel spacing, and the filter curves until it gives a better performance signal to the existing WDM system. As a result BER of $1.0x\ 10^{-12}$ was obtained corresponding to receive power level of -24.27 dBm with isolation of 23.22 dB. When the wavelength spacing is reduced to 1 nm, the isolation is only 11.30 dB and BER increased to $5.49x10^{-7}$ with a received power of -15.39 dBm.

The author would like to acknowledge this work was supported by the High Ministry of Education Malaysia. The work has been done in 2003 and yet not been published. The author would like to thanks to Late Prof Dr Mohamed Khazani Abdullah and his team at Universiti Putra Malaysia previously in 2003. Most of the work was done in Universiti Putra Malaysia previously.

References

1. Elnegaard K. N., Stordahl K., Ims L. A.; *International Symposium on Services and Local access, ISSLS 2000*, Stockholm, Sweden, 18-23 June 2000.F. De Lillo, F. Cecconi, G. Lacorata, A. Vulpiani, EPL, **84** (2008).
2. G. Held "Best Practice Series Network Design Principles and Applications", Auerbach Publications, Florida, (2000).
3. Roka, R.; *Mobile Future and Symposium on Trends in Communications, 2003, SympoTIC '03. Joint First Workshop* Oct. 2003, pp160 – 162.(2003).
4. RBN, "Characteristic of CWDM: Roots, Current status and Future Opportunity", White Paper, Redfren Broadband Networks, San Francisco,(2002).
5. Djafar K. Mynbaev, and Loell L. Scheiner.: "Fiber-Optic Communication Technology." Prentice-Hall, New Jersey,(2001).
6. Leo, C.J.; Ramana, P.V.; Sudharsanam, K.; *Electronics Packaging Technology, 2003 5th Conference (EPTC 2003)* , pp 647-651, (2003).
7. Pastor-Satorras, Romualdo, Alexei Vázquez, and Alessandro Vespignani, Phys. Rev. Let. **87**,25 (2001): 258701.
8. de Laat, Cees, Erik Radius, and Steven Wallace. Future Generation Computer Systems, **19**,6 pp 999-1008, (2003).

Quality of polarization entanglement in spontaneous parametric down conversion

N Ali[1,2,4*], *S* Soekardjo[5], *S.* Saharudin[3] and *M. R. B.* Wahiddin[6]

[1]School of Microelectronics Engineering, Universiti Malaysia Perlis (UniMAP), 02600 Arau, Perlis, Malaysia
[2]Semiconductor Photonics & Integrated Lightwave Systems (SPILS), School of Microelectronic Engineering, Universiti Malaysia Perlis, Pauh Putra Main Campus, 02600 Arau, Perlis, Malaysia
[3]Nano Photonics and Nano Electronics, Mimos Berhad, Technology Park Malaysia, 57000 Kuala Lumpur
[4]Advanced Communication Engineering, Centre of Excellence-School of Computer and Communication Engineering, Universiti Malaysia Perlis
[5]Department of Physics, Kulliyyah of Science, International Islamic University Malaysia, Jalan Sultan Ahmad Shah, Bandar Indera Mahkota 25200 Kuantan, Pahang.
[6]Dept. of CS and Deputy Rector (Research and Innovation) at the International Islamic University Malaysia (IIUM)

Abstract. We experimentally demonstrated a high degree of polarization entanglement known as entanglement visibility through spontaneous parametric down conversion process pumped by a femtosecond laser. The entangled-photon pair was obtained using two type-I BBO crystal. The down-converted photons from these crystals demonstrates a high visibility of 98.7% (Θ2 = 0°) and 90% (Θ2 = 22.5°). These results are in agreement with the theory which expects high visibility from such arrangement.

1 Introduction

Entanglement, a counterintuitive phenomenon of quantum mechanics plays a crucial role in many interesting application in quantum mechanics such as in the development of quantum computation, quantum information, and quantum cryptography [1]. It is also the main point of the debate in foundation issues and interpretation of quantum mechanics since firstly pointed out by Einstein, Podolsky and Rosen in their renowned paper [2]. Entangled state of spin-1/2 particle has been utilized in Ekert's proposal [3] to realize quantum cryptographic protocol based on the notion of quantum non-locality (i.e. Bell theorem). Such cryptographic protocol was experimentally demonstrated for the first time using polarization-entangled photon [4]. Entanglement is also an efficient alternative source for single-particle quantum key distribution protocols (see for instance: [5]).
Furthermore, it has been proved that the protocol using entangled photon has the potential to extend the length of communication distance in the presence of realistic experimental imperfection [6]. These significant examples strongly highlight the importance of entangled photon generation. In today's technology, the most accessible and controllable source of entanglement comes from spontaneous parametric down conversion (SPDC) through stochastic process in the interaction between light and nonlinear optical materials [7]. It is necessary to mention here that the study on optimization of the process will contribute much to the development of unconditional secure communication.
Spontaneous parametric down conversion process in nonlinear optical materials is utilized in order to create one of four Bell states which are maximally entangled. The states can be mathematically written as follows:

$$\left|\Phi_{12}^{\pm}\right\rangle = \frac{1}{\sqrt{2}}\left(\left|H\right\rangle_1 \otimes \left|V\right\rangle_2 \pm \left|V\right\rangle_1 \otimes \left|H\right\rangle_2\right),$$

$$\left|\Psi_{12}^{\pm}\right\rangle = \frac{1}{\sqrt{2}}\left(\left|H\right\rangle_1 \otimes \left|H\right\rangle_2 \pm \left|H\right\rangle_1 \otimes \left|H\right\rangle_2\right)$$

where, $\left|H\right\rangle\left(\left|V\right\rangle\right)$ denotes Horizontal (Vertical) polarization of the photons. We use the notation ⊗, as a tensor product describing a composite system of two spatially separated photons.
In this letter, we report experimental results on the entanglement polarization from type-I phase matching down conversion processes by using femtosecond-pulse laser pumping. We shall demonstrate that highly entanglement polarization states with sufficiently large flux were obtained when an appropriate temporal compensation was given in the pump pulses. The resulted will be shown to be consistent with the theory of visibility which incorporates the coincidence rate as a function of the polarization analyzer angle.

* Corresponding author: norshamsuri@unimap.edu.my

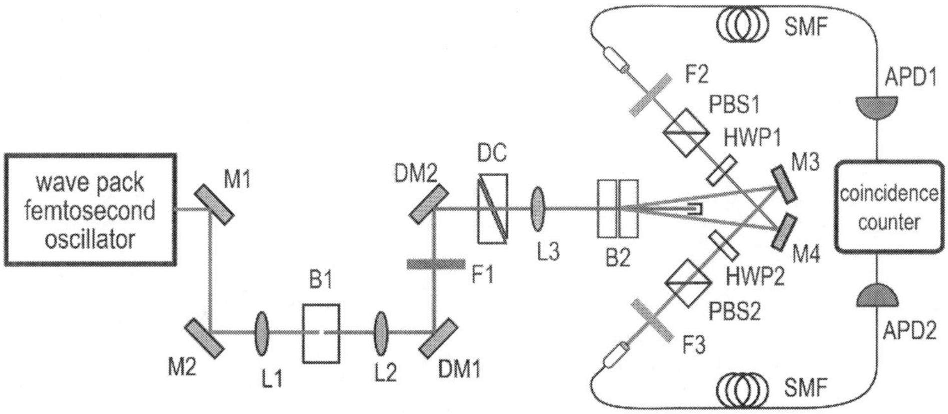

Fig. 1. Schematic of experimental setup used for generate polarization-entangled photon from spontaneous parametric down-conversion processes in the two type-I β-BBO (β− BaB_2O_4 - Beta Barium Borate) crystals. M, mirror; L, lens; B, β-Barium Borate crystal; F, color filter; DM, dichroic mirror; DC, delay compensator; HWP, half-wave plate; PBS, polarizing beam splitter; SMF, single mode fiber; APD, avalanche photo diode detector

2 Experimental setup

An experimental setup for generation of polarization entangles state, by spontaneous parametric down-conversion (SPDC) processes in a nonlinear crystal, is shown in Fig 1.
The configuration uses an active mode-locked femtosecond Sapphire oscillator emitting pulses of light at 794 nm. This radiation was frequency doubled by using 1 mm type-I β-BBO (β−BaB2O4 - Beta Barium Borate) crystal (B1), cut for collinear second harmonic generation (SHG) processes, to provided 200 fsec pulses (FWHM) at center-wavelength of 397 nm, with a repetition rate of 80 MHz and maximum power of ~300 mW. A pair of dichroic mirrors (M3 and M4) and a color glass filter (F1) were used to remove the residual fundamental wavelength coming from the femtosecond sapphire oscillator.

The cascaded type-I BBO crystal (B2) was placed at the waist position of the sequence femto second pulse collimated by lenses L2 and L3. Each crystal was designed in certain cut-angle to achieve phase matching condition for non-collinear frequency degenerate down conversion. The crystal are rotated with respect to other in such a way that horizontally polarized pump light generates pair of vertically polarized photon in the first crystal, and vertically polarized pump light generate pairs of horizontally polarized photon in the second crystal [7]. In order to generate a pair of polarized entangled photon we used pump beam polarized at 45° with respect to the crystal axis. By neglecting losses from passing through the first crystal, this polarized pump photon will be equally likely to down-convert in either crystal. Nevertheless, because of the combined effects of group velocity dispersion and birefringence in the two crystals, the space-time components of the two-photon state associated with the polarized states and are expected to be temporally displaced after the crystals. As a result, information about "which-polarization" the emitted photon pair has, as well as "which crystal" each pair originated from, may be available from the arrival time of the photon at the detector. On the other hand, the effective polarization entanglement requires the suppression of any distinguishing information in the other degrees of freedom that can provide potential information about the emitted pair have.

To eliminate this distinguishing space-time information, a polarization dependent optical delay line for the pump was inserted before the crystal, which is denoted as the delay compensator (DC) in Figure 1. The delay compensator was used as the pre-compensator for group velocity mismatch between ordinary and extraordinary rays in the BBO crystal, so that the state associated with $|HH\rangle$ created in the first crystal relative to the state associated with $|VV\rangle$ created in second crystal overlapped with them temporally. As the result, a two-photon Bell state $|VV\rangle + |HH\rangle$, can be directly created.

To experimentally demonstrate the polarization entanglement of the collected photon pairs, their polarization correlations in two conjugate bases were measured. This was done by directing the down-converted light into adjustable polarization analyzers, each consisting of a polarizing beam-splitter cube (PBS) preceded by a rotatable half-wave plate (HWP). The residual violet light coming from the first BBO crystal was blocked using long-pass filters. After passing through the analyzers, the photons were coupled into single-mode fibers by using the aspheric coupling lenses

DOI: 10.1051/epjconf/201716201033

with focal length of $f = 7.5$ mm, the desired spectral bandwidth and the numerical aperture of the optical fiber, and then detected with passively quenched silicon avalanche photon diodes (Si-APDs). The detectors were linked to a time-to-amplitude converter for a record of coincidence counts. To keep the experimental setup compact, two mirrors (M3 and M4) were used to fold the paths of the fluorescence beams.

3 Results and Discussions

The experiments were carried out to confirm whether the polarized-entangled photon state $|\Psi^+\rangle$ was successfully prepared. The first measurement was performed by setting the azimuth angle of HWP in path-2 θ_2 to be 0° and the HWP in path-1 rotated. During the second measurement, the azimuth angle of HWP in path-2 θ_2 was set to 22.5°, with the azimuth angle of HWP in path- 1 rotated. This setup corresponds to the projection measurement onto $|L\rangle \otimes |\theta\rangle$ or $|R\rangle \otimes |\theta\rangle$ where $|L\rangle = (|H\rangle + i|V\rangle)/\sqrt{2}$ and $|R\rangle = (|H\rangle - i|V\rangle)/\sqrt{2}$, while $|\theta\rangle = \cos\theta|H\rangle + i\sin\theta|V\rangle$. Since the density matrix of the states approximately described by $\hat{\rho} = p|\psi^+\rangle\langle\psi^+| + q|\psi^-\rangle\langle\psi^-|$ where p and q are probabilities of finding our system in the state $|\psi^+\rangle$ and $|\psi^-\rangle$, respectively, the state should exhibit interference in coincidence rates for the two detector in proportion to $R_c(\theta_1, \theta_2) \propto \cos^2(\theta_1 - \theta_2)$.

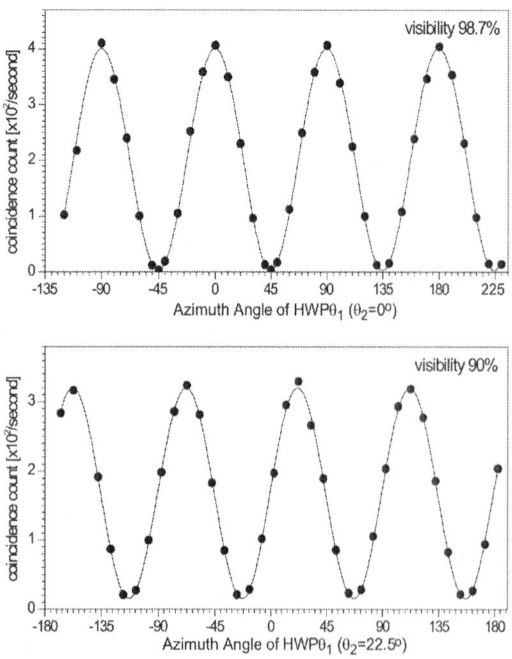

Fig. 2: Plot of coincidence rate as a function of idler polarization analyzer angle θ_1. (a) Signal polarization analyzer was set at $\theta_2 = 0°$. The solid curve is a best fit, with visibility of V = 98.7%. (b) Coincidence rate for the signal at polarization analyzer at $\theta_2 = 22.5°$, with the visibility of 90%.

To evaluate the quality of a polarization entangled-photon state, we use the polarization visibility equation $V = (R_{cMax} - R_{cMin})/(R_{cMax} + R_{cMin})$. The equation states that one should observe high-visibility modulation in polarization correlation measurement for certain values of θ_1 and θ_2. The visibility was determined by a fit of the data with cosine-squared function of coincidence rates. Figure 2 shows the coincidence rate as a function of the HWM angle in path-2 and the evaluated visibility values.

The solid lines in Figure 2 are coincidence rates curve fitted to the measured data (dark circles). In this experiment, the values of visibility for the azimuth angles HWP in path-2, θ_2 of 0° and 22.5° were 98.7% and 90%, respectively. Since the visibility of an observed interference pattern gives another convenient measure for the entanglement, this result clearly demonstrate that the high degree of polarization entanglement has been achieved in this experiment setup.

3 Summary

We have demonstrated a scheme to prepare pulse polarized-entangled photon pairs by femtosecond pulse-pumped SPDC in a cascade of two type-I crystals. The highly polarized entangled photon pairs were successfully obtained by using the pre-compensator optical component which appropriate to temporally delayed the orthogonal polarization components for the pump.

We are grateful to Professor S. Chwirot, K. Banazsek, P. Kolenderski and M. Karpinski for the hospitality and helpful supports during experimental visit to Torun Poland and to R. S. Said for his contributions. This research is part of MIMOS - National Laboratory for Atomic Molecular and Optical Physics (FAMO) Poland research collaborations.

References

1. M.A. Nielsen and I.L. Chuang, Quantum Computation and Quantum Information, Cambridge University Press, Cambridge (2000).
2. A. Einstein, B. Podolsky, and N. Rosen, Phys. Rev. **47**, 777 (1935)
3. A. Ekert, Phys. Rev. Lett. **67**, 661 (1991).
4. D.S. Naik et. al., Phys. Rev. Lett. **84**, 4733 (2000).
5. C.H. Bennett, G. Brassard, and N.D. Mermin, Phys. Rev. Lett. **68**, 557 (1992)
6. E. Waks, A. Zeevi, and Y. Yamamoto, Phys. Rev. A **65**, 052310 (2002)
7. P.G. Kwiat, E. Waks, A.G. White, I. Appelbaum, and P.H. Eberhard, Phys. Rev. A **60**, R773 (1999).

Implementation of CWDM technique in access network

N. Ali[1,2,4*], *S.* Shaari[1], *N. A. M.* Ahmad Hambali1[1,2], *Liza* M. A[3] and *C. B. M.* Rashidi[4]

[1]School of Microelectronics Engineering, Universiti Malaysia Perlis (UniMAP), 02600 Arau, Perlis, Malaysia
[2]Semiconductor Photonics & Integrated Lightwave Systems (SPILS), School of Microelectronic Engineering, Universiti Malaysia Perlis, Pauh Putra Main Campus, 02600 Arau, Perlis, Malaysia
[3]University Kuala Lumpur British Malaysian Institute, Batu 8, Jalan Sungai Pusu 53100, Gombak Selangor
[4]Advanced Communication Engineering, Centre of Excellence-School of Computer and Communication Engineering, Universiti Malaysia Perlis

Abstract. The Coarse Wavelength Division Multiplexing (CWDM) system presented in this paper has several distinct characteristics from the standard CWDM system. One of them is the channel spacing whereby it uses smaller channel spacing compared to the standard system. The other is that, the system uses a broad spectrum laser source. The source is filtered by using 'Bragg Grating' before being transmitted. The filtration gives higher OSNR compared to standard CWDM setup. It was found that the system can transmit through each channel up to 2.5 Gbps with the probability of error of 10^{-12}. The maximum distance that can be supported by this system is about 27 km, for the data rate of 155 Mbps.

1 Introduction

Access network mainly comprises of the voice network, which is traditionally based on copper twisted pair cable. This is still a viable option to a certain extend in the network today even if the data loading in the network increases since the technology in communication system is also improving. As the speed increases further, copper cables can no longer be used and need to be replaced by fiber cables as feeder line to the copper cables, which are then connected to the customer. This is the reason that in the latest trend [1] the link between telco exchanges and the customer is connected all the way by fiber. In existing architecture the access distributions are using copper cable with coverage of 4 km in radii. The coverage span limits the connection from the access node to the end user. The optical connection is used when the coverage of copper cable cannot be supported and a high bandwidth line is needed. Optical lines will increased the price of service but by providing high bandwidth to the end user the price will be economical. The problem arises in the optical lines transmission where, the lines are not shareable by the other users in order to connect to medium and lower class users. The low-bandwidth users that are far away from the access node; i.e. unreachable by the copper cable; cannot be connected. One of the solutions available is FTTx system. FTTx is beneficial because it can support not only low bandwidth but also high bandwidth and long distance. However, the cost incurred by FTTx is prohibitive and not practical for medium-bandwidth users. One of them is FTTH (Fiber to the Home)

Many studies have been carried out to reduce the cost of FTTx i.e. FTTH system such as Yaojun et. al. [2]. The architecture proposed to maximize the utilization of the optical lines in order to reduce the cost of implementation. The CWDM implementation in this architecture is only available at the feeder of access network, and it does not cover up to the distribution of access network. The distributions use FTTx system for higher end user and copper to lower end user. The architecture does not cover medium bandwidth user and also the lower bandwidth user, which are located far away from the access node. Even though the utilization of the access node using CWDM can reduce the cost but it still cannot support the optical lines for medium and lower bandwidth user. Most of the time the connection to the end user using optical lines is dedicated for high end user only. In this architecture, the connection to the end user has to use a new scheme since the network architecture cannot support existing scheme or data type. The author proposed modulator on each user, which increased the cost of implementation. In the architecture proposed, the distribution of the channel uses power splitting which reduces the cost but it has to place erbium-doped amplifier (EDFA) to increase the power reduction during splitting which increases the level of noise and also raises the cost of system. The wavelength spacing used is 3.6 nm, which is larger compared to our architecture (2 nm).

Our architecture network has a new multiplexing device that can integrate the two physical types of transmitted data: optical and electrical. This device will

* Corresponding author: norshamsuri@unimap.edu.my

 DOI: 10.1051/epjconf/201716201034

work on the cascaded network architecture where the access network will consist of an optical network and a copper network. The system can transmit up to 2.5 Gbps on each channel with probability of errors of 10^{-12}. The existing scheme can be integrated into the architecture since the system is transparent of data format.

2 WORKING PRINCIPLE AND SYSTEM DESCRIPTION

In the existing architecture users are connected by the access distribution network. The network normally starts from the exchange terminal and in urban areas they are connected from the curb nearby. The network architecture is normally in point-to-point form where the connection is done directly to access node. The connections are implemented using a copper cable and for high-end lines they use optical lines. Most paper and research today are more concentrated on the FTTx system, which provides the optical line to the each house. The limitations on the FTTx system are that the components and equipment used may incur a high cost of implementation.

In the existing multiplexer system, the technique used to increase the bandwidth is TDM technique. The conventional method is by using the electrical signal, which uses a coaxial or twisted pair cable. The optical solution available requires a pair of fiber for a single connection to the end user. The connection is made from the exchange point to the customer premises and fixed at certain data rate only. Inflexibility of existing solution pushes the telecommunication company using existing E1 architecture and increases the availability of exchange point to the place where it does not cover before. As the number of users increases the number of port has to be increased and introduce complexity in network. It also raises the price since it needs a number of ports in order to reach many users.

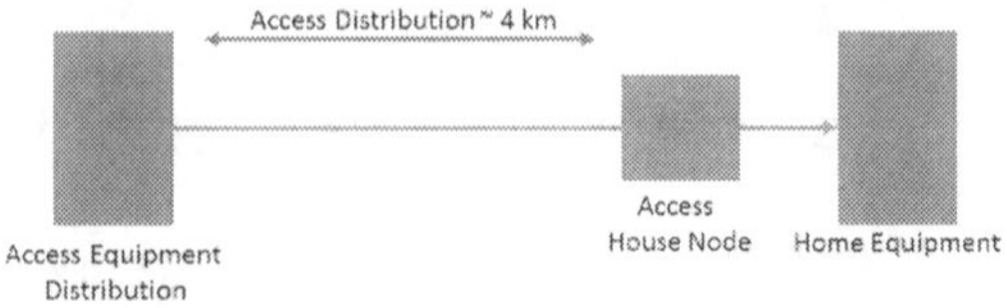

Fig. 1. Existing network architecture in access distribution node

The proposed multiplexer uses the CWDM technique to multiplex the optical signal into a single connection, thus increasing the capacity and optimizing the fiber usage. The interfacing of electrical signal is done at the media converter where it converts signals to optical signals independent of data rates and data formats and then multiplexed to the network optically. The signals received by the end user are demultiplexed according to the specific channel and are converted to electrical signal. The end user equipment may consist of a media converter only or a multiplexer depending on the service used. The system also supports electrical multiplexing such as in the fiber multiplexer. The multiplexed electrical signal is converted using media converter before transmitting into the proposed multiplexer. Figure 2 below shows the implementation of multiplexed access network into the access network today where the distribution lines from access network as shown in Figure 1 is converted into architecture in Figure 2.

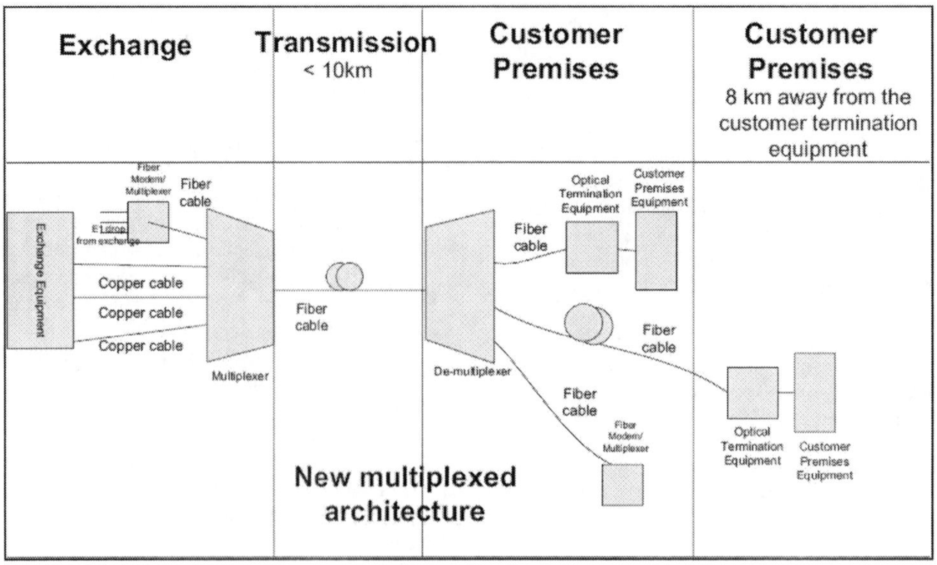

Fig. 2. Multiplexed access network architecture.

The architecture of this system is expandable architecture where it can be expanded to nearby customers by optical or electrical connection depending on the distance of the other users as shown in Figure 2. The existing point-to-point architecture has been changed to the bus architecture in the new design. The new architecture provides the transmission media with either optical or electrical or both. This architecture is capable of supporting existing equipment used in the old system (existing architecture) either optical devices or electrical devices. The utilization of fiber in the network also can be increased since more than one signal can be transmitted in the same fiber. The nearby users can be connected easily by extending the network. The existing limitation faced by the previous network such as distance reach, limited number of users, cost incurred as the number of user increases, and electrical problems, can be eliminated by implementing this new network architecture.

In this architecture, the distance reachable problem is solved by extending the distance using optical cables. The connection to the end user can also be maintained in electrical domain since the proposed multiplexer supported both (optical and electrical) type of signals. The number of user can be easily added by using the fiber multiplexer or single connection of E1 services to the nearby users. Adding the number of users does not require new installation of new fiber. The usage of fiber can be maximized since all connection will be multiplexed into a fiber. This indirectly reduces the cost incurred in the transmission. The electrical problem is not a main issue in the proposed multiplexer. If the electrical problem arises at a certain user, it will not disturb other users in the network. The electrical problem of the system is also reduced since the proposed multiplexer is mainly in optical domain where it is immune to the electrical problems.

The architecture proposed is not only capable to multiplex all optical signals but also electrical signals. The multiplexer has an electrical port for all electrical termination in the access network. The converter has design where it can receive all type electrical formats. The special feature on this multiplexer is it can convert all electrical signal with format independent up to 600 Mbps. This will be able all electrical signal including the 2 wire and 4 wire format for telephony system that can be converted to electrical signal. Thus, all networks at distribution access node can be combined into a single network until up to end user or to the curb. This technique will reduce the complexity in the existing network and also cost of implementation.

3 Results and Analysis

Existing techniques that convert the electrical into optical depends on the format of the signal such as fiber multiplexer, media converter and EO converter. For example, fiber multiplexer only converts the E1 electrical signal and media converter converts the Ethernet signals. Furthermore, the price of individual equipment is considered expensive compare the purpose's multiplexer. The comparison of price is shown in Table 1. As clearly shown in the Table 1, the price of individual equipment for the conventional converter only is about same as complete proposed multiplexer. It clearly shows that the proposed multiplexer is considered economical compared to the conventional system since it provides better flexibility and new features to the network architecture and also competitive prices.

Table 1. Price comparison between conventional equipment and proposed system.

Proposed Multiplexer	Conventional Equipment
Converter RM 2,000 (Development price not a sale price)	EO Convertor RM 5,000 [4]
Multiplexer RM 2, 000 for 2 channel (Development price not a sale price)	Media Convertor RM 2,220 [5]
	Fiber Multiplexer RM 8,000 [6]
	Array wave guide/ Multiplexer RM 15,000 [7]

The experiment setup to analyze the performance of the new CWDM system is shown in Figure 3.

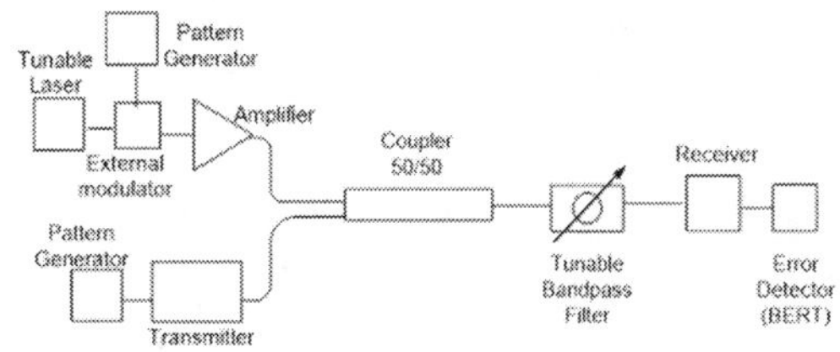

Fig. 3. Experimental setup to test the proposed system in access network.

This setup is a point-to-point transmission where the distance of the transmission is based on power budget calculation. The main blocks of the experiment are the transmitter, coupler, tuneable band-pass filter and the receiver. Coupler acts as the multiplexer while the tuneable bandpass filter acts as the de-multiplexer.

The main signal is modulated by the external modulator, while the adjacent signal is modulated from the 2.5 Gbps transmitter. The main signal has to be amplified since the external modulator attenuates the transmitted power by 8.3 dB. Both signals are then, multiplexed by coupler before being filtered by the tuneable bandpass filter to recover back the main signal. The performance of this setup is determined by the BER value detected by BER tester at the receiving side. The BER calculation is based on the 95% confidence level at $1e^{-12}$ error rates with 2 minutes requirement for every data collected. The CWDM portion in the architecture is tested for various data rate from 1.2 Gbps up to 2.5 Gbps. The system is tested on the wavelength spacing variation towards the performance of the signal. It found that the BER is better when the wavelength spacing 0.58

nm. Figure 4 shows the performance of the system in term of BER towards variation spacing.

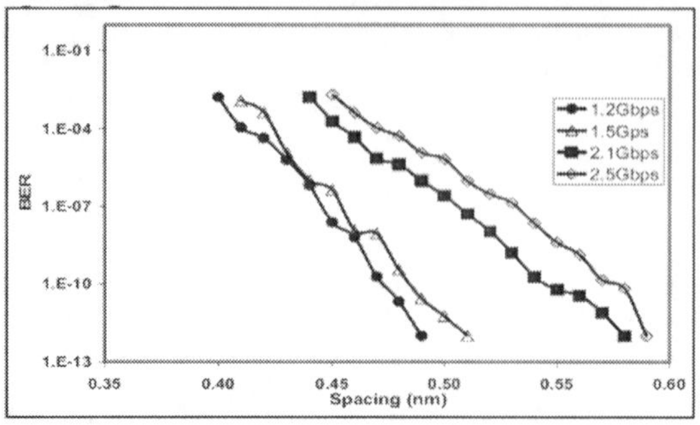

Fig. 4. The BER of system modulated adjacent signal performance towards wavelength spacing at different data rate.

The graph also shows that the wavelength-spacing can be reduced more when the data rate transmitted is reduced. This shows that the architecture can provide a better performance even though the transmitted signal is at 2.5 Gbps. The graph shows that decreasing the wavelength spacing will increase the system BER. When the spacing reduces, the overlapping spectra between channels increases. This will increase the cross-talk and introduce error into the channel, thus increasing the BER. It is important that to find the optimize spacing since it provide the number of channel can be transmitted in single core while reducing the price of services.

The system is also tested on the coverage span as shown in Figure 5. The system shows a better performance even though the data rate transmitted at 2.5 Gbps. It is found that the system can reach up to 15.4 km distance when system at 2.5 Gbps. As the data rate decreased, the graph shifted to the right, towards higher attenuation indicating that the system's sensitivity is better (higher) at lower data rate. For example, the system could attenuate up to 2.1 dB only for a stable performance at 2.5 Gbps system but the system could attenuate up to 4.2 dB when the data rate was 155 Mbps.

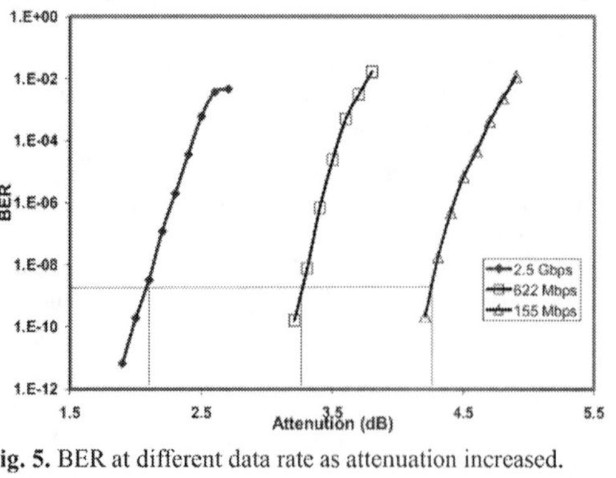

Fig. 5. BER at different data rate as attenuation increased.

Since this is a WDM system, in order to transmit higher data rates and longer distances, each channel can just carry 155 Mbps and can be multiplexed to obtain higher bandwidth such as 622 Mbps or 2.5 Gbps whereas transmitting the data of 2.5 Gbps in a single channel, will only support a short transmission distance. Thus, the system can support higher bandwidth with multiple channels. The system can reach up to 21.5 km at 155Mbps data rate. This bandwidth is already sufficient to support the needs at the end user.

4 Conclusions

The implementation of the CWDM in the access will resolve the current problem arises today especially on the bandwidth and distance reach. The architecture can be used as an intermediate solution before implementing full optical lines to the home. It was found that the system can transmit each channel up to 2.5 Gbps with the probability of error of 10^{-12}. The maximum distance that can be supported by this system is about 27 km, for the data rate of 155 Mbps.

The author would like to acknowledge this work was supported by the High Ministry of Education Malaysia. The work has been done in 2003. The author would like to thanks to Late Prof Dr Mohamed Khazani Abdullah and his team at Universiti Putra Malaysia previously in 2003. Most of the work was done in Universiti Putra Malaysia previously.

References

1. R. Nelson., 2003, .Fibre in the access network – where are the opportunities in Asia?.,http://www.ovum.com/Fibre in theaccess network - where are the opportunities in Asia.htm
2. Q. Yaojun, Q. Jiang, P. Hongtu, C. Shuqiang, G. Kejian, *WCC. ICCT 2000 Communication Technology Proceedings*, **2**, pp. 1201 - 1204 (2000)
3. A.A.M. Saleh, J. M. Simmons, Journal of Lightwave Technology, **17**, Issue 12, pp. 2431 - 2448 (1999)
4. Nvision, SD 4171 SDI E/O Converter. http://www.nvision1.com/Prod
5. Omnitron iConverter GX/T -media converter,http://reviews.cnet.com/Omnitron_iConverter_GX_T_media_converter/4505-243_730643524.html
6. RDMux 200. Fiber Multiplexer, http://www.raddirect.com/Product-Fiber-Multiplexer-RDmux200.html
7. JDSU Filter, 100 GHz, Add/Drop Module, http://www.jdsu.com/index.cfm?productid=280 &pagepa th=Products.

Investigation of AWG demultiplexer based SOI for CWDM application

Nurjuliana Juhari[1], *P Susthitha* Menon[2], *Sahbudin* Shaari[2], *Abang Annuar* Ehsan[2]

[1]School of Microelectronic Engineering, Universiti Malaysia Perlis (UniMAP), 02600 Arau, Perlis and [2]Institute of Microengineering [2]Nanoelectronics (IMEN), Universiti Kebangsaan Malaysia (UKM), 43600 UKM Bangi, Selangor, Malaysia

Abstract. 9-channel Arrayed Waveguide Grating (AWG) demultiplexer for conventional and tapered structure were simulated using beam propagation method (BPM) with channel spacing of 20 nm. The AWG demultiplexer was design using high refractive index (n~3.47) material namely silicon-on-insulator (SOI) with rib waveguide structure. The characteristics of insertion loss, adjacent crosstalk and output spectrum response at central wavelength of 1.55 μm for both designs were compared and analyzed. The conventional AWG produced a minimum insertion loss of 6.64 dB whereas the tapered AWG design reduced the insertion loss by 2.66 dB. The lowest adjacent crosstalk value of -16.96 dB was obtained in the conventional AWG design and this was much smaller compared to the tapered AWG design where the lowest crosstalk value is -17.23 dB. Hence, a tapered AWG design significantly reduces the insertion loss but has a slightly higher adjacent crosstalk compared to the conventional AWG design. On the other hand, the output spectrum responses that are obtained from both designs were close to the Coarse Wavelength Division Multiplexing (CWDM) wavelength grid.

1 Introduction

Planar lightwave circuits (PLC) are waveguide devices that are fabricated on flat substrates, commonly on silicon or glass. PLCs are also widely utilized in wavelength-division multiplexing (WDM), time-divison multiplexing (TDM) systems and fiber-to-the-home (FTTH) access networks [1]. WDM systems have some key components that have the function to multiplexer/demultiplexer (mux/demux) signal that can increase transmission capacity on a single optical fiber. These components are such as the Arrayed Waveguide Grating (AWG), Echelle gratings, ring resonators and lattice form filters. Actually, the research on mux/demux that focused on grating based and phased array (PHASAR) based devices were developed since 1990 [2]. Silica was the first material that was widely used in developing AWGs. In the late 1990's, a 64-channel 50 GHz InP based AWG was developed [3] and this was followed by other materials such as polymer, Al_2O_3, SiGe, $LiNbO_3$ and silicon-on-insulator (SOI) materials. Since silicon is widely used in the microelectronics industry, integration between silicon photonics and electronic devices became more interesting. Based on sophisticated semiconductor technology, [4] it is predicted that silicon photonics would provide inexpensive integrated electronics-photonics platform. Therefore, the Array Waveguide Grating (AWG) developed on a SOI-based rib waveguide have great potential especially for WDM systems.

In this work, we designed a novel SOI-based AWG mux/demux with channel spacing catering for Coarse Wavelength Division Multiplexing (CWDM) networks. Two different designs of the AWG device were developed; one with a conventional waveguide and the other with a tapered design. The insertion loss and adjacent crosstalk values of both designs were analyzed and compared with. The output transmission peak from this AWG device was then verified using standard International Transmission Unit (ITU) wavelength grid for WDM/CWDM system.

2 Fundamental theory of arrayed waveguide gratings (AWG)

Fig. 1 (a) shows the schematic of an rib optical waveguide (cross-section) while Fig. 1 (b) shows the effective refractive index, n_{eff} value of the rib waveguide at transverse mode, TE00 (single mode). The measurements were executed using RSOFT Fullwave software where the mathematical equations are based on finite difference method which gives n_{eff} of 3.31715. Layout of the AWG device is depicted in Fig. 2 and the circle signage show the tapered design located at the phase array and at the input/output region respectively. In an AWG, the fundamental mode is the correct operation that allows light to propagate through a waveguide with less dispersion. The design of the SOI-based rib waveguide structure with fundamental mode condition can be designed where the geometry of the waveguide is determined using Eq. (1) and (2) [6]. Silicon waveguide which is used as the core have a refractive index, n ~ 3.5 and is surrounded with SiO_2 as a

* Corresponding author: nurjuliana@unimap.edu.my

cladding with n ~1.5 which gives a step index value of Δn~ 2. In this design, the value of core width W, core height H, and etch depth h, are W= 1 μm, H= 0.6 μm and h= 0.1 μm respectively.

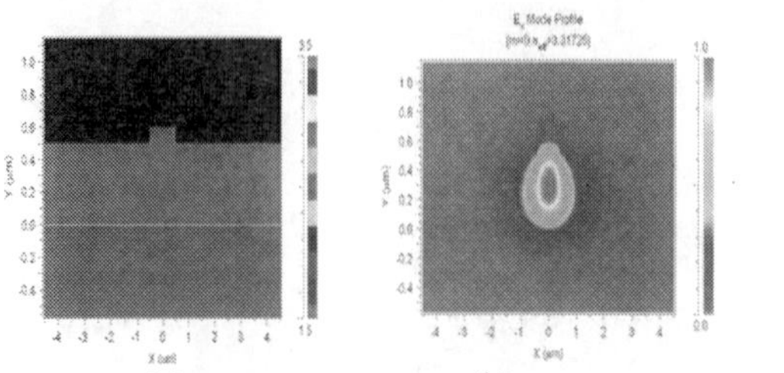

(a) (b)

Fig. 1. (a) Cross section (b) Single mode pattern of 1 μm core width of SOI-based rib waveguide

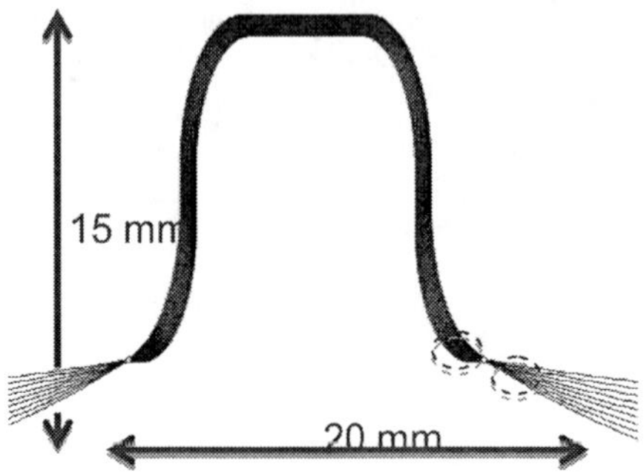

Fig. 2. Schematic layout of conventional 9-channel AWG and circle sign show the tapered design at phase array tapered designed at I/O port region.

$$\frac{W}{H} = \alpha + \frac{\frac{h}{H}}{\sqrt{1-\left(\frac{h}{H}\right)^2}} \qquad (1)$$

$$\frac{h}{H} \geq 0.5 \qquad (2)$$

Basically, this AWG consist of 9 x 9 input and output waveguides, two free propagation regions (FPR) at length of 263.0688 μm and 34 arrayed waveguides produced. Some important parameters are the input/output waveguide separation, D, arrayed waveguide separation, d, and the radius of curvature, f, at both FPRs. AWG works based on the interferometer principle. Light which carries different wavelength in one optical fiber propagating at the input waveguide is then diffracted at different diffraction order in the FPR. This light is then coupled into the arrayed waveguide. The arrayed waveguide is constructed based on Eq. 3 at central wavelength λ_c, 1.55 μm where ΔL is the difference in the optical path length, m is the order of the phased array, and N_g is the modal effective index of the waveguide.

$$\Delta L = m\frac{\lambda_c}{N_g} \qquad (3)$$

The phase difference depends on the input wavelength. It can be determined when the light beam is channeledback into the output star coupler. The constructive interference in the second slab is then propagated through the output channel that brings about different wavelengths. To determine the channel spacing obtained at 20 nm, Eq. 4 is used

$$\Delta\lambda = \frac{n_s d D \lambda_0}{N_c f \Delta L} \qquad (4)$$

where n_s is effective index at the FPR region, d is arrayed waveguide separation, D is waveguide separation, λ_0 is the channel spacing, N_c is the group index, f is the curvature radius and ΔL is path length difference. A variety of loss mechanisms occur in AWGs but generally they are mainly contributed by the loss coming from the junction between the free propagation region and the waveguide array [7].

3 Methodology

The design and simulation was conducted using PHASAR software where the calculation is based on the beam propagation method (BPM) algorithm. BPM simulation gives accurate data such as loss and radiation in the arrayed waveguide, coupling of waveguide and phase error. Numerical method based on (effective index method) EIM is a popular technique in analyzing integrated optical devices meanwhile (S-matrix method) SMM used to analyze the phased array when the decoupling process occurs. The material used is SOI which have an effective index between core/cladding at 1550 nm of 3.31725/1.5. The port separation of the input/output adjacent channels is designed to be 250 μm to enable pigtailing to fiber ribbons [8]. The tapered design for width waveguide separation and length at input/output port is set at 5 μm and 700 μm respectively. All the result of the design parameters are tabulated in Table 1. The difference in the parameters between the conventional and the tapered AWG is only the size of waveguide width and the length of the taper while other parameters remain the same. The simulation results presents the value of adjacent crosstalk, channel spacing, bandwith at -3dB level and insertion loss.

Table 1. Design parameters of 1 μm core width of conventional and tapered AWG device

Parameter	Conventional	Tapered AWG

 DOI: 10.1051/epjconf/201716201035

	AWG Value	Value
I/O and array waveguide effective index	3.2858	3.2858
Free Propagation Region (FPR)	263.0688 μm	263.0688 μm
Number of Arrays	34	34
Modal index (I/O & array waveguide)	3.2559	3.2559
Free Spectral Range (FSR)	32235.484 GHz	32235.484 GHz
Optical path length, ΔL	2.83 μm	2.83 μm
Phasar order, m	6	6
Channel spacing, Δλ	20 nm	20 nm
Central wavelength, λ_0	1.55 μm	1.55 μm
Taper width and length (I/O)	-	2.3 μm/75.74 μm
Taper width and length (I/O)	-	2.3 μm/75.74 μm

4 Results and discussions

The AWG device was compared between two design structures i.e. the conventional and tapered design. Both devices were designed using SOI material with a rib waveguide structure. The thickness of the Si layer is 0.6 μm on top of a 2 μm thick buried oxide (BOX). Fig. 1 (a) shows the cross section of the SOI-based optical waveguide designed in RSoft software. Fig. 1 (b) shows the single-mode transmission obtained within the optical waveguide with core width of 1 um. The AWG is being designed as a multiplexer in CWDM networks where the channel spacing is 20 ± 5 nm with operating wavelength of 1.55 μm in the E, S, C and L optical windows. To fulfill the complete 34 arrays of the design structure, the bending radius is adjusted until four arc-straight-arc shaped is obtained as in Fig. 2 (a). The phase array loss was determined in the range of 0.006-0.005 dB. The rib waveguide design structure makes the device become non-compact. Therefore, the total device size for both AWG designs is 20 mm x 15 mm where this design is much bigger as compared to previously developed AWG designs [9-10].

Table 1 shows the value of all design parameters. The ΔL, m and Δλ are calculated using Eq. 3 and Eq. 4. With phasar order of 6 at λ_0=1.55 μm, the corresponding dispersion is 12.4 nmGHz^{-1} at a FPR of 263.0688 μm. Hence at TE mode, the single mode for 1.0 μm core width with slab height of 0.1 μm produced a 3.2858 effective index value. The idea to reduce insertion loss of phased array can be implemented by introducing tapered waveguide at the ends of the phased array. However, the insertion loss does not reduce too much, therefore the tapered design is added at input/output star coupler. The angle tapers, α at width 2.5 μm tapering waveguide produced the length, L_{taper} with the value of 75.74 μm. Table 2 and Table 3 shows the output parameter with -3dB bandwith for conventional and tapered design of 9 channel AWG device. Channel spacings of 20 nm are successfully produced where these designs are applicable for CWDM networks. However, for the conventional design, the channel spacing between channel 4 and 5 is 21.7 nm. The same scenario also occurs at the tapered AWG design when the 21.7 nm spacing is produced at between channel 7 and channel 8. The highest value of adjacent crosstalk is obtained at channel 5 for the tapered AWG which is much higher compared to the conventional AWG which gives the value of -22.087654 dB and -20.144659 dB respectively.

The same phenomena occurs at the lowest adjacent crosstalk where the minimum crosstalk for the conventional AWG is still less compared to the tapered AWG. Therefore the tapered and conventional AWG design has not much effect to the adjacent crosstalk. In comparison, these adjacent crosstalk values for both designs still presents a good performance when compared to previous designs [11] where the suggested crosstalk was indicated in the range of ~ 35-37 dB. In comparison, a fabricated Si nanowire-based AWG waveguide produced a general crosstalk with value of -8 dB [12] which shows that the developed designs have better crosstalk. The idea of the tapered design is to reduce the insertion loss especially at the phase arrayed waveguide. From the results in Table 3, it is found that the minimum insertion loss for a tapered design is about 50% lower as compared to the conventional design. This loss is due to the imperfect light capture loss [13] and diffraction loss [14] at the star coupler and array interface.

Fig. 3 (a) and (b) show the spectral response output of a 9-channel conventional AWG device where the simulations was set from 1440 nm until 1635 nm at 121 iteration steps. The wavelength for the nine output of the tapered design with λ_1, λ_2, λ_3, λ_4, λ_5 λ_6, λ_7, λ_8, and λ_9 are located at 1454 nm, 1474 nm, 1494 nm, 1514 nm, 1535.667 nm, 1555.667 nm, 1575.667 nm, 1595.667 nm and 1615.667 nm while for the conventional design λ_1, λ_2, λ_3, λ_4, λ_5 λ_6, λ_7, λ_8, and λ_9 are located at 1454 nm, 1474 nm, 1494 nm, 1514 nm, 1534 nm, 1554 nm, 1575.667 nm, 1595.667 nm and 1615.667 nm respectively. The value for each peak of the output response is based on the channel spacing as tabulated in Table 2 and Table 3. The results of peak wavelength for both designs are found to be slightly different. The reason is due to the channel spacing between some of the channels where the spacing is not exactly 20 nm. However, if compared with ITU Grid [15] the wavelengths are still within the CWDM network [16].

Table 2. Output parameter calculated with -3 dB bandwidth for 9 channel tapered AWG.

Channel	Peak (dB)	Crosstalk (dB)	Spacing (nm)	Peak Wavelength from simulation (nm)	Peak Wavelength: G.694.2 CWDM Wavelength Grid (nm)
1	-4.17	-17.23	0.020	1454.000±5	1451
2	-3.35	-18.39	0.020	1474.000±5	1471
3	-3.03	-20.08	0.020	1494.000±5	1491
4	-2.78	-20.98	0.022	1514.000±5	1511
5	-2.66	-22.09	0.020	1535.667±5	1531
6	-2.77	-21.69	0.020	1555.667±5	1551
7	-3.14	-19.98	0.020	1575.667±5	1571
8	-3.88	-19.29	0.020	1595.667±5	1591
9	-4.86	-17.44	-	1615.667±5	1611

Table 3. Output parameter calculated with -3 dB bandwidth for 9 channel conventional AWG

Channel	Peak (dB)	Crosstalk (dB)	Spacing (nm)	Peak Wavelength from simulation (nm)	Peak Wavelength: G.694.2 CWDM Wavelength Grid (nm)
1	-7.62	-16.96	0.020	1454.000±5	1451
2	-7.06	-17.71	0.020	1474.000±5	1471
3	-6.80	-18.85	0.020	1494.000±5	1491
4	-6.72	-19.69	0.020	1514.000±5	1511
5	-6.64	-20.14	0.020	1534.000±5	1531
6	-6.94	-19.27	0.020	1554.000±5	1551
7	-7.26	-19.34	0.022	1575.667±5	1571
8	-8.02	-18.54	0.020	1595.667±5	1591
9	-8.905	-17.99	-	1615.667±5	1611

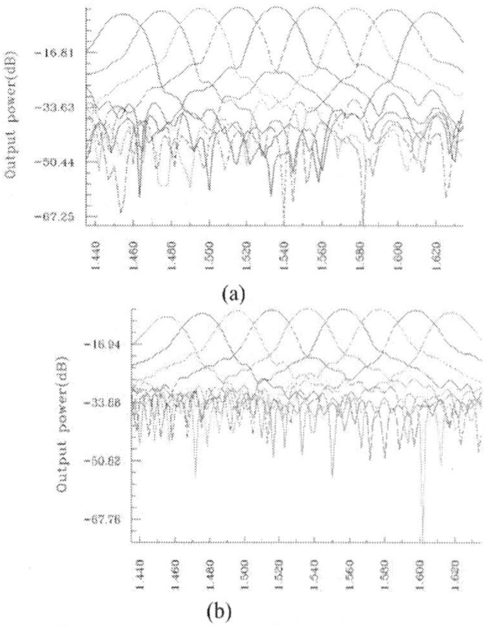

Fig. 3 : Output spectral response of 9 channel SOI-based CWDM AWG for (a) tapered AWG and (b) conventional AWG

5 Conclusions

Conventional and tapered 9 channels AWG-based rib waveguide structure SOI have been successfully designed to obtain a standard wavelength transmission in the optical windows of E, S, C and L for CWDM network. The conventional AWG produced a minimum insertion loss of 6.64 dB whereas the tapered AWG design reduced the insertion loss to 2.66 dB. The lowest adjacent crosstalk value of -16.96 dB was obtained in the conventional AWG design and this was much smaller compared to the tapered AWG design where the lowest crosstalk value is -17.23 dB. The output spectral response for 9-channel AWG for both designs satisfy the ITU grid of a CWDM network. In summary, a tapered AWG design significantly reduces insertion loss and has an equivalent adjacent crosstalk as compared to a conventional AWG.

The author would like to thank the Malaysian Ministry of Education and Universiti Kebangsaan Malaysia (UKM) for sponsoring this work under research grant FRGS/2/2013/SG02/UKM/02/4. The payment for this paper is supported by FRGS grant (9003-00405).

* Corresponding author: nurjuliana@unimap.edu.my

References

1. K. Okamoto, Laser Photonics Rev. 1, 14–23, (2012).
2. M.K Smit, C.V. Dam, IEEE J. Sel. Topics Quantum Electron., **2**, 236–250, (1996).
3. P. Cheben, Optical Waveguide From Theory to Applied Technologies CRC Press Taylor & Francis Group, U.S, pp.173-230, (2007).
4. K.Yamada, Silicon Photonics II, Springer-Verlag, Berlin Heidelberg, pp.129 -187, (2011).
5. B. Chomycz, Planning Fiber Optics Networks, New York: McGraw Hills Company, (2009).
6. O. Powell, Single-mode Condition for Silicon Rib Waveguides, J. Lightw. Technol., **10**, 1851– 1855, (2002).
7. X.J.M. Leijtens, B.Kuhlow, M.K. Smit, Filters in Fibre Optics, Springer, Berlin Germany, pp. 125-187, (2006).
8. H.C.Woei, N.A.Rahman, S. Shaari, ICSE2004, 473-476, (2004).
9. Q. Fang, F. Li, Y. Liu, Opt. Commun **2**, 155–158, (2006).
10. T. Fukazawa, F. Ohno, T. Baba, Jpn. J. Appl. Phys., **5B**, 673–675, (2004).
11. I P Kaminow, C.R. Doerr, C. Dragone, T. Koch, U. Koren, IEEE J. Sel. Areas Commun., **14**, 780-789, (1996).
12. Z. Lei, A. Jumming, Z. Jiashun, S. Shijiao, W. Yuanda, H. Xiongwei, J. Semicond., **32**, 1-4, (2011).
13. K. Okamoto, Fundamental of Optical Waveguide, Elsevier, USA, (2005).
14. S.L. Tsao and Y.H. Lin, Proc. SPIE, (2004),.
15. H.J. Thiele, M. Nebeling, Coarse Wavelength Division Multiplexing Technologies and Application. Taylor & Francis Group, USA, (2007).
16. N. Juhari, P.S. Menon, A.A. Ehsan, 4th Int. Conf. Photonics, 230 – 233, (2013).

Digital fringe projection for hand surface coordinate variation analysis caused by osteoarthritis

Wan Mokhdzani Wan Nor Haimi*, *Cheek Hau* Tan[1], *Vithyacharan* Retnasamy[1], *Rajendaran* Vairavan[1], *Zaliman* Sauli[1], *Nor Roshidah* Yusof, *Nor Azura Malini Ahmad* Hambali[1], *Muhammad Hafiz Ab* Aziz[1] and *Ahmad Syahir Ahmad* Bakhit[1]

[1]School of Microelectronic Engineering, Universiti Malaysia Perlis, Malaysia

Abstract. Hand osteoarthritis is one of the most common forms of arthritis which impact millions of people worldwide. The disabling problem occurs when the protective cartilage on the boundaries of bones wear off over time. Currently, in order to identify hand osteoarthritis, special instruments namely X-ray scanning and MRI are used for the detection but it also has its limitations such as radiation exposure and can be quite costly. In this work, an optical metrology system based on digital fringe projection which comprises of an LCD projector, CCD camera and a personal computer has been developed to anticipate abnormal growth or deformation on the joints of the hand which are common symptoms of osteoarthritis. The main concept of this optical metrology system is to apply structured light as imaging source for surface change detection. The imaging source utilizes fringe patterns generated by C++ programming and is shifted by 3 phase shifts based on the 3 steps 2 shifts method. Phase wrapping technique and analysis were applied in order to detect the deformation of live subjects. The result has demonstrated a successful method of hand deformation detection based on the pixel tracking differences of a normal and deformed state.

1 Introduction

Osteoarthritis (OA) is a relatively prevalence musculoskeletal disorder problem which frequently affects the joints in chronic conditions and it is the most common type of arthritis. Hand OA is a joint degeneration disease due to the failure reparation of joint damage which causes stiffness, pain and impaired movement[1,2]. According to the World Health Organization (WHO), the Department of Chronic Diseases and Health Promotion estimated that 10% to 15% of all adults aged over 60 have a certain degree of osteoarthritis, with OA more common among women compared to men[3]. Knees, hips, hands, big toes and spine are common body parts where OA can develop. A population-based analysis based on National Health Examination Survey (NHES) data found that the prevalence of hand OA was 29.5% among those aged 25 years old and onwards[4]. The probability of getting hand OA was higher compared to the feet OA even though both of these body parts has equal numbers of joint. This is due to the cartilage inside the joints of the hand are used more regularly than the feet to cover and protect the end tips of bones where they meet to form a joint.

In general, OA can be characterised by three types of diagnosis which is clinical, pathological and radiological[5]. Pathological diagnosis are carried out in certain focal areas within synovial joints that are located with hypertrophy of bone and thickening capsule and are inspected by x-rays. In the clinical analysis, the OA syndrome can be classified by joint pain, tenderness, restriction of movement, crepitus, occasionally effusion and variety degrees of local inflammation where significant physical symptoms of OA can be seen.

Radiographic diagnosis is executed with X-rays which can detect changes of the interphalangeal joints such as focal narrowing, marginal osteophyte changes and sclerosis. The X-ray utilizes high energy level that allows the beam to penetrate through the body and create an image. Since X-rays applies ionizing radiation, cells can be damaged and develop into cancer. A short duration and small dosage have been know to avoid cells being damaged namely diagnosis such as the mammogram [6, 7].

Several other methods of hand OA diagnostic has also been developed throughout the years such as the Magnetic Resonance Imaging (MRI), Micro CT Scanning and ultrasonography. The MRI diagnostic method has long being applied in medical image diagnosis. It applies a magnetic field and pulses of radio wave energy to create image of organs and bones inside the body which also provides different image information compared to X-rays and CT scan. However, there is a limitation to the MRI in terms of low sensitivity, low specificity and also cost ineffectiveness [8].

* Corresponding author: wanmokhdzani@unimap.edu.my

DOI: 10.1051/epjconf/201716201036

The Micro-CT scanning method is a non-invasive,whole volume imaging technology that provides high contrast between bone and soft tissues and produces quantitative data on bone stereology[9]. It is the advanced version of X-ray because micro-CT scanning offers higher resolution than a plain X-ray. Although both diagnostic tools use X-rays and ionizing radiation as imaging sources, micro-CT scanning enables one to view in 3D microscopy.

The Ultrasonography method appears favourable to assess OA in both clinic and clinical trials. It possesses higher resolutions than conventional radiographs, does not involve any ionising radiation, and allows multi-planar dynamic imaging of joints[10].

The method of hand OA diagnostic proposed in this work is by using the digital fringe projection based on structured light projection. Digital fringe projection has previously been applied on human subjects for analysis on face wrinkles, dental analysis, fingerprint imaging, human skin and wound dimension analysis [11-16]. Based on the literature reviews carried and to the best of the author's knowledge, hand OA detection using digital fringe projection has been quite limited. The digital fringe projection technique applied in this work utilizes the phase shifting method. Human hand were tested whereby two states of the hand were assessed, normal state and deformed state of the hand under the fringe projected image with phase shifting.

2 Theory

In this work, the basic C++ language is applied to generate three fringe patterns with different phase. The phase angle is shifted every $2\pi/3$ radians respectively to the three different fringe patterns. The three-step technique of phase shifting method is applied to obtain phase map. The mathematical descriptions of these three fringe patterns are as[16]:

$$I_1 = A(i,j) + B(i,j)cos\phi(i,j) \quad (1)$$

$$I_2 = A(i,j) + B(i,j)cos[\phi(i,j)+2\pi/3] \quad (2)$$

$$I_3 = A(i,j) + B(i,j)cos[\phi(i,j)+4\pi/3] \quad (3)$$

Where $A(i,j)$ is the uniform background intensity, $B(i,j)$ is the amplitude of sinusoidal waveform and $\phi(i,j)$is the phase angle at the point (i,j). The three equations are also known as the resultant intensity patterns.

The phase wrapped mapping can be obtained by evaluating $\phi(i,j)$simultaneously with Equation 1, 2 and 3 which leads t0 Equation 4:

$$\phi\ (i,j)=tan^{-1}[\sqrt{3}[I_3(i,j)-I_2(i,j)]\ /\ 2I_1(i,j)-\ I_3(i,j)-I_2(i,j)] \quad (4)$$

The significance of ϕ given by equation (4) is within $\pm\ \pi$. Nevertheless, by examining the signs of the numerator and denominator, the significance of ϕ between 0 to 2π can be resulted [17].

3 Methodology

3.1 Experimental Setup

The general fringe projection system setup used for the hand deformation inspection is shown in Figure 1. A standard LCD projector ((Panasonic PT-LW80NTEA) is utilized to project computer generated fringe patterns and images are recorded with a CCD camera(monochrome (AVT Marlin F-146C) 1392 x 1040 resolution). The LCD projector and digital CCD camera are mounted on top of the experimental surface under test. The sinusoidal fringe patterns were programmed using C++ in Microsoft Visual Studio 2010. The hand used as the test subject was placed on the reference plane and fringe visibility was best observed at a height of 490 mm above the hand sample after a series of fringe visibility analysis based on camera distance from the hand sample.

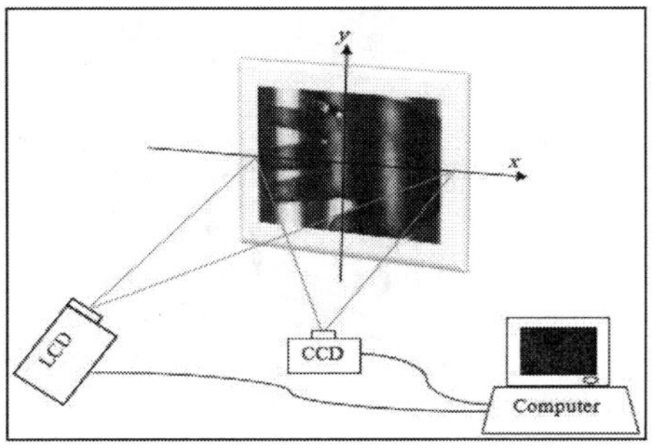

Fig. 1. Digital Fringe Projection Experimental Setup

3.1 Analysis Method

In this work, a human hand test subject was tested in two different states, normal positioning of the hand and a slightly deformed finger of the hand. The method adopted for analysis of hand OA are as follow:

1. The fringe patterns generated by C++ coding was shifted three times by $2\pi/3$ radians and projected onto the actual hand model in the normal state and deformed state as shown in Figure 2. Image of the hand after fringe projection were recorded with the CCD camera. This process was repeated for five times to ensure each phase shifting algorithms were done in static condition.

2. The three phase shifted recorded images were then applied in the phase mapping process using the UU imaging software to obtain the wrapped phase map [18].

3. A total of 106 digital markers with an equal distance of 50 pixels were plotted onto the wrapped phase map as illustrated in Figure 3. The digital markers were only marked onto the region of actual hand and not the reference plane. Blue markers indicated the centre gridline of the image while the red markers were

DOI: 10.1051/epjconf/201716201036

directed to the region of actual finger to be analyzed. The gray scale values and RGB values for each red markers were used as reference.

4. The resultant grayscale and RGB values can be determined from wrapped phase map by conducting deformed state phase shifting algorithms.

5. Comparisons were then made between the normal state and deformed state by constructing a line plot from specified digital markers. If a gray scale or RGB variation happens, pixel tracing of Y-axis will be conducted to detect the equal gray scale value and the new coordinate point location was obtained.

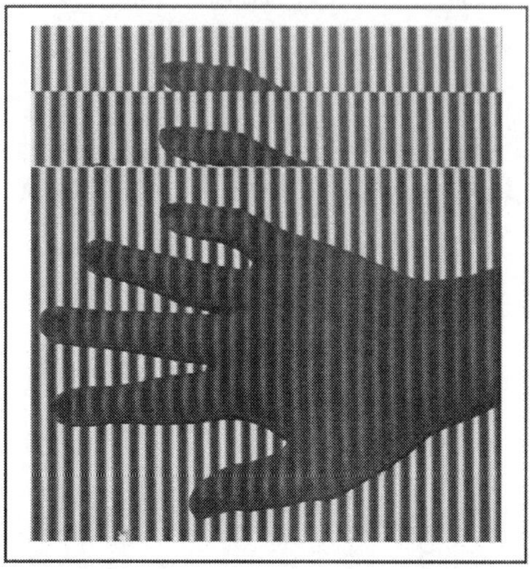

Fig. 2. Three phase shifted images of the hand

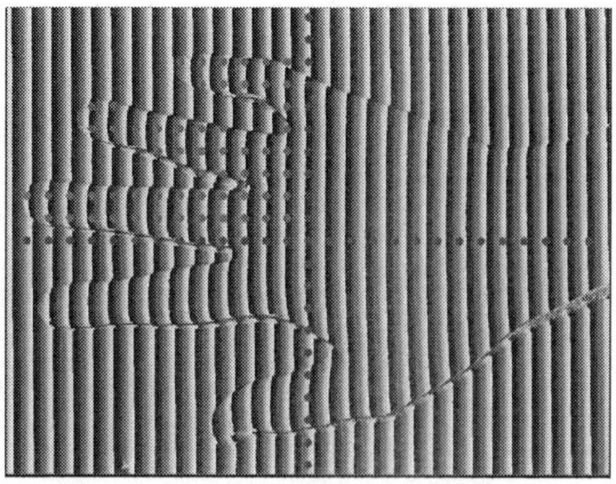

Fig. 3. Phase Map Image of the hand

4 Results & Discussion

The main purpose in this work was to apply digital fringe projection in detecting surface abnormalities of the hand or early stages of hand osteoarthritis. The fringe projection technique adopted the phase shifting technique of the generated fringes. In the analysis, the shifted fringe pattern was projected onto the hand sample in the normal and deformed state whereby a phase map was produced.

Based on the 106 digital markers from the fringe image, the coordinate for each marker was determined from the fringe image. The gray scale value for each coordinate marker was located from the phase map image of the normal hand condition as shown in Figure 4. The area of interest for the hand OA analysis are the little finger, ring finger and the middle finger demonstrated by the red coordinate markers. Coordinate markers were then plotted for the deformed hand condition with the same area of interest as the normal state phase map image. Comparisons were made for both the gray scale values and an early variation of gray scale value can be observed circled in yellow as shown in Figure 5.

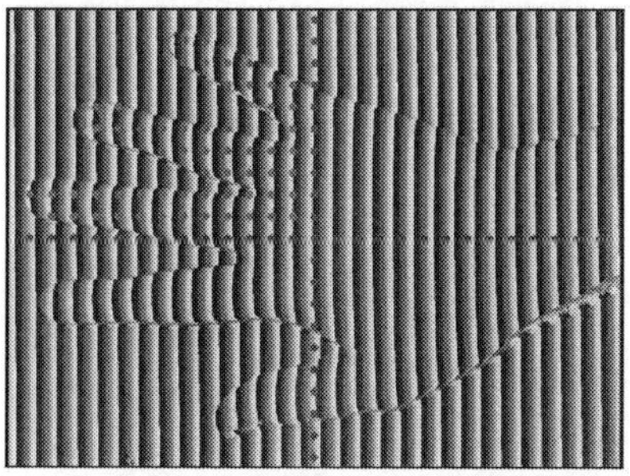

Fig. 4. Coordinate Markers of the normal state of the hand

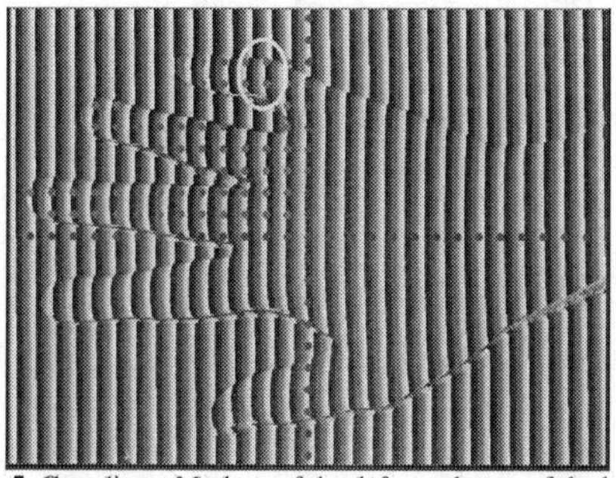

Fig. 5. Coordinate Markers of the deformed state of the hand

Pixel tracing was then conducted and plotted to investigate the surface change of the deformed hand as shown in Figure 6. A total 59 out of the 106 red coordinate markers were plotted onto the wrapped phase map image of the hand test subject shown in Figure 6. The solid blue line signifies the surface plot for the normal hand condition, while the red dashed line signifies the surface plot for the deformed hand. It can be seen in that the plotting for both hand conditions are almost identical but for a small number of coordinate location. Based on the phase map image of the deformed hand in Figure 5, the main difference of the pixel coordinate is situated at the little finger location.

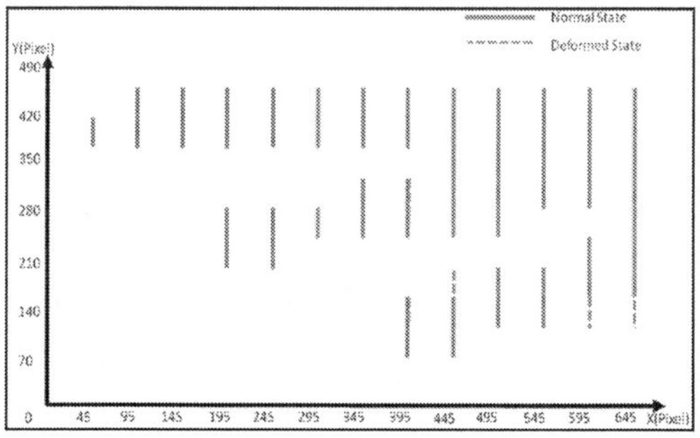

Fig. 6. Hand surface variation plotting of the normal and deformated state of the hand dinate Markers of the normal state of the hand

Evidently the coordinate variation occurs at four main coordinates from the normal hand condition at (x_1,y_1) = (395,70), (x_2,y_2) = (445,70), (x_3,y_3) = (595,170) and (x_4,y_4) = (645,190). The variation of the deformed coordinates from the normal hand coordinates are shown in Table 1.

Table 1. Pixel tracing comparison of deformed and normal hand state

Pixel Coordinates		
X-axis	Y-Axis (normal)	Y-Axis (deformed)
395	70	120
445	70	120
595	170	120
645	190	120

The results obtained in this work demonstrate that the application of digital fringe projection can be used in the analysis of hand osteoarthritis or abnormalities. The method applied for hand osteoarthritis detection can be another added technique where its main advantage is that it is a non-contact and free of ionizing radiation compared to the current existing methods. Previous similar technique utilizing digital fringe projection on early breast carcinoma has also been successfully demonstrated [16] and this work has successfully demonstrated another early detection of a medical condition, hand osteoarthritis detection. Extensive analysis will still need to be conducted to ensure a more accurate and efficient application of digital fringe projection on hand osteoarthritis detection.

5 Conclusions

The preliminary result of the digital fringe projection system has successfully been proven to be an effective surface measurement for early hand osteoarthritis detection. The phase shift technique and the pixel tracing method have been fully utilized to evaluate the changes of joints and finger deformation. Further improvement can be done by testing the fringe projection system on real patients with osteoarthritis and classify the stages of the patients osteoarthritis which will require a databank of patients medical osteoarthritis history. The fringe shifting and phase mapping method used in this work can also be applied to other medical conditions where physical changes occur.

References

1. R. W. Moskowitz, *Hosp. Pract.*, **14**, 7, pp. 75–80, 85–87, (1979).
2. Doherty, Michael, Hans Bijlsma, Nigel Arden, Nicola Dalbeth, and David J. Hunter, eds. *Oxford Textbook of Osteoarthritis and Crystal Arthropathy*. Oxford University Press, (2016).
3. R. C. Lawrence *et al.*, Arthritis Rheum., **41**, **5**, pp. 778–799, (1998).
4. World Health Organization In *Summary meeting report*, pp. 5-7. (2004).
5. EUMUSC, Eur. Musculoskelet. Heal. Surveill. Infromation Netw., pp. 1–13, (2014).
6. P. Information, "Plain Radiography / X-rays," no. June, (2014).
7. J. F. Holmes and R. Akkinepalli, "Computed Tomography Versus Plain Radiography to Screen for Cervical Spine Injury : A Meta-Analysis," no. May, pp. 902–905, (2005).
8. S. B. Paper, W. Oa, G. Burden, D. Study, E. U. M. States, and U. Nations, 6.12 Osteoarthritis, **12**, pp. 6–8, (2010).
9. H. Fuller, R. Fuller, R. Maria, and R. Pereira, *Rev. Bras. Ortop. English Ed.*, **55**, 4, pp. 352–362, (2015).
10. H. I. Keen *et al.*, "The development of a preliminary ultrasonographic scoring system for features of hand osteoarthritis," pp. 651–656, (2007).
11. Luebberding, S., Krueger, N., & Kerscher, M., Dermatologic Surgery, **40**, 1, 22–32 (2014).
12. Munera, N., Lora, G. J., & Garcia-Sucerquia, J., Dyna, 79, 171, 65–73 (2012).
13. Huang, S., Zhang, Z., Zhao, Y., Dai, J., Chen, C., Xu, Y. Xie, L., Optics and Lasers in Engineering, **52**, 123–130 (2014).

 DOI: 10.1051/epjconf/201716201036

14. Rohr, M., & Schrader, A, In Non Invasive Diagnostic Techniques in Clinical Dermatology , Springer. pp. 55–64) (2014).
15. Saito, M. T., Yoshimura, E. M., Palacios, F. F., Lino, A. C. L., Palacios, F., & Sousa, M. V. P., SPIE BiOS, **8572**, 1–7 (2013).
16. Vairavan, Rajendaran, Vithyacharan Retnasamy, Mukhzeer Mohamad Shahimin, Zaliman Sauli, Lai Siang Leng, Wan Mokhzani Wan Norhaimi, Prema Boshani Retnasamy, Othman Abdullah, and Supap Kirtsaeng. *Proc. of SPIE Vol*, **10043**, pp. 1004314-1. (2017).
17. Retnasamy, V., & Ratnam, M. MThe International Journal of Advanced Manufacturing Technology, **38**(11–12), 1172–1180 (2008).
18. Chen, L., In *International Conference on Optics in Precision Engineering and Nanotechnology (icOPEN2013)* (p. 87690U–87690U). International Society for Optics and Photonics (2013).

DOI: 10.1051/epjconf/201716201037

Doping concentration effect on performance of single QW double-heterostructure InGaN/AlGaN light emitting diode

N. Syafira Abdul Halim[1,*], *M.Halim* A. Wahid[1], *N. Azura* M. Ahmad Hambali[1], *Shanise* Rashid[1], and *Mukhzeer* M.Shahimin[2]

[1]Semiconductor Photonics & Integrated Lightwave Systems (SPILS), School of Microelectronic Engineering, Universiti Malaysia Perlis, Pauh Putra Campus, 02600 Malaysia.
[2]Department of Electrical and Electronic Engineering, Faculty of Engineering, National Defence University of Malaysia (UPNM), Kem Sungai Besi, 57000 Kuala Lumpur.

Abstract. Light emitting diode (LED) employed a numerous applications such as displaying information, communication, sensing, illumination and lighting. In this paper, InGaN/AlGaN based on one quantum well (1QW) light emitting diode (LED) is modeled and studied numerically by using COMSOL Multiphysics 5.1 version. We have selected $In_{0.06}Ga_{0.94}N$ as the active layer with thickness 50nm sandwiched between 0.15µm thick layers of p and n-type $Al_{0.15}Ga_{0.85}N$ of cladding layers. We investigated an effect of doping concentration on InGaN/AlGaN double heterostructure of light-emitting diode (LED). Thus, energy levels, carrier concentration, electron concentration and forward voltage (*IV*) are extracted from the simulation results. As the doping concentration is increasing, the performance of threshold voltage, Vth on one quantum well (1QW) is also increases from 2.8V to 3.1V.

1 Introduction

Wide band gap gallium nitride (GaN) based semiconductor device has become a great potential of electronic and semiconductor research community due to its extensive applications for optoelectronic devices in the blue and ultraviolet (UV) wavelength region. Gallium nitride (GaN) based LEDs have played a progressively imperative roles in countless promising reals, for instance, an exterior automobile lighting, traffic signals, backlighting sources for solid-state lighting and liquid-crytal displays (LCD) [1]. Thus, the wavelength range determines the LED applications for instance blue LED with 430nm-470nm is used for medical field while other range operates in different fields.

The improvement in these devices obtained by enhancing the device structure and quality is still on research progress which look forward to upgrading its efficiencies and performances. Efficiency droop effect [2], high forward voltage [3], current-crowding [4], low hole concentration in p-GaN [5] are such an obstacle that is still under research progress. Thus, several solutions have been introduced to overcome these problems. Besides, doping concentration also lead to a serious issues in GaN. It has been widely observed by many research communities that an electron mobility increases when the free electron concentration is decreased, and then decreases after a critical electron concentration is reached. The threshold voltage is typically in the low $1x10^{17}cm^{-3}$ range and is dependent upon the threading dislocation density. In this context, the improvement of increasing doping concentration in gallium nitride (GaN) LED structure has a great importance for better performances in GaN LED structure.

The modelling, design and to understand more into a complex nitrides device structure needs multiphysics tool [6]. Furthermore, during last few recent years, various method of numerical modelling and simulation of gallium nitride (GaN) based light-emitting diode (LED) have been published and develop [7-11]. Thus, for enhancing the device performance of GaN LED, typically the doping concentration plays an important roles to increase the great performance of light-emitting diode itself.

In this work, a study is performed to evaluate an effect of doping concentration on InGaN/AlGaN double heterostructure gallium nitride (GaN) LED. Fundamental relations such as the band diagram, carrier concentration, electron concentration, current-voltage characteristics (*IV*) are explored with different values of doping concentration.

2 LED device modeling

Single quantum well (1QW) InGaN/AlGaN LED is investigated in multiphysics tool to depict the behaviour of LED. One quantum well (1QW) is basically designed with 3 domains namely, n-AlGaN from coordinate 0µm to 0.15µm, first quantum well (QW) of InGaN from

* Corresponding author: syafiraatiqah@gmail.com

0.15μm to 0.155μm and p-AlGaN region from 0.155μm to 0.305μm respectively.

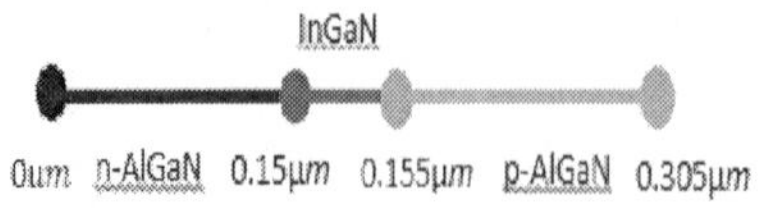

Fig. 1. Device geometry for 1QW InGaN/AlGaN double heterostructure LED.

The device structure can be modeled using 1D line in Comsol. The lower AlGaN layer represents n-doped and the upper AlGaN layer denotes p-doped. Meanwhile, InGaN layer which is located in the centre between two AlGaN layers is undoped. This constructs a P-I-N diode structure with the central (intrinsic) undoped region, InGaN layer that emits light.

In this study, the boundary conditions referred to p and n metal contact boundary conditions are applied to InGaN/AlGaN double heterostructure LED. Moreover, the boundary conditions play an imperative roles as this feature is added to the bounds of device structure for applying voltages and currents to the device. Next, the boundary conditions are vital to toggle amid the continuous quasi-Fermi levels and thermionic emission heterojunction continuity models. In addition, optical transitions are applied in this model to calculate the value of spontaneous emission from LED. This is important as it will assist and make it straightforward to calculate the process such as atom or molecule transition from an excited energy state to a lower energy state and emits a quantum in the form of photon.

Besides, the number of doping concentration is applied for all domains of p and n-type AlGaN. The range of doping concentration is 10^{16} cm^{-3} to 10^{19} $cm^{-3.}$ This doping concentration range is a possible range for light-emitting diode devices to effectively operate in various applications.

Next, Auger recombination feature is added to the structure as it functionally to include a non-radiative recombination mechanism. Thus, this is the only possible non-radiative recombination mechanism, in fact a various technique may be present such as lattice or impurity scattering and phonon interactions. Besides, the purpose to display a proof model, Auger recombination is adequate to monitor the behaviour such as the efficiency is reduced as the current densities are increased. In addition, trap-asisted recombination feature is added to all domains in the LED structure.

In addition, mesh is applied onto the entire geometry of 1D of InGaN/AlGaN double heterostructure LED. The first distribution is developed for domain 1 and 3 only. Meanwhile, second distribution is built for domain 2 only. The mesh is important for a device structure as it can evaluate the transition area of electron and photon.

Afterwards, two types of studies which are preliminary study and bias voltage are evaluated and contemplated in this work for resolving the problems. Firstly, the preliminary study is mainly focused on ramping the doping and analyse the data transition from the continuous quasi-Fermi level heterojunction to the thermionic emission layer. This work includes two study modules whereby the first module solves the case of zero applied bias, by using a low value of doping and the continuous quasi-Fermi level heterojunction continuity model. The resolution from this module is used as an indicator for the second study module which changes to the thermionic emission continuity model and ramps the doping up to the preferred level.

Secondly, the biased voltage study is evaluated and represented as voltage sweeps across the device in the range from 0V to 3.3V. This feature extracts an energy level diagram for the device along with the carrier concentration distribution and the current-voltage characteristics *(IV)* relation.

Furthermore, the performance of single quantum well (1QW) InGaN/AlGaN double heterostructure LED is compared based on energy levels, carrier concentrations, current-voltage characteristics *(IV)* at different values of doping concentration in a range from $1x10^{16\ cm-3}$ to $1x10^{19}$ cm^{-3}.

2 Simulation result and discussion

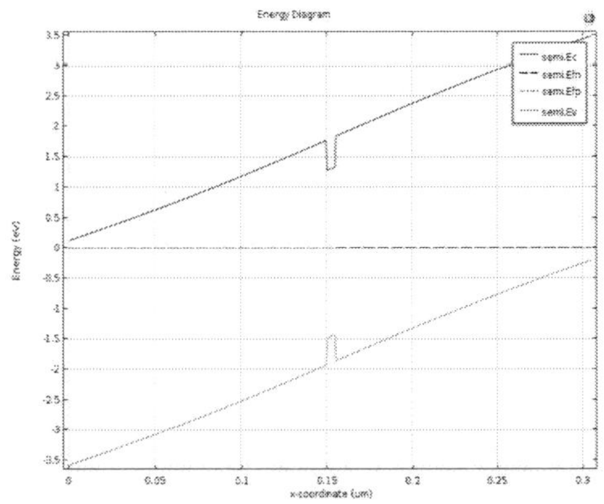

(a)

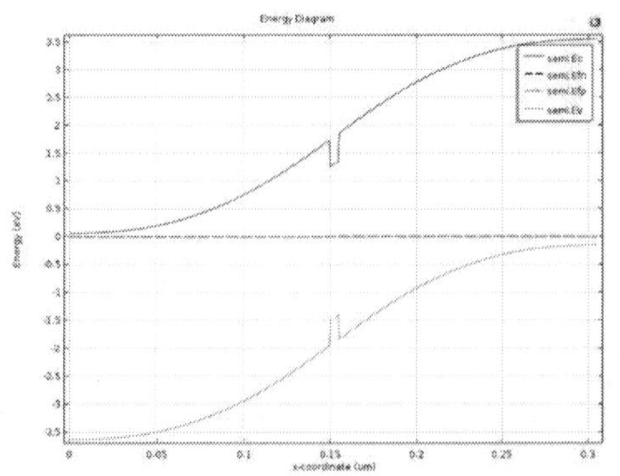

(b)

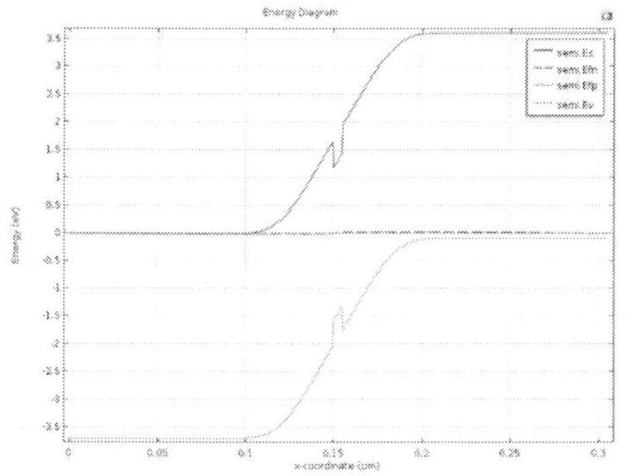

(c)

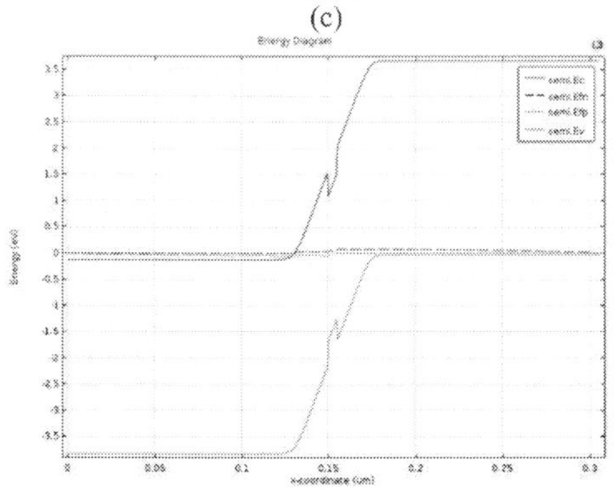

(d)

Fig. 2. Energy band diagram at 0V for doping concentration (a) $1x10^{16}$ cm-^{3} (b) $1x10^{17}$ cm-^{3} (c) $1x10^{18}$ cm-^{3} (d) $1x10^{19}$ cm-^{3}.

Fig. 2 depicts the energy band diagram at 0V for doping concentration of $1x10^{16}$ cm-^{3}, $1x10^{17}$ cm-^{3}, $1x10^{18}$ cm-^{3} and $1x10^{19}$ cm-^{3}. Based on the band diagram illustrated in Fig.2, the semi.Ec, semi.Ev, semi.Efn, semi.Efp are the designation for conduction band, valence band, quasi-Fermi level in region n and p respectively. It shows that the potential barrier is formed between the left side n-doped region and p-doped region. This occurs due to the potential well in both conduction and valence band are lower than the band gap of InGaN layer. Fig. 2(a) depicts a linear relation of energy diagram for both conduction and valence band at 0V for doping concentration of $1x10^{16}$ cm-^{3}. Meanwhile, as doping concentration increases from $1x10^{16}$ cm-^{3} to $1x10^{19}$ cm-^{3}, the trend changes from linear to curvature relation. This is because the number of holes and electron is increased due to the number of doping concentration which also increased in the recombination process. In addition at zero bias, the quasi-Fermi level is away from the band edges of InGaN layer for all modules.

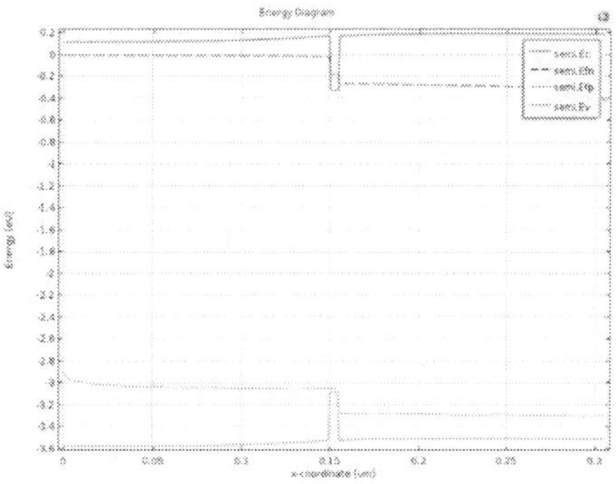

(a)

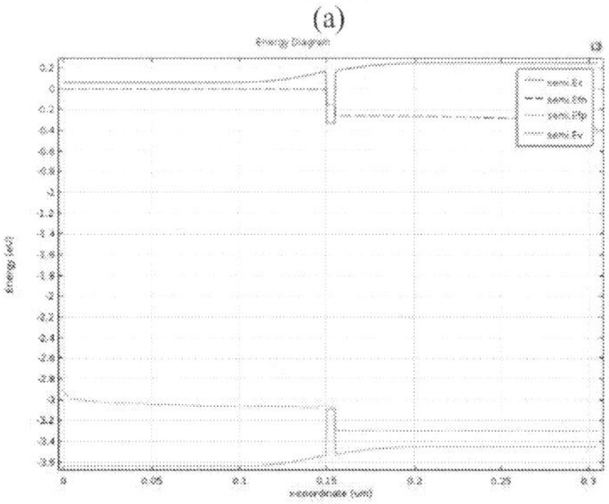

(b)

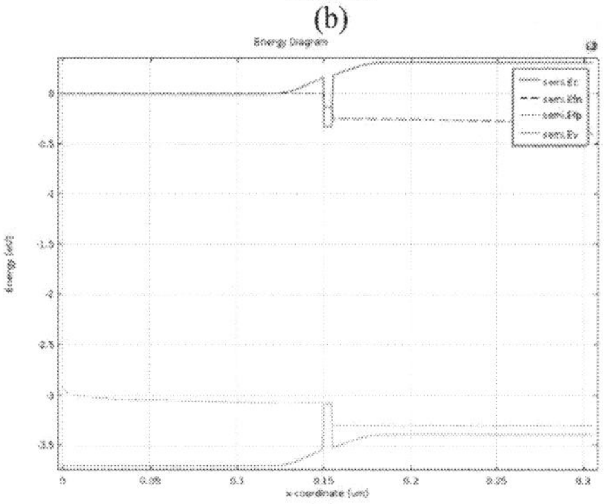

(c)

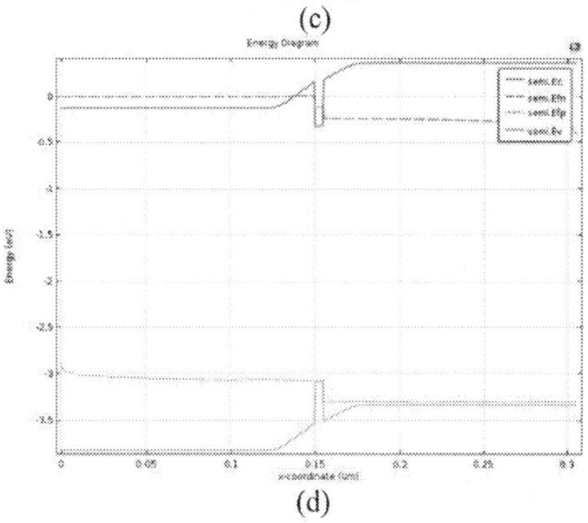

(d)

Fig. 3. Energy band diagram at 3.3V results for doping concentration (a) $1x10^{16}$ cm-^{3} (b) $1x10^{17}$ cm-^{3} (c) $1x10^{18}$ cm-^{3} (d) $1x10^{19}$ cm-^{3}.

Fig. 3 displays that under voltage bias 3.3V, the holes will be injected into one quantum-well (1QW).

Besides, the height of potential barrier between n-region and p-region is decreased when forward biased at 3.3V.

Fig. 3(a) depicts that the level of conduction and valence bands is far enough from the Fermi-level in the p and n-type materials. As the number of doping concentration is increased, the trend of semi.Ec and semi.Ev is close enough to the Fermi-level.

In all models, the hole quasi-Fermi energy levels are very near to the valence band edge. Thus, it can be predicted that both conduction and valence bands are populated with electrons and holes respectively at aforementioned forward bias condition in all modules. This is due to the materials that are brought together to form a junction.

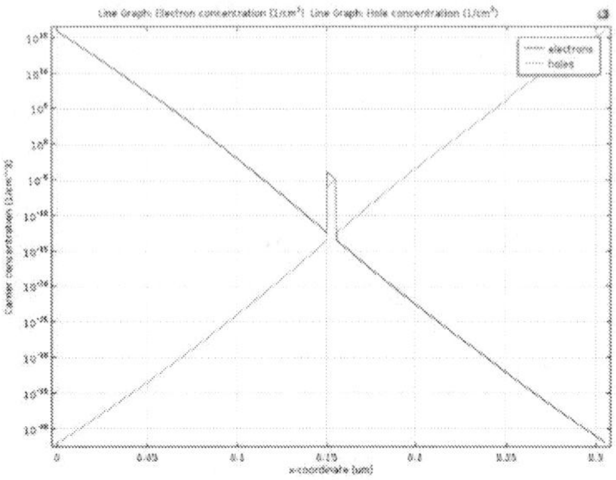

(a)

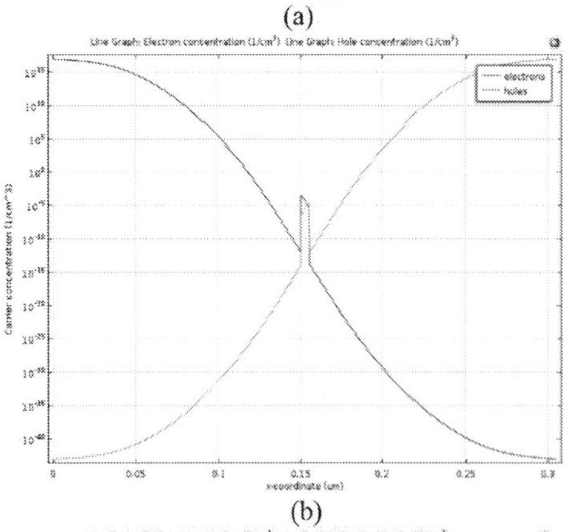

(b)

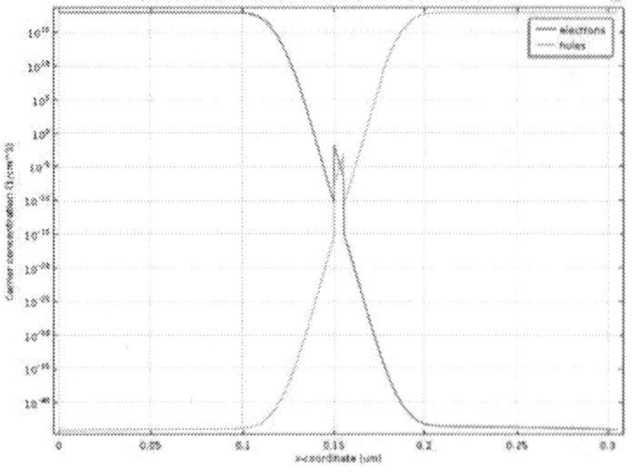

(c)

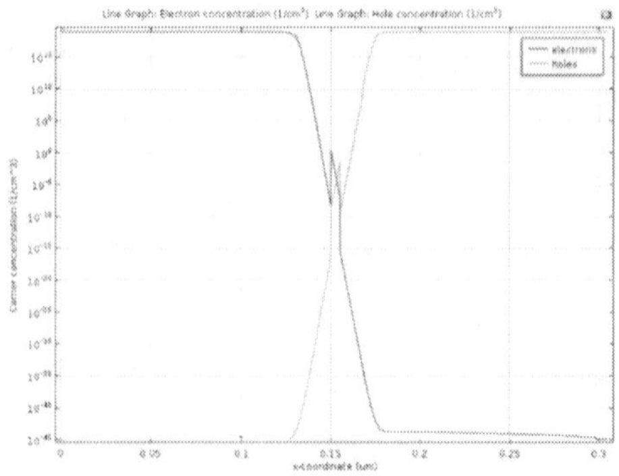

(d)

Fig. 4. Carrier concentration at zero bias (a) $1x10^{16}$ cm^{-3} , (b) $1x10^{17}$ cm^{-3} , (c) $1x10^{18}$ cm^{-3} , (d) $1x10^{19}$ cm^{-3}

Fig. 4 demonstrates the carrier concentration involving both holes and electrons for different values of doping concentration in a range from $1x10^{16}$ cm^{-3} to $1x10^{19}$ cm^{-3}. As illustrated in Fig. 4, as the number of doping concentration increases, the combination peak of holes and electron reaches approximately at the same level.

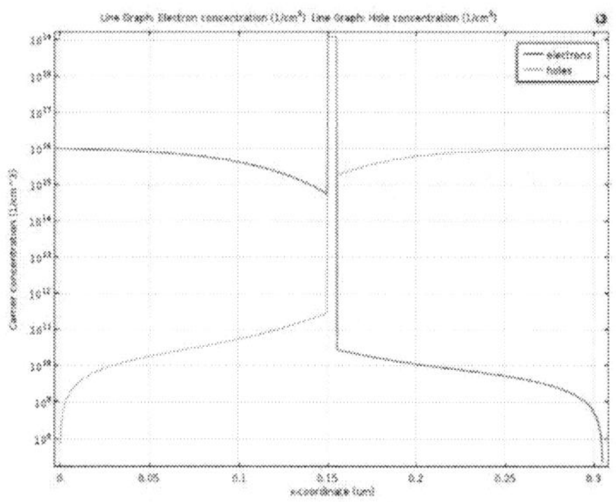

(a)

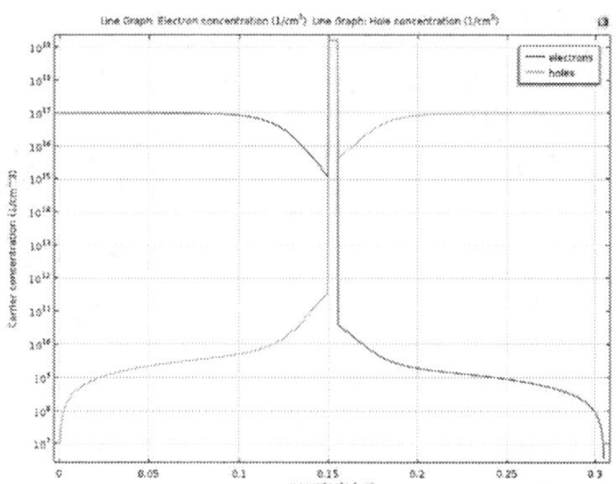

(b)

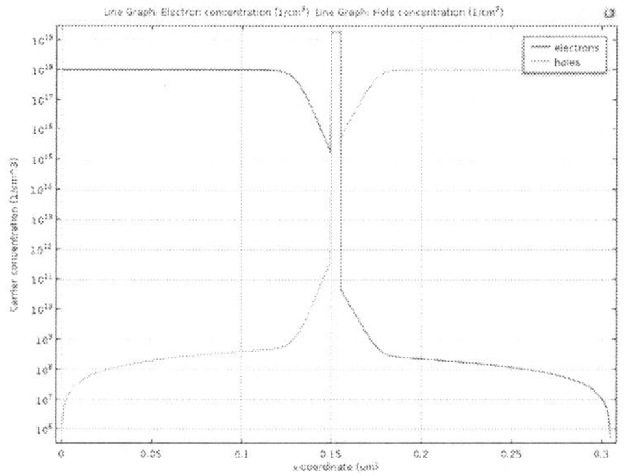

(c)

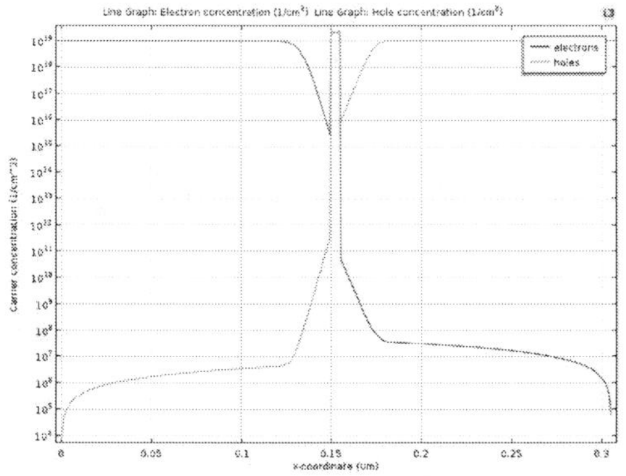

(d)

Fig. 5. Carrier concentration at 3.3V forward bias (a) $1x10^{16}$ cm-^{3}, (b) $1x10^{17}$ cm-^{3}, (c) $1x10^{18}$ cm-^{3}, (d) $1x10^{19}$ cm-^{3}

As illustrated in Fig. 5, the carrier concentration versus device coordinate plot showed that the population of carriers in InGaN layer increased drastically with the increased bias of LED along with a different number of doping concentrations. This occurs as doping concentration is increased gradually from $1x10^{16}$ cm-^{3} to $1x10^{19}$ cm-^{3} whereby the number of holes and electrons also increased.

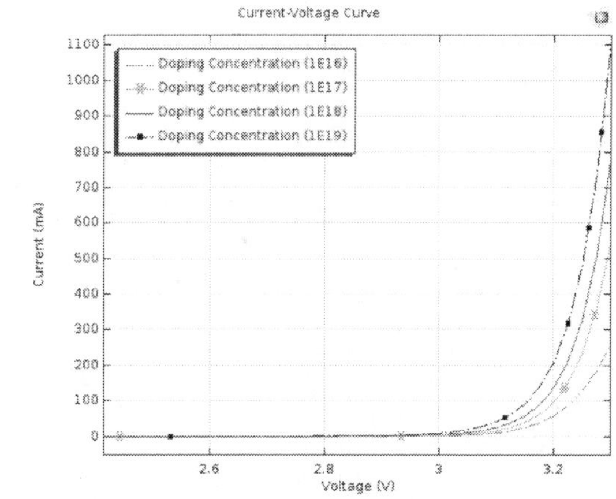

Fig. 6. Current-voltage (*IV*) characteristics for different value doping concentration.

Fig. 6 indicates the current-voltage (*IV*) characteristics for different values of doping concentration. In addition, the light-emitting diode is biased in voltage ranges of 0V-3.3V. Besides, the graph depicts that the value of threshold voltage, Vth is increased with a low value of doping concentration, $1x10^{16}$ cm-^{3}. Meanwhile, the minimum turn on voltage is needed for LED to operate at a high doping concentration which is $1x10^{19}$ cm-^{3}. This is happen due to the minimum energy needed to turn-on voltage as both electron and holes are increased along with an increment of doping concentration.

3 Conclusion

In this paper, the InGaN/AlGaN double heterostructure light-emitting diode (LED) is successfully developed. The band diagram, carrier concentration, electron concentration, current-voltage characteristics (*IV*) are studied with their relation at different values of doping concentration. It is observed that energy band diagram under zero bias caused the potential barrier to be created between the left side n-doped and p-doped regions. Meanwhile, under 3.3V forward bias, potential barrier between p and n-doped region is decreased as the forward bias is applied. This is caused by the different materials which are brought together to form a junction. Besides, as the number of doping concentration increases, the peak of holes and electrons combination reaches approximately zero. Under forward bias of 3.3V, the population of carriers in InGaN layer increased drastically through the increment of LED's drive voltage along with different levels of doping concentration. The current-voltage (*IV*) characteristics extract that the threshold voltage, Vth is higher at lower doping concentration.

The authors would like to thank the School of Microelectronics Engineering, Universiti Malaysia Perlis (UniMAP) especially SPILS for their support in this work. Partial of this research is

 DOI: 10.1051/epjconf/201716201037

funded by FRGS 9003-00461 under Ministry of Higher Education, Malaysia.

References

1. He An-He et al., Chinese.Physics B, vol.**19**, no. 6, (2010).
2. J. Piprek, 13th International Conference on Numerical Simulation of Optoelectronic Devices (NUSOD), Vancouver, BC, pp. 107-108, (2013).
3. Yoshitaka Taniyasu and Makoto Kasu, Diamond & Related Materials, **4**, (2008).
4. Tae Hoon Seo, Applied physics Express, **5**, 115101, (2012).
5. Guofeng Yang et. al, Superlattices and Microstructures, **83**, pp.1–8, (2015).
6. Joachim Piprek, "Nitride Semiconductor Devices- and Simulation", Wiley-Vch Verlag GmbH & Co. KGaA, Weinheim, (2007).
7. Friedhard R¨omer and Bernd Witzigmann, Optics Express, A1443, vol. **22**, no. S6, (2014).
8. Chia-Lung Tsai and Wei-Che Wu, Materials, vol.**7**, pp.3758-3771, (2014).
9. Kwang Jae Lee et al., Optics Express, A1164 vol. **22**, no. S4, (2014).
10. Xing-ming Long, Rui-jin Liao and Jing Zhou, Advances in Opto Electronics, vol. 2012, article ID 495981, pp. **1-7**, (2012).
11. D. Oriato, Alison B. Walker and W. N. Wang, VLSI Design, vol. **13**, no. 1-4, pp. 295-299, (2001).

 DOI: 10.1051/epjconf/201716201038

Effect of InGaN thickness on assisted trap recombination and behaviour of InGaN/AlGaN double heterostructure LED

Shanise Rashid[1,*], *M.Halim* A. Wahid[1], *N. Azura* M. Ahmad Hambali[1], *N.Syafira* Abdul Halim[1], and *Mukhzeer* M.Shahimin[2]

[1]Semiconductor Photonics & Integrated Lightwave Systems (SPILS), School of Microelectronic Engineering, Universiti Malaysia Perlis, Pauh Putra Campus, 02600 Malaysia.
[2]Department of Electrical and Electronic Engineering, Faculty of Engineering, National Defence University of Malaysia (UPNM), Kem Sungai Besi, 57000 Kuala Lumpur.

Abstract. This work is dedicated to the study of InGaN based LED on the thickness variation effect. The operating voltage, total emission rate, efficiency droop and spontaneous recombination rate was improved by increasing the thickness of InGaN layer of InGaN/AlGaN electron blocking layer (EBL). Based on optical and electrical results, the thicker InGaN layer could have lower operating voltage and higher total emission rate. LED with 5nm thick InGaN layer also could prevent electron leakage into p-region and improve hole injection efficiency. As a result it decreases 60% of efficiency droops and reduces the trap assisted recombination in InGaN/AlGaN active light emitting region. This indicates that InGaN layer may decrease the non-radiative recombination.

1 Introduction

Nitrides semiconductors, especially the gallium nitride (GaN) based light emitting diode (LED) with an InGaN/AlGaN double heterostructure, are among the most promising optoelectronic devices to be developed in recent years [1-3]. With wide bandgap of GaN, thus will emit light from close to the UV range to blue, it can be used in various lighting devices with colour converting material [4]. Recent works based on microsphere array [5,6] and photonic crystal [7] showing great potential to enhance the light extraction efficiency (LEE) in GaN based LEDs. Although the output power of current LEDs is already high, the output efficiency of these LEDs is still required further improvement to achieve feasible solid state lighting (SSL). Besides that, the carrier injection efficiency has the key functions in the light emission efficiency of GaN based LEDs. Meeyard et al. [8, 9] reported asymmetry during carrier transport because of the much lower concentration and mobility of holes in p-GaN and p-AlGaN compared with electrons. The asymmetry in carrier transport reduces the light emission efficiency of LEDs and enlarges the efficiency droop.

Recently, it has been suggested to redesign the electron blocking layer (EBL) because it can lead to weak electron confinement and can act as a potential barrier for holes injection [10]. Yu et al. [3] experimentally and theoretically showed that EBL with thick InGaN layer have improved hole injection efficiency and increasing emission properties.

In this study, we investigate the thickness effect of InGaN layer on optical and electrical properties of the active light emitting region of InGaN/AlGaN LEDs.

2 Methodology

The active double heterojunction region, from which the desired emitted light, is modelled stand on the very basis of the GaN based blue LED structure that once has been fabricated by Nakamura et al. [11]. This device consists of p-GaN layer, an active region of 0.15µm thickness of $Al_{0.15}Ga_{0.85}N$ layer and 2nm thickness of $In_{0.06}Ga_{0.94}N$ layer. Note that only the active light emitting region of GaN based blue LED is modelled. The lower AlGaN layer is n-doped while the upper AlGaN layer is p-doped. The InGaN layer is undoped, thus creates PIN diode structure with the light emitting InGaN layer in the central intrinsic region. The device geometry has lateral dimension of 200x200 µm which is a standard size of LED device. The structure of the LED device is shown in Figure 1.

Modelling tool (1D analysis of LED structure) is used to simulate InGaN/AlGaN based LEDs, in which Auger recombination factors (Auger recombination

* Corresponding author: shanise605@gmail.com

factor for electrons and holes = $1.7x10^{-30}$ cm^6s^{-1}) and carrier lifetimes for trap assisted recombination (electron and hole lifetime = $1x10^8$ s) have been included in the materials [12-13]. All the parameters are based on GaN properties from semiconductor module material library but values of electron affinity and bandgap energy have been altered to reflect In and Al alloy composition. Quantum confinement effect is not included within the thin InGaN material in this model. This is because quantum dots take many geometries, thus Schrodinger equation has to be solved which is not the interest of the study at the moment, hence it will take longer computation time.

In order to apply voltage and current bias to the device, metal contact boundary conditions are applied to the bounds of geometry as illustrated in Figure 1(a). This is to enable toggling between thermionic emission heterojunction continuity model and continuous quasi-Fermi levels.

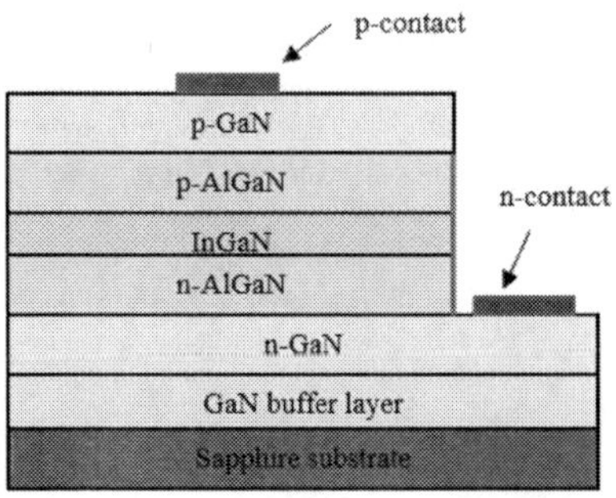

(a)

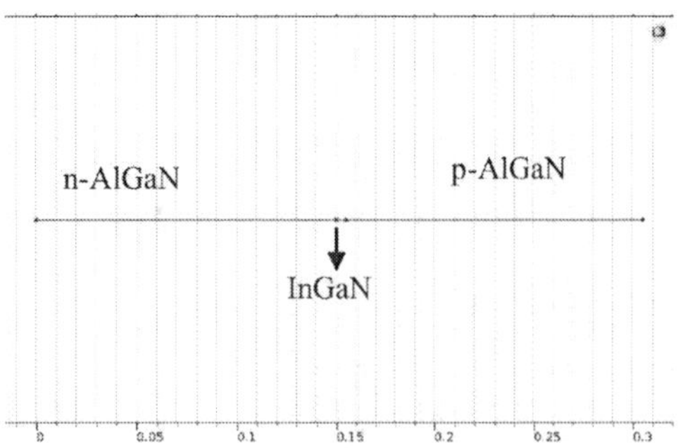

(b)

Fig. 1. Geometry of LED device. (a) Structure of modelled layers. The focused section is highlighted in red line. (b) Double heterostructure region of full device in 1D model.

The AlGaN thickness and Al% in InGaN/AlGaN EBL is fixed at 0.15μm and 15%, respectively. The InGaN In% in InGaN/AlGaN is fixed at 6%. The InGaN thickness in InGaN/AlGaN EBL for 2nm is set as reference LED (ref-LED) and others are variably set at 3,4 and 5 nm and labelled as LED I, LED II and LED III respectively.

Thicknesses of double heterojunction region and its characteristics of GaN LED are investigated, as well as reduction of trap assisted recombination in the active region. Only the thicknesses have been varied while other parameters that can affect the simulation such as bandgap and value of electron affinity of InGaN and AlGaN are fixed as summarized in Table 1.

Table 1. Material properties.

Property	Unit, V
Bandgap of AlGaN	3.7
Bandgap of InGaN	2.8
Electron affinity of AlGaN	4.1
Electron affinity of InGaN	4.6

Three investigations are performed in this work which are preliminary mode, voltage biased and current biased studies. Preliminary study is used to connect the doping and transition level from the continuous quasi-Fermi level heterojunction continuity model to suitable level of thermionic emission model. Then, voltage sweep is used to apply bias voltage ranges from 0V to 3.3V across the device. Current injection study is adopted to converge initial solution from voltage driven and to sweep several orders of magnitude of 1μA to 700 μA through the device.

3 Results and discussion

Figure 2 shows the forward current-voltage (I-V) characteristic curves of all fabricated LEDs with different thicknesses of InGaN layer. Forward voltage (V_f) of ref-LED is 3.19V while LEDs I to III are 3.186V, 3.183V and 3.182V respectively. The V_f of LEDs I, II and III are lower than ref-LED is recorded. This reduction might be due to the decrement of effective potential barrier on the hole carrier between p-GaN and p-AlGaN/InGaN active double heterojunction region.

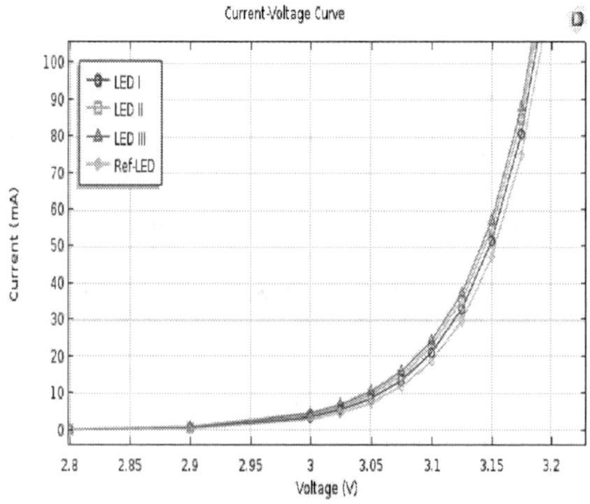

Fig. 2. I-V characteristics of all LEDs.

Increasing InGaN thickness of InGaN/AlGaN electron blocking layer would also increase the total emission rate and shrink the efficiency droops of LED as illustrated in Figures 3 and 4. Figure 3 depicts the total emission rate from InGaN layer as a function of applied current. The visible effect is known as LED droop. This LED droop caused by a non-radiative carrier loss mechanism that has little effect at low currents, but become dominant at high currents. In general, carrier losses can occur either inside or outside the active region. The results show that the total emission rate of InGaN layer increases with thickness. At 100mA, LED III shows the highest total emission rate, $1.2x10^{17}$ (s^{-1}) follows by LED II and LED I, $1.12x10^{17}$ (s^{-1}) and $1.03x10^{17}$ (s^{-1}) respectively. Ref-LED demonstrates the lowest total emission rate of $1x10^{17} (s^{-1})$.

The efficiency droop is even more apparent in Figure 4 which portrays the internal quantum efficiency (IQE) as a function of current density. Not all electron hole recombination is radiative. IQE is the proportion of all electron-hole recombination in the active region that are radiative, producing photons. There is a peak in the IQE at current density around 30 (A/cm^{-2}) with applied current in the 10-30mA range and there is rapid drop off for densities in access 100 (A/cm^{-2}). The IQEs for all LEDs are almost the same at low current density. However, when the current increases to 1600 (A/cm^{2}), the efficiency droop appears and the difference is clearly visible. It is also observed that the IQE of ref-LED decreases almost 73% when the current increases to 1600 (A/cm^{2}), while IQE for LED I decreases by 63%, follows by LED II 64% and LED III decreases only 60%. This demonstrates that the structure with thicker InGaN layer can improve the efficiency droop significantly, which suggests that the proposed thickness of InGaN layer can mitigate the problem of injection current efficiently.

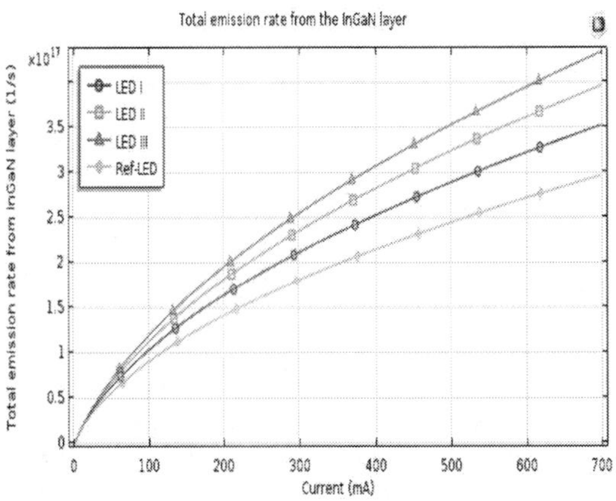

Fig. 3. Total emission rate as a function of current of all LEDs. The sublinear curve is a characteristic of LED droop.

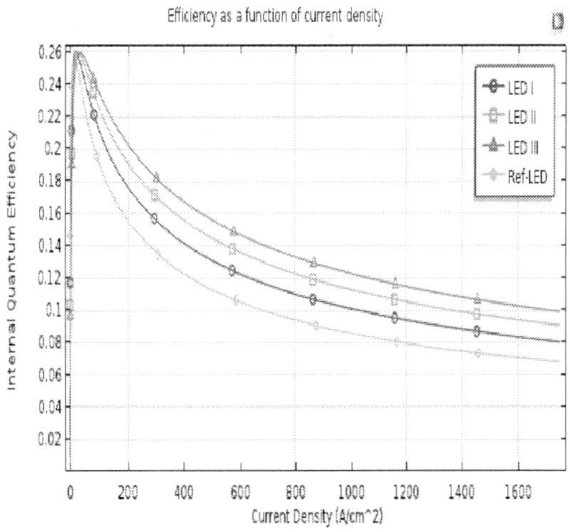

Fig. 4. Efficiency as a function of current density of all LEDs.

Based on droop efficiency that increases with thickness of InGaN layer, this indicates that InGaN layer may decrease the non-radiative recombination. Non-radiative recombination consists of Shockley Read Hall (SRH) or called as trap-assisted recombination process and Auger recombination process. Auger recombination is smaller than SRH recombination which is constant with Kuo's result that the Auger recombination can be ignored under the case that the carrier density is below $1x10^{20}$ cm^{-3} [14]. Semiconductor diode would deviate from ideal behaviour when generated-recombination processes took place in the depletion region as described by Shockley [15]. It is confirmed that SRH generated-recombination in the depletion region could control the feature of semiconductor diodes. The performance of LED is further explored to support those theories.

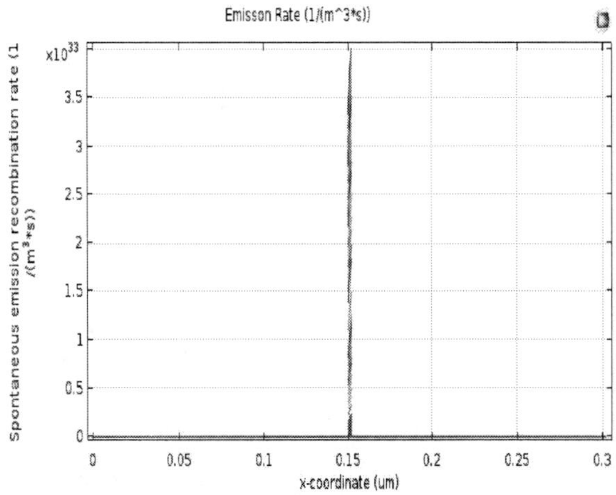

(a) 2nm (ref-LED) thickness of InGaN layer.

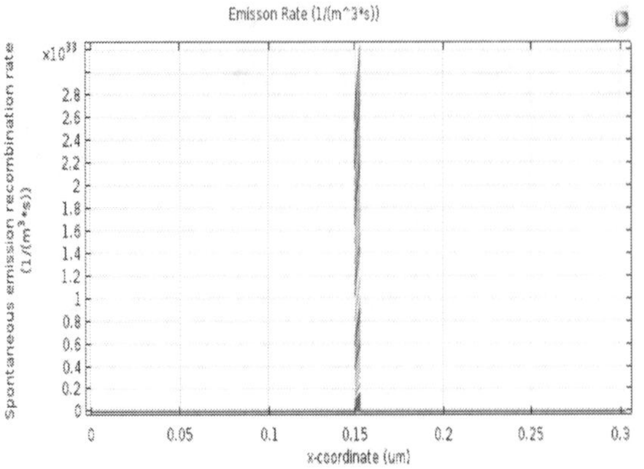

(b) 3nm thickness of InGaN layer

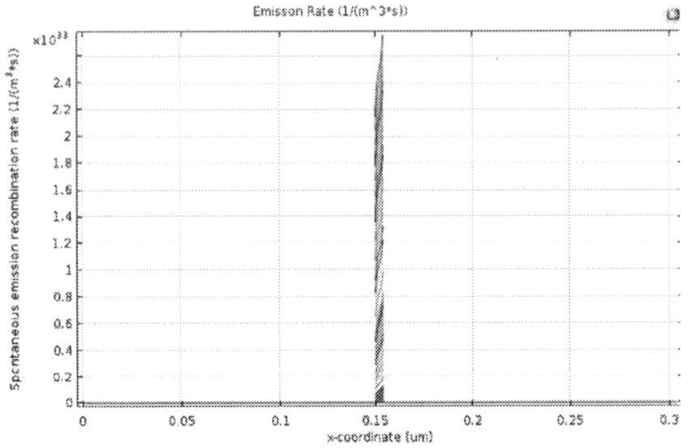

(c) 4nm thickness of InGaN layer

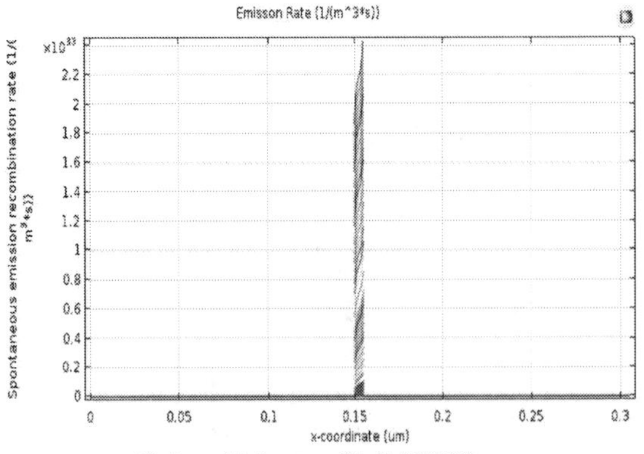

(d) 5nm thickness of InGaN LED

Fig. 5. Emission rate of all LED samples in the selected range of injected currents.

Figure 5 depicts the emission rate in the active region of LEDs with different InGaN thicknesses. The emission from InGaN layer in all samples is significantly larger than in the cladding material due to InGaN layer is the only region of the device which has a high concentration of both holes and electrons. The results show that as thickness of InGaN layer increases, spontaneous emission recombination rate decreases.

The spontaneous emission recombination rate of LED chips with 5nm GaN layer is minimum among all samples which is 2.42×10^{33} ($1/(m^3*s)$) compared to 2nm with highest spontaneous emission rate, 4.01×10^{33} ($1/(m^3*s)$). The optical and electrical behaviour of LED is greatly improved by inserting InGaN layer of appropriate thickness which is beneficial to improve the depletion region quality, thus decreases the spontaneous recombination rate.

4 Conclusion

Characteristics of InGaN/AlGaN LED with undoped InGaN EBL have been investigated with different thicknesses of InGaN that sandwiched between AlGaN. The results indicated that InGaN layer thickness affected LEDs' performance greatly. It changes the effective potential height of electrons and holes respectively. Therefore, LED III with 5nm InGaN layer of InGaN/AlGaN EBL showed highest total emission rate with less efficiency droops with the injection current, improved hole injection efficiency and reduce the non-radiative SRH recombination.

The authors would like to thank the School of Microelectronics Engineering, Universiti Malaysia Perlis (UniMAP) especially SPILS for their support in this work. Partial of this research is funded by FRGS 9003-00461 under Ministry of Higher Education, Malaysia.

References

1. S. Nakamura, Mukai, T., Senoh, M., Jpn. J. Appl. Phys. 30 L1705, (1991).
2. S. Nakamura, IEEE K. Sel. Topics ., vol. **3**, no. 2, pp. 435-442, (1997).
3. C. C. Yu, C. F. Chu, J. Y. Tsai, H. W. Huang, T. H. Hsueh, C. F. Lin, and S. C. Wang, Jpn. J. Appl. Phys. 2, Lett., vol. **41**, no. 8B, pp. L910–L912, (2002).
4. R. Mueller-Mach, G. Mueller, M. R. Krames, H. A. Hoeppe, F. Stadler, W. Schnick, T. Juestel, and P. Schmidt, Phys. Stat. Sol. (a) **202**, No. 9, 1727-1732, (2005)
5. P. Zhu, G. Liu, J. Zhang, and N. Tansu, J. Displ. Technol. **9(5)**, 317–323, (2013).
6. X. Li, P. Zhu, G. Liu, J. Zhang, R. Song, Y. Ee, P. Kumnorkaew, J. F. Gilchrist, and N. Tansu, J. Displ. Technol. **9(5)**, 324–332 (2013).
7. J. Jewell, D. Simeonov, S. C. Huang, Y. L. Hu, S. Nakamura, J. Speck, and C. Weisbuch, Appl. Phys. Lett. **100(17)**, 171105, (2012).
8. D. S. Meyaard, G.-B. Lin, Q. Shan, J. H. Cho, E. F. Schubert, H. Shim, M. H. Kim, and C. Sone, Appl. Phys. Lett. **99(25)**, 251115, (2011).
9. G.-B. Lin, D. Meyaard, J. H. Cho, E. F. Schubert, H. Shim, and C. Sone, Appl. Phys. Lett. **100(16)**, 161106, (2012).
10. S.H. Han, D.Y. Lee, S.J. Lee, C.Y. Cho, M.K. Kwon, S.P. Lee, D.Y. Noh, D.J. Kim, Y.C. Kim, S.J. Park, Appl. Phys. Lett. **94**, 231123, (2009).

11. S. Nakamura,T. Mukia, and M. Senoh, Appl. Phys. Lett. **64**, 1687, (1994).
12. Y.C. Shen, G. O. Mueller, S. Watanabe, N. F. Gardner, A. Munkholm, and M. R. Krames, Appl. Phys. Lett. **91**, 141101 (2007).
13. Q. Dai et al., Appl. Phys. Lett. **94**, 111109, (2009).
14. S. H. Yen, M. C. Tsai, M. L. Tsai, Y. J. Shen, T. C. Hsu, and Y. K. Kuo, Appl. Phys., A Mater. Sci. Process. **97(3),** 705–708 (2009).
15. W. Shockley, Bell Syst. Tech. J. **28**, 435–489 (1949).

Preparation of quantum state *(review)*

N Ali[1,2,4*], *N. R.* Yusof[1,2], *S* Soekardjo[5], *S.* Saharudin[3] and *R.* Endut[4]

[1]School of Microelectronics Engineering, Universiti Malaysia Perlis (UniMAP), 02600 Arau, Perlis, Malaysia
[2]Semiconductor Photonics & Integrated Lightwave Systems (SPILS), School of Microelectronic Engineering, Universiti Malaysia Perlis, Pauh Putra Main Campus, 02600 Arau, Perlis, Malaysia
[3]Nano Photonics and Nano Electronics, Mimos Berhad, Technology Park Malaysia, 57000 Kuala Lumpur
[4]Advanced Communication Engineering, Centre of Excellence-School of Computer and Communication Engineering, Universiti Malaysia Perlis
[5]Department of Physics, Kulliyyah of Science, International Islamic University Malaysia, Jalan Sultan Ahmad Shah, Bandar Indera Mahkota 25200 Kuantan, Pahang.

Abstract. We reviewed experimental results and publications prepared by the Quantum Laboratory, Mimos Berhad. The complexity of the setups lies mainly in preparing the quantum states. Optics is chosen as the medium of this quantum system. The two methods - fiber based and free space systems are different from each other in terms of experimental setups, components configuration, and selections.

1 Introduction

The communications today are secured by the public key cryptography which is based on the mathematical complexity. As the computer performance improves through technology development, the length of the encryption key has to be increased in order to maintain confidentiality and security of the encrypted massages as it transfers to the other party. On the other hand, the speed of the computer can overcome the mathematical complexity of the encryption system [7], [8]. The main threat of this encryption system is the invention of the quantum computer, which potentially has the computing power billions of times faster than the current technology to process and calculate any digital information [1].

Quantum Computer works in quantum bits (qubits) [9] value as opposed to the current computer that uses binary bits. Quantum mechanic's law is different with the classical computer. A qubit can exist in a normal classical bit and it also can exist in the superposition state which can be both values at the same time. Thus, a computer working in qubit rather than standard bits can make a calculation using both values simultaneously. There is also the term qubyte which is a group of eight qubits. A quantum computer can have power beyond today's classical supercomputer and can just resolve calculation as fast as 1 billion times faster than Pentium III PC [2].

As a preventive measure, people try to maintain the security of messages by securing the encryption key which is breakable very fast by a quantum computer for current public key cryptography. Symmetric key cryptography will assure the messages sent are secured since the keys are only shared by the sender and receiver. On the other hand, the system is difficult to be implemented since it is difficult to send the key securely into the other place via a public communication channel. A trusted party is needed to transfer the key to the other end securely. By introducing Quantum Key Distribution (QKD) the key can be shared between two parties securely via a normal communication link. The security of this system lies in the quantum mechanical principles and will result in securely encrypted messages.

In the experimental implementation of a QKD system, there are two mainly used techniques for the optical quantum system which are the free space system and the fiber based system. In the free space system, the quantum information is sent via polarization of photon since the polarization states are stable in the free space region, even at the quantum level. This is one of the advantages of transmission in the free space region, but the limited line-of-sight limits the expected range of transmission. The second technique is the fiber based system whereby the photon is transmitted through the optical fiber. In the fiber based system, the photon cannot be coded using the state of polarization because it is susceptible in the fiber. Due to this limitation, the system use phase coding method in the fiber which can be retained throughout the transmission.

In this paper, we will go through a theoretical review of the protocol that was used for both setups (fiber based and free space). Then we will explain the experimental setup on the fiber based followed by the free space. After that, this paper will discuss the experimental setup which will explain how the preparation of quantum state affect the experimental setup, and lastly conclusion.

* Corresponding author: norshamsuri@unimap.edu.my

2 Theoretical review

The system is working based on BB84 protocol. In this protocol, the key is transmitted in one of four states and decoded by two bases. The important part is that the protocol uses two conjugate bases. The sender (Alice) encodes the key by two orthogonal states of the quantum system. She will randomly change the two orthogonal states that refer to each basis. This means that, for each encoded key, she will choose the state from the two conjugate bases. The state of this protocol can be any quantum state, but for this explanation, we will use polarization state as an example. One basis consists of two polarization states i.e $|H\rangle$ and $|V\rangle$, or $|45\rangle$ and $|135\rangle$. Where $|H\rangle$ is a horizontal state, $|V\rangle$ is a vertical state. Both states are in rectilinear basis. $|45\rangle$ is diagonal state and can be represented as $|D\rangle$ and $|135\rangle$ represents anti diagonal state and also can be represented as $|A\rangle$. Both states are in diagonal basis. The protocol encoded as below:

Horizontal basis

$$|H\rangle \Rightarrow "0" \quad (1)$$

$$|V\rangle \Rightarrow "1" \quad (2)$$

Diagonal basis

$$|D\rangle \Rightarrow "0" \quad (3)$$

$$|A\rangle \Rightarrow "1" \quad (4)$$

$$|D\rangle = \frac{1}{\sqrt{2}}(|V\rangle + |H\rangle) \quad (5)$$

$$|A\rangle = \frac{1}{\sqrt{2}}(|V\rangle - |H\rangle) \quad (6)$$

These four states satisfy the following relationship;

$$\langle H|V\rangle = \langle A|D\rangle = 0 \quad (7)$$

$$\langle H|H\rangle = \langle A|A\rangle = \langle D|D\rangle = \langle V|V\rangle = 1 \quad (8)$$

$$[\langle H|A\rangle]^2 = [\langle H|D\rangle]^2 = [\langle A|V\rangle]^2 = [\langle D|V\rangle]^2 = \frac{1}{2} \quad (9)$$

A deterministic result will be produced only when the measurement performed in the basis is identical to the preparation basis. Any measurement in the rectilinear basis on a photon prepared in the diagonal basis will yield an outcome with equal probabilities and vice versa for diagonal basis. This measurement should be initiated by pre agreeing on the key (0 or 1) carried by respective photon states as shown in equation (1) - (4) above. Since Alice is the sender, she will send random polarization $|H\rangle$, $|V\rangle$, $|D\rangle$, or $|A\rangle$ to Bob and Bob will decode the key by randomly and independently chooses either diagonal basis or rectangular basis. Statistically, Bob will only have 50% probability of measuring the photon with the same bases as Alice prepared i.e. 50% chance of deterministic measurements by Bob. He will only know if his outcomes are deterministic when he compares his measurement basis with Alice's through a normal (classical) communication channel. Whenever the basis is identical, Bob will keep the bit and whenever it is not identical he will discard the measured bit. This classical channel can be intercepted by an eavesdropper but she can only know the basis sent, not the information on the code.

In this experiment, the quality of the interference pattern is measured by measuring the visibility. The ratio of the size or amplitude of oscillations to the sum of the powers of the individual waves is defined as the visibility. The sum of the intensities (or powers) of the two interfering waves equals the average of the fringes and can be written as,

$$Visibility_{Real} = \frac{amplitude}{average} \quad (10)$$

Alternatively, the above equation can be written as follows;

$$Visibility_{Real} = \frac{max - min}{max + min}$$

where max- the maximum of the oscillations
min- the minimum of the oscillations

When the two waves/particles have the same polarization, then the predicted visibility will be;

$$Visibility_{ideal} = \frac{2\sqrt{I_1 I_2}}{I_1 + I_2}$$

where I_1, I_2 - Intensities of the optical waves

3 Experimental setup

3.1. Fibre Based QKD

The set-up for the experiment is shown in Figure 1. It represents one interferometer in a complete plug and plays QKD scheme (Alice and Bob's setups will produce only one interferometer). In this setup, the idea is to modulate in the phase of the photon since the state of polarization is scrambled throughout the fiber distance. The light pulse from the LD is guided out at the common port to the bi-directional optical ratio coupler (ORC). The optical pulse will then take the path of either one of the ORC arm. One part of the ORC arm is connected to a phase modulator (PM) while the other (shorter in length) is connected to a polarization maintaining optical fiber with a length of approximately 2 meters. The longer arm has 8 meters extra length from the shorter arm. In this setup, all fiber is polarization maintaining fiber (PM fiber) which all the polarization state is maintained. Because the
PM is a polarization dependent device; the connection of our rotation polarization is after a pass through the phase modulator in order to ensure only the pulse with correct polarization state enters the PM. The light pulses from both arms are then recombined using a polarization beam coupler/splitter (PBCS) where the output is connected to a 10 km optical fiber link. Due to the unbalanced nature

of the interferometer used, the two superposition optical pulses that recombined and left the PBS will be delayed about 40 ns apart. The optical power (intensity) of the two pulses is attenuated (to an average power of -108.93 dBm) by the VOA to achieve a low number of photons per pulse. In optical fiber QKD scheme, 'single photon' is approximated by light pulses with Poisson photon-number distributions characterized by small values of μ, the mean number of photons/pulse. This action is achieved by attenuating the optical pulses such that the μ value is in the range of 0.1 to 1.

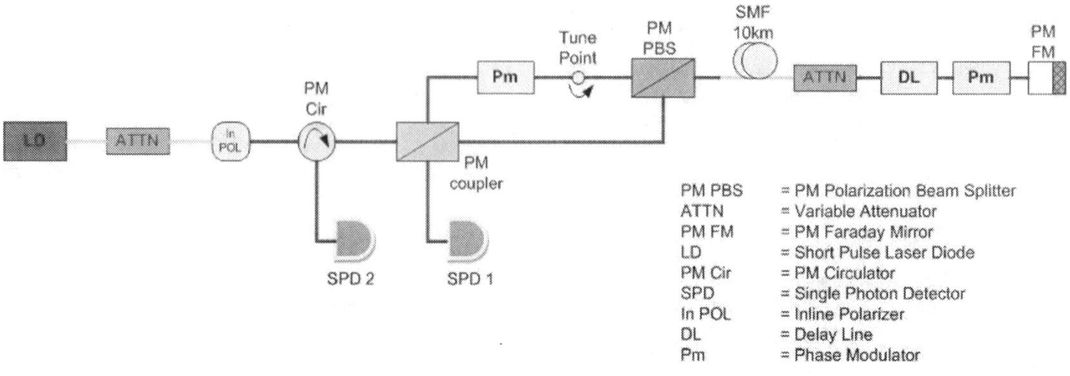

Fig. 1. Schematic of the experimental setup used for Plug and Play configuration QKD interferometer.

The attenuated optical pulse then travels through a standard 10 km telecommunication grade optical fiber (SMF) before it reaches Alice. At Alice, the pulses have to go through the variable attenuator (ATTN) again in order to make sure that the light that reflects back to Bob has a μ value between 0.1 to 1.0. Then the pulses are delayed by a few meter of single mode fiber before being modulated at Alice's modulator. Then the modulated pulses are reflected back by a Faraday mirror (FM) and sent back to Bob. The two pulses will go to the PBS and they will take opposing routes as they come since the polarization of both pulses is already flipped by the FM. Bob will modulate the basis at Bob's modulator, which is placed in the long arm. The interference of the two pulses will determine the readings on the detectors.

3.2 Free sSpace QKD

The cascaded type-I BBO crystal (B2) was placed at the waist position of the sequence femto second pulse collimated by lenses L2 and L3. Each crystal was designed in certain cut-angle to achieve phase matching condition for non-collinear frequency degenerate down conversion. The crystal is rotated with respect to other in such a way that horizontally polarized pump light generates a pair of vertically polarized photon in the first crystal, and vertically polarized pump light generate pairs of a horizontally polarized photon in the second crystal [3]. In order to generate a pair of polarized entangled photon $|\psi\rangle \propto |VV\rangle + |HH\rangle$ we used pump beam polarized at 45o with respect to the crystal axis. By neglecting losses from passing through the first crystal, this polarized pump photon will be equally likely to down-convert in either crystal. Nevertheless, because of the combined effects of group velocity dispersion and birefringence in the two crystals, the space-time components of the two-photon state $|VV\rangle$ associated with the polarized states and $|HH\rangle$ is expected to be temporally displaced after the crystals. As a result, information about "which-polarization" the emitted photon pair has, as well as "which crystal" each pair originated from, may be available from the arrival time of the photon at the detector. On the other hand, the effective polarization entanglement requires the suppression of any distinguishing information in the other degrees of freedom that can provide potential information about the emitted pair.

To eliminate this distinguishing space-time information, a polarization dependent optical delay line for the pump was inserted before the crystal, which is denoted as the delay compensator (DC) in Figure 2. The delay compensator was used as the pre-compensator for group velocity mismatch between ordinary and extraordinary rays in the BBO crystal so that the state associated with $|HH\rangle$ created in the first crystal relative to the state associated with $|VV\rangle$ created in second crystal overlapped with them temporally. As the result, a two-photon Bell state $|VV\rangle+|HH\rangle$ can be directly created.

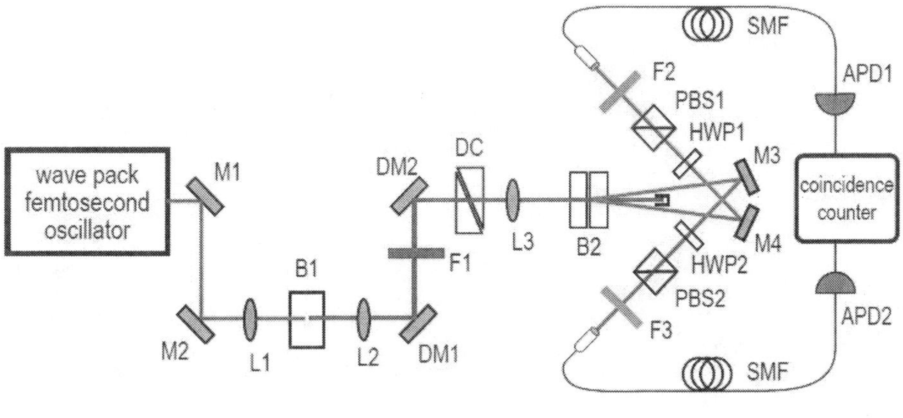

Fig. 2. Plot of coincidence rate as a function of idler polarization analyzer angle θ_1. (a) Signal polarisation analyzer was set at $\theta_1 = 0^o$ and the visibility achieved is V = 98:7%. (b) Coincidence rate for the signal at polarisation analyzer at $\theta_2 = 22.5^o$, with the visibility of 90%

To experimentally demonstrate the polarization entanglement of the collected photon pairs, their polarization correlations in two conjugate bases were measured. This was done by directing the down-converted light into adjustable polarization analyzers, each consisting of a polarizing beam-splitter cube (PBS) preceded by a rotatable half-wave plate (HWP). The residual violet light coming from the first BBO crystal was blocked using long-pass filters. After passing through the analyzers, the photons were coupled into single-mode fibers by using the aspheric coupling lenses with focal length of f= 7.5 mm, the desired spectral bandwidth and the numerical aperture of the optical fiber, and then detected with passively quenched silicon avalanche photon diodes (Si-APDs). The detectors were linked to a time-to-amplitude converter for a record of coincidence counts. To keep the experimental setup compact, two mirrors (M3 and M4) were used to fold the paths of the fluorescence beams.

4 Discussion

4.1. Fibre based QKD

Since the light pulse is phase modulated, the setup should be able to translate between phase and codes according to BB84 protocol. The setup should correctly translate the protocol when Alice codes into the system while the coded message should be correctly translated when Bob is doing the measurement. In order to translate the data in phase, there should be a reference pulse that can interfere with the other pulse that has been phase code. The interference results will depend on the phase difference between the two pulses. The interferometer is used in order to locate the interference and the interference should occur at the Bob setup since Bob will collect the decoded data. In order to have the reference pulses for interference, the unbalanced interferometer is built, and then the pulses will travel by superposition of the two pulses.

The problem is, when the pulses arrived at the Bob setup there will be two pulses that will not interfere with any pulses. They are the pulses that travel through both setups through the short path (SS) and pulses that travel in both setups through the long path (LL). Due to this problem, the interference becomes worsen since there are lost due to the unbalanced interferometer. In addition, the best interference result occurs when the pulses are in the similar polarization state. Due to this limitation, the polarization of the light is used as the compensation method to make sure that the traveling pulses between the setup only pass through the interferometer by these two paths either long-short(LS) arms or short-long(SL) arms. The pulses have to travel from Bob barely without any modulation and then reflected by Faraday mirror purposely to swap the two pulses' polarization states so that these two pulses will follow one of the two paths, LS or SL. The signal is modulated and attenuated into single photon level only after the reflection at Alice. Then, the pulses will go through the Bob setup where Bob will insert the bases phase modulation in order to know which state was sent by Alice. The preparation of the QKD state actually reflected the design of optical setup that needs to be used.

3.2 Free space QKD

Spontaneous parametric down conversion (SPDC) process in nonlinear optical materials is utilized in order to create one of the four Bell states which are maximally entangled. The states can mathematically be written as follows:

$$|\Phi_{12}^{\pm}\rangle = \frac{1}{\sqrt{2}}(|H\rangle_1 \otimes |V\rangle_2 \pm |V\rangle_1 \otimes |H\rangle_2),$$
$$|\Psi_{12}^{\pm}\rangle = \frac{1}{\sqrt{2}}(|H\rangle_1 \otimes |H\rangle_2 \pm |V\rangle_1 \otimes |V\rangle_2),$$

$|H\rangle$ ($|V\rangle$) denotes Horizontal (Vertical) polarisation of the photons. We use the notation $\otimes$, as a tensor product describing a composite system of two spatially separated photons. Since the medium is free space the polarisation states are used as the coding method. SPDC process will be able to produce the entangled photon in polarisation state. The value of the measurement can only be revealed by measuring the signal state. This is an ideal random generator for QKD system where both Alice and Bob only know what they are measuring by comparing the bases that they used.

In the QKD system, the entangled photon source of this system needs to be placed somewhere between Alice and

Bob. This means that it is also possible to locate it near (next) to Alice's setup. Alice will measure one of the detector arms by setting the azimuth angle of HWP in path-2 θ_2 to be rectilinear or diagonal bases which correspond to the state of $|V\rangle$, $|D\rangle$, $|H\rangle$ or $|A\rangle$ that Alice measured. In Bob's measurement, while the azimuth angle of HWP in path-2 θ_2 was set to Alice measurement bases, the azimuth angle of HWP in path-1 was rotated to the respective bases. This setup corresponds to the projection measurement onto $|L\rangle \otimes |\theta\rangle$ or $|R\rangle\otimes|\theta\rangle$ where $|L\rangle = (|H\rangle + i|V\rangle)/\sqrt{2}$, $|R\rangle = (|H\rangle - i|V\rangle)/\sqrt{2}$, and $|\theta\rangle = \cos\theta\,|H\rangle + i\sin\theta|V\rangle$. Since the density matrix of the states approximately described by $\hat{\rho} = p|\Psi^+\rangle\langle\Psi^+| + q|\Psi^-\rangle\langle\Psi^-|$, where p and q are probabilities of finding our system in the state $|\Psi^+\rangle$ and $|\Psi^-\rangle$, respectively. The state should exhibit interference in coincidence rates for the two detector in proportion to $R_C(\theta_1, \theta_2) \propto \cos^2(\theta_1 - \theta_2)$.

5 Conclusion

The preparation of the photon state is the main characteristic that determines experimental setup of the QKD system. The preparation is not only to address the issue of how to produce the state but also how to measure the state so that it can translate the key from one end to another end while maintaining the quantum properties. The visibilities of the interference pattern actually the main important characteristic since it will translate to a good data coding, decoding, and reduced errors during preparation and measurement. Both two setups presented produced good visibilities (90%~98%) which come from good optical setup preparation and configuration.

References

1. Quantum Computer Could Solve Problems In A Few Months That Would Take Conventional Computers Millions Of Years (http://www:sciencedaily:com=releases=2001=09=010913074828:htm), Accessed on 24th March 2010
2. Beyond the PC: Atomic QC Quantum computers could be a billion times faster than Pentium III(http://www:amd1:com=quantum□computers:html) Accessed on 14 January 2010Luigi T. De Luca, *Propulsion physics*, EDP Sciences, Les Ulis, (2009)
3. P.G. Kwiat, E. Waks, A.G. White, I. Appelbaum, and P.H. Eberhard, Phys. Rev. A 60, R773 (1999).
4. Norshamsuri Ali, Hafzulfika, Salim Ali Al-Kathiri, Abdulla Al-Attas, Suhairi Saharudin, and Mohamed Ridza Wahiddin, *International Conference on Cryptography, Coding and Information Security*, (2008).
5. Norshamsuri Ali, Hafizulfika, and Salim Ali Al-Kathiri, The NCTT-MCP, (2008).
6. Suryadi, Norshamsuri Ali, Suhairi Saharudin, and Mohamed Ridza Wahiddin, MIMOS R&D Symposium, (2006).
7. Alex Biryukov, Orr Dunkelman, Nathan Keller, Dmitry Khovratovich, and Adi Shamir Key Recovery Attacks of Practical Complexity on AES Variants With Up To 10 Rounds (http://eprint.iacr.org/2009/374.pdf)
8. Alex Biryukov and Dmitry Khovratovich, Related key Cryptanalysis of the Full AES-192 and AES-256 (https://cryptolux.org/mediawiki/uploads/1/1a/Aes-192-256.pdf)
9. C. Monroe, D. M. Meekhof, B. E. King, W. M. Itano, and D. J. Wineland, Demonstration of a Fundamental Quantum Logic Gate (http://tf.nist.gov/general/pdf/140.pdf)

 DOI: 10.1051/epjconf/201716201040

Biophotonics sensor acclimatization to stem cells environment

Mukhzeer Mohamad Shahimin[1,*]

[1]Department of Electrical and Electronic Engineering, Faculty of Engineering, National Defence University of Malaysia (UPNM), Kem Sungai Besi, 57000 Kuala Lumpur.

Abstract. The ability to discriminate, characterise and purify biological cells from heterogeneous population of cells is fundamental to numerous prognosis and diagnosis applications; often forming the basis for current and emerging clinical protocols in stem cell therapy. Current sorting approaches exploit differences in cell density, specific immunologic targets, or receptor-ligand interactions to isolate particular cells. Identification of novel properties by which different cell types may be discerned and of new ways for their selective manipulation are clearly fundamental components for improving sorting methodologies. Biophotonics sensor developed by our team are potentially capable of discriminating cells according to their refractive index (which is highly dependable on the organelles inside the cell), size (indicator to cell stage) and shape (in certain cases as an indicator to cell type). The sensor, which already discriminate particles efficiently, is modified to acclimatize into biological environment, especially for stem cell applications.

1 Introduction

Recent derivation of human embryonic stem cell open new opportunities for regenerative medicine as well as understanding basic aspect of embryonic development and diseases such as cancers [1]. Many ideas that are now discussed have long history and much has been underpinned by the earlier studies of teratocarcinomas and embryonal carcinoma stem cells, which present a malignant surrogate for the normal stem cells of the early embryo.

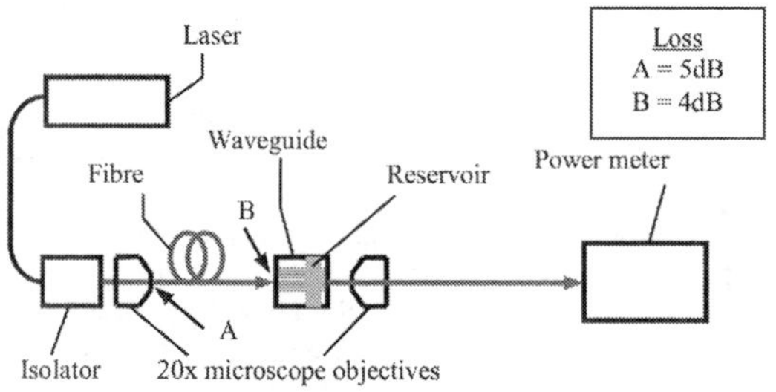

Fig. 1. Experimental set up for particle/cell trapping and propulsion

Although the potential of embryonic stem cell and embryonic carcinoma cells to differentiate into a wide range of tissue is now well attested, little is understood of the key regulatory mechanism to control their differentiation. The culture environment, morphological; all aspect plays important role in this process. Hence, there is a need for an approach to provide a pure population of stem cells that is free from mechanical (fluid sheer stress, cyclic stretch and pressure), electrical (field induced) or chemical (need for labelling) induced cellular response. Such an approach will be able to provide the effective characterization and study of different stem cell populations and ultimately clearer strategies for regenerative medicine. Optical trapping and propulsion is seen as a potential candidate as a sorting technique that avoids detrimental effects on the stem cells.

A biophotonics sensor comprises of caesium ion exchanged optical waveguides with various parameters have been developed to efficiently discriminate polystyrene particles according to their size [2]. In this paper, the system is modified to adapt into biological environment, especially for stem cell applications.

There are several characteristics needed to be investigated in order to ensure the compatibility of the system for optical trapping and propulsion of stem cells. For this purpose, characteristics of a mammalian cell type were listed. The cells should be non-adherent, at least pseudo-spherical and, if possible, have a high refractive index. Lymphoblastoma cells were seen as a good candidate not only because the cells are non-adherent and pseudo-spherical in shape, but also they are readily available and easy to culture. The investigation of optical trapping and propulsion on lymphoblastoma was carried out first, as shown in this paper, before applying the technique to teratocarcinoma cells, the cell type that is targeted for the biophotonics sensor application.

* Corresponding author: mukhzeer@upnm.edu.my

2 Experimental procedures

In order to optically propel particles and cells, a solution of particles or cells is placed in a reservoir on top of the waveguide, as illustrated in Fig. 1. The trapping and propulsion is powered by a diode pumped ytterbium doped fibre laser (IPG Photonics) operating at 1064nm. The wavelength is used as the laser source is readily available and it was proven to work for polystyrene particles propulsion on caesium ion-exchanged channel waveguides [3]. The laser source, which is linearly polarised and producing a continuous wave, is capable of supplying power up to 5W with a Gaussian output beam of 1.6mm diameter. The laser source is directed to a built-in isolator in order to provide a collimated beam and to avoid the backscatter reflection that can damage the laser source. Output from the isolator is coupled into a single mode polarisation maintaining (PM) fibre (Fibercore, HB980-T) via a 20x objective lens (loss ≈ 5dB).

The PM fibre is used to adjust the input field to the optical waveguide into different polarisation modes. The PM fibre is aligned in three dimensions to give maximum output, using a fibre holder with a nanopositioner (Thorlabs - MAX313), and set to the correct polarisation using a nanorotator (Thorlabs - HFR003) attached to the fibre holder. Scattering due to Fresnel reflection is minimised by cleaving the fibre using a fibre cleaver (PK FK11). A coupling efficiency, between the fibre and the waveguide input facet, of approximately 4dB is achieved using this setup. An optical microscope was used to place the PM fibre in close proximity to the waveguide input facet. The coupling was optimised by monitoring the output of the waveguide.

Each of the experimental data was compared with a theoretical model. The theoretical model was simulated by adapting the Arbitrary Beam Theory (ABT) developed by [4, 5]. The simulation was made using the same model as in [6, 7]. Unless stated otherwise, the waveguide parameters used in the simulation were a 4μm channel width, a substrate index of 1.50, a waveguide index of 1.54 and a particle index of 1.59 dispersed in water (index of 1.33). The wavelength used in the simulation was 1064nm with an input power of 500mW. Note that the model assumed that there is no power loss due to the Fresnel scattering, modal mismatch or propagation loss along the channel waveguide. Furthermore, it is also assumed that the propulsion of particles is not affected by any non-optical forces.

3 Results and discussion

3.1 Particle/cell size matching

With a view to apply the existing system in the trapping and propulsion of particles to mammalian cells, certain characteristics of the mammalian cells can be investigated using particles of a similar property. One of the characteristics that can be investigated using particles is the cell size. The diameter of the cell used varies depending on its type and the state of the cell cycle. Lymphoblastoma cells can vary from approximately 8μm to 12μm, while teratocarcinoma cells vary from 15μm to 23μm. In order to evaluate the effect of varying cell size, six different polystyrene particle sizes, ranging from 1μm to 20μm diameter, were used. The propulsion velocity of each particle size was measured using a 12 hour caesium ion-exchanged waveguide [2].

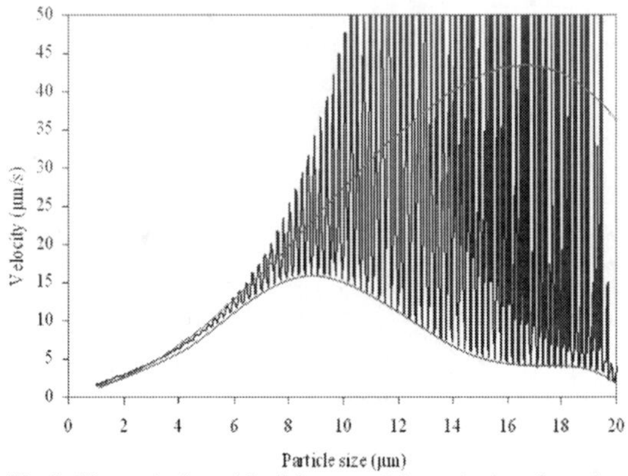

Fig. 2. Theoretical model of the propulsion velocity of varying particle sizes at TM mode. The red line indicates the best fit between the two adjacent maxima and minima in the theoretical model and the green line is the estimated non-resonance propulsion estimation

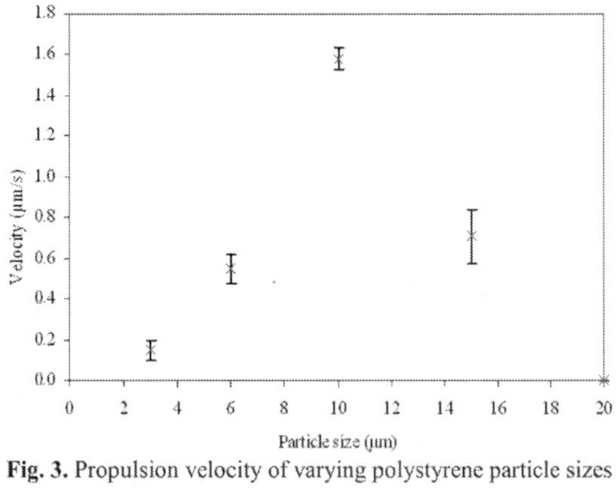

Fig. 3. Propulsion velocity of varying polystyrene particle sizes

Fig. 2 illustrates the theoretical model of the propulsion velocity for varying particle sizes. The theoretical plot shows velocity resonance, an effect that is observed in total scattering and extinction cross-sections for Mie particles [8, 9]. These resonances originate from the electromagnetic modes of a sphere and are known as Morphology Dependent Resonances (MDRs) [8, 9]. With increasing radius, the loss for the internal field decreases, the internal field accumulates and resonances occur [7]. Note that the resonances become narrower with particle size. This indicates that the condition for resonance is harder to satisfy for larger particle size. There are a few limitations for the theoretical model to be met experimentally. The model assumes a lossless particle where the particle is assumed to be a perfect sphere and of high quality factor, Q. These particle characteristics are not met with the particles used in this experiment. Furthermore, the laser is assumed to be an ideal laser of a single frequency linewidth.

The experimental data is illustrated in Fig. 3. The propulsion velocity was observed to increase with particle size. However, the 15μm particles show reduction in velocity and no propulsion observed for 20μm particles. An increase in the particle size increases the polarizability of the particle and the optical force. The scattering image from a 20μm particle proves that there was optical force acting on the particle although no propulsion was observed. There was also no apparent propulsion for 1μm particles observed in the experiments, mostly due to the limitation of the lenses available to monitor and track the 1μm particles. However, lateral trapping of the particles can be detected. The increment in the intensity line profile indicates an increasing number of 1μm particles trapped in the illuminated channel with increasing time. As indicated in previous papers [10-12], the Brownian motion is more apparent as the size of the particles decreases. The 1μm particles were seen to move randomly in Brownian motion. However the motion was directed to the illuminated channel, indicating a strong lateral trapping.

There are some discrepancies between experimental velocities and the predicted velocity in the theory. Apart from the reasons stated previously, the standard deviation of the particles used was observed to be higher than the distance between two adjacent maxima and minima of the velocity resonance. For example, the standard deviation for a 10μm particle is 0.763μm while the difference between maximum and minimum near 10μm particle size is 0.06μm in the theoretical plot. In addition, small deviations from spherical symmetry are known to affect the quality factor of the narrowest resonances [13]. Therefore this may indicate that the particles that were used in the experiments are not perfectly spherical. Hence, the experiments did not observed a wide range of propulsion velocities as expected from the theory. In addition, estimation of mammalian cell propulsion using the theoretical model is also invalid. Mammalian cells were observed to be hemispherical on surface [2], which indirectly increases the frictional force due to a larger contact area. However, assuming non-resonance propulsion estimation, as illustrated in Fig. 2, the trend of particles propulsion observed in the experiments follows the theory. This might be a better approximation to estimate the propulsion velocity according to size. The theory should also taken into consideration the reduction of laser power due to modal mismatch, scattering and Fresnel reflection.

3.2 Refractive index variation

Apart from differences in size, density, structure and surface properties, cells also have a low refractive index compared to the polystyrene particles. A variation in the refractive index changes the propulsion behavior of the cells. The refractive index of the polystyrene particles used in all experiments is n=1.59. Exact values of the refractive index of lymphoblastoma cells or teratocarcinoma cells on the other hand could not be found in the literature. However, as most of the cell structure consists of cytoplasm, it is fair to assume that the refractive index will be close to n≈1.39, as indicated in [14, 15]. Prior to manipulation of cells, propulsion of particles with the refractive index close to the cell index was investigated to predict the propulsion behavior of a low index material.

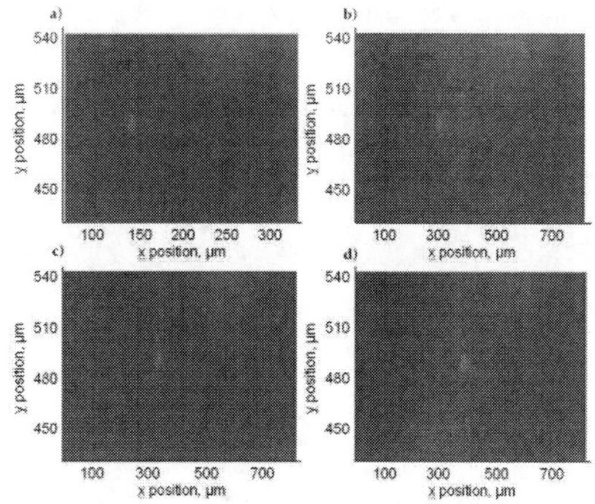

Fig. 4. Propulsion of PMMA particles a) Image taken at t = 0s, b) t = 66s, c) t = 133s and d) t = 200s

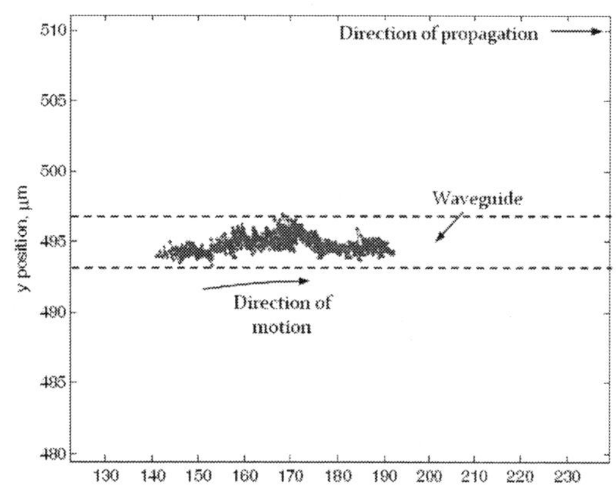

Fig. 5. Output from the particle tracking software. PMMA particle propulsion is presented in ■ coloured line

The lowest refractive index available commercially is obtained from silica spheres, which is approximately 1.46. The refractive index is the closest to the assumed refractive index of mammalian cells; unfortunately the particle is quite dense with the density of 2.5gcm^{-3}. A high density particle has a larger mass (hence, more friction force) and preliminary propulsion experiments using silica spheres show no motion at all [2]. Hence, 10μm diameter polymethylmethacrylate (PMMA)

 DOI: 10.1051/epjconf/201716201040

particles with a refractive index of 1.489 were chosen [16]. The same optical and waveguide setup was used to carry out trapping and propulsion analysis of PMMA particles. A series of images of PMMA particles propulsion is shown in Fig. 4. The images were compressed (in the x-direction) in order to accommodate the distance travelled by the particle. The sequence of images in Fig. 4 shows that the PMMA particle moved with a mean velocity of 0.23±0.07μm/s. As illustrated in Fig. 5, Brownian motion was more pronounced in the PMMA particle propulsion vis–à–vis the observation for the polystyrene particles. This is due to the lower refractive index of the PMMA particles. This investigation was extended by varying the input power supplied to the 4μm nominal width channel waveguide.

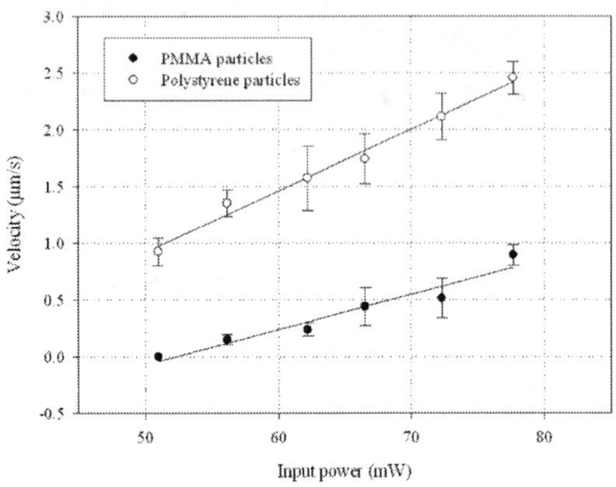

Fig. 6. Propulsion velocity of 10μm PMMA particles with varying input power. Propulsion velocity of polystyrene particles of the same size is also plotted for comparison

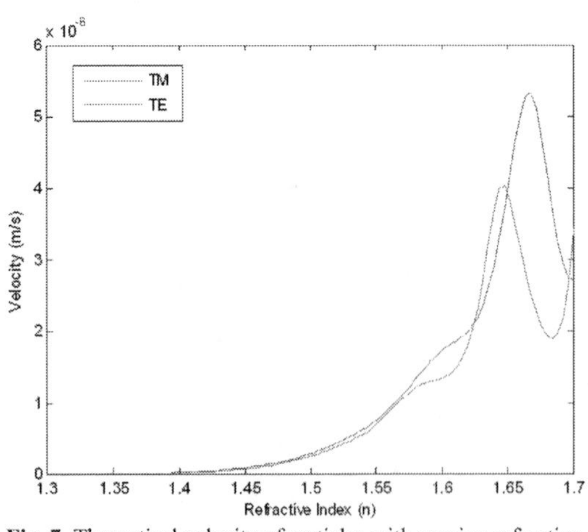

Fig. 7. Theoretical velocity of particles with varying refractive index

Fig. 6 shows the velocity of 10μm PMMA particles on the 12 hour caesium ion-exchanged waveguide. The input power was varied from about 400mW up to 620mW by adjusting the current supplied to the laser (corresponds to 50mW to 78mW input power to the waveguide). Propulsion of a 10μm polystyrene particle on the same waveguide is also plotted on the same graph for comparison. The propulsion for PMMA particles is clearly much less than for the polystyrene particle in all ranges of input power tested, as expected from the theoretical model in Fig. 7. PMMA particles were also observed to be unable of propulsion at about 50mW input power. This may indicate the minimum threshold power needed for PMMA particle's propulsion. At similar input powers, polystyrene particles propel, on average, at a velocity of 0.93μm/s. PMMA and polystyrene particles propel with a mean velocity of 0.24μm/s and 1.58μm/s respectively with an input power of approximately 66mW.

In order to understand the effect of different refractive indices of particles, a simulation using the same model as in [6, 7] was carried out. The same optical and waveguide arrangement, as described earlier in this section, was set in the simulation. The simulated velocity for polystyrene particles was normalized to the experimental values in order to have a more realistic estimation of the propulsion velocity of PMMA particles. The predicted particle velocity against the particle's refractive index is illustrated in Fig. 7. The simulation indicates that higher propulsion velocity is achieved at the TM mode compared to the TE mode for the particles under consideration (refractive index less than n=1.6) as predicted in previous papers [10-12]. The velocity for the PMMA particle is expected to be around 0.2266μm/s due to a lower effective index which approximately the same as observed experimentally. The observations on varying particle size and refractive index suggest that the propulsion of mammalian cells of interest, lymphoblastoma and teratocarcinoma cells was indeed feasible. The size range of the mammalian cells overlaps the size range of particles that could be propelled on the caesium ion-exchanged waveguide. The theoretical model, as illustrated in Fig. 7, also shows the predicted velocity for cells of refractive index near 1.39, which is about 6.273nm/s.

3.3 Effect of surface functionalization

Mammalian cells involved in this project are lymphoblastoma and two types of teratocarcinoma cells; TERA1 and NT2 cells. Although lymphoblastoma grows in suspension and is less likely to adhere to the surface, teratocarcinoma cells adhere to the surface and form a monolayer in culture. Thus, a method to reduce adhesion and promote the dominance of a double layer repulsive force is by functionalizing the waveguide surface. A 12 hour caesium ion-exchanged waveguide was PEG-functionalized, as detailed in [2]. Prior to the functionalization step, propulsion experiments of 8μm, 10μm and 12μm polystyrene particles were carried out on the waveguide for comparison. Fig. 8 shows results from the propulsion experiments on both PEG-functionalized and plain surfaces.

It can be observed in the propulsion experiments that particles of all sizes tested show a faster propulsion on the PEG-functionalised surface. From Fig. 8, propulsion of polystyrene particles is on average 25% faster than propulsion on a plain surface. Apart from preventing cell adhesion [17, 18], the faster propulsion observed also indicates a reduction in the frictional force that is opposing the particle motion, as discussed in [12]. This is beneficial in terms of applying the system to propel mammalian cells since a low refractive index is expected to reduce the propulsion velocity dramatically. Any minimization of forces that hinder the propulsion (frictional force and cell adhesion) is definitely seen as an advantage.

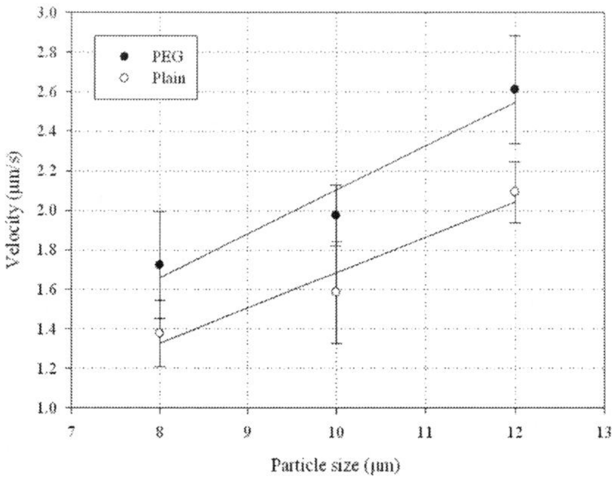

Fig. 8. Propulsion of 8μm, 10μm and 12μm polystyrene particles on PEG-functionalized and plain waveguide surfaces

4 Conclusion

The optical trapping and propulsion of polymer particles, namely polystyrene and PMMA were presented in this paper. Biological cells not only vary in terms of size and refractive index but also their surface characteristics. Hence, the propulsion of polymer particles of varying sizes and refractive indexes were investigated initially on plain and functionalized surfaces of caesium ion-exchanged waveguides. It was found that as the refractive index of the particles decreased, the propulsion velocity was also observed to decrease, as demonstrated by the PMMA particles (propulsion rate of $3.9 \times 10^{-3} \mu ms^{-1} mW^{-1}$). The investigation of different surfaces showed that the propulsion velocity of particles on a PEG-functionalized surface increased on average by 25%.

The work carried out here determined the optimum optical, waveguide parameters and surface characteristics to be further exploited for trapping and sorting of teratocarcinoma cells and ultimately stem cells on caesium ion-exchanged channel waveguides. These studies on biophotonics sensor is expected to provide a fast, reliable and high throughput discriminating mechanism that is practical and applicable to real environment in the future.

References

1. J. A. Thomson, Science, **282**, pp. 1827-1827 (1998)
2. M. M. Shahimin, "Propulsion of polymer particles on caesium ion-exchanged channel waveguides for stem cell sorting applications," in *School of Electronics and Computer Sciences*. PhD Thesis Southampton: University of Southampton, pp. 1-206 (2009)
3. K. Grujic, O. G. Hellesø, J. S. Wilkinson, and J. P. Hole, Optics Communications, **239**, pp. 227-235 (2004)
4. J. P. Barton, D. R. Alexander, and S. A. Schaub, Journal of Applied Physics, **64**, p. 1632 (1988)
5. J. P. Barton, D. R. Alexander, and S. A. Schaub, Journal of Applied Physics, **66**, p. 4594 (1989)
6. H. Jaising, K. Grujic, and O. G. Hellesø, Optical Review, **12**, p. 4 (2005)
7. H. Y. Jaising and O. G. Hellesø, Optics Communications, **246**, pp. 373-83 (2005)
8. C. F. Bohren and D. R. Huffman, *Absorption and scattering of light by small particles*, (John Wiley & Sons, New York, 1983)
9. Y. Panitchob, G. Senthil Murugan, M. N. Zervas, P. Horak, S. Berneschi, S. Pelli, G. Nunzi Conti, and J. S. Wilkinson, Optics Express, **16**, pp. 11066-11076 (2008)
10. M. M. Shahimin, O. G. Hellesø, J. S. Wilkinson, and T. Melvin, *Proceedings of 25th Regional Conference on Solid State Science and Technology* (2009)
11. M. M. Shahimin, O. G. Hellesø, J. S. Wilkinson, and T. Melvin, *Proceedings of Nanomaterials Synthesis and Characterisation Conference* (2009)
12. M. M. Shahimin, O. G. Hellesø, J. S. Wilkinson, and T. Melvin, *Proceedings of Malaysian Technical Universities Conference on Engineering and Technology* (2009)
13. M. I. Mishchenko and A. A. Lacis, Applied Optics, **42**, pp. 5551-5556 (2003)
14. S. R. Goodman, *Medical cell biology*. (Alabama: J B Lippincott Company, 1994)
15. E. D. P. DeRobertis, F. A. Saez, and E. M. F. J. DeRobertis, *Cell biology* (6th ed. Philadelphia: W B Saunders Company, 1975)
16. Microparticles GmbH Berlin, *Poly(methyl methacrylate) Particles: Physical and Chemical Properties* (2008)
17. S. Mandal, J. M. Rouillard, O. Srivannavit, and E. Gulari, Biotechnology Progress, **23**, p. 972 (2007)
18. A. Wolter, R. Niessner, and M. Seidel, Analytical Chemistry, **79**, p. 4529 (2007)
19. J. P. Hole, J. S. Wilkinson, K. Grujic, and O. G. Hellesø, *Conference on Lasers and Electro-Optics (CLEO 2005)* (2005)

EPJ Web of Conferences 162, 01041 (2017) DOI: 10.1051/epjconf/201716201041
InCAPE2017

The influence of design parameters on the performance of FBAR in 10–14 GHz

*N.I.M.*Nor[1,*], *R.A.M.* Osman[1], *M.S.* Idris[2], *N.* Khalid[1], *M.* Mohamad Isa[1], *N.* Ahmad[1], *Siti S.* Mat Isa[1], *Muhammad M.* Ramli[1], and *S.R.* Kasjoo[1]

[1]School of Microelectronic Engineering, Universiti Malaysia Perlis, Pauh Putra Campus 02600 Arau, Perlis, Malaysia
[2]School of Material Engineering, Universiti Malaysia Perlis, Jejawi Campus 02600 Arau, Perlis, Malaysia

Abstract. This research presents the analysis of the influence of design parameters on the performance of film bilk acoustic wave resonator (FBAR) working from 10 GHz to 14 GHz. The analysis is done by implementing one-dimensional (1-D) modellings, which are 1-D Mason model and Butterworth Van Dyke (BVD) model. The physical parameters such as piezoelectric materials and its thickness, and size of area affecting the characteristics of the FBAR are analyzed in detail. Zinc oxide (ZnO) and aluminum nitride (AlN) are chosen as the piezoelectric materials. The resonance area is varied at 25μm×25μm to 35μm×35μm. From the analysis, it is found that as the frequency increases, the thickness of the piezoelectric material decreases. Meanwhile, the static capacitance increases as the frequency increases. It is also found that as the area increases, the electrical impedance and static capacitance also increases.

1 Introduction

High demand for wireless communications technology and video has led to demand for more channels and wider bandwidth. Conventional frequency bands (below 6 GHz) are too congested. Therefore, to meet this demand, the study on the receiving system that works at frequency higher than 10 GHz has increased. Film bulk acoustic wave resonator (FBAR) filters and FBAR diplexers designed using microelectromechanical system (MEMS) technology have been widely used for WiFi and WiMAX applications. Such MEMS components have shown better performance and a higher integration level, and the same performance is expected to be achieved in Ku-band transceivers by using FBAR filters. FBARs operating in the frequency range of 5 GHz to 20 GHz have been reported to show a very high quality (*Q*) factor, good power handling and small size. The most common piezoelectric materials used for development of FBAR are aluminum nitride (AlN) and zinc oxide (ZnO) [1-4]. FBARs based on lead zirconate titanate (PZT) and cadmium sulphide (CdS) are also found in the literature [5-6]. CdS has low acoustic impedance and an electromechanical coupling coefficient (~2.4%). PZT has good performance in terms of the electromechanical coupling coefficient (k^2_{eff}) values of 19.8%, but has high intrinsic losses at high frequencies [1]. Thus, PZT is mostly used for low-frequency FBAR devices or applications that do not require a high *Q* factor [7]. For higher frequency applications, ZnO and AlN are the most suitable piezoelectric materials. These materials have the same hexagonal wurtzite structure. Although the coupling coefficient of ZnO is higher than AlN, AlN is superior to ZnO due to its moderate mechanical coupling factor, higher acoustic velocity and higher *Q* value. This makes AlN suitable to fabricate bulk acoustic wave (BAW) resonators/filters in several gigahertzes (GHz). Furthermore, AlN is compatible with CMOS technology and more easily manufactured compared to ZnO [7-8].

Modelling is a fundamental step in analyzing the performance of an FBAR. Several 1-D models have been proposed in order to characterize the electrical behavior of the FBAR. The 1-D Mason model is mainly used to represent the electrical behavior of BAW resonators and has been widely employed in work related to it [9-10]. The Krimholtz-Leedom-Matthaei (KLM) and Butterworth Van Dyke (BVD) model also provide a very good approach for characterizing the electrical behavior of the FBAR at fundamental modes and higher harmonics [11-12]. Therefore, in this work, the both models are implemented in the simulation in order to characterize the performance of the FBAR.

* Corresponding author: izza@unimap.edu.my

2 Design and Modelling of FBAR

2.1 FBAR Concept

FBAR is one of the technologies for fabricating BAW devices, the first FBAR device being released by Lakin and Wang in 1981. Figure 1 shows the acoustic wave propagation in FBAR through its active layer structure, which is usually a piezoelectric material. The acoustic wave causes the deformation of the piezoelectric material. Therefore, the actuation and detection mechanisms involved in FBAR operation are due to the piezoelectric and inverse piezoelectric effects. By using these principles, a voltage applied to the resonators electrodes induces strain of the acoustic layer, and vice versa, and following a mechanical strain of the acoustic layer a voltage can be read out of the electrodes [13].

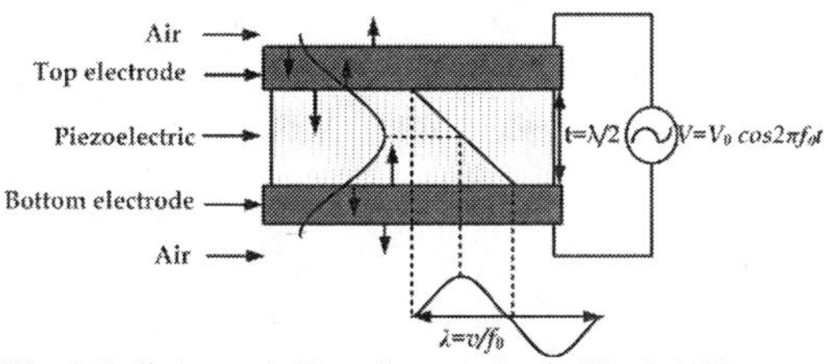

Fig. 1. Bulk Acoustic Wave Propagation in FBAR [13].

The resonance frequency of an FBAR operating in fundamental, longitudinal mode is determined mainly by the thickness (t) of the piezoelectric layer given as:

$$\theta = \frac{2\pi f_0 t}{v^D} \tag{1}$$

where θ is the phase, v^D is acoustic velocity, f is the frequency of acoustic wave propagating through the bulk acoustic layer and t is the thickness of the piezoelectric film. The acoustic velocity can be calculated using the approximation given as :

$$v^D = \sqrt{\frac{c_{33}^D}{\rho}} \tag{2}$$

where c^D_{33} is the elastic stiffness at constant electric displacement and ρ is the density of the piezoelectric material.

At resonance (f=f_0), the acoustic phase of the film is θ=π, under these conditions, the f_0 can be calculated as in [13]:

$$f_0 = \frac{v_L}{2t} \tag{3}$$

From Equation 3, it follows that resonance occurs when the film thickness is equal to half the wavelength of the acoustic wave. As shown in Fig. 1, the acoustic wave is confined by the reflecting electrode surfaces at the thin film interface where t must be half the acoustic wavelength, ignoring the electrode loading effects.

2.1 1-D Mason Model

The Mason model has been widely used in deriving solutions for the wave propagation through the acoustic layer by using the network theory approach. Fig. 2 shows the thickness excitation of the piezoelectric layer, ignoring the electrodes. The piezoelectric layer can be seen as a three-port component. Two ports are the mechanical ports presented by the forces (F) and the displacement velocity (v). The other port is the electrical port given by voltage (V) and current (I).

The electrical input impedance of a single piezoelectric layer is given as:

$$Z_{in} = \frac{1}{jwC}\left(1 - k_t^2 \frac{\tan(kd/2)}{kd/2}\right) \tag{4}$$

where d is the thickness of the piezoelectric layer. F_1 and F_2 are the forces on the top and bottom surface of the resonator, respectively. v_1 and v_2 represent the acoustic velocities of the top and bottom surface plane of the resonator, respectively. V is the external electric voltage and I is the current. C is the static capacitance, ω=$2\pi f$ is the angular frequency, $k = \omega\ v_L$ where v_L is the longitudinal acoustic wave velocity and h=e/ε_s where e is piezoelectric coefficient and ε_s is the permittivity of the piezoelectric layer. k_t^2 can be computed using equation 2; while C_0 is given in equation 6 where A is the area.

$$k_t^2 = \frac{e_{33}^2}{c_{33}^D \varepsilon_{33}^s} \tag{5}$$

$$C_o = \frac{\varepsilon A}{d} \tag{6}$$

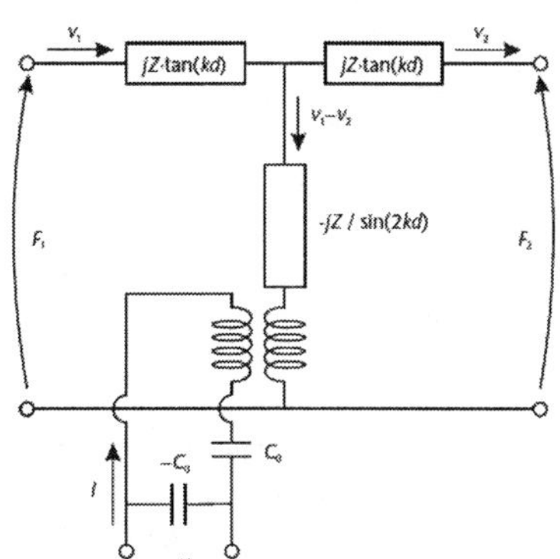

Fig. 2. Three-Port Mason"s Model Equivalent Circuit [7]

2.2 BVD Model

The Mason model may be simplified to the six-lumped element model under the assumption that FBARs have very thin electrodes. The Butterworth- Van Dyke (BVD) model is a common lumped element equivalent circuit model used by the crystal filter to simplify the transcendental functions that totally characterize the resonators used as filter elements. This BVD model, as shown in Fig. 3 (a), comprises of a series motional inductor (L_m), motional capacitor (C_m) and motional resistor (R_m) resonator in parallel with static capacitance (C_o). The C_o is the electrical capacitance between the two electrodes through which the electric fields are applied. The motional components (C_m, L_m and R_m) represent the electromechanical response of a piezoelectric material. BVD model is the most convenient model to use but if the loss from the electrodes is to be considered then the modified Butterworth Van Dyke Circuit (MBVD) as shown in Fig. 3 (b) is adopted instead. The MBVD includes the dielectric loss (R_o) of the piezoelectric material and electrical losses (R_s) of the electrodes [14]. This model provides a better method, both simple and accurate, for characterizing electrical behavior of the FBAR at fundamental modes and higher harmonics and designing bandpass filters [11-14].

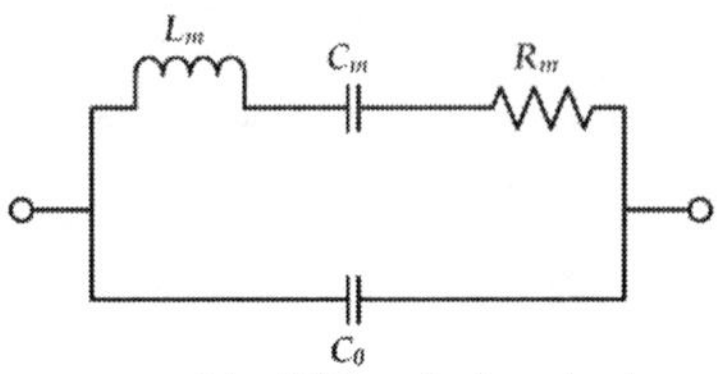

(a) BVD equivalent circuit

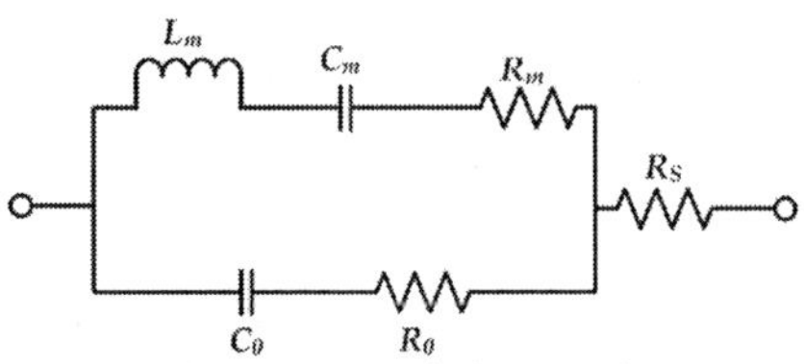

(b) MBVD equivalent circuit

Fig. 3. Equivalent Circuit for Acoustic Resonator

Several methods can be found in the literature to calculate the Q factor, for example, calculations based on the BVD) model or evaluation from the phase angle (imaginary part) of the electrical impedance given as:

$$Q_{s/p} = \frac{f_{s/p}}{2}\left[\frac{d\angle Z_{in}}{df}\right]_{f_s/f_p} \quad (7)$$

where f_s and f_p are the resonance frequencies and Z_{in} is the electrical impedance of the FBAR.

3 Results and Discussions

3.1 Influence of Different Piezoelectric Materials

In this 1-D modelling, the effect of electrodes on f_s and f_p is ignored. By referring to the estimation given in [7] for the purpose of this analysis, the bottom and top electrodes, Ru is set to 50 nm. Fig. 4 shows the relationship between resonance frequency and piezoelectric thickness. The simulation result indicates that as the thickness of the AlN and ZnO increases, the series resonance frequency decreases. This suggests that a thicker piezoelectric film reduces the influence of the electrode thickness upon the resonance frequency. This occurrence is supported by the notion that the acoustic path is proportional to the piezoelectric film thickness as modelled in equation 4. The piezoelectric film thickness is directly proportional to the acoustic path and inversely proportional to the resonance frequency. As seen from the graph too, the thickness of AlN is thicker than ZnO at the same working frequency. For example, FBAR working at 10 GHz, the thickness of AlN required is 0.553μm. Meanwhile, the thickness of ZnO is 0.318μm. This is due higher acoustic velocity of AlN which is 11050 m/s compared to ZnO that has 6350 m/s. This is true as given in equation 3.

Fig. 5 shows the relationship between different piezoelectric materials and their thicknesses on static capacitance, C_0. The results show as the frequency increases, the static capacitance gradually increases too for both AlN and ZnO. This can be verified by using equation (6), where the static capacitance is inversely proportional to the thickness of the piezoelectric material.

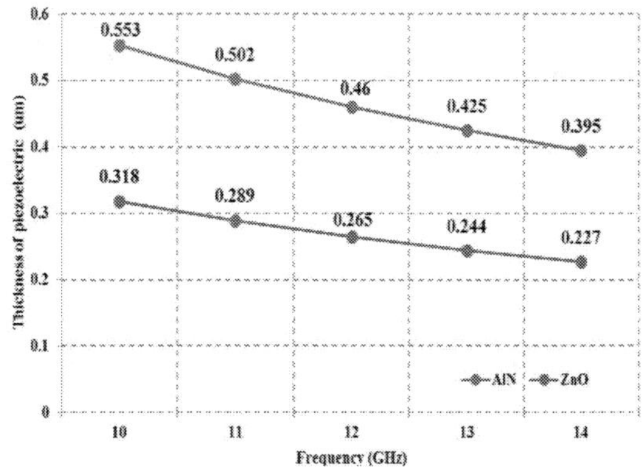

Fig. 4. Influence of Different Piezoelectric Materials and Thickness on Frequency

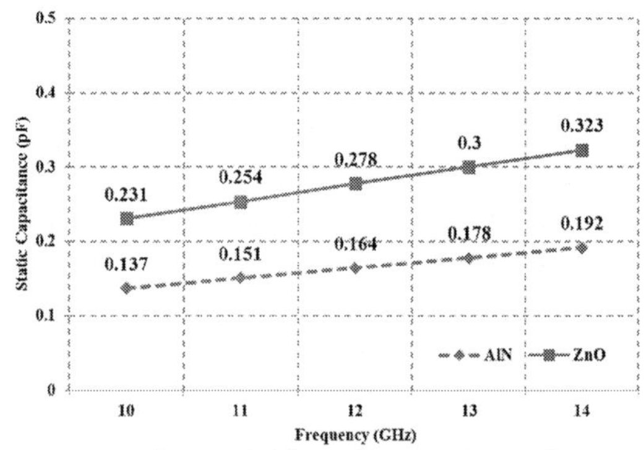

Fig. 5. Influence of Different Piezoelectric Materials

3.2 Influence of Different Area Size

Table 1 and Table 2 summarizes the influence of the resonance area on the electrical impedance, Z_{in} of FBAR using AlN and ZnO respectively. The area is set to 25x25 μm², 30x30 μm² and 35x35 μm² for each frequency. It is observed that as the resonance area increases, there is no significant change in the resonance frequencies, f_s and f_p. However the electrical impedance

DOI: 10.1051/epjconf/201716201041

of the FBAR decreases as the resonance area increases. According to equation 4, this is true as the electrical impedance is inversely proportional to resonance frequency.

Table 1. Influence of Area Size on Impedance, Z_{in} using AlN.

Frequency (GHz)	Area (μm^2)	f_s (GHz)	f_p (GHz)	Z_{in} (Ω)
10	25x25	09.712	09.984	0.374
	30x30	09.700	09.980	0.259
	35x35	09.710	09.980	0.191
11	25x25	10.704	11.000	1.801
	30x30	10.700	11.000	1.252
	35x35	10.700	11.000	0.923
12	25x25	11.680	12.000	0.772
	30x30	11.700	12.000	0.536
	35x35	11.700	12.000	0.393
13	25x25	12.640	12.992	0.062
	30x30	12.700	13.000	0.043
	35x35	12.600	13.000	0.032
14	25x25	13.600	13.976	0.191
	30x30	13.600	14.000	0.132
	35x35	13.600	14.000	0.097

Table 2. Influence of Area Size on Impedance, Z_{in} using ZnO.

Frequency (GHz)	Area (μm^2)	f_s (GHz)	f_p (GHz)	Z_{in} (Ω)
10	25x25	9.620	9.980	1.009
	30x30	9.630	9.990	0.765
	35x35	9.640	10.000	0.560
11	25x25	10.600	11.000	0.483
	30x30	10.600	11.000	0.304
	35x35	10.600	11.000	0.224
12	25x25	11.500	12.000	0.253
	30x30	11.600	12.000	0.175
	35x35	11.600	12.000	0.129
13	25x25	12.500	13.000	0.492
	30x30	12.500	13.000	0.343
	35x35	12.500	13.000	0.251
14	25x25	13.500	14.000	0.036
	30x30	13.500	14.000	0.025
	35x35	13.500	14.000	0.019

Fig. 6 and Fig. 7 shows the influence of different area size on the static capacitance of FBAR using AlN and ZnO respectively. From the graphs, it is clear that value of static capacitance is proportional to the size of area for both materials. Therefore, as the area size increases, the static capacitance increases too. This is verified by using equation 6. As seen from the graphs too, it is found that the values of static capacitance is higher for FBAR using ZnO compared to FBAR using AlN at the same working frequency. This is due to the permittivity value of ZnO is slightly lower than AlN. The permittivity of AlN and ZnO is 9.5 and 9.2 respectively.

From the analysis in section 3.1 and 3.2, the finding shows that the resonance frequency of FBARs is determined by the thickness and phase velocity of the piezoelectric layer. Therefore, for FBARs operating at frequencies higher than 10 GHz, the thickness of the piezoelectric layer is in the hundred nanometre (nm) scale. Furthermore, the properties and thickness of the piezoelectric materials have a significant influence on the performance of the FBAR in terms of resonance frequency, static capacitance and the electrical impedance of the FBAR.

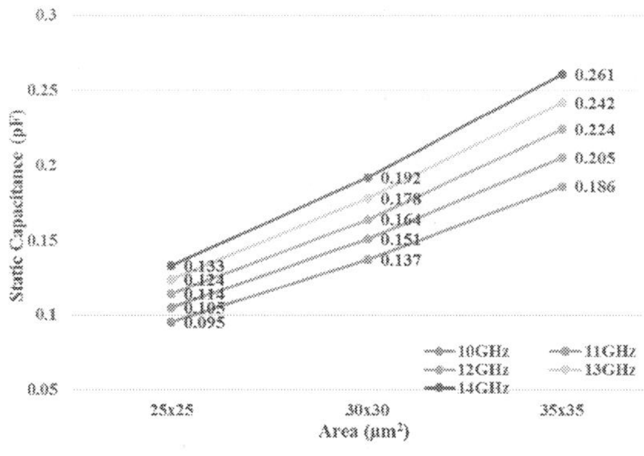

Fig. 6. Influence of Different Piezoelectric Materials

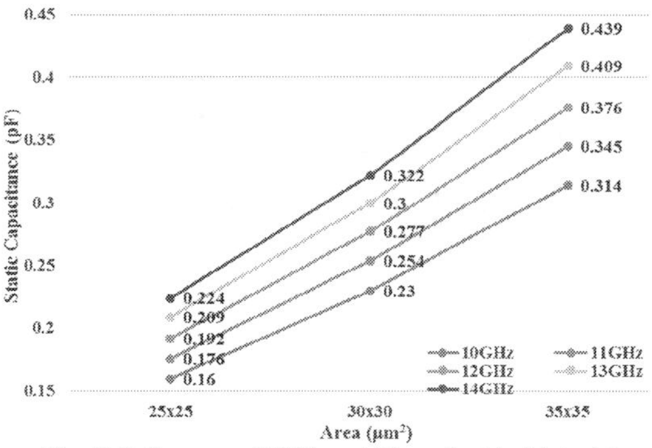

Fig. 7. Influence of Different Piezoelectric Materials

4 Conclusion

This work presented the design and analysis of FBAR using 1-D modelling. The influence of various geometrical parameters on FBAR performance were analysed and explored to find suitable solutions for designing FBAR operates at frequency higher than 10 GHz with a high *Q* factor and wide bandwidth. However, the presence of spurious modes at the frequency of interest due to lateral waves cannot be predicted by 1-D modelling, therefore, the three-dimensional (3-D) simulation tool will be beneficial in future work.

References

1. Klee, C. Metzmacher, et al, Materials Chemistry and Physics, **79**, pp. 143-146 (2003).

2. W.-K. Liu, K.-W. Tay, S.-C. Kuo, et al, Science in China Series G: Physics Mechanics and Astronomy, **52**, pp. 226-232 (2009).
3. R. Kubo, H. Fujii, H. Kawamura, et al, IEEE Symposium on Ultrasonics, **1**, pp. 166-169 (2003).
4. Q. X. Su, P. Kirby, E. Komuro, et al, IEEE Transactions on Microwave Theory and Techniques, **49**, pp. 769-778 (2001).
5. E. K. Kim, T. Y. Lee, Y. H. Jeong, et al, Thin Solid Films, **496**, pp. 653-657 (2006).
6. J. G. Gualtieri, J. A. Kosinski, and A. Ballato, IEEE Transactions on Ultrasonics, Ferroelectrics and Frequency Control, **41**, pp. 53-59 (1994).
7. K.-y. Hashimoto, *RF Bulk Acoustic Wave Filters for Communications,* Artech House (2009).
8. R. LANZ, *Piezoelectric Thin Films for Bulk Acoustic Wave Resonator Applications: From Processing to Microwave Filters,PhD*, ÉCOLE POLYTECHNIQUE FÉDÉRALE DE LAUSANNE (2004).
9. J. V. Tirado, *Bulk Acoustic Wave Resonators and Their Application to Microwave Devices*, PhD, Universitat Antonoma de Barcelona, (2010).
10. D. S. Shim and D. A. Feld, Int. J. RF Microw. Comput.-Aided Eng., **21**, pp. 486-495 (2011).
11. K. M. Lakin, IEEE Transactions on Ultrasonics, Ferroelectrics and Frequency Control, **52**, pp. 707-716, (2005).
12. H. Jin, S. R. Dong, J. K. Luo, et al., Electro. Let., **47**, pp. 424-426, (2011).
13. H. Campanella, *Acoustic Wave and Electromechanical Resonators*: Artech House (2010).
14. J. D. Larson, III, P. D. Bradley, S. Wartenberg, *et al.*, IEEE Ultrasonics Symposium, pp. 863-868, (2000).
15. M. Hara, T. Yokoyama, M. Ueda, *et al.*, IEEE Ultrasonics Symposium, pp. 1152-1155, (2007).

 DOI: 10.1051/epjconf/201716201042

Correlation of film thickness to optical band gap of Sol-gel derived $Ba_{0.9}Gd_{0.1}TiO_3$ thin films for optoelectronic applications

Yen Chin Teh[1,2] , *Ala 'eddin* A. Saif [1,2*], *Zul Azhar* Zahid Jamal[1], and *Prabakaran* Poopalan[1,2]

[1]School of Microelectronic Engineering, University Malaysia Perlis, Pauh Putra Campus, 02600 Arau, Perlis Malaysia

[2]AMBIENCE UniMAP, Kompleks Kilang SME Bank, Kawasan Perindustrian Kuala Perlis, 02000 Kangar, Perlis Malaysia

Abstract. $Ba_{0.9}Gd_{0.1}TiO_3$ thin films have been fabricated on SiO_2/Si and fused silica by sol-gel method. The films are prepared through a spin coating process and annealed at 900 °C to obtain crystallized films. The effect of film thickness on the microstructure and optical band gap has been investigated using X-ray diffractometer, atomic force microscope and ultraviolet-visible spectroscopy, respectively. XRD patterns confirm that the films crystallized with tetragonal phase perovskite structure. The films surface morphology is analysed through amplitude parameter analysis to find out that the grain size and surface roughness are increased with the increase of films thickness. The transmittance and absorbance spectra reveal that all films exhibit high absorption in UV region. The evaluated optical band gap is obtained in the range of 3.67 - 3.78 eV and is found to be decreased as the thickness increase.

1 Introduction

Barium titanate ($BaTiO_3$) is a ferroelectric material that has gained much attention in recent decades for its excellent electrical and optical properties, and the simplicity of its crystal structure that can adapt variety types of dopant [1-2]. The incorporation of different element as dopant into $BaTiO_3$ has been extensively investigated in order to improve its original properties and extends the areas of application of barium titanate [3-4]. In spite of the fact that Gd^{3+} possessed a strong influences on electrical properties with high refractive index which makes it a good candidate for optoelectronic applications [4-6].

In optoelectronics, the responsivity of the material toward light wavelength is important to define its application. As a result, many researchers have focused on tuning the material's band gap based on desired applications through variety of parameters. The thickness of perovskite oxide films plays a role as one of parameters that can influence on many properties of the films, such as transport properties, optical band gap and the electronic structure [7]. Chen *et al.* [8] reported that film quality is highly dependent on the film thickness of sputtered BFO thin films, whereby thinner film displays greater leakage current with less light absorption yet demonstrated a great photovoltaic effect. Fasasi *et al.* [9] observed changes in band gap energy by altering the thickness of Zn-$BaTiO_3$ thin films. Thus, a better understanding of the thickness effect on the film properties is essential to optimize the optoelectronic device for better performance. Among the thin films fabrication technique, Sol-gel method possessed a great advantages for its simplicity in producing high homogeneity films at low processing temperature with low cost equipment [10]. In the present work, multi-layered of $Ba_{0.9}Gd_{0.1}TiO_3$ (BGT) thin films are fabricated on SiO_2/Si and fused silica substrate by sol-gel technique. The effect of different thicknesses on the microstructure and optical properties of BGT films are investigated using XRD, AFM and UV-Vis spectroscopy. The correlation between the optical band gap and microstructure at different film thicknesses are also discussed in some details.

2 Experimental Procedures

Pre-calculated amount of barium and gadolinium acetates were dissolved in heated glacial acetic acid at constant stirring, and followed by refluxing until a clear and transparent solution was obtained. At room temperature condition, a titanium solution was prepared by mix-stirred a stoichiometric amount of titanium (IV) isopropoxide with 2-methoxyethanol. This solution was then added into Ba:Gd solution and continued in stirring at room temperature. The stirred mixture was refluxed at 120 °C to obtain a clear and homogeneous solution. The final solution is filtered and used as stock solution. The BGT thin films were fabricated by spin coating

* Corresponding author: alaeddinsaif@gmail.com

technique. The stock solution was spun-coated onto SiO_2/Si and fused silica substrates at 4000 r.p.m for 20s, and then followed by heating at 200 °C for 10 minutes to vaporize the organic solvent which formed a thin film. The deposition and heating processes were repeated until desired thickness was achieved. The final as-deposited thin films were annealed at 900 °C for 1 hour in atmosphere ambient to form crystallized films. Four samples with different thicknesses were prepared, and denoted as BGT1, BGT2, BGT3 and BGT4. In order to measure the film thickness, the films were partially etched using diluted hydrofluoric acid to create a step profile. This step profile was measured using stylus profilometer. The obtained average thicknesses of fabricated sample BGT1 to 4 were 121.215 nm, 219.556 nm, 363.890 nm and 490.398 nm, respectively. The crystal structure of the BGT films was confirmed by using X-ray diffractometer (XRD, XRD 6000, Shimadzu) with Cu-kα radiation source (λ = 1.5406 Å). Surface morphologies were characterized using a non-contact mode of atomic force microscopy (AFM, SPA 400, SII Nanotechnology, Inc) at different location away from sample edges. The optical transmittance and absorbance spectra were measured by using UV-Vis spectrometer (Perkin Elmer, LAMBDA 950) at room temperature.

3 Results and discussion

Fig. 1 illustrates the XRD pattern of BGT films, deposited on SiO_2/Si and annealed at 900 °C. The orientation peaks (marked as P) appear in the pattern confirms the existence of perovskite structure in the films. A minor secondary phase corresponding to fresnoite (marked as f) is detected within the diffraction range. This secondary phase is attributed to the inter-diffusion between BGT films and substrate at high annealing temperature [11]. The appearance of secondary phase at 900 °C is consistent with Huang *et al.* [12] results for Gd_2O_3 deposited on silicon substrate. The determined lattice constant, a and c, from the perovskite peaks are 3.9786 ± 0.0056 Å and 4.0248 ± 0.0189 Å, respectively. This indicates that the film is crystallized with tetragonal perovskite structure phase.

AFM analysis has been employed to study the surface morphology of BGT films deposited on fused silica. Fig. 2 shows the AFM three-dimensional micrographs of different thicknesses of BGT films with scanned area of 5 µm x 5 µm. These micrographs reveal the microstructure nature of the films and the behaviour of grain structure with the film thickness. It is apparently seen from the fig. that dense and smooth surfaces are obtained with grains distributed uniformly over the surface area. The grain size of distributed grain on the surface are measured and listed in Table 1. The grain size is found to be increased as the thickness increases. This increment is attributed to the grain growth during heating and annealing processes [13].

In order to further analyse the surface roughness of BGT films, amplitude parameter analysis has been performed at different point on the sample surface and the results are compiled in Table 1. From the table, both average roughness and RMS roughness increased with the increase of thickness. This increase in roughness could be due to the surface grain growth [14]. The surface asymmetry and flatness can be evaluated from the parameter of skewness and kurtosis [15]. Apparently, from the table all the samples show positive skewness surface and kurtosis surface of less than 3 indicating that all the film surfaces contain disproportionate number of peaks [16].

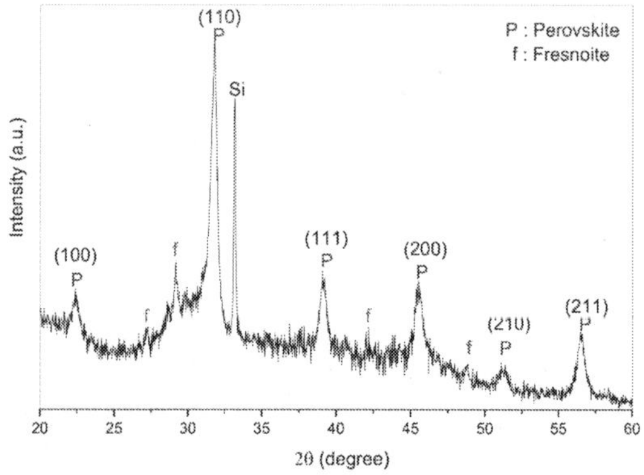

Fig. 1. XRD pattern of $Ba_{0.9}Gd_{0.1}TiO_3$ thin films.

Fig. 3 shows the transmittance and absorbance spectra of BGT films at different thicknesses, recorded at room temperature. It is found that the transmittance of the BGT films in the wavelength region of 350 nm to 800 nm reduces as the thickness increases as shown in fig. 3(a). This transmittance reduction has satisfies the Beer-Lambert law which is given by [17]

$$T = \exp^{(-\alpha d)} \quad (1)$$

where the intensity of transmitted light, T, decreases exponentially with thickness of sample, d, and absorption coefficient, α. As incident light shines through the sample, the total incident of light reduces as a result of scattering from the films. This light scattering could be due to the grain size increment and higher surface roughness [18].

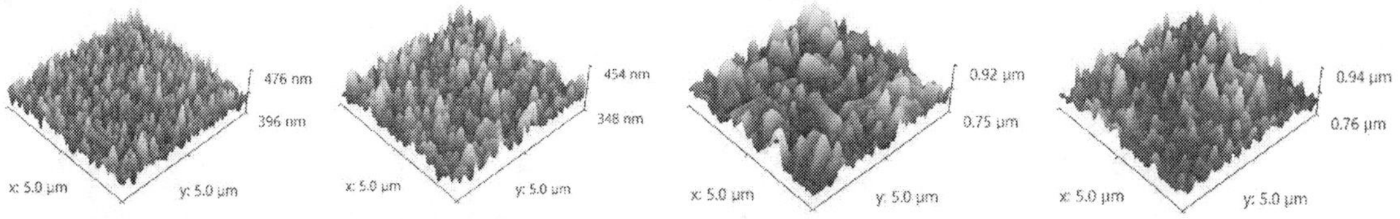

(a) (b) (c) (d)

Fig. 2. Three-dimensional AFM micrograph of BGT films fabricated at different thickness of (a) 121.215 nm, (b) 219.556 nm, (c) 363.890 nm and (d) 490.398 nm .

Table 1 Amplitude parameters of BGT films at different thicknesses.

Sample	BGT1	BGT2	BGT3	BGT4
Grain size (nm)	160.522 ±3.202	164.594 ±3.486	174.794 ±3.693	187.4545 ±4.193
Average Roughness (nm)	7.839 ±0.078	11.782 ±0.085	19.542 ±0.405	15.122 ±0.410
RMS Roughness (nm)	9.7936 ±0.092	14.590 ±0.080	24.185 ±0.466	19.500 ±0.514
Skewness	0.2614 ±0.041	0.601 ±0.018	0.558 ±0.010	0.971 ±0.059
Kurtosis	0.21556 ±0.121	0.166 ±0.030	0.243 ±0.059	1.208 ±0.179

The absorbance spectra reveal the light absorption behaviour in the wavelength range of 200 nm to 800 nm. It is clearly seen that relatively low in absorbance found in the range of 350 nm to 800 nm as shown in fig. 3 (b). However, BGT films apparently have high absorption in the UV wavelength region of 200 – 350 nm as shown in fig. 3(b). It is also observed that the absorbance in this region is proportionally increased with the film thickness which could be due to film crystallinity improvement. A significant shift in cut off wavelength towards longer wavelength as the thickness increase can be observed clearly from both transmittance and absorbance spectra, which attributed to the thickness effect [19].

(a)

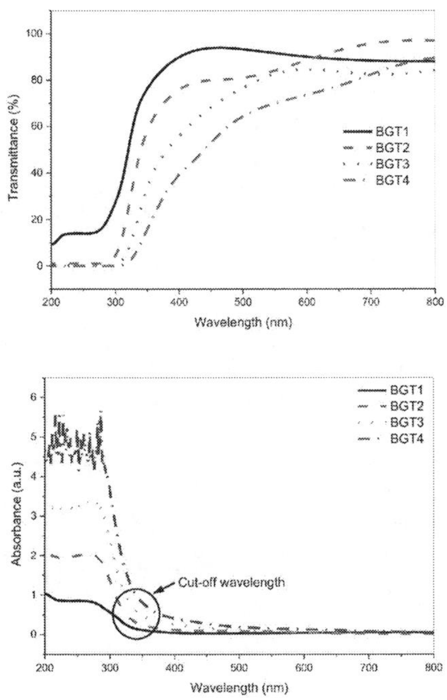

(b)

Fig. 3. Optical spectra, (a) transmittance and (b) absorbance, of BGT films fabricated at different thicknesses.

The optical interband transition and the value of band gap can be determined using Tauc relation which given as [20]

$$(\alpha h\nu) = B(h\nu - E_g)^m \quad (2)$$

where $h\nu$ is the photon energy, B is a constant, $m = 1/2$ and 3/2 for allowed direct and forbidden direct transitions, and $m = 2$ and 3 for allowed indirect and forbidden indirect transitions respectively [21]. Absorption coefficient, denoted as α can be calculated from transmittance spectra which given as

$$\alpha = 1/d \ln (1/T) \quad (3)$$

Thus, the optical band gap is deduced from the intercept of x-axis at zero y-axis from the plot of $\alpha h\nu^{1/m}$ against $h\nu$ through extrapolation of the linear part of the plot.

(a)

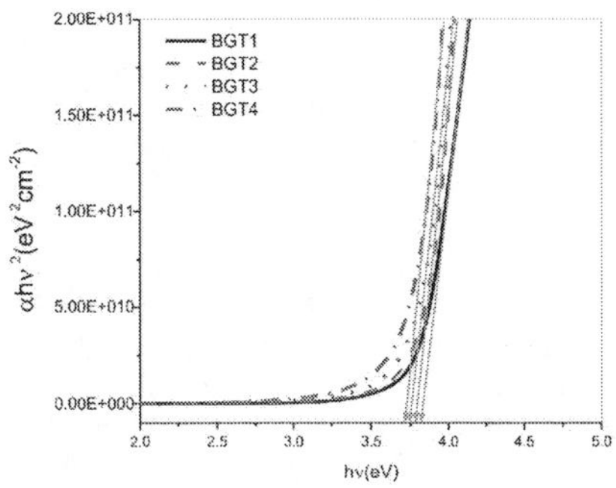

(b)

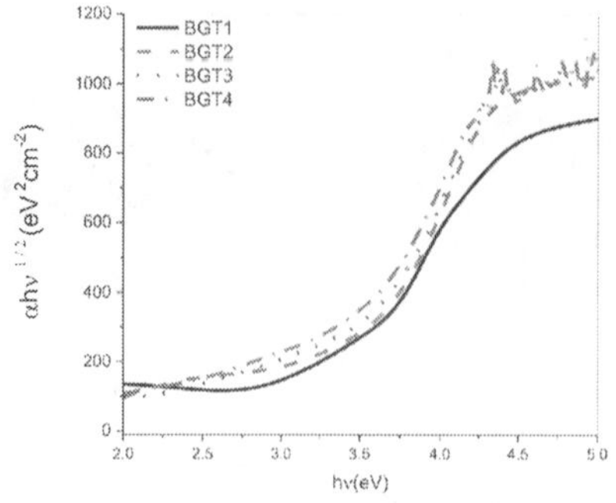

Fig. 4. Tauc plot of (a) $(\alpha hv)^2$ and (b) $(\alpha hv)^{1/2}$ as a function of hv of BGT films at distinct thicknesses.

In order to determine the interband transition of BGT films at different thicknesses in this work, the $\alpha hv^{1/m}$ with direct and indirect transition are plotted against hv as illustrated in fig. 4 (a) and (b). By comparing both graphs, it is observed that the best fit of straight line is obtained for m = 2, indicating that direct-allowed transition takes place between valence bands and conduction bands for BGT films. The value of optical band gap for BGT1, BGT2, BGT3 and BGT4 films are found to be 3.78 eV, 3.73 eV, 3.69 eV and 3.67 eV, respectively, which are decreasing as the thickness increased. The decrement in band gap can be linked to the increment of grain size as discussed in AFM results, which is attributed to the quantum size effect, whereby the smaller the crystal size, the larger the energy band gap [22]. Similar reduction in optical band gap as the thickness increased has been reported by Zhou *et al.* [23] for BFO thin films.

4 Conclusions

Gadolinium doped barium titanate thin films have been fabricated on SiO_2/Si and fused silica by sol-gel method. The phase structure of the crystallized films obtained from XRD result confirms the existence of perovskite structure with tetragonal phase. AFM micrographs illustrate the distribution behaviour of grain over the surface as the thickness increases. The average grain size is found to be increased with the increase of film thickness, which is resulted from the grain growth mechanism during heating and annealing processes. Similar increment is observed for surface roughness with thickness increase. From the skewness and kurtosis, the films surfaces are found to be having disproportionate number of peaks and spread out distribution. Both the transmittance and absorbance spectra illustrate that all the films exhibit high absorption in the wavelength region of 200 - 350 nm. The optical band gap that determined from transmittance spectra is found to be reduced as the film thickness increases, which due to quantum size effect as correlated with the grain size of the films. Therefore, the obtained results illustrate the possible integration BGT thin films into optoelectronic devices depending on the application needs.

This work is supported by the Fundamental Research Grant Scheme (FRGS) with the Grant number 9003-0479, funded by the Ministry of Higher Education (KPT).

References

1. Z. H. Dughaish, *Acta Phys. Pol. A*, **123**, 87 (2013)
2. Y. Huan, X. Wang, J. Fang, L. Li, *J. Eur. Ceram. Soc.*, **34**, 1445 (2014)
3. A. Rani, J. Kolte, S. S. Vadla, and P. Gopalan, Ceram. Int., **42**, 8010 (2016)
4. J. P. H. Lara, M. P. Labra, F. R. B. Hernández, J. A. R. Serrano, E. O. A. Davila, P. Thangarasu, and A. H. Ramirez, *Mater. Res.*, 1 (2017)
5. A. Y. Fasasi, B. D. Ngom, J. B. Kana-Kana, R. Bucher, M. Maaza, C. Theron, U. Buttner, *J. Phys. Chem. Solids*, **70**, 1322 (2009)
6. Y. Li, Y. Hao, X. Wang, X. Yao, *Ferroelectrics*, **407**, 134 (2010)
7. X. He, K.-J. Jin, C. Ge, C. Wang, H.-B. Lu, G.-Z. Yang, *J. Epl*, **102**, 37007 (2013)
8. M. Chen, J. Ding, J. Qiu, N. Yuan, *Mater. Lett.*, **139**, 325 (2015)
9. A. Y. Fasasi, M. Maaza, E. G. Rohwer, D. Knoessen, C. Theron, A. Leitch, U. Buttner, *Thin Solid Films*, **516**, 6226 (2008)
10. M. N. Kamalasanan, S. Chandra, P. C. Joshi, A. Mansingh, *Appl. Phys. Lett.*, **59**, 3547 (1991)
11. T. M. Stawski, W. J. C. Vijselaar, O. F. Göbel, S. A. Veldhuis, B. F. Smith, D. H. A. Blank, J. E. Ten Elshof, *Thin Solid Films*, **520**, 4394 (2012)
12. M. R. S. Huang, C. P. Liu, J. C. Wang, Y. K. Chen, C. S. Lai, Y. C. Fang, L. Shu, *Thin Solid Films*, **520**, 5579 (2012)
13. Y. C. Teh, A. A. Saif, Z. A. Z. Jamal, P. Poopalan, *Adv. Mater. Res.*, **911**, 251 (2014)
14. A. A. Saif, N. Ramli, P. Poopalan, *Jordan J. Phys.*, **3**, 61 (2010)
15. M. Raposo, Q. Ferreira, P. A Ribeiro, *Mod. Res. Educ. Top. Microsc.*, 758 (2007)
16. G. Yildirim, S. Bal, M. Gulen, A. Varilci, E. Budak, M. Akdogan, *Cryst. Res. Technol.*, **47**, 195 (2012)
17. W. Leng, C. Yang, H. Ji, J. Zhang, J. Tang, H. Chen, *J. Mater. Sci. Mater. Electron.*, **17**, 1041–1045 (2006)
18. S. Bobby Singh, H. B. Sharma, H. N. K. Sarma, S. Phanjoubam, *Physica B: Condensed Matter*, **403**. 2678 (2008)
19. A. Singh, Z. R. Khan, P. M. Vilarinho, V. Gupta, R. S. Katiyar, *Mater. Res. Bull.*, **49**, 531 (2014)
20. S. Kumari, N. Ortega, A. Kumar, J. F. Scott, R. S. Katiyar, *AIP Adv.*, **4**, 37101 (2014)
21. J. Ahmad, H. Minami, S. Alam, J. Yu, Y. Arai, H. Uwe, *Chinese Phys. Lett.*, **25**, 4421 (2008)
22. K. Suzuki K. Kijima, *Jpn. J. Appl. Phys.*, **44**, 2081 (2005)
23. Y.-E. Zhou, X.-Y. Tan, B.-F. Yu, L. Liu, S.-L. Yuan, W.-H. Jiao, *Chinese Phys. Lett.*, **31**, 037304 (2014)

Critical analysis of stability and performance of organometal halide perovskite solar cells via various fabrication method *(Review)*

Suriati Suhaimi[1], *Zaliman* Sauli[1], *Nor Azura Malini* Ahmad Hambali[2], *Mohamad Halim* Abd Wahid[2], *Vithyacharan* Retnasamy[1], *Mukhzeer* Mohamad Shahimin[3,*]

[1] School of Microelectronic Engineering (SoME), Universiti Malaysia Perlis (UniMAP), Pauh Putra Main Campus, Jalan Arau-Changlun, 02600 Arau, Perlis, Malaysia
[2] Semiconductor Photonics & Integrated Lightwave Systems (SPILS), Tun Abdul Razak Laser Laboratory (TAREL),School of Microelectronic Engineering, Universiti Malaysia Perlis (UniMAP), Kampus Pauh Putra, Arau, Perlis, Malaysia.
[3] Department of Electrical and Electronic Engineering, Faculty of Engineering, National Defence University of Malaysia (UPNM), Kem Sungai Besi, 5700 Kuala Lumpur, Malaysia

Abstract. Organometal halide perovskite solar cells (Omh-PSCs) have attracted attention due to its unique electrical and optical properties. Ideally, the Omh-PSCs should remain free from degradation under normal operating conditions for several years, preferably tens of years. In order to produce high power conversion efficiency with low potential of degradation, different fabrication methods have been developed. The reported stability of perovskite films can vary significantly and reported to decay substantially up to 20% of its original performance. A thorough understanding of fabrication process upon the stability of the device is regarded as crucial to pave the way for future endeavors. This review summarized and highlighted the recent research of fabrication methods that gave an impact to the stability of perovskite devices.

1 Introduction

As a new type of photovoltaic technology, perovskite solar cell will become the ultimate low cost technology from now on [1]. A new generation of mixed organic–inorganic halide perovskites hit the alluring prospects on both views; the overall cost to manufacture as well as the power conversion efficiencies [2]. Dye-sensitized solar cell (DSSCs) [3], organic solar cell [4] and hybrid solar cell [5] are emerging solar cell that made strides especially low cost fabrication technology using solution processing at low temperature and outlying large scale application. Perovskite solar cells initial research is correlated with DSSCs work, which is the perovskite materials that absorbed on mesoporous TiO_2 employed as a dye-sensitizer in DSSCs. However, only half or less power conversion efficiencies (PCE) achieved as compared to the best commercial Si PV cells [6].

Most studies of light absorber layers are focusing on methylammonium (MA) lead halide $MAPbX_3$ (X ¼ halogen) and mixed-halide perovskites [7]. Recently, solar cells based on halide perovskites have been widely explored due to their high PCE, straightforward and low manufacturing cost [8]. Besides that, organometal halide perovskites have been attractive due to very excellent electrical and optical properties such as: (i) in mixed halide perovskite, the diffusion lengths is up to 1 [9], [10], (ii) good ambipolar (can be n- or p-type) transport of charge [11-12], (iii) absorption in the visible range is efficient [15–17], and (iv) synthesis method straightforward [16].

Over a few years under normal operating state, organometal halide perovskite solar cells (Omh-PSCs) must stay free from degradation, in order for Omh-PSCs to grow economically feasible which preferably tens of years. The assessment and indication of reliability of the devices are tough as the stability is correlated to the fading process of single solar cells [17]. Thus, for practical applications the stability is critical. The goal of this review is to look up a comprehensive view of the fabrication method that affects the stability of perovskite devices.

2 Structure and Working Principle

To assemble the perovskite solar cells, the various architectures of the device has been identified. General configuration of conventional PSC device is depicted in Fig.1 (a). Between a hole transporting material (HTM) and an electron transporting material (ETM) perovskite layer is sandwiched together with metal electrode substrates of indium tin oxide (ITO)/fluorine tin oxide

*Corresponding author: mukhzeer@upnm.edu.my

(FTO). Structure of employing mesoporous metal oxide materials (TiO_2) and planar heterojunction structure without mesoporous materials [18-19] are two major architecture of perovskite solar cells devices .

Perovskite solar cells work by following processes: (i) the incident light absorbs by perovskite layer, (ii) ETMs and HTMs are extracted and transported the generated electron and hole, respectively as illustrated in Fig.1 (b). At the end, the electrode collected the charge carriers and forming complete cycle of the perovskite solar cells. Principally, the transportation and extraction of the hole and electron must be efficient. The highest occupied molecular orbital (HOMO) of HTMs must be greater than the maximum valence band (VBM), while, the minimum conduction band (CBM) of ETMs must be lower than the CBM of perovskite materials [15], which is important to be fully grasped.

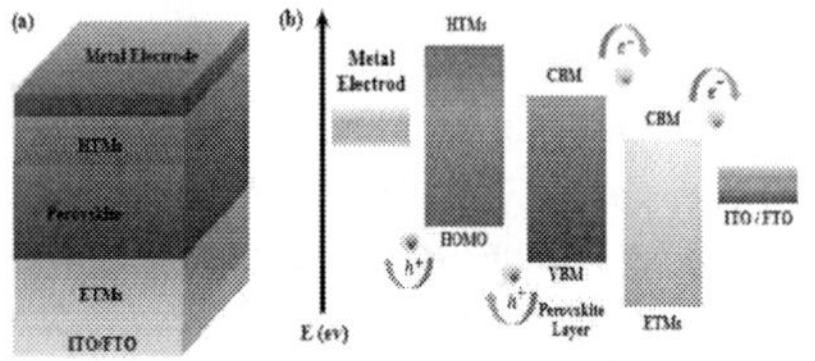

Fig.1. (a) A schematic illustration of perovskite device structure, (b) Working principle of ideal perovskite solar cells based on (a) [20].

3 Fabrication Method

Different methods that have been developed with regard to fabricate perovskite solar cell devices. There are two common methods that have been used to fabricate perovskite solar cells which are (i) one step coating and (ii) two step coating. However, on the side of producing high PCE with low potential of degradation, other additional method is introduced.

3.1. Spin coating

In order to grow quality perovskite thin films, a whole new field of technology has been developed. The one-step solution deposition which is used as direct deposition of perovskite precursor solution on top of mesoscopic, or planar oxide [7] is the most typical deposition procedure. The one-step spin coating has been extensively used in fabricating perovskite solar cells due to the easy fabrication method. However, the one-step spin coating in ambient air is not widely investigated because this approach crucially needs to be carried out in the glove box, in order to neglect the influence of moisture. [21]. Fig.2 shown a schematic illustration of the fabrication in present of $TiO_2/CH_3NH_3PbI_3$ photovoltaic cells. Previous work are reported about the fabrication process in details except for the mesoporous TiO_2 layer [22].

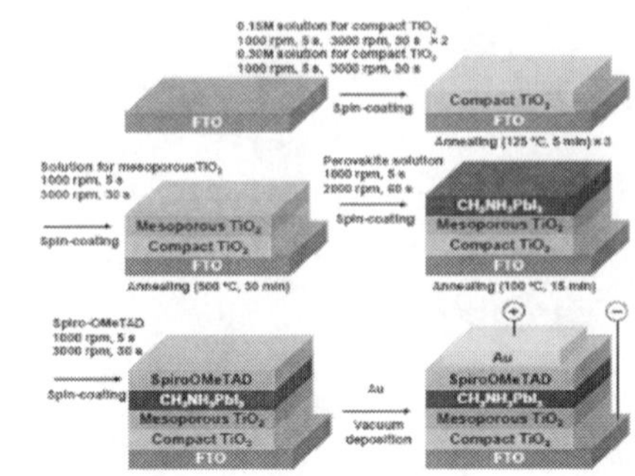

Fig.2. Schematic illustration of the fabrication $TiO_2/CH_3NH_3PbI_3$ photovoltaic cells using spin coating deposition method [22].

3.2 Doctor Blade

As been reported, the doctor blade method that fabricate in ambient air condition with planar structure has benefit in stability and photovoltaic performance of perovskite solar cell [23]. Based on the study, scanning electron microscope (SEM) shows the large domains of crystalline and the thin film formation on film substrate are uniformly covered, as depicted in Fig.3. The light absorption that produced the film prepared by using doctor blade technique proved that the dropping PV performance using spin coating technique in ambient air are related to the perovskite degradation [23].

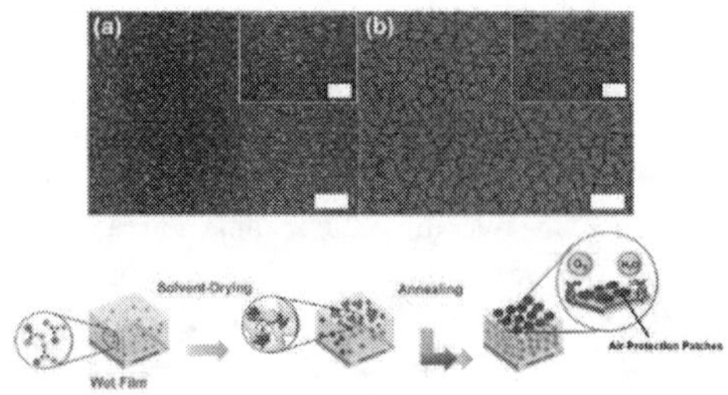

Fig.3. Perovskite thin films SEM image using (a) spin coating and (b) doctor blade method [23].

3.3. Sequential Deposition

Gratzel and co-workers was first used the sequential deposition method to deposit the absorber layer using $MAPbI_3$ [24]. Generally, to achieve high reproducibility with an excellent performance of the perovskite, sequential deposition method is introduced to fabricate the perovskite sensitized mesoscopic solar cells. The best device for solution processed achieved PCE about 15%, that conclude to be a new record as amongst the highest PCE for organic or hybrid inorganic solar cells [25]. One-step or sequential deposition method are the common technique to deposit $MAPbI_3$ perovskite thin films. As widely known, the sequential deposition method give first-string control over the growth of perovskite, and under near equilibrium condition the formation of crystalline might occur, as the resulting structure of $MAPbI_3$ is as fascinating as one-step method [24].

3.4 Hybrid Chemical Vapor Deposition

Another method has also been used to deposit $MAPbI_3$ film is hybrid chemical vapor deposition. The PCE about 11.8% obtained by using hybrid CVD and the same efficiency demonstrated although after 1100 hour the device stored in dark and dry N_2 gas condition [26]. Preparation of perovskite film using vapor-based deposition technique is one of the promising approaches that gives an advantage for fabrication of large scale perovskite module fabrication and able to prevent fast reaction rate. Recently, a few works applying different vapor-assisted techniques have been reported, with the concern to the high-quality absorber layer that faces up the reduction of the response rate for a pin-hole-free layer [27].

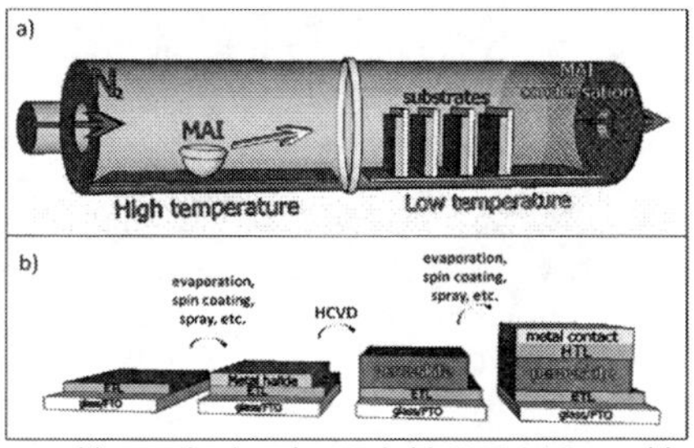

Fig.4. Perovskite synthesis using Hybrid CVS (a) Deposition of MAI substrate onto metal halide seeded in HCVD furnace. (b) The full solar cell structure of each layer fabricated in HCVD furnace. Overall layer of solar cell; (i) FTO layer, (ii) ETL, (iii) perovskite layer (iii) HTL and (iv) metal contact [26].

Fig.4 (a) illustrated one example of HCVD deposition process that used $PbCI_2$ and MAI. The substrates FTO glass substrate with TiO_2 compact layer pre-deposited with $PbCI_2$ and MAI is loaded into the furnace with two separate temperature control zones. Fig.4 (b) shown a schematic diagram of overall cell and layers. The titanium dioxide on FTO thin film coated by spray pyrolysis is the electron transport layer (ETL) of the cell. The following layer is deposited with a metal halide layer and HCVD perovskite synthesis. A film of Spiro-MeOTAD is the hole transport layer (HTL) that spun onto the surface of perovskite followed by deposition via evaporation of top electrode with gold. By using hybrid CVD, the perovskite solar cells showed a stability higher to 11.8% which can encourage further study in the stability of perovskite especially under ambient conditions [26].

3.5 Alternating Layer-by-Layer

Alternating layer-by-layer vacuum deposition is one of the efficient method that has been formulated to fabricate perovskite solar cells. This method obtained 91% from its first value after 62 days being stored under an ambient condition with less degradation of <9% [28]. The best efficiency achieved is 16.03% and average efficiency recorded by the planar device is 15.37% with a minimum standard deviation of $\pm 0.37\%$ [28]. A factor of producing good stability was believed due to the dense formation of perovskite thin film with overall coverage surface that able to prevent penetration of moisture [29].

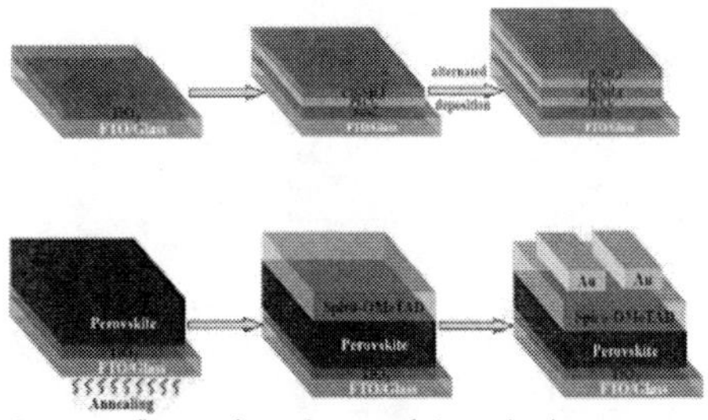

Fig.5. Process flow using alternated layer-by-layer vacuum deposition of perovskite solar cells [28].

A layer of $PbCI_2$ was evaporated onto substrate/TCO/ TiO_2 in a vacuum chamber via alternating precursor layer deposition as illustrated in Fig.5. The process begins with thermally sublimed the layer of CH_3NH_3I onto the surface of the $PbCI_2$ film, while the thickness of CH_3NH_3I film is controlled to secure the sufficient quantity for deposited $PbCI_2$. To achieve high PCE, both of the processes are repeated until the acceptable thickness is generated. Then, the sample is annealed for 2 hours at temperature 20°C in a glove box condition. The sample color switched from taupe to dark color right after annealing, which reveals that perovskite film is formed through the interaction reaction between $PbCI_2$ and CH_3NH_3I. Consequently, both of the spiro-MeOTAD and gold which act as HTM and electrode were deposited on the perovskite film, separately [28]. Therefore, the fabrication process is found to be very reproducible.

4 Effect of Fabrication Method

The fabrication method is the one of the critical factors that gives an impact to the device stability. The reported stability of perovskite films is shown to vary quite significantly. In one case, perovskite is reported to decay as much as 20% substantially from its initial performance over the 6 days course in the N_2 environment, and in others, it is reported to be stable.

Table 1. Reliability of perovskite solar cells using various fabrication method.

Fabrication Method	Composition/Structure	PCE (%)	Reliability	Degradation	Ref.
Spin Coating	FTO /c-TiO_2/ Al_2O_3/$MAPbI_{3-x}Cl_x$/ Au	~12	1000 hour		[30]
	FTO /c-TiO_2/m-TiO_2/$MAPbI_{3-x}Cl_x$/ Au	5.9	Without encapsulation. 50% of initial PCC after 5 hour	Photocurrent degradation	[30]
		7.1	With encapsulation, 10% of initial PCC after 5 hour		
		6.2	With encapsulation and UV Filter: No degradation		
	FTO /c-TiO_2/ $MAPbI_{3-x}Cl_x$/ spiro/Au	10.2	15 days PCE dropped one order of magnitude	PbI_2 formation due to interaction between perovskite with moisture based on XRD measurement	[31]
	ITO/ PEDOT:PSS/ $MAPbI_{3-x}Cl_x$ / $PC_{61}BM$ /C_{60}-bis /Ag	1.23	Decomposed after 5 days	Perovskite degradation	[23]
	FTO /c-TiO_2/m-TiO_2/$MAPbI_{3-x}Cl_x$/spiro/ Au (Precursor: $PbCl_2$)	11.1	15 days, 20% μ_o	Lead precursor effect the stability device, film morphology and cell performance	[32]
	FTO /c-TiO_2/m-TiO_2/$MAPbI_{3-x}Cl_x$/spiro/ Au	10.8	15 days: degradation is slightly higher than cells with $PbCl_2$ precursor	Different lead precursor	[32]
Blade Coating	ITO/PEDOT:PSS/ $MAPbI_{3-x}Cl_x$ / $PC_{61}BM$/ C_{60}-bis /Ag	9.29	PCE of 95% after 268 hour	Enhance surface morphology, uniformity of perovskite film and improve crystalline size to be larger	[23]
	ITO/PEDOT:PSS/ $CH_3NH_3PbI_xCl_{3-x}$/ $PC_{61}BM$/ Bis-C_{60} / Ag	10.44	Maintain >90% of original PCE after being stored for 10 days in inert atmosphere	Control humidity to enhance performance	[33]
Sequential Deposition	FTO/ TiO_2/ $CH_3NH_3PbI_3$ / Spiro-MeOTAD /Au	13.0	Maintain >80% of initial PCE after period of 500 hour.	Decrease both fill factor and open circuit voltage due to change of shunt resistance	[25]
Hybrid Chemical Vapor Deposition	FTO/c-TiO_2/ perovskite/spiro/ gold	11.8	1100 hour: No degradation	Perovskite reaches temperature above 160°C	[26]
Alternating layer-by-layer	FTO/c-TiO_2/ $MAPbI_{3-x}Cl_x$/spiro/ gold	16.03	62 days: 9% decrease in μ_o	Fabrication of dense perovskite film with overall coverage of surface that prevents moisture penetration resulted in high stability	[34]

As summarized in Table 1, the different fabrication methods with different structure and architecture of perovskite solar cell resulted in the different PCE which reflect the stability and reliability of the device. Based on the review, different fabrication process produced different stability and reliability of the device. Thus, its shows that the fabrication method also affected the PCE performance of the perovskite solar cells.

5 Recommendation and Outlook

The previous researcher have been investigating myriad of approaches in order to enhance the stability of the perovskite solar cells such as using different structures of the metal oxide films and different hole transport materials (HTM) that related to the hydrophobicity of the HTM, which is important to protect perovskite from atmospheric moisture. In this review, we critically analyzed the progress of various methods for fabrication PSCs and have discussed a number of degradation issues existed. However, a key challenge to fabricate PSCs, is the difficulty in comparing the stability testing results from the other group because of the different optimum condition that are utilized in each experiment such as humidity, temperature, and encapsulation technique. Therefore, for stability testing, it is necessary to standardize the required optimum condition. Regardless, to improve the stability of the PSC devices, the development of new technology in encapsulation is recommended. Specify standards testing devices need to be introduced such as device hysteresis, mechanical stability, stability under thermal stress and exposure to light, oxygen and moisture for every fabrication method. In summary, research on PSCs will eventually lead to the ultimate device that can rival the conventional silicon-based photovoltaics. However, it is highly critical to overcoming the stability issues that correlated with both the perovskite material and the fabrication method of the device to meet the required of long term stability. When the device is exposed to the high intensity of moisture or at high temperature, the perovskite solar cell shows the inevitable degradation. Thus, it is crucial to have a comprehensive understanding of the fabrication method that tangled in the deterioration of perovskite material to evolve robust strategy to resolve these issues.

FRGS (9003-00396), RACE (9017-00002), ERGS (9010-00004) and UPNM-STG (1. UPNM/2016/GPJP/4/ST/1) are acknowledged for the funding used for this investigation.

References

1. H. J. Snaith, J. Phys. Chem. Lett., **4**, no. 21, pp. 3623–3630 (2013)
2. N.-G. Park, J. Phys. Chem. Lett., **4**, no. 15, pp. 2423–2429, Aug. (2013)
3. S. Suhaimi, M. M. Shahimin, Z. A. Alahmed, J. Chyský, and A. H. & Reshak, Int. J. Electrochem. Sci., **10**, pp. 2859–2871 (2015)
4. R. Vairavan, M. M. Shahimin, and N. Juhari, *2011 IEEE Colloquium on Humanities, Science and*

Engineering. pp. 454–458 (2011)
5. A. H. Reshak, M. M. Shahimin, N. Juhari, and R. Vairavan, Curr. Appl. Phys., **13**, no. 9, pp. 1894–1898 (2013)
6. P. K. Nayak and D. Cahen, Adv. Mater., **26**, no. 10, pp. 1622–1628 (2014)
7. S. Casaluci *et al.*, J. Power Sources, **297**, pp. 504–510 (2015)
8. J. Qian, B. Xu, and W. Tian, Org. Electron. physics, Mater. Appl., **37**, pp. 61–73 (2016)
9. S. D. Stranks *et al.*, Science (80-.)., **342**, no. 6156, p. 341 LP-344 (2013)
10. G. Xing *et al.*, Science, **342**, no. 6156, pp. 344–347, (2013)
11. E. Edri *et al.*, **14**, no. 2, pp. 1000–1004, (2014)
12. C. Zuo, H. J. Bolink, H. Han, J. Huang, D. Cahen, and L. Ding, Adv. Sci., **3**, no. 7, p. 1500324–n/a (2016)
13. J. H. Noh, S. H. Im, J. H. Heo, T. N. Mandal, and S. Il Seok, Nano Lett., **13**, no. 4, pp. 1764–1769, (2013)
14. G. E. Eperon, S. D. Stranks, C. Menelaou, M. B. Johnston, L. M. Herz, and H. J. Snaith, Energy Environ. Sci., **7**, no. 3, pp. 982–988 (2014)
15. S. De Wolf *et al.*, J. Phys. Chem. Lett., **5**, no. 6, pp. 1035–1039, (2014)
16. H.-S. Kim *et al.*, " *Sci. Rep.*, **2**, p. 591, (2012)
17. T. A. Berhe *et al.*, Energy Environ. Sci., **9**, no. 2, pp. 323–356 (2016)
18. J. Shi *et al.*, ACS Appl. Mater. Interfaces, **6**, no. 12, pp. 9711–9718, (2014)
19. L. Liu *et al.*, J. Am. Chem. Soc., **137**, no. 5, pp. 1790–1793, (2015)
20. X. Tong, F. Lin, J. Wu, and Z. M. Wang, Adv. Sci., **3**, no. 5, p. 1500201–n/a (2016)
21. G. Wang *et al.*, RSC Adv., **6**, no. 49, pp. 43299–43303 (2016)
22. T. Oku, T. Matsumoto, A. Suzuki, and K. Suzuki, Coatings, pp. 646–655 (2015)
23. J. H. Kim, S. T. Williams, N. Cho, C.-C. Chueh, and A. K.-Y. Jen, Adv. Energy Mater., **5**, no. 4, p. 1401229–n/a (2015)
24. H. Li *et al.*, Sol. Energy, **126**, pp. 243–251, (2016)
25. J. Burschka *et al.*, Nature, **499**, no. 7458, pp. 316–319, (2013)
26. M. R. Leyden, L. K. Ono, S. R. Raga, Y. Kato, S. Wang, and Y. Qi, J. Mater. Chem. A, **2**, no. 44, pp. 18742–18745 (2014)
27. M. H. Li, P. S. Shen, J. S. Chen, Y. H. Chiang, P. Chen, and T. F. Guo, *2016 23rd International Workshop on Active-Matrix Flatpanel Displays and Devices (AM-FPD)*. pp. 256–257 (2016)
28. D. Yang, Z. Yang, W. Qin, Y. Zhang, S. (Frank) Liu, and C. Li, J. Mater. Chem. A, **3**, no. 18, pp. 9401–9405, (2015)
29. M. Shahbazi and H. Wang, Sol. Energy, **123**, pp. 74–87, (2016)
30. T. Leijtens, G. E. Eperon, S. Pathak, A. Abate, M. M. Lee, and H. J. Snaith, Nat. Commun., **4**, pp. 1–8, (2013)
31. A. Mei *et al.*, Science (80-.)., **345**, no. 6194, p. 295 LP-298 (2014)
32. F. K. Aldibaja, L. Badia, E. Mas-Marza, R. S. Sanchez, E. M. Barea, and I. Mora-Sero, *J. Mater. Chem. A*, **3**, no. 17, pp. 9194–9200 (2015)
33. Z. Yang, C. Chueh, F. Zuo, J. H. Kim, P. Liang, and A. K. Jen, Advanced Energy Materials, **5** no. 13 (2015)
34. J. Yang, B. D. Siempelkamp, D. Liu, and T. L. Kelly, ACS Nano, **9**, no. 2, pp. 1955–1963, (2015)

Optimization of intrinsic layer thickness, dopant layer thickness and concentration for a-SiC/a-SiGe multilayer solar cell efficiency performance using Silvaco software

Wong Wei Yuan[1], *Mohd Natashah* Norizan[1,2,3,*], *Ili Salwani* Mohamad[1,2,3], *Nurnaeimah* Jamalullail[1,2], and *Nor Hidayah* Saad[1]

[1]School of Microelectronic Engineering, Universiti Malaysia Perlis, Pauh Putra Campus, 02600 Arau, Perlis, Malaysia.
[2]Centre of Excellence Geopolymer and Green Technology (CeGeoGTech), Universiti Malaysia Perlis, Perlis, Malaysia.
[3]Advanced Multi-Disciplinary MEMS-Based Integrated NCER Centre of Excellence (AMBIENCE), Universiti Malaysia Perlis, Perlis, Malaysia.

Abstract. Solar cell is expanding as green renewable alternative to conventional fossil fuel electricity generation, but compared to other land-used electrical generators, it is a comparative beginner. Many applications covered by solar cells starting from low power mobile devices, terrestrial, satellites and many more. To date, the highest efficiency solar cell is given by GaAs based multilayer solar cell. However, this material is very expensive in fabrication and material costs compared to silicon which is cheaper due to the abundance of supply. Thus, this research is devoted to develop multilayer solar cell by combining two different layers of P-I-N structures with silicon carbide and silicon germanium. This research focused on optimising the intrinsic layer thickness, p-doped layer thickness and concentration, n-doped layer thickness and concentration in achieving the highest efficiency. As a result, both single layer a-SiC and a-SiGe showed positive efficiency improvement with the record of 27.19% and 9.07% respectively via parametric optimization. The optimized parameters is then applied on both SiC and SiGe P-I-N layers and resulted the convincing efficiency of 33.80%.

1 Introduction

In modelling multilayer Silicon (Si) solar cell using silicon carbide (SiC) based solar cell layer, amorphous silicon (a-Si) alloy is considered as an important material of the multilayer solar cell based on many researches [1–4]. a-Si alloy is a material that used to reduce the cost of solar cells in the future [5,6]. The material is easy to dope compared to crystalline Si, hence the energy used in doping process can be saved [7,8]. By combining a-Si and intrinsic layer formation, it has high optical absorption coefficient over the wavelength of solar spectrum that enable to fabricate thin film solar cell that can fully absorb the incoming light in its very thin thickness layer [9]. Hence, it can be used to design low cost thin film solar cell [10].

1.1 P-I-N solar cell structure

In an a-Si alloy, the alloy needed to be hydrogenated because of amorphous silicon has different bond angles and bond length in single units of a-Si cell compared to crystalline Si [11]. Thus, a-Si has possibility to form dangling bond defect, which the certain bond will be totally missing Si atom. Therefore, hydrogen dilution is needed to prevent loss bonding of neighbouring Si atoms.

Figure 1 shows the structure of P-I-N solar cell with its doping density. The top doping is p-layer while the bottom doping is n-layer of the solar cell. Two layers of Indium Tin Oxide (ITO) which is colourless and transparent are coated on the bottom and top of the solar cell to act as a contact.

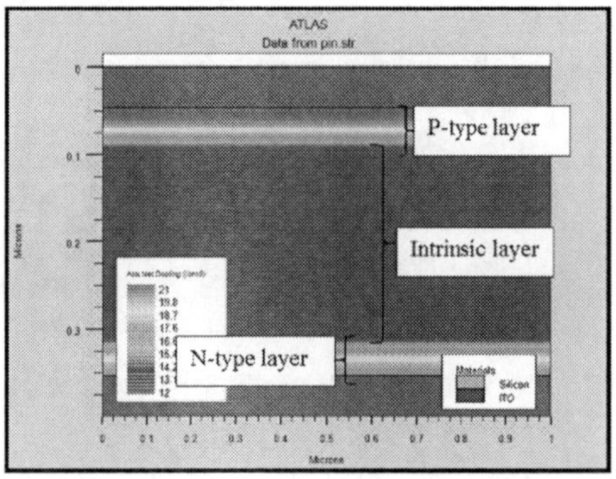

Fig. 1. Structure of P-I-N Solar Cell with its doping density.

Combination of a-Si and intrinsic layer produced an a-Si layer with low dopant concentration which is important to form P-I-N structure of the solar cell. This

* Corresponding author: mohdnatashah@unimap.edu.my

design has high optical absorption coefficient over the wavelength of solar spectrum that enable to fabricate thin film solar cell that can fully absorb the incoming light in its very thin thickness layer [12]. The band gap energy of this combination of materials is high enough to cover relatively more high light intensity in solar spectrum compared to crystalline Si [13]. The intrinsic layer also has much stronger internal electric field to separate electrons and holes efficiently, which avoid additional energy losses [14].

1.2 Operation of P-I-N solar cell

The operation of solar cell consists of two main steps. In the first step, photons are striking on the Si layer surface and absorbed at these layers to create free electrons and holes[15]. The energy of the photon must higher or equal to the band gap energy of these Si layers to ensure absorption of the photon. When the energy of photon is equal to the energy of the band gap, single electron hole pair will be generated without loss of energy. When the energy of photon is larger than the energy of the band gap, the excess energy of photon minus the band gap energy (E_{ph}-E_g) will rapidly transform into heat after generation of the single electron hole pair. When the energy of photon is lesser than the band gap energy, the photon will either be reflected or absorbed elsewhere [16]. This low energy of photons will be absorbed due to impurities or defect introduced within the gap.

In the second step, these electron hole pairs have to be isolated by guiding the electron to one side and the holes to the other side of the solar cell device [16]. Thus, an internal electric field created by intrinsic layer can separate them by a potential difference as shown in Figure 2. Figure 2 is not only an energy band diagram that shows the band gap energy of p-doped, n-doped layer and intrinsic layer of silicon, but it shows the larger separation of p-doped layer energy band and n-doped layer energy band which can separate electron at conduction band of n-doped layer recombination to valence band of p-doped layer. Hence, majority carriers of p-doped layer and n-doped layer which were electrons and holes that have opposite electric charges will move in opposite directions within an electric field to create a voltage for external use [16].

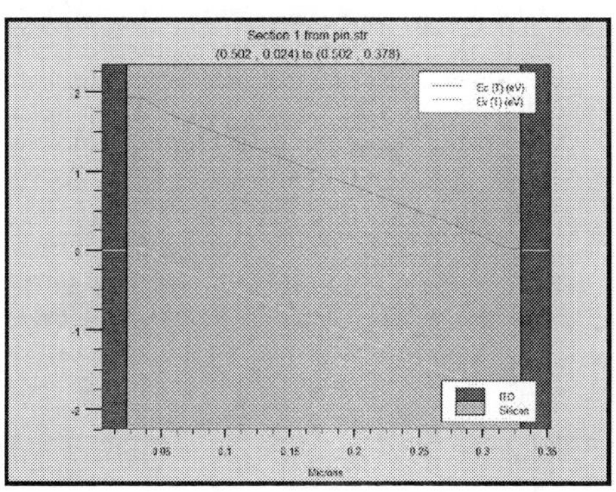

Fig. 2. Energy band diagram in P-I-N solar cell.

1.3 Design of P-I-N multilayer SiC solar cell

In all layers of multilayer silicon alloy solar cell shown in Figure 3, p-doped a-H is hydrogenated amorphous material that doped with a p-type dopant, an intrinsic layer of a-H is hydrogenated amorphous material that doped lightly with n-type dopant, and N-doped a-H is hydrogenated amorphous material that doped with n-type dopant.

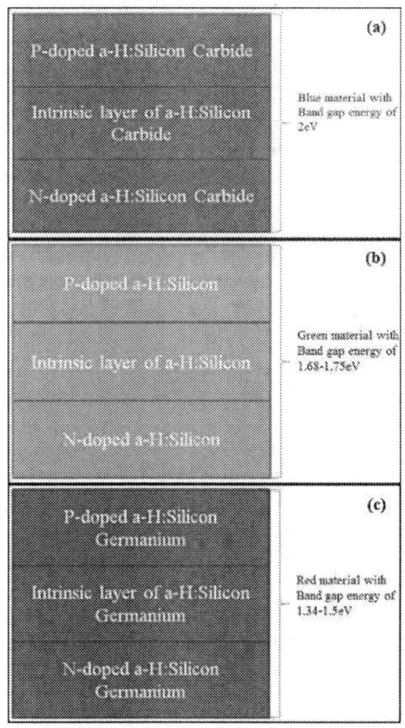

Fig. 3. (a) Top, (b) middle and (c) bottom layer of multilayer Si alloy solar cell.

The material for top layer is silicon carbide, middle layer is silicon, and bottom layer is silicon germanium. The blue material with band gap energy of 2eV is able to absorb short wavelength with high light intensity which will not only contribute high voltage but also high current with its high light intensity [17]. Secondly, green material (a-Si) with energy band gap in between 1.68eV and 1.75eV is prefect material to be the second layer only if third layer that absorbs red photon exist.

2 Details of the simulated solar cell

In order to simulate a-SiC/SiGe multilayer solar cells, a commercial simulation software application, Silvaco TCAD was used. The program computes the generating functions of electron-hole pairs from the optical parameters of each layer [18-20]. In this study, optical constants have been used, which are the refractive index and the extinction coefficient as functions of the wavelength for the layers used in the solar cell, of all individual layers as input optical parameters in the reference.

2.1. Baseline parameter of single layer solar cell

Since amorphous silicon carbide and amorphous silicon germanium are amorphous silicon alloy, amorphous silicon is the most basic material that can be alloyed with other elements such as carbon (C) and germanium (Ge) [5]. Thus, baseline values of the parameters used to optimize the efficiency of every layer solar cell can be referred from the parameter values of efficiency optimized amorphous silicon single layer solar cell. The parameters of interest are intrinsic layer thickness, p-doped layer thickness, n-doped layer thickness, p-doped layer concentration and n-doped layer concentration. Other than that, the parameters that are constant were set by simulating to default value, only one parameter can be changed to find the maximum efficiency in order to optimize this single layer solar cell efficiency.

In Table 1, values in a row no. 1 until row no. 5 are the increment of P type dopant concentration to give the increment result of the efficiency of P-I-N solar cell. The values of n-type dopant were increased as shown by the values in row no. 5 until row no.8 until a maximum efficiency was achieved as shown in row no. 7 which is 4.42 percent. This showed that increment of p-type and n-types dopant concentration will increase the efficiency, but depend on the increment of its opposite layer dopant concentration.

For an example, when p-type concentration increases until a value of efficiency stagnate, increment of its opposite type layer dopant concentration will actually increase the efficiency of P-I-N solar cell. Therefore, increment of opposite type layer dopant concentration after maximizing its dopant concentration will increase the efficiency. However, row of no. 9 explained that the efficiency does not increase but decrease to 4.21 percent. This proven that the maximum efficiency from manipulation of P type and N type dopant concentration is 4.42 percent. In conclusion, the baseline value for P type and N type dopant concentration were 1e21 and 1e21 respectively.

Table 1. Efficiency optimization with p-doped layer and n-doped layer.

No.	P- Concentration (cm^{-3})	N- Concentration (cm^{-3})	Efficiency (%)
1	4×10^{17}	1×10^{18}	4.134
2	1×10^{18}	1×10^{18}	4.213
3	1×10^{19}	1×10^{18}	4.33
4	1×10^{20}	1×10^{18}	4.35
5	1×10^{21}	1×10^{18}	4.35
6	1×10^{21}	1×10^{19}	4.36
7	1×10^{21}	1×10^{20}	4.38
8	1×10^{21}	1×10^{21}	4.42
9	1×10^{22}	1×10^{21}	4.21

For the intrinsic layer, the thickness can be changed in between 0.15 micron and 0.3 micron [15]. In Table 2, the thickness of intrinsic layer is increased to 2 microns to achieve efficiency of 10.05%, otherwise the efficiency is lower than 10.05%. Therefore, the baseline value for the intrinsic layer thickness is chosed to be 2 μm.

Table 2. Efficiency optimization with intrinsic layer thickness.

No.	Thickness (μm)	Efficiency (%)
1	0.25	4.36
2	0.3	4.42
3	0.35	4.826
4	0.4	5.16
5	0.5	5.96
6	0.6	6.63
7	0.7	7.17
8	0.8	7.63
9	0.9	7.65
10	1	8.51
11	1.5	9.42
12	2	10.05
13	2.5	10.005

The default values of n-doped layer and p-doped layer thickness were not changed since the efficiency of solar cell will be decreased when the thickness were changed as shown in Table 3 and 4. Therefore, baseline values of n-doped layer and p-doped layer thickness were 0.05 microns.

Table 3. Efficiency optimization with p-doped layer thickness.

No.	Thickness (μm)	Efficiency (%)
1	0.06	9.76
2	0.05	10.05
3	0.04	10.007

Table 4. Efficiency optimization with n-doped layer thickness.

No.	Thickness (μm)	Efficiency (%)
1	0.06	10.003
2	0.05	10.05
3	0.04	10.045

2.2 Simulation of a-SiC/a-SiGe multilayer solar cells

Figure 4 (a) shows the baseline structure with the meshing, thickness of the material region and electrodes; doping statement in Figure 4 (b) specified the doping concentration in each region. The meshing used in x-axis is spaced every 0.2 micron to 1 micron and automatically spaced in the y-axis. Meshing statement "qty.m" and "qtx.m" is used to declare the structure in that dimension of a tunnel junction.

Material thickness of each region is listed in Figure 4 (b) using "y.min" and "y.max". The materials of each region was listed inside the coding using "mat". The tunnel junction is deposited in between a-SiC top layer solar cell and a-SiGe bottom layer solar cell. The material used for tunnel junction is SiC, which is the same material used in the top layer of solar cell. The thickness and the doping concentration of p-type layer and n-type layer are both 0.05 micron and 1×10^{15} cm^{-3} respectively.

 DOI: 10.1051/epjconf/201716201044

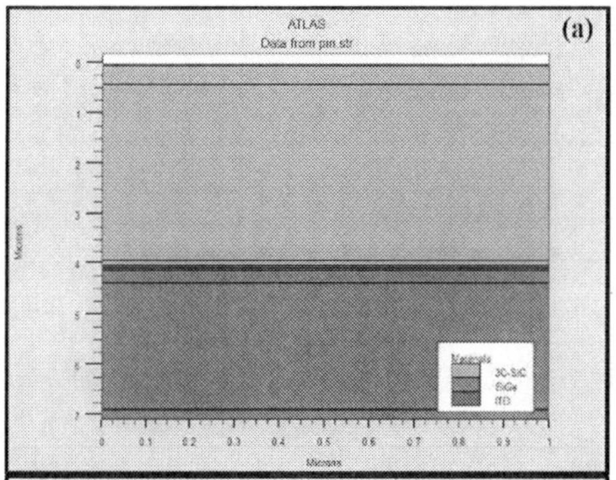

(b)

```
mesh auto

x.m l=0.0 s=0.25
x.m loc=1.0 s=0.25

qtx.m loc=0.0 s=0.25
qtx.m loc=1.0 s=0.25

qty.m l=4.05   s=0.0002
qty.m l=4.1  s=0.0001
qty.m l=4.15   s=0.0002

region num=1 mat=3C-SiC y.min=0.05 y.max=0.45 name=a-SiC:H
region num=2 mat=3C-SiC y.min=0.45 y.max=3.95 name=a-SiC:H
region num=3 mat=3C-SiC y.min=3.95 y.max=4.05 name=a-SiC:H
region num=4 mat=3C-SiC y.min=4.05 y.max=4.1 name=a-SiC:H
region num=5 mat=3C-SiC y.min=4.1 y.max=4.15 name=a-SiC:H
region num=6 mat=SiGe y.min=4.15 y.max=4.4  x.comp=0.3 name=a-SiGe
region num=7 mat=SiGe y.min=4.4 y.max=6.9 x.comp=0.3 name=a-SiGe
region num=8 mat=SiGe y.min=6.9 y.max=7.1 x.comp=0.3 name=a-SiGe

elec num=1 name=anode   y.max=0.05             material=ITO
elec num=2 name=cathode y.min=7.1 y.max=7.15 material=ITO

doping region=1 uniform p.type conc=1e29
doping region=3 uniform n.type conc=1e25
doping region=4 uniform n.type conc=1e15
doping region=5 uniform p.type conc=1e15
doping region=6 uniform p.type conc=1e10
doping region=8 uniform n.type conc=1e21
```

Fig. 4. (a) Baseline structure of multilayer solar cell and (b) coding of multilayer solar cell structure specification group statements.

3 Results and discussions

The optimized results shown in Table 5 were obtained by doing a parametric study on every parameter. Parameters that effect the efficiency increment of a-SiC single layer solar cell obtained from this study were n-type dopant concentration and intrinsic layer thickness while for a-SiGe single layer solar cell is n-type layer thickness.

Table 5. Optimized results for a-SiC and a-SiGe single layer solar cell.

Parameter	a-SiC	a-SiGe
p-type dopant concentration (cm^{-3})	1 x 10^{29}	1 x 10^{10}
n-type dopant concentration (cm^{-3})	1 x 10^{25}	1 x 10^{21}
n-type layer thickness (micron)	0.1	0.2
p-type layer thickness (micron)	0.4	0.25
Intrinsic layer thickness (μm)	3.5	2.5

In Figure 5 (a) and (c), open-circuit voltage (V_{oc}) were 1.85 V same for both (a-SiC and SiC/a-SiGe) and at this time there were no current flows through the circuit and it is the maximum voltage that the solar cell can generate. At Short-circuit current (I_{sc}) 1.59 mA (a-SiC) and 1.97 mA (SiC/a-SiGe), this was where current flow through the circuit when the electrodes are short-circuited. The I_{sc} of a solar cell depends on the photon flux density incident on the solar cell, which is determined by the spectrum of incident light.

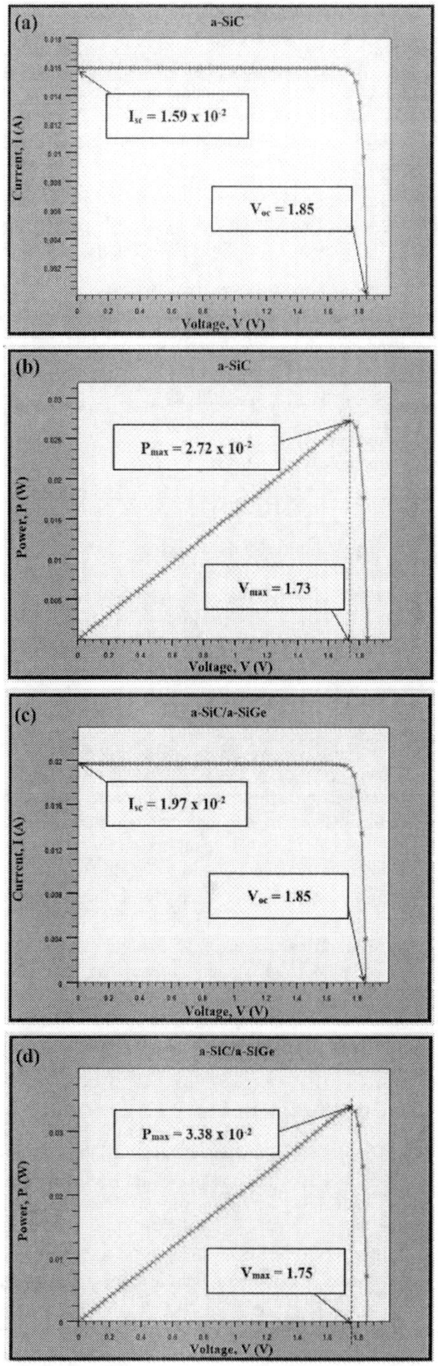

Fig. 5. Characteristic curves (a) I-V and (b) P-V of a-SiC solar cell and (c) I-V and (d) P-V of a-SiC/a-SiGe solar cell.

In Figure 5 (b) and (d), maximum power output (P_{max}) also known as the maximum power point is 2.72 mW for an a-SiC and 3.38 mW for a SiC/a-SiGe solar cell as shown at the peak of the P-V curve. P_{max} is a key parameter since solar cells are used to produce electrical energy. Solar cell produces no power when it is in an open-circuit or short-circuit state. The maximum voltage (V_{max}) is 1.73 V (a-SiC) and 1.75 V (SiC/a-SiGe) when a solar cell reaches it's maximum power. The maximum current (I_{max}) of the solar cell is 1.57 mA (a-SiC) and 1.93 mA (SiC/a-SiGe) calculated using the following formula $I_{max} = P_{max}/V_{max}$.

Table 6. Optimized simulation results of a-SiC single layer solar cell and a-SiC/a-SiGe multilayer solar cell.

Photovoltaic Parameter	a-SiC	a-SiC/a-SiGe
Short Circuit Current, I_{sc} (mA)	1.59	1.97
Open Circuit Voltage, V_{oc} (V)	1.85	1.85
Maximum Voltage, V_{max} (V)	1.73	1.75
Maximum Current, I_{max} (mA)	1.57	1.93
Fill Factor, FF	0.92	0.93
Maximum Power, P_{max} (mW)	2.72	3.38
Input Power, P_{in} (kW/m^2)	0.10	0.10
Efficiency η (%)	27.20	33.80

The simulated values such as I_{sc}, V_{oc}, V_{max}, I_{max}, P_{max} were collected from Figure 5 and summarized in Table 6. Fill factor (FF) of 0.92 (a-SiC) and 0.93 (a-SiC/a-SiGe) were calculated by using the equation $FF=P_{max}/(I_{sc}\times V_{oc})$. The efficiency of optimized a-SiC/a-SiGe multilayer solar cell is 33.8 %, while the efficiency of optimized a-SiC single layer solar cell is 27.2 % and this shows an improvement of 6.6%

The n-type dopant concentration of 1 x 10^{25} cm^{-3} is considered highly doped concentration. High n-doped concentration allowed more light to pass through the intrinsic layer to improve the open-circuit voltage and short-circuit. This result supported by A. Kosarian and P. Jelodarian in their research results which shown that high n-doped concentration will increase open-circuit voltage and short-circuit current [21].

The intrinsic layer of 3.5 µm play very important role to separate electrons and holes that was generated to prevent electron recombine back to holes at valence band. Thus, it improved a-SiC/a-SiGe multilayer solar cell from avoiding additional energy losses.

4 Conclusions

This work demonstrated the optimization of intrinsic layer thickness, dopant layer thickness and concentration for a-SiC/a-SiGe multilayer solar cell, which improved the efficiency performance. Efficiency of 33.8% of the double layers a-SiC/a-SiGe solar cell which improved 6.6% in efficiency from single layer a-SiC solar cell has been achieved through this research by optimization through parametric study. The results proved that correct material combination can contribute to the increment of solar cell efficiency when compared with the single layer a-SiC solar cell simulated with only 27.2% in efficiency.

The author would like to acknowledge the support from the Fundamental Research Grant Scheme (FRGS) Grant under a grant number of 9003-00339 and Research Acculturation Grant Scheme (RAGS) under a grant number of RAGS/1/2015/SG0/UNIMAP/03/3 from the Ministry of Higher Education (MoHE) Malaysia.

References

1. I. Afdi Yunaz, S. Miyajima, and M. Konagai, in *2010 35th IEEE Photovolt. Spec. Conf.* (IEEE, 2010), pp. 000317–000322.
2. S. De Wolf, A. Descoeudres, Z. C. Holman, and C. Ballif, Green **2**, 7 (2012).
3. Kyung Hoon Jun, Sung Won Kwon, and Koeng Su Lim, in *Conf. Rec. Twenty-Eighth IEEE Photovolt. Spec. Conf. - 2000 (Cat. No.00CH37036)* (IEEE, 2000), pp. 924–927.
4. J. Burdick and T. Glatfelter, Sol. Cells **18**, 301 (1986).
5. I. A. Yunaz, A. Yamada, and M. Konagai, Jpn. J. Appl. Phys. **46**, L1152 (2007).
6. M. A. Green, Prog. Photovoltaics Res. Appl. **9**, 123 (2001).
7. J. Poortmans and V. Arkhipov, in *Sol. Cells* (2006), p. 471.
8. A. Scuto, M. Foti, C. Gerardi, A. Battaglia, and S. Lombardo, *IEEE Int. Reliab. Phys. Symp. Proc.* **2016–Septe**, 3C31 (2016).
9. J. Zhu, Z. Yu, G. F. Burkhard, C.-M. Hsu, S. T. Connor, Y. Xu, Q. Wang, M. McGehee, S. Fan, and Y. Cui, Nano Lett. **9**, 279 (2009).
10. S. Kim, J. W. Chung, H. Lee, J. Park, Y. Heo, and H. M. Lee, Sol. Energy Mater. Sol. Cells **119**, 26 (2013).
11. W. G. J. H. M. Van Sark, in *Thin Film. Nanostructures* (2002), pp. 1–215.
12. X. Meng, G. Gomard, O. El Daif, E. Drouard, R. Orobtchouk, A. Kaminski, A. Fave, M. Lemiti, A. Abramov, P. Roca i Cabarrocas, and C. Seassal, Sol. Energy Mater. Sol. Cells **95**, S32 (2011).
13. P. Sánchez, O. Lorenzo, A. Menéndez, J. L. Menéndez, D. Gomez, R. Pereiro, and B. Fernández, Int. J. Mol. Sci. **12**, 2200 (2011).
14. L. Li, P. A. Salvador, and G. S. Rohrer, Nanoscale **6**, 24 (2014).
15. S. M. Zahari, R. A. M. Osman, M. N. Norizan, I. S. Mohamad, A. Malini, and S. Taking, Appl. Mech. Mater., pp. 1182–1186, 2015.
16. A. Shah, *Thin-Film Silicon Solar Cells* (EPFL Press, 2010).
17. A. A. Boussettine, Y. Belhadji, and A. Benmansour, Energy Procedia **18**, 693 (2012).
18. Silvaco, *ATLAS User's Manual* (2016).
19. M. N. Norizan, S. M. Zahari, I. S. Mohamad, R. A. M. Osman, M. M. Shahimin, and S. A. Z. Murad, *IOP Conf. Ser. Mater. Sci. Eng.*, **209**, (2017).

 DOI: 10.1051/epjconf/201716201044

20. S. M. Zahari, M. N. Norizan, I. S. Mohamad, R. A. M. Osman, and S. Taking, *AIP Conf. Proc.*, 1660, May, (2015).
21. A. Kosarian and P. Jelodarian, in *2011 Third Int. Conf. Comput. Intell. Model. Simul.* pp. 417–422. (IEEE, 2011).

Recent improvements on TiO_2 and ZnO nanostructure photoanode for dye sensitized solar cells: A brief review

Nurnaeimah Jamalullail[1,2], *Ili Salwani* Mohamad[1,2,3,*], *Mohd Natashah* Norizan[1,2,3], *Norsuria* Mahmed[2,4], and *Bibi Nadia* Taib[1]

[1]School of Microelectronic Engineering, Universiti Malaysia Perlis, Pauh Putra Campus, 02600 Arau, Perlis, Malaysia.
[2]Centre of Excellence Geopolymer and Green Technology (CeGeoGTech), Universiti Malaysia Perlis, Perlis, Malaysia.
[3]Advanced Multi-Disciplinary MEMS-Based Integrated NCER Centre of Excellence (AMBIENCE), Universiti Malaysia Perlis, Perlis, Malaysia.
[4]School of Material Engineering, Universiti Malaysia Perlis, P.O. Box 77, D/A Pejabat Pos Besar, 01000, Kangar, Perlis, Malaysia.

Abstract. Dye sensitized solar cell (DSSC) is a promising candidate for a low cost solar harvesting technology as it promised a low manufacturing cost, ease of fabrication and reasonable conversion efficiency. Basic structure of DSSC consists of photoanode, dye, electrolyte and counter electrode. Photoanode plays an important role for a DSSC as it supports the dye molecules and helps in the electron transfer that will determine the energy conversion efficiency. This paper emphasizes the various improvements that had been done on the TiO_2 and ZnO photoanode nanostructures synthesized through thermal method. For overall comparisons, ZnO nanoflowers photoanode had achieved the highest energy conversion efficiency of 4.7% due to its ability of internal light scattering that had increased the electron transportation rate. This has made ZnO as a potential candidate to replace TiO_2 as a photoanode material in DSSC.

1 Introduction

The energy sources that the world primarily depending today is harvested from the non-renewable energy sources which is from fossil fuel [1][2], coal [3][4] and nuclear [5][6]. This energy sources will be run out any time from now. Besides, the process of utilizing the non-renewable energy left us vulnerable to air pollutants because of the emission of carbon dioxide (CO_2) [7][8]. The increase of carbon dioxide will lead to the global warming and greenhouse effect [9][10]. Thus, in order to maintain the energy demand and the environmental problem, the world now is urging to find highly effective and carbon-free sources from renewable energy.

Renewable energy is defined as a natural energy that is inexhaustible and abundant such as solar, wind and tidal. Among all those type of renewable energy, solar is the most commonly used as the energy source due to the ability that can be deployed nearly everywhere on earth [11]. Solar energy falls on the surface of the earth everyday on the rate of 120 petawatts which is able to satisfy more than 20 years of energy demand [12][13]. The technology that had been used to convert solar energy into electrical energy is called the photovoltaic effect that is utilized by solar cell.

There are many types of solar cell such as silicon solar cell, thin film solar cell and dye sensitized solar cell (DSSC). Each type of solar cell has its own speciality but the advantages of simplest fabrication process , low cost and the reasonable efficiency has made DSSC as a promising candidate in low power appliances such as outdoor lamp, gadget charger and if it is assembled in a big scale it can be used as smart window power generator [14][15]. The first fabricated DSSC was discovered by O'Regan and Gratzel with the efficiency of 7% [16]. Since then, DSSC has attracted a lot of attention in researches and developments areas.

Basic DSSC structure consists of working electrode that is sandwiched together with semiconductor materials, dye or sensitizer, electrolyte and counter electrode as illustrated in Fig. 1. Photoanode is composed of working electrode that had been coated with semiconductor materials. The semiconductor material plays an important role in electron injection and transportation which will determine the performance of the DSSC [14-16]. As the important component of DSSC, the nanostructure of the semiconductor materials for the photoanode has been widely studied.

1.1 Working principle

Dye sensitized solar cell working principle is applied from photosynthesis process and diffusion is a major mechanism for electron to travel through the components [16-18]. The working principle of dye sensitized solar cell involved several steps as shown in Fig. 1. During the day, the sun will release energy called photon. Photon will strike a dye molecule and bring it to the excited

* Corresponding author: ilisalwani@unimap.edu.my

state. The excited dye molecule is then injects an electron into the semiconductor material, leaving a hole in the molecule. The electron will diffuse through the semiconductor material, reaching both working and counter electrode. On the counter electrode, electron will regenerate the oxidized dye through the iodide (I_3^-) and form triiodide ($3I^-$). Meanwhile, the triiodide ($3I^-$) will recover its missing electron (e^-) from cathode and regenerating the iodides. The process will continue with the presence of sunlight.

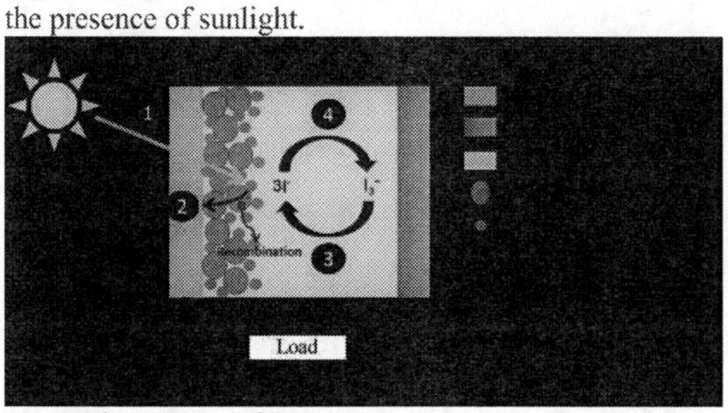

Fig. 1. The structure of DSSC.

2 Improvements in Photoanode Semiconductor Materials

In order to perform as a good photoanode semiconductor materials, the selected materials must have wide energy and direct bandgap [21][22], high electron mobility [23][24] and ideal surface morphology [25][26]. In DSSC, diffusion is a major route for electron to travel through [16-18]. Upon dye excitation caused by the photon, the electron from the excited dye is injected into the semiconductor material. The reduction of the oxidized sensitizer dye by the electrolyte is in competition with the unfavorable recombination of the injected electron with the oxidized dye [27] as illustrated in Figure 1. Thus, it is obvious that the electron transportation is a very important for the energy conversion process. Henceforth, the selection of the photoanode need to be taken seriously so that the electron can be transferred efficiently.

2.1. Titanium Dioxide (TiO_2)

TiO_2 is the most commonly used semiconductor material for photoanode [28][29]. TiO_2 is a semiconductor material with wide energy bandgap that is existed in three crystalline phase; rutile [26][31], anatase [31][32] and brookite [31][33]. This type of material is not only differs in the crystalline phase but also in the energy bandgap as rutile have the energy bandgap of 3.02 eV, 3.2 eV for anatase and 2.96 eV for brookite [30][31]. Anatase is the most commonly used in DSSC [34][35]. However, TiO_2 has been suffering from high recombination losses and low electron mobility properties [31][36]. In DSSC, the recombination takes place between the injected electron in the semiconductor and the oxidized dye that will lead to inefficiency [27].

Thus, in order to enhance the electron mobility in the TiO_2, several studies in surface area, porosity, grain size, particle diameter, orientation, and crystal phase of the nanostructures had been studied. Recent studies have involved the fabrication of TiO_2 anatase as nanoparticles [37][38], nanowires [34-36], nanotubes [37-40] and nanorods [41-43]. Taleb et. al demonstrated that the aggregation of TiO_2 nanoparticles (Fig. 2 (a)) strongly influences the performance of DSSC by achieving 4.3% of energy conversion efficiency [46]. According to the research, the high energy conversion efficiency is achieved due to the large surface area for dye attached, enhanced light scattering and improved connectivity for charge transport and thus lowering the recombination process.

La et al. in his previous research have developed a method to synthesize anatase TiO_2 nanowires with the energy conversion efficiency of 4.12% by employing salinization treatment to the TiO_2 nanowire [41]. Through the salinization treatment, it can prevent the detachment of dyes from the surface of TiO_2 nanowires which allowing more sunlight energy to be absorbed and converted. In other hand, TiO_2 nanorods structure also had been synthesized with the aim to provide direct pathways for electron transportation and thus increase the energy conversion rate with the energy conversion efficiency of 1.86% [47]. However, the energy conversion efficiency recorded is still lower due to numbers of the grain boundary for the electrons to pass through. Meanwhile, TiO_2 nanotubes also had been studied, according to Liu et al. in his research, the energy conversion efficiency recorded for TiO_2 nanotubes photoanode is 4.59% [48]. TiO_2 nanotubes provides hollow structure that gives large surface area for the dye attachement and allow the electron to move in one direction. Ameen et. al in had developed highly dense and well-defined TiO_2 nanoflowers photoanode grown by the hydrothermal process [49]. The fabricated DSSC with TiO_2 nanoflowers photoanode reveal that the surface morphology is in the form of uniform clover-like petals and had accomplished a reasonably good energy conversion efficiency of 3.64%. The efficiency improvement on TiO_2 nanostructure photoanode is illustrated in Table 1.

2.2 Zinc Oxide (ZnO)

ZnO is a semiconductor with hexagonal crystalline wurzite structure with direct wide energy bandgap (3.37 eV) [50][51]. ZnO is also a favorable semiconductor material for photoanode due to large free excitation binding energy (60 meV) and high electron mobility (155 cm^2 $V^{-1}.S^{-1}$) [52][53]. Up to now, ZnO has been considered as a promising candidate to replace TiO_2. Interestingly, ZnO crystal structure enables various morphological changes including nanoparticles [54][55], nanowires [40][51], nanorods [56][57], nanotubes [40][58], nanoflowers [59][60] and etc. The variety of nanostructures can boost the electron mobility and enhance the energy conversion efficiency of DSSC.

To achieve high energy conversion efficiency performance, the photoanode needs to possess a large surface area for the dye attached and good electron transport capability [61][62]. Pandey et al. in his research have achieved energy conversion efficiency of 3.35% through ZnO nanoparticles photoanode [63]. ZnO nanoparticle film provides a large enough surface area; however, electron transport is difficult because of the need for electrons to hop across neighboring nanoparticles and thus slow down the electron mobility that leads to inefficiency [64]. In other hand, Jung et al. have fabricated DSSC with ZnO nanowires photoanode that improved light scattering and facilitating electron with energy conversion efficiency of 0.26% [65]. However, small surface area for dye attach and too many path for the light scattering direction causing it to slow down the electron mobility and leads to inefficiency. Rouhi et al. research have synthesized ZnO nanorods photoanode in his research and achieved energy conversion efficiency of 0.93% [66]. Based on the theory, nanorods will provide direct pathways for the electron to pass through the semiconductor materials. However, since the nanorods mainly lie down on the glass, electrons still need to face many grain boundaries that slow down their speed to the electrode. Meanwhile, Han et. al had develop ZnO nanotubes photoanode with the energy conversion efficiency of 1.18% [67]. This research compares the performance of DSSC with ZnO nanorods and nanotubes where the nanotubes was modified from the nanorods structure. The research shows significant improvement with nanotubes that achieved higher energy conversion efficiency compared to nanorods. Next, Wahyuono et. al in his research has reported energy conversion efficiency of 4.7% [68] through the fabrication of DSSC with ZnO nanoflowers photoanode which recorded as the highest efficiency compared to others that had been stated before. In the research, ZnO nanoflowers have shown a suitable structure with improved inner light scaterring compared to TiO_2 nanoflowers structure

The energy conversion efficiency of fabricated DSSC with ZnO photoanode is comparable with TiO_2 photoanode which has made ZnO as a potential candidate to replace TiO_2 as a photoanode material in DSSC. The summary of the electrical performance on improvements in ZnO nanostructures compared to TiO_2 nanostructures is summarized in Table 1 and illustrated in Figure 2.

Table 1 presents the electrical data recorded based on previous works that had been done for both TiO_2 and ZnO nanostructure. According to the data, the highest efficiency recorded is from ZnO nanoflowers photoanode with the energy conversion efficiency of 4.7%. ZnO nanoflowers can work as a good photoanode due to its ability of internal light scattering that had increased the electron transportation rate. This has made ZnO as a potential candidate to replace TiO_2 as a photoanode material in DSSC.

Table 1. Summary of performance on improvements in nanostructures of photoanode.

Photoanode	Nanostructures	J_{sc} (mA/cm^2)	V_{oc} (V)	FF	Efficiency (%)	Ref
TiO_2	Nanoparticles	15.00	0.62	0.55	4.30	[46]
	Nanowires	8.31	0.76	0.65	4.12	[41]
	Nanorods	4.15	0.76	0.59	1.86	[47]
	Nanotubes	8.65	0.76	67.7	4.59	[48]
	Nanoflowers	9.60	0.67	0.57	3.60	[49]
ZnO	Nanoparticles	8.82	0.62	0.58	3.35	[63]
	Nanowires	1.13	0.73	0.26	0.26	[65]
	Nanorods	4.90	0.43	0.44	0.93	[66]
	Nanotubes	3.24	0.68	0.58	1.18	[67]
	Nanoflowers	19.60	0.57	0.42	4.70	[68]

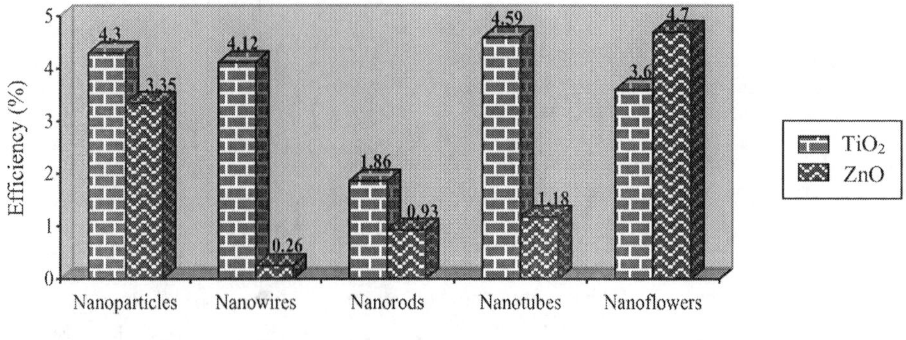

Fig. 2. Efficiency improvements in nanostructures of photoanode.

3 Conclusions

Many studies have been devoted to enhance the performance of the DSSC especially in the nanostructure of photoanode materials since photoanode is an important component that provides the electron transportation for the energy conversion process. This paper emphasizes the various improvements that had been done on the TiO_2 and ZnO photoanode nanostructures (nanoparticles, nanowires, nanorods, nanotubes and nanoflowers) synthesized through thermal method where all of the above works have shown some significant improvement in the performance of the DSSC. In conclusion, ZnO nanoflowers photoanode had achieved the highest energy conversion efficiency of 4.7% due to its ability of internal light scattering that had increased the electron transportation rate. This is also supported by the chemical properties of ZnO itself where it has the high electron mobility (155 cm^2 $V^{-1}.S^{-1}$) and large free excitation binding energy (60 meV). Hence, the development and innovation of the photoanode material should be continued to enhance the performance of DSSC.

The author would like to acknowledge the support from the Research Acculturation Grant Scheme (RAGS) under a grant number of RAGS/1/2015/SG0/UNIMAP/03/3 from the Ministry of Higher Education Malaysia.

References

1. FS-UNEP, *Global Trends in Renewable Energy* (2016).
2. International Energy Agency, *Key CO2 Emissions Trends* (2016).
3. World Energy Council, *World Energy Resources 2016* (2016).
4. International Energy Agency, *Key World Energy Statistics* (2016).
5. S. Sanaeepur, H. Sanaeepur, A. Kargari, and M. H. Habibi, Int. J. Sustain. Energy **33**, 203 (2014).
6. International Energy Agency, *Energy Technology Perspectives 2017 Executive Summary* (2017).
7. OECD/IEA, *The Future of Trucks - Implications for Energy and the Environment* (2017).
8. U. F. Akpan and G. E. Akpan, Int. J. Energy Econ. Policy **2**, 21 (2012).
9. S. Mekhilef, A. Safari, W. E. S. Mustaffa, R. Saidur, R. Omar, and M. A. A. Younis, Renew. Sustain. Energy Rev. **16**, 386 (2012).
10. P. Taylor and M. Balat, 37 (2007).
11. M. Mazalan, M. M. Noh, Y. Wahab, M. N. Norizan, and I. S. Mohamad, in *2013 IEEE Conf. Clean Energy Technol.* (IEEE, 2013), pp. 187–191.
12. E. Zakhidov, M. Zakhidova, A. Kokhkharov, A. Yarbekov, V. Kuvondikov, S. Nematov, E. Normatov, and A. Saparbayev, Turkish J. Biol. **39**, 276 (2015).
13. J. Rogers and W. Laura, *Solar Power on the Rise* (2014).
14. J. Kim and J. Park, Build. Environ. **46**, 1899 (2011).
15. I. S. Mohamad, S. S. Ismail, M. N. Norizan, S. A. Z. Murad, and M. M. A. Abdullah, IOP Conf. Ser. Mater. Sci. Eng. **209**, 12028 (2017).
16. O. Brian and M. Gratzel, Nature **353**, 737 (1991).
17. M. K. Nazeeruddin, E. Baranoff, and M. Grätzel, Sol. Energy **85**, 1172 (2011).
18. S. Sharma, S. Bulkesh, S. K. Ghoshal, and D. Mohan, Renew. Sustain. Energy Rev. **70**, 529 (2017).
19. G. Paruthimal Kalaignan and Y. Soo Kang, J. Photochem. Photobiol. C Photochem. Rev. **7**, 17 (2006).
20. K. Park, Q. Zhang, J. Xi, and G. Cao, Thin Solid Films **588**, 19 (2015).
21. K. Jasim, Sol. Cells - Dye. Devices 171 (2011).
22. Z. Abdin, M. A. Alim, R. Saidur, M. R. Islam, W. Rashmi, S. Mekhilef, and A. Wadi, Renew. Sustain. Energy Rev. **26**, 837 (2013).
23. J. Gong, J. Liang, and K. Sumathy, Renew. Sustain. Energy Rev. **16**, 5848 (2012).
24. E. Ramasamy and J. Lee, J. Phys. Chem. C **114**, 22032 (2010).
25. W. J. Lee, E. Ramasamy, D. Y. Lee, and J. S. Song, J. Photochem. Photobiol. A Chem. **194**, 27 (2008).
26. S. Sardar, S. Ghosh, H. Remita, P. Kar, B. Liu, C. Bhattacharya, P. Lemmens, and S. K. Pal, RSC Adv. **6**, 33433 (2016).
27. X. Tang, Y. Wang, and G. Cao, J. Electroanal. Chem. **694**, 6 (2013).
28. M. Nebi and D. Peker, J. Phys. Conf. Ser. **766**, 12026 (2016).
29. M. M. Rashad, A. E. Shalan, M. Lira-Cantú, and M. S. A. Abdel-Mottaleb, Appl. Nanosci. **3**, 167 (2013).
30. S. Perera, N. Zelenski, and E. G. Gillan, Chem. Mater. **18**, 2381 (2006).
31. J. Zhang, P. Zhou, J. Liu, and J. Yu, Phys. Chem. Chem. Phys. **16**, 20382 (2014).
32. T. Luttrell, S. Halpegamage, J. Tao, A. Kramer, E. Sutter, and M. Batzill, Sci. Rep. **4**, 4043 (2014).
33. A. Di Paola, M. Bellardita, and L. Palmisano, Catalysts **3**, 36 (2013).
34. G. Cheng, M. S. Akhtar, O. Yang, and F. J. Stadler, Electrochim. Acta **113**, 527 (2013).
35. M. Grätzel, J. Photochem. Photobiol. C Photochem. Rev. **4**, 145 (2003).
36. S. Chai, T. Lau, J. Dayou, C. S. Sipaut, and R. F.

Mansa, Int. J. Renew. Energy Res. **4**, (2014).
37. P. Agarwala, P. Makkar, S. Sharma, and R. Garg, J. Mater. Eng. Perform. **23**, 3703 (2014).
38. X. Wei, G. Zhu, J. Fang, and J. Chen, Int. J. Photoenergy **2013**, (2013).
39. M. Adachi, Y. Murata, J. Takao, J. Jiu, M. Sakamoto, and F. Wang, J. Am. Chem. Soc. **126**, 14943 (2004).
40. J. Maçaira, L. Andrade, and A. Mendes, Renew. Sustain. Energy Rev. **27**, 334 (2013).
41. M. La, Y. Feng, C. Chen, C. Yang, and S. Li, **10**, 1563 (2015).
42. S. Y. Ho, C. Su, C. C. Cheng, S. Kathirvel, C. Y. Li, and W. R. Li, Nanoscale Res. Lett. **7**, 147 (2012).
43. D. Regonini, G. Chen, C. Leach, and F. J. Clemens, Electrochim. Acta **213**, 31 (2016).
44. Y. C. Tu, J. F. Lin, W. C. Lin, C. P. Liu, J. J. Shyue, and W. F. Su, CrystEngComm **14**, 4772 (2012).
45. J. Liu, J. Huo, M. Zhang, and X. Dong, Thin Solid Films **623**, 25 (2017).
46. A. Taleb, F. Mesguich, A. Hérissan, C. Colbeau-justin, X. Yanpeng, and P. Dubot, Sol. Energy Mater. Sol. Cells **148**, 1 (2015).
47. M. Iraj, F. D. Nayeri, E. Asl-Soleimani, and K. Narimani, J. Alloys Compd. **659**, 44 (2016).
48. X. Liu, J. Lin, Y. H. Tsang, X. Chen, P. Hing, and H. Huang, J. Alloys Compd. **607**, 50 (2014).
49. S. Ameen, M. S. Akhtar, H. Seo, and H. Shin, 3020 (2014).
50. K. S. Kim, H. Song, S. H. Nam, S. M. Kim, H. Jeong, W. B. Kim, and G. Y. Jung, Adv. Mater. **24**, 792 (2012).
51. T. Marimuthu, N. Anandhan, R. Thangamuthu, M. Mummoorthi, and G. Ravi, J. Alloys Compd. **677**, 211 (2016).
52. A. Al-Kahlout, Thin Solid Films **520**, 1814 (2012).
53. I. S. Mohamad, M. N. Norizan, M. K. F. M. Hanifiah, I. A. M. Amin, and M. M. Shahimin, AIP Conf. Proc. **1660**, (2015).
54. I. Journal, O. F. Electrochemical, M. Mohamad, S. Universiti, J. Chysk, and W. Bohemia, **10**, 2859 (2015).
55. S. Rani, P. Suri, P. K. Shishodia, and R. M. Mehra, Sol. Energy Mater. Sol. Cells **92**, 1639 (2008).
56. L.-T. Yan, F.-L. Wu, L. Peng, L.-J. Zhang, P.-J. Li, S.-Y. Dou, and T.-X. Li, Int. J. Photoenergy **2012**, 1 (2012).
57. A. Omar and H. Abdullah, Renew. Sustain. Energy Rev. **31**, 149 (2014).
58. A.-E. Marwa, M. Nafiseh, Z. Lei, P. T. Joseph, M. Donald, S. Saurabh, and K. T. Leung, Nanoscale 1 (2015).
59. H.-Y. Chen, D.-B. Kuang, and C.-Y. Su, J. Mater. Chem. **22**, 15475 (2012).
60. Q. Zhang and G. Cao, Nano Today **6**, 91 (2011).
61. C. Cavallo, F. Di Pascasio, A. Latini, M. Bonomo, and D. Dini, **2017**, (2017).
62. V. Sugathan, E. John, and K. Sudhakar, Renew. Sustain. Energy Rev. **52**, 54 (2015).
63. P. Pandey, M. R. Parra, F. Z. Haque, and R. Kurchania, J. Mater. Sci. Mater. Electron. **28**, 1537 (2017).
64. T. H. Lee, H. J. Sue, and X. Cheng, Nanoscale Res. Lett. **6**, 517 (2011).
65. J. Jung, J. Myoung, and S. Lim, Thin Solid Films **520**, 5779 (2012).
66. J. Rouhi, M. H. Mamat, C. H. R. Ooi, and S. Mahmud, 1 (2015).
67. J. Han, F. Fan, C. Xu, S. Lin, M. Wei, X. Duan, and Z. L. Wang, **405203**, (2010).
68. R. A. Wahyuono, C. Schmidt, A. Dellith, J. Dellith, M. Schulz, M. Seyring, M. Rettenmayr, J. Plentz, and B. Dietzek, Open Chem. **14**, 158 (2016).

FET-biosensor for cardiac troponin biomarker

Mohd Khairuddin Md Arshad[1,2,*], *Mohamad Faris Mohamad* Fathil[1], and *Uda* Hashim[1,2]

[1]Institute of Nano Electronic Engineering, Universiti Malaysia Perlis, 01000 Malaysia
[2]School of Microelectronic Engineering, Universiti Malaysia Perlis, 02600 Malaysia

Abstract. Acute myocardial infarction or myocardial infarction (MI) is a major health problem, due to diminished flow of blood to the heart, leads to higher rates of mortality and morbidity. The most specific markers for cardiac injury are cardiac troponin I (cTnI) and cardiac troponin T (cTnT) which have been considered as 'gold standard'. Due to higher specificity, determination of the level of cardiac troponins became a predominant indicator for MI. Currently, field-effect transistor (FET)-based biosensors have been main interest to be implemented in portable sensors with the ultimate application in point-of-care testing (POCT). In this paper, we review on the FET-based biosensor based on its principle of operation, integration with nanomaterial, surface functionalization as well as immobilization, and the introduction of additional gate (for ambipolar conduction) on the device architecture for the detection of cardiac troponin I (cTnI) biomarker.

1 Introduction

Cardiac biomarkers are the indicators, which have been pre-dominantly used in the detection of acute myocardial infarction (AMI). It is logical to use protein quantification in a blood sample for this purpose as stated by Rosalki et al. [1]. The myocyte is the major cell in the heart, and the heart's purpose is to pump blood. When myocytes essentially cannot be regenerated due to heart cells die, then cardiac function has a high probability of being damaged. When the cell dies, the biomarker proteins (i.e. myoglobin, creatine-kinase MB, C-reactive protein and cardiac troponin are most commonly used) inside the cell will be released, with proteins in the cytoplasm leaving the cell more rapidly than the ones in membranes or fixed cell elements.

For MI, cardiac troponin T (cTnT) and cardiac troponin I (cTnI) are regarded as more sensitive and specific [2]. Both are released from the death cell within 2–4 h and 3–4 h, respectively, after the onset of MI symptoms [3]. In principle, cTnT and cTnI remain in the blood stream approximately more than 10 days, reaches to peak approximately 1–2 days [4] after myocardial injury. Because of its prolonged release in the blood, these biomarkers are useful in diagnosing sub-acute myocardial infarction [2]. In normal patients, the level of cTnI concentration is around 1 pg/ml, but increased to 100 ng/ml in MI patients [5]. Even the concentration as low as 10 pg/ml can be related to heart failure. **Fig. 1** shows general information regarding MI.

Biosensors can be used to detect and quantify the target molecules involved with cardiac biomarker interaction. Biosensors are integrated diagnostic devices, which merge biological or biologically-derived sensing element associated with a physicochemical transducer [6]. Generally, surface of a suitable transducer of a biosensor is immobilized with a biological receptor material (DNA, RNA or antibody), which enables conversion of biochemical signal into quantifiable electronic signal [7].

To overcome the limitations of conventional immunoassay-labelled method such as enzyme-linked immunosorbent assay (ELISA), which is lack of portability, late detection time, and high complexity of fabrication process [8], the electrical detection of bio-molecular interaction development is highly beneficial because it is suitable to become the low cost portable sensor and can be used non-specialized personnel [9]. Electrical detection is conducted by transducing the molecular binding event into a usable electrical signal [10]. In order to sense chemical and biological species which is very small in size, researchers have intensely studied nanostructures, such as nanowires (NWs), nanobelts, carbon nanotubes (CNTs), graphene, and nanoparticles for biosensing, due to comparable size between sensor and target. The electrical detection has become the main interest in recent year, which was evidenced by higher number of publications related to diagnostic MI.

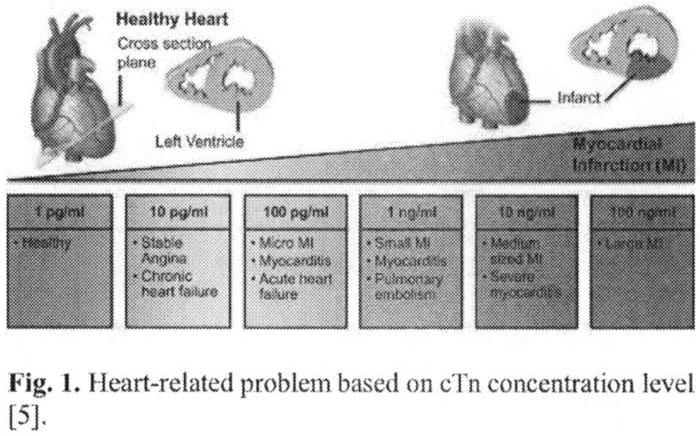

Fig. 1. Heart-related problem based on cTn concentration level [5].

* Corresponding author: mohd.khairuddin@unimap.edu.my

2 Field-effect transistors biosensor

Field-effect transistor (FET)-based biosensors are currently at the centre of interest for biological molecules (biomolecules) detection due to anticipated qualities, i.e. label-free, fast electrical detection, low power operation, easy to move from one place to another, low cost for mass production, and both sensor and measurement system are compatible to be integrated into a single chip [11]. These kind of biosensor is a FET gated by biomolecules, which change the gate charge distribution, hence influencing the conductance of the FET channel [12].

2.1. Principle of operation.

The FET-based biosensor (**Fig. 2**) is a device that enables the detection of biomolecules by having surface's channel, which is directly exposed to the surrounding, producing difference surface potential on the surface of a transducer upon biological interaction, thus contributes to modulation of the current flow inside the semiconductor channel, located between drain and source regions [13] enabling biological recognition. The application of biomolecules to the gate electrode generates an electric field, which controls the current flow between source and drain region [14]. The type of semiconductor (n- or p-type) determines the type of carriers (electrons or holes) of the conduction current.

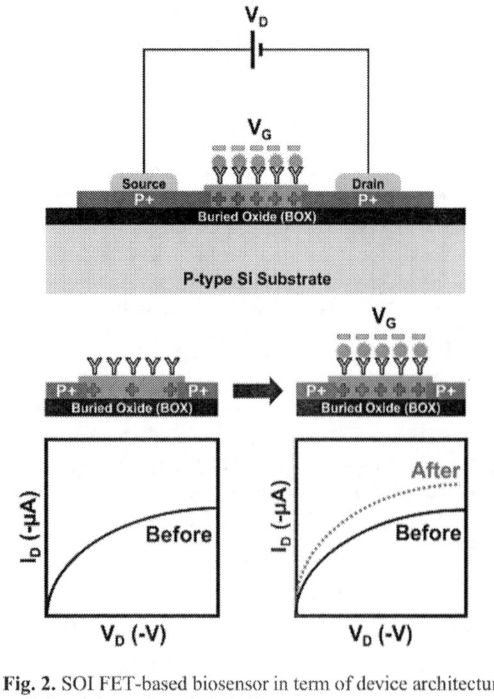

Fig. 2. SOI FET-based biosensor in term of device architecture and electrical characteristic [13].

2.2. Integration with nanomaterial as transducer

In the recent years, nanostructured metal-oxides (NMOs) materials (i.e. ZnO, TiO2, SnO2, and etc.) have grabbed considerable attention as transducer in development of biosensor. They exhibited unique non-toxic nano-morphology with functional biocompatibility, and interesting catalytic properties [15]. The materials also offered appropriate biomolecules immobilization micro-environment by owning strong adsorption capability, hence improved electron transfer between the biomolecule's active site and electrode and enhanced biosensing characteristic [16]. NMOs also offer high surface area for higher biomolecules loading and help biomolecules to maintain its bioactivity by providing a suitable microenvironment [17]. The NMOs' unique characteristic provided possibility of 1) utilizing electronic signal transduction as interface for biomolecules detection and 2) next generation of novel bioelectronics devices [15]. Selection of an appropriate NMOs for immobilization of the required bio-receptor is important in fabrication of efficient biosensor. The biosensor's performance is also influenced by immobilized surface between NMOs bind with a biomolecule. This can be done by having an effective NMOs immobilized surface, which aid a biomolecule to maintain its activity with high stability by exhibiting a biocompatible microenvironment. Commonly, NMOs bind to the biomolecules by physical adsorption or covalent binding. Weak interactions of NMO to biomolecule (e.g. van der Waals, electrostatic, physisorption) are some examples of the physical adsorption, which rely on the surface morphology, reaction medium and net surface charge. On the other hand, covalent binding of NMO to a biomolecule relies on the available functional groups on the NMO's surface. This can be arranged through suitable chemical reactions. Therefore, the use of NMOs in development of biosensor can enhanced sensitivity and limit of detection, in addition to simple and inexpensive fabrication deposition process.

2.2 Surface functionalization and immobilization

The specific detection of biomolecules is possible by utilizing specific biological receptors (bio-receptors), i.e. DNA, RNA, aptamers, or Antibodies (Abs), which are immobilized at the surface of the transducer through covalent binding with suitable chemical linkers. The biomolecules, which can be positively or negatively charge, captured by the bio-receptors and influence the changes in electrical behaviour of the transducers. The changes, whether increase or decrease the conductance of the transducer's channel (depending on the n- or p-type of semiconductor materials), allow more or less current flows from source to drain region [11].

The covalent bonding as in **Fig. 3** is the most frequently used technique for immobilization of cardiac biomarkers as bio-receptors in FET-based biosensor. This technique alters the biosensor's surface for acquiring a reactive functional group to enable attachment of the specific bio-receptor. Proteins comprised of amino acid side chain, which owned nucleophilic functional groups (i.e. amino, carboxylic, imidazole, thiol, hydroxyl, and etc.). These functional groups are used for the coupling, which requires the utilization of reagent such as glutaraldehyde (GA), carbodiimide, succinimide esters, maleinimides, or periodate for covalent immobilization [18]. The advantages of using covalent binding for bio-receptors' immobilization on the surface of the transducer is the enhancement in bio-receptors' uniformity, density, and distribution, in addition to the reproducibility and homogeneity of the surfaces [19]. This technique also possible to reduce or remove some common complications i.e. instability, diffusion and aggregation, or inactivation of biomolecules [19].

The high selectivity of a biosensor is possible through immobilization of the bio-receptor, which is the cTnI monoclonal antibody (Mab-cTnI), on the sensor to allow specific interaction of compounds [20]. Mab-cTnI is a sophisticated biomolecule, built from hundreds of singular amino acids arranged in highly ordered sequence and highly specific to cTnI biomarker [19]. It is produced by the immune system which identifies and remembers an cTnI biomarker that occupies the human body. The produced Mab-cTnI, commonly illustrated as Y-shape, is a complementary match to the cTnI biomarker, which is identical to the lock-and-key mechanism of enzyme-substrate binding) [21], hence suitable to be utilized as exceptional bio-receptors in bio-sensing applications.

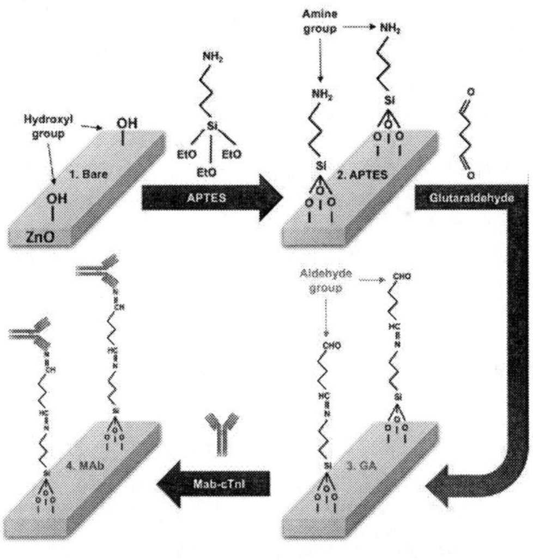

Fig. 3. Overall stages of the surface functionalization and immobilization on the ZnO-NPs thin film [22].

2.3 Integration with substrate-gate biasing

Several gating approaches on FET-based biosensor have been demonstrated in the literature [23–29], which are possible to enhance their performance in biomolecules detection with higher sensitivity. Jieun Lee et al. [26] and M. Nuzaihan et al. [28] utilized a single liquid and molecular gate control, respectively, where modulation of surface charge through the nanowires (gate) occurred during interaction between bio-receptor and target molecule, thus influences the conductivity of the nanowires. In addition, different device architecture was presented by Ahn et al. [30], a double-gate nanowire FET, where the sensitivity is improved by controlling the conduction channel in the silicon nanowire (SiNW) through the application of bias symmetric/asymmetric on double-gate, which are located at both sides of the nanowire. Different approach or device architecture of double gate FET [23,24,29,31] was presented utilizing the back-gate, which provides low V_T values, suggests a more effective back-gate coupling.

Fathil et al. [22] demonstrated various substrate-gate voltage (V_{SG}) biases effect on the electron concentration of the ZnO-FET-based biosensor. **Fig. 4** shows the horizontal cut near (a) the BOX/substrate interface and (b) near the surface of *p-n*-p junction.

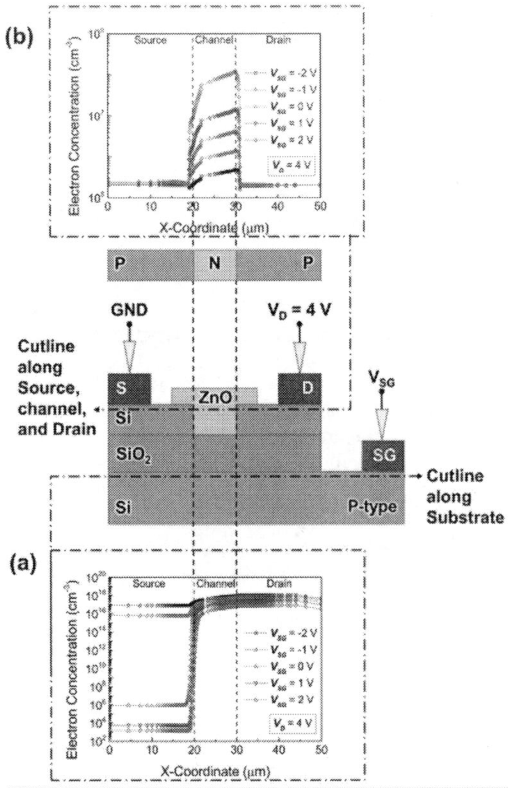

Fig. 4. Electron concentration on the FET biosensor at (a) BOX/Si substrate interface; (b) near the surface of the source, gate and drain region for different V_{SG} bias [22].

As shown in electron concentration cut in **Fig. 4**(a), at negatively V_{SG} bias (V_{SG}= −1 and −2 V), accumulation of high electron concentration occurs, thus introducing the inversion space charge condition in the p-type Si substrate along the BOX/Si substrate interface. This result causes the accumulation of holes, producing a hole conduction layer along the source, gate, and drain region (**Fig. 4**(b)), and allows more current flow between source and drain. On the contrary, as positively V_{SG} bias (V_{SG}= 1 and 2 V) are given to the substrate-gate electrode, the electron concentration along the BOX/Si substrate interface has become lesser, especially at the location beneath the source region, hence reduced the hole concentration inside hole conduction layer at the gate region. The electrical characterization of the FET biosensors, respectively, as a function of I_D with various V_{SG} (V_{SG} = -

 DOI: 10.1051/epjconf/201716201046

2, -1, 0, 1, and 2 V) as in **Fig. 5**. The results show the biosensor exhibits a linear characteristic during forward biased, and shift of V_T to the left with increases of I_D for negatively V_{SG}. Opposite trend can be observed for positively V_{SG}.

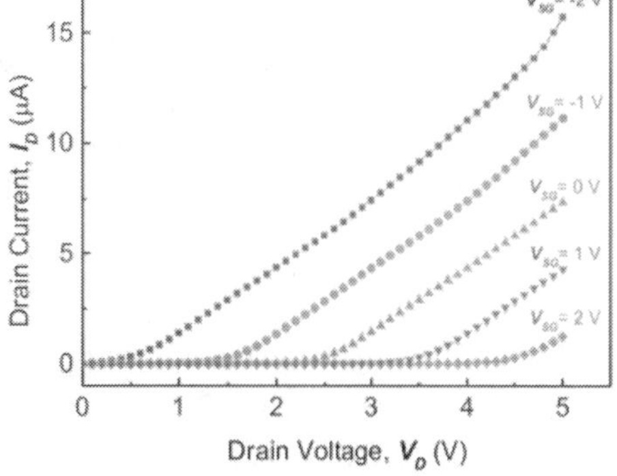

Fig. 5. Different V_{SG} bias effect in term of I_D against V_D characteristic on the ZnO-FET biosensor with substrate-gate coupling [22].

3 Detection of cTnI cardiac biomarker

Fathil et al. [22] demonstrated a FET-based biosensor's example for the detection of cTnI biomarker by using zinc oxide (ZnO)-FET biosensor. The device exhibited a significant decrease in measured I_D as the concentrations of cTnI biomarker were increased from 1 ng/ml, 10 ng/ml, 100 ng/ml, 1 μg/ ml to 10 μg/ml as shown in **Fig. 6**.

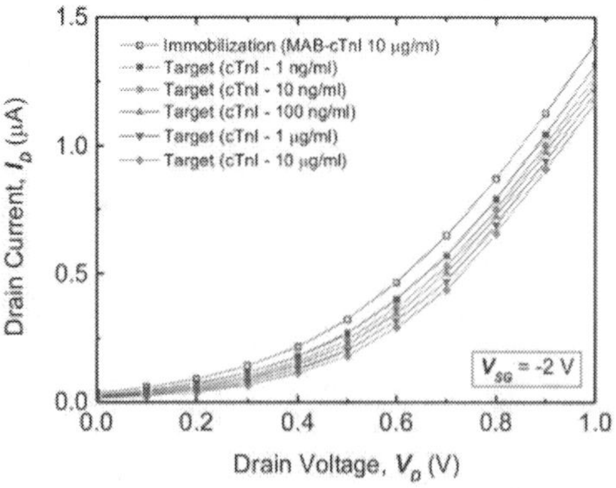

Fig. 6. I_D against V_D characteristic at different cTnI biomarker concentrations.

This event occurs due to the fact that cTnI biomarkers have high positive charge (pI 9.87) [32], which repulse the positively charge, decreased the hole concentration, and thus diminished the hole conduction layer of the n-type ZnO-NPs thin film between p-type source and drain region, hence reduced the measured I_D. The sensitivity of our biosensor [28], which is 35.3 nA·(g/ml)$^{-1}$ with limit of detection (LOD) of 3.24 pg/ml [22]. The specificity of the biosensor was also confirmed with 10 μg/ml of cTnT biomarker as non-specific protein (control), and 10 μg/ml of cTnI target biomarker. In comparison, a huge change in I_D (~17%) was displayed by 10 μg/ml of cTnI biomarker compared to 10 μg/ml cTnT non-complementary biomarker (~1%). Only cTnI biomarker, which exhibited a specific binding to the MAb-cTnI as bio-receptor allows the relative changes in measured I_D [22].

4 Conclusion and future perspective

cTnI detections in recent years through the mean of electrical label-free based biosensors have been rising due to their ability to directly interpret interactions between cTnI and the transducers' surface into readable electrical signals. FET-based biosensors, have become favourable method among researchers due to ultra-sensitivity and high specificity, which are most likely reduce false-negative results in terms of AMI determination and gives more accurate cTnI levels in real time. Simplicity in the detection of cTnI without the need for labels addition has enabled a simple and inexpensive fabrication process, thus reducing the complexity of the device detection. Although with all the promising privileges presented by the electrical label-free detection biosensors, they are not intended as a substitute for the existing technologies for AMI detection. Rather, the biosensors should be used side by side with the present methods, such as ECG, to aid and offer quantitative outcomes for the determination of cardiac biomarker levels in patients; thus, allowing doctors to justify their decision making in the diagnostics of AMI with accurately, quickly, and wisely.

With the rapid advancement of technology, there is hope that, in the future, an optimized, stable and portable biosensor can be fabricated as a plug-in accessory to a smartphone or tablet that serves as a cTn level reader; thus, enabling the diagnosis process to be performed anywhere and at any time as in **Fig. 7**.

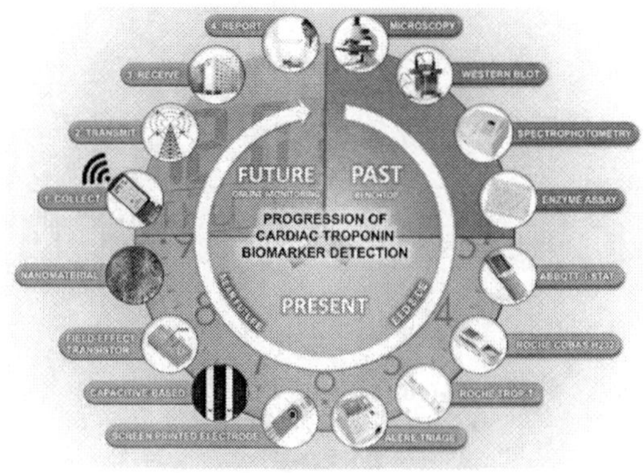

Fig. 7. Progression of cTn biomarkers devices detection from the past, present, and future perspective, a transition from benchtop towards translational research [33].

The results could not only be displayed to the user but also transmitted to the hospital for further analysis by physicians. For the patient, this will save significant travelling and waiting time at the hospital for a medical check-up, while from the doctors' point of view, these tests could reduce the number of patients with this condition falling under false-negative trips to the emergency department. Based on the results, doctors can provide suitable advice to the patient through voice conversation via smartphone depending on the level of cTn in the patient's blood or can even dispatch an ambulance to the patient's present location if the cTn level is above the reference value at an early stage. Hence, the probability to save lives due to MI can be increased significantly.

The authors are grateful to the Department of Higher Education, Ministry of Higher Education, (KPT) for funding this research through the Fundamental Research Grant Scheme (FRGS) with the grant number 9003-00536. The authors also would like to acknowledge all the team members at the Institute of Nano Electronic Engineering (INEE), Universiti Malaysia Perlis (UniMAP), for their guidance and help.

References

1. S. B. Rosalki, R. Roberts, H. A. Katus, E. Giannitsis, and J. H. Ladenson, Clin. Chem. **50**, 2205 (2004)
2. A. S. Jaffe and J. Ordonez-Llanos, Rev. Española Cardiol. (English Ed. **63**, 763 (2010)
3. E. Burcu Bahadır and M. Kemal Sezgintürk, Talanta **132**, 162 (2015)
4. R. S. Wright, J. L. Anderson, C. D. Adams, C. R. Bridges, D. E. Casey, S. M. Ettinger, F. M. Fesmire, T. G. Ganiats, H. Jneid, A. M. Lincoff, E. D. Peterson, G. J. Philippides, P. Theroux, N. K. Wenger, J. P. Zidar, and A. K. Jacobs, Circulation **123**, 2022 (2011)
5. S. Agewall, E. Giannitsis, T. Jernberg, and H. Katus, Eur. Heart J. **32**, 404 (2011)
6. M. Mascini and S. Tombelli, Biomarkers **13**, 637 (2008)
7. A. Qureshi, Y. Gurbuz, and J. H. Niazi, Sensors Actuators B Chem. **171–172**, 62 (2012)
8. T. Kong, R. Su, B. Zhang, Q. Zhang, and G. Cheng, Biosens. Bioelectron. **34**, 267 (2012)
9. P. Estrela, A. G. Stewart, F. Yan, and P. Migliorato, Electrochim. Acta **50**, 4995 (2005)
10. G. Zhang and Y. Ning, Anal. Chim. Acta **749**, 1 (2012)
11. D. Sarkar, W. Liu, X. Xie, A. C. Anselmo, S. Mitragotri, and K. Banerjee, ACS Nano **8**, 3992 (2014)
12. U. Brand, L. Brandes, V. Koch, T. Kullik, B. Reinhardt, F. Rüther, T. Scheper, K. Schügerl, S. Wang, X. Wu, R. Ferretti, S. Prasad, and D. Wilhelm, Appl. Microbiol. Biotechnol. **36**, 167 (1991)
13. S. Liu and X. Guo, NPG Asia Mater. **4**, e23 (2012)
14. P. Bergveld, Sensors Actuators B Chem. **88**, 1 (2003)
15. P. R. Solanki, A. Kaushik, V. V. Agrawal, and B. D. Malhotra, NPG Asia Mater. **3**, 17 (2011)
16. Q. Xu, C. Mao, N. Liu, J. Zhu, and J. Sheng, Biosens. Bioelectron. **22**, 768 (2006)
17. S. P. Singh, S. K. Arya, P. Pandey, B. D. Malhotra, S. Saha, K. Sreenivas, and V. Gupta, Appl. Phys. Lett. **91**, 63901 (2007)
18. A. F. Collings and F. Caruso, Reports Prog. Phys. **60**, 1397 (1997)
19. R. Monošík, M. Streďanský, and E. Šturdík, Acta Chim. Slovaca **5**, 109 (2012)
20. A. Chaubey and B. D. Malhotra, Biosens. Bioelectron. **17**, 441 (2002)
21. J.-Y. Yoon, *Introduction to Biosensors* (Springer International Publishing, Cham, 2016)
22. M. F. M. Fathil, M. K. Md Arshad, A. R. Ruslinda, S. C. B. Gopinath, M. Nuzaihan M.N., R. Adzhri, U. Hashim, and H. Y. Lam, Sensors Actuators B Chem. **242**, 1142 (2017)
23. E. Buitrago, M. F.-B. Badia, Y. M. Georgiev, R. Yu, O. Lotty, J. D. Holmes, A. M. Nightingale, H. M. Guerin, and A. M. Ionescu, Sensors Actuators B Chem. **199**, 291 (2014)
24. P. Dai, A. Gao, N. Lu, T. Li, and Y. Wang, Jpn. J. Appl. Phys. **52**, 121301 (2013)
25. Z. Iskierko, M. Sosnowska, P. S. Sharma, T. Benincori, F. D'Souza, I. Kaminska, K. Fronc, and K. Noworyta, Biosens. Bioelectron. **74**, 526 (2015)
26. J. Lee, J. Jang, B. Choi, J. Yoon, J.-Y. Kim, Y.-K. Choi, D. Myong Kim, D. Hwan Kim, and S. Choi, Sci. Rep. **5**, 12286 (2015)
27. J. Li, M. Sun, X. Wei, L. Zhang, and Y. Zhang, Biosens. Bioelectron. **74**, 423 (2015)
28. M. N. M. Nuzaihan, U. Hashim, M. K. Md Arshad, S. R. Kasjoo, S. F. A. Rahman, A. . Ruslinda, M. F. M. Fathil, R. Adzhri, and M. M. Shahimin, Biosens. Bioelectron. **83**, 106 (2016)
29. J.-K. Park, H. Jang, J. Park, and W. Cho, Solid. State. Electron. **97**, 2 (2014)
30. J.-H. Ahn, S.-J. Choi, J.-W. Han, T. J. Park, S. Y. Lee, and Y.-K. Choi, Nano Lett. **10**, 2934 (2010)
31. P. G. Fernandes, R. A. Chapman, O. Seitz, H. J. Stiegler, H.-C. Wen, Y. J. Chabal, and E. M. Vogel, IEEE Electron Device Lett. **33**, 447 (2012)
32. S. Eriksson, Clin. Chem. **49**, 1095 (2003)
33. M. F. M. Fathil, M. K. Md Arshad, A. R. Ruslinda, M. Nuzaihan, S. C. B. Gopinath, R. Adzhri, and U. Hashim, Anal. Chim. Acta (2016)

Gate recess study for high thermal stability pHEMT devices

M. Mohamad Isa[1], *N.* Ahmad[1], *Siti S.* Mat Isa[1], *Muhammad M.* Ramli[1], *N.* Khalid[1], *N.I. M.* Nor[1], *S.R.* Kasjoo[1] *and M.* Missous[2]

[1]School of Microelectronic Engineering, Pauh Putra Campus, University Malaysia Perlis (UniMAP), 02600 Arau, Perlis, MALAYSIA
[2]School of Electrical and Electronic Engineering, University of Manchester, M13 9PL, UK

Abstract. Gate formation is a crucial steps, especially in FET fabrication process. At this steps, the characteristics are very much influenced by the processing parameters, particularly in the processing temperature. In this paper, we report the thermal stability study and sidewall etch to reduce the off-state Schottky's gate leakage on 1 µm gate pHEMT device. In our study, we found that low sintering temperatures of 200 °C is preferable and sidewall etching of 10 minutes has reduces the gate leakage by almost 5 times as compared with the devices with no sidewall etching. The optimised processing recipe is proposed for low off-state Schottky's gate leakage, where low leakage has significant influence in the device performances, especially for future high speed and low noise applications.

1 Introduction

III-V compounds have a number of desirable material characteristics which make them superior to Silicon (Si) in high-frequency applications. The primary characteristic is high carrier electron mobility. This contributes to making the materials very responsive to rapid changes in applied electric fields. Furthermore, since the mobility of electrons far exceeds that of the holes, electrons are invariably used in high-frequency devices.

Another useful property of III-V compounds is the higher band gap energy compared to Si, allowing the synthesis of insulating and semi-insulating materials that are essential in high-frequency circuits. The rich variety of available materials also allows for lower band gap materials with superior transport characteristics and very high mobilities [1].

High Electron Mobility Transistor (HEMT), one of Field Effect Transistor (FET) family, is an example of a high-speed device that takes the advantage of this materials. Using advanced epitaxial growth tools such Molecular Beam Epitaxy (MBE) [2-4], the advancement in devices properties have been proved by the invention of new high-speed and low-noise devices such as pseudomorphic HEMT (pHEMT) [5].

Temperature is an important process control parameter, where sufficient thermal budget is necessary to improve the metal-semiconductor contact and therefore increase the device performances such as better threshold voltage (V_{th}) roll-offs [6-8].

In FET devices, the gate formation is one of the complicated whilst important steps in devices' fabrication process. Here, the gate biasing is used to control the current conduction in the channel. At gate patterning steps, the thermal budget is more trivial, where additional thermal energy could damage the channel and consequently degrading the device characteristics.

Previously, we have reported the elimination of dome effect [9] and high selectivity etching process [10] using Succinic Acid as a better gate recess etchant for pHEMT devices. The gate recess etching using Succinic has been demonstrated to improve the Schottky contact quality and considerably lower the gate leakage by a factor of ten. However, at a higher reverse gate voltage (V_{GS}), the gate still suffers higher leakage, which reduces the gate stability at high V_{GS} values. Therefore, it is essential that the Schottky gate contact is both mechanically and thermally stable, so that the metal gates can survive during the high-temperature process and during harsh device operation.

2 Sample Preparation

For these reasons, another set of samples was prepared for the gate's thermal stability study. This time, several sizes of 2-1-2 µm Source-Gate-Drain spacing 1 µm Schottky gate length were patterned on XMBE #178 samples. The epitaxial layer for the samples is shown in Figure 1.

$In_{0.53}Ga_{0.47}As$	Cap layer (50 Å)
$In_{0.25}Al_{0.75}As$	Schottky layer (300 Å)
	δ
$In_{0.25}Al_{0.75}As$	Spacer layer 1 (100 Å)
$In_{0.7}Ga_{0.3}As$	Channel layer (150 Å)
$In_{0.52}Al_{0.48}As$	Buffer layer (4500 Å)
InP	Substrate

Fig. 1. Epitaxial layer for XMBE #178 (thickness not to scale)

* Corresponding author: muammar@unimap.edu.my

As illustrated in the figure, the fabricated devices have 50 Å InGaAs cap layer, and a wide band gap $In_{0.25}Al_{0.75}As$ material as the Schottky barrier layer so that thermal stability studies on the Schottky Diode were performed from the advantage of a higher barrier height. As an extension to the previous experiments, three sets of gate recess etching time were also tested in this experiment, namely 3, 4 and 5 minutes. All samples were prepared using the succinic acid composition listed in Table 1. All etching processes were done using agitation during the etching period.

Table 1. Succinic Acid (etchant)compositions [1]

Parameter	Composition/value
Succinic acid granulates	10 g
Hydrogen Peroxide (H_2O_2)	5 ml
pH	5.5
DI water (H_2O)	50 ml

A 50 nm of Ti and 400 nm of Au (hereafter denoted as 50/400nm of Ti/Au) metal evaporation took place after the samples were etched. After liftoff, the sintering studies and device's Schottky diode measurements were carried out to examine the gate thermal stability together with the quality of the each gate recess step. For comparison purposes, four sets of different temperature parameters were used in this test: Presintering (25 °C), 200 °C, 250 °C, and 300 °C were used. Each sample was sintered on a hot plate set to the desired temperature for 5 minutes. The Schottky diode measurements were performed using Agilent's B1500 Semiconductor Device Analyser.

3 Thermal Stability Study

Figure 2 shows the gate I-V characteristic for 50 μm total gate width at different sintering temperature after 3 minutes of gate etching time. In this section and hereafter, at least four devices from random locations were measured. The results shown in the figure are only the characteristics for the best device as they were all uniform within variation of 10%.

The almost flat reverse current indicates the advantage of the used high band gap material as the Supply layer. The forward current for Room Temperature (Pre-sintering) and all other sintering temperature peaks at 100 mA/mm at the forward voltage of 1 V. When reverse biased, the graph clearly shows interesting diode characteristics. For pre-sintering and low temperature (200 ° C), the leakage current at a reverse voltage -8 V is about -20 μA/mm whereas at 250 °C the leakage increases almost doubles to -130 μA/mm and becomes worst at 300 °C where the leakage is -1500 μA/mm. Even though heat does not change the forward current, but too much of heat clearly damages the Schottky as seen by the much larger leakage current at 250 °C and 300 °C.

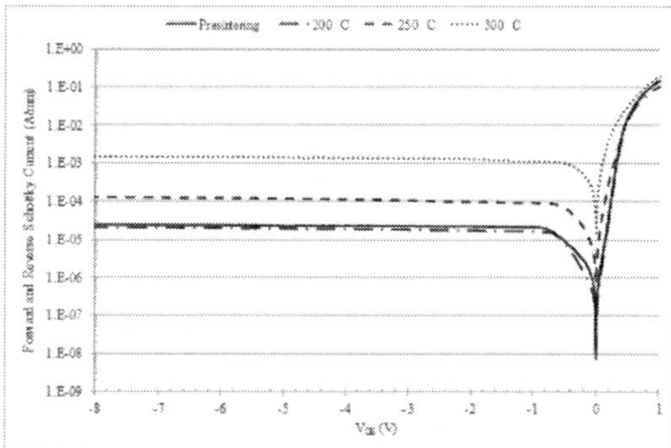

Fig. 2. Sintering studies for 3 minutes succinic etch time.

Previously the presence of Aluminium oxide layer between the gate metal and InAlAs interface has been reported [11]. Because Aluminium will react instantaneously with oxygen to form oxides, this process becomes more prevalent at higher composition of Al atoms in the InAlAs layer and will accelerate at high curing temperatures. This result in an unavoidable oxide layer, which will pull down the barrier heights (ϕ_B) causing the large leakage is observed.

In this experiment, the barrier heights is used to indicate the quality of the gate contact and was deduced to be 0.7 eV after 200 °C sintering temperature and reduced to 0.6 eV and 0.5 eV at sintering temperatures of 250 °C and 300 °C, respectively. The ϕ_B is extracted from the semi log plot of the forward current over the applied V_{GS} values [12].

Similar trends were observed for both the 4 minutes and 5 minutes gate recess etching times. The peak forward gate current is consistent at 100 mA/mm for all etching times. The Schottky leakage was increased from 10 μA/mm before sintering to 30 – 60 μA/mm after sintering at temperatures of 200 °C and 250 °C. At 300 °C, the leakage was considerably higher than 100 μA/mm for both etching times. The Schottky diode data for both the 4 and 5 minutes etching times are illustrated in Figure 3 and Figure 4 respectively.

From the measured data, it is clear that a low sintering temperature is preferred for better Schottky and lower leakage current. At sintering temperatures lower than 250 °C, small leakage current (less than 100 μA/mm) can be achieved in this epitaxial structure. For sintering temperature of 300 °C and for all gate recess etching times, the leakage gets smaller as the etching time was increased. This shows that longer etching time will produce a more stable Schottky contact with lower Schottky gate leakage. It can be concluded from the results that 5 minutes gate recess etching and 200 °C sintering temperature can produce a thermally stable Schottky contact.

DOI: 10.1051/epjconf/201716201047

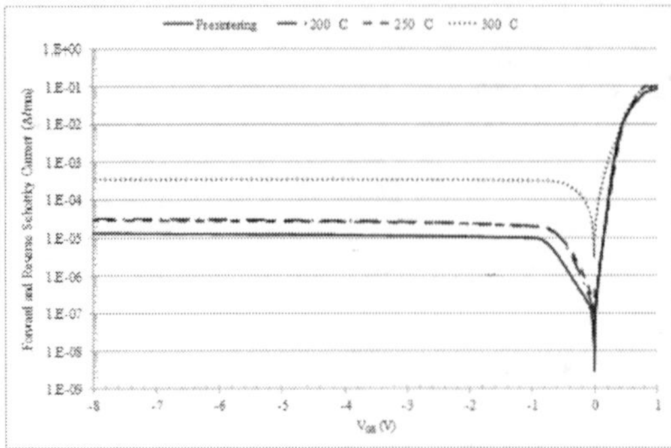

Fig. 3. Sintering studies for 4 minutes succinic etch time.

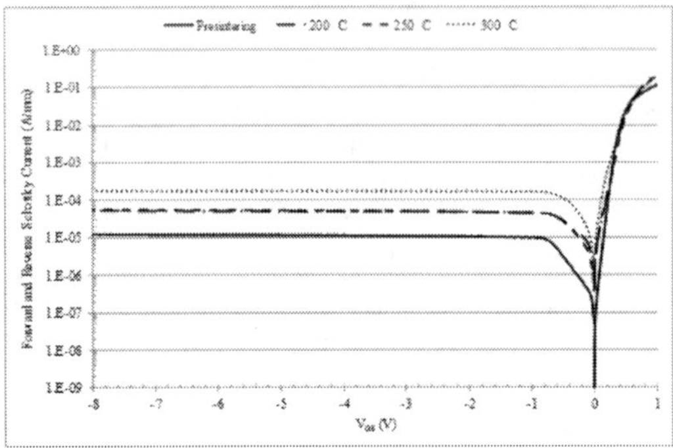

Fig. 4. Sintering studies for 5 minutes succinic etch time.

4 Sidewall Spacer

In our pHEMT device process, the devices were isolated using a MESA isolation technique. Here, the pyramid like MESA structure was realised by non-selective orthophosporic solution wet etchant Hydrogen Peroxide (H_3PO_4) with 3:1:50 chemical composition of H_3PO_4:H_2O_2:H_2O. This structure however, will expose and short-circuited the narrow band gap InGaAs channel to the Schottky gate metal after gate metal evaporation and lift off. The gate metal running up the MESA side is depicted in Figure 5. The gate metal in contact with the InGaAs channel will form a complete sidewall-gate leakage path from gate to channel [13-15].

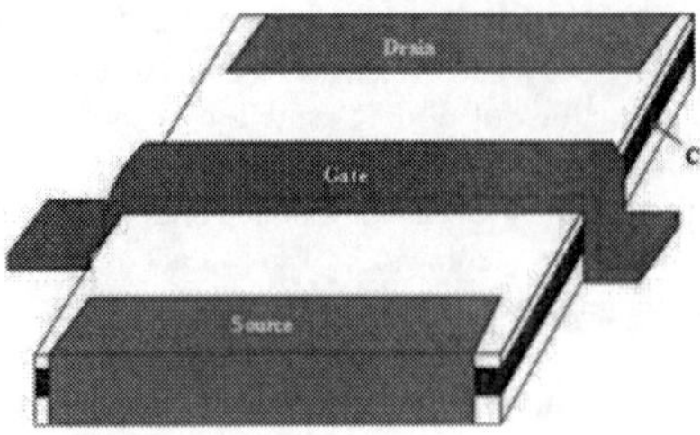

Fig. 5. Graphical representation of MESA side wall. The Schottky gate running up the MESA will in contact with barrow band gap channel layer which complete the sidewall leakage path [14]

Although the metal-channel contact area is very small several orders smaller than the gate area, the low Schottky barrier height of metals with narrow band gap InGaAs channel potentially results in a notable leakage whose path runs from the gate to the channel. The sidewall leakage current has been known to be the main reason of excessive gate leakage current and severely degrades the gate breakdown voltage. Both effects are not desirable in low noise and power device applications. In addition, sidewall leakages were found to worsen with higher doping, increased channel thickness, and increased x (x > 0.53) in InxGa1-xAs channel [13].

Bahl et al has proposed and demonstrated a new and simple one step technique, which is self-aligned to the MESA and requires no additional masks. The technique is not gate-length dependent, and works for MESAs in all crystallographic directions on the (100) surface [60].

Bahl et al had selectively recessed the exposed InGaAs channel using Succinic Acid:H_3PO_4 solution without removing the MESA mask after orthophosporic etch. As a result, the high selectivity of the InGaAs channel etchant formed a cavity which the flowing gate metallization did not enter the cavity and remained isolated from the channel layer. The sidewall structure with the channel cavity is illustrated in Figure 6.

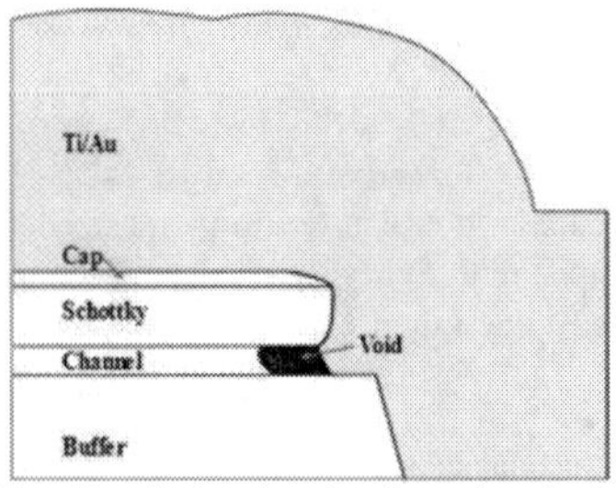

Fig. 6. Sidewall etch showing the channel cavity isolate the gate from InGaAs channel [14]

The same technique was used in the next experiment in our study. The same succinic acid composition as listed in Table 1 was used in this experiment. The aim of this experiment was to find out a suitable sidewall etching time to produce the lowest possible gate leakage current as well as a thermally stable gate.

The sample was cleaned and the MESA pattern was transferred using conventional contact photolithography processes. After hard baking the resist mask, the MESA structure was etched down to the buffer layer using the orthophosporic acid solution, for about 150 nm etch depth. Using the same MESA mask, the sample was etched for different etching times of 5 and 10 minutes. To complete the comparisons, the Schottky diode measurements were also taken with no sidewall etching. As a continuation to previous sintering study, the same sintering temperatures were used here as well. Since 200 °C was found to be the optimal sintering temperature, only this curing temperature is presented here. The sintering time was set to be 5 minutes. A higher sintering temperature or longer sintering time has proven to cause higher leakage, and will not be shown in

DOI: 10.1051/epjconf/201716201047

this section. For comparison purpose, the data before heat treatment for all side wall etching cases were also measured.

Figure 7 shows the Schottky diode characteristic before heat treatment. From the forward bias characteristics, the ideality factor (η) and ϕ_B for all devices were comparable. For the reverse biased characteristic, all devices had almost flat reverse currents, where the gate leakages are well below 100 μA/mm, until at least -8 V reverse voltage.

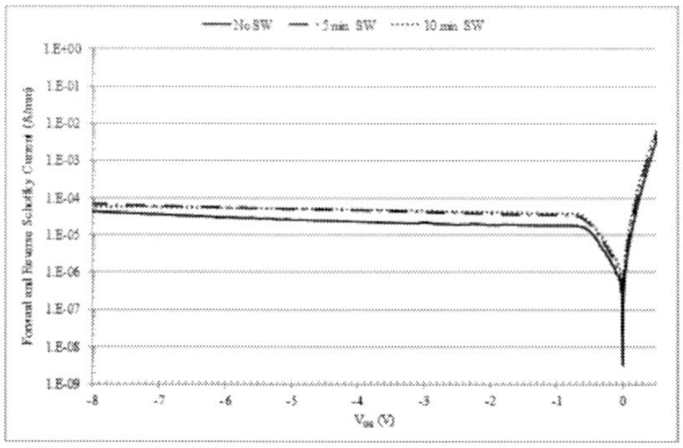

Fig. 7. Schottky diode characteristic for 1 x 50 μm gate for all sidewall etch condition before heat treatment

By inspecting the leakage current right after gate evaporation and lift-off of 50/100 nm Ti/Au gate metallisation, the leakage of the device without sidewall etching is considerably lower than the leakage of the other two samples with 5 and 10 minutes side wall etching. The gate leakage at V_{GS} = -8 V are 48 μA/mm for no sidewall etching, 57 μA/mm for 5 minutes sidewall etching and 56 μA/mm corresponding for 10 minutes sidewall etching. However, a curing temperature is always performed during the process flow to improve the adhesion between metal and semiconductor.

Figure 8 and Figure 9 show the diode characteristics of the devices for all etching time after sintering at 200 °C and 250 °C for 5 minutes. At 200 °C, the gate leakage for the device with no sidewall etching was consistent with that of the device before sintering. The gate leakage without sidewall is ~ 50 μA/mm. However, the gate leakage of the device with 5 minutes and 10 minutes sidewall etching show considerable improvement with the leakage reduced to 15 μA/mm, five times lower than before sintering. The forward currents are comparable with those of the devices without sintering, demonstrating a negligible change in η and ϕ_B.

However, a decreasing trend can be observed for a sintering temperature of 250 °C. The 5 and 10 minutes sidewall etching have a considerably higher gate leakage current, almost ten times higher. The diodes with no sidewall etching exhibited the worst leakage of 215 μA/mm at VGS = -8 V. For $V_{GS} \geq 0$ V, a less steep slope and higher saturation current (IS) can be observed in the forward bias region implying a falling in ideality factor and ϕ_B. The samples with 5 and 10 minutes sidewall etching also showed lower leakage and better thermal stability as compared to the devices with no sidewall etching.

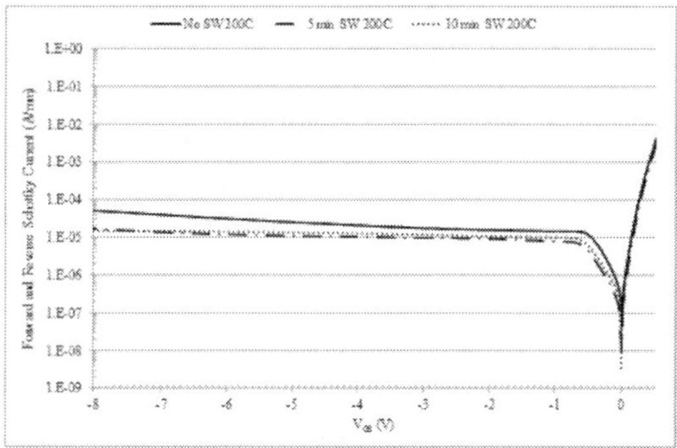

Fig. 8. Schottky diode characteristic side wall etch condition at sintering temperature of 200 °C for 5 minutes

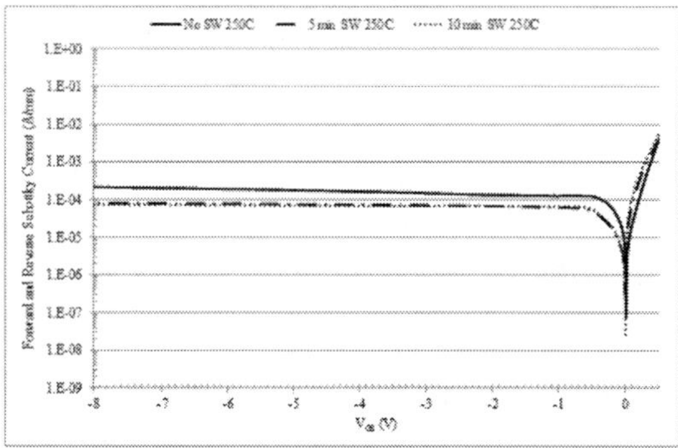

Fig. 9. Schottky diode characteristic for side wall etch condition at sintering temperature of 250 °C for 5 minutes

From this work, two important findings can be observed:

a. The Schottky diodes are thermally stable at low sintering temperatures (T < 200 °C).
b. At sintering temperature 200 °C the sidewall etching reduces the gate leakage by almost 5 times compared with the gate leakage of the devices with no sidewall etching.

Since the off-state Schottky gate leakage of the devices with the 5 and 10 minutes sidewall etching does not show significant changes in gate leakage current, a 10 minutes etching time was chosen for all subsequent processing for better sample uniformity. Similarly, and in agreement with previous experiments, excessive thermal energy can damage the metal-semiconductor junction, and therefore should be avoided. A sintering temperature of 200 °C is found to be optimum.

5 Conclusions

In this work, the thermal stability studies were conducted on 1 μm gate length devices, in which it has been found that high processing temperatures are not favoured and can deteriorate the Schottky barriers. The adoption of a sidewall etching technique, which isolates the metal gate and the channel, can significantly reduce the gate leakage during the sintering process. The optimal thermal budget in all the thermal stability studies was found to be 200 °C, with a sidewall etching time of 10 minutes.

The optimised chemical composition and thermal budget were applied in the fabrication of novel low noise and high breakdown pHEMTs for future high speed and low noise applications.

This research is funded by the Malaysian Ministry of Higher Education (MoHE) under the Fundamental Research Scheme (FRGS) with Grant number 9003 -00421.

References

1. W. R. Frensley and N. G. Einspruch, *VLSI Electronics: Microstructure Science,* (San Diego: Academic Press, 1995).
2. G. C. Osbourn, J. App.Phy., **53**, pp. 1586-1589 (1982).
3. G. Bauer and F. Schäffler, Physica Status Solidi (a), **203**, pp. 3496-3505 (2006).
4. T. F. Kuech, et al., *IEEE In. Con. IPRM 09*, pp. 63-64 (2009).
5. E. Carey and S. Lidholm, *Millimeter-Wave Integrated Circuits*, (Boston, Springer Science, 2005).
6. F.V Seggern, I. Keskin, E. Koos, R. Kruk, H. Hahn and S. Dasgupta, ACS Appl. Mater. Interfaces, **8**, pp. 31757 (2016).
7. B. Nasr, Zhirong, Z. Arger, D. Wang, R. Kruk, H. Hahn and S. Dasgupta, J. Mater. Chem. C, **1**, 2548-2552 (2013).
8. F. Ootsuka; A. Katakami; K. Shirai; T.Watanabe; H. Nakata; M. Kitajima; T. Aoyama; T. Eimori; Y. Nara; Y. Ohji and M. Tanjyo, IEEE Trans. on Elect. Dev., **55** (2008).
9. M. Mohamad Isa, N. Ahmad, F. Packeer, M., J. Eng. Tech., (2015).
10. M. Mohamad Isa *et. al*, *Proc. ICED 2016* (2016).
11. S.Fujita *et al.*, J. of App Phy. , **73**, pp. 1284-1287 (1993).
12. D. K. Schroder, *Semiconductor Material and Device Characterization,* John Wiley & Sons, New Jersey, (2006).
13. S. R. Bahl, *et al.*, Electron Devices, IEEE Transactions , **39**, pp. 2037-2043 (1992).
14. S. R. Bahl and I. A. del Alamo, Electron Device Letters, IEEE, **13**, pp. 195-197 (1992)
15. S. R. Bahl and J. A. d. Alamo, *Proc. 2nd Int. Conf. IPRM .*, (1990).

Low voltage low power FGMOS based current mirror

D. Nurulain[1,*], *F.A.S.* Musa[1], *M.* Mohamad Isa[1], *N.* Ahmad[1] and S.R.Kasjoo[1]

[1]School of Microelectronic Engineering University Malaysia Perlis, 02600 Arau, Perlis, Malaysia

Abstract. This paper presents the comparison of a conventional current mirror with the one utilizing floating gate MOSFET transistors (FGMOS) to achieve low power (LP) and low voltage (LV) design. The device structure has been simulated with 0.1μ CMOS technology and 1.2V voltage supply by using SAED 90nm PDK with the Synopsys Custom Designer tool. The FGMOS circuit has shown to have low power consumption of 9.62mW, smaller threshold voltage of 0.2V and Iout of 20 mA. The improvement of 40.1% from conventional current mirror has shown the LV and LP capability of FGMOS transistor.

1 Introduction

LV and LP operations are the main issues for both digital and analog integrated circuit (IC), due to the growing demand especially for portable electronic applications. Current mirror is one of the commonly used in almost all classification of analog and mixed mode designs. Here, the power consumption is influenced by the voltage source where low supply voltages significantly reduce the power utilization and hence expand the battery lifetime of the portable devices such as tablets, cell phones etc. While having LV and LP operation, it maintains the same speed, design accuracy and supply current, comparable with conventional MOS transistor designs [1].

Thus, the greatest challenge here is to design and develop high performance analog circuit without sacrificing the circuit performance and at the same time significantly reducing the circuit's power consumption. Some of the design techniques used by the other designers are level shifter, bulk driven transistors, self-cascode structures, floating gate and MOSFET that is operating in sub-threshold region [1].

Floating gate technique is an attractive option for LP and LV design. This technique simplifies the signals processing configuration chain and reduce the complexity of the circuits. Eventually, it can work below the normal operational limits of supply voltage levels for a particular technology and therefore consumes less power than minimum power required for a CMOS circuit technology without compromising on device performance [1-2].

The paper will examine the advantage of floating gate in analog circuit, particularly a current mirror. The paper structure is as follow. As an introduction, a simple current mirror is discussed in section 2, followed by characteristic of FGMOS presented in section 3, implememntation of current mirror by FGMOS in section 4, section 5 discusses about the comparison results and conclusion in section 6.

2 Current mirror (CM)

LV and LP design structures are frequently used in almost all mobile electronic devices, which generally have a mixed of digital-analog circuit, embedded with analog sub-sections [3]. A current mirror (CM) is an essential part of any anlog signal processing structure in CMOS based analog IC. A CM is a circuit that will copy the reference current source through one reference active device by controlling the current in another active device thus, keeping the output current constant regardless of circuit loading [4]. A high performance of CM should have low input impedance, at the same time giving a high output impedance [4-5].

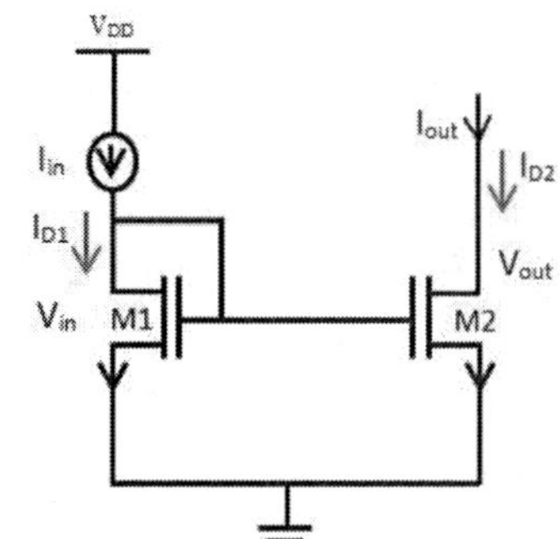

Fig. 1. Simple NMOS current mirror [6]

Figure 1 shows the simplest form of n-type Metal Oxide Semiconductor (NMOS) current mirror which consists of reference transistor M1 and output transistor M2, supplied with supply voltage,V_{DD}. Transistor M1 is connected in diode load configuration. Here, the drain-gate voltage of M1 is at the same potential. Referring to

* Corresponding author: nurulaindolah@gmail.com

transistor operation region, the diode configuration makes the V_{DS} always greater than V_{GS} - V_{TH}, thus the transistor is always in saturation region. The output of the CM is at the drain of transistor M2. Here, M2 also works in the saturation region and both transistors have infinite output resistance. I_{D2} is controlled by V_{GS2}, which is equal to V_{GS1}, $V_{GS1}=V_{GS2}$.

Assuming all transistors are similar and if the transistors are at identical size, $\left(\frac{W}{L}\right)_2 = \left(\frac{W}{L}\right)_1$, therefore,

$$I_{out} = I_{D2} = I_{D1} \quad (1)$$

Thus, the current flows in the drain of transistor M1 is mirrored to the drain of M2 as shown by Equation (1). Applying kirchhoff's current law (KCL) and Equation (1) at the drain of M1 yield to [6]

$$I_{D1} = I_{in} \quad (2)$$

where I_{in} is the reference current. Here, the gate current, I_{G1} = 0A. Therefore, from (1) and (2), it is found that,

$$I_{in} = I_{out} \quad (3)$$

Thus, the output current, I_{out} is mirrored to the reference current, I_{in}.

The other current amplification can be obtained by modifying the size of $(W/L)_2$. CM can be used as current amplification, level shifting, biasing and loading in the circuit. Therefore, CM can be used in various applications such as analog IC operational amplifiers, current conveyers, analog-to-digital and digital-to-analog data converter [5].

3 Floating gate MOSFET (FGMOS)

The threshold voltage of the MOS can be altered by using FGMOS configuration [3]. FGMOS has multiple input gate terminal where the threshold voltage can be controlled and tuned by the values of capacitors and the applied bias voltage [1]. The multiple input gates are isolated to the main gate terminal by an insulator. Figure 2(a) and (b) show the symbol of N-input FGMOS transistor and its equivalent circuit respectively [2].

In both figures, V_i (for i=1,2,...N) are the control input voltages. The N-inputs are only capacitively connected to the floating gate (FG) since it is completely surrounded by highly resistive material. In this way, at DC analysis, the FG is seen as a floating node.

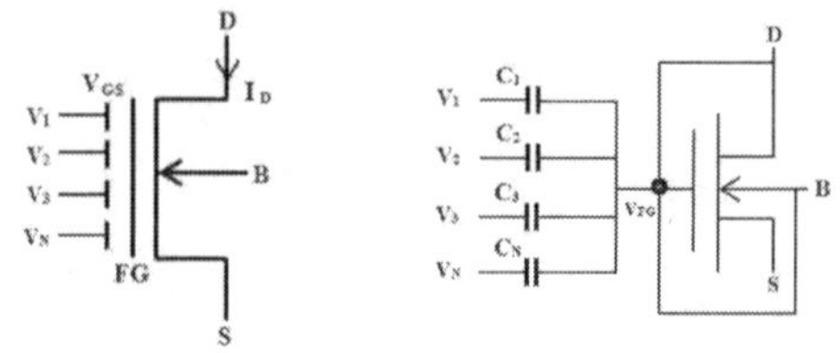

Fig. 2. (a) Symbol of FGMOS (b) FGMOS equivalent circuit [7]

The voltage on FG (V_{FG}) is given in Equation (4) [2].

$$V_{FG} = \frac{\sum_{i=1}^{N} C_i V_i + C_{FGD} V_D + C_{FGS} V_S + Q_{FG}}{C_T} \quad (4)$$

where C_i is the set of input capacitors associated with effective inputs. The total floating gate capacitance (C_T) is given by,

$$C_T = C_1 + C_2 + C_3 + C_{FGD} + C_{FGS} + C_{FGB} \quad (5)$$

C_{FGD}, C_{FGS} and C_{FGB} are the overlap capacitance of floating gate with drain, source and bulk respectively. V_{DS} is drain to source voltage, V_{BS} is the bulk to source voltage and Q_{FG} is the residual charge trapped in the silicon oxide interface during fabrication process. Removal of Q_{FG} can be done by method suggested in [8]. Therefore neglecting Q_{FG} (4) can be written as

$$V_{FG} = \frac{\sum_{i=1}^{N} C_i V_{is} + C_{FGD} V_{DS} + C_{FGB} V_{BS}}{C_T} \quad (6)$$

The drain current (I_D) of the FGMOS transistor operating in saturation region is given by

$$I_D = \frac{\mu_o Cox}{2} \frac{W}{L} \left(V_{FG} - V_{TH}\right)^2 \quad (7)$$

Assuming C_i>> C_{FGD}, C_{FGB} and Q_{FG} = 0, the I_D of the FGMOS transistor is represented by [2]

$$I_D = \frac{\beta}{2}\left(V_{FG} - V_{TH}\right)^2 = \frac{\beta}{2}\left(k_i V_{iS} - V_{TH}\right)^2 \quad (8)$$

where $\beta = \mu_o C_{ox} \frac{W}{L}$, V_{TH} is threshold voltage and k represents the number of metals available in the technology, where $k_N = \frac{C_N}{C_T}$.

The equivalent threshold voltage $V_{TH,eq}$ is lower than the conventional MOS, and given by [9]

$$V_{TH,eq} = \frac{V_{TH} - k_N V_N}{k_1} \tag{9}$$

and V_{TH} depended on V_N, k_N and k_1.

4 Current mirror floating gate (CMFG)

The conventional CM in Figure 1 is modified to obtain LP and LV current mirror by utilizing floating gate technique. The implementation of FGMOS current mirror is shown in Figure 3 where one of the gates is used for gate biasing and the other one is used to modify the threshold voltage and consequently decreasing the input voltage of the current mirror. The minimum input voltage now becomes $V_{TH,eq}$, which can be lowered by using value of k_N and V_N.

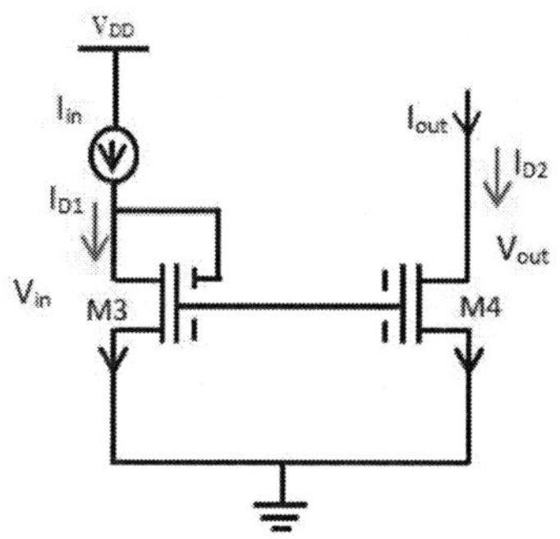

Fig. 3. FGMOS current mirror

The configuration of CM and CMFG is same but different at the gate of transistor where the CMFG have multiple input gates. The multiple input gates of transistor M3 and M4 are called FGMOS. The CMFG consists of reference transistor M3 connected with supply voltage,V_{DD} and the output is at the drain of transistor M4. The transistor consist of four resistances and four capacitances at the floating gate [1].

In the design, width and length (W/L) are the ratio of transistor M1 and M2 of conventional CM, at the same time M3 and M4 of CMFG also used the same ratio. Both current mirrors are simulated with the ratio of;

$$\left(\frac{W}{L}\right)_1 = \left(\frac{W}{L}\right)_2 = \left(\frac{20u}{0.1u}\right) \tag{10}$$

5 Results and discussion

In this section, the simulation results between conventional CM and CMFG are presented. Simulation has been carried out using 0.1µ CMOS technology by SAED 90nm PDK with the Synopsys Custom Designer tool.

Figure 4 shows the simulated drain current,I_D against V_{DS} for CM and CMFG. It shows when same V_{DD}, 1.2V is supplied on both circuit resulting input current of 20mA for CM while the input current for CMFG is 8.02mA. Based on power (P) equation (P=IV), the supply voltage is directly proportional to the consumed power which is the power for CM is 24mW while the power for CMFG is 9.62mW. Therefore, reducing the required voltage supply will contribute to low power consumption.

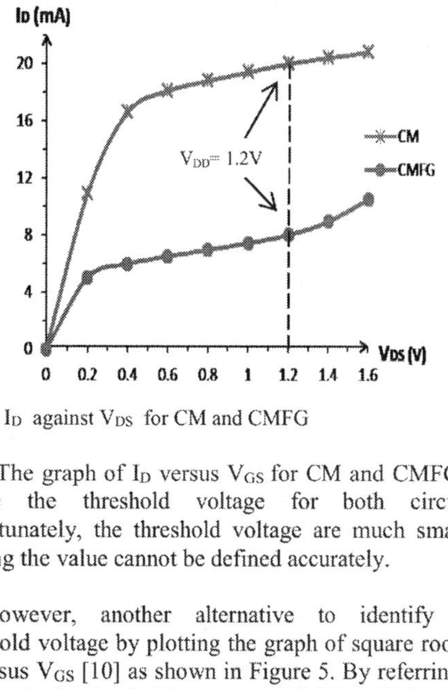

Fig. 4. I_D against V_{DS} for CM and CMFG

The graph of I_D versus V_{GS} for CM and CMFG to define the threshold voltage for both circuits. Unfortunately, the threshold voltage are much smaller causing the value cannot be defined accurately.

However, another alternative to identify the threshold voltage by plotting the graph of square root of I_D versus V_{GS} [10] as shown in Figure 5. By referring to the graph, the threshold voltage of CM is 0.45 V while CMFG is 0.2 V, which is much smaller than the threshold voltage of CM.

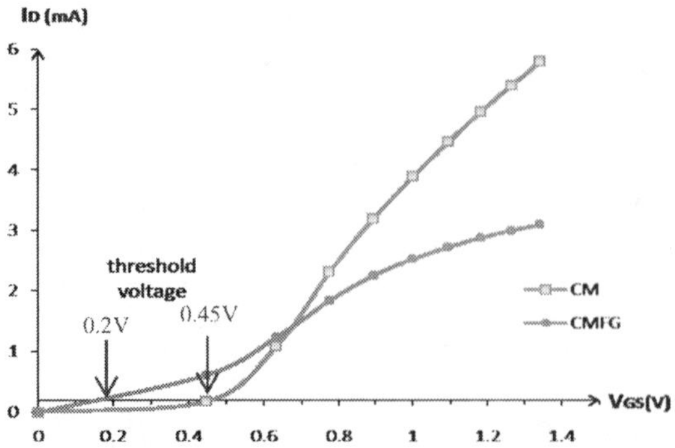

Fig. 5. Square root of I_D vs. V_{GS} for CM and CMFG

Figure 6 shows the graph represents current-voltage characteristic between I_{in} and the output voltage, V_{out} for both CM and CMFG. The same current which is 20mA was supplied to both circuits at the reference transistor, I_{D1}. The CMFG has the V_{out} of 1.7 V which is lower than the V_{out} of CM, 4.4 V.

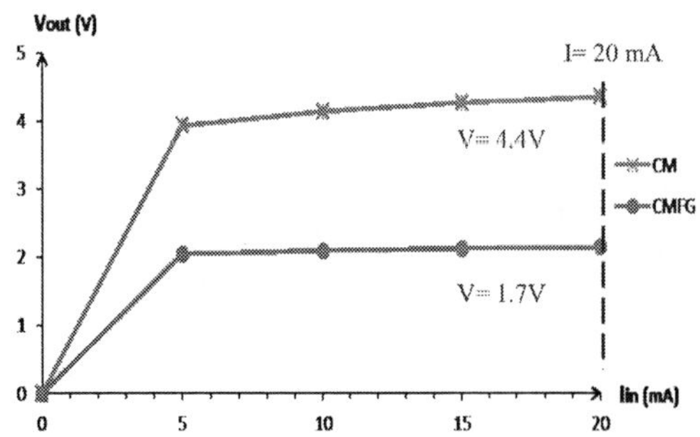

Fig. 6. Vout versus Iin for CM and CMFG

6 Conclusion

In this paper, simulation of CMFG and conventional CM have been presented. Firstly, the same voltage is applied to both circuits to identify its power consumption and then the same input current is applied to determine its output voltage. It can be seen that the improvement percentage of power for CMFG is 40.1% of conventional CM. At the same time, at I_{in} = 20 mA, the percentage of output voltage for CMFG is decreased about 22.3% from conventional CM. In conclusion, the LP and LV design can be enhanced by replacing CMOS with FGMOS, due to its unique advantages, which is consequently reduced supply voltage and power consumption. These circuits operate with a supply voltage of 1.2V with 0.1 μm CMOS technology and the power consumption is 9.6 mW. It has been realized that using CMFG is making them suitable for LV and LP applications.

However, this study is not sufficient to conclude that FG design is much better than the conventional MOS design. Optimisation on the device's width – length dimentions for example, is another parameter to be explored in the next future work. The parameter will apply with current mirror in a complete circuit.

Table 1. Comparison between CM and CMFG

Parameters	CM	CMFG
V_{DD} (V)	1.20	1.20
I_D (mA)	20.0	8.00
P (mW)	24.0	9.62
Vth (V)	0.40	0.20

This research is fully funded by the Fundamental Research Scheme (FRGS) with Grant number 9003 -00421, sponsored by the Malaysian Ministry of Higher Education (MoHE).

References

1. N. Singh and H. Kumar, **5**, 9, pp. 2840–2843, (2016)
2. R. Srivastava, M. Gupta, and U. Singh, Analog Integr. Circuits Signal Process., **78**, 1, pp. 245–252.
3. S. Jamuar, S. Sharma, and S. Rajput, J. Act. Passiv. Electron. Devices, **3**, pp. 109–124, (2008)
4. A. Anand, A. Dash, and B. S. Patro, **4**, no. 2, pp. 39–50, (2013)
5. C. Rana, **vol. 10**, no. 6, pp. 263–271, (2017)
6. B. Tiwari and J. K. Dhanoa, Int. J. Eng. Commun. Technol., **6**, 3, pp. 14–17, (2015)
7. R. Pandey and M. Gupta, Inf. MIDEM, **43**, 3, pp. 173–178, (2013)
8. E. Rodriguez-Villegas, M. Jimenez, and R. G. Carvajal, IEEE Trans. Circuits Syst. II Express Briefs, **54**, 2, pp. 156–160, (2007)
9. F. Khateb, N. Khatib, and D. Kubánek, Microelectronics J., **42**, 5, pp. 622–631, (2011).
10. A. Ortiz-conde, F. J. García-sánchez, J. Muci, A. Terán, and J. J. Liou, Microelectron. Reliab., (2012)

[11] N. Singh and R. Srivastava, Int. J. Adv. Res. Electron. Commun. Eng., **5**, 6, pp. 1683–1686, (2016)

 DOI: 10.1051/epjconf/201716201049

Graphene transfer process and optimization of graphene coverage

Syarifah Norfaezah Sabki[1*], *Shafiq Hafly* Shamsuri[1], *Siti Fazlina* Fauzi[1], *Meghashama Lim* Chon-Ki[1] *and Noraini* Othman[1]

School of Microelectronic Engineering, Universiti Malaysia Perlis (UniMAP), Kampus UniMAP Pauh Putra, 02600 Arau, Perlis, Malaysia.

Abstract. Graphene grown on transition metal is known to be high in quality due to its controlled amount of defects and potentially used for many electronic applications. The transfer process of graphene grown on transition metal to a new substrate requires optimization in order to ensure that high graphene coverage can be obtained. In this work, an improvement in the graphene transfer process is performed from graphene grown on copper foil. It has been observed that the graphene coverage is affected by the pressure given to the top of PDMS to eliminate water and air between graphene and SiO_2 (new substrate). This work experimented with different approaches to optimize the graphene coverage, and stamping method has proven to be the best technique in obtaining the largest graphene coverage. This work also highlights the elimination of impurities from graphene after the transfer process, known to be PMMA residues, which involved immersion of graphene in acetone. This method has improved the graphene conductivity.

1 Introduction

Since its first discovery by the two physicists from Manchester University in 2004 [1], graphene has become a trending in research and applications [2-5]. In recent years, many applications have been explored using graphene as the base material due to its remarkable characteristics, particularly mechanical and electrical properties. This makes graphene very practical and feasible to be incorporated in electronics and RF applications [6-11]. In order to incorporate graphene in devices for applications, the right transfer process plays a major role in order to produce well function devices.

Graphene grown on transition metals have been incorporated in many electronic applications due to its quality and coverage of monolayer graphene. The transfer method of this graphene on transition metal is important in order to ensure a high coverage of graphene is transferred and graphene quality is sustained.

There is several graphene transfer methods have been carried out in graphene researches. One of the method is using a polymeric layer which are polymer supported transfer (PST) [12]. Another well known transfer method is known as direct transfer [13] and pick up transfer [14]. All the three methods use polymethil methacrylic (PMMA) as the support layer for graphene. So far, these three methods have successfuly transferred graphene on to substrates [12, 13, 14]. Among all the methods, the graphene transfer using polymeric layer have been popular due to its easy manage and better yield. However graphene transferred by using this method suffers crack and impurities which is undesirable for a device. Therefore this paper discusses about a new approach that can be applied to optimize the graphene coverage and eliminate impurities from the graphene layer after it is transferred to a new substrate surface.

2 Methodology

2.1 Transfer process

In the graphene transfer, the process started with spin coating the graphene on copper (Cu) surface with polymethil methacrylate (PMMA), followed with drops of polymethyl siloxane (PDMS). Both polymers are cured at ~100°C to solidize and harden. The following step is to etch the Cu foils in aqueous iron chloride ($FeCl_3(aq)$), leaving graphene attached to PMMA and PDMS. PMMA is used because this polymer can be easily dissolved in acetone. PDMS works as a protective layer and it eases the sample handling.

After the Cu etching process, the graphene on PMMA/PDMS is cleaned in distilled water and left on the water surface to be scooped using a spatula, and finally placed on a targeted substrate. In this case of study, silicon dioxide (SiO_2) is used as the substrate. Three graphene on copper samples are transferred based on this method. The process followed with giving a significant pressure to the top of the PDMS, which we believe affects the coverage of graphene that will be obtained on SiO_2. Applying pressure to the top of the PDMS is important to eliminate excess water and air because during the transfer, water and air is trapped between both medium, which could lead to non-uniform coverage of graphene.

To optimize the graphene coverage, three different approaches in applying pressures to the top of PDMS are performed. The approaches are illustrated in **Fig. 1**. The first approach (**Fig. 1(a)**) is performed by giving pressure to the top of PDMS using finger tip and the pressure is given repeatedly in 10s. This sample is marked as sample 1. The second approach (**Fig. 1(b)**) is performed by applying a constant pressure for ~10s using fingertip and the sample is marked as sample 2. The final step is performed by applying a constant

* Syarifah Norfaezah Sabki: faiesns@gmail.com

DOI: 10.1051/epjconf/201716201049

pressure (also at ~10s) using a wooden block with very flat surface and similar surface area as the graphene area (**Fig. 1(c)**), marked as sample 3.

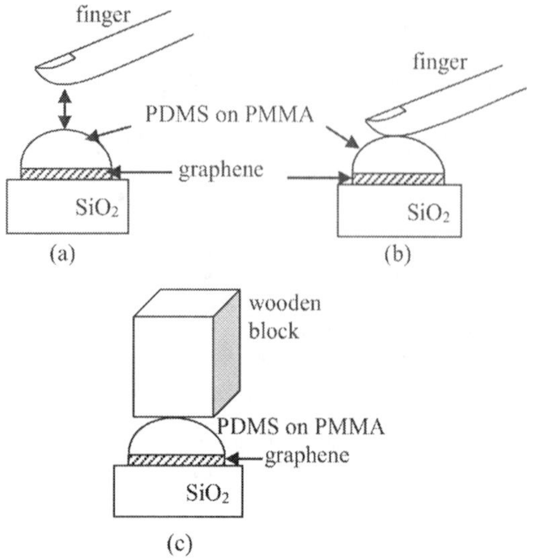

Fig. 1. Approaches taken to eliminate water and air between graphene and SiO_2 to optimize the graphene coverage (a) finger tip is pressed repeatedly in 10s (sample 1), (b) finger tip is pressed constantly for 10s (sample 2), (c) wooden block with flat surface is pressed constantly for 10s (sample 3).

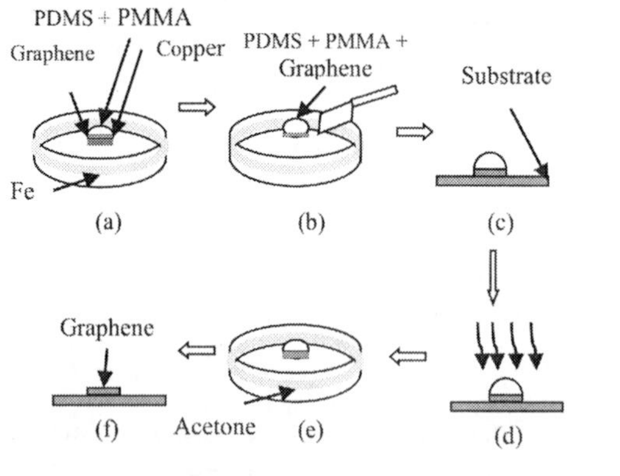

Fig. 2. Transfer process. (a) PDMS + PMMA + graphene + Cu are etched in $FeCl_3$. (b) PDMS + PMMA + graphene are transferred onto substrate using spatula. (c) PDMS + PMMA + graphene are being pressed down onto new substrate (SiO_2) surface. (d) The sample is baked in convection oven. (e) PDMS + PMMA + graphene are immersed in acetone. (e) PMMA is dissolved, leaving graphene on SiO_2.

After all the three different approaches are performed, all the samples are then annealed in a furnace at ~100°C for ~10 minutes to ensure the samples are completely dry and van der Waals force between graphene and SiO_2 surface is formed. The samples are then immersed in acetone to dissolve the PMMA, leaving graphene attached on SiO_2. The transfer process is illustrated in **Fig. 2**.

2.2 Characterization of the transferred graphene layer.

The surface characteristics of the graphene layer are observed using atomic-force microscopy (AFM), SPA-400 from SII Instrument. AFM images are taken by using tapping mode with the scale of 5 × 5μm and scan speed of 1 Hz to obtain high resolution topographic images. The image of the transferred graphene layer on SiO2 is captured by using High Power Microscope (HPM) ECLIPSE L300N from Nikon. The magnification of 5 × 0.15 A is used to inspect the transferred graphene layers quality and coverage. For further investigation, the electrical test on the transferred graphene layers is carried out by using two points probe, Semiconductor Parameter Analyser (SPA) 4200-SCS from KEITHLEY. The electrical test result is analysed by using KITE software with the voltage sweep -10V to +10 V and step of 0.5.

3 Results and Discussions

3.1 Surface characteristics of the transfered graphene

As shown in **Fig. 1**, three transferred graphene samples (Sample 1, 2, and 3) are investigated to observe of which sample produced the best graphene quality after the transfer process. Based on sample 1 which is pressed with finger tip repeatedly produced graphene patches that are not uniform across the substrate surface as shown in **Fig. 3(a)**. The repeated pressure given to the top of PDMS cause disturbance between graphene and the substrate surface, which lead to high degree of graphene edge defects and non-uniform graphene coverage. This result is undesirable for electronic applications, where a more controlled surface coverage is required.

Therefore, sample 2 is prepared by pressing finger tip on the top of PDMS firmly and constantly for about ten seconds to ensure that the graphene is well attached to the substrate surface. The optical microscopy image in **Fig. 3(b)** shows that there is a larger coverage of graphene compared to sample 1. However a small amount of graphene patches are still observed.

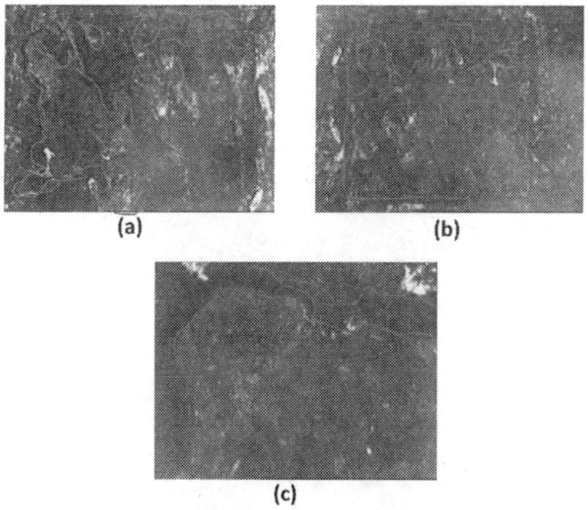

Fig.3. Optical microscopy image of graphene on SiO_2 after transfer process. The size of the image is equivalent to the size of actual area of the transferred graphene, 1 cm x 1 cm. (a) Sample 1- finger tip is pressed repeatedly in 10s, (b) Sample 2 - finger tip is pressed constantly for 10s, (c) Sample 3 - wooden block with flat surface is pressed constantly for 10s (Stamping method).

Graphene on Sample 3 is transferred by pressing the top of PDMS using a wooden block that has a flat surface with dimension of 1 cm x 1 cm, which is exactly similar to the size of the overall graphene sample. The optical microscopy image of the graphene after the transfer process as in **Fig. 3(c)** shows that a large coverage of graphene is obtained. The graphene coverage in sample 3 is the largest compared to sample 1 and 2. This shows that, by pressing the graphene sample with a flat wooden block of the same size as the graphene sample yielded the best graphene coverage. Since the action of pressing a wooden block is similar to stamping process, therefore we named this process as 'stamping method'. Note that we have tried different sizes of wooden block and we observed that smaller surface area of the wooden block reduced the graphene coverage. Hence a larger surface area of the tool is required because a more stable pressure can be applied.

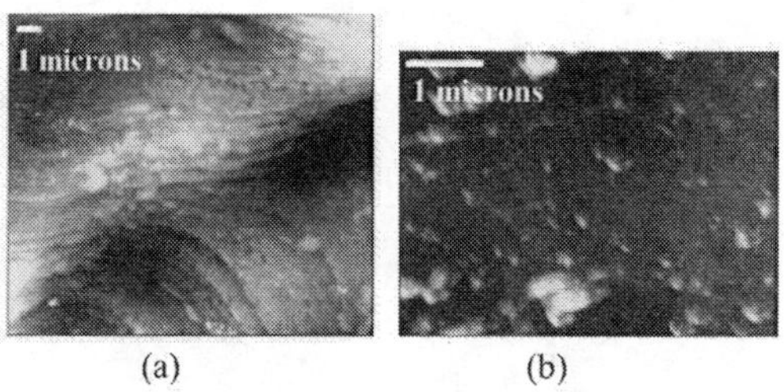

Fig.4. AFM images of (a) untreated graphene on copper foil before transfer process (b) graphene transferred to SiO_2 (Sample 3 – by stamping method using wooden block), immersed 1 time in acetone for 30 min.

Fig. 4(a) above shows graphene on Cu foil of thickness 0.25 μm, with copper orientation in (111). The step edges of Cu can be seen clearly. After the transfer process, the Cu step edges is no longer present, however graphene cracks can be seen as in **Fig. 4(b)**. It is also obvious that there are impurities on the graphene layer, as represented by the spots with lower contrast. Possible impurities are dusts or polymer residues that are also transferred during the transfer process. However, it is suggested that dusts are already removed with only 1 time immersion in acetone, because only van der Waals force is present. To remove polymer residues requires more immersion time or duration in acetone.

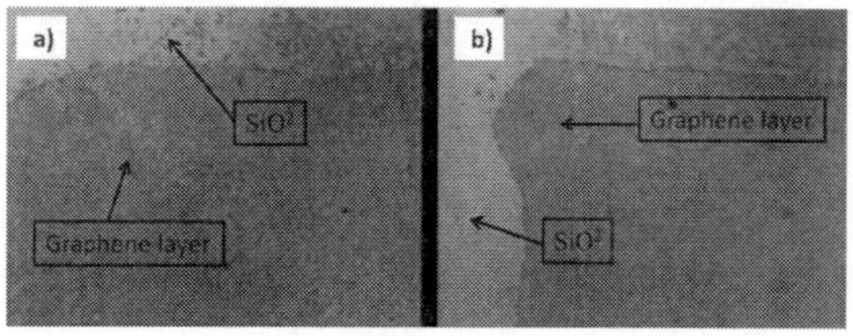

Fig.5. Optical microscopy images of transferred graphene layer on SiO^2 (a) 1 time immersion in acetone for 30 min and (b) 3 times immersion in acetone with immersion of 30 min each time.

Not only having outstanding electrical properties, graphene is also well known by its optical properties which can be used as high transparent electrode in nanoelectronic devices. The transparency of the graphene layer can be seen in **Fig. 5**, where **Fig.5 (a)** shows that graphene layer that is transferred in this work is slightly high in contrast, which may be suggested by the presence of impurities. The possible residues may be suggested by PMMA that is used in the transfer process, where it may not be totally dissolved. There should be no PDMS residues on the graphene surface because PDMS does not dissolve in acetone and it is not attached directly on graphene. Dissolving PMMA in acetone is one typical technique that is easy to conduct, low cost and feasible. Another method that can be used is to anneal the sample with PMMA in ultra high vacuum (UHV) system, which will evaporate and remove the PMMA physically [15]. However the process requires high cost, complicated system and difficult to conduct.

As shown in **Fig. 5(b)**, after further cleaning, better surface area is obtained, in which the transparency is improved. The image of graphene on SiO_2 in **Fig. 5(b)** shows that graphene layer is more apparent. Such sample surface is obtained after 3 times immersion in acetone, with each immersion; the sample is left in acetone for about 30s. It is also suggested that the immersion of the sample in acetone can be done longer than 2 hours to obtain the same result.

It is important to ensure all the polymer residues are eliminated from the graphene after the transfer process to reduce the effect of graphene defects and to optimize the graphene quality. High amount of PMMA residues may affect the performance of graphene layer for any electronic application, since PMMA is non conductive. It has been investigated that polymer residue such as PMMA provides scattering sites that may alter the electronic structure of graphene especially its mean free paths of carrier charges [16].

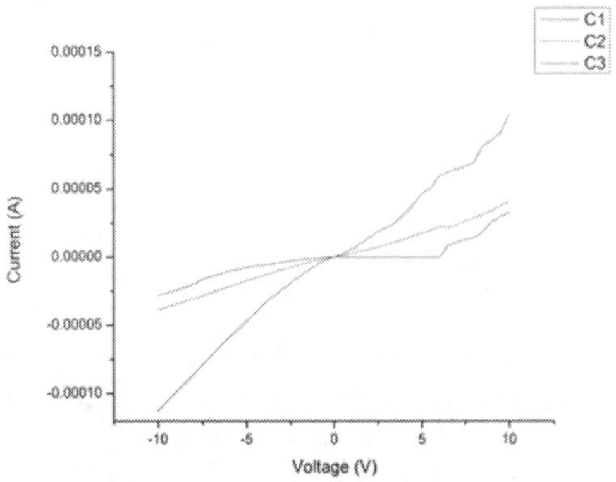

Fig.6. I-V curve of transferred graphene layer obtained from sample 1 (C1), sample 2 (C2) and sample 3 (C3).

To observe the quality of the obtained graphene layer after the transfer process, IV measurement is performed. The IV characteristics in **Fig. 6** is referred to sample 3 (stamping method using wooden block) that has undergone 3 times immersion in acetone (30 min each immersion). The result of the first immersion is represented by curve C1, followed with second immersion represented by curve C2 and the third immersion is represented by curve C3. It is observed that C1 shows the lowest conductivity, and C3 shows the highest conductivity. Even though the obtained conductivity is quite low compared to conductivity of ideal monolayer graphene, this IV characteristics justify that number of immersion in acetone to dissolve PMMA residues improves the quality of the graphene layer hence improving its conductivity. The overall low conductivity range in **Fig. 6** may be suggested by the cracks and defects that present in the graphene layer as observed in **Fig. 4**. Note also the average resistance of C1, C2 and C3 are ~$1.424E^{+9}$, ~$2.962E^{+5}$ and $1.173E^{+5}$ respectively.

4 Conclusions

It can be concluded that graphene transfer by using stamping method, specifically using a tool with flat and large surface area (where this work used a wooden block), lead to a large graphene coverage after the transfer process because this method implies a stable pressure during the process. This result is crucial to ensure optimized graphene coverage, which is required for electronic applications. This work has also pointed out the importance of eliminating impurities from the graphene layer in the transfer process, where the impurities are significantly came from PMMA residues. However, this problem can easily be overcomed by cleaning the graphene layer, by immersing the samples repeatedly in solvent, particularly acetone.

References

1. Novoselov, K. S., Geim, A. K., Morozov, S. V, & Jiang, D., Science **306**, Issue 5696, pp. 666-669, (2004).
2. Yue, W., Jiang, S., Xu, S., Ma, Y., & Bai, C. Sensors and Actuators B: Chemical, **214**, pp. 204–210, (2015).
3. Arefinia, Z., & Asgari, A. Solar Energy Materials and Solar Cells, **137**, pp.146–153, (2015).
4. Li, Z., Zhang, W., Guo, J., Yang, B., & Yuan, J. Vacuum, **117**, 35–39. (2015).
5. Valentini, L. Materials Letters, **148**, 204–207. (2015).
6. Ramachandran, R., Saranya, M., Velmurugan, V., Raghupathy, B. P. C., Jeong, S. K., & Grace, A. N,Applied Energy, **153**, 22–31. (2015).
7. Kwon, K.-W., Choi, J.-W., Jung, M. W., You, Y.-H., Chang, R., Seo, J. H., ... Hwang, J.-H. Ceramics International, **41**,7, 8360–8366. (2015).
8. Bolotin, K. I., Sikes, K. J., Jiang, Z., Klima, M., Fudenberg, G., Hone, J.,Stormer, H. L.,Solid State Comm. **146**,9-10, 351–434. (2008).
9. Lemme, M., & Echtermeyer, T. Electron Device Letters, **28**,4, 282–284. (2007).
10. Lin, Y., Chiu, H., Jenkins, K. A., Member, S., Farmer, D. B., Avouris, P., Abstract, A. Dual-Gate Graphene FETs With f T of 50 GHz, **31**,1, 2009–2011. (2010).
11. Liao, L., & Duan, X. Materials Today, **15**(7-8), 328–338. (2012).
12. Reina, A., et al., Nano Letters, **9** (1) ,p, 30-35 , (2009).
13. Regan, W..,et al Applied Physics Letters **96**,p. 113102,(2010).
14. Ren, Y., et al, Nano, **7**, (2012), p.1150001.
15. A. Pirkle, J. Chan, A. Venugopal, D. Hinojos, C. W. Magnuson, S. McDonnell, L. Colombo, E. M. Vogel, R. S. Ruoff, and R. M. Wallace, Appl. Phys. Lett ,**99**,pp3-5, (2011).
16. J. W. Suk, W. H. Lee, J. Lee, H. Chou, R. D. Piner, Y. Hao, D. Akinwande, and R. S. Ruoff, Nano Lett, **13**, 1462 (2013).

Structural and electrical properties of Barium Titanate ($BaTiO_3$) and Neodymium doped $BaTiO_3$ ($Ba_{0.995}Nd_{0.005}TiO_3$)

Tuan Amirah Tuan Sulong[1, 3*], *Rozana Aina* Maulat Osman[1, 3], *Mohd Sobri* Idris [2, 3] and *Zul Azhar* Zahid Jamal[1, 3]

[1]School of Microelectronics Engineering, University Malaysia Perlis, 02600, Arau, Perlis, Malaysia
[2]School of Materials Engineering, University Malaysia Perlis, 01000, kangar, Perlis, Malaysia
[3]COE Frontier Material Research Centre, School of Material Engineering, University Malaysia Perlis, 01000, Perlis, Malaysia

Abstract. Barium titanate ($BaTiO_3$) and Neodymium (Nd) doped $BaTiO_3$ with composition $Ba_{0.995}Nd_{0.005}TiO_3$ were prepared using conventional solid state reaction method to study the dielectric properties of materials. Pure phase samples were found at final heating temperature of 1400 °C for overnight. X-ray diffraction analysis reveals the changes in the lattice parameter and unit cell volume of the pure perovskite tetragonal structure with space group (*P4mm*). Electrical analysis is carried out to investigate the dielectric properties, conductivity behaviour and dielectric loss of $BaTiO_3$ and $Ba_{0.995}Nd_{0.005}TiO_3$. $Ba_{0.995}Nd_{0.005}TiO_3$ have a broaden dielectric peaks with high permittivity of 8000 and reasonably low loss tan δ which is about 0.004 (1 kHz).

1 Introduction

Dielectric ceramics are widely used in advance microelectronics technologies such as capacitor and microwave communication [1]. A high dielectric constant ceramic had received a lot of attention in reducing the size of the microelectronics circuits where high permittivity and low dielectric loss are required for many electronics application. $BaTiO_3$ composition modified by doping have been extensively studied for high dielectric materials applications.

In previous work reported by Shaikh & Vest [2], $BaTiO_3$ structure are very sensitive to the doping level and will change the electrical behaviour of pure $BaTiO_3$. In order to improve the properties of $BaTiO_3$, some modifications on $BaTiO_3$ structure was made and rare earth materials were of interest due to a huge improvement in the permittivity as reported by Abdul Hamid et. al. [3] in Lanthanum (La) doped $BaTiO_3$ and Samarium (Sm) doped $BaTiO_3$ by Ganguly et al. [4].

Neodymium (Nd) doped $BaTiO_3$ was previously studied by other researcher, however, in this study, we present a simple study about the physical behaviour and electrical behaviour change by small amount of Nd doping to the A-site perovskite of $BaTiO_3$. The trivalent Nd is very interesting to study because it previously reported that small amount of Nd will exhibit a PTCR effect and change the electrical behaviour from insulating to semiconducting [5]. Instead of that, in term of structure, the radius of Nd^{3+} (1.27Å) is very close to the radius of Ba^{2+} (1.61Å) and small size of Ti^{4+} (0.605.Å) at the center of the octahedral site towards one of the other corner oxygens. In view of the radius, it is possible for Nd^{3+} to enter into the A-sites of $BaTiO_3$ perovskite and effect the properties of $BaTiO_3$ ceramics. In this work, the phase analysis, structural analysis and electrical behaviour of $BaTiO_3$ and $Ba_{0.995}Nd_{0.005}TiO_3$ was presented. The incorporation location of small amount of doping Nd doped to $BaTiO_3$ was identified using Rietveld Refinement analysis.

* Corresponding author: rozana@unimap.edu.my

2 Experimental Procedure

High purity of raw materials $BaCO_3$ (99%), TiO_2 (99%) and Nd_2O_3 (99%) from Sigma-Aldrich were used as starting materials. Stoichiometric amount of materials were mixed and grounds using pestle and mortar. Samples were pressed uniaxially using handpress into pellets and heated at final temperature of 1400 °C for 12 hours with heating rate 5°C/min. The phase purity of the samples were analyzed using the X-ray Diffraction (XRD: D2-ADVANCE, Bruker, Germany) from 2θ, 20°-130°. Structural studied obtained were refined using the General Structure Analysis Software (GSAS). For impedance measurements, pressed pellets were sintered at 1400°C for 6 hours. Silver paste was used as electrode. Samples was painted with silver paste on both pellets' surfaces. Pellets were connected to the Platinum (Pt) leads of a conductivity jig which was mounted inside a horizontal tube furnace. Impedance data were recorded on a stepwise heating cycle using HIOKI IM 3522 Impedance Analyzer instrumentation in frequency range 10 Hz to 100 kHz.

 DOI: 10.1051/epjconf/201716201050

3 Result and Discussion

Figure 1 shows the XRD patterns of $BaTiO_3$ and $Ba_{0.995}Nd_{0.005}TiO_3$ respectively. The reflection patterns of $Ba_{0.995}Nd_{0.005}TiO_3$ remain similar as $BaTiO_3$ sample, except with the presence of broad peak at 2θ: 22° and 51° respectively. All samples of the $BaTiO_3$ and Nd-doped $BaTiO_3$ were indexed with tetragonal phase with the unit cell a = 3.9937 Å, c= 4.0342 Å and a= 3.9964 Å, c = 4.0288 Å, volume = 64.38 Å and volume = 64.34 Å respectively. The space group belong to P4mm. Figure 2 shows the variation of lattice parameters and unit cell volume of $BaTiO_3$ and $Ba_{0.995}Nd_{0.005}TiO_3$. Introducing the small amount of Nd content into pure $BaTiO_3$ at A—sites perovskite structure, resulting the changes as shown in Figure 2(a) and 2(b). A decrement in the lattice parameter and unit cell volume was observed. The decrement in lattice parameter and unit cell volume means that the lattice is shrinkage and the size of the unit cell become smaller. This effect is might be due to the size incompability of Nd with ionic radii of Nd^{3+} and Ba^{2+}, thus causes an overall shrinkage of unit cell and volume.

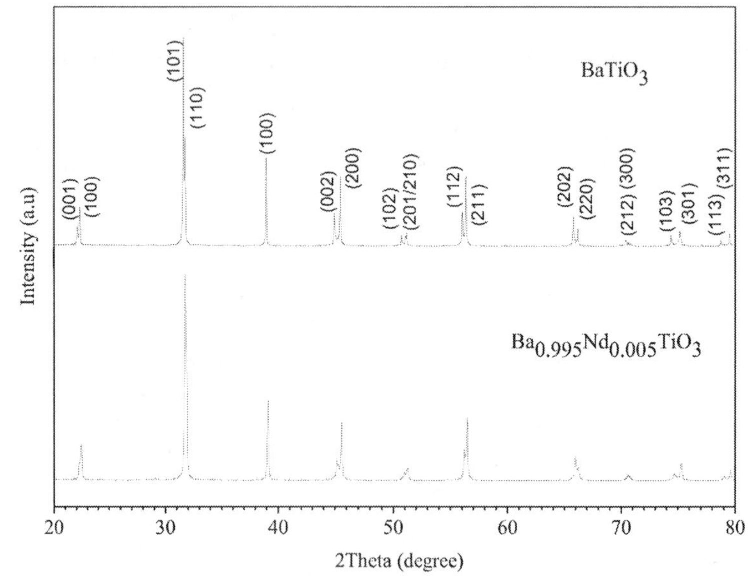

Fig.1. Indexed XRD patterns of $BaTiO_3$ and $Ba_{0.995}Nd_{0.005}TiO_3$

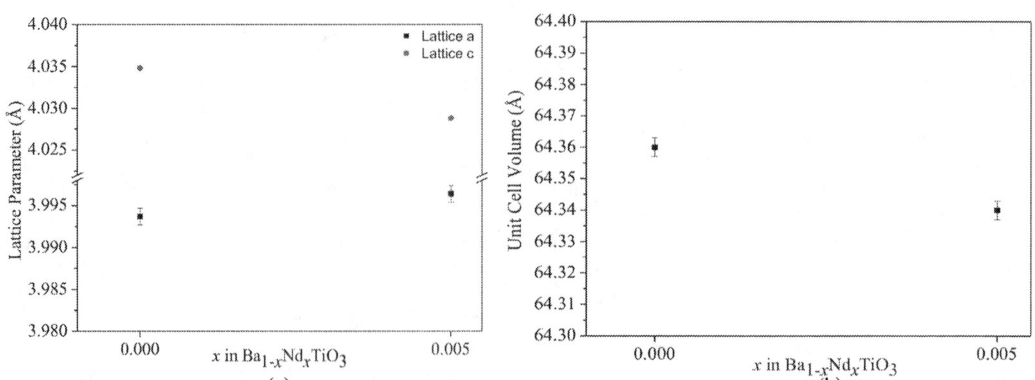

Fig. 2. Variation in: (a) lattice parameter and (b) unit cell volume of $BaTiO_3$ and $Ba_{0.995}Nd_{0.005}TiO_3$.

XRD data for $Ba_{1-x}Nd_xTiO_3$ ($0 \leq x \leq 0.015$) were collected and the structure was refined using the atomic coordinates of tetragonal perovskite (ICSD No: 245945) as astarting parameters. The cation site occupancies were varied with Ba and Nd on A-sites and Ti on B-sites. The optimized parameters for refinement were detailed in [6,7]. The refinement result, Table 1, shown the substitution of Nd into $BaTiO_3$. Reliability refinement parameters show relatively lower R_p, R_{wp} and χ^2. This result give some ideas to claim that substitution

 DOI: 10.1051/epjconf/201716201050

mechanism, suggest by the refinement result are mostly reliable. Fitting parameters (R_p and R_{wp}) indicate acceptable agreement between refined and observed XRD patterns for $BaTiO_3$ and $Ba_{0.995}Nd_{0.005}TiO_3$ ceramics with both tetragonal structure. Some variations in occupancy values were showed due to the replacement of Ba atoms by Nd atoms which indicates that (Ba or Nd) atoms located in the same A-site occupy the same site in a fractional percentage. Note that there are small variations in atomic positions related to the titanium and oxygen atoms while the barium and neodymium atoms keep their fixed positions within the structure. Figure 3(a–d) shown the difference profile of (a) $BaTiO_3$ and (b) $Ba_{0.995}Nd_{0.005}TiO_3$ after refinement. Results obtained by using the Rietveld refinement method indicate acceptable agreement between observed XRD patterns and fitted theoretical results. This indicate the success of the Rietveld refinement method which displays small differences near zero in the intensity scale.

Table 1. Refinement data for $BaTiO_3$ and $Ba_{0.995}Nd_{0.005}TiO_3$.

Doping Concentration	$x = 0$	$x = 0.005$
Ti, z	0.5171 (3)	0.5126 (2)
Ba (1a), U_{iso}	0.3841 (8)	0.0052 (7)
Nd (1a), U_{iso}	-	0.0076 (9)
Ti (1b), U_{iso}	0.5740 (7)	0.0050 (9)
01(1b), U_{iso}	0.8923 (9)	0.0064 (5)
O2 (2c), U_{iso}	0.2377 (5)	0.0118 (9)
R_{wp}	14.04	13.89
R_p	10.44	11.04
χ^2	3.994	4.468

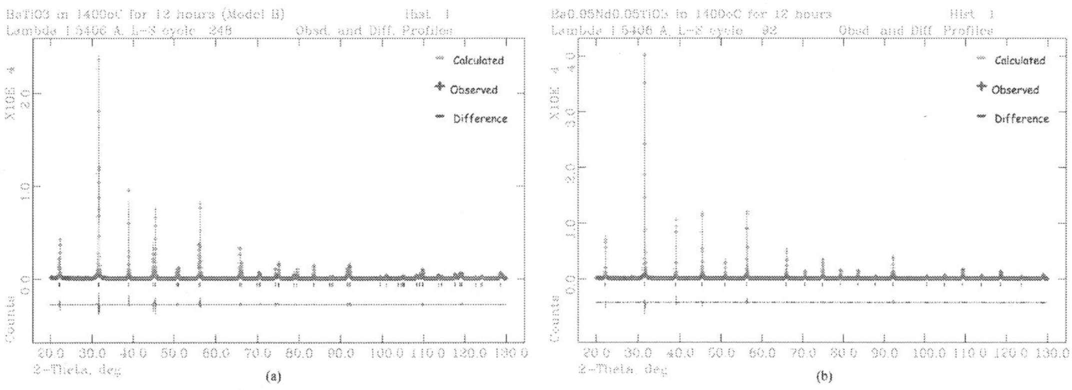

Fig. 3. Structural refinement plot for a) $BaTiO_3$ and b) $Ba_{0.995}Nd_{0.005}TiO_3$

Permittvity at fixed frequency of 1 kHz, 10 kHz and 100 kHz for $BaTiO_3$ and $Ba_{0.995}Nd_{0.005}TiO_3$ sintered at1400 °C for 6 hours are shown in Figure 4 respectively. Data show sharp maximum permittivity for all samples in accordance with literature. Permittivity at T_c was remarkedly decreased for Nd-doped $BaTiO_3$ as compared to undoped $BaTiO_3$. Maximum permittivity for pure $BaTiO_3$ about 8500, at curie temperature (T_c) of 120 °C, while for Nd-doped $BaTiO_3$, permittivity at T_c are 8000, 5000 and 2000 at different frequency. At room temperature, the permittivity of Nd-doped $BaTiO_3$ show remarkable value as compared to permittivity of $BaTiO_3$. Small amount of Nd shows a very interesting result by shifting the T_c towards lower temperature, and slight decreament in permittivity change at T_c. This might be related to the high dipole polarisation of rare-earth based-Nd, microstructure and grain size of the sample. The permittivity reaches a peak at the Curie point and falls of at higher temperatures in accordance with the Curie–Weiss law [8]. This phenomena were strongly related with the shifting of maximum permittivity at T_c of Nd-Doped $BaTiO_3$ at tetragonal phase to cubic phase above T_c. Generally, in heavily doped $BaTiO_3$ ceramics grain size is usually suppressed and the permittivity of $BaTiO_3$ ceramics attains the maximum when the grain size is 1μm or less[9]. Factors that influence permittivity include type of dopant, doping mechanism, processing conditions and grain size [10]. On the other hand, high permittivity may be associated with sample geometry and, in particular, with thin layer effects associated with grain boundaries, surface layers or sample-electrode contacts [11].

DOI: 10.1051/epjconf/201716201050

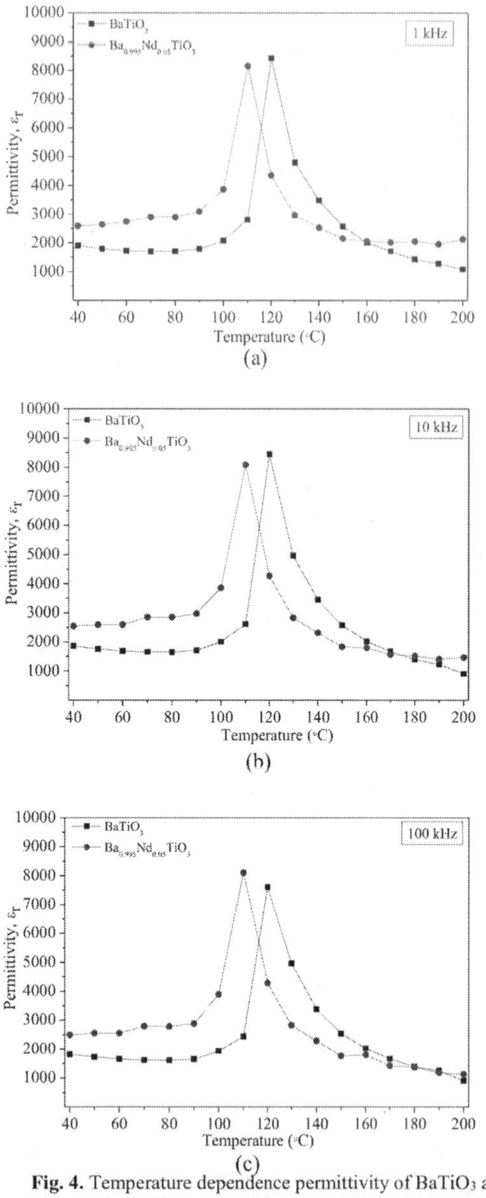

Fig. 4. Temperature dependence permittivity of $BaTiO_3$ and $Ba_{0.995}Nd_{0.005}TiO_3$ at (a) 1 kHz, (b) 10 kHz and (c) 100 kHz respectively.

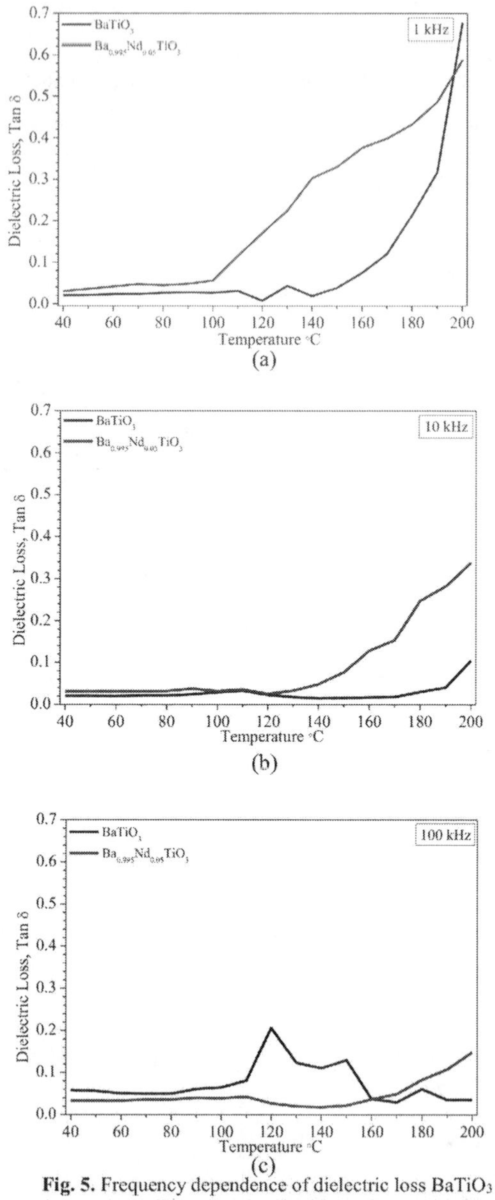

Fig. 5. Frequency dependence of dielectric loss $BaTiO_3$ and $Ba_{0.995}Nd_{0.005}TiO_3$ of (a) 1 kHz, (b) 10 kHz and (c) 100 kHz respectively.

Figure 5 shows the temperature dependence in dielectric loss of $BaTiO_3$ and $Ba_{0.995}Nd_{0.005}TiO_3$ at a frequency of 1 kHz. The dielectric loss of the $Ba_{0.995}Nd_{0.005}TiO_3$ is very low at much lower temperature which is below 0.04 loss. The dielectric loss of $BaTiO_3$ and $Ba_{0.995}Nd_{0.005}TiO_3$ at frequency of 1 kHz is very low below T_c and increase a little bit to 0.06 about T_c. However the loss was very minimal as frequency increased in Figure 5(b) and 5(c). Minimum dielectric loss as low as 0.2 was mantained at lower temperature, however, rapidly increased at high temperature. By doping Nd, below T_c, the dielectric loss was bound from 0.04 to 0.004. Figure 6 (a) and (b) shows the frequency dependence conductivity of $BaTiO_3$ and $Ba_{0.995}Nd_{0.005}TiO_3$ at frequency of 1 kHz, 10 kHz and 100 kHz sintered at 1400°C for 6 hours. The conductivity of $BaTiO_3$ changes with the frequencies of 1 kHz to 100 kHz. At 1 kHz, the conductivity is about $5.4x10^{-8}$ Scm^{-1}. However a small increase of 10^{-7} Scm^{-1} at 10 kHz and increased again at 100 kHz to 10^{-6} Scm^{-1}. The conductivity at T_c for $BaTiO_3$ is higher compared to other temperatures and much higher at higher frequency. The conductivity of $Ba_{0.995}Nd_{0.005}TiO$ tend to increase from 10^{-8} to 10^{-6} Scm^{-1} approaching T_c at 110°C.

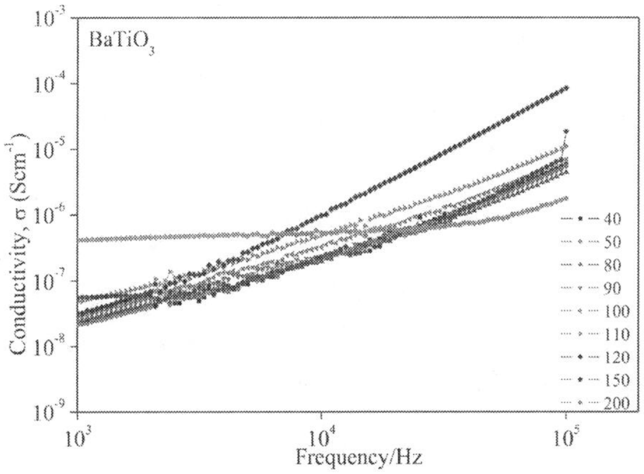

(a)

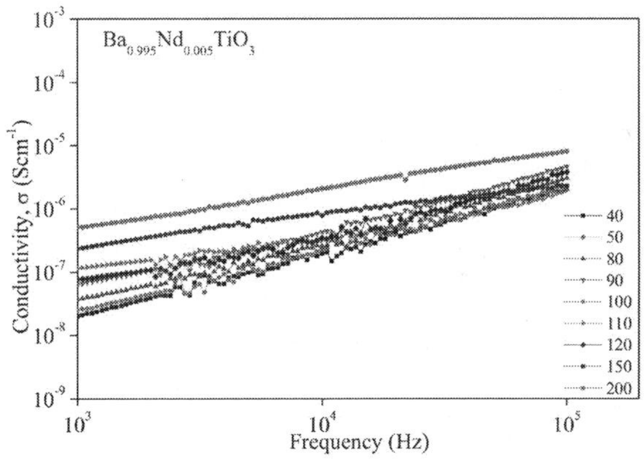

(b)

Fig. 6. Frequency dependence of conductivity of (a) $BaTiO_3$ and (b) $Ba_{0.995}Nd_{0.005}TiO_3$

The range of conductivity at low frequency of 1 kHz and 100 kHz were about 9.0 x 10^{-7} Scm^{-1} and $1x10^{-5}$ Scm^{-1} respectively. The conductivity difference considered high, with two orders of magnitude. This is in agreement with the exact conductivity of a semiconductor material, correlated with the observed properties and related phenomenon responsible [10]. Therefore, conductivity was greatly affected by temperature and frequency of range 1 kHz to 100 kHz. Data collection was limited to frequencies < 1 MHz, due to increases in inductive effects that influenced the impedance data [11].

4 Conclusion

The effect of Nd incorporation to the A-sites of perovskite $BaTiO_3$ was studied. This effect is seen by observing the lattice and volume change in the structure of $BaTiO_3$ with percentage 0.68% of shrinkage. However, the structure was still mantain with tetragonal unit cell and the structural site of Nd was confirmed by rietveld analysis with Nd at A-sites. The permittivity maximum was at T_c with 8500 and reduced to 8000 by doping at low frequency. However at room temperature, the substitution of Nd show a very high permittivity compared to pure $BaTiO_3$ at three different frequencies. The conductivities of $BaTiO_3$ and $Ba_{0.995}Nd_{0.005}TiO_3$ increased by increasing the frequency and temperature.

The authors acknowledge the financial support from the Ministry of Education, Malaysia for funding this project (FRGS Grant No: 9003-00496) and Universiti Malaysia Perlis (UniMAP) for the opportunities of performing this research.

DOI: 10.1051/epjconf/201716201050

References

1. X. Hao, Journal of Advanced Dielectrics, **3**, 1, 1-14, (2013).
2. A. S. Shaikh and R. W. Vest, Journal of American Ceramic Society, **69**, 9, 689-694, (1986).
3. N.F. Abdul Hamid, R. A. M. Osman and M. S. Idris, Applied Mechanics and Materials, **815**, 131-135, (2015).
4. M. Ganguly, S. K. Rout, W. S. Woo, C. W. Ahn and I. W. Kim, Physica B, **411**, 26-34, (2013).
5. N. Hirose, J. M. S. Skakle and A.R. West, Journal of Electroceramics, **3**, 3, 233-238, (1999).
6. M. S. Idris and R. A .M. Osman, Structure Refinement Strategy of Li-based Complex Oxides Using GSAS-EXPGUI Software Package, Advanced Materials Research, 795, 479-482, (2013).
7. T.Q. Tan, M. S. Idris, R. A. M. Osman, M. V. Reddy and B. V. R. Chowdry, Solid State Ionics, **278**, 43-48, (2015).
8. W. Heywang, Journal of Material Science, **6**, 1214-1226, (1971).
9. Z. Yao, H. Liu, Y. Liu, Z. Wu, Z. Shen, Y. Liu and M. Cao, Material Chemistry and Physics, **109**, 475-481, (2008).
10. A. R. West, Basic Solid State Chemistry, 2nd edition. Chichester: John & Wiley &Sons Ltd., (2014).
11. A. R. West, T. B. Adams, F.D. Morrison and D. C. Sinclair, Journal of the European Ceramic Society, **24**, 1439-1448, (2004).

Dielectric and microstructural properties of $BaTiO_3$ and $Ba_{0.9925}Er_{0.0075}TiO_3$ ceramics

Fatin Adila Ismail[1,2*], *Rozana Aina* Maulat Osman[1,2], *Mohd Sobri* Idris[1,3], *Sanna* Taking[2] and *Zul Azhar* Zahid Jamal[2]

[1]CoE Frontier Materials Research, School of Material Engineering, Universiti Malaysia Perlis, 01000 Kangar, Perlis, Malaysia
[2]School of Microelectronic Engineering, Universiti Malaysia Perlis, 02600 Arau, Perlis, Malaysia
[3]School of Material Engineering, Universiti Malaysia Perlis, 02600 Arau, Perlis, Malaysia

Abstract. $BaTiO_3$ and $Ba_{0.9925}Er_{0.0075}TiO_3$ ceramics were investigated regarding their dielectric and microstructure properties via conventional solid state reaction method. The phase pure samples were obtained when heated at 1400°C for overnight. The effect of Er^{3+} doped into $BaTiO_3$ on dielectric properties and microstructural properties was investigated for composition of $BaTiO_3$ and $Ba_{0.9925}Er_{0.0075}TiO_3$. The analysis was made by X-ray Diffraction (XRD), Scanning Electron Microscopy (SEM) and Impedance Analyzer. The XRD patterns of $BaTiO_3$ and $Ba_{0.9925}Er_{0.0075}TiO_3$ are phase pure and identical with tetragonal perovskite structure with space group of P4mm. The lattice parameters and unit cell volume of $BaTiO_3$ increased by doping with Erbium as the crystallite size decreased. Measurements of dielectric properties were carried out as a function of temperature up to 200°C at different frequencies. $Ba_{0.9925}Er_{0.0075}TiO_3$ exhibit the high value of dielectric constant (ε=6179) at Curie temperature (T_C) of 120°C. SEM analysis of $BaTiO_3$ and $Ba_{0.9925}Er_{0.0075}TiO_3$ ceramics showed that the grain sizes of $BaTiO_3$ and $Ba_{0.9925}Er_{0.0075}TiO_3$ were ranged from 3.3μm-7.8μm and 2.2μm-4.7μm respectively.

1 Introduction

The Barium titanate ($BaTiO_3$) – based ferroelectric ceramics have gained importance in electronic industry as they have high dielectric constant and offer wide range of applications such as capacitors, sensors, actuators, power transmission devices, memory devices and high energy storage devices [1-2]. $BaTiO_3$ has a perovskite structure and due to this perovskite structure, $BaTiO_3$ has intrinsic capability to host different dopant ions with different sizes. The influence of dopant ion depends on the substitutional site (Ba^{2+} or Ti^{4+}) as well as on the defects itself. The ionic radius of the dopant ion is essential for the determination of the substitutional site of host material. The ions with small ionic radii such as La^{3+} and Ce^{3+}will substitute into A- site whereas the ions with bigger ionic radii such as Lu^{3+} and Yb^{3+} possibly substitute at B site of $BaTiO_3$ structure. The ions that have amphoteric behaviour such as Y^{3+}, Dy^{3+}, Ho^{3+} and Er^{3+} can occupy both cation lattice site in the $BaTiO_3$ structure [3-4]. The influence of the dopant incorporation into $BaTiO_3$ as well as the influence of some transition metal and large rare earth ions in the structural and electrical properties of $BaTiO_3$ ceramics has been extensively investigated. Among them, Erbium (Er^{3+}) has been widely used due to its great technological interest. It has been previously reported by Leyet et. al. [5] that with the Erbium cation incorporation into $BaTiO_3$ crystalline lattice according to the formula $Ba_{1-x}Er_xTiO_3$, increase in dielectric constant is observed. This suggests that the incorporation of Er^{3+} into $BaTiO_3$ crystalline lattice can be greatly improve the dielectric properties of $BaTiO_3$.The aim of this work is to investigate the structural and electrical properties of Er doped $BaTiO_3$ as a function of the Er^{3+} concentration as well as to correlate the structure properties of pure $BaTiO_3$ and $Ba_{0.9925}Er_{0.0075}TiO_3$.

2 Experimental

$BaTiO_3$ and $Ba_{0.9925}Er_{0.0075}TiO_3$ compositions were prepared via conventional solid state reaction method. The chemicals used were high purity of $BaCO_3$ (99.99% Pure, Sigma-Aldrich), USA), TiO_2 (99.99%, Sigma-Aldrich) and Er_2O_3 (99%, Sigma-Aldrich) powders. $BaCO_3$, TiO_2 and Er_2O_3 were weighed using A&D Weighing HR250AZ pharma balance. Powder samples were mixed and ground using pestle and mortar. Samples were ground for half an hour by hand using acetone as the medium for mixing. Samples were then pressed into pellets shape by using uniaxial press with 1 tonne pressured. Pellets were then heated in air at high temperature between 1000°C and 1400°C at 5°C/min for overnight. The crystal structures of the samples were confirmed by using X-ray Diffraction (Bruker,D2-Phaser) with Cu Kα radiation (λ = 1.5406). Samples

* Corresponding author: fatinadilaismail@gmail.com

 DOI: 10.1051/epjconf/201716201051

were measured in range of 2θ = 20° - 80° using step size of 0.02. The sintering temperature for electrical measurement was 1300°C for 3 hours with heating and cooling rate of about 5°C/min in air. For electrical measurements, the surfaces of sintered pellets were painted with silver conductive paint. The electrical characterization is measured using Impedance Analyser (Hioki IM3570) from room temperature to 200°C at 1kHz, 10 kHz and 100 kHz. The microstructure was studied by using Scanning Electron Microscopy (SEM). The SEM micrographs were obtained using a JEOL JSM-6460LA SEM. The grains size distribution was measured using SEM image analysis software, Smile View.

3 Results and Discussions

3.1 XRD Analysis

Fig. 1 shows the X-ray Diffraction (XRD) patterns of $BaTiO_3$ and $Ba_{0.9925}Er_{0.0075}TiO_3$ samples heated at 1400°C for overnight. The XRD patterns show similar XRD pattern with an early reported $BaTiO_3$ tetragonal perovskite structure [6, 7]. The XRD patterns are the same and show the single phase tetragonal perovskite structure with space group of P4mm. The lattice parameters and unit cell volume for the sample $BaTiO_3$ composition sintered at 1400°C were a = 3.9918 Å and c = 4.0338 Å and unit cell volume, V = 64.28 Å^3. The lattice parameters and unit cell volume for the sample $Ba_{0.9925}Er_{0.0075}TiO_3$ composition sintered at 1400°C were a = 3.9940 Å and c = 4.0362 Å and unit cell volume, V = 64.38 Å^3. Table 1 shows the structure parameters of $BaTiO_3$ and $Ba_{0.9925}Er_{0.0075}TiO_3$ obtained. An increase in lattice parameters of $BaTiO_3$ as doped with Er can be observed. The average crystallite size values of pure $BaTiO_3$ and Erbium doped $BaTiO_3$ ceramics, revealing there is a decrease in the size suggesting the effect of dopant on crystallite size of $BaTiO_3$. The crystallite sizes were evaluated through the Scherrer's equation. It is commonly determined by measuring the Bragg's peak width at the half maximum intensity and by using the Scherrer's formula (Eq. 1), where d is the crystalline size, λ is the wavelength of the x-ray, ß is the peak width at the half maximum intensity and theta is the Bragg's angle [7].

$$d = \frac{0.9\lambda}{\beta cos\theta} \quad (1)$$

Table 1. Structural parameters of $BaTiO_3$ and $Ba_{0.9925}Er_{0.0075}TiO_3$ prepared samples and their calculated crystallite sizes.

Sample	*a (Å)*	*c (Å)*	*V(cm^3)*	*Crystallite size (Å)*
$BaTiO_3$	3.9918(3)	4.0338(5)	64.28(2)	1068.44
$Ba_{0.9925}Er_{0.0075}TiO_3$	3.9940(3)	4.0362(5)	64.38(1)	931

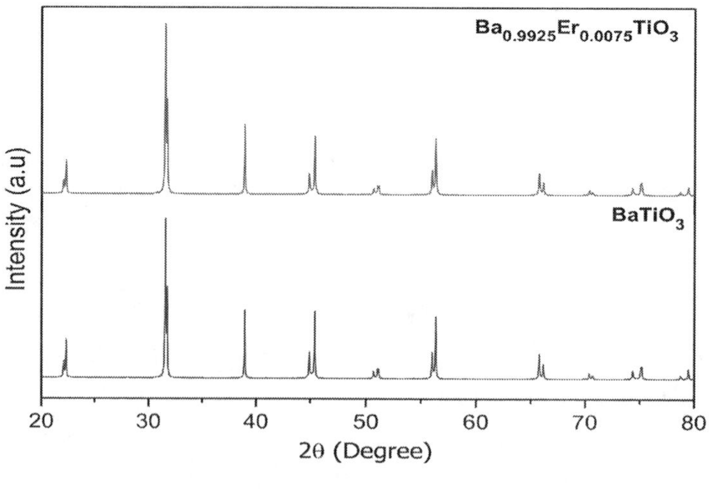

Fig. 1. XRD pattern of $BaTiO_3$and $Ba_{0.9925}Er_{0.0075}TiO_3$ heated at 1400°C

3.2 Electrical Analysis

The dielectric behavior and dielectric loss as a function of temperature for the sintered pellets of $BaTiO_3$ and $Ba_{0.9925}Er_{0.0075}TiO_3$ at 1 kHz is shown in Fig. 2. It can be observed that by doping, the permittivity increase from 5194 to 6502 for $BaTiO_3$ and $Ba_{0.9925}Er_{0.0075}TiO_3$. The Curie temperature, T_C increased by doping which is from 110°C to 120°C for $BaTiO_3$ and $Ba_{0.9925}Er_{0.0075}TiO_3$ respectively. Increase in T_C may be due to the misfit between the grain core and the grain shell give rise to stresses and shift the Curie point to higher temperature as suggested by Hwang et. al [8]. Some of rare earth doping that increase the T_C are Yb doped $BaTiO_3$ and Lu doped $BaTiO_3$ which is quite similar to the current Er doped $BaTiO_3$ system. Based on Fig. 2, an increase in dielectric constant from 5194 to 6502 for $BaTiO_3$ and $Ba_{0.9925}Er_{0.0075}TiO_3$ can be observed at 1 kHz. The increase in dielectric constant can be due to the modifications in the grain size, resulting from the incorporation of the erbium in the $BaTiO_3$ structure [5].

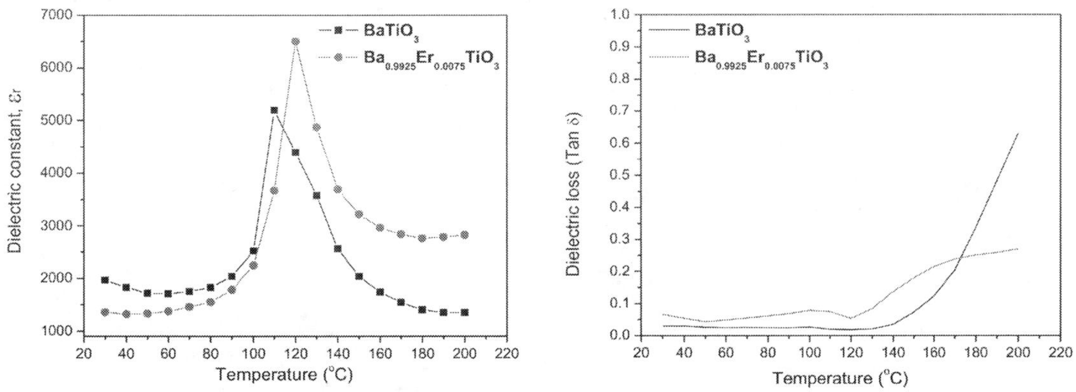

Fig. 2. Dielectric constant and dielectric loss of $BaTiO_3$ and $Ba_{0.9925}Er_{0.0075}TiO_3$ at 1 kHz

As for the dielectric loss, it can be seen that the dielectric loss of $BaTiO_3$ and $Ba_{0.9925}Er_{0.0075}TiO_3$ were less than 0.1 at temperature below 120°C. Much higher dielectric loss which is more than 0.1 observed at higher temperature above 120°C up to 200°C. Table 2 summarizes the dielectric constant values and T_C of $BaTiO_3$ and $Ba_{0.9925}Er_{0.0075}TiO_3$ at different frequencies. The dielectric constant values are decreased as the frequency increased.

Table 2. Dielectric constant and T_C of $BaTiO_3$ and $Ba_{0.9925}Er_{0.0075}TiO_3$ at different frequencies.

Sample	Dielectric constant	Curie temperatre, T_C	Frequency
$BaTiO_3$	5194	110°C	1kHz
	5072		10kHz
	4880		100kHz
$Ba_{0.9925}Er_{0.0075}TiO_3$	6502	120°C	1kHz
	6179		10kHz
	5891		100kHz

$Ba_{0.9925}Er_{0.0075}TiO_3$ shows the highest dielectric constant at 1 kHz which increased by almost 1000 as compared to the $BaTiO_3$. The T_C of $Ba_{0.9925}Er_{0.0075}TiO_3$ increased by 10°C from T_C of $BaTiO_3$. However, as the frequency increased, the dielectric constant values decreased. The decreasing dielectric constant values across the frequencies may be due to space charge polarization as reported by Ganguly et al. [9]. Fig. 3(a) shows the variation of capacitance value for undoped $BaTiO_3$ and Er doped $BaTiO_3$ at x=0.0075. The capacitance values increases from about ~$3x10^{-9}$ Fcm^{-1} to ~$4x10^{-9}$ Fcm^{-1} by doping Er into $BaTiO_3$. These results are consistent with the results of dielectric constant. Fig. 3(b) shows the conductivity values of $BaTiO_3$ and $Ba_{0.9925}Er_{0.0075}TiO_3$ at specific temperatures from 30°C to 200°C. The conductivity was between ~10^{-7} Scm^{-1} and ~10^{-6} Scm^{-1}.The conductivity of $BaTiO_3$ shows the trend for insulator behavior which increased from room temperature to T_C and decreased after T_C. For $Ba_{0.9925}Er_{0.0075}TiO_3$, the conductivity of samples shows

 DOI: 10.1051/epjconf/201716201051

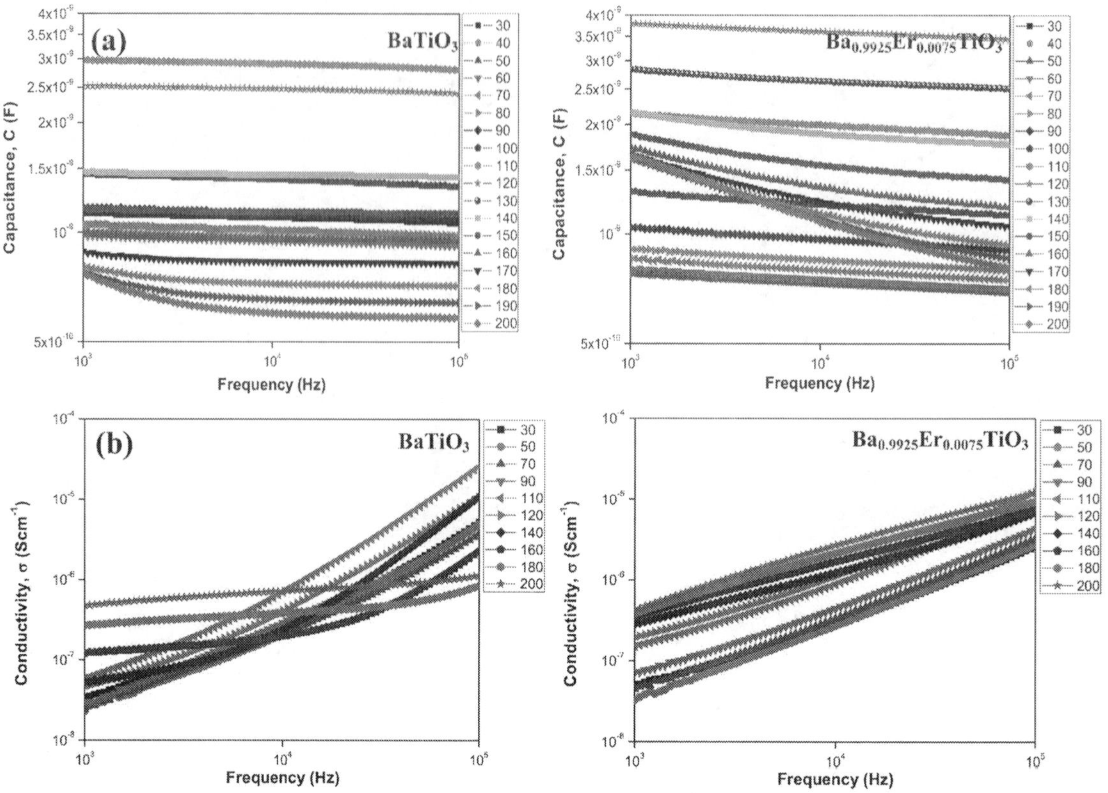

Fig. 3. a) Capacitance and b) conductivity of $BaTiO_3$ and $Ba_{1-0.9925}Er_{0.0075}TiO_3$ as a function of frequencies at different temperature

the typical ceramic behavior which increased as the temperature increased.

3.3 Microstructural Analysis

Fig. 4 shows that the surface morphology of sintered $BaTiO_3$ and $Ba_{0.9925}Er_{0.0075}TiO_3$ by using SEM. The relation between dielectric constant and the grain size of sintered $BaTiO_3$ and $Ba_{0.9925}Er_{0.0075}TiO_3$ were observed. These micrographs were captured by 1000x magnification using Secondary Electron (SE) mode.

From Fig. 4(a) and 4(b), the irregular shape of grains can be seen for both samples. However, there is difference in the grain size as shown in grain size distribution graph in Fig. 5(a) and 5(b). The grain size of $BaTiO_3$ was in between 3.3μm-7.8μm and the grain size for $Ba_{0.9925}Er_{0.0075}TiO_3$ was ranged from 2.2μm-4.7μm. The grain size of $Ba_{0.9925}Er_{0.0075}TiO_3$ is much smaller as compared to $BaTiO_3$. Therefore, the dielectric constant of $Ba_{0.9925}Er_{0.0075}TiO_3$ sample is much higher than $BaTiO_3$ samples.

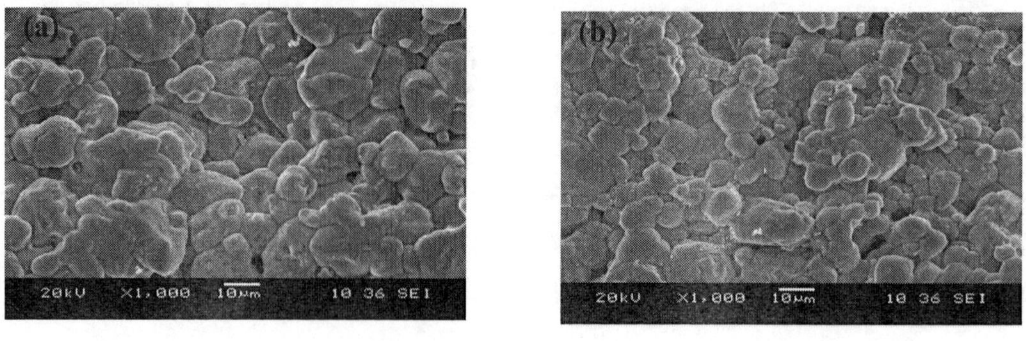

Fig. 4. Scanning electron micrographs of sintered (a) $BaTiO_3$ and (b) $Ba_{1-0.9925}Er_{0.0075}TiO_3$ at 1000x magnification

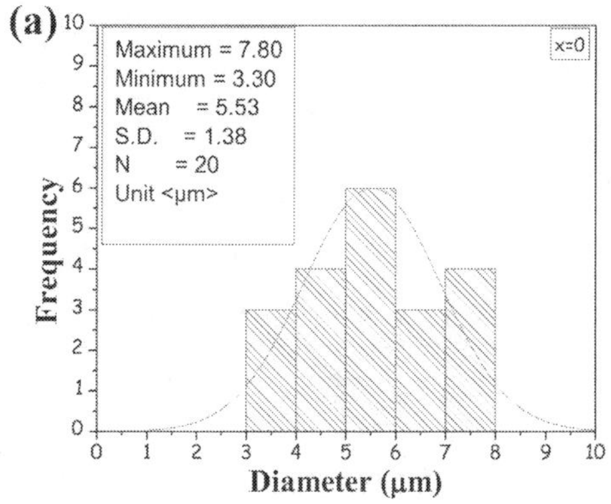

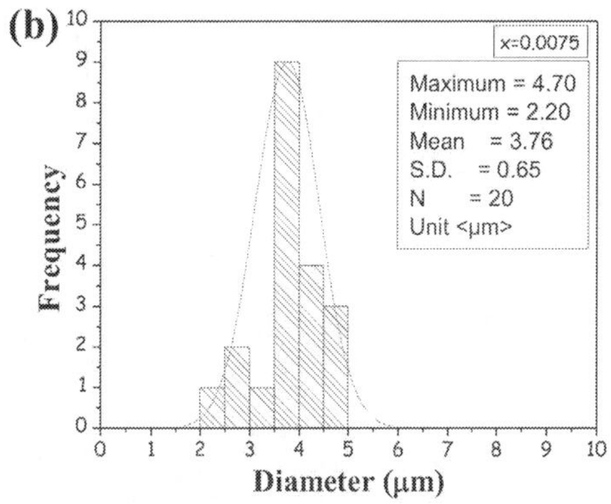

Fig. 5. Grain size distribution of sintered (a) $BaTiO_3$ and (b) $Ba_{1-0.9925}Er_{0.0075}TiO_3$ at 1000x magnification

4 Conclusions

Doping study using Erbium doped $BaTiO_3$ was studied with the effect of small amount of Er to the A-site of $BaTiO_3$. The dielectric behaviour of $BaTiO_3$ improved by doping. The highest dielectric constant observed at Er doped $BaTiO_3$ with dielectric constant maximum at T_C about 120°C which is 6502. As for dielectric loss, the value is lower than 0.1. Besides, the capacitance and conductivity also improved significantly. The capacitance increased from ~3x10^{-9} Fcm^{-1} to ~4 x 10^{-9} Fcm^{-1}. The conductivity was between ~10^{-7} Scm^{-1} and ~10^{-6} Scm^{-1}. The effects of Erbium doping on the microstructural properties were verified as the grain size decreased as Er doped into $BaTiO_3$. The grain size of $BaTiO_3$ was in between 3.3 μm - 7.8 μm which is larger as compared to the grain size for $Ba_{0.9925}Er_{0.0075}TiO_3$ ranged from 2.2 μm - 4.7 μm. Finally, it can be concluded that by doping Er into $BaTiO_3$, the dielectric properties were improved and can be correlated with microstructure properties.

The authors are grateful to Universiti Malaysia Perlis (UniMAP), for giving opportunities to do this research project, Ministry of High Education (MOHE) for funding on Exploratory Research Grant Scheme, FRGS (Grant No: 9010-00032).

References

1. G. H. Haertling, Journal of the American Ceramic Society, **82(4)**, 797-818 (1999).
2. A. J. Moulson and J. M. Herbert, *Electroceramics: Materials, Properties, Applications* (1990).
3. M. T. Buscaglia, , M. Viviani, V. Buscaglia, C. Bottino and P. Nanni, Journal of the American Ceramic Society, **85(6)**, 1569-1575(2002).
4. P. Yongping, Y. Wenhu, and C. Shoutian, Journal of Rare Earths, **25**, 154-157(2007).
5. Y. Leyet, R. Peña, Y. Zulueta, F. Guerrero, J. Anglada-Rivera, Y. Romaguera, and J. P. Cruz, Materials Science and Engineering: B, **177(11)**, 832-837(2012).
6. N. A. Hamid, R. A. M. Osman, M. S. Idris and T.Q. Tan, Materials Science Forum, **819** (2015).
7. Y. Leyet, L. Aguilera, A. Perez-Rivero, F. Guerrero and M. I. B. Bernardi, Rev. Cub. Quim, **22**,62-72 (2010).
8. J. H. Hwang, S. K. Choi and Y. H. Han, Japanese Journal of Applied Physics, **40(8)**, 4952 (2001).
9. M. Ganguly, S. K. Rout, P. K. Barhai, C. W. Ahn and I. W. Kim, Phase Transitions, **87(2)**, 157-174 (2014).

Physical and electrical properties of $SrTiO_3$ and $SrZrO_3$

Norhizatol Fashren Muhamad[1, 2*], *Rozana Aina* Maulat Osman[1, 2], *Mohd Sobri* Idris[2, 3] and *Mohd Najib* Mohd Yasin[1]

[1]School of Microelectronic Engineering, Universiti Malaysia Perlis, Pauh Putra Campus, 02600 Arau, Perlis, Malaysia.
[2]CEO Frontier Material Research Centre, School of Material Engineering, Universiti Malaysia Perlis, 01000 Kangar, Perlis, Malaysia.
[3]School of Materials Engineering, Universiti Malaysia Perlis, jejawi, 02600 Arau, perlis, Malaysia.

Abstract. Perovskite type oxide strontium titanate ($SrTiO_3$) and strontium zirconate ($SrZrO_3$) ceramic powder has been synthesized using conventional solid state reaction method. The powders were mixed and ground undergone calcinations at 1400°C for 12 h and sintered at 1550°C for 5h. X-ray Diffraction exposes physical properties $SrTiO_3$ which exhibit cubic phase (space group: pm-3m) at room temperature meanwhile $SrZrO_3$ has Orthorhombic phase (space group: pnma). The electrical properties such as dielectric constant (εr), dielectric loss (tan δ), and conductivity (σ) were studied in variation temperature and frequency. High dielectric constant of $SrTiO_3$ and $SrZrO_3$ were observed at 10 kHz for both samples about 240 and 21 respectively at room temperature. The dielectric loss of $SrTiO_3$ and $SrZrO_3$ is very low loss value approximately 0.00076 and 0.67512 indicates very good dielectric.

1 Introduction

Ceramic with perovskite crystal structure have long been investigated due to their excellent ferroelectric, paraelectric properties and chemically stable-structure. Its applicable in displays, Multilayer Ceramic Capacitor (MLCC), electronic/piezoelectric devices, sensors, actuators, transducers, wireless communications [1-5]. The ideal Perovskite structure, which have the general formula ABX_3 are consisting of two different cations (A and B) in equal ratio and an anion (X), which is usually oxygen.

$SrTiO_3$ and $SrZrO_3$ are such materials that have been current interest to study due high dielectric constant and its a lead free material. At room temperature, $SrTiO_3$ has a simple cubic perovskite structure with space group (pm-3m) meanwhile $SrZrO_3$ has orthorhombic structure with space group (pnma) but both composition undergoes a series of phase transitions by varying the temperature. $SrTiO_3$ undergoes phase transitions from high to low symmetry [6], while low to high symmetry for $SrZrO_3$ [7].

$SrTiO_3$ and $SrZrO_3$ is a perovskite dielectric material applied to many application fields such as integrated microelectronic and microwave device [8]. These features are attributed by their unique properties such as high dielectric constant, low dielectric loss, tunability, high breakdown strength and low leakage current density [9-11]. Besides, the melting point of $SrTiO_3$ and $SrZrO_3$ is 2080°C and 2600°C respectively, making it applicable at high temperatures. In 2001, Shende et al. [11] claim that $SrTiO_3$ and $SrZrO_3$ can be used in high-voltage applications because of its high breakdown strengths which these materials are not expected to experience electromechanical failure mechanisms that may result in dielectric breakdown. Parida et al. [12] in 2012 report that $SrTiO_3$ and $SrZrO_3$ has high permittivity and low loss make it suitable for microwave antenna application. Based on Wang et al [13] study, high dielectric constant and favourable bias stability inside $SrTiO_3$ are widely required for the application of high energy storage density dielectrics.

Numerous attempts have been explored to further improve the properties of $SrTiO_3$ based ceramics. Among them, doping was considered as an effective approach for altering their properties. ZrO_2 was selected as cation substitute the B-type site of $SrTiO_3$, because Zr^{4+} ions in $SrTiO_3$ can stabilize the charge of Ti^{4+} and suppress the oxygen dissociation when sintered at high temperature [13]. $SrTiO_3$–$SrZrO_3$ solid solution is one of them, which shows a super lattice structure due to its cell enlargement as a result of tilting of BO_6 (B = Ti, Zr) octahedral [14]. This type of disorder perovskite solid solution offers exciting new possibilities both in the investigation of fundamental physical behaviour and in the exploitation of novel properties for various applications.

In this study, the electrical properties of $SrTiO_3$ and $SrZrO_3$ prepared via solid state reaction method were investigated as respect to its physical structure properties. So, the system is expected to provide access to the desired phase or properties at convenient temperatures.

2 Experimental procedure

The Perovskite-type oxide were prepared by mixing high purity $SrCO_3$ (99.9% Aldrich), TiO_2 (99.0% Sigma Aldrich), and ZrO_2 (99% Aldrich) in the appropriate stoichiometric ratios. Sample was mixed and ground using pestle and mortar with acetone as mixing medium. The obtained powders were pressed into pellets of diameter 13mm and heated at high temperature 1400°C for 12h with heating rate 5°C/m in the furnace. The samples were sintered for 1550°C for 5h in air followed by characterization.

X-ray Diffraction (XRD) were conducted at room temperature to determine the crystalline structure of all samples by using X-Ray Diffractometer (Model: D2 PHASER from Bruker AXS) in a 2θ range from 10°

* Corresponding author: norhizatolfashren@gmail.com

to 80°, using Cu Kα1 radiation. The electrical properties such as dielectric constant (εr), dielectric loss (tan δ), conductivity(σ) as well as impedance (Z) were studied by making use of impedance-capacitance-resistance (LCR) meter (Model: HIOKI 3522-50 LCR HiTESTER) at temperature 30-600°C. For this purpose, sintered pallets were coated with silver paste as electrode. Samples are measured over the frequency range of 10 Hz – 100 kHz with different temperature.

3 Results and Discussion

The structures of the oxide $SrTiO_3$ and $SrZrO_3$ were first studied using powder X-ray diffraction data within 10° to 80° value of 2θ in range as shown in **Figure 1(a) and 1(b).** All the peaks are sharp and there is no unwanted peak is found in XRD pattern, representing the crystal is single phase without any impurities and identical to the pdf reported [$SrTiO_3$=PDF00-035-0734] and [$SrZrO_3$= PDF01-076-9442]. From the diffraction pattern, the lattice parameter along with the average crystallite size for both the samples has been calculated as shown in **Table 1**. $SrTiO_3$ was indexed Cubic phase (space group: P m -3 m) with unit cell with lattice constant formulated by a = 3.9000Å which are in contrast to the orthorhombic structure of the $SrZrO_3$ with unit cell with lattice constant formulated by a = 5.7915Å, b=5.8129Å, c=8.2000Å.

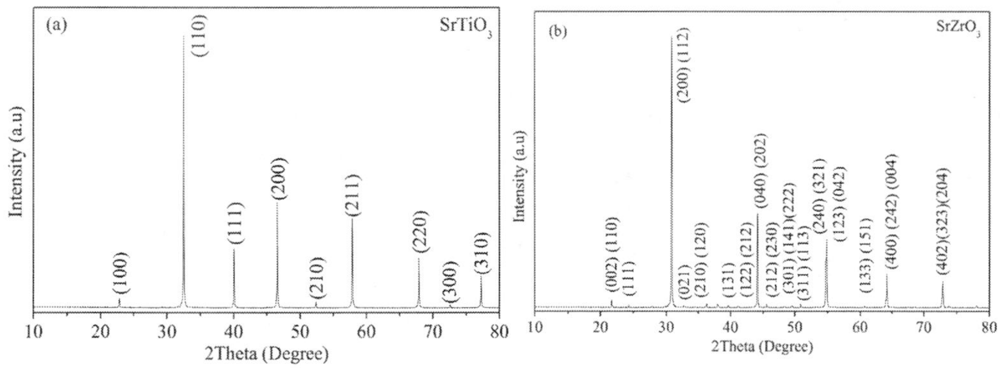

Fig. 1. Index XRD pattern of (a) $SrTiO_3$ (b) $SrZrO_3$

Table 1 Latice parameter,crystallite size and space group of $SrTiO_3$ and $SrZrO_3$

	$SrTiO_3$	$SrZrO_3$
a **(Å)**	3.9000 (1)	5.7915 (1)
b **(Å)**	-	5.8129 (1)
c **(Å)**	-	8.2000 (1)
V (Å^3)	59.5 (1)	276.1 (1)
Crystallite Size	973.14	796.86
Space Group	P m -3 m	P n m a

The Dielectric Constant (εr) was calculated from the measured capacitance (Cp) obtain by LCR meter between 30 - 600°C temperatures. The variations of dielectric constant with temperature at some selected frequencies for $SrTiO_3$ and $SrZrO_3$ ceramics are shown in **Figure 2(a) and 2(b).** For $SrTiO_3$ ceramics, the dielectric constant decreases gradually up to a certain temperature and increases rapidly with increasing temperature. Similar phenomenon of the dielectric constant was observed in $SrZrO_3$ ceramics. There is no curie peak detected for both sample in the whole measurement temperature range. The dielectric constant of $SrTiO_3$ much higher compared to $SrZrO_3$ .dielectric constant is decrease in from $\varepsilon_r = 240$ ($SrTiO_3$) to $\varepsilon_r = 21$ ($SrZrO_3$) is due to the decrease in ionic polarization [15].

Figure 3(a) and 3(b) shows the Capacitance against frequency, the spectrum displace toward higher frequency. $SrTiO_3$ exhibit capacitance of average 10^{-9} ~ 10^{-10} F, meanwhile $SrZrO_3$ show to have capacitance value in range of 10^{-10} ~ 10^{-11} F, Thus the range of capacitance probably belonged to grain boundary [16]. The highest capacitance value is obtain at temperature 600°C which give 2.9×10^{-9} F for $SrTiO_3$ and 1.3×10^{-10} F for $SrZrO_3$ at 1 kHz. The capacitance related to the ability of $SrTiO_3$ and $SrZrO_3$ sample to collect and store energy in the form of an electrical charge.

Figure 4(a) and 4(b) shows the dielectric loss (tan δ) of $SrTiO_3$ and $SrZrO_3$. The dielectric loss for the $SrTiO_3$ measured at below temperature 200°C shows very small loss which about 0.00191 but by increasing the temperature from 200°C to 600°C, the dielectric loss value fluctuate and increased at about 1. However, for $SrZrO_3$ below 300°C the dielectric loss is less than 1 but increasing after 300°C. The dielectric loss of $SrZrO_3$ ceramics seems to remain similar trend but it gives much higher of loss as compared to $SrTiO_3$.

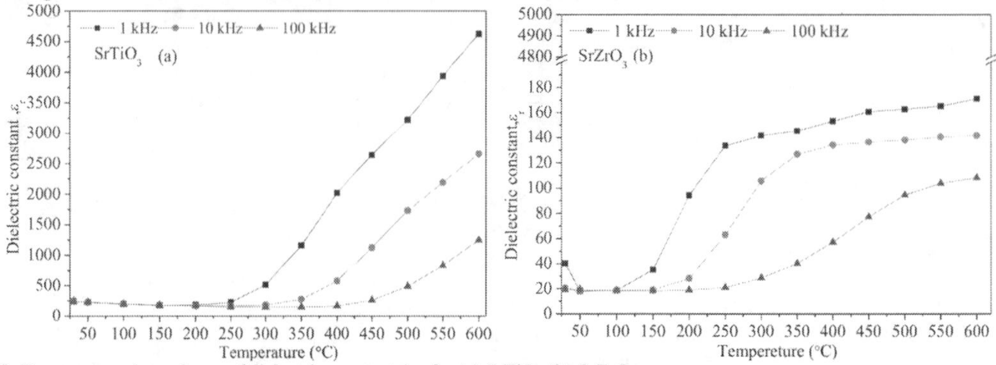

Fig. 2. Temperature dependence of dielectric constant (ε_r) for (a) $SrTiO_3$ (b) $SrZrO_3$

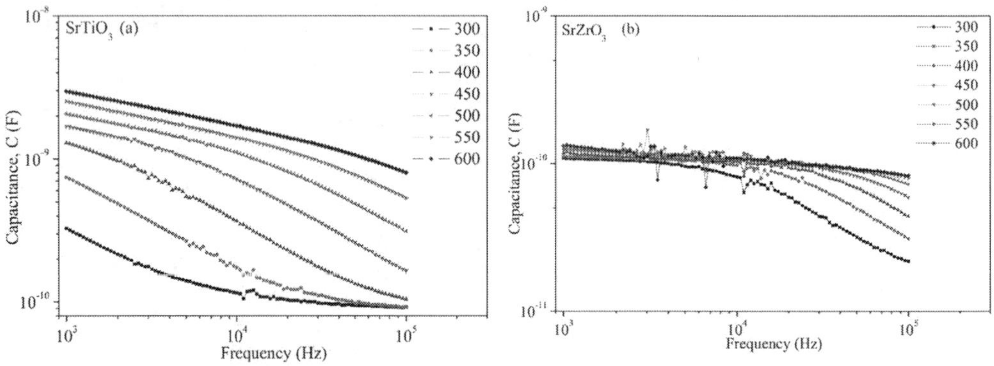

Fig. 3. Frequency dependence of capacitance for (a) $SrTiO_3$ (b) $SrZrO_3$

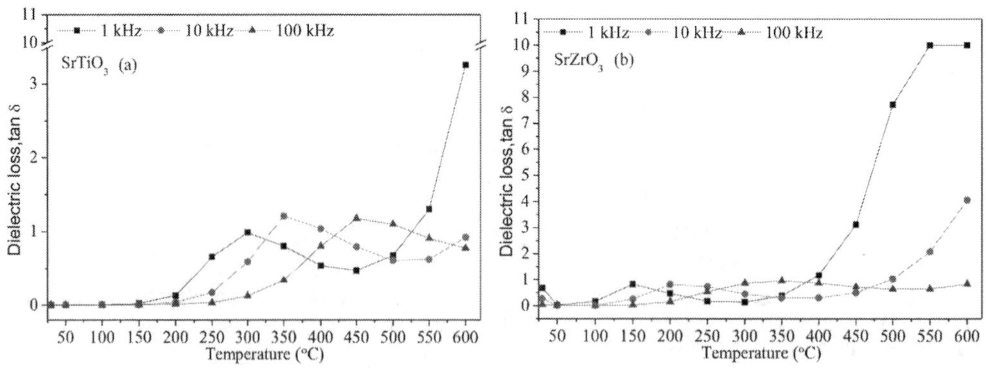

Fig. 4. Temperature dependence of Dielectric loss (tan δ) for (a) $SrTiO_3$ (b) $SrZrO_3$

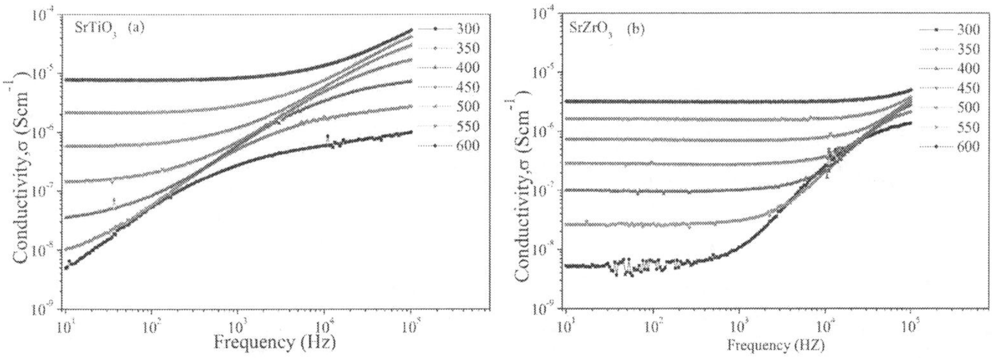

Fig. 5. Frequency dependence of conductivity (σ) for (a) $SrTiO_3$ (b) $SrZrO_3$

The frequency dependent conductivity for $SrTiO_3$ and $SrZrO_3$ in the temperature range of 300–600°C was shown in **Figure 5(a) and 5(b)**. The conductivity observed for $SrZrO_3$ slightly lower as compared to pure $SrTiO_3$. For $SrTiO_3$ it shows that the conductivity increased by increasing temperature which behave like typical dielectric ceramic. The dc conductivity was about $4.9x10^{-9}$ Scm^{-1} to $7.7x10^{-6}$ Scm^{-1} and $5.3x10^{-9}$ Scm^{-1} to $5.2x10^{-6}$ Scm^{-1} for $SrZrO_3$. $SrZrO_3$ show the conductivity behaviour at frequency 10 Hz and 10 kHz. This phenomenon are related to the conductivity of grain boundaries as reported by Liu [17]

4 Conclusions

$SrTiO_3$ and $SrZrO_3$ was phase pure after heating at 1400°C for 12h. $SrTiO_3$ exhibit cubic structure with space group pm-3m however $SrZrO_3$ exhibit orthorhombic structure with space group pnma. This is proven by the change in unit cell and volume of the samples. In term of the dielectric properties, $SrTiO_3$ have much higher value of dielectric constant which is 240 compare to $SrZrO_3$ is 21 at 10 kHz. The dielectric loss of $SrTiO_3$ is 0.00191 and $SrZrO_3$ is 0.26547. Besides, at 400°C the conductivity of $SrTiO_3$ is $7.8x10^{-6}$ Scm^{-1} and $3.2x10^{-6}$ Scm^{-1} for $SrTiO_3$. Overall the properties of $SrTiO_3$ is much better than $SrZrO_3$.

This project was financial supported by Ministry of Education, Malaysia under FRGS Grant No: 9003-00496 and 9003-00429. The author would also like to express their gratitude to University Malaysia Perlis (UniMAP) for giving opportunity to conduct this research.

References

1. M. T. Sebastian, Dielectric Materials for Wireless Communication, Elsevier, Amsterdam, Netherlands, pp. 49–77, (2008).
2. A. A. Vives, Piezoelectric Transducers and Application, 2nd edition, Springer, Berlin, pp. 509, (2008).
3. K. Uchino, Ceram. Int., 21, 309–315 (1995).
4. S. S. Park,Integrated Ferroelectrics, **74**, 87-94 (2005).
5. N. Zakaria, R. A. M. Osman, M. S. Idris, *AIP Conference Proceedings*, 1756, 090006, (2016).
6. F. W. Lyte, Appl. Phys., **35**, 7, 2212-2215 (1964).
7. R. E. A. McKnight, C. J. Howard, M. A. Carpenter, J. Phys. Condens. Matter, **21**, 015901 (2009).
8. T. A. T. Sulong, R. A. M. Osman, and M. S. Idris, AIP Conference Proceedings, 1756, 070003 (2016).
9. J. Zhao, X. Wu, L. Li, X. Li, Solid-State Electron, **48**, 2287–2291 (2004).
10. J. Xie, H. Hao, H. Liu, Z. Yao, Z. Song, L. Zhang, Q. Xu, J. Dai, M. Cao, Ceram. Int, **42**, 12796–12801 (2016).
11. P. K. Petrov, E. F. Carlsson, P. Larsson, M. Friesel, G. I. Zdravko, J. Appl. Phys. **84**, 3134–3140 (1998).
12. R. V. Shende, D. S. Krueger, G. A. Rossetti Jr, S. J. Lombardo, J. Am. Ceram. Soc., **84**,7, 1648–50 (2001).
13. S. Parida, S. K. Rout, V. Subramanian, P. K. Barhai, N. Gupta, V. R. Gupta, J. Alloy. Compd, **528**, 126–134 (2012).
14. Z. Wang, M. cao, Z. Yao, Z. song, G. Li, W.Hu, H. Hoa, H. Liu, Ceram. Int, **40**, 14127–14132 (2014).
15. T. K. Y. Wong, B. J. Kennedy, C. J. Howard, B. A. Hunter, T. Vogt, J. Solid State Chem., **156**, 255-263 (2001).
16. T. Tsurumi, T. Teranishi, S. Wada, H. Kakemoto, M. Nakada, J. Akedo, ,J. Cer. Soc. Japan, 774-781(2006).
17. J. T. S. Irvine, D. C. Sinclair, and A. R. West, , Adv. Mater., 2,3, 132-138 (1990).
18. L. N. Liu, C. C. Wang, X. H. Sun, G. J. Wang, C. M. Lei, T. Li, J. Alloy. Compd., **552**, 279–282 (2013).

DOI: 10.1051/epjconf/201716201053

Structure and electrochemical properties of Zn and Co dual-doped ($Li_2Co_{1-x}Zn_xMn_3O_8$) as cathode material for rechargeable lithium-ion batteries

Tze Qing Tan[1, 2], *Rozana Aina Maulat* Osman[1, 3], M. V. Reddy[4], *Shing Fhan* Khor[5] and *Mohd Sobri* Idris[1, 2*]

[1] Centre for Frontier Materials Research, Universiti Malaysia Perlis, 01000 Kangar, Perlis, Malaysia.
[2] School of Materials Engineering, Universiti Malaysia Perlis, Jejawi, 02600 Arau, Perlis, Malaysia.
[3] School of Microelectronics Engineering, Universiti Malaysia Perlis, Pauh Putra, 02600 Arau, Perlis, Malaysia.
[4] Advanced Batteries Lab, Department of Physics, National University of Singapore, Singapore 117542, Singapore.
[5] School of Electrical System Engineering, Universiti Malaysia Perlis, Pauh Putra, 02600 Arau, Perlis, Malaysia.

Abstract. Spinel Zn doped $Li_2CoMn_3O_8$ (or also known as $LiCo_{0.5}Mn_{1.5}O_4$) yielding formula $Li_2Co_{1-x}Zn_xMn_3O_8$ ($0 \leq x \leq 1$) were produced via conventional solid state method. XRD results and the variation of cell lattice and volume showed that the solid solution limit of these compositions was at x=0.6. Impurities were detected when the amount of Zn was beyond 60 %. The discharge capacities deteriorate as Zn content was increased. However, these Zn doped samples exhibited excellent cycle-ability (99.9% capacity retention) throughout 50 charging and discharging cycles which indicated that doping of Zn could possibly stabilised the spinel structure.

1 Introduction

The significant finding of the first Li-ion cells by Sony in the early 1990s has strongly motivated the researches for better performance active materials in order to provide energy storage for some major applications, such as portable electronic devices and other potential applications. Considering cathode materials contributed the most influence on the performance of current Li-ion batteries, a wide variety of studies on cathode materials have been done for the past few decades to fully understand the factors that influence the electrochemical performance of lithium ion batteries and at the same time reducing the cost and simplifying the fabrication process.

An alternative spinel $LiMn_2O_4$ based cathode materials have been proposed as one of the most anticipating cathode materials after the discovery of lithium intercalation-deintercalation reversibility in $LiMn_2O_4$ by Thackeray and co-workers [1]. In addition, it is low cost, thermally stable and benevolent to the environment. Spinel $LiMn_2O_4$ with space group Fd-3m comprises lithium ions on tetrahedral 8a sites and manganese ions on octahedral 16d sites while oxygen ions in 32e sites [2]. However, this spinel suffered from capacity fading which becomes a huge hindrance from feasible use as cathode materials. The capacity fading especially at elevated temperature was due to the existence of cooperative Jahn-Teller distortion when more than 50% of Mn^{3+} appeared [3]. One of the many solutions to reduce this fading is by replacing Mn partially by other elements (M) to increase the Mn valence in the spinel phase, and hence reducing the possibility of the presence of Jahn-Teller distortion. Such doping includes Ru [4], Gd [5], Cu [6], Co [7], Al [8], Mg [9,10], Zn [11], Ni [12] and others. Recently, there are studies focusing on dual substitutions to improve the structural stability and consequently its electrochemical performance [13–15]. On the basis of available literatures, it can be deduced that an interesting composition $LiM_{0.5}Mn_{1.5}O_4$ where exactly ¼ of the Mn was substituted can favour an ordering of 16d cations. Moreover, it is reported previously that substitution of Zn in $LiM_{0.5}Mn_{1.5}O_4$ resulted in a superstructure primitive cubic symmetry [11]. In addition, there are reports about Li-Zn cation inversion where partial Zn ions were located at Li ions sites [16,17].

Hence, in this study, we prepared $Li_2Co_{1-x}Zn_xMn_3O_8$ by conventional solid-state reaction and characterized its structure as a function of Zn content. Structure and morphology of the compounds were characterised by X-ray diffraction (XRD) and scanning electron microscopy (SEM). Furthermore, this study also reports the electrochemical performance of Zn doped $Li_2CoMn_3O_8$.

2 Experimental

Zn doped $Li_2CoMn_3O_8$ ($Li_2Co_{1-x}Zn_xMn_3O_8$) was produced via conventional solid state method [18] using acetate salts as starting materials. Stoichiometric amounts of lithium, cobalt, manganese, and zinc acetates were mixed and ground in acetone to aid homogenous mixing and grounding. Then, the compounds were heated at 400 °C for 12 hours with slow heating rate to

* Corresponding author: sobri@unimap.edu.my

remove acetates. After that, the compounds were heated at 800 °C for 12 hours in air. A Bruker D2 Phaser benchtop X-ray diffractometer equipped with LYNXEYE 1D-detector with Cu-K_{α} radiation was used. The diffraction patterns were recorded using step size of 0.02° and exposure time 1 second per step. Least-square refinements of XRD data were performed using GSAS/EXPGUI [19,20] software to determine the structural parameters.

Electrochemical characterisation of all the synthesised compounds were performed by using a 2016 coin cell (20 mm outside diameter and 1.6 mm thickness). The cell was composed of synthesised compound as a cathode and a metallic lithium as an anode separated by a Celgard 2502 membrane separator with a solution of 1M $LiPF_6$ in ethylene carbonate (EC) + dimethyl carbonate (DMC) + diethyl carbonate (DEC) (with mixing ratio of 1:1:1 Vol. %) as the electrolyte. The cathodes were prepared by mixing active material ($Li_2Co_{0.5}Zn_{0.5}Mn_3O_8$), super P carbon black and the binder poly(vinylidene fluoride) (PVdF: Kynar 2801) in weight percent ratio of 70:15:15 with small amount of N-methyl pyrrolidone (NMP) to form slurry. Then, the obtained slurry was coated on aluminium foil current collector and dried at 80 °C overnight. Then, the prepared electrode was kept in a vacuum oven for 24 h at 70 °C before being used in the Ar-filled glove box (MBraun, Germany) which was maintained at O_2 and H_2O level of less than 2 ppm. The cells were aged for at least 8 h before measurement. Galvanostatic charge-discharge cycling was carried out at room temperature by using a computer-controlled Bitrode multiple battery testers (Model SCN, Bitrode, USA) at a constant current density of 15 mA g^{-1} (C/10).

3 Results and discussion

Fig. 1 shows the XRD patterns of $Li_2Co_{1-x}Zn_xMn_3O_8$ ($0 \leq x \leq 1.0$) that were prepared at 800 °C in air. For the parent composition $Li_2CoMn_3O_8$ (x=0), the XRD pattern was matched with the ICDD database (No. PDF card: 01-070-4214). Furthermore, its reflections could be indexed by the cubic spinel structure with the space group of Fd-3m. For the Zn-doped $Li_2CoMn_3O_8$, the XRD patterns for $Li_2Co_{1-x}Zn_xMn_3O_8$ ($0 \leq x \leq 0.4$) were reasonably similar, except a small shoulder at the (400) reflection getting stronger as the amount of Zn was increased. However, the XRD patterns for x=0.6, 0.8 and 1.0 exhibited numerous additional reflections that could be indexed by the primitive symmetry with the space group of $P2_13$. But three reflections were unable to be indexed (denoted as *) and it was believed that they belong to $ZnMn_3O_8$.

In this case, it is speculated that for $x \geq 0.2$, the presence of relatively higher intensity of 220 reflection (2θ= ~31°) compared to the parent $Li_2CoMn_3O_8$ was probably due to Li-Zn cation inversion occurred whereby a partial of Zn ions are located at tetrahedral 8 (a) sites instead of octahedral 16 (d) sites as proposed by Ohzuku *et al.* [16] This occurrence of Li-Zn cation inversion is plausibly because of the strong tetrahedral preference of Zn^{2+}. This partial occupancy of immobile Zn ions in the Li site will probably obstruct the diffusion of Li ions.

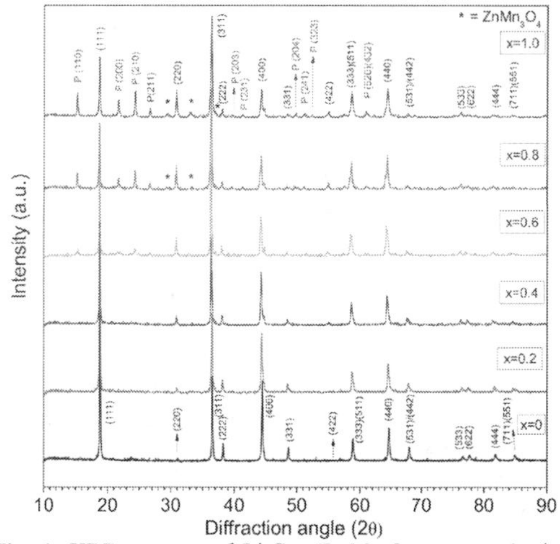

Fig. 1. XRD patterns of $Li_2Co_{1-x}Zn_xMn_3O_8$ compounds that were synthesised at 800 °C.

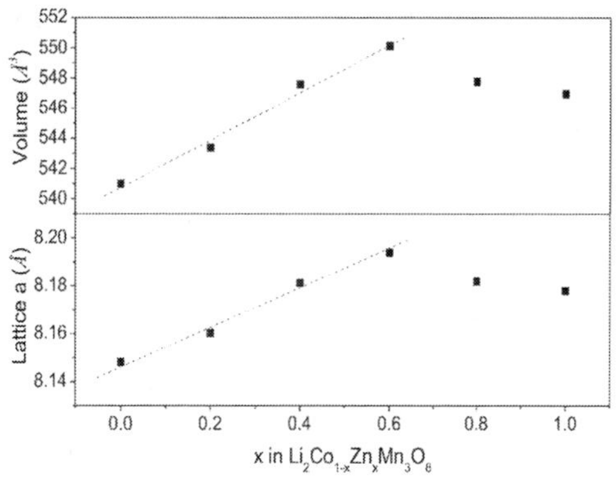

Fig. 2. Variations of structural parameters *a* and unit cell volume as a function of Zn content.

The variations of structural parameters and unit cell volume obtained by the least-square refinements of XRD data were shown in Table 1 and Fig. 2. The structural parameter and unit cell volume of the parent composition $Li_2CoMn_3O_8$ (x=0) obtained from least-square refinements were a=8.1481(1) Å and V=540.96(1) Å^3, respectively. The structural parameters and unit cell volume increased linearly within the error (3 e.s.d.s) until x=0.6, followed by slight reduced for x=0.8 and above. The changes of structural parameters obeyed the Vegard's law as the impurities were observed in the XRD patterns for x=0.8 and 1.0. Thus, the solid solution limit for the $Li_2Co_{1-x}Zn_xMn_3O_8$ was x=0.6.

The increase of structural parameters is probably due to the octahedral substitution of Zn^{2+} ions (0.74 Å) which have larger ionic radii compared to Co^{3+} (0.545 Å) ions [21]. However, the structural parameters decreased beyond x=0.6 was possibly due to the increase of Li/Zn cation inversion as mentioned previously whereby the ionic radii of Zn^{2+} ions at tetrahedral sites is 0.60 Å which is smaller than Zn^{2+} at octahedral sites.

Table 1: Structural parameters, unit cell volume and crystallite size of $Li_2Co_{1-x}Zn_xMn_3O_8$ ($0 \le x \le 1.0$).

x in $Li_2Co_xZn_{1-x}Mn_3O_8$	Space group	Lattice parameter, a (Å)	Unit cell volume (Å^3)	Crystallite size (Å)
x = 0.0	Fd-3m	8.1481(1)	540.96(1)	661.3
x = 0.2	Fd-3m	8.1602(1)	543.39(1)	604.7
x = 0.4	Fd-3m	8.1812(1)	547.59(1)	594.1
x = 0.6	P2$_1$3	8.1938(1)	550.12(2)	597.9
x = 0.8	P2$_1$3	8.1821(2)	547.76(2)	510.8
x = 1.0	P2$_1$3	8.1780(3)	546.95(3)	493.7

The average crystallite sizes of $Li_2Co_{1-x}Zn_xMn_3O_8$ were calculated by Dybye-Scherrer equation from the line width of the strongest reflection plane which refers to the (311) plane. The average crystallite size for all the samples obtained from the calculations are listed in Table 1. The parent composition $Li_2CoMn_3O_8$ (x=0) exhibited the largest crystallite size (661.3 Å) and the size decreased as Zn content increased whereby x=1 possessed the smallest size of 493.7 Å.

Microstructural images observed by SEM analysis of spinel $Li_2Co_{1-x}Zn_xMn_3O_8$ ($0 \le x \le 1$) compounds synthesised at 800 °C are shown in Fig. 3. The images show typical solid state samples with random particle size distribution. Obviously, all the samples exhibit well developed particles of regular shape. The micrographs were taken at an appropriate magnification to emphasize the morphology of the samples. It is noticeable that all the particles are in micron size (< 5 μm). From observation, the primary particle size increases as the Zn content increases.

Fig. 4 depicted the galvanostatic charge-discharge profile for 1st, 2nd, and 50th cycle of $Li_2Co_{1-x}Zn_xMn_3O_8$ compounds that were synthesised at 800 °C. The cycling tests were carried out at a current rate of 15 mA g^{-1} in the voltage window of 2.5-4.6 V vs. Li. All the cells possessed ~ 3.0 V of open circuit voltage and a flat voltage profile at ~4.0 V where the oxidation and reduction reaction occurs (except for x=0.8 and 1.0).

The discharge capacities vs cycle number plots were shown in Fig. 5. The discharging capacities deteriorated as doping of Zn increased. The initial discharging capacity of undoped $Li_2CoMn_3O_8$ is ~115.7 mAh g^{-1} which was comparable to the theoretical capacity value (~147 mAh g^{-1}). The discharge capacity faded about 11.11 % after 50 cycles. Furthermore, the initial discharging capacity decreased from ~115.7 mAh g^{-1} to ~ 68.1 mAh g^{-1}, ~ 63.7 mAh g^{-1} and ~48.4 mAh g^{-1} when 20 %, 40 % and 60 % of Co was replaced by Zn, respectively. The discharging capacity dropped to ~15.6 mAh g^{-1} and ~7.7 mAh g^{-1} when 80 % and 100 % of Zn was doped. Although the discharging capacities continue to decrease as more Co was replaced by Zn, but the discharge capacities can be retained approximately 99.9 % after 50 cycles for all the Zn doped samples. It is speculated that the capacities were affected by the immobile Zn ions in the Li site which obstructed the diffusion of Li ions. Nonetheless, the doping of Zn stabilized the structure during charging and discharging cycle.

Fig. 3. SEM images of $Li_2Co_{1-x}Zn_xMn_3O_8$ ($0 \le x \le 1$) compounds at magnification of 5000×.

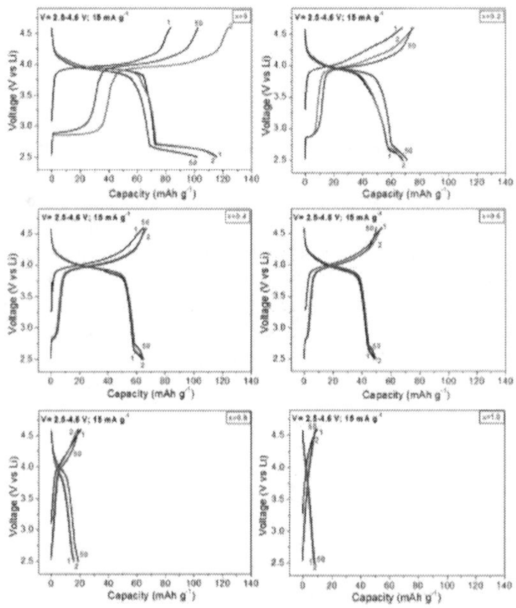

Fig. 4. The galvanostatic charge-discharge profile for 1st, 2nd, and 50th cycle of $Li_2Co_{1-x}Zn_xMn_3O_8$ ($0 \le x \le 1$) compounds that were synthesised at 800 °C.

* Corresponding author: sobri@unimap.edu.my

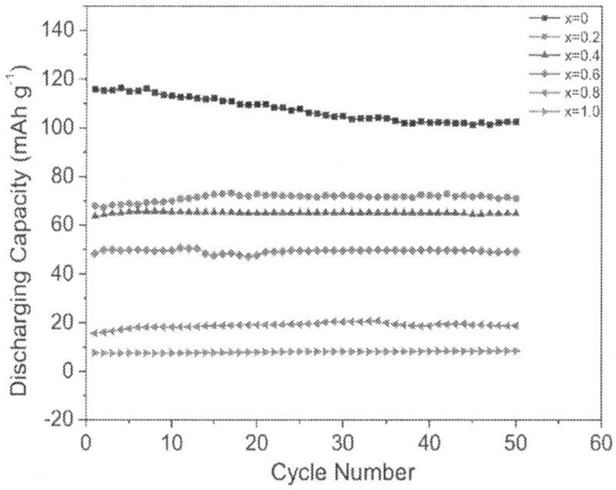

Fig. 5. The discharge capacities vs cycle number plots of $Li_2Co_{1-x}Zn_xMn_3O_8$ ($0 \le x \le 1$) compounds that were synthesised at 800 °C.

4 Conclusion

Zn doped $Li_2CoMn_3O_8$ ($Li_2Co_{1-x}Zn_xMn_3O_8$) were synthesised via simple conventional solid state route. In this systematic study, the relationship of the Zn content and electrochemical performance of Zn doped $Li_2CoMn_3O_8$ ($Li_2Co_{1-x}Zn_xMn_3O_8$) was established. Li/Zn cation inversion is believed to impede the diffusion of Li ion and hence influenced the discharge capacity. However, Zn was found to stabilise the spinel structure as these Zn-doped compositions demonstrated a stable cycle-ability in Li-ion batteries. Further investigation will be done to determine or estimate the amount of Li-Zn cation inversion.

The authors would like to acknowledge the Ministry of Higher Education Malaysia for the financial support through the Fundamental Research Grant Scheme 2014 (FRGS Grant No.: 9003-00429). We acknowledged the support of the National University of Singapore in providing the Advanced Batteries Research facilities used in this work.

References

1. M. M. Thackeray, W. I. F. David, P. G. Bruce, and J. B. Goodenough, Mater. Res. Bull. **18**, 461 (1983).
2. J. M. Tarascon, J. Electrochem. Soc. **138**, 2859 (1991).
3. G. Amatucci, A. Du Pasquier, A. Blyr, T. Zheng, and J.-M. Tarascon, Electrochim. Acta **45**, 255 (1999).
4. M. V Reddy, S. S. Manoharan, J. John, B. Singh, G. V. Subba Rao, and B. V. R. Chowdari, J. Electrochem. Soc. **156**, A652 (2009).
5. S. C. Han, S. P. Singh, Y. -h. Hwang, E. G. Bae, B. K. Park, K.-S. Sohn, and M. Pyo, J. Electrochem. Soc. **159**, A1867 (2012).
6. H.-W. Chan, J.-G. Duh, and H.-S. Sheu, J. Electrochem. Soc. **153**, A1533 (2006).
7. C. H. Shen, R. S. Liu, R. Gundakaram, J. M. Chen, S. M. Huang, J. S. Chen, and C. M. Wang, J. Power Sources **102**, 21 (2001).
8. L. Xiao, Y. Zhao, Y. Yang, Y. Cao, X. Ai, and H. Yang, Electrochim. Acta **54**, 545 (2008).
9. B. Deng, H. Nakamura, Q. Zhang, M. Yoshio, and Y. Xia, Electrochim. Acta **49**, 1823 (2004).
10. G.-M. Song, W.-J. Li, and Y. Zhou, Mater. Chem. Phys. **87**, 162 (2004).
11. Y. J. Lee, S.-H. Park, C. Eng, J. B. Parise, and C. P. Grey, Chem. Mater. **14**, 194 (2002).
12. C. A. Kim, H. J. Choi, J. H. Lee, S. Y. Yoo, J. W. Kim, J. H. Shim, and B. Kang, Electrochim. Acta **184**, 134 (2015).
13. D.-L. Fang, J.-C. Li, X. Liu, P.-F. Huang, T.-R. Xu, M.-C. Qian, and C.-H. Zheng, J. Alloys Compd. **640**, 82 (2015).
14. R. Thirunakaran, R. Ravikumar, S. Gopukumar, and A. Sivashanmugam, J. Alloys Compd. **556**, 266 (2013).
15. H. Zhang, Y. Xu, D. Liu, X. Zhang, and C. Zhao, Electrochim. Acta **125**, 225 (2014).
16. T. Ohzuku, S. Takeda, and M. Iwanaga, J. Power Sources **81–82**, 90 (1999).
17. P. Strobel, A. Ibarra-Palos, M. Anne, C. Poinsignon, and A. Crisci, Solid State Sci. **5**, 1009 (2003).
18. T. Q. Tan, M. S. Idris, R. A. M. Osman, M. V. Reddy, and B. V. R. Chowdari, Solid State Ionics **278**, 43 (2015).
19. A. C. Larson and R. B. Von Dreele, Los Alamos Natl. Lab. Rep. LAUR **748**, 86 (2004).
20. B. H. Toby, J. Appl. Crystallogr. **34**, 210 (2001).
21. R. D. Shannon, Acta Crystallogr. Sect. A **32**, 751 (1976).

EPJ Web of Conferences 162, 01054 (2017) DOI: 10.1051/epjconf/201716201054
InCAPE2017

Structural and electrical properties of $Li_2AlMn_3O_8$

Wan Masku Wan Mohamed [1,2], *Rozana Aina* Maulat Osman [1,3], *Mohd Sobri* Idris [1,2] *and Nazuhusna* Khalid [3] *and Nurul Izza* Mohd Nor[3]

[1]Centre for Frontier Materials Research, Universiti Malaysia Perlis, 01000 Kangar, Perlis, Malaysia.
[2]School of Materials Engineering, Universiti Malaysia Perlis, Jejawi, 02600 Arau, Perlis, Malaysia.
[3]School of Microelectronic Engineering, Universiti Malaysia Perlis, Pauh Putra, 02600 Arau, Perlis, Malaysia.

Abstract. The spinel $Li_2AlMn_3O_8$ was synthesized using a conventional solid-state method at 800°C for 12 h in air. The samples were analyzed using X-ray Diffraction (XRD) and Electrochemical Impedance Analyzer to study the structural and electrical properties of the samples. Well defined reflections were obtained with additional weak peaks of Al_2O_3. However, small values of χ^2 indicating a good agreement between calculated and observed pattern. A single semicircle from the complex plane plot (Z'' vs. Z'), and the capacitance value at the top point of semicircular arc was determined as a grain boundary region. The conductivity of $Li_2AlMn_3O_8$ at room temperature is about 2.99 x 10^{-5} S cm^{-1}. Meanwhile, the Arrhenius plot related from conductivity shows the activation energy also fairly similar at frequency range from 10^1 to 10^5 Hz.

1 Introduction

Lithium ion batteries (LIB) have been widely used in electronic products in the past decade. Inexpensive and environmentally benign spinel cubic structure of $LiMn_2O_4$ has Fd3m symmetry in which the Li^+ and the Mn^{3+}/Mn^{4+} are located in the 8a tetrahedral and 16d octahedral sites, respectively while O^{2-} ions was occupy the 32e sites. The 8a sites share a face with the vacant 16c octahedral sites, offers 3D conduction pathways for Li^+ ions .[1]

However, this composition has a bottleneck related to capacity fades issues on charge-discharge cycling. Widespread studies have been focused towards various strategies to improve the cycleability of $LiMn_2O_4$. Among these, one effective approach is to substitute small amount of dopant ions place of Mn ions. It is believed that, the structure of $LiMn_2O_4$ become more stable since the dopant ions are occupied in the 16d sites of the Mn ions in the spinel lattice.[2]

The substitution of aluminium as a non-transition metal in $LiMn_2O_4$ has a potential of reducing strain during repeated cycling and facilitates a higher discharge capacity retention.[3] In this paper, the studied of the new composition of spinel structure, $Li_2AlMn_3O_8$ was reported. Much attention has been done to investigate the effect of non-transition metal that has fixed oxidation state in the 16d site of the spinel structure. Result of the structural and electrical property of $Li_2AlMn_3O_8$ was reported. The structural property has been done using XRD, coupled with Rietveld refinement. Meanwhile, to meet the needs of analyzing the electrical properties of $Li_2AlMn_3O_8$, electrochemical impedance analyzer has been used. The electrical properties was discussed with a set of impedance data measured between 303K and 26K.

2 Experimental

$Li_2CoMn_3O_8$, $Li_2NiMn_3O_8$, and $Li_2AlMn_3O_8$ samples were prepared by the conventional solid-state. Reagent grade $Li(CH_3COO)\cdot 2H_2O$, $Al(NO_3)_3\cdot 9H_2O$ and $Mn(CH_3COO)_2\cdot 4H_2O$ were used as raw materials. The powder samples from conventional solid state synthesis were prepared using a stoichiometric amount of raw materials were mixed and ground by using pestle and mortar. After that, the samples were slowly heated to remove acetates at 400°C for 12 h. Then, samples were further heated at 800°C for 12 h in the air with intermittent grinding.

Bruker D2 Phaser benchtop X-ray diffractometer equipped with LYNXEYE 1D-detector with Cu-Kα radiation) was used to analyzed the purity of the final products at room temperature with a scan rate about 1.0 s/step and 0.02 of step size. The detector is the fast detector with 192 silicon strips. Rietveld refinements were carried out by using GSAS/EXPGUI software to determine the structural parameters.

For electrical measurements, a pellet was cold-pressed uniaxially at 4 tons, sintered at 950°C for 12 hours in order to increase the mechanical strength, slowly cooled to 600°C and maintained at 600°C for 12 hours before slowly cooled to room temperature. In/Ga paste electrodes were coated on opposites sides of sintered pellet. Impedance measurements were carried out over the frequency range 10 Hz-100 kHz, using

* Corresponding author: author@e-mail.org

Hioki Impedance Analyzer IM3570 at the temperature range between 303K and 268K.

3 Results and Discussion

The powder XRD pattern for $Li_2AlMn_3O_8$ that were prepared at 800 °C in air was shown in Fig. 1. The XRD pattern of $Li_2AlMn_3O_8$ shows very well defined reflections; refer to PDF Card 00-048-0261. All the peaks indexable in the Fd-3m space group with a cubic lattice. The lattice belongs to face-centered cubic, since the planes agreed with rule for reflection to be observed, which (h, k, l) either all odd or all even. However, a few extra peaks could be observed at 2θ=37.49°, 45.65° and 66.50°, indicating the presence of mixed phases. The additional peaks in these diffractogram are ascribed to Al_2O_3.

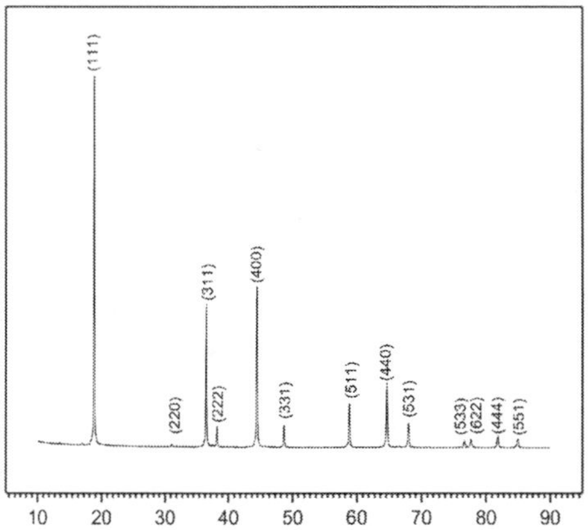

Fig. 1. XRD pattern of $Li_2AlMn_3O_8$ heated in air at 800°C for 12 h.

Replacing the transition metal with Al is not easy. It was reported that, at calcination temperatures above 600 °C, the inessential phases such as Al_2O_3, β-$LiAlO_2$ and γ-$LiAlO_2$, can easily form.[3] Manganese was expected to become $Mn^{3+/4+}$ in the composition of $Li_2AlMn_3O_8$, since Al has fixed valence states as Al^{3+}. Al^{3+} mostly occupies in octahedral site (16d) and was predicted to stabilize the spinel structure. However, the presence of additional peaks of Al_2O_3 was indicated Mn oxidize as Mn^{4+}.

The purpose of obtaining the details of crystallographic properties of sample was accomplished with structure refinement using Rietveld analysis. The refinement were performed regarding to the steps described elsewhere.[4–6] The spinel structure of $Li_2CoMn_3O_8$ was used as the structural model with a space group Fd-3m. Li ions were located at 8a-site (0.125,0.125,0.125), transition metal ions (Al and Mn) located at 16d-site (0.5,0.5,0.5) and oxygen at the 32e-site (0.2629,0.2629,0.2629). The initial structure model assumed that there is no cation disorder between Li and transition metal ions. Thus, the occupancy of Li ion at 8a-site was fixed at 100%, while transition metals at 16d-site were fixed with Al, Mn1 and Mn2 as 0.25, 0.25 and 0.5, respectively. The detail of crystallographic properties of $Li_2AlMn_3O_8$ was summarized in Table 1.

Table 1. Refined structural data for $Li_2AlMn_3O_8$ synthesized at 800°C in air for 12 h

Composition	$Li_2AlMn_3O_8$
a (Å)	8.1548(1)
Volume (Å^3)	542.31(1)
Oxygen, XYZ	0.2625(1)
8a Li occ.	1.00
16d Al occ.	0.25
16d Mn occ.	0.25
16d Mn occ.	0.50
32e O occ.	1.00
8a Li U_{iso}	0.01(1)
16d Al U_{iso}	0.009(1)
16d Mn U_{iso}	0.009(1)
16d Mn U_{iso}	0.009(1)
32e O U_{iso}	0.011(1)
R_{wp}	8.42 %
R_p	6.62 %
χ^2	2.138

The disagreements between observed and calculated patterns were represented as R_{wp}, R_p, and χ^2. Small values of χ^2 indicating a good agreement between calculated and observed pattern. The small weak peaks of Al_2O_3 were excluded and negligible during the refinement.

Fig. 2 shows a set of typical impedance data measured for $Li_2AlMn_3O_8$. An impedance complex plane, which is imaginary vs. real at 303K was plot in Fig. 2(a). The value of capacitance calculated from the relation ωRC=1 at the maximum point of semicircle is 5.88 x 10^{-8} F. Meanwhile, Z", M" spectroscopic plots from the relation of Z* and M* at 303K was shown in Fig. 2(b). The plots shows Z" and M" peak at low frequencies and high frequencies, respectively. A capacitance vs. frequency plots (shown in Fig. 2(c)) shows the frequency and temperature dependence. The value of capacity was decreased with increasing the frequency, with at three orders of magnitude while decreased with decreasing of temperature between 303K and 268K. Other than that, the conductivity of $Li_2AlMn_3O_8$ also shows clearly frequency dependence as shown in Fig. 2(d). A clear plateau was observed at the intermediate measured frequency (between 10^2 and 10^3 Hz). The plot also shows temperature dependent conductivity since the conductivity was decreased with decreasing the temperature from 303K to 268K. The conductivity at room temperature is about 2.99 x 10^{-5} S cm^{-1}, while an Arrhenius plots in Fig. 2(e) shows average of activation energy at frequency between 10 Hz and 100 kHz is about 0.3 eV.

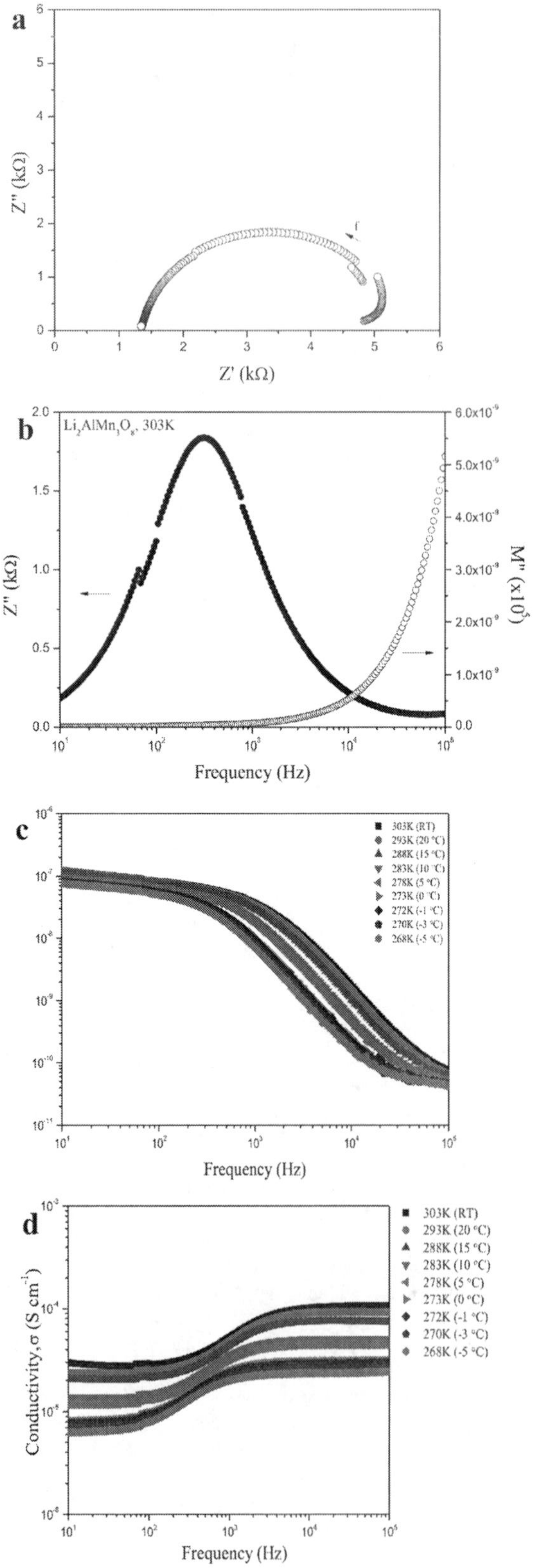

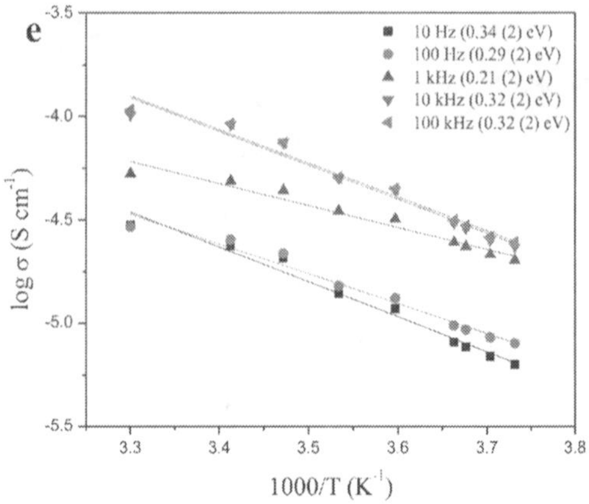

Fig. 2. Typical IS data set of $Li_2AlMn_3O_8$ (a) an impedance complex plane at 303K, (b) Z", M" spectroscopic plots at 303K, (c) capacitance spectroscopic plot, (d) conductivity spectroscopic plot and (e) Arrhenius plot.

The magnitude and direction of the complex number can be expressed by real, Z' and imaginary, Z' components through the relation;

$$Z^* = Z' - jZ'' \text{ where } j = \sqrt{-1} \qquad (1)$$

From the top point of semicircular arc (Fig. 2 (a)), the capacitance value was attributed to grain boundary regions of the sample.[7] The complex electric modulus, M*, is related to the complex impedance Z*, through the relation;

$$M^* = j\omega C_o Z^* \qquad (2)$$

Where $M' = \omega C_o Z''$ and $M'' = \omega C_o Z'$; $C_o = \varepsilon_o A/l$ is the vacuum capacitance of the empty measuring cell with electrode area, A and separation, l. The permittivity of free space, $\varepsilon_o = 8.854 \times 10^{-14}$ Fcm^{-1}.[8] Non-ideal RC element of $Li_2AlMn_3O_8$ also observed from the Z", M" vs. frequency plot (Fig. 2(b). The Z" spectra are broadened on the low frequency side of the peak maximum, while the M" spectra are broadened on the high frequency side. These two separated peak maxima contributed to the non-ideal RC element, with an additional parallel constant phase element (CPE). An accurate fit experimental data for a range of electrically homogeneous and heterogeneous materials also was provided by insertion of a CPE into equivalent circuits.[9,10] The frequency dependence capacitance of this composition also lies into grain boundary region. At low frequency, capacitance almost constant contributes more to grain boundary. Other than that, the capacitance was declined with increasing of frequency. The limiting high frequency capacitance corresponds to the bulk capacitance. However, the bulk capacitance was modified by the grain boundary capacitance.[10]

The electrical conductivity and lithium-ion diffusivity are most important properties responsible for the electrochemical performance of the LIB cathode material. The conductivity can be written as;

$$\sigma = 1/R*l/A \quad (3)$$

where l is the thickness and A the area of the electrode deposited on the sample. Low conductivity may lead to a poor rate capability because it provides a kinetic limitation and induces polarization during lithium intercalation and deintercalation. Using Eq. (3), the conductivity of $Li_2AlMn_3O_8$ was obtained. The plot shows frequency dependent conductivity of of $Li_2AlMn_3O_8$. The frequency dependence conductivity spectrum exhibits three different regions. The three different regions are; (a) low frequency dispersed, (b) an intermediate plateau and (c) conductivity dispersion at high frequencies. In the low and intermediate frequency region, conductivity is almost found to be frequency independent and equal to DC conductivity.[10] The intermediate frequency region is obtained due to the transportation of mobile ions response to applied electric field.[11] The intermediate frequency conductivity was clearly seen on $Li_2AlMn_3O_8$ (Fig. 2(d)), with existence of clear plateau between 10^2 and 10^3 Hz. In high-frequency region, the conductivity increases with frequency and become more sensitive. It is caused by the situation known as hopping frequency. It is shifted towards higher frequency with a rise in temperature. The conductivity increases due to long range movement of charge carriers and the material show the relaxor properties in high frequency. The difference of conductivity in the low-frequency range is due to the polarization belongings. The accumulation of charge occurs at very low frequencies, hence drop the conductivity. An Arrhenius law that was related with conductivity can be presented as;

$$\sigma = \sigma_o \exp[-E_a/kT] \quad (4)$$

where σ_o is a pre-exponential factor and characteristics of the materials, Ea, k, and T are activation energy for conduction, Boltzmann's constant and the absolute temperature, respectively.[12] The Arrhenius plot from Eq. (4) in shows the activation energy also fairly similar at frequency range from 10^1 to 10^5 Hz.

4 Conclusions

The composition of spinel $Li_2AlMn_3O_8$ was synthesized with the aim of investigates the structural and electrical properties. XRD studies shows that, this composition had existence of extra peaks belongs to Al_2O_3 at 2θ=37.49°, 45.65° and 66.50°. Meanwhile, the refinement data shows a good agreement between calculated and observed pattern. A single semicircle from the complex plane plot (Z" vs. Z'), and the capacitance value at the top point of semicircular arc was determined as a grain boundary region. Non-ideal RC element also observed from the Z", M" vs. frequency plot, with the presence of two separated peak maxima. The conductivity of $Li_2AlMn_3O_8$ at room temperature is about 2.99 x 10^{-5} S cm^{-1}. An Arrhenius plot related from conductivity shows the activation energy also fairly similar at frequency range from 10^1 to 10^5 Hz with the average about 0.3 eV.

This research was financially supported by the Fundamental Research Grant Scheme (FRGS) (Grant No.: 9003-00429) and MyBrain15 scholarship from Ministry of Higher Education Malaysia (MOHE).

References

1. M.M. Thackeray, W.I.F. David, P.G. Bruce, J.B. Goodenough, Mater. Res. Bull. **18** (1983)
2. B. Hwang, R. Santhanam, D. Liu, J. Power Sources. **101** (2001).
3. G.T. Fey, C. Lu, T.P. Kumar, Mater. Chem. Phys. **80** (2003).
4. M. Sobri Idris, a. R. West, J. Electrochem. Soc. **159** (2012).
5. T.Q. Tan, M.S. Idris, R.A.M. Osman, M.V. Reddy, B.V.R. Chowdari, Solid State Ionics. **278** (2015)
6. J.J. Biendicho, A.R. West, Solid State Ionics. **203** (2011).
7. J.T.S. Irvine, D.C. Sinclair, A.R. West, Adv. Mater. **2** (1990).
8. D.C. Sinclair, D.C. Bol. La Soc. Esp. Ceram. Y Vidr. **65** (1995).
9. J.J. Biendicho, A.R. West, Solid State Ionics. **226** (2012).
10. A.R. West, D.C. Sinclair, N. Hirose, J. Electroceramics. **1** (1997).
11. T. Badapanda, R.K. Harichandan, S.S. Nayak, A. Mishra, S. Anwar, Process. Appl. Ceram. **8** (2014).
12. S. Lanfredi, a. C.M. Rodrigues, J. Appl. Phys. **86** (1999).

Characterization of Zinc Oxide (ZnO) piezoelectric properties for Surface Acoustic Wave (SAW) device

Mohd Rosydi Zakaria[1,2*], *Shazlina* Johari[3], *Mohd Hafiz* Ismail[3], and *Uda* Hashim[2,3]

[1]Center of Diploma Studies, Universiti Malaysia Perlis, 02100 Padang Besar, Perlis, Malaysia.
[2]Institute of Nano Electronic Engineering, Universiti Malaysia Perlis, 01000 Kangar, Perlis, Malaysia.
[3]School of Microelectronic Engineering, Universiti Malaysia Perlis (UniMAP), Perlis, Malaysia

Abstract. In fabricating Surface Acoustic Wave (SAW) biosensors device, the substrate is one of important factors that affected to performance device. there are many types of piezoelectric substrate in the markets and the cheapest is zinc Oxide substrate. Zinc Oxide (ZnO) with its unique properties can be used as piezoelectric substrate along with SAW devices for detection of DNA in this research. In this project, ZnO thin film is deposited onto silicon oxide substrate using electron beam evaporation (E-beam) and Sol-Gel technique. Different material structure is used to compare the roughness and best piezoelectric substrate of ZnO thin film. Two different structures of ZnO target which are pellet and granular are used for e-beam deposition and one sol-gel liquid were synthesize and compared. Parameter for thickness of ZnO e-beam deposition is fixed to a 0.1kÅ for both materials structure and sol-gel was coat using spin coat technique. After the process is done, samples are annealed at temperature of 500°C for 2 hours. The structural properties of effect of post annealing using different material structure of ZnO are studied using Atomic Force Microscopic (AFM) for surface morphology and X-ray Diffraction (XRD) for phase structure.

1 Introduction

Surface Acoustic Wave (SAW) sensors have been successfully enhanced in its technology almost in every type of application including temperature, pressure, telecommunication and biochemical sensing. Relevantly, material of substrate will give an effect towards the sensing performance in biosensor field [1,2]. Today, many technology SAW has been used in Biosensor detection. SAW biosensor basically consists of three main parts which is piezoelectric substrate, a pair of metallic InterDigital Transducers (IDTs) and sensing area [1-3]. Operation of SAW biosensor normally depends on the output frequency response for characterization. It is operated by generating input IDT and output IDT will then detect the acoustic waves by employing the piezoelectric effect. Surface acoustic wave is generated from the input IDT and transducer mechanical wave into electrical signals by the output IDT [4]. Piezoelectric layer is very important in fabricating the SAW biosensor. This thin film is one of the important factors that improve the performance of the SAW device in term of sensitivity and accuracy in detection [5].

Piezoelectricity is an electrical charge which produces when certain material experience mechanical stress. Alternately, the mechanical force will be created when electricity applied on the same material. The commonly used piezoelectric substrate materials are Quartz (SiO_2), Lithium Niobate ($LiNbO_3$), Lithium Tantalate ($LiTaO_3$), and zinc oxide (ZnO) [4,5]. Each of them has different properties such as cost, piezoelectric coupling coefficients, attenuation, temperature sensitivities and propagation velocity.

However, in this study ZnO is chosen as it is a material with great advantages for piezoelectric substrate. ZnO is a n-type semiconductor material and used for several sensor devices considered its sensitivity and easy doping method to improve sensing performance [6]. It is characterized with its direct wide band gap of 3.37 ev at temperature room, high excitation binding energy of 60 mev allows it to have some specific behaviour in physical and chemical properties [1,4,6,7]. This II-IV compound semiconductor is crystallized in hexagonal wurtzite structures [6,8]. With these bunches of outstanding characteristics, ZnO is believed to an exhibition of strong piezoelectric properties [1,4,8]. ZnO is said to be high in piezoelectric coupling factor and a good bonding to varying substrate such as silicon, silicon oxide and silicon nitride [8].

Thus, in this research ZnO thin film is deposited onto silicon oxide surface to ensure good bonding and prevention of eroded during further fabrication step such as lift-off process. There are various technologies that can be used for ZnO deposition onto different substrate material. Currently, variety deposition techniques available are Pulse Laser Deposition (PLD) [2,6], Chemical Vapor Deposition (CVD), RF Sputtering [2,6,9,10], Sol-gel method [6] and Electron beam evaporation [9,10]. The substrate temperature, coating

* Corresponding author: rosydizakaria@unimap.edu.my

conditions, annealing and thin film thickness will eventually effect on the optical and physical characteristics of ZnO thin films grown [11]. In this study, ZnO thin film is evaporated by reactive electron beam technique in oxygen environment and prepared by sol-gel technique to compare the best roughness for SAW device. The analysis will be done by characterized using AFM and XRD.

2 Experimental Details

2.1 Silicon Wafer sample preparation

Pure Silicon substrate wafer (100) is first cleaned using Buffered Oxide (BOE) to clear any unwanted native oxide. Then, the wafer is checked using Film Spectrophotometer to ensure zero oxide on the surface of the wafer. Thermal oxidation process is done onto bare wafer in which silicon will react with oxygen under high temperature to form a layer of silicon dioxide. Usually the thickness of silicon dioxide depends on the oxidation rate whereby oxidation type such as dry or wet oxidation, temperature and flow rate in the furnace is considered. Wet oxidation is used in this process as it is aiming to grow high quality insulator with parameter of 1000°C for 45 minutes using oxidation furnace. For resistivity improvement, SiO_2 was growth to a range of 3000-5000μm respectively. ZnO target is then deposited onto SiO_2 by using different structures which is pellet and granular and sol gel ZnO liquid.

The deposition is set by e-beam evaporator Korean Coating Material & Component (KCMS) KES-200D with a parameter of 0.1kA for thickness deposition. An electron source in a vacuum will generate electron beams in order to irradiate any evaporate material. Then, the evaporated material will be heated and evaporated to ensure evaporated material forms a thin film on subjects samples. The chamber gauge is evacuated to a pressure of 3.5^{-5} to 4.5^{-5} Torr and the high voltage of about 7.3kV. The film growth rate of deposition is varied along the time within the range 0.01Ås^{-1} to 0.10Ås-1 with a duration of 2 hours. However the parameter is depending on the substrate temperature, oxygen pressure and filament current of the electron beam [12].

This is followed by thin film characterization using Atomic Force Microscopy (AFM) and X-ray diffraction (XRD) for surface morphology and composition studies.

2.2 Pellet source for E-Beam Evaporator

To perform an e-beam technique, ZnO pellets firstly needs to be prepared. The conventional sintering technique is used to form undoped ZnO pellets using zinc oxide powders with high purity of ≈99%. ZnO powder was first balanced for 1g on the scales with up to ±0.02g uncertainty. ZnO powder is then milled in an agate mortar by using a pestle and mortar in the Sustainable Engineering Research Center. Manual Hand Pellet Presser was used to form pellets disks with approximate pressure of 500psi. Then, the pellets were sintered at various temperature of 700°C-1000°C in ambient air for 3 hours. Lastly, the pellets is quenched to normal room temperature to freeze the structure formed. The two pellets of 1cm x 1cm diameter size sintered pellets used for one deposition of ZnO thin film using e-beam. Figure 1 defines the undoped ZnO pellet formed.

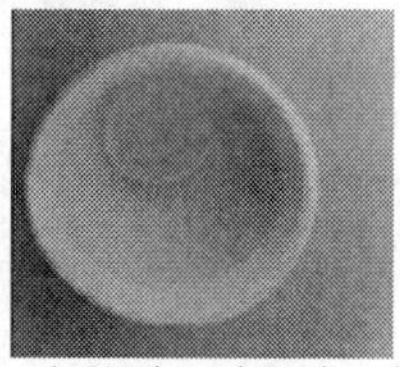

Fig. 1. Undoped ZnO pellet

2.3 Granular sources for E-Beam Evaporator

As for granular source, which is provided directly from the manufacturer, it is characterized by a small size of approximately of 1-3mm and with high purity of 99.99%. An amount of ZnO pellet is loaded into boat for E-beam evaporation. Figure 2 shows an amount of ZnO pellet granular source.

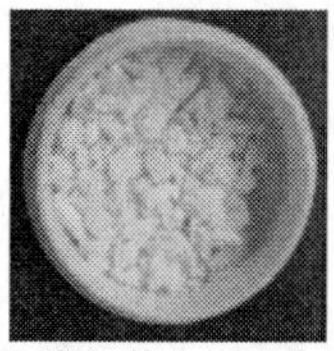

Fig. 2. ZnO pellet granular source

2.4 Sol Gel Zinc Oxide Preparation

Zinc oxide sol-gel solution was prepared before spin coat onto substrate. A 880mg of zinc acetate dihydrate (ZAD) is measured using an electronic mass scale and mix into a beaker of 20 ml of dehydrated 2-Methoxyethanol solvent. The mixture stirred at 1000 RPM on a magnetic stirrer for 20 minutes at 60 °C. After that, a total of 244 μl of diethanolamine (DEA) is added drop by drop every 10 minutes into the mixture in the beaker on the magnetic stirrer for the next 2 hours. Finally, the mixture is left for 24 hours at room temperature for aging time before the solution is used. Steps for ZnO preparation as shown in Figure 5.

The ZnO is deposited using a conventional deposition process where the ZnO solution will be dropped on the silicon oxide surface using a dropper. The substrate will be spun using spinning to get a uniformity form. The zinc oxide sol-gel is dropped and coat uniformly onto the sample using the photoresist spin coater. After that, the sample will be pre-heat on the hot plate for 10 minutes to remove the sol-gel solvent followed by cooling down on a ceramic plate. The coating process is repeated until the desire number thin film layers achieved. Next, the samples are annealed at 500 °C for 2 hours with nitrogen gas to decrease the sheet resistance of the piezoelectric layer as well as to improve the surface quality of the zinc oxide. Since more

 DOI: 10.1051/epjconf/201716201055

zinc oxide sol-gel solution is needed to coat the samples for 25 layers, 40 ml of zinc oxide sol-gel solution is prepared instead of 20 ml. The target is piezoelectric zinc oxide thin film with a thickness of 0.5 μm. Figure 3 shows steps for ZnO sol-gel technique.

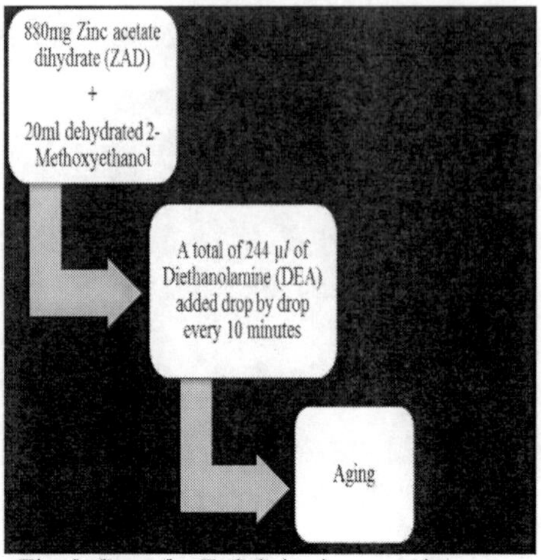

Fig. 3. Steps for ZnO Sol-grl preparation.

3 Results and Discussion

3.1 AFM Analysis

The surface morphology of deposition using different structures of ZnO is analyzed using Scanning Probe Microscope SPA400 (AFM). Figure 4(a) shows a 3D image of ZnO thin film using pellet source. Cone-like topography can be seen clearly from the certain expected of good uniformity scan area. The 3D image illustrates the light and darker color of the non-uniformity with highest peak is reach up to 100nm. Average roughness Ra of the sample is 20.39 nm and a peak to valley roughness P-V of 185.5nm shows that the surface of the sample got a more significant deviation between the distance of the highest peak and the lower valley of the grains. The small grain size of the sample is equivalent to 546nm with RMS roughness of equal to 25.71 nm. Crystalline structure and rough surface of ZnO sample is slightly better after annealing process.

Figure 4(b) defines the 3D image of ZnO thin film using granular source for deposition of e-beam. The highest peak is 150nm with average roughness of 28.74 nm. Grain size is equivalent to 1029.926 nm with P-V of 246.9 nm. Sample defines a slight roughness and crystalline structure using ZnO granular source after annealing process is far better compared to the ZnO pellet source as it is good in uniformity.

When coated for 25 times and anealing, the sample surface exhibit bigger ZnO grains based on the 3D images with 40 nm peak in Figure 4(c). Surface analysis confirms that average roughness Ra increases with the increasing number of coating in average 19.3 nm with P-V 106 nm. ZnO grain size also has been observed to increase with the increase in thickness of the film with more sol-gel coatings. The grain size has increase from average 170 nm. R. Gayen et al. [13] also reported that there an increase in grain size with an increase in zinc oxide thin film layer

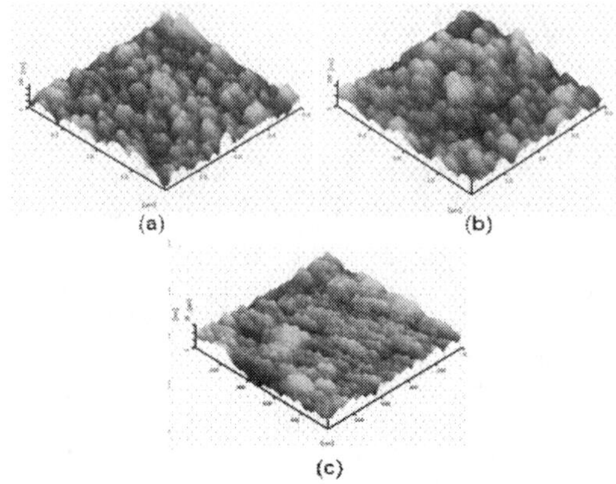

Fig. 4. The AFM 3D image of ZnO thin film a)using pellets and b) using granular source c) sol-gel

Surface analysis for all three techniques are summarized in Table 1. From results shows that the ZnO granular source has better quality compare to others samples in term of surface roughness and thickness. In reference [14], A. Zaki has claimed that, smooth surface morphology, sharp interface and perfect c-axis texture of piezoelectric film can achieve good SAW performance. Therefore, the reduction in surface roughness can enhance the performance of piezoelectric substrate.

Table 1. Comparison result of AFM for ZnO

Characteristic	ZnO Pellet Source	ZnO granular source	Sol gel (25 coatings)
Root mean square (rms) (nm)	25.71	4.55	23.29
Peak to valley height(P-V)(nm)	185.5	246.9	106
Average roughness (Ra) (nm)	20.39	28.74	19.37
Thickness (nm)	100	200	20

3.2 X-ray Diffraction (XRD) Studies

To confirm the thin film coating of the ZnO sol-gel in consistency with crystalline species of the ZnO material, the samples were tested using diffractometer. D2 Phaser desktop diffractometer from Bruker AXS were used and

DOI: 10.1051/epjconf/201716201055

operated at 30 kV and 10 mA with a Cu radiation source. 2θ/θ scan type is employed with scan angle 20° to 60°. All diffractograms were analyzed using DIFFRAC.EVA software from Bruker AXS and undergo material matching and correction processes such as smoothing and baseline correction. The characterization of XRD focuses on the range of 30° to 40° as it shows the most eligible range for ZnO and matched the reference pattern perfectly. Figure 5 illustrates a comparison of XRD result for e-Beam deposition for using ZnO pellets and granular sources after annealing process. Result granular source (line blue) shows the highest peak intensity up to 654 with at an angle of 34.43° with preferred (111) crystal axis orientation whereas the weak peak responded to an angle of 32.25° and 36.60° respectively. For preparation ZnO pellet source shows an average counts through angle 30° to 40° and sudden peak at angle 34.39° with a value of 466. Others weak peak are none clearly seen and have an average intensity along the tested of 2θ. Both of data within the range angle of 30°-40° matched perfectly referring to a PDF database of Cui Shouxin, et.al (2009) [15]. Using granular source indicates the greater intensity compared to the pellets with same parameter and condition applied. Since samples had undergone annealing process, when temperature increases thus pressure also increases. After annealing, wurtzite ZnO is compressed which reduced the height and shifted down energy [15].

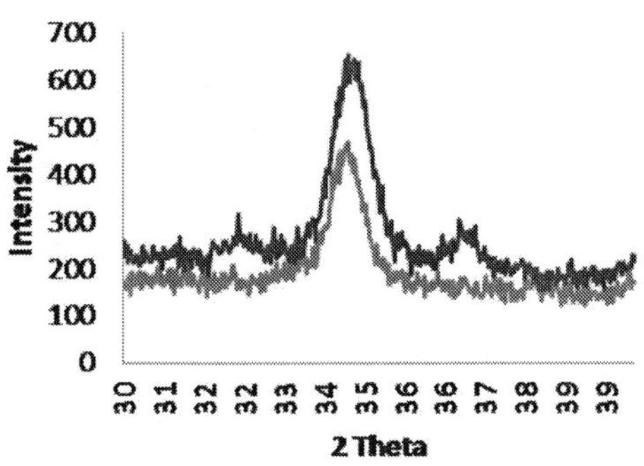

Fig. 5. Comparison of XRD using ZnO pellets and granular source

For the Sol-Gel analysis the samples exhibit 2θ peak at 31.8°, 34.3° and 36.4°, and 47.6°. However, the strongest peak intensity of the X-Ray diffraction was observed at 34.3° that related to preferred (002) c-axis ZnO crystal orientation. Other weak peak intensity corresponds to ZnO crystal orientation are for (100) at 31.8°, (101) at 36.4° and (102) at 47.6°. Therefore, the ZnO sol-gel coatings has been confirmed to produce preferable c-axis orientation for piezoelectric thin film with the presence of peaks of reflections of the ZnO wurzite structure with lattice constant a= 3.242 and c=5.176. A review on zinc oxide sol-gel done by L. Znaidi [16] also reported that a similar finding on the preferred zinc oxide thin film with (002) crystal orientation.

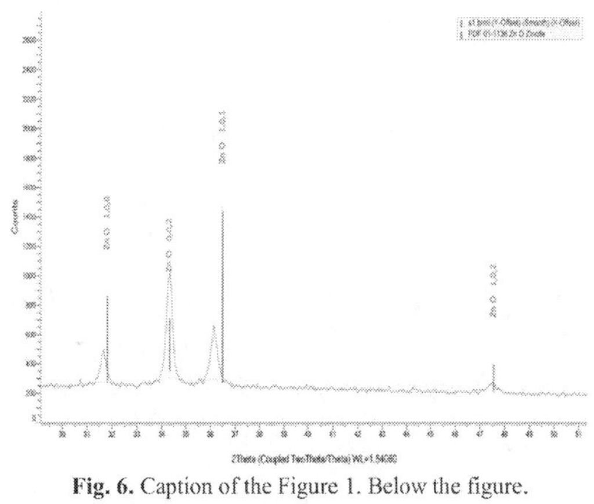

Fig. 6. Caption of the Figure 1. Below the figure.

3.3 UV-VIS Analysis

UV-Vis transmittance spectra analysis of the ZnO prepared by granular deposition and sol-gel coating are shown in Figure 7 after heat treated with annealing 500 C temperatures. The range of the wavelength is taken from 300nm to 800 nm. The ZnO thin film absorbs UV light in the wavelength from range 300-400 nm for both results. From the graph, the potential cut-off wavelength takes place from 370nm – 385nm and no other peak was observed in the spectrum confirms that the synthesized are pure ZnO and its wide-band-gap is about 3.3eV. [17,18] The band gap energy can be evaluate by substituting the value of the absorption peak at a given wavelength in the following equation [19]

$$E_g = h\nu_g = hc/\lambda_g \qquad (1)$$

Where h = 4.14×10^{-15} eVs, c = 2.99×10^8 m/s, and λ is referred in Figure 6.

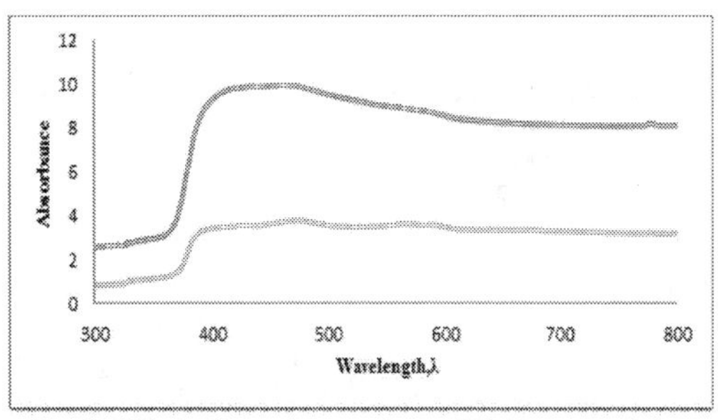

Fig. 7. UV-vis absorption spectrum of ZnO nanoparticles.

3.4 SEM Analysis

Figure 8 shows the SEM images of the granular source samples that undergo different annealing temperature. All the results taken were measured by the average applied voltage 15kV and the beam diameter around 1μm while the magnification range applied ranging from 10000X to 50000X in order to observe the difference surface morphology of the thin films. Figure shows, the mean size and also size distribution of particles are increasing with increasing annealing temperature. The important results concern the smoothest surface grain size of ZnO. The results are in agreement with the previous work reported [20], demonstrated that the annealing effect will improve the structure of ZnO form nanostructure ZnO thin film

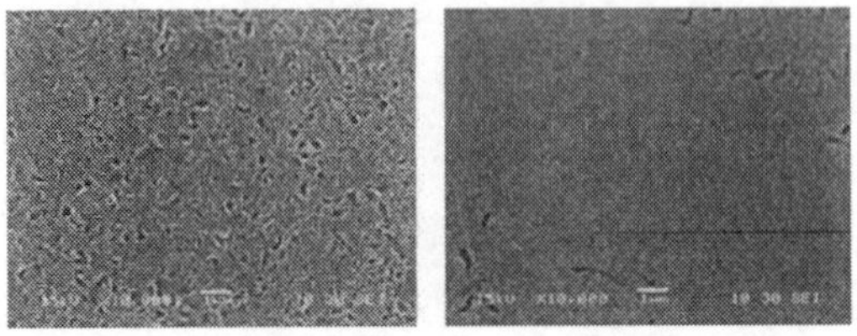

Fig. 8. SEM image of granular deposition ZnO a) before annealing b) after anealing.

4 Conclusion

Nanostructured ZnO thin films have been deposited using E-Beam and 25 layers coated by sol-gel technique. The samples were characterized by XRD, SEM, AFM and UV-VIS. The XRD analysis confirmed that the thin film was ZnO nanostructure. The thin Film produces preferable c-axis orientation for piezoelectric thin film with the presence of peaks of reflections of the ZnO website structure. From SEM analysis shows that the average particle size of ZnO increases with increasing annealing temperatures. UV-VIS analysis shows the range wavelength in range 300nm and 400nm with band gap 3.3eV. From the result, granular source shows the best layer in term of roughness and absorbance. These ZnO Thin films can be used in the SAW Biosensor application.

The authors wish to thank Universiti Malaysia Perlis (UniMAP), for giving the opportunities to do this research in the Micro Fabrication Cleanroom School of Microelectronic Engineering and Institute of Nanoelectronics Engineering (INEE). The appreciation also goes to MOSTI and we are grateful for the funding provided by FRGS 9003:00460

References

1. M.-I. Rocha-Gaso, C. March-Iborra, A. Montoya-Baides, and A. Arnau-Vives, Sensors (Basel), **9**, no. 7, pp. 5740–69 (2009)
2. K. Länge, B. E. Rapp, and M. Rapp, Anal. Bioanal. Chem., **391**, no. 5, pp. 1509–19 (2008)
3. O. Tigli and M. E. Zaghloul, *53rd IEEE Int. Midwest Symp. Circuits Syst.*, pp. 77–80 (2010)
4. C. K. Campbell, *Proceedings of the IEEE*, **77**, no. 10, pp. 1453–1484 (1989)
5. G. Salah, A. Telia, et al., *3rd International Conference on Control, Engineering & Information Technology (CEIT)* (2015)
6. N, Nayan, T. Shuhana, et al., *IEEE Regional Symposium on Micro and Nanoelectronics (RSM)* (2013)
7. M. Yusoff, Y. Yusof et.al., *IEEE Colloqium on Humanities, Science and Engineering (CHUSER)*, (2012)
8. I. Sayago, M. Aleixandre, *2t al. Spanish Conference on Electron Devices* (2005)
9. V. D. Falco, M. E. L. Sabino, et al., *33rd IEEE. Photovoltaic, Specialists Conference, PVSC.* 08. (2008)
10. J. P. Atanas, R. Al Asmar, et al. Sensor and Actuators A physical 127(1), 49-55 (2006)
11. M. R. Khanlary, and S. Isazadeh., Micro & Nano Letters, IET 6(9), 767-769
12. R. Al Asmar, R. D. Zaouk, et al., Microelectronic Engineering 83(3): 393-398 (2006)
13. R. N. Gayen, K. Sarkar, S. Hussain, R. Bhar, and A. K. Pal, Indian Journal of Pure & Applied Physics, **49**, pp. 470–477 (2011)
14. A. Zaki, H. Elsimary, and M. Zaghloul, Microelectronics J., **38**, no. 3, pp. 426–429 (2007)
15. S. Cui, W. Feng, et al. Journal of Alloys and Compounds, 306-310 (2009)
16. L. Znaidi, Mater. Sci. Eng. B, **174**, no. 1–3, pp. 18–30 (2010)
17. M. P. Santos and E. J. P. Santos, Thin Solid Films, **516**, no. 18, pp. 6210–6214 (2008)
18. Y. Q. Fu, J. K. Luo,and X. Y. Du, Sensors Actuators B Chem., **143**, no. 2, pp. 606–619 (2010)
19. B. W. Chieng, Y.y Loo, Material Letters, **73**, pp. 78-82 (2012)
20. R. H. Prasad, U. Hashim, K. L. Foo, and M. Shafiq, Advanced Materials Research, **832**, pp. 68–72 (2014)

DOI: 10.1051/epjconf/201716201056

Hydrothermal growth of titania nanowires for SAW device sensing area

Mohd Rosydi Zakaria[1,4], *Sh.Nadzirah* S. Ayub[3,4], *Mohd Hafiz* Ismail[2],*Shazlina* Johari[2], and *Uda* Hashim[2,4]

[1]Center of Diploma Studies, Universiti Malaysia Perlis, 02100 Padang Besar, Perlis, Malaysia.
[2]School of Microelectronic Engineering, Universiti Malaysia Perlis (UniMAP),02600 Arau, Perlis, Malaysia
[3]NANO-SciTech Centre, Institute. of Science University Technology (UiTM), 40450 Shah Alam, Selangor, Malaysia
[4]Institute of Nano Electronic Engineering,Universiti Malaysia Perlis (UniMAP)01000 Kangar, Perlis Malaysia.

Abstract. Synthesis of titania or titanium dioxide (TiO_2) is attracted to energy and environmental applications. Here, the growth of nanostructure TiO_2 nanowires on Si (100) substrates by using the two-step method. Different seed layers of TiO_2 were deposited by spin coating and annealing, followed by the growth of TiO_2 nanowires by using the hydrothermal method. The sol-gel technique was used in preparing the TiO_2 solution for the thin film deposition purpose. Acetic acid, hydrochloric acid and tris (2-aminoethyl) amine were used as a stabilizer to synthesize three different TiO_2 seed layers. The aim of this study was to understand the role of polycrystalline size on thin film towards the diameter of nanowires grown as a sensing area in Surface Acoustic Wave (SAW) Biosensor. The morphology and structure of the thin film and TiO_2 nanowires were characterized using X-Ray diffraction (XRD), scanning electron microscope (SEM), field emission scanning electron microscope (FESEM) and atomic force microscopy (AFM).

1 Introduction

Surface Acoustic Wave devices have become very important in sensor application such as communication, chemical gas sensing and has been used for many years [1,2,3]. Nowadays, these sensors have been applied and used for biological sensor (biosensor) application to detect the DNA and protein [3,4]. This project is important because current biosensor have limitation in the term of low sensitivity and low accuracy in biomolecular detection [5]. For this reason these types of sensor were chosen for general biomedical applications due to its advantages in high sensitivity, ease of integration into electronic circuits, and all around robustness [6,7]. SAW Bio-sensors is one alternative offer a great potential for achieving this goal. SAW Bio-sensing can be achieved by efforts in three significant components called (1) Interface/substrate (2) Transducer (IDT) and (3) Sensing area as shown in Fig.1. The sensitivity of the sensor for surface modifications is increased, by the choice of the sensing area material, and the design of the IDT as well as by the structure of the sensor and the transducers [8]. In SAW biosensor device, sensing area also receives the biggest research interest. It is not only determines the performance quality of the device but also contributes to ultimate detection limit, accuracy and sensitivity [9]. Hence, makes SAW biosensing is highly interdisciplinary in biomolecular selection. Sensing area is typically a biological complex to determine the specificity and contributes to sensitivity. Conventional SAW Biosensor commonly used gold material layer as a sensing area for providing a medium for the transducer to come into contact with the sensing material through strong binding [9,10]. However, this technique has several limitations such as may cause contamination problems and put restrictions, e.g. temperature limitation, for the subsequent processes [11]. From the studies, metal oxide materials such as titanium dioxide (TiO_2) and zinc oxide (ZnO) thin films is another alternative and have been extensively studied for various applications such as solar cells, gas sensors and protective coating [12]. Recently, more researcher focus on the preparation and characterization of nanowire on SAW sensing area, due the sensitivity in the biosensor detection [13].

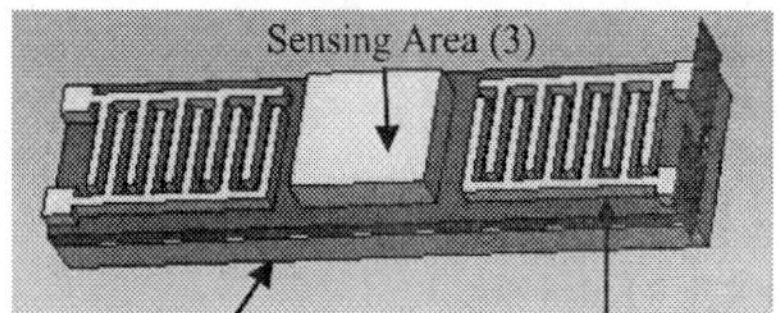

Fig.1. Surface Acoustic Wave (SAW) full structure.

TiO_2 or also known as a titanium (IV) oxide or titania is the promising nanostructure metal oxide material have excited tremendous interest due to its various functional properties. TiO_2 is an important photoactive material with existing and potential uses in a wide variety of environmental and energy applications, such as photocatalytic oxidation of contaminations [14]. TiO_2

* Corresponding author: rosydizakaria@unimap.edu.my

 DOI: 10.1051/epjconf/201716201056

also been reported as one of the most suitable semiconductor materials in a wide range of technological applications, such as photo-catalyst, because of its photo-catalytic activity through the irradiation of light at the surface [15]. Nanostructure TiO_2 with different morphologies has attracted great on a function of the shape and structure of TiO_2 nanomaterials. From the review of this metal oxide material found that that TiO_2 are suited for medical application because it is biocompatible and very stable formation [16]. Other than that, the Ti (Titanium) are chosen because it occurs naturally as an oxide meanwhile TiO_2 will used as a main material due to its good biocompability that will allow this material to be as biomolecules detection in biological systems. TiO_2 are providing a high chemical stability [17], high resistance to deteriorate in acid or alkali. This material is chosen to form the nanowires due to the advantages which are non-poisonous [18], safe, easy to make and inexpensive [18,19]. Due to these advantages the TiO_2 nanowire was chosen as a sensing material medium in the fabrication of SAW biosensor devices. To prepare the TiO_2 film, many methods can be used such as chemical vapor deposition (CVD), electron beam evaporation, ion sputtering and also the sol-gel method [20,21].

2 Experimental details

2.1 Sampel and substrate preparation

The (100) Si wafer substrates were cleaned to remove any contamination and particle on the wafer surface. Three cleaning procedures were used which is used Pirahna, RCA and Diluted HF. The piranha is a general cleaning procedure that used to clean organic residues of substrates. After cleaning process, the wafer was going through in the oxidation process. The oxidation process is an adding process, which is adding oxygen to a silicon wafer to form a silicon dioxide on the wafer surface. In this process, dry oxidation was used due to its lower growth rate and better film quality with the thickness of 195 Å.

The ZnO layer then growth on the silicon substrate as a piezoelectric material. The sample then continued with aluminum deposition for IDT formation. After fabrication of IDT was completed, the fabrication process for sensing area were taking part. Fig. 2 shows the samples of the conventional SAW device after IDT formation was completed.

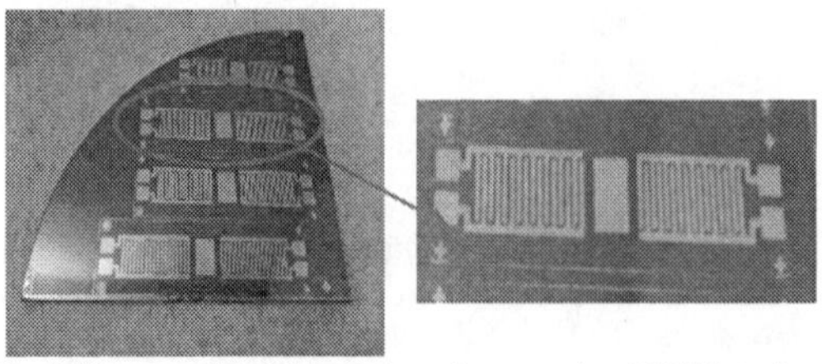

Fig.2. The completed fabrication of conventional SAW device.

2.2 TiO_2 Film Deposition

In this part, there were two main processes which are TiO_2 solution preparation and process for thin film coating. For the solution preparation, it used three different catalysts consist of acetic acid, tris (2-aminoelthyl) amine and hydrochloric acid (HCl). In this project, these three solutions were prepared by using the sol gel technique. Samples were labeled as Sample 1, Sample 2 and Sample 3 for acetic acid, HCl and tris (2-aminoelthyl) amine respectively. For the sample 1, there are three components used to produce TiO_2 thin film namely, titanium (IV) isopropoxide TIP (Ti $(C_3H_7O)_4$) as a titanium precursor, ethanol (C_2H_5OH) as a solvent and acetic acid (CH_3COOH) as a stabilizer with a fixed ratio of 1:9:0.1. Acetic acid were used to stabilize the solution and to slow down the hydrolysis process [22]. Acetic acid stabilizer was replaced by HCl and tris (2-aminoelthyl) amine for Sample 2 and 3 respectively. .

After the solution had been prepared, the next step was the coating process where the solutions was doped onto the SiO_2 substrates. A layer of TiO_2 film was spin coated on the substrates by using a spinner with certain parameters. Then, the sample was going through for annealing process certain time and temperature before being cool at room temperature. Finally, this TiO_2 thin film is characterized by using SEM for thickness and XRD for phase structure. Table 1 summarizes the process steps for all solution preparation of TiO_2 film.

Table 1. Process step for preparing TiO_2 film coating

Sample	1	2	3
Precursor	Titanium Isopropoxide TIP (Ti $(C_3H_7O)_4$)		
TiO_2 solution stabilizer	Acetic acid	Hydrochlori c acid (HCl)	Tris-2aminoethyl amine
Solvent	Ethanol (C_2H_5OH)	Ethanol (C_2H_5OH)	Isopropyl alcohol
Ratio	1:9:0.1	1:9:0.1	1:9:0.1

2.3 TiO_2 nanowire growth

For nanowires growth, titanium butoxide ($Ti[O(CH_2)_3CH_3]_4$), hydrochloric acid (HCl) and distilled water were mixed together. In a typical of hydrothermal synthesis, Solutions were mixed under magnetic-stirrer in a Teflon-lined stainless steel autoclave for 10 min. Then, 0.51mL $Ti[O(CH_2)_3CH_3]_4$ was added into the mixture as a small droplet. The solutions then were stirred continuously for another 30 min. In the hydrothermal growth process, the precoated substrates were immerged horizontally with the coated side was kept face down inside the Teflon-liner containing the growth precursors. The autoclaved that contains the substrates was sealed and be annealed in the vacuum oven at 175°C for 4h. After that, the autoclaved was cooled at the room temperature for a few hours. The substrates were finally washed with distilled water to

 DOI: 10.1051/epjconf/201716201056

remove any residual and followed by air-drying [23]. Steps were repeated for 1 and 3 h annealing process.

The TiO_2 nanowires then had been characterized by using PAN analytical X'Pert Pro X-ray diffractometer with Cu Kα radiation at λ = 1.5406 Å. The X-ray diffraction (XRD) pattern was recorded in the range of 20° to 60° operating at 45 kV and 20 mA. Morphological studies were performed using field emission scanning electron microscopy (FESEM, HITACHI SU8020) and Atomic Force Microscopy (AFM) SII Sciko Instrument INC SPI3800N Probe station. While the nanowires growth were analysed by Scanning Electron Microscopy (SEM, JEOL,JSM6460LA).

3 Results and Discussion

Fig. 3 shows the 3D AFM images of TiO_2 thin film of (a) Sample 1, (b) Sample 2 and (c) Sample 3 which were deposited on SiO_2 substrate. It is clearly observed that the grain size of the TiO_2 thin films deposited on SiO_2 substrates increases significantly from Sample 1, 2 and 3. The grains growth have spherical shaped for Sample 1 and 2 while Sample 3 has bigger and elongated grains. Thus the Root Mean Square (RMS) surface roughness were increase from 1.337, 2.09 and 2.29 nm for Sample 1, 2 and 3 respectively. The increment in surface roughness can be related to the increase in crystallite size.

(a)

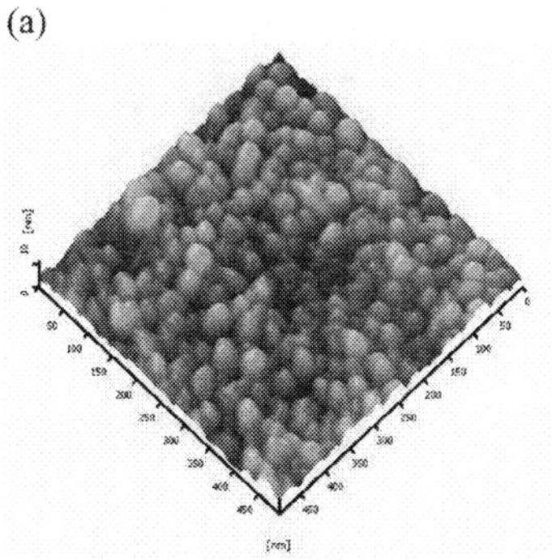

(b)

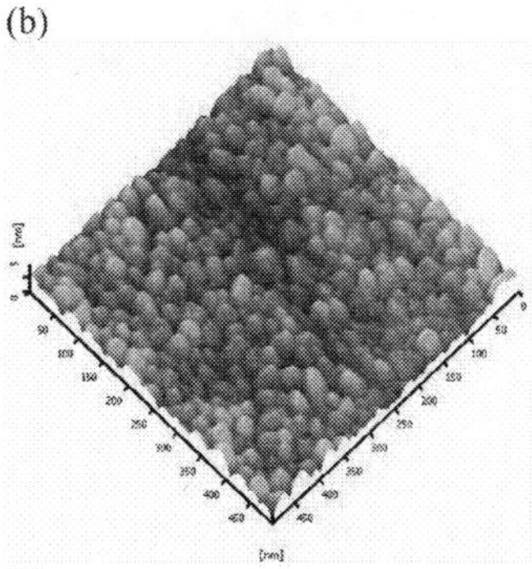

(c)

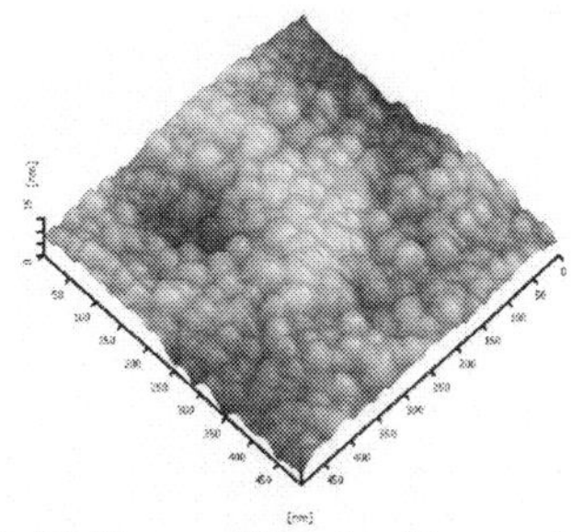

Fig. 3. 3D AFM image of TiO_2 thin film annealed at 600 °C for (a) Sample 1, (b) Sample 2 and (c) Sample 3.

Further confirmation of the morphological TiO_2 can be seen through FESEM image as shown in Fig. 4. It can be seen that nanoparticles were uniformly growth in the range of 19, 31 and 21 nm at 600 °C annealing temperature.

(a)

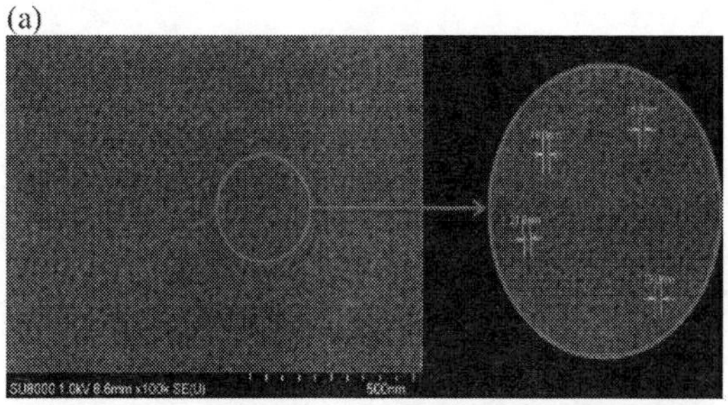

(b)

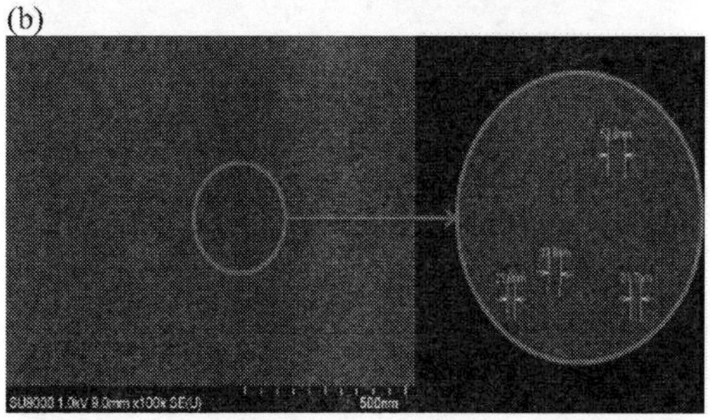

(c)

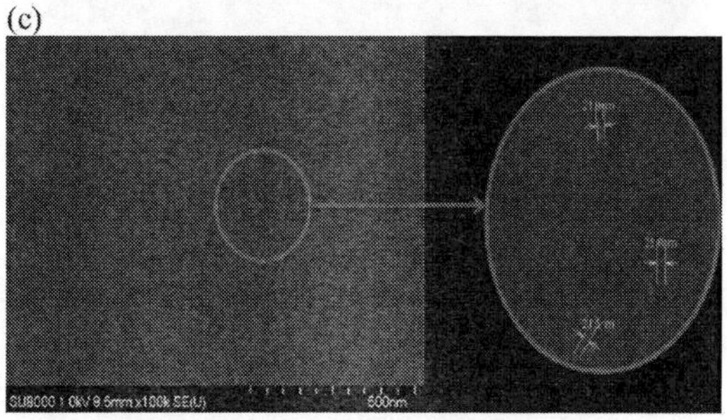

Fig. 4. 2D FESEM image of TiO_2 thin film of (a) Sample 1, (b) Sample 2 and (c) Sample 3.

The crystalline structure of the nanoparticles was studied by XRD as shown in Fig. 5. The diffraction peaks were indexed according to JCPDS File No. 21-

 DOI: 10.1051/epjconf/201716201056

1272 for anatase and JCPDS File No. 21-1276 for rutile. There are anatase (101), (004) and rutile (210) peaks occur when acetic acid was used as stabilizer namely as Sample 1. Only one anatase (101) peak occurs for Sample 2 which used hydrochloric acid as stabilizer and Sample 3 has (101), (111) and (210) rutile peaks. The average crystalline size can be calculated using the Scherer equation.

$$r = 0.9\,\lambda / B \cos\theta \qquad (1)$$

here λ is the X-ray radiation and B is the full-width at half-maximum of the diffraction peak. The calculated crystallite size values of TiO_2 films annealed at 600 °C were 21.56, 33.23 and 28.67 nm for Sample 1, Sample 2 and Sample 3 respectively. These values were interrelated with measured nanoparticles trough FESEM images. Different size of nanoparticles from multiple seed layers were expected to synthesis different shape and size of nanowires grow. Further confirmation can be seen through SEM images.

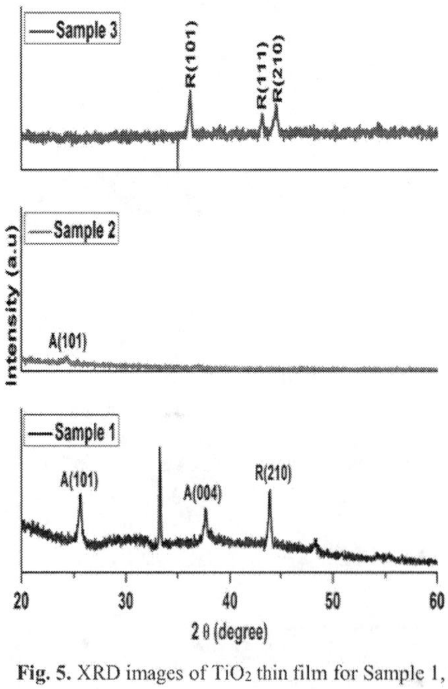

Fig. 5. XRD images of TiO_2 thin film for Sample 1, 2 and 3.

Several trends of nanowires growth on different seed layers can be seen in Fig.6-8. There were few nanowires growth on Sample 1 and Sample 3 as seed layers even though only an hour was spent; regarding to the acidic environment of HCl as stabilizer [23,24]. The length of nanowires was 2.76 +/- 0.1 and 4.6 +/- 0.2 μm for Sample 1 and 3 respectively; while no nanowires were observed on Sample 2. When the growth is carried out for longer time (3h), most nanowires appear perpendicular to the TiO_2 seed layers with square top. The average diameter and length as determined from SEM were 0.17 +/- 0.1 and 3.76 +/- 0.2; 0.3 +/- 0.1 and 5.14 +/- 0.2 also 0.23 +/- 0.1 and 6.5 +/- 0.2 μm for Sample 1, 2 and 3 respectively.

Finally, after 4 h, the diameter and length of nanowires growth on Sample 2 were 0.36 +/- 0.1 and 9.2 +/- 0.2 μm. It was slightly increased however, the pattern changes from square top to semisphere. While a white uniform film consists of nanowires begin to peel-off on the Sample 1 and 3. The film looks like a white paper which can be easily handled by a tweezer and possible to transfer on other substrate in the form of a film. This observation was almost similar to Liu et.al, who studied the TiO_2 nanorods growth [24]. According to them, a competition between crystal growth and dissolution caused the wires tends to peel-off. The uniform wires growth tends to peel-off from substrate [23]. Thus, 3 h is the most suitable time for TiO_2 nanowires to grow.

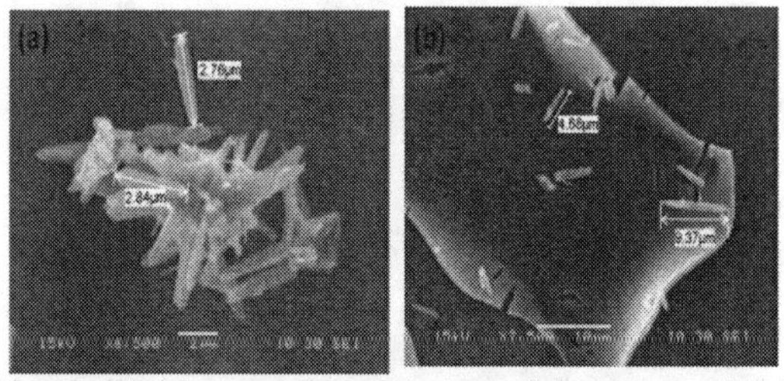

Fig. 6. SEM images of nanowires on (a) Sample 1 and (b) Sample 3 were growth for 1h.

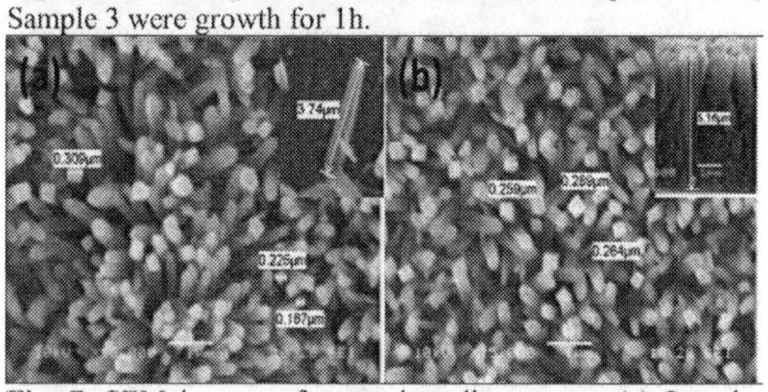

Fig. 7. SEM images of nanowires diameter on (a) Sample 1 and (b) Sample 2, were growth for 3h; the inset is the diameter.

Fig. 8. SEM images of the (a) diameter and (b) length of 4h nanowires growth on Sample 2.

Therefore, it is important to get the smallest particle size since it will be affected the diameter of nanowires grown on it.

DOI: 10.1051/epjconf/201716201056

4 Conclusion

In summary, we have successfully synthesis TiO_2 thin film and nanowires with various parameters, including the stabilizer used for thin film preparation using inexpensive sol-gel spin-coating technique, the nanowires growth time applied. From the XRD analysis, high crystalline tetragonal anatase and rutile structures of TiO_2 nanoparticles were synthesized in average diameters for acetic acid, HCl and tris(2-aminoethyl) amine stabilizers were 19, 31 and 21 nm respectively. From these seed layers, nanowires were growth hydrothermally at different time. SEM image s obviously show that nanowires were only growth at 3h with 0.17 +/- 0.1 and 3.74 +/- 0.2; 0.3 +/- 0.1 and 5.14 +/- 0.2 also 0.23 +/- 0.1 and 6.5 +/- 0.2 μm for Sample 1, 2 and 3 respectively. However, when it further growth to 4h, the nanowires tend to peel-off.

The authors wish to thank Universiti Malaysia Perlis (UniMAP), for giving the opportunities to do this research in the Micro Fabrication Cleanroom School of Microelectronic Engineering and Institute of Nanoelectronics Engineering (INEE). The appreciation also goes to MOSTI and we are grateful for the funding provided by FRGS 9003:00460

References

1. David A. Powell, et al., Encyclopedia of Sensors, 5I-M, pp. 245-263 (2006)
2. Lange, K., B. Rapp, and M. Rapp, Surface acoustic wave biosensors: a review, Anal Bioanal Chem., 391: 1509-1519 (2008)
3. Y. Hur, et al., Sensors and Actuators A: Physical. 120(2): p. 462-467 (2005)
4. Gronewold and T.M. A., Analytica Chimica Acta. 603(2): p. 119-128 (2007)
5. Cular, S., et al, IEEE Sensors Journal, 8(3): p. 314-320 (2008)
6. Chou, F.-F. and J.-S. Shih, Sensors and Actuators B: Chemical. 129(1): p. 176-183 (2008)
7. K. Lange, et al., Anal. Chem. 75: p. 5561-5566 (2003)
8. Francis, L.A., J.-M. F, and P. Bertrand, Sensors and Actuators A: Physical. 123-124(0): p. 360-369 (2005)
9. Tigli, O. and M.E. Zaghloul, in Circuits and Systems (MWSCAS) (2010)
10. T.K. Shing, H.H. Pan, I.C. Chen, C.I. Kuo., Tamkang J. Sci. and Eng. 7 (3) p. 135 (2004)
11. S. Venkatachalam, H. Hayashi, T. Ebina and H. Nanjo, Optoelectronics-Advanced Materials and Devices (2013)
12. Watson, S., et al., Journal of Nanoparticle Research. 6(2): p. 193-207 (2004)
13. Peng, W., et al., Sensors and Actuators A: Physical. 184(0): p. 34-40 (2012)
14. Sakulkhaemaruethai, S., et al., Materials Letters. 59(23): p. 2965-2968 (2005)
15. Wu, G., M. Tian, and A. Chen., Journal of Photochemistry and Photobiology A: Chemistry. 233(0): p. 65-71 (2012)
16. Srimala Sreekantan et al, IOP Conf. Ser.: Mater. Sci. Eng. 21 012002 (2011)
17. K. Hashimoto, H. IRIE, and A. Fujishima, Japanese Journal of Applied Physics, **44** (No. 12): p. pp. 8269–8285. (2005)
18. U. Diebold, Surface Science Reports,. 48(5-8): p. 53-229. (2003)
19. AN Banerjee., Nanotechnol Sci Appl. 4:35–65. (2011)
20. Sonawane, R.S., S.G. Hegde, and M.K. Dongare, Materials Chemistry and Physics. 77(3): p. 744-750. (2003)
21. Ding, Z., et al., Catalysis Today. 68(1-3): p. 173-182. (2001)
22. Fan, Q., et al., Journal of Physics D: Applied Physics, p. 2683 (2000)
23. Wu, J., et al., Journal of Materials Research. (2011) 26(13): p. 1646-1652.
24. Liu, B. and E.S. Aydil, Journal of the American Chemical Society. 131(11): p. 3985-3990. (2009)

Study of zinc oxide thin film characteristics

Shazlina Johari[1], *Nazalea Yazmin* Muhammad[1] and *Mohd Rosydi* Zakaria[2]

[1]School of Microelectronic Engineering, Universiti Malaysia Perlis, Pauh Putra Kampus, 02610 Arau, Perlis, Malaysia
[2]Institute of Nano Electronic Engineering (INEE), Universiti Malaysia Perlis (UniMAP), Perlis, Malaysia

Abstract. This paper presents the characterization of ZnO thin films with the thickness of 8nm, 30nm, and 200nm. The thin films were prepared using sol-gel method and has been deposited onto different substrate of silicon wafer, glass and quartz. The thin films were annealed at 400, 500 and 600°C. By using UV-Vis, the optical transmittance measurement were recorded by using a single beam spectrophotometer in the wavelength 250nm to 800nm. However, the transmittance in the visible range is hardly influenced by the film thickness, substrate used and annealed temperature and the averages are all above 80%. On surface morphology observed by AFM and FESEM, the results show that the increase of film thickness and annealed temperature will increase the mean grain size, surface-to-volume ration and RMS roughness. Besides that, higher annealing temperature cause the crystalline quality to gradually improve and the wurtzite structure of ZnO can be seen more clearly. Nonetheless, the substrate used had no effect on surface morphology, yet the uniformity of deposition on silicon wafer is better than glass and quartz.

1 Introduction

ZnO is considered as a versatile material with a very wide range of properties that include good transparency, high electron mobility, high thermal conductivity, wide bandgap, and strong room temperature luminescence [1-2]. Besides that, ZnO have attracted the interest of many research group due its application in light-emitting diodes (LED), laser diodes, field emission devices, solar cells gas sensors, surface acoustic wave, transparent contacts and thin film transistor (TFT) [1, 3-7, 8]. It was reported that ZnO thin films are highly transparent in the visible wavelength, also making it suitable for optoelectronic devices such as solar cells [1, 4, 6]. Moreover, it is chemically and mechanically stable, non-toxic, and high abundant. Thin film ZnO can be prepared by many methods and technique such as molecular beam epitaxy, atomic layer epitaxy, chemical vapor deposition, spray pyrolysis, sol-gel method, pulsed laser deposition, magnetron sputtering, electron beam evaporation, spin coating, thermal evaporation, electrochemical deposition, reactive evaporation [4 - 6]. Each of the technique has their own advantage and disadvantage but among all of these techniques, sol-gel method attracts much attention because it is low in cost, simple deposition equipment, easier adjustment of composition, able to carry out doping at molecular level, and easy fabricate on large-area films [5].

There are many factors that affect the performance of the optical and structural properties of ZnO thin film such as annealing temperature of thin film, the preparation method of the thin film, substrate that are used and the thickness of thin film. In terms of the annealing temperature, the thin films have been reported to be typically annealed at different temperature ranging from 200°C to 800°C while thickness of ZnO usually are varied from 40nm to 360nm. Furthermore, substrate that are commonly used with ZnO are glass, quartz and silicon. With regards to the preparation technique, ZnO thin films can be prepared using two methods, which are chemical deposition and physical deposition. For the chemical deposition, it contains spin coating process, spray pyrolysis process and dip coating process while for the physical deposition, it is divided into two parts, which are evaporation and sputter deposition

Although many works have been done to investigate the properties of ZnO thin film, it is still questionable which method and factor influence the performance of the thin film the most. These factors include ZnO solution concentration, preheating temperature, post-annealing temperature, annealing atmosphere, film thickness and so on [5]. Between all these factors, the influence of film thickness on structural, optical and electrical properties of ZnO thin films (especially undoped ZnO thin films) derived from sol–gel method was less studied. Moreover, there are still many differences between the results that have been reported. In order to obtain high-quality optoelectronic devices based on ZnO thin films, it is very significant to deeply investigate the effect of the film parameters such as film thickness.

This paper aims to investigate the structure, morphology and optical properties of ZnO using Field

* Corresponding author: shazlinajohari@unimap.edu.my

Emission Scanning Electron Microscopy (FESEM), Atomic force microscopy (AFM) and ultraviolet-visible spectrometry (UV-Vis) techniques and determine the best thin film parameter in terms of thickness, annealed temperature and substrate used that could produce optimum surface morphology and high transmission rate.

2 Methodology

Initially the substrate (quartz, glass or silicon wafer) was cleaned with chemical (acetone + ethanol + isopropanol) using ultrasonic machine for 10 minutes to remove the contaminants on the surface of the substrate. Zinc acetate dehydrate had to be dissolved in mixture of 2-methoxyethanol and mono-ethanolamine. After that, the mixture solution was stirred by using a magnetic stirrer at 60°C for 2 hours using a hot plate and stirrer. The solution is then left for 24 hours at room temperature. Solution was coated on a substrate (quartz, glass, and silicon) at 3000 rpm for 60 seconds. Then precursor thin films (sample) were heated at 150°C for 10 minutes and after that, coating and heating process were repeated for 10 times depending on the thickness of the solution (ZnO). The films were inserted into the furnace and annealed in ambient atmosphere at several temperatures (400°C, 500°C and 600°C). Samples are characterized using Atomic Force Microscope (AFM), UV-visible Spectrophotometer and Field Emission Scanning Electron Microscopy (FESEM) to observe the structural and optical properties. Then, observation was made in each different parameter in every characterization equipment and identify the differences.

Atomic Force Microscopy (AFM) is essential for the visualization and measurement of nanometre scale structures. This equipment is a combination of the principle of the scanning tunnelling microscope and the stylus profilometer. Meanwhile, ultraviolet–visible spectroscopy or ultraviolet-visible spectrophotometry (UV-Vis or UV/Vis) refers to absorption spectroscopy or reflectance spectroscopy in the ultraviolet-visible spectral region. This means, it uses light in the visible and adjacent ranges. FESEM is microscope that works with electrons (particles with a negative charge) instead of light and the electrons are liberated using a field emission source. In addition, FESEM is an advance technology of SEM because it have a better resolution and more focus beam [9].

3 Results and Discussion

First, observation was made to investigate the effect of annealing temperature which was varied from 400 °C, 500 °C and 600°C towards the thin film. The morphology and topography of the thin films were observed and compared using AFM and FESEM. Figure 1 shows the topography of the thin film with different annealing temperature. The thickness of film is set to be constant to approximately 8 nm.

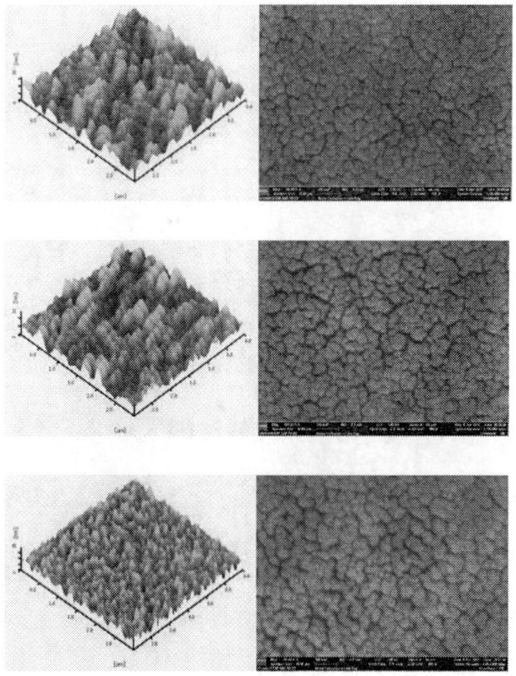

Fig. 1. AFM and FESEM images of ZnO with 30 nm thickness deposited on Silicon substrate annealed at different temperature of. (a) 400°C (b) 500°C (c) 600°C

The RMS roughness of the grain are 0.7337nm, 0.8209nm, 0.8344nm for 400°C, 500°C and 600°C respectively. Therefore, the RMS roughness increases as the annealed temperature increases. The RMS roughness increases due to increases in grain size and the crystalline quality improved as the annealed temperature increases. For all samples annealed at all the varied temperature, the peak of the grain are sharp and pointed and the grain boundaries is clear. Besides that, the uniformity is good and the surface is smooth because the RMS roughness is less than 1nm. This trend is similar and comparable with the work reported by [7], where they also confirmed that annealing temperature has direct influence over the roughness of the grain structure.

The film thickness was also varied in deposition process using a spin coater. The layer of ZnO was build up to 8nm (one layer), 30nm (three layers) and 200 nm (six layers). The differences in film thickness of ZnO was observed using AFM and FESEM to study the morphology and topography of the ZnO thin film. Figure 2 shows the morphology surface of the ZnO thin film scanned by FESEM. The magnification used to scan the thin film was 50k times and 100k times. In one layer of ZnO, the structure size of the ZnO was very small, smooth and had a good uniformity.

On the other hand, in three layers of ZnO, the size of structure was bigger than one layer of ZnO and the quality of crystalline was improved. The layer of overlapped ZnO can be seen in the three layers FESEM image because of the repetition in deposition process. Lastly, in six layers of ZnO, the overlapping of ZnO was

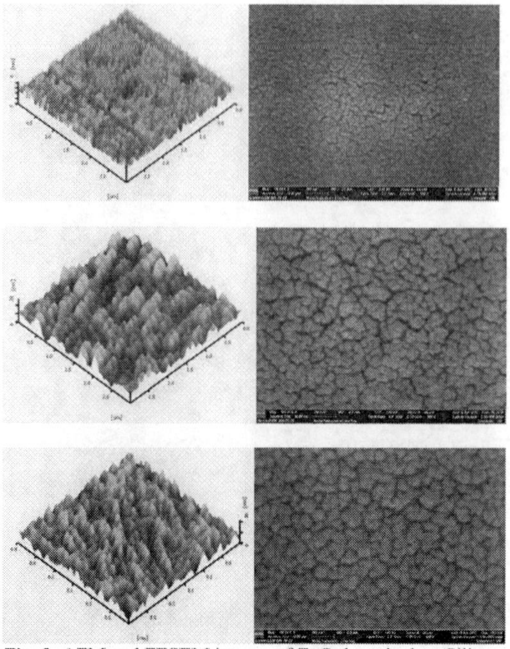

Fig. 2. AFM and FESEM images of ZnO deposited on Silicon substrate annealed at 500°C with different thickness of (a) 8nm (b) 30nm (c) 200 nm

clear and the uniformity decreases. Despite that, the quality of crystalline in six layers was greater compared to three layers. Moreover, the size of the ZnO structure also increase as the thickness increases.

We also compared the physical properties of ZnO deposited on different substrate, as shown in Figure 3 where it shows the result of 30nm ZnO annealed at 400°C with different substrate. The morphology and topography of these films were observed to differentiate it. It can be seen that the differences between the substrate were obvious. The thickness of film were 8nm, 60nm and 6nm for Silicon wafer, glass and quartz respectively. Meanwhile, surface analysis shows that silicon wafer, glass and quartz have 0.8044nm, 5.617nm and 69.68nm in RMS roughness respectively. Silicon had a smooth and uniform structure. That is why silicon RMS roughness and the thickness value was small. Moreover, the peak of the grain were sharp and pointed and the grain boundaries was clear. The growth mode is vertical growth and the grain size was small. Next, for glass substrate, the uniformity was decreased, which formed a thicker layer of ZnO than in silicon and quartz but the surface was smooth because the RMS roughness was only 5.617nm. The peak of the grain were round and the surface of the film appeared flat, which make the grain boundaries less clear. However, the growth mode of glass was still in vertical direction. Lastly, in quartz, the thickness was small but the RMS roughness were high. The surface was very rough which make it difficult for AFM to scan the entire surface clearly. In addition, quartz have an issue with uniformity but the growth mode was in lateral direction because the crystalline quality improved and at the same time, resulting in the grain to became rough.

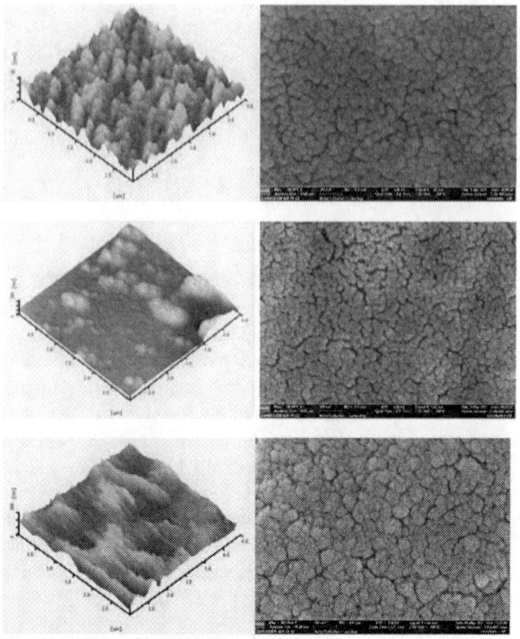

Fig. 3. AFM and FESEM images of ZnO annealed at 400°C with thickness of 30 nm on different substrates (a) Silicon (b) Glass (c) Quartz

In terms of optical properties obtained, ZnO deposited on quartz substrate achieved the highest transmittance percentage, which is 90%. Meanwhile, for ZnO deposited on glass, the transmittance was more than 80% but it did not reach 90%. Nonetheless, the curve of both graph were similar. At wavelength of 400nm, the transmittance percentage of ZnO using glass substrate increase slightly until 800nm. On the other hand, the transmittance percentage for ZnO when using quartz as substrate, increase dramatically and maintained the transmittance percentage until 700nm. Therefore, it can be concluded that, quartz was a better substrate having a high transparency in visible region.

The optical properties are investigated by varying the film thickness and the annealing temperature, as depicted in Figure 4. The high transmittance in visible region was good for the optical performance of the thin film. From the graph, in terms of annealing temperature parameter, it have proved that 400°C have the best transmittance. Then thickness was varied at 400°C to study the effect of thickness. The highest transmittance percentage was one layer of ZnO with the percentage more than 80%. Then, it followed by six and three layers. This proved that as the film thickness decreases, the increases of wavelength would increase the transmittance percentage too. Nevertheless, the result was not consistent because in some of the results, it seems like the film thickness did not affect the transmittance.

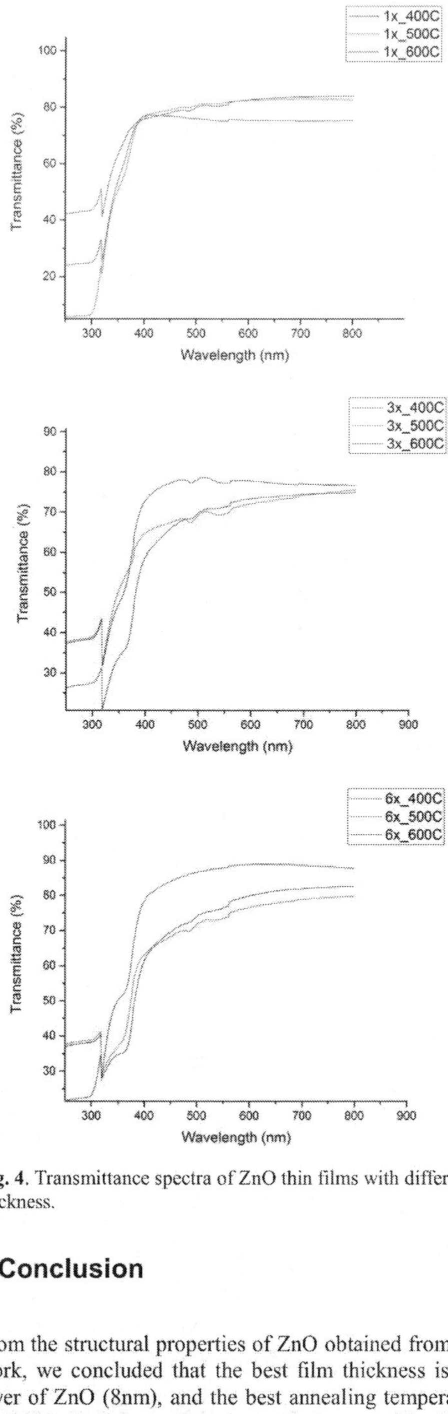

Fig. 4. Transmittance spectra of ZnO thin films with different thickness.

4 Conclusion

From the structural properties of ZnO obtained from this work, we concluded that the best film thickness is one layer of ZnO (8nm), and the best annealing temperature is 600°C. Furthermore, the best substrate used is silicon wafer. This, however, depends on the potential application of thin film. Nonetheless, for the optical properties, parameter that achieved the highest transmittance percentage is when one layer ZnO is deposited on quartz substrate and annealed at 400°C. In order to have a high quality of thin film, many criteria have to be investigated to find the best results that suits the requirement of thin film. For the future studies, the other parameters that can be tested and studied are the reliability of the ZnO thin film, by using a different type of material to dope with the ZnO. and by using different type of method to prepare the ZnO. If all the parameters have been optimized, ZnO can have a number of applications, making it a very versatile material. It has already made its way into MEMS devices and its nano characteristics and the uses are continuing to be explored. There are a lot of potential for zinc oxide, and other uses for this material will continue to be discovered and explored.

We are grateful for the funding provided by FRGS 9003:00405 and the authors would like to thank School of Microelectronic (SoME), Universiti Malaysia Perlis for their technical assistance and supports.

References

1. A. Janotti and C. G V. d. Walle, Rep. Prog. Phys., **72**, pp. 29 (2009)
2. R. Kumar, G. Kumar, O. Al-Dossary and A. Umar, Mater. Express, **5**, No 1 (2015)
3. S.Sanjeev and D.Kekuda, "*IOP Conf. Series: Materials Science and Engineering*, **73**, (2015)
4. J.Lv, W.Gong, K.Huang, J.Zhu, F.Meng, X.Song, and Z.Sun, Superlattices and Microstructures, **50**, pp. 98-106 (2011)
5. L.Xu, X.Li, Y.Chen, and F.Xu, Applied Surface Science, **257**, pp. 4031-4037 (2011)
6. Y. Aoun, B. Benhaoua, S. Benramache, and B. Gasmi, " Optik 126, pp. 5407-5411 (2015).
7. S. S. Shariffudin, N. T. C. Ibrahim, M. S. P. Sarah, and H. Hashim, *IEEE-ICSE2016 Proc.*, (2016)
8. S.M.S, IJRSET, **2**, pp. 19-24 (2015)
9. H. Yao, and K. Kimura, Modern Research and Educational Topics in Microscopy (2007)

DOI: 10.1051/epjconf/201716201058

Niobium Pentoxide thin films employ simple colloidal suspension at low preparation temperature

M. K. Abood[1*], *M. H. A.* Wahid[2], *E. T.* Salim[1*], *J. A.* Saimon[1]

[1]Laser Science Branch, University of Technology, 10066 Baghdad, Iraq
[2]Semiconductor Photonics & Integrated Lightwave Systems (SPILS), School of Microelectronic Engineering, University Malaysia Perlis, Pauh Putra Campus, Arau 02600 Malaysia

Abstract. In this work a nano-colloidal suspension is used to prepare Nb_2O_5 thin films. The effect of different substrates on structural properties of niobium pentoxide thin film deposited by spin coating technique on silicon and quartz substrates are presented. We observed that the obtained structure is monocline in both substrates. The diffraction peaks in both substrates ensured the successful formation of Nb_2O_5 thin films with a clear polymorphous structure. However, the structure became more crystalline with additional distinguished peaks on silicon substrate comparing to quartz substrate. The extracted structural parameters from X- Ray diffraction show that the grain size of the thin films on quartz is smaller than silicon with the values of 16.47 nm and 20.98 nm respectively. The stress measurement records the values of 0.19 and 0.00719 for the thin films deposited on silicon and quartz substrates respectively. Effects of film thickness depicted increment in the absorbance and reduction in the band gap. Energy gaps of 2.7, 2.58 and, 2.5 eV are measured as a result of increasing the film thicknesses of 325, 420 and 450 nm respectively.

1 Introduction

Niobium pentoxide (Nb_2O_5) is an n-type semiconductor with a broad band gap of 2.8-3.4 eV [1, 2]. It has abundant polymorphic shapes which give height to a pleasant series of structural phases [3] that rely on composition mechanisms, pressure and temperature. Thus, it enables existent various physical characteristics like a semiconductor or as a dielectric material, relying on the similarity. The phases are mostly established on NbO_6 octahedral groups, shaping different arrangements from the rectangular blocks or columns [4-9]. It is supposed that numerous of the present phases must be metastable or, fully, stabilized through impurities [5]. An appropriate incorporation of chemical, electronic and optical characteristic produces this oxide which helpful in different applications [6-9].

From the technological viewpoint, the Nb_2O_5 is significantly noted on account of its energy absorption existence in the ultraviolet region. Hence, this possibly be utilized to safeguard materials from that irradiation [10-12]. In addition, Nb_2O_5 films have been exploited greatly as optical interference filters, solar cell, gas sensors materials, electrochromic devices [13-16], and etc.

Different physical or chemical growing procedures have been advanced to present Nb_2O_5 thin films and nanostructures such as pulsed laser deposition, reactive RF magnetron sputtering, atomic layer deposition, and sol–gel [17-20].

2 Methodology

The required niobium pentoxide colloidal solution was obtained using Nb_2O_5 powder (ultra-purity, 99.99%); hydrofluoric acid and ammonia without further purification. The chemicals were prepared with the following molar ratio 1:1:4:8 of DIW, ethanol, HF, and Nh_3OH respectively, and added with 0.2 g Nb_2O_5 powder. Firstly, the hydrofluoric acid and Nb_2O_5 were mixed together with continuous heating and stirring for 1 hour until a transparent solution was obtained. After that ammonia, water and ethanol were added with constant stirring for 1 hour in water bath at 100°C. A milky like colloidal solution was observed at this stage. The chemical reactions can be simplified as follows [21, 22].

$$Nb_2O_5 + 10HF \rightarrow 2H_2[NbOF_5] + 3H_2O$$
$$2H_2[NbOF_5] + 10NH_4OH \rightarrow Nb_2O_5 \downarrow + 10NH4F + 7H_2O$$

A spin coating technique was employed to achieve homogeneous Nb_2O_5 thin films. The colloidal solution was deposited on quartz and silicon substrates at a

* Corresponding author: mkabood@gmail.com

DOI: 10.1051/epjconf/201716201058

spinning speed of 2500 rpm in 1 min. Its structural property is recorded by using x-ray diffraction technique (Shimadzu6000). Later, three different thicknesses T_1=325, T_2=420 and T_3= 450 nm were deposited on quartz substrate for the study of optical properties by uv-visible spectrophotometer (T60 UV-Vis). Morphological characterization was performed using scanning electron microscope (AA-3000).

In this work, preparation and characterization of niobium pentoxide thin films at low temperature without post heat treatment using new raw material is established. It is the first of its kind in film's fabrication in our knowledge. The effect of the substrate types on structural properties of the Nb_2O_5 deposited thin film was also investigated.

3 Results and discussion

3.1 Structural studies

The crystal structure and orientation of the Nb_2O_5 films was inspected by XRD patterns. Figure 1 shows the XRD output of the prepared sample from deposition of the chemical colloidal suspension without ammonia.

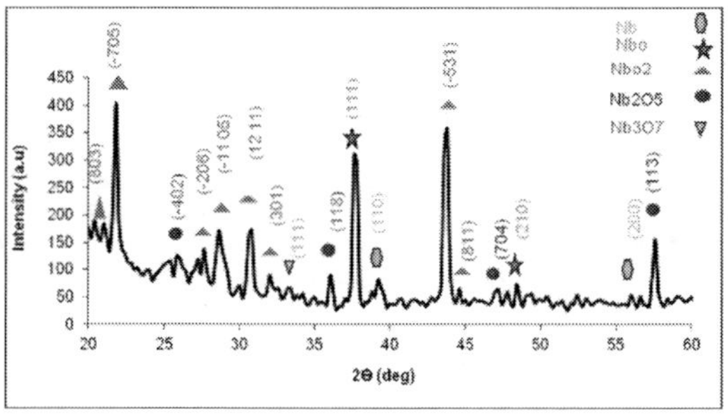

Fig. 1. X-ray diffraction spectrum for the prepared thin films before adding ammonia

The diffraction peaks indicate amorphous structure. It is also proved the existence of niobium metal in addition to different niobium oxidation degree at different phases Nb, NbO, NbO_2, Nb_2O_5 and Nb_3O_7. However, niobium monoxide NbO is the dominated one. The main diffraction peaks belong to different niobium oxide phases are located at 21.80, 28.60, 30.80, 37.80, 43.80 and 57.60 which identify the (-705), (-11 05), (12 11), (111), (-531) and (113) respectively.

When the ammonia was added the structure totally changed as shown in Figure 2. All the diffraction peaks of the Nb_2O_5 films deposited on quartz substrate were corresponded to the peaks of standard Card (00-037-1468). While the diffraction peaks of the thin film deposited on silicon substrate matched to the standard Card (00-030-0872). The diffraction peaks in both cases ensured the successful formation of Nb_2O_5 thin films with a clear polymorphous structure, but the structure became more crystalline with more distinguished peaks when deposited on silicon substrate comparing with quartz substrate.

The dominated diffraction peaks for quartz and silicon substrates were referred to (301) and (111) diffraction plane respectively, which insure the formation of monoclinic crystalline structure in both cases. The impurities (NbO) were present in the two substrates represented by (011) and (101) diffraction planes for quartz and silicon respectively. However, its intensity was monitored to be smaller in silicon substrate.

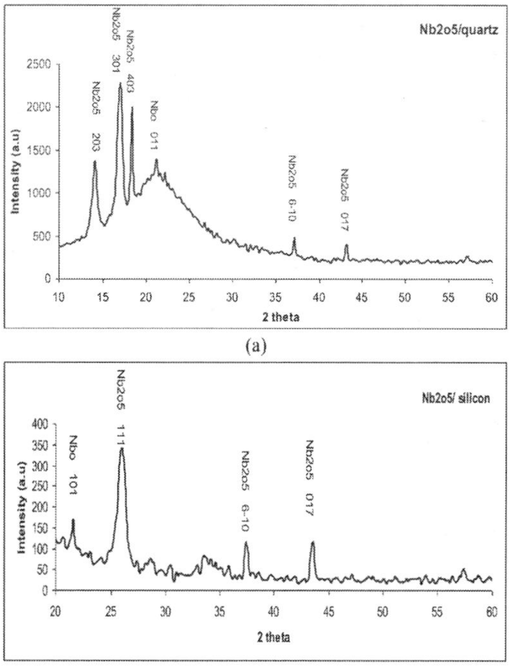

(a)

(b)

Fig. 2. X-ray diffraction spectrum for the prepared thin films deposited on (a) quartz and (b) silicon substrates

The parameters which discerned the structure of deposited thin film on silicon and quartz were characterized base on the analysis of XRD results and calculated using the values summarized in Table 1. In addition, the following equations were applied in the analyses:

$$D = \frac{K\lambda}{\beta \cos\theta} \quad (1)$$

$$\delta = \frac{1}{D^2} \quad (2)$$

$$\varepsilon = \frac{\beta}{4 \tan \theta} \quad (3)$$

where D is grain size in (nm), K is constant depended on the used X-Ray wave length, β is the full width at half maximum, Θ is the diffracted angle, δ is the density of dislocation which refers to the number of dislocations in a unit volume of a crystalline material and ε is the stress [23-30].

The structural parameters in Table 1 indicate that the thin film deposited on quartz substrate has grain size smaller than silicon substrate. Furthermore, the density of dislocations or deformations is larger in quartz substrate about 4.1×10^{-3} compare to 2.51×10^{-3} for silicon substrate. Besides, the average value of stress is smaller for quartz substrate than silicon with values of 0.00719 and 0.19 respectively. This is attributed to the existence of preferred growth orientation in the silicon substrate [31].

Table 1. Structural parameters extracted from X-ray diffraction method

	2Θ (deg)	d spacing (A°)	hkl	D (nm)	δ $*10^{-3}$ (line/m^2)	ε
silicon	21.20	4.10	101	26.42	1.43	0.21
	26.00	3.4	111	16.08	3.87	0.28
	37.40	2.39	601	19.06	2.75	0.15
	43.60	2.07	017	22.36	2.00	0.11
average				20.98	2.51	0.19
quartz	14.2	6.2	203	13.9	5.1	0.0057
	17.0	5.2	301	12.0	7.0	0.0109
	18.4	4.8	403	21.0	2.3	0.0052
	37.2	2.4	610	17.18	3.3	0.0074
	43.2	2.0	017	18.23	3.0	0.0067
average				16.47	4.1	0.00719

3.2 Optical studies

Optical properties of the deposited thin film on quartz substrate were observed using uv-visble spectrophotometer as shown in Figure 3.

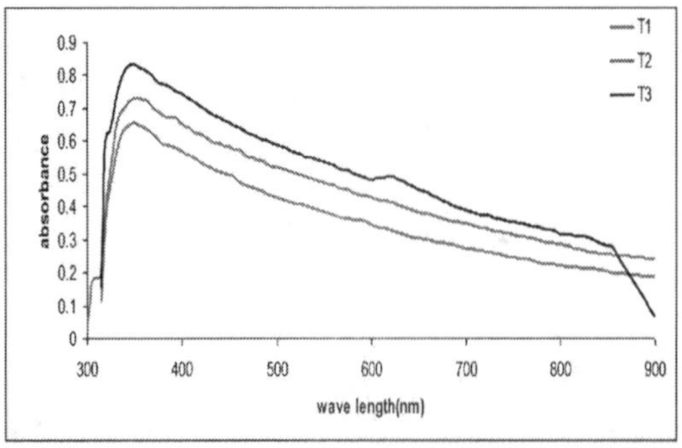

Fig. 3. Relation between the wave length and the absorbance of the prepared thin films deposited on quartz substrate with three different thicknesses

It is clearly depicted that the absorbance intensity of the deposited thin films increases with the thickness. This is due to the increment density of Nb_2O_5 particles which absorbed the photons, hence the absorbance increases. It was similarly recorded in Ref. [33] that gave good agreement with our observation.

Energy gap estimation using absorption spectrum of the deposited thin films with different thicknesses is shown in Figure 4. It indicates that the energy gap decreases with thickness from 2.7, 2.58 and 2.5 eV to T_1, T_2 and T_3 respectively. This is due to the improvement in the crystals, changes in the morphology of the deposited films, atomic distances variation and the existence of structural defects in the films [32].

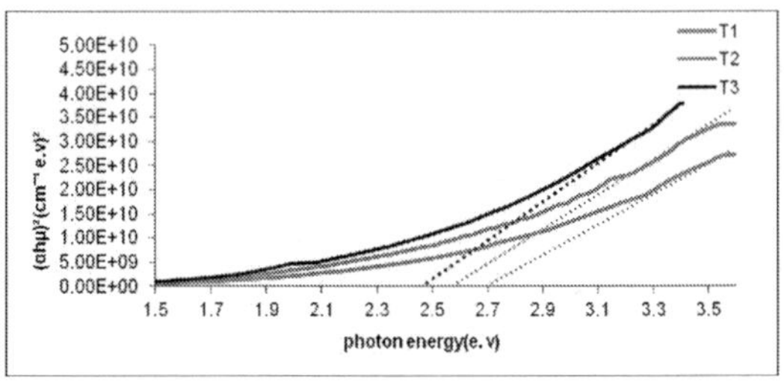

Fig. 4. Energy gap estimation for Nb_2O_5 thin films deposited on quartz substrate with three different thicknesses

3.3 Morphological studies

The morphological properties for Nb_2O_5 thin film deposited on quartz substrate are shown in SEM micrograph in Figure 5. The formation of microstructure with an earth rock like structure along the films surface is visibly noted. Such structure is a normal characteristic of films derived from sol-gel method [34].

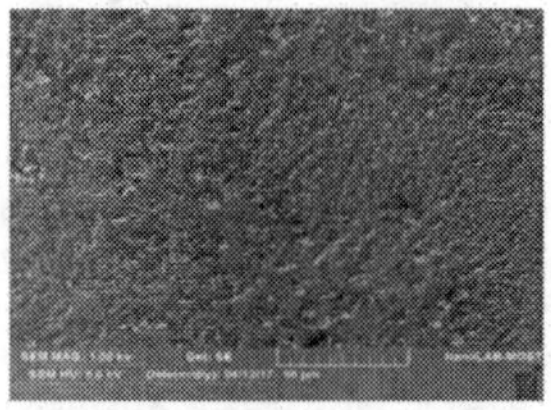

(a)

(b)

Fig. 5. Scanning electron microscope micrograph for thin film deposited on quartz substrate (a) 50 um scale and (b) 10 um scale

Conclusion

Nb_2O_5 thin films were successfully prepared using niobic acid based colloidal solution as a raw material at room temperature. The films were fabricated on quartz and silicon substrates using spin coating method. Structural parameters extracted from X- Ray diffraction showed many niobium oxide phases resulting from the complete dissolved of the raw material in the chemical reaction. A clear formation of Nb_2O_5 thin films was observed at a certain amount of ammonia. The thin films deposited on quartz have smaller grain size and stress but they have higher density of dislocation. Optical properties revealed an increase in the absorbance and a decrease in the band gap as a result of film thickness increment.

References

1. H. Luo, W. Song, P. G. Hoertz, K. Hanson, R. Ghosh, S. Rangan, M. K. Brennaman, J. J. Concepcion, R. A. Binstead, R. A. Bartynski, R. Lopez, T. J. Meyer Chem. Mater. **25**, 122, (2013).
2. Y. Zhou, Z. Qiu, M. Lu, A. Zhang, and Q. Ma Journal of Luminescence **128**, 1369, (2008).
3. G. C. Vezzoli, Phys. Rev. B **26**, 3954, (1982).
4. H. Sch¨afer, R. Gruehn and F. Schulte, Angew. Chem., Int. Ed. Engl. **5**, 40–52, (1966).
5. S. Tamura, J. Mater. Sci. 7, 298–302, (1972).
6. J. Z. Ou, R. A. Rani, M. H. Ham, M. R. Field, Y. Zhang, H. Zheng, P. Reece, S. Zhuiykov, S. Sriram, M. Bhaskaran, et al., ACS Nano **5**, 4045–4053, (2012).
7. X. Fang, L. Hu, K. Huo, B. Gao, L. Zhao, M. Liao, P. K. Chu, Y. Bando, D. Golberg, Adv. Funct. Mater. **21**, 3907–3915, (2011).
8. S. Qi, R. Zuo, Y. Liu, Y. Wang, Mater. Res. Bull. **48**, 1213–1217, (2013).
9. C. Nico, T. Monteiro, M. P. F. Gracca, Prog. Mater. Sci. **80**, 1–37, (2016).
10. A. G. S. Prado, L. B. Bolzon, C. P. Pedroso, A. O. Moura, L. L. Costa Appl. Catal., B **82**, 219, (2008).
11. A. M. Ferrari-Lima, R. G. Marques, M. L. Gimenes, N. R. C. Fernandes-Machado Catal. Today **254**, 119, (2015).
12. H. A. J. L. Mourão, V. R. Mendonça, A. R. Malagutti, C. Ribeiro Quim. Nova **32**, 2181, (2009).
13. I. Sieber, H. Hilderbrand, A. Friedrich, P. Schmuki, Electrochem. Commun. **7**, 97, (2005).
14. P. Viswanathamurthi, N. Bhattarai, H.Y. Kim, D.R. Lee, S.R. Kim,M.A. Morris, Chem. Phys. Lett. **374**, 79, (2003).
15. R. Ghosh, M.K. Brennaman, T. Uher, M.R. Ok, E.T. Samulski, L.E. McNeil et al., ACS Appl Mater Interfaces **3**, 3929-3935, (2011).
16. M. Schmitt, S. Heusing, M.A. Aegerter, A. Pawlicka, C. Avellaneda, Sol. Energy Mater. Sol. Cells **54**, 9, (1998).
17. Z. Weibin, W. Weidong, W. Xueming, C. Xinlu, Y. Dawei, S. Changle, P. Liping, W. Yuying and B. Li Surf. Interface Anal. **45**, 1206–1210, (2013).
18. O. D. Coskun and S. Demirela Appl. Surf. Sci. **277**, 35–39, (2013).
19. T. Blanquart, J. Niinisto, M. Heikkila, T. Sajavaara, K. Kukli, E. Puukilainen, C. Xu, W. Hunks, M. Ritala and M. Leskela Chem. Mater. **24**, 975–980, (2012).
20. M. R. Joya, J. J. B. Ortega, A. M. R. Paez, J. G. da S. Filho and P. de T. C. Freire, Metals **7**, 142, (2017).
21. O S Ayanda and F A Adekola, Journal of Minerals & Materials Characterization & Engineering **10**, 245, (2011).
22. N Izabela and Z Maria Chemical Reviews **99**, 3603, (1999).
23. E.T. Salim, M.A. Fakhri, H. Hasan, Int. J. Nanoelectronics and Materials **6**, 121-128, (2013).
24. Zaid T Salim, U Hashim, MK Md Arshad, Makram A Fakhri, Materials Research Bulletin **86**, 215-219, (2017).
25. M. A. Fakhri, U. Hashim, E. T. Salim, Z. T. Salim, Journal of Materials Science: Materials in Electronics **27** (12), 13105–13112, (2016).
26. E. T Salim, M. S Al Wazny, M.A Fakhry, Modern Physics Letters B, **27**, (2013).
27. Fakhri, M.A., et al., Applied Physics B, **121**(1), p. 107-116, (2015).
28. E.T.Salim, Indian Journal of Physics **87**, 349, (2013).
29. M. A. Muhsien, E. T. Salem, I. R. Agool, H. H. Hamdan, Applied Nanoscience **4** (6), 719–732, (2014).
30. A. Kadhim, E. T. Salim, S. M. Fayadh,A. A. Al-Amiery, A. Amir H. Kadhum, and A. B. Mohamad, The Scientific World Journal, (2014).
31. Fan, L. and R. Muller, Solid-State Sensor and Actuator Workshop, Technical Digest, (1988).
32. M. A. Fakhri, Y. Al-Douri, U. Hashim, E. T. Salim, Advanced Materials Research **1133**, 457-461, (2016).
33. Akaltun, Y., et al., Optics Communications, **284(9)**, p. 2307-2311, (2011).
34. W-K Kim and et al. Optics Express **17**, 2638, (2009).

 DOI: 10.1051/epjconf/201716201059

Effects of geometrical specification of microengineered surfaces to Casimir-Polder potential

A.F Khaidir[1*], *S.N.* Sabki[2], and *M.N.A.* Halif[1]

[1]Institute of Engineering Mathematics, Universiti Malaysia Perlis (UniMAP), Pauh Putra Campus, 02600, Arau, Perlis
[2]School of Microelectronic Engineering, Universiti Malaysia Perlis (UniMAP), Pauh Putra Campus, 02600, Arau, Perlis

Abstract. Cold atoms can be trapped and guided using nanofabricated wires on microengineered surfaces, achieving the scales required by quantum information proposals on the consequences of Casimir-Polder (CP) force to its stability and lifetime. The atoms are found to be less attractive to the thin films compared to dielectric half plate, where the attractive potential drops even further when the thickness of the surface is lower than both thermal wavelength and the spacing length. Amiably, for the spherical surface, the CP potential decays even further as radial thickness decreases. We approximate the CP potential using the analytical expression obtained from the pairwise summation (PWS) calculated by taking a sum of single atom with the jumbled atoms inside the surface, treating them as two-bodies system as a whole. The means of PWS are to calculate the possibilities of geometrical effect: planar and curved surface on the CP potential.

1 Introduction

The implementation of such nanofabricated wires (atom chips) implanted in microengineered surface over a coherent matter waves of cold atoms for the purpose of trapping and manipulating near the micron distance away from the surface has been underway generating interest to facilitate study of the attractive power of law's effect over the stability and the lifetime of Bose-Einstein Condensate (BEC) atoms [1-3]. The transition from various laser cooling techniques, spatially-varying magnetic trap, evaporative cooling to the placement of microelectronic chip near the 'atomic conveyer belt' or atom chips has seen the new challenges to nanotechnological developments during the trapping process of BEC atoms [4]. Apparently, the precise control of the delivery of atoms during the microtrap process is also necessary to minimize further inevitable noise and decoherence leading into improvisation of atom chips in recent years includes an integrated coherent matter wave circuits and a two-photon transition integrated atom chips for deterministic positioning of the atom cloud [5]. The atom chip has been used recently to probe quantum zero dynamics [6] and to prepare any arbitrary family of internal states of rubidium (^{87}Rb) BEC atoms yielded in atom-chip-based microtrap in order to increase the lifetime of coherence phase-space densities [7].

Nonetheless, the geometry specification of designated atom chips also needs to be taken in consideration. This is because the geometrical surface has shown greater possibilities in affecting the attractive potential between the two closed dielectric surfaces, none other than the Casimir-Polder (CP) potential which is studied over the years [8, 9]. CP potential derives from the Casimir's effect's between the two-body systems [10] because of the occurring quantum electromagnetic dipole fluctuations [11] that involves the evanescence waves from the reproduction and annihilation of virtual photons surrounding it [12]. The simulation of this CP potential is shown intuitively on the reflective patterns of atom clouds notably the dynamical excitations and formation of vortex rings when they loaded closer to the surface, probing the possibilities of trapping the micron-atom-surface-distance through quantum reflection [13,14]. Under a very tight confining magnetic field, the CP potential overwhelms the evaporative cooling process causing trapping instabilities as many coherent BEC atoms escape during the process [15]. Therefore, coherency cannot be maintained for obtaining pure condensate of cold atoms trapped near microengineered surface of atom chips. Tuning the CP potential as good as maintaining coherency serves as a part of problem-solving to the main goal of this research, this is by specifying the type of geometrical surface that may weaken the CP potential.

2 A pairwise summation (PWS) method to estimate Casimir-Polder (CP) potential

Multiple attempts are done to compute the CP potential either numerically or by analytical method. The computation of CP potential using conventional exact scattering method is daunted challenging as the $\mathfrak{R}$-operators of the involved bodies need to be assessed first, and there is no known method to find the $\mathfrak{R}$-operator for material surfaces of arbitrary shape [16]. Therefore, any approximation tool like the one we introduced here, a pairwise summation method (PWS) [17] is preferable due to its functionality to choose different geometrical surface at any best without going through the trouble like finding the complex matrix operators. The PWS tool is noted for underestimating the lateral atom-surfaces forces as it covers most of the

* Corresponding author: dauswq@gmail.com

geometric corrections for the calculation of dispersive force between the ground state atoms and corrugated surface [18]. The idea of PWS is to sum up the interactions between a single atom and every atom inside the surface, where the single atom is treated as giant coherence single matter waves in ground state energy level interacting with the surface with dielectric permittivity. The PWS method ignores the Van der Waals (vdW) many-body interactions inside the surface because they are negligible compared to the interaction of approaching atom with atoms made up the surface. Using PWS method, the atom-surface interaction can be expressed as volume integral of an atom over all atoms building up the surface,

$$U_m(\vec{r}) = -n \int \frac{C_m}{|r|^m} d^3\vec{r} \tag{1}$$

where $d^3\vec{r}$ is a volume element of the surface, n is a density of atoms that made up the surface and C_m is a interaction coefficient in which m = 6, 7 defined for non-retarded London-dispersion potential and retarded CP potential, respectively. Also, their atom-atom interaction constants may be written as follow [19],

$$C_6 = \frac{3\,\hbar}{16\,\pi^3\,\varepsilon_0{}^2} \int_0^\infty \alpha_{atom}(i\omega)\ \alpha_{surface}(i\omega)\ d\omega \tag{2}$$

$$C_7 = \frac{23\,\hbar\,c}{64\,\pi^3\,\varepsilon_0{}^2}\ \alpha_{atom}(0)\ \alpha_{surface}(0) \tag{3}$$

where α are the polarizabilities of respective dielectric bodies, ω is a mode defined in imaginary term, i, $\hbar$ is Plank constant divided by 2π and ε_0 is a vacuum permittivity. In this research, we study the interaction of ^{87}Rb-atoms with silicon-based[19],

Table 1. Values of C_6 and C_7 coeeficients for a ^{87}Rb atom in front of silicon bodies.

C_m	Si
C_6 (10^{-48} J m^6)	3.23
C_7 (10^{-55} J m^7)	5.12

As both the outer atoms and its nearby surfaces holding atoms together are treated as single macroscopic bodies, their densities, n are valid enough to be taken into account and related with the surface's permittivity, ε by Clausius-Mossotti's relations [19],

$$\frac{n\,\alpha(\omega)}{3\,\varepsilon_0} = \frac{\varepsilon\,(\omega)-1}{\varepsilon\,(\omega)+2} \tag{4}$$

Lastly, the total estimated CP potential, $U_p(z)$ can be obtained using simple proposition below [20],

$$U_p(z) = (U_6{}^{-1}(z) + U_7{}^{-1}(z))^{-1} \tag{5}$$

where m = 6, 7 defined for non-retarded and retarded potential respectively. In the following section, we will apply this Eqn. (5) to compute interaction between an atom and a flat or planar surface: finite (half-plate) and infinite thickness (half-space), and also a spherically curved surface as indicated in Fig. 1.

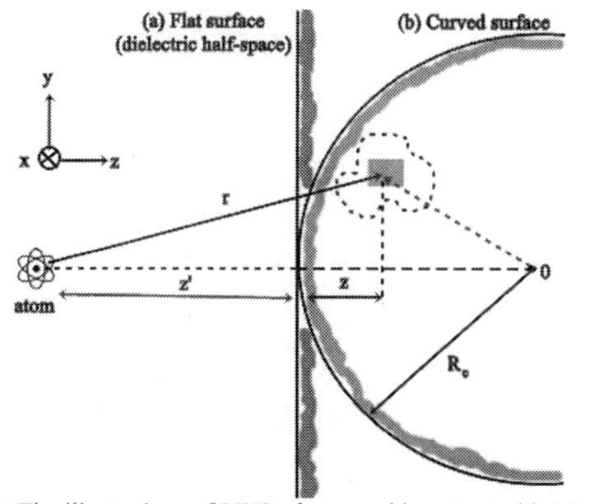

Fig. 1. The illustrations of PWS of approaching atom with (a) a planar surface, and (b) curved surfaces: spherical and cylindrical surface with length, L in x-axis. The z' is longitudal interspace distance while z is defined as the thickness length for (a) a planar surface, or as radial length for (b) a curved surface with a maximal R_C. As the atom are brought away from the surface for z' >> z, atom's perception of curvature gradually changes to flat conception, we will expect the CP potential of curved surface at larger distance are approximately closer to the one calculated for a planar surface. These probe the capability of geometrical surface in affecting the CP potential.

2.1. Atom with nearby planar surface

Let us begin with the application of PWS method to simple case of an infinite planar surface, or dielectric half-space. This means a surface having dielectric permitivity, $\varepsilon\ (\omega)$ that occupy a half-space of coordinates, defined as $z<0$. Let make assumption that atom is located at (0, 0, z + z'), disregarding the x and y coordinates due to translational symmetry. We use the cylindrical coordinates to transform PWS computation of CP potential into much simpler expressions,

$$U_m(z) = -2\pi\, C_m \int_0^z dz' \int_0^\infty \frac{\rho\, d\rho}{[(z+z')^2+\rho^2]^{\frac{m}{2}}} \tag{6}$$

where m = 6, 7. For limit z → ∞, the Eqn. (6) becomes the dielectric half-space. While for any arbitrary z, the potential is computed for different thickness suited for dielectric semi half-plate, or thin film. Eqn. (6) was to be compared with the one computed using exact scattering method [21],

$$U_{m'}(z) = -\frac{C_{m'}}{z^{m'}} \tag{7}$$

where C_3 and C_4 are both coefficients calculated from the exact scattering method, canonically different from interaction coefficients we have defined before. Both $U_3(z)$ and $U_4(z)$ from Eqn. (7) are the London-dispersion potential (non-retarded) and CP potential (retarded) respectively, serve as comparisons to $U_6\ (z)$ and $U_7\ (z)$ obtained by PWS method in Eqn. (6). After

DOI: 10.1051/epjconf/201716201059

combining PWS for both regimes using Eqn. (5), the crossover between them, where it can be determined by the effective transition wavelength λ, are obtained through the coefficients containing the polarizabilities,

$$\lambda = \frac{3\,C_3}{4\,C_4} = \frac{3\,C_7}{4\,C_6} \tag{8}$$

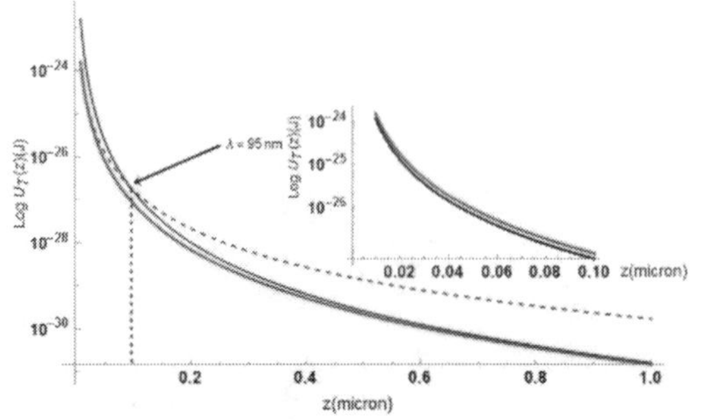

Fig. 2. The top-right graph shows the both attractive potentials of [87]Rb-atoms decays with the increasing distance from the silicon surface which are computed using PWS method (red color) and exact scattering method (black color) respectively. The bottom graph shows the three different potentials: the non-retardation potential $U_3(z)$ represents the short-ranged interaction (blue-dashed), retardation potential $U_4(z)$ represents the long-ranged interaction (blue) and the total pairwise summation (PWS) is calculated based on Eqn. (5) (red-bolded). The effective transition wavelength, λ was obtained through the intersection point between the non-retardation potential and retardation potential, and was compared with exact scattering method in Eqn. (8).

It is verified in Fig. 2 that λ obtained via PWS method is 95nm , just 5% error compared to the one calculated from exact scattering method which is λ = *100nm* [21]. By this initial result, the PWS method is still good approximation tool at least for determining the atom's behavior when it approaches different geometrical surface. The error is probably caused by ignoring the CP potential generated through the interatomic interaction inside the surface. Later, we extend to the different case of thickness, a slab of finite thickness (half-plate). The PWS for this case are formulated as follows,

$$U_{HP}(z) = U_P(z0 + z) - U_P(z) \tag{9}$$

for any $z \geq \lambda$,

$$U_{HP}(z) = C'_{m=3,4}\{\frac{1}{(z+z0)^m} - \frac{1}{z^m}\} \tag{10}$$

for any $d < z < \lambda$,

$$U_{HP}(z) \approx -C'_{m=3,4}\frac{m\,d}{z^{m+1}} \tag{11}$$

where C_3 a any $z \ll d$ (using the Taylor Series over the predecessor Eqn. 10) where z_0 is the thickness of the dielectric half-plate. m = *3, 4* refers to C'_3 and C'_4, respectively. This shows that as the thickness $d << z$, the inverse power-law reduces which cause the non-retardation $U_{3;F} \propto -\frac{1}{z^4}$ and retardation $U_{4;F} \propto -\frac{1}{z^5}$ as compares with those thickness when approaching the half-plate. The behavioral change of CP potential with regards to the slab thickness is shown in Fig. 3.

The inverse power-law reduction has deviated the CP potential of thickness, $d = 1nm$ so much from its

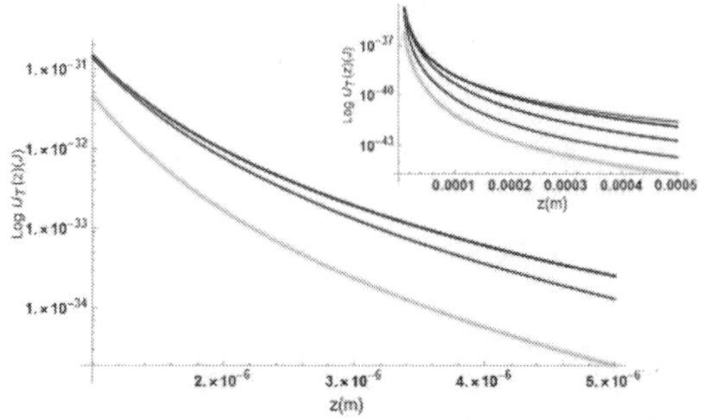

Fig. 3. Comparison of total PWS potentials for each thickness, d: 100 μm (black) ,10 μm (blue), 100nm (purple) and 1nm (green) for the interaction of [87]Rb-atoms with Si thin slabs.

neighboring $d = 100nm$ for extremely smaller scale of z. From here, it also can be concluded that as thin films become much thinner, CP Potential decays subsequently. But, as d become larger, it almost approaches with the one of half-plate (red color) in Figure 3. These results agree with actual CP potential calculated for thin slabs of dielectric half-plate [22].

2.2. Atom with nearby spherical surface

We now focus on the case of spherical surface to which we study how the curvature play a part in shaping CP potential critically as compared with planar one. The PWS for this case is given as follow,

$$U_m(z') = \pi\, C_m \int_{z'-R}^{z'+R} \frac{R^2-(r-z\prime)^2}{r^{m-1}}\,dr \tag{12}$$

where R is a radius of curvature and $z = z' + R$ by referring to Fig.1.

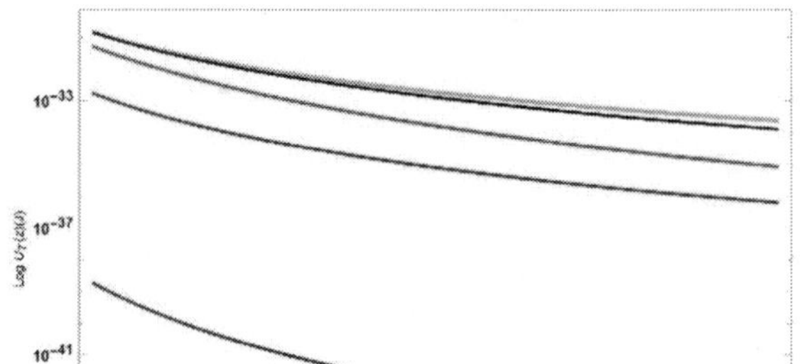

Fig. 4. Comparison of total pairwise potentials ([87]Rb-atoms and silicon surface) for both spherical surfaces (curved) and the planar surface (thin slabs) made up of 100 μm (green, pink) ,1 μm (red, black) and 1nm (blue, purple) thickness respectively. A thin film with thickness, d = 1nm doesn't reduce the CP potential as much as a spherical surface does with similar R_s=1nm.

As we may observe in Fig. 4, the spherical surface serves as the best criteria for lowering the CP potential when compared with planar surface. Coincidently, both spherical and planar surfaces agree that as their thickness increase, they behave like a dielectric half-plate.

3 Discussions

Based on the CP potential obtained for the interaction of an atom with the thin slabs of varied thickness and an atom with spherical surface, it clearly shows that geometrical surfaces play vital roles in determining the attractive CP potential. When interpreting CP potential of atom-surface interaction, atom's perspective towards the nearby surfaces has eventually changed to dielectric half-space perfection as they are located far away from the source, proven for both case of planar and spherical surfaces. Reducing the attractive CP potential by making the silicon slabs much thinner will help to increase the stability and lifetime of 87Rb-atoms trapped nearby the microengineered surface.

The next work we wish to extend to a cylindrical surface: infinitely long cylinder and finite length, L. The first case involves the CP computation of infinitely long cylindrical surface with varied thickness, R_c. The second case attributed to the limit of radial thickness, as $R_c \ll z$, whether the close like CP behavior may be observed as in dielectric half-plate (thin films). Lastly, the third case that we will consider when we calibrate the finite length, L cylindrical surfaces. The shrinking of L will change the topological structure of cylindrical surface into a disk, as compared to the affined sphere-like surface. The studies of curved surfaces' effects over CP-potential will fuel up our interests to manipulate atoms over nanofabricated wires forming the microengineered surface (atom chips) with the best condition that it may achieve peculiarly in BEC experiment.

References

1. R. Folman, P. Krüger, D. Cassettari, B. Hessmo, T. Maier, and J. Schmiedmayer Phys. Rev. Lett. **84**, 4749 (2000)
2. J. Reichel. Applied Physics B, **74(6)**:469-487 (2002)
3. M. N. A. Halif, R. Messina and T. M. Fromhold. J. Phys. Conf. Ser. 286 012045 (2011)
4. Hansel, W; Hommelhoff, P; Hansch, T W; Reichel, J. Nature; London 413.6855 (2001)
5. R. Roy. *An integrated atom chip for the detection and manipulation of cold atoms using a two-photon transition.* A thesis submitted to National University of Singapore (2015)
6. S. Gleyzes, and J.M. Raimond. Comptes Rendus Physique **17**:7 (2016)
7. C. Lovecchio, F. Schäfer, S. Cherukattil, M. Ali Khan, I. Herrera, F. S. Cataliotti, T. Calarco, S. Montangero, and F. Caruso. Phys. Rev. A **93**, 010304(R) (2016)
8. Salem R, Y Japha, J Chabé, B Hadad, M Keil, K A Milton and R Folman. New J. Phys. **12** 023039 (2010)
9. M. Keil, O. Amit, S. Zhou, D. Groswasser, Y. Japha, and R. Folman. Journal of Modern Optics, 1–46 (2016)
10. H. B. G. Casimir and D. Polder, Phys. Rev. **73**, 360 (1948)
11. D. Meschede, W. Jhe, and E. A. Hinds, Phys. Rev. A **41**, 1587 (1990)
12. C. Eberlein and S. T. Wu. Phys. Rev. A **68**, 033813 (2003)
13. T. A. Pasquini, Y Shin, C. Sanner, M. Saba, A. Shirotzek, D. E. Pritchard, and W. Ketterle, Phys. Rev. Lett. **93**, 223201 (2004)
14. R. G. Scott, A. M. Martin, T. M. Fromhold, and F. W. Sheard, Phys. Rev. Lett. **95**, 073201 (2005)
15. Y. J. Lin, I. Teper, C. Chin, and V. Vuletić. Phys. Rev. Lett. **92**, 050404 (2004)
16. G. Bimonte Phys. Rev. D 95, 065004 (2017)
17. H. Hamaker. Physica **4**, 1058 (1937)
18. R. Messina, D. A. R. Dalvit, P. A. Maia Neto, A. Lambrecht, and S. Reynaud, Phys. Rev. A **80**, 022119 (2009)
19. R. Messina, M. N. A. Halif, B. Kaczmarek, T. M. Fromhold (Unpublished)
20. F. Shimizu, Phys. Rev. Lett. **86**, 987 (2001)
21. S. T. Wu and C. Eberlein. Proc. R. Soc. Lond. A, 455 (1999)
22. A. M. C. Reyes and C. Eberlein. Phys. Rev. A **80**, 032901 (2009)

Frequency and voltage dependent electrical responses of poly(triarylamine) thin film-based organic Schottky diode

Khairul Anuar Mohamad[1,*], *Hang Tak* Hoh[1], *Afishah* Alias[1], *Bablu Kumar* Ghosh[1], and *Hisashi* Fukuda[2]

[1]Faculty of Engineering, Universiti Malaysia Sabah, Jalan UMS, Kota Kinabalu, Sabah, 88400 Malaysia
[2]Faculty of Engineering, Muroran Institute of Technology, 27-1 Mizumoto, Muroran, Hokkaido, 050-8585 Japan

Abstract. A metal-organic-metal (MOM) type Schottky diode based on poly (triarylamine) (PTAA) thin films has been fabricated by using the spin coating method. Investigation of the frequency dependent conductance-voltage (*G-V-f*) and capacitance-voltage (*C-V-f*) characteristics of the ITO/PTAA/Al MOM type diode were carried out in the frequency range from 12 Hz to 100 kHz using an LCR meter at room temperature. The frequency and bias voltage dependent electrical response were determined by admittance-based measured method in terms of an equivalent circuit model of the parallel combination of resistance and capacitance (*RC* circuit). Investigation revealed that the conductance is frequency and a bias voltage dependent in which conductance continuous increase as the increasing frequency, respectively. Meanwhile, the capacitance is dependent on frequency up to a certain value of frequency (100 Hz) but decreases at high frequency (1 – 10 kHz). The interface state density in the Schottky diode was determined from *G-V* and *C-V* characteristics. The interface state density has values almost constant of 2.8 x 10^{12} $eV^{-1}cm^{-2}$ with slightly decrease by increasing frequencies. Consequently, both series resistance and interface trap density were found to decrease with increasing frequency. The frequency dependence of the electrical responses is attributed the distribution density of interface states that could follow the alternating current (AC) signal.

1 Introduction

Organic semiconductor raises intense attention in producing thin film application devices like Schottky diodes, field-effect transistors, and solar cells, as a potential candidate for future emerging electronics analogous to conventional technology [1,2]. Organic semiconductor materials have advantages such as an easy fabrication process, low cost, mechanical flexibility, large area electronic applications, low temperature process and more versatile than inorganic semiconductor, which allowed a manipulation and the molecular level and building of eco-friendly and sustainable future electronics ranging from flexible and wearable electronics to energy-harvesting devices. Recently, solution-processed organic semiconductor, especially polymer, gained significant attention as new functional organic materials applied in the development and advancement in high performance and stability with reasonable electrical properties to meet the challenge due to their unique electronic and optical properties [3-5]. Among the family of polymer, poly(triarylamine) (PTAA) is an amorphous p-type polymer semiconductor, which offers a lot of advantages for device technologies such as low-cost materials, simple fabrication techniques and enables thin films at room temperature [6,7], but have not yet demonstrated the high performance device with excellence electrical properties such as mobility [8,9], but is known as polymer binder that enable to form uniform polymer thin films in large areas [10,11]. Thus, metal/organic Schottky diodes are useful ways to investigate the influence of conduction mechanism to the electrical properties.

Despite the investigation of the organic-based devices in DC measurements, very little is known about their characteristics in AC operation on conduction or transport mechanism, charge carrier injection, relaxation process, disordered molecule and interfacial layers between molecules. Organic semiconductors showed strongly frequency-dependent electrical response in a wide frequency range, especially at low and intermediate frequencies due to the interface states or traps and surface polarization [12-15]. The formation of an interfacial layer between metal and organic semiconductor, a variation of the semiconductor surface and series resistance between contact interface cause deviations of the ideal diode behavior [16]. Therefore, parameters such as interface states and series resistance play an important role in the determination of reasonable electrical characteristics of the metal-organic-metal (MOM) type diode. The information of such parameters can be obtained from the conductance and capacitance methods such as conductance-voltage-frequency (*G-V-f*) and capacitance-voltage-frequency (*C-V-f*) measurements [17-20]. These measurements are common tools to characterize the quality of the insulator/organic interfaces and various measurement techniques are available for determining the interface trap density [21-25]. The interface trap density and series resistance usually depend on frequency and cause frequency dispersion [17,26-29].

* Corresponding author: khairul@ums.edu.my

In this work, an investigation of the frequency dependent electrical responses of ITO/PTAA/Al Schottky diode using the conductance-frequency (*G-f*), the conductance-voltage (*G-V*), the capacitance-frequency (*C-f*) and the capacitance-voltage (*C-V*) measurements are presented to elucidate information on the electrical parameters of the organic Schottky diode. The characterization and analysis of the ac response will give important information for the conduction process of the thin films.

2 Experiments

PTAA (Lumtec) without further purification, as shown in Figure 1, was dispersed in chloroform to generate 0.1 wt% polymer solution. PTAA was soaked overnight until all solid materials were dissolved at a room temperature. Prior to thin film deposition, the glass as a substrate was soaked overnight in Decon90 and ultrasonically cleaned with ethanol, acetone and distilled water followed by drying with nitrogen gas. Then, a PTAA thin film was deposited by spin-coating at 500 rpm for 30 s, and followed further sintering at 2000 rpm for 50 s from a syringe with a filter to form an approximately 100-nm-thick thin film. Aluminum (Al) metal with a thickness of 150 nm was deposited on the spin-coated thin films through a designated mask in a vacuum chamber for top contact using sputtering technique. The area dimension of the diode was 0.5 x 0.5 cm^2. The schematic diagram of the fabricated MOM type Schottky diode structure is shown in Figure 1. The frequency dependent conductance (*G-f*) and capacitance (*C-f*) were measured over a frequency range of 12 Hz – 100 kHz using a precision LCR meter (GW Instek LCR-821). The voltage dependent conductance (*G-V*) and capacitance (*C-V*) were carried out at the various bias voltages from 0.2 – 5.0 V with an AC oscillating wave. All measurements were recorded at the room temperature in an ambient condition.

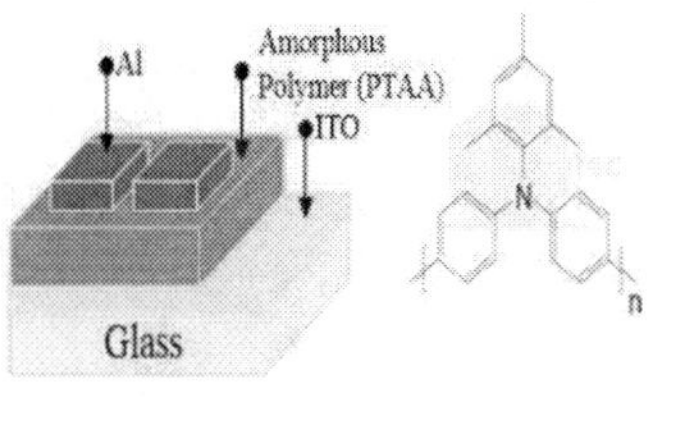

Fig. 1. Molecular structures of PTAA and schematic diagram of the ITO/PTAA/Al metal-organic-metal (MOM) type Schottky diode structure.

3 Results and discussion

3.1. Frequency dependent conductance and capacitance

The frequency dependent conductance and capacitance characteristics provide important information on the conduction process and these characteristics were obtained based on the following complex admittance equation in terms of an equivalent circuit model of the parallel combination of resistance and capacitance (*RC* circuit) [24,25]:

$$Y = 1/Z = G + j\omega C = G + j(2\pi f)C \quad (1)$$

where Y is the admittance, Z is the impedance, G is the conductance, ω is the angular frequency, f is the frequency and C is the capacitance. Figure 2 shows the conductance characteristic as a function of frequency (*G-f*) with various bias voltages (0.2 – 5.0 V) of the ITO/PTAA/Al Schottky diode. The *G-f* characteristics indicated that conductance has remained almost constant up to a certain value of the low frequency. As the frequency was increased, the conductance of the PTAA-based diode showed a continuous increase over the applied bias voltages, hence the conductance is frequency dependent characteristics. The behavior of the *G-f* characteristics is related to the long-range ordering which involving a translational motion with a sudden hopping [16]. The origin of the conductance stayed in the relaxation phenomena and when a mobile charge carrier hopped from origin to a new position, the charge remained in the state of displacement between two molecules, which then forming the hopping conductivity [30].

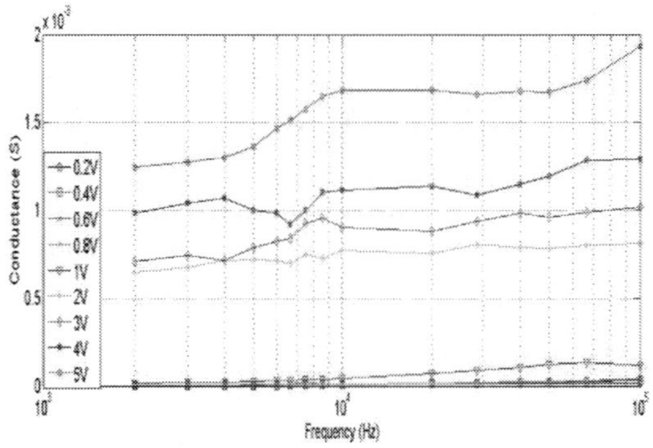

Fig. 2. Conductance characteristic as a function of frequency (*G-f*) with various bias voltages (0.2 – 5.0 V) of the ITO/PTAA/Al Schottky diode.

Figure 3 shows the capacitance characteristics as a function of frequency (*C-f*) of ITO/PTAA/Al Schottky diode in the bias voltage range from 0.2 to 5.0 V. The *C-f* characteristics indicated a decrease of the capacitance values as the frequency was increased (< 100 Hz) and it remained constant at the high frequency range (> 10 kHz). The capacitance is defined by the quality of trap charge in the band of a device [31]. The high values of capacitance at low frequencies (< 100 Hz) were attributed to the excess capacitance from the contribution of the interface trap charges, which depends on the frequency and the relaxation time of charge at interface state or the interface trap [32]. However, a continuous contribution of interface trap charges was limited by the high frequency region (> 10 kHz). As the small portion of interface charges, known as space charge capacitance,

 DOI: 10.1051/epjconf/201716201060

cannot follow the signal and negligibly small, only the interface state capacitance is contributing to the total capacitance and yield from the geometrical capacitance value [33].

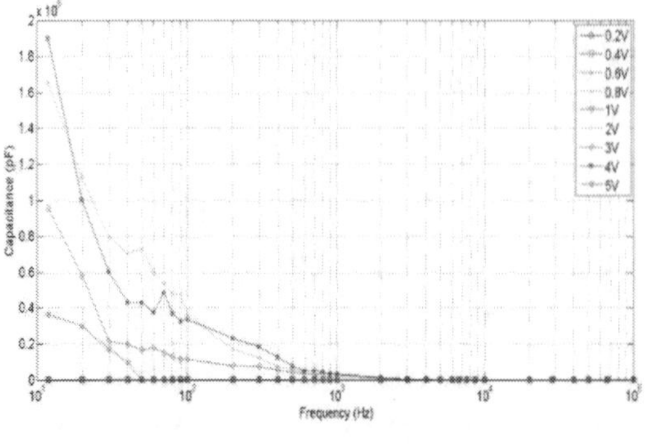

Fig. 3. Capacitance characteristics as a function of frequency (*C-f*) of ITO/PTAA/Al Schottky diode in the bias voltage range from 0.2 to 5.0 V.

3.2 Voltage dependent conductance and capacitance

Figure 4 and Figure 5 show the conductance (*G-V*) and capacitance (*C-V*) characteristics as a function of bias voltage at different frequencies for ITO/PTAA/Al Schottky diode, respectively. The *G-V* characteristic indicated that the conductance values increase from 0.2 to 5.0 V at a frequency of 2 kHz, 10 kHz and 100 kHz, respectively as shown in Figure 4. Furthermore, the conductance value also increases with increasing frequency at a given bias voltage. On the other hand, the capacitance value of the diode within the range of 0.2 V and 5.0 V increases with decreasing frequency at a given voltage as shown in *C-V* of Figure 5. At low frequency, capacitance exhibited a high value which indicated that the organic diode is turned on and the charges are trapped due to the threshold voltage at approximately 1.0 V and the charge had enough energy to pass through the depletion region at full turn-on voltage of 4.0 V.

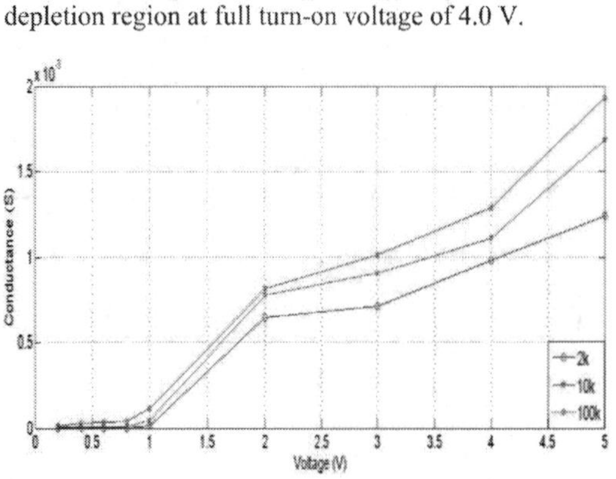

Fig. 4. Conductance characteristics as a function of bias voltage (*G-V*) at different frequencies for ITO/PTAA/Al Schottky diode.

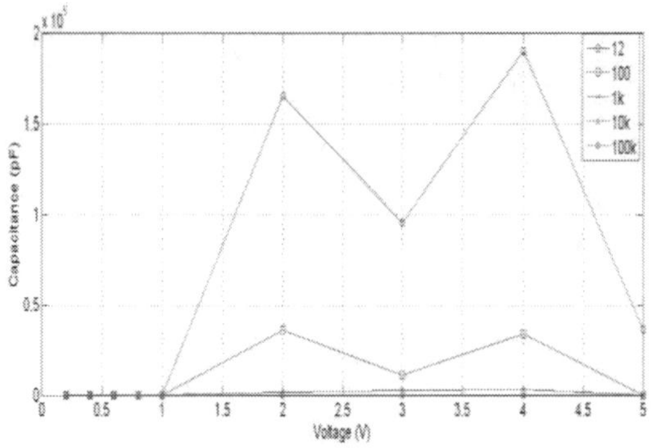

Fig. 5. Capacitance characteristics as a function of bias voltage (*C-V*) at different frequencies for ITO/PTAA/Al Schottky diode.

This phenomenon is attributed to the existence of interface state traps, interfacial PTAA layer and series resistance effect of the Schottky diode [17]. In fact, that the charges at the interface states can follow the AC signal at low frequencies, the total excess capacitance is contributed by the interface charges in addition to the space charge capacitance. Thus, the capacitance is strongly dependent on the frequency at this low range. As the voltage is built-in in the diode, the capacitance achieves the highest value in the range between 2.0 to 4.0 V and this capacitance peak region is attributed to the imperfect ohmic contacts that add a contact resistance to the series resistance or only the interface states [34,35]. Furthermore, the values of the capacitance at the peak region decrease with increasing frequency. This is attributed to the interface states are limited by the high frequencies. In addition, both *G-V* and *C-V* characteristics show three regions which indicated the inversion, accumulation, and depletion regions. The capacitance and conductance values were changed at the accumulation and depletion regions, but remained constant in the inversion region.

3.3 Frequency dependent series resistance and state density

Comparing the real and imaginary part of the admittance of the parallel *RC* circuit in Equation (1), the series resistance (R_S) at strong accumulation region of the device is given by [24,25]:

$$R_S = G_m / G_m^2 + (\omega C_m)^2 \qquad (2)$$

where the value of G_m and C_m are the measured conductance and capacitance as a constant bias voltage at various frequencies. Figure 6 and Figure 7 show the series resistance as a function of frequency (R_S-*f*) and bias voltage (R_S-*V*) for the ITO/PTAA/Al diode, respectively. These significant characteristics revealed the effects of the series resistance in the application of admittance-based measured methods. The R_S-*f* characteristics indicated that the series resistance is depends on frequency and can be differentiated into two regions [36]. The value of series resistance decreases

DOI: 10.1051/epjconf/201716201060

with the increasing frequency and remains almost constant in the higher frequency region for each bias voltage as shown in Figure 6. This attributed to the interfacial polymer layer and the interface states between polymer layer and contact interface, which cannot follow the ac signal and consequently cannot contribute to the diode. On the other hand, the values of series resistance are high at a low bias voltage ranging from 0.2 to 2.0 V due to intensive interface states at low frequency as shown in the R_S-V characteristics of Figure 7. However, series resistance almost disappears or negligibly small as the bias voltage increases further from 2.0 to 5.0 V which attributed to the trapped charges in interface state with low energy that unable to pass through the depletion region [33], thus this contributed to high values of capacitance in the voltage range.

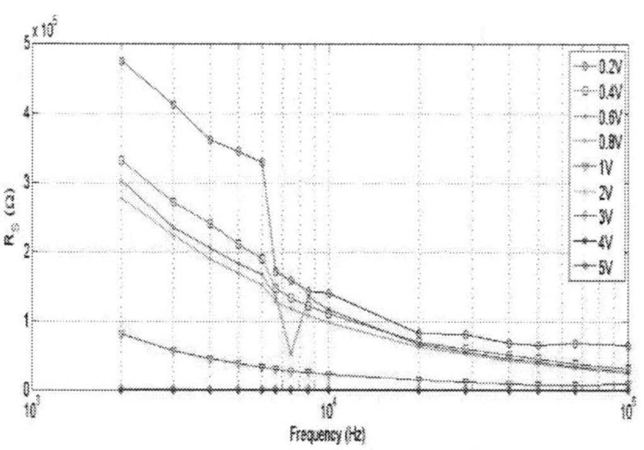

Fig. 6. Series resistance as a function of frequency (R_S-f) for the ITO/PTAA/Al diode.

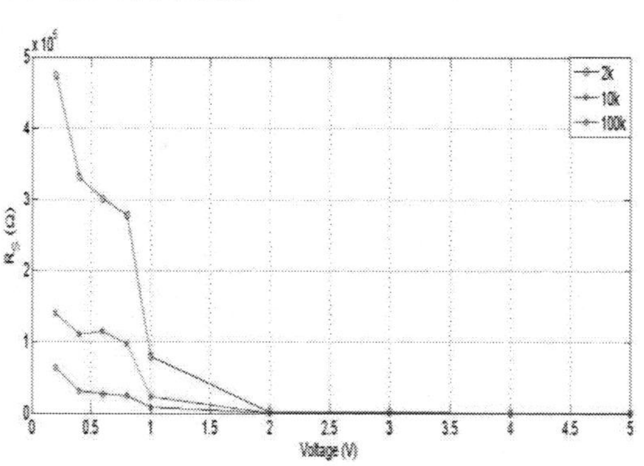

Fig. 7. Series resistance as a function of bias voltage (R_S-V) for the ITO/PTAA/Al diode.

In addition, frequency dependent of interface state density (N_{SS}) was obtained using the Hill-Coleman method, where the interface states density values can be calculated by using the following equation [23]:

$$N_{SS} = 2/gA \cdot G_{m,\max} / (G_{m,\max} C_{ox})^2 + (1 - C_m/C)^2 \quad (3)$$

where, A is the area of the diode, G_m and C_m are the measured conductance and capacitance, respectively, and C is the capacitance of polymer layer and can be obtained from G-V and C-V measurements in strong accumulation regions of a constant bias voltage of 4.0 V at constant frequency of 10 kHz by using the following relation [18]:

$$C = C_{ma}\left[1 + G_{ma}^2 / (\omega C_{ma})^2\right] \quad (4)$$

The interface states density as a function of frequency (N_{SS}-f) is shown in Figure 8. The N_{SS}-f characteristic indicated that the interface states density values almost constant at 2.8 x 10^{12} $eV^{-1}cm^{-2}$ with slightly decrease with increasing frequencies. At low frequency, the interface states density depends on frequency, showing the high value of capacitance, which attributed to the excess capacitance resulting from the interface states. However, it is almost independent of frequency at high frequencies. Thus, the diode is frequency dependent interface states density, which similar behavior and comparable to that to the other organic Schottky diodes [18,26,37], and currently further investigations to enhance the electrical responses of PTAA-based Schottky diode by adding the small molecule such as soluble pentacene to form a blend PTAA-pentacene system is currently undergoing [38].

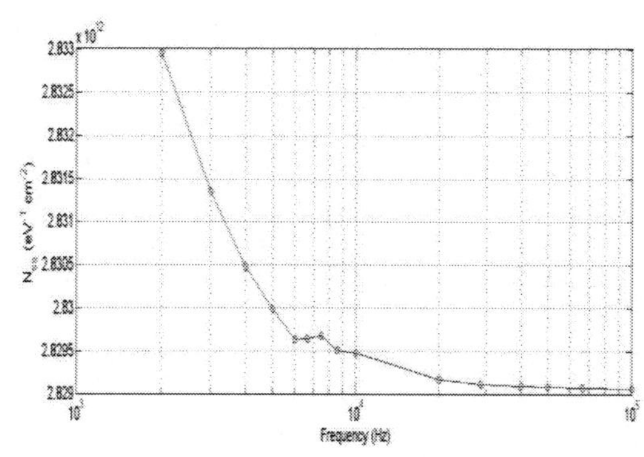

Fig. 8. Interface state density as a function of frequency (N_{SS}-f) for the ITO/PTAA/Al diode.

3 Conclusions

In summary, we have successfully fabricated a metal-organic-metal (MOM) type Schottky diode based on poly(triarylamine) (PTAA) thin films by using the spin coating method. The frequency and voltage dependence conductance and capacitance characteristics of metal-organic-metal (ITO/PTAA/Al) Schottky diode were investigated in the frequency range of 12 Hz to 100 kHz and bias voltage range of 0.2 to 5.0 V, respectively at room temperature. The frequency dependent characteristics of the Schottky diode show that both conductance and capacitance were sensitive to frequency and voltage, especially at low frequency region, which attributed to the interface states at the polymer interface and series resistance. Thus, this behavior for both conductance and capacitance can follow an AC signal at low frequencies but cannot follow at high frequencies. Moreover, series resistance is dependent both frequency and bias voltage and have a significant effect on electrical characteristics of ITO/PTAA/Al MOM type Schottky diode. Moreover, the interface state density

 DOI: 10.1051/epjconf/201716201060

was found almost constant with slightly decrease with increasing frequency. This development of Schottky diode based on amorphous PTAA thin film has been realized and further improvements in the electrical responses should be possible by improving the thin film morphology and structure as well as the fabrication conditions for possible applications.

The authors would like to thank Artificial Intelligence Research Unit (AiRU) at Universiti Malaysia Sabah for the facilities provided.

References

1. X. Ye, H. Lin, X. Yu, S. Han, M. Shang, L. Zhang, Q. Jiang, J. Zhon, Synth. Met. **209**, 337 (2015)
2. J. Takeya, T. Uemura, K. Sakai, Y. Okada, Thin Solid Films **554**, 19 (2014)
3. M. U. Hassan, Y. C. Liu, K. U. Hasan, H. Butt, J. F. Chang, R. H. Friend, Nano Energy **21**, 62 (2016)
4. P. S. K. Amegadze, Y. Y. Noh, Thin Solid Films **556**, 414 (2014)
5. T. Ohe, M. Kuribayashi, R. Yasuda, A. Tsuboi, K. Nomoto, K. Satori, M. Itabashi, J. Kasahara, Appl. Phys. Lett. **93**, 053303 (2008)
6. Y. Xu, T. Minari, K. Tsukagoshi, R. Gwoziecki, R. Coppard, F. Balestra, G. Ghibaudo, Org. Electron. **12**, 2019 (2011)
7. C. Celle, C. Suspene, J. Simonato, S. Lenfant, M. Ternisien, D. Vuillaume, Org. Electron. **10**, 119 (2009)
8. K. A. Mohamad, H. T. Hoh, B. K. Ghosh, A. Alias, I. Saad, J. Adv. Res. Appl. Mech. **13**, 24 (2015)
9. E. M. Barea, G. Garcia-Belmonte, M. Sommer, S. Huttner, H. J. Bolink, M. Thelakkat, Thin Solid Films **518**, 3315 (2010)
10. Y. H. Kim, J. E. Anthony, S. K. Park, Org. Electron. **13**, 1152 (2012)
11. J. H. Kwon, S. I. Shin, J. Choi, M. H. Chung, T. Y. Oh, K. H. Kim, M. J. Cho, K. N. Kim, D. H. Choi, B. K. Ju, Org. Electron. **10**, 729 (2009)
12. B. Arredondo, B. Romero, G. D. Pozo, M. Sessler, C. Veit, U. Wurfel, Sol. Energy Mater. Sol. Cells **128**, 351 (2014)
13. M. Ullah, D. M. Taylor, R. Schwodiauer, H. Sitter, S. Bauer, N. S. Sariciftci, T. B. Singh, J. Appl. Phys. **106**, 114505 (2009)
14. K. P. Chandra, K. Prasad, R. N. Gupta, Physica B **388**, pp. 118 (2007)
15. J. H. Ahn, J. U. Lee, T. W. Kim, Curr. Appl. Phys. **7**, 509 (2007)
16. H. Dogan, N. Yildirim, I. Orak, S. Elagoz, A. Turut, Physica B **457**, 48 (2015)
17. B. Baris, Physica B **426**, 132 (2013)
18. S. Zeyrek, E. Acaroglu, A. Altindal, S. Birdogan, M. M. Bulbul, Curr. Appl. Phys. **13**, 1125 (2013)
19. T. Tunc, S. Altindal, I. Uslu, I. Dokme, H. Uslu, Mater. Sci. Semicond. Process. **14**, 139 (2011)
20. T. Kilicoglu, M. E. Aydin, Y. S. Ocak, Physica B 388, 244 (2007)
21. K. K. Hung, Y. C. Cheng, J. Appl. Phys. **62**, 4204 (1987)
22. S. Kar, S. Varma, J. Appl. Phys. **58**, 4256 (1985)
23. W. A. Hill, C. C. Coleman, Solid State Electron. **23**, 987 (1980)
24. E. H. Nicollian, A. Goettzberger, Bell Sys. Tech. J. **46**, 1055 (1967)
25. E. H. Nicollian, A. Goettzberger, Appl. Phys. Lett. **7**, 216 (1965)
26. M. S. P. Reddy, J. H. Lee, J. S. Jang, Synth. Met. **185-186**, 167 (2013)
27. W. G. Osiris, A. A. M. Farag, I. S. Yahia, Synth. Met. **161**, 1079 (2011)
28. H. Xio, S. Huang, Mater. Sci. Semicond. Process. **13**, 395 (2010)
29. S. Sonmezoglu, S. Senkul, R. Tas, G, Cankaya, and M. Can, Thin Solid Films **518**, 3475 (2010)
30. A. Cherif, S. Jomni, H. Saghrouni, W. Belgacem, K.Khirouni, L. Beji, J. Alloys Compd. **685**, 28 (2016)
31. Y. S. Lee, J. H. Park, S. C. Jong, J. Korean Phys. Soc. **42**, 294 (2003)
32. C. Bilkan, A. Gumus, S. Altindal, Mater. Sci. Semicond. Process. **39**, 484 (2015)
33. S. Karatas, A. Turut, Microelectron. Reliab. **50**, 351 (2010)
34. E. J. Miller, X. Z. Dang, H. H. Wieder, P. M. Asbeck, E. T. Yu, G. J. Sullivan, J. M. Redwing, J. Appl. Phys. **87(11)**, 8070 (2000)
35. X. Wuand, E. S. Yang, J. Appl. Phys. **65**, 3560 (1989)
36. V. S. Reddy, S. Das, S. K. Ray, A. Dhar, J. Phys. D Appl. Phys. **40**, 7687 (2007)
37. C. Tozlu and A. Mutlu, Synth. Met. **211**, 99 (2016)
38. K. A. Mohamad, H. T. Hoh, A. Alias, B. K Ghosh, I. Saad, H. Fukuda, *Ext. Abstr. Solid State Devices and Materials*, 313 (2016)

 DOI: 10.1051/epjconf/201716201061

Annealing dependence on flexible p-$CuGaO_2$/n-ZnO heterojunction diode deposited by RF sputtering method

Mui Li Lam[1], *Muhammad Hafiz* Abu Bakar[1], *Wai Yip* Lam[1], *Afishah* Alias[1,*], *Abu Bakar* Abd Rahman[1], *Khairul Anuar* Mohamad[2], and *Katsuhiro* Uesugi[3]

[1]Faculty of Science and Natural Resources, Universiti Malaysia Sabah, Jalan UMS, Kota Kinabalu, Sabah, 88400 Malaysia
[2]Faculty of Engineering, Universiti Malaysia Sabah, Jalan UMS, Kota Kinabalu, Sabah, 88400 Malaysia
[3]Faculty of Engineering, Muroran Institute of Technology, 27-1 Mizumoto, Muroran, Hokkaido, 050-8585 Japan

Abstract. In this work, p-$CuGaO_2$/n-ZnO heterojunction diodes were deposited by RF powered sputtering method on polyethylene terephthalate (PETP, PET) substrates. Structural, morphology, optical and electrical properties of $CuGaO_2$/ZnO heterojunction was investigated as a function of annealing duration. The structural properties show the ZnO films (002) peak were stronger at the range of 34° while $CuGaO_2$ (015) peak is not visible at 44°C. The surface morphology revealed that RMS roughness become smoother as the annealing duration increase to 30 minutes and become rougher as the annealing duration is increased to 60 minutes. The optical properties of $CuGaO_2$/ZnO heterojunction diode at 30 minutes exhibit approximately 75% optical transmittance in the invisible region. The diodes exhibited a rectifying characteristic and the maximum forward current was observed for the diode annealed for 30 minutes. The diodes show an ideality factor range from 43.69 to 71.29 and turn on voltage between 0.75 V and 1.05 V.

1 Introduction

Transparent conducting oxide (TCO) thin films have been extensively studied because it's have very interesting properties in various electrical and optical application such as gas sensor, light-emitting diodes, solar cells, thin-film transistors and varistor [1,2]. ZnO have become a promising TCO materials due to its non-toxic, inexpensive and abundant material compared to ITO that once was a commonly used materials [3-6]. Besides that, ZnO have a wide direct band gap of 3.3 eV, high electrical conductivity, high optical transmission and high excitation binding energy of 60 meV, which makes the $CuGaO_2$/ZnO heterojunction a demanding candidate in optoelectronic applications [7,8]. In order to fabricate p-n junction with n-ZnO, delafossite Copper Gallium Oxide ($CuGaO_2$) is proposed as p-type material, due to its smaller lattice mismatch with ZnO [9].

A flexible diode is the last missing jigsaw piece needed to achieve a lightweight, slim, flexible, high durability and cheaper devices. Therefore, plastic substrate is used in order to opens new application of $CuGaO_2$/ZnO heterojunction thin films as transparent films that utilize curved surfaces, which means the plastic device are flexible and bendable to a large angle [10,11]. Thin films diode on plastic or flexible substrates have many merits due to the fact that they can be lightweight, flexible, unbreakable, low cost and easily folded than other substrates like silicon and glass [12,13]. Contributions of annealing towards heterojunction thin films are to suppress or eliminate detrimental surface defects so that it is unable to trap carries again [14]. Annealing also can enhance device performance by the improvement of the active layer crystallinity, morphology and optical absorption [15,16]. However, the annealing duration effect the p-$CuGaO_2$/n-ZnO heterojunction diodes and has not been investigated yet.

This paper describes the deposition of p-$CuGaO_2$/n-ZnO heterojunction diodes by RF powered sputtering method on polyethylene terephthalate (PETP, PET) substrates. The motivation behind the present work is to clarify the dependence of annealing duration on the characteristics of p-$CuGaO_2$/n-ZnO heterojunction diodes. More specifically, the turn on voltage, saturation current and ideality factor of the p-$CuGaO_2$/n-ZnO heterojunction diodes were examined with respect to the annealing duration applied after the deposition.

2 Experiments

$CuGaO_2$/ZnO heterojunction thin films were prepared on commercially available indium tin oxide (ITO) coated polyethylene terephthalate (PETP) substrates by RF powered magnetron sputtering. The targets used are 3 inches diameter ceramic target of undoped ZnO and $CuGaO_2$ with 99.99 % purity. The substrates were ultrasonically cleaned for 5 minutes in distilled water, followed by ethanol and lastly acetone. After that, the substrates were rinsed again with distilled water and then dried with nitrogen gas.

A RF sputtering was used to prepare ZnO and $CuGaO_2$ layers, and a DC sputtering was used for the Cu contact. The deposition of the ZnO films was carried out at 200 °C and 9 mtorr with a thickness of 500 nm for 25

* Corresponding author: afishah@ums.edu.my

minutes. The sputtering power was adjusted to maintain at 100 W and substrate rotation was 5 rpm, while the argon gas flow was kept at 25 sccm. The sputtering process of $CuGaO_2$ layer was performed in an argon atmosphere with flow rate 10 sccm at room temperature for 30 minute. The sputtering was carried out at RF power of 100 W and base pressure of 2.5 x 10^{-5} torr. Then, the films were annealed in a furnace at 250 °C and 5°C/min at various annealing duration (10 minute to 60 minute) to investigate the effect of annealing duration to obtain the properties of $CuGaO_2$/ZnO heterojunction. After that, Cu electrode was sputtered on top of the $CuGaO_2$ layer through shadow masks. The sputtering process of Cu contact was performed in an argon atmosphere with 10 sccm flow rate at room temperature for 10 minutes.

The surface roughness was measured using a Nanolab 550 Profilometer. The structural and crystalline properties of the heterojunction diodes were characterized using X-Ray Diffraction Philips Expert Pro. The optical properties were analyzed by using UV-Vis Spectrometer Lambda EZ210 in the wavelength range from 340 nm to 840 nm and the electrical properties were investigated using a Keithley 2400 source meter.

3 Results and discussion

3.1. Structural and optical properties

Figure 1 shows XRD patterns of as-deposited and annealed $CuGaO_2$/ZnO/PETP films. The diffraction intensity peak of (002) plane at 34° are high indicated that the ZnO films were preferentially oriented in (002) direction, agreed with the results of Chang *et al.* and Sayago *et al* [17,18]. An obvious increment in the intensity of ZnO (002) peak noticed from 10, 20 and 30 minutes annealing duration indicating better crystallinity. An increase in the diffraction intensity observed with long annealing duration due to the enhanced films quality [19]. However, the intensity of ZnO (002) peak drops at 40 minutes and continues to decrease in longer annealing duration of 60 minutes indicate the films crystallinity deteriorates. Therefore, it is considered that the diffraction intensities of the ZnO films were strongly affected by the annealing duration for the films. From this analysis, it can be concluded that variations in the annealing durations are responsible for deviations in the crystallinity where there exists an optimum value of annealing time beyond which the film crystallinity deteriorates [20]. $CuGaO_2$ (015) peak is not visible at 44°C due to insufficient thermal energy for the $CuGaO_2$ thin films to become crystalline. There is no $CuGaO_2$ (015) peak visible for thin film annealed at 250°C probably because the thin films is still in amorphous state at this temperature [21].

Figure 2 shows the root mean square (RMS) roughness of $CuGaO_2$/ZnO heterojunctions diode in various annealing duration. The inset shows the profilometer 3D images for 30 minutes annealing duration. The RMS value of $CuGaO_2$/ZnO heterojunction diode surface with no annealing treatment was 4.08 nm. After undergoing annealing duration from 10, 20 and 30 minutes, the RMS roughness become smooth compared to without annealing treatment $CuGaO_2$/ZnO heterojunction diode with the value of RMS roughness are 3.58 nm, 3.39 nm and 3.20 nm, respectively. The RMS roughness values shows decreasing trend with increasing annealing duration [20]. The surface become rougher as the annealing duration is increased from 40, 50 and 60 minutes with RMS roughness of 4.70 nm, 5.39 nm and 5.44 nm, respectively. The inset 3D images for $CuGaO_2$/ZnO heterojunctions annealing for 30 minutes shows not only lower roughness compared to the other heterojunction thin films.

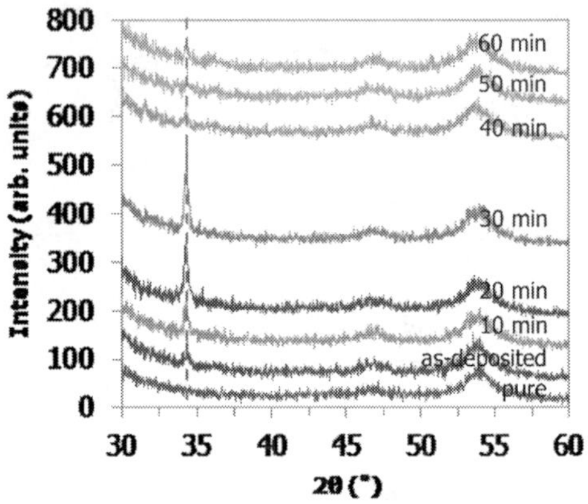

Fig. 1. (Black and White) XRD patterns of the $CuGaO_2$/ZnO/PETP films annealed for 0, 10, 20, 30, 40, 50 and 60 minutes in the furnace at 250 °C.

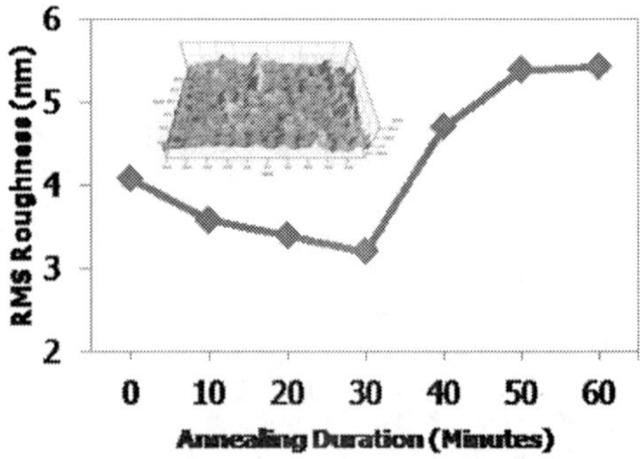

Fig. 2. (Black and White) The measured $CuGaO_2$/ZnO heterojunction diode RMS roughness for various annealing duration. The inset shows the profilometer 3D images for 30 minutes annealing duration.

Optical transmittance measurement has been done in order to investigate the transparency properties of the $CuGaO_2$/ZnO heterojunction diode. Figure 3 shows the optical transmittance versus wavelength spectra of the heterojunction thin film diode at 30 minutes annealing time that exhibit approximately 75% in the invisible region between 500 and 700nm. This result can be compared to an optical transmission in the visible that greater than 70% transmittance of thin film based p-

$CuCrO_2$:Mg/n-ZnO heterojunction and 80% transmittance of *p*-$CuCrO_2$:Mg/*n*-ZnO heterojunction in the visible region [22,23].

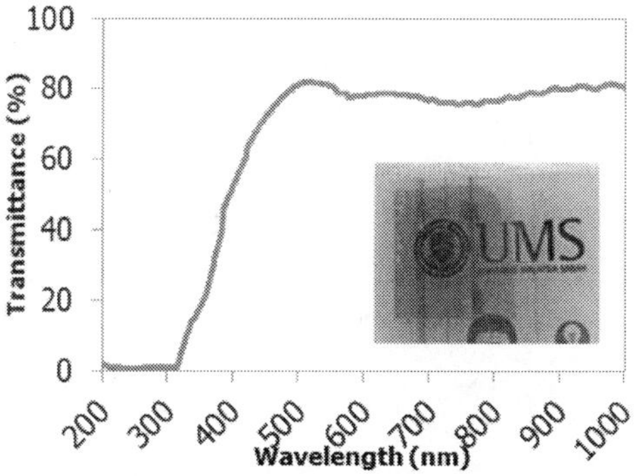

Fig. 3. (Black and White) Optical transmittance spectra of a typical p-$CuGaO_2$/n-ZnO heterojunction diode onto an ITO/PET substrate for 30 minutes annealing time. The inset shows a photograph of a $CuGaO_2$/ZnO/ITO/PET diode device to illustrate its level of transparency

3.2 Electrical diode characterization

Figure 4 illustrates the schematic diagram of the $CuGaO_2$/ZnO heterojunction structure and current-voltage (I-V) characteristics annealed at various times at 250 °C. The diodes exhibited a rectifying characteristic with turn on voltage, V_T between 0.75 V and 1.05 V. The maximum forward current was observed for the diode annealed for 30 minutes. This indicated that 30-minute annealing time was the suitable annealing duration to obtain a non-linear behavior of a diode. As shown in Figure 4, the saturation current (I_0), barrier height (φ_b) and series resistance ranging from 1.18×10^{-5} A to 3.51×10^{-7} A, 0.840 eV to 0.963 eV and 1693 Ω to 13560 Ω were obtained, respectively. In this region, an ideality factor range from 20.9 to 25.6 was extracted from the slope of the linear region of the forward bias I-V characteristics. The effect of annealing duration on the ideality factor calculated forward bias I-V characteristic was observed is in the range of those reported for ITO/ poly(3,4,ethylenedioxythiophene):poly(styrenesulfonate) (PEDOT:PSS)/poly(2-methoxy-5-(2-ethlhexyloxy))-1,4-phenylenevinylene)(MEH-PPV)/Alq3/Lif/Au Schottky diode [24], and comparable to that those reported for ITO/MEH-PPV/A21 [28]. Large values of ideality factor were obtained due to the presence of recombination in the junction deposition or at junction interface [25], besides some structural defects and lattice mismatch between ZnO and $CuGaO_2$ [19]. The improvement of the device performance is attributed to the smoother surface [26]. One possible reason for this might be the rougher surface that provides better surface adsorption, resulting in degradation of electrical properties [27]. Besides that, smooth $CuGaO_2$/ZnO heterojunction interface would provide a more ideal interface for carrier exchange [28].

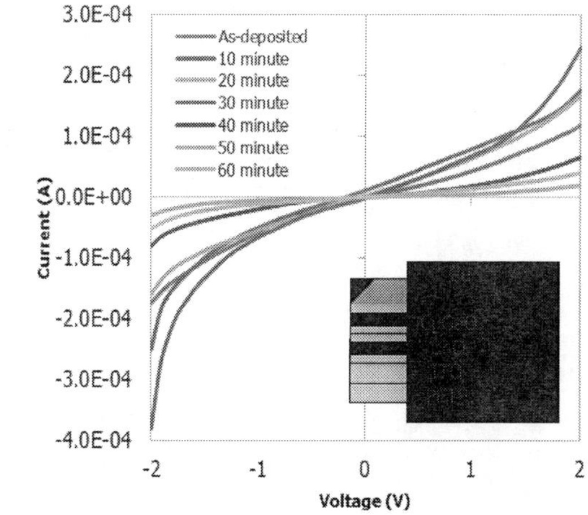

Fig. 4. (Black and White) The schematic diagram of the $CuGaO_2$/ZnO heterojunction and I-V characteristics annealed at various times at 250 °C.

In addition, as shown in Figure 4, the current for forward bias I-V characteristic annealed more than 30 minutes decreases due to coarsening of the morphology at longer annealing durations. These result consistent with the RMS value of p-n device that increase from 4.70 nm to 5.44 nm for 40 minute to 60 minute. The degradation of the rectification characteristics at longer annealing durations is due to the interfacial migration of carrier by the excessive heat treatment which usually increases the roughness of device surface [29,30]. For 10, 20 and 30 minute, the RMS roughness of the p-n device become smoother indicated the improvement on the flow of device current. RMS roughness of device decreased as annealing duration increased up to 30 minutes and its roughness increased as annealing time increased up to 60 minutes. 30 minutes is an optimum annealing duration for ideal $CuGaO_2$/ZnO heterojunction diode due to the maximum forward current found for that annealing durations. Besides that, the value of p-n device RMS roughness at that annealing duration also the smallest indicate that the device have the smoother surface morphology.

3 Conclusions

$CuGaO_2$/ZnO heterojunction diodes have been successfully fabricated on flexible substrate by using RF powered magnetron sputtering. The effect of annealing duration on the structural, morphology, optical and electrical properties of $CuGaO_2$/ZnO heterojunction diodes was investigated. The XRD pattern exhibits that ZnO films (002) peak were stronger at the range of 34° while $CuGaO_2$ (015) peak is not visible at 44°C due to insufficient thermal energy. The surface RMS roughness become smoother as the annealing duration increase to 30 minutes and become rougher as the annealing duration is increased to 60 minutes. Optical transmittance of $CuGaO_2$/ZnO heterojunction diode annealed at 30 minutes approximately 75% in the invisible region. The diodes exhibited a rectifying

DOI: 10.1051/epjconf/201716201061

characteristic with turn on voltage between 0.75 V and 1.05 V. The maximum forward current was observed for the diode annealed for 30 minutes. The diodes show an ideality factor range from 20.9 to 25.6 and barrier height (φ_b) range from 0.840 eV to 0.963 eV. This work has open up new opportunities for development in transparent and flexible optoelectronics device.

This work was financially supported by the Ministry of Higher Education under grant no. FRGS-0323-TK-1/2013. And the authors would like to thanks Universiti Malaysia Sabah for the facilities provided.

References

1. M. I. Medina-Montes, S. H. Lee, M. Perez, L. A. Baldenegro-Perez, M. A. Quevedo- Lopez, B. Gnade, R. Ramirez-Bon. J. Electron. Mater. **40**, 6 (2011)
2. T. Singh, T. Lehnen, T. Leuning, D. Sahu, S. Mathur. Appl. Surf. Sci. **289**, 27 (2014)
3. K. A. Sierros, D. A. Banerjee, N. J. Morris, D. R. Cairns, J. Kortidis, G. Kiriakidis. Thin Solid Films. 519, 325 (2010)
4. M. Bouderbala, S. Hamzaoui, B. Amrani, A. H. Reshak, M. Adnane, T. Sharaoui, M Zerdali. Physica B. 403, 3326 (2008)
5. Y. C. Lin, M. Z. Chen, C. C. Kuo, and W. T. Yen. Colloids Surf. A: Physicochem. Eng. Aspects. **337**, 52-56 (2009)
6. T. P. Rao, M. C. S. Kumar, S. A. Angayarkanni, M. Ashok. J. Alloys. Comp. **485**, 413 (2009)
7. S. S. Lin, and J. L. Huang. Surf. Coat. Technol. **185**, 222-227 (2004)
8. C. Koidis, S. Logothetidis, A. Laskarakis, I. Tsiaoussis, N. Frangis. Micron **40**, 130 (2009)
9. T. Omata, H. Nagatani, I. Suzuki, M. Kita, H. Yanagi, and N. Ohashi. J. Am. Chem. Soc. 136 (9), 3378 (2014)
10. E. Fortunato, P. Nunes, A. Marques, D. Costa, H. Aguas, I. Ferreira, M. E. V. Costa, M. H. Godinho, P. L. Almeida, J. P. Borges, and R. Martins. Surf. Coat. Technol. **151-152**, 247-251 (2002)
11. F. L. Wong, M. K. Fung, S. W. Tong, C. S. Lee, and S. T. Lee. Thin Solid Film. **466**, 225-230 (2004)
12. J. J. Kim, J. H. Lee, J. Y. Lee, D. J. Lee, B. R. Jang, and H. S. Kim. J. Korean Phys. Soc. **55**, 1910-1914 (2009)
13. C. Nunes de Carvalho, A. Luis, G. Lavareda, E. Fortunato, and A. Amaral. Surf. Coat. Technol. **151-152**, 252-256 (2002)
14. M. Y. Soomro, I. Hussain, N. Bano, O. Nur, and M. Willander. Superlattices Microstruct. **62**, 201 (2013)
15. S. Karak, S. K. Ray, and A. Dhar. Sol. Energy Mater. Sol. Cells. **94 (5)**, 837 (2010)
16. J. H. Huang, C. Y. Yang, Z. Y. Ho, D. Kekuda, M. C. Wu, F. C. Chien, P. Chen, C. W. Chu, and K. C. Ho. **10(1)**, 28 (2009)
17. H. P. Chang, F. H. Wang, J. C. Chao, C. C. Huang, And H. W. Liu. Current Apply. Phys. **11**, S186 (2001)
18. I. Sayago, M. Aleixandre, L. Ares, M. J. Fernandez, J. P. Santos, J. Gutierrez, M. C. Horrillo. Appl. Surf. Sci. **245**, 278 (2005)
19. N. F. Habubi, R. A. Ismail, W. K. Hamoudi, and H. R. Abid. Surf. Rev. Lett. **22-1**, 1550027 (2015)
20. K. S. Rahman, N. A. Khan, M. N. Imamzai, and M. Akhtaruzzaman. *IEEE Proc. Regional Symposium on Micro and Nanoelectronics* (2015)
21. M. H. Abu Bakar, L. M. Li, K. A. Mohamad, S. Sulaiman, S. Salleh, and A. Alias., J. Adv. (2015)
22. T. W. Chiu, K. Tonooka, and N. Kikuchi. Vacuum. **83-3**, 614-617 (2008)
23. K. Tonooka, and N. Kikuchi. Thin Solid Films. **515-4**, 2415-2418 (2006)
24. M. H. Suhail, I. M. Ibrahim, Z. Ahmad, S. M. Abdullah, and K. Sulaiman. J. Appl. Innov. Eng. Management. **2-1**, 130-136 (2013)
25. R. A Ismail, K. Z. Yahya, and O. A Abdulrazaq. Surf. Rev. Lett. **12**, 299 (2005)
26. W. Qi-Hui. Crit. Rev. Solid State Mater. Sci. **38**, 318 (2013).
27. D. K. Min, and H. B. Kim. Superlattices. Microstructures. **85**, 50-58 (2015)
28. O. Dhibi, A. Ltaief, N. Chaaben, and A. Bouazizi. Microelectronic Eng. 129, 24-30 (2014)
29. X. Tian, Z. Pan, H. Zhang, Y. Xie, X. Zeng, C. Xiao, G. Hu, and Z. Wei. Mater. Lett. **97**, 71-73 (2013)
30. S. K. Sharma, S. Heo, B. Lee, H. Lee, C. Kim, and D. Y. Kim. Current Appl. Phys. **14**, 1696-1702 (2014)

Study of low dimensional SiGe island on Si for potential visible Metal-Semiconductor-Metal photodetector

Alhan Farhanah Abd Rahim[1,*], *Nur'Amirah* Zainal Badri[1], *Rosfariza* Radzali[1] and *Ainorkhilah* Mahmood[2]

[1]Faculty of Electrical Engineering, Universiti Teknologi MARA Cawangan Pulau Pinang,13500 Permatang Pauh, Penang, Malaysia
[2]Department of Applied Sciences, Universiti Teknologi MARA Cawangan Pulau Pinang,13500 Permatang Pauh, Penang, Malaysia

Abstract. In this paper, an investigation of design and simulation of silicon germanium (SiGe) islands on silicon (Si) was presented for potential visible metal semiconductor metal (MSM) photodetector. The characterization of the performances in term of the structural, optical and electrical properties of the structures was analyzed from the simulation results. The project involves simulation using SILVACO Technology Computer Aided Design (TCAD) tools. The different structures of the silicon germanium (SiGe) island on silicon substrate were created, which were large SiGe, small SiGe, combination SiGe and bulk Ge. All the structures were tested for potential Metal Semiconductor Metal (MSM) photodetector. The extracted data such as current versus voltage characteristic, current gain and spectral response were obtained using ATLAS SILVACO tools. The performance of SiGe island structures and bulk Ge on Si substrate as (MSM) photodetector was evaluated by photo and dark current-voltage (I-V) characteristics. It was found that SiGe islands exhibited higher energy band gap compared to bulk Ge. The SiGe islands current-voltage characteristics showed improved current gain compared to bulk Ge. Specifically the enhancement of the islands gain was contributed by the enhanced photo currents and lower dark currents. The spectral responses of the SiGe islands showed peak response at 590 nm (yellow) which is at the visible wavelength. This shows the feasibility of the SiGe islands to be utilized for visible photodetections.

1 Introduction

Recent integrated circuit (IC) technology used millions of transistors on a small of silicon die [1]. Silicon is most commonly used semiconductor for the technology [2]. Low dimensional structures can be defined as structures and components that will improve physical, chemical, and biological properties that result in new phenomena and processes due to their nanoscale size [3]. Hence making SiGe into small sizes are expected to tailor the energy band gap of the SiGe [4].

At the complementary metal oxide semiconductor (CMOS) and bipolar CMOS (BiCMOS) technologies, silicon germanium is an important material for the fabrication of SiGe heterojunction bipolar transistors and strained Si metal oxide semiconductor (MOS) transistors [5]. Besides, the combination between the silicon germanium and the silicon will produce a heterojunction that also a benefit to silicon homojuction [6]. Silicon Germanium quantum dots have a way to reach high performance hybrid photonics circuitry within CMOS platforms [7]. For the improved light emission properties at telecommunication wavelengths with higher activation energies and improved ratio of radiative to non-radiative recombination, it used the self-assembled of germanium islands on silicon [8]. Si and SiGe is indirect semiconductor which unable to emit and detect light at visible spectrum [9]. Germanium has very favourable properties as a semiconductor, when compared to silicon whereby germanium has a much higher mobility of electrons and holes than silicon [10].

This has motivated the needs of the SiGe islands on Si to investigate its optical performance towards the visible spectrum compared with the bulk Ge using SILVACO (TCAD). Silvaco is an attractive comprehensive TCAD tool which capable for simulating the fabrication and characteristic of the semiconductor device [11]. Then, using SILVACO tools, the optimum SiGe island growth condition can be obtain while the expected performance of the device can be foresee and the time for fabrication can be optimized [9]. In this project, different sizes of SiGe islands and bulk Ge (for comparison) on Si substrate were simulated. Then, the structures were subsequently tested for potential visible MSM photodetector. Finally, the analysis and comparison of structural, optical and electrical characteristic between SiGe island and bulk Ge devices on Si substrate MSM photo detector were presented.

* Corresponding author: alhan570@ppinang.uitm.edu.my

2 Methodology

2.1. Silvaco Technology Computer Aided (TCAD)

The project started with design and simulation of SiGe islands on Si using SILVACO tools. The project involved the construction of the four different structures of SiGe islands on silicon substrate. The four SiGe structures were varied in term of their island diameters. Table 1 shows the structures which are large SiGe, small SiGe, combination SiGe and bulk Ge. Fig. 1 shows the flowchart of the steps undertaken for the project that involved the simulation of the structures.

Table 1. Size of SiGe islands and Bulk Ge on Si substrate.

Structure	SiGe island width (μm)
Large SiGe	0.5
Small SiGe	0.1
Combination SiGe	0.5 & 0.1
Bulk Ge	12

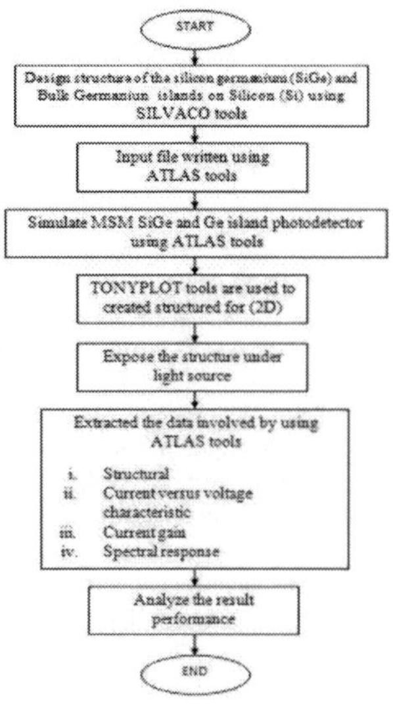

Fig. 1. Flowchart for simulation of SiGe islands on Si using ATLAS Silvaco TCAD tools.

In ATLAS, each file must contain five groups of statements in the correct order [12]. Failure to follow the order will usually cause termination of the program and an error message, then lead to incorrect operation of the program [13]. Table 2 shows ATLAS command group that involved in the simulation of the structures to characterize the device performance.

Table 2. ATLAS commands with the primary statements [13].

Group	Statements
Structure Specification	Mesh, Region, Electrode, Doping
Material Model Specification	Material, Models, Contact, Interface
Numerical Method Selection	Method
Solution Specification	Log ,Solve, Load, Save
Result analysis	Extract, Tonyplot

The process simulation started with the first part where the structures were constructed and simulated for four different island sizes using ATLAS tools. In the ATLAS input file, the Si substrate was first initialized, and the mesh was defined by setting the x- and y-mesh. Fig. 2 shows the fine mesh defined at the islands region and the wetting layer regions (underneath the island) whereas coarse mesh were tabulated around the substrate.

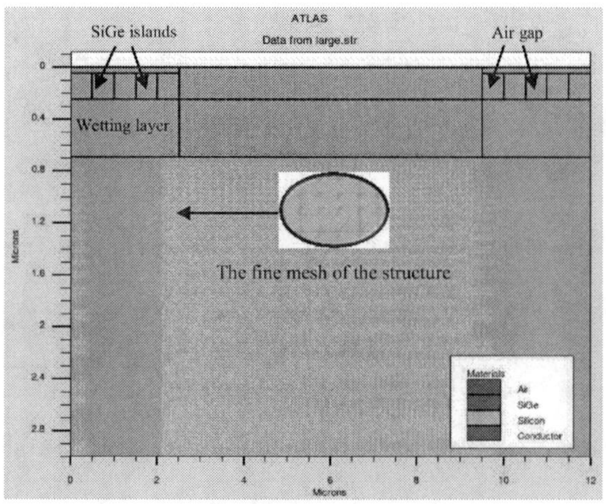

Fig. 2. Mesh defination for large structure of islands.

The construction of the SiGe islands MSM photodetector is the second part in the simulation of the project. During the second part, the region and contact statements were set. The example coding written for the islands are shown in Table 3.

Table 3. Region and contact specification for MSM photodetector for large structure of islands in ATLAS.

	ATLAS input file coding
Defining region and materials	region num=4 x.min=0.5 x.max=1.0 y.min=0.05 y.max=0.25 material=SiGe
Defining contact (anode & cathode)	electr name=anode x.min=2.0 x.max=2.5 y.min=0.05 y.max=0.25 electr name=cathode x.min=9.5 x.max=12.0 y.min=0.0 y.max=0.1

The next step of the simulation is setting up the material and physical model in the input file. In Silvaco TCAD tools, there are wide number of physical model such as models for recombination, photogeneration, mobility and lifetime [14]. The project used AUGER recombination model for the physical model as shown in Table 4. Precise selection for type of physical model is really important for accurate simulation process [13].

Table 4. Physical model coding

Type of physical model	Physical model coding in ATLAS input file
AUGER MODEL	physical models model conmob fldmob srh auger print

Then, to check the photo response of the photodetector, the light source was exposed to the sample by defining the beam power and its respective wavelength. Table 5 shows the solution specification for the biasing of the device and the setting of the light source that had been applied to the SiGe islands on Si MSM photodetector. The photodetector was biased from 0.5 V to 5 V. Fig. 3 shows the light source (power beam) shined onto the large island of the SiGe structure. Nickel (Ni) contact has been used by defining the work function of 5.01 eV. The Ni contact was chosen in order to obtain Schottky behaviour for the photodetector.

Table 5. Solution specification and beam power setting

	ATLAS input file
Solution specification and beam power coding	method newton trap solve init log outfile=I-Vpho.log solve b1=750 vanode=0.5 vstep=0.1 vfinal=5.0 name=anode Tonyplot I-Vdark.log -overlay I-Vpho.log -I-Vpho.set

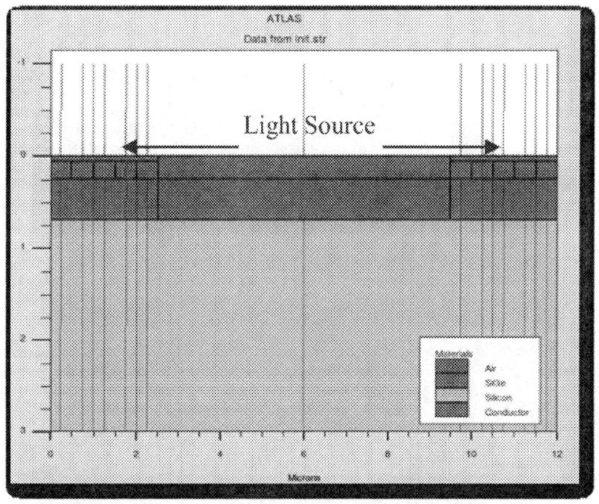

Fig. 3. Light source applied to the structure.

Lastly, the current-voltage (I-V) characteristic was obtained from the simulation, then the performance of photodetection was being evaluated by extracting the current gain which is the ratio of the photo current (Ip) over the dark current (Id). The energy band gap (Eg) and spectral response of the photodetector were also being characterized.

3 Result and discussion

3.1 Structural characteristics of the SiGe islands

The structure of different islands was created from the simulation. Fig. 4 shows the notation of the region and material in the structures.

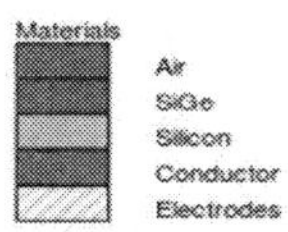

Fig. 4. Label identification of the region and material of the structure.

Fig. 5 shows the structure of the large SiGe, small SiGe and combination SiGe islands. The bulk Ge also included to compare the characteristic of the properties with the islands. The dark green colour represents SiGe islands and Ge bulk.

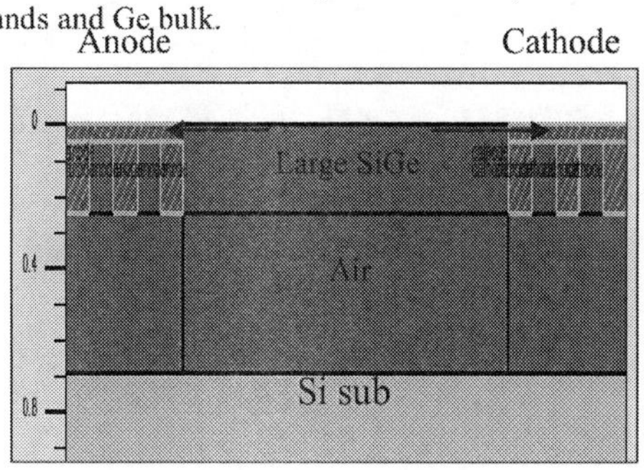

(a)

DOI: 10.1051/epjconf/201716201062

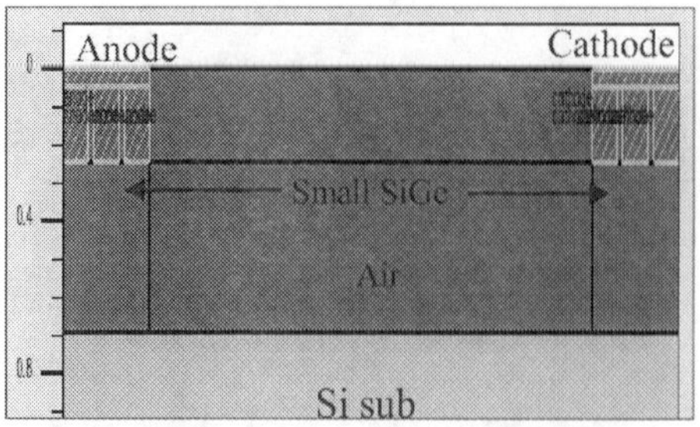

(b)

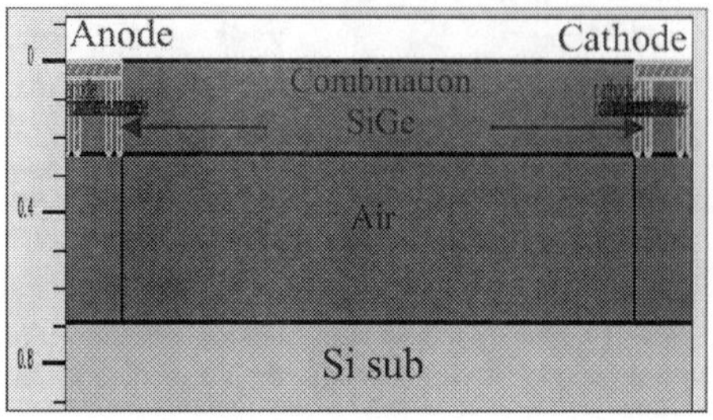

(c)

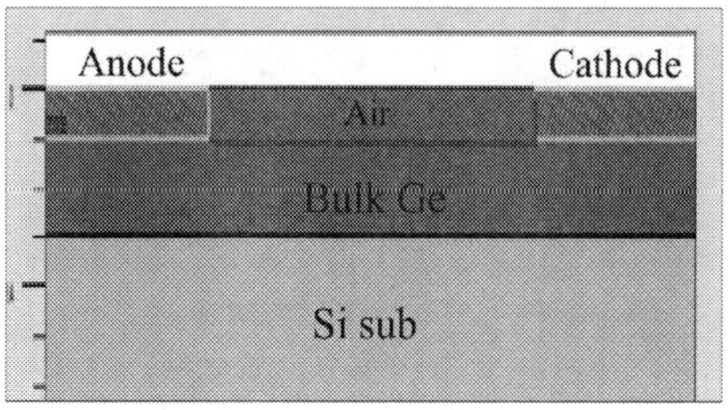

(d)

Fig. 5. Final structures for (a) large SiGe (0.5µm width), (b) small SiGe (0.1µm width), (c) combination SiGe (0.5µm – 0.1µm width) islands on Si substrate and (d) bulk Ge (12 µm width on Si substrate.

3.2 Bandgap of the structure

The band gap for the SiGe islands and the bulk Ge were extracted from the simulation as shown in Fig. 6. It can be seen that altering the island sizes has no cause to the changes in the band gap of the SiGe structures. However, alloying Si to Ge which is SiGe (Eg = 1.5 eV) does increase the band gap compared to Ge bulk (Eg = 0.8 eV). This agrees with the quantum confinement effect which stated that decreasing crystallite size will increase the band gap [15]. In addition, the increase in the bandgap indicate the potential for visible photodetection by the SiGe island on Si substrate whereby a change from indirect semiconductor to direct semiconductor will lead to the enhance photon absorption.

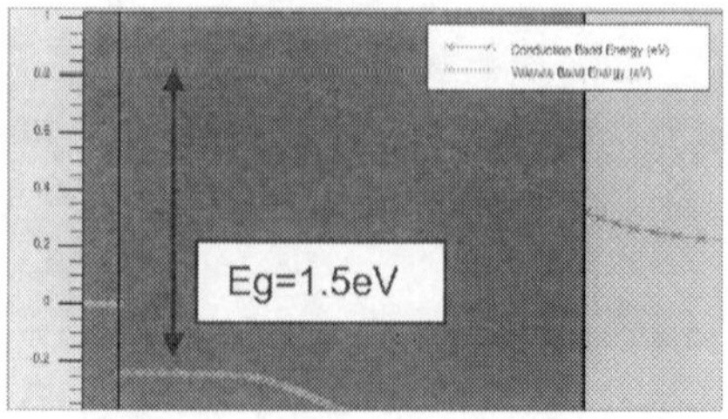

(a)

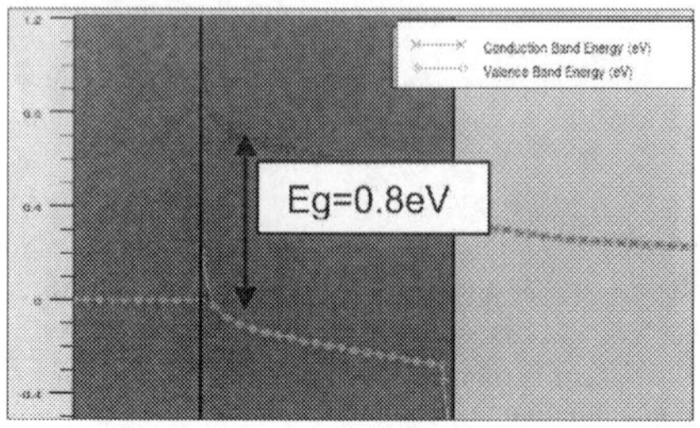

(b)

Fig. 6. Extracted bandgap of (a) SiGe islands and (b) bulk Ge.

3.3 I-V Characteristic and Gain (Ip/Id) of the structures

The metal finger contact was deposited on the SiGe islands and also bulk Ge to form metal-SiGe islands-metal photodetector and to analyse the potential of the structures for optoelectronic application. Fig. 7 presents the three SiGe island samples, then bulk Ge was included as the control sample to compare for the current-voltage (I-V) characteristics of the MSM photodetectors measured under the illumination of light (Ip) and under dark (Id). The current of the SiGe islands MSM photodetector increases with the bias voltage and saturated gradually as the voltage reach 1V. It can be seen that SiGe islands for all different sizes produced lower dark current compared to bulk Ge. In addition, the SiGe island devices also produced higher photocurrents compared to the bulk Ge photodetector. This showed that upon photons illumination, all the different size of the SiGe islands exhibited tremendous responses, producing significant free carriers for the enhanced current conduction than the bulk Ge.

DOI: 10.1051/epjconf/201716201062

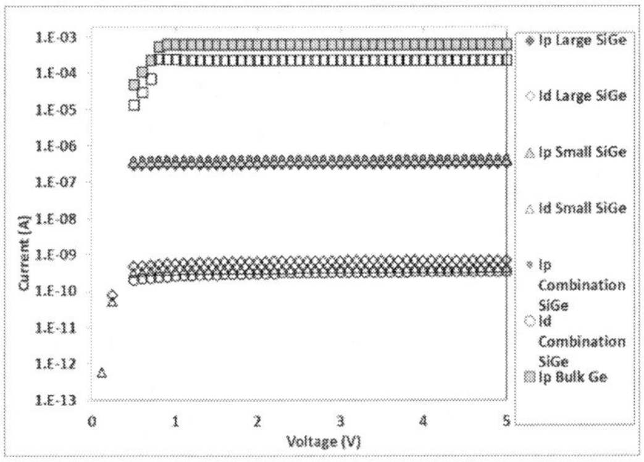

Fig. 7. I-V characteristic SiGe islands and bulk Ge.

Fig. 8 shows current gain for the SiGe islands and bulk Ge, to analyze the performance for photodetection. The current gain is the ratio of photo current-to-dark current. The result of the current gain shows that the three SiGe islands obtained higher current gain compared to bulk Ge. The higher photo currents from the SiGe islands were attributed to the optical phonon confinement of the low dimensional SiGe and the higher SiGe mobility than the bulk Ge.

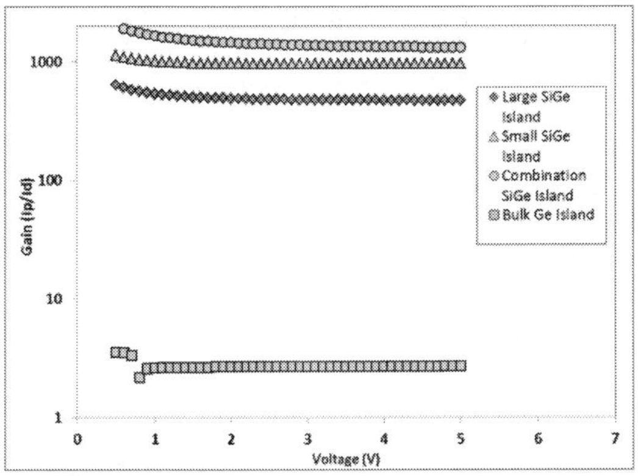

Fig. 8. Current gain of the three SiGe islands and bulk Ge.

3.4 Spectral Response

Fig. 9 shows the spectral response of the SiGe islands and bulk Ge from deep UV to infrared regions. Interestingly, all the SiGe islands devices exhibited significant response towards yellow regions (visible light). Specifically, the three SiGe islands devices showed peak current at the same wavelength of 0.59 μm, while bulk Ge at 0.94μm respectively. Therefore, the SiGe Island on Si devices show potential for photodetection at visible (yellow) spectrum, while the bulk Ge shows no response over the visible spectrum.

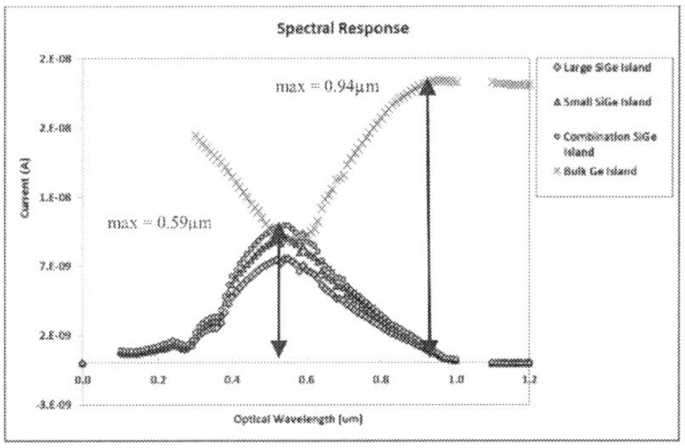

Fig. 9. Spectral response of SiGe islands and bulk Ge.

3.5 Schottky Barrier Height

Schottky barrier height which will probe the quality of the Schottky contacts was extracted to further analyze the performance of the MSM photodetector. The quality of the interface between the deposited metal (Ni) and the semiconductor surface (SiGe islands and Ge bulk) and the nature of the current transport across this interface has been extracted to determine the performance and reliability of the Schottky contacts. The thermionic emission theory is the model used to study the forward I-V characteristics of a Schottky diode [8]. According to this theory, the diode current, Id is described by the following equation [8],

$$I_d = I_o exp(qV_d / nkT) [1-exp(-qV_d / kT)] \quad (1)$$

where V_d is the voltage across the diode, n is the ideality factor (which may depend on temperature and independent on voltage), k is the Boltzman constant, and I_o is the saturation current given by:

$$I_o = AA^{**}T^2 exp [-q\varphi_B / k] \quad (2)$$

where q is the electron charge, T is the temperature, A is the contact area, A** is the effective Richardson constant and φ_B is the Schottky barrier height [8]. The saturation current, Io is mainly determined by the barrier height φ_B and the effective Richardson constant A** [8]. Equation (1) at V>3kT/q, can be simplified to:

$$I_d = I_o exp(qV_d / nkT) \quad (3)$$

The theoretical value of A** can be calculated using A** = 4πm*qk2/h3. The plot of ln I_d versus V_d will give a straight line with a slope of q/(nkT), and the intercept with y-axis will yield I_0 in which Schottky barrier height, φ_B can be obtained using $I_o = AA^{**}T^2 exp [-q\varphi_B / k]$ (2). The Schottky barrier height for dark current of SiGe islands and bulk Ge on Si substrate were listed as shown in Table 6. It was observed that the SiGe islands exhibited lower dark currents compared to bulk Ge. This in agreement with the larger schottky barrier height (SBH) of the SiGe islands compared to bulk Ge. The lowest dark current of the SiGe islands could be attributed to the high barrier height of the metal contact than the bulk Ge that have higher of dark current. In addition, the lower dark currents of the SiGe islands also contributed to the higher resistance for the devices.

Table 6. Data extracted from (I-V) characteristics graph for dark current situation.

	Large SiGe	Small SiGe	Combination SiGe	Bulk Ge
SBH in dark (eV)	0.931	0.937	0.948	0.845
Id at 5V (A)	7.01E-10	4.24E-10	3.55E-10	2.32E-04
Ip at 5V (A)	3.30E-07	4.06E-07	4.67E-07	6.31E-04
Rs (Ω)	7.E+10	1.E+10	1.E+10	2.16E+04
Gain at 5V	470.76	957.55	1315.49	2.73

4 Conclusion

As a conclusion, the structure of the silicon germanium (SiGe) islands and bulk Germanium (Ge) on Silicon (Si) have been successfully designed and simulated for potential visible metal semiconductor metal (MSM) photodetector using SILVACO TCAD tools. It is shown that the SiGe islands exhibited higher energy band gap (1.5eV) compared to bulk Ge (0.8eV). It is believed that the quantum confinement effect of the low dimensional SiGe island structure is responsible for the enhanced photo response and lower dark current. The SiGe islands produced higher photocurrents and lower dark currents compared to the bulk Ge. Moreover, SiGe Island on Si devices exhibited photoresponse at 0.59 μm of visible (yellow) spectrum but bulk Ge at 0.94 μm which is at infrared (red) spectrum region. Hence, the SiGe islands on the Si substrate provide better performance than Ge bulk for the potential visible metal semiconductor metal (MSM) photodetector.

The author wishes to thank Universiti Teknologi MARA for financial assistance and members of the IC Design Laboratory, and Department of Applied Science, Universiti Teknologi Mara Pulau Pinang, for endless technical assistance. Support from the Fundamental Research Grant Scheme (600-IRMI/FRGS 5/3(0107/2016) is gratefully acknowledged.

References

1. Y. An, A. Behnam, E Pop, and A.Ural. Appl. Phys. Lett. **102**, 013110 (2013).
2. S. Bhushan, S. Sarangi, G. K. S, and A. Santra. J. of Semicond. Tech. and Sc. **13**(4), 367-380 (2013).
3. N. U. R. Ershadiah, B. Abdul. Thesis / Undegraduate Project Paper, Falculty of Electrical Engineering, Universiti Teknologi Malaysia (2011).
4. L. Nataraj. D.Eng., University of Delaware, **183**, 3440493 (2010)
5. M. A. A. Hamid and F.Sulaiman. Int. J. of Elec. and Electronic Sys. Res **5**, 67-73 (2012).
6. D. V. Shenai, , R. L. DiCarlo Jr, M. B. Power, A. Amamchyan, R. J. Goyette and E. Woelk. Journal of Crystal Growth **298**, 172–175(2007).
7. V. V. Chandroday, N. Tiwari, J. S. Parmar, and G. S. Mittal. Int. J. of Eng. and Tech. Res. (IJETR) **2**, 187-190 (2014)
8. P. Behera and S. K. Mohapatra. Int. J. of Adv. Sci. and Tech. **71**, 59-66 (2014)
9. A. Baharin and M. R. Hashim. Semicond. Sci. Technol. **22**, 905–910 (2007)
10. M. A. Bilya and M. H. Sani. International Journal of Science and Research (IJSR) **5 (2)**, 1137-1140 (2016)
11. H. Batwani, M. Gaur, and M. J. Kumar. The Int. J. for Comp. and Math. in Elec. and Electronic Eng. **28**, 353–371 (2009).
12. M. Kaifi, M. J. Siddiqui, T. A. Abbasi, and M. U. Khan. J. of Electronic and Electrical Eng., **1**, 9-12 (2010).
13. D. S. Software. ATLAS User' s Manual, 567–1000 (2006)
14. V. Subramanian, S. Member, and K. C. Saraswat. IEEE Trans. on Electron Dev. **45**, 1690-1695 (1998).
15. S. Hu. Theses, Rochester Institute of Technology (2012)

Parametric studies on effects of defected ground structure (DGS) for dual band bandstop microstrip filter

Farah Shazuani Mahmud, *Muhammad Shahrazel* Razalli, *Hasliza* A Rahim, *Wee* Fwen Hoon *and Mohd Zaizu* Ilyas

Bioelectromagnetics Research Group (BioEM), School of Computer and Communication Engineering, Universiti Malaysia Perlis (UniMAP), 02600 Arau, Perlis, Malaysia.

Abstract. A dual-band bandstop microstrip filter is proposed and designed by using microstrip resonator with Defected Ground Structure (DGS). The DGS are placed in the ground plane to obtain high attenuation of the bandstop filter response. The proposed is designed by using R04003C printed circuit board with dimension of 38 mm x 20 mm. Simulated result shows an insertion loss of -12.31 dB, a return loss at –1.75 dB for center frequency at 2.4 GHz and insertion loss of -22.93 dB, a return loss of –0.78 dB for center frequency at 5.8 GHz The results of simulated in different parameters of DGS is analysed and discussed.

1 Introduction

Recently, many applications such as microwave or wireless communication systems are widely used bandstop filters. Bandstop filters used to block a band of frequencies which is determined by the upper and lower cutoff frequencies and allows all other frequencies. To understanding the bandstop filter, there are numerous approaches and designs have been determined [1]. The proposed bandstop filter in [2] using a simple layout configuration with high attenuation performances. The size of configuration is reduced by using method of dual band filter [3-4]. In the recent past, the researchers have been utilized several kinds of design and method of dual band bandstop filter [5-7].

Defected Ground Structure (DGS) was popular in bandstop filter to accomplish the S-parameter performances and smaller size [8-9]. DGS has affected the ground plane structures periodic or non-periodic cascaded. The shield current distribution in the ground plane is disturbed cause of the DGS. The disturbance will produce the additional inductance and capacitance to create resonant at some frequency. The DGS in the microstrip line applied in the ground planes and it gives a band rejection feature from the resonant characteristics. Meanwhile, in order to enhance the stop band characteristics, the DGS is recently used. Since it have band-rejection properties.

There are several types of DGS geometries that have been used for filter such as L-shaped, U-shaped, V-shaped, I-shaped hairpin DGS, square, rectangular, circular, square, dumbbell, spiral, hexagonal, cross shaped and interdigital [10-11]. Depending on the DGS shape and dimensions in the ground plane, the disturbed resulting in a controlled excitation and propagation of the electromagnetic waves through the substrate layer [12]. In this paper, the design of dual band bandstop microstrip filter is developed by placed the slot DGS in the ground plane. A rectangular shaped DGS is varying the dimension of width and length are analyzed and discussed in this paper. The dual band bandstop filter is simulated by using Electromagnetic Wave simulator.

.

2 Bandstop DGS filter

The configuration of dual band bandstop microstrip filter is illustrated in Figure 1. Figure 1 show the dual-band bandstop microstrip resonator filter without DGS. The ground plane is etched in rectangular pattern. The slot DGS pattern consists of length, l and width, w, as illustrated in Figure 2. Figure 3 shows the dual band bandstop filter with DGS. The configuration is designed by using R04003C printed circuit board with dielectric constant of 3.38 and a thickness of 0.508mm. The specifications of bandstop filter are shown in Table 1.

Table 1. Bandstop filter specifications

Physical Parameter	Specifications
Center Frequency, f_0	2.4 GHz and 5.8 GHz
Insertion loss, S_{21}	$\leq$ -10 dB
Return loss, S_{11}	$\geq$ -1 dB

*Corresponding author: farahshazuani@ymail.com

DOI: 10.1051/epjconf/201716201063

Based on the orientation or configuration from Figure 1, the layout is combined both resonators in one single transmission line to obtain dual band bandstop filter. In obtaining an easy layout, miniature and compact size, [13] the layout is designed as illustrated in Figure 1.

The slot defected ground structure (DGS) as shown in Figure 2 are added to the design layout in Figure 3. The insertion loss (S_{21}) and return loss (S_{11}) of the bandstop filter are improved the performances by using DGS [14].

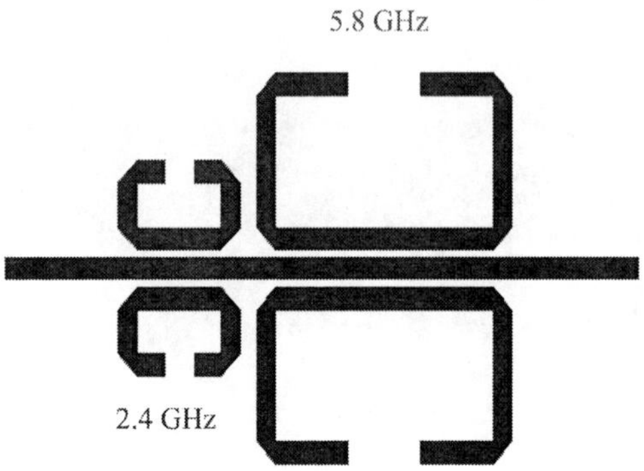

Fig. 1. Dual band bandstop microstrip filter without DGS

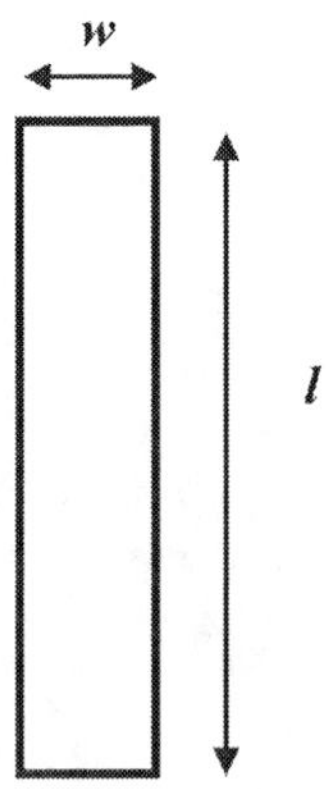

Fig. 2. Rectangular shaped DGS pattern

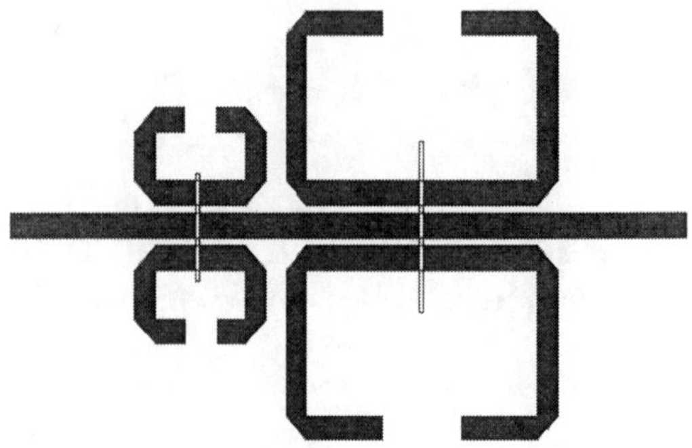

Fig. 3. Dual band bandstop microstrip filter with DGS

3 Results and Discussion

Based on the structure in Figure 1, the simulation result shows in Figure 4. There are two center frequencies of bandstop filter. The first resonant frequency is at 2.40 GHz and the second resonant frequency delivers at 5.80 GHz. The insertion loss, S_{21} in the stop bands are -8.45 dB and -10.42 dB. The return loss, S_{11} shows at -3.23 dB and -3.11 dB on it.

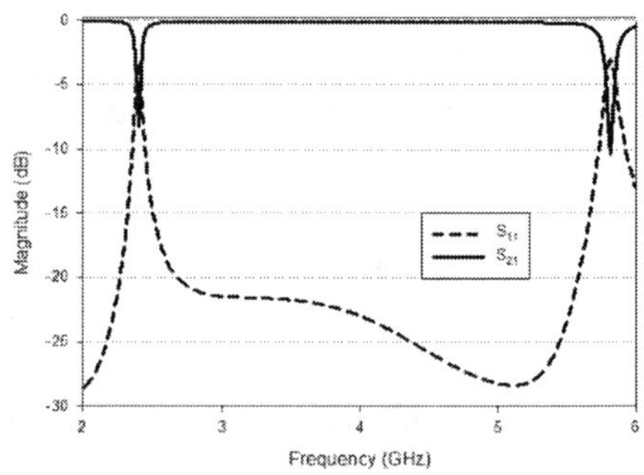

Fig. 4. Simulation result for dual band bandstop microstrip filter without DGS

However, the resonant frequency is affected and varied due to the improper length and width of the slot DGS in the ground structure. The center frequency are shifted when applied the slot DGS in the ground structure.

From the studies, the DGS width is found slightly or less impact to the variation of the resonant frequencies. This is due to the lack of tightness of the current density in the DGS width direction [15]. However, the DGS length is significantly and gives a huge impact to the resonant frequencies. Moreover, the DGS length controlled the resonant frequency and the DGS width has little impact on it. Figure 4 and Figure 5 shows the simulation S-parameter results of DGS length and width.

Initially, both DGS widths are set at 0.3 mm for both resonators and the DGS length are varying with

 DOI: 10.1051/epjconf/201716201063

different length that are 5mm, 7mm and 10mm. Referring to Figure 4(a) and (b) shows the simulation result for insertion loss, S_{21} and return loss, S_{11} by using different length of DGS.

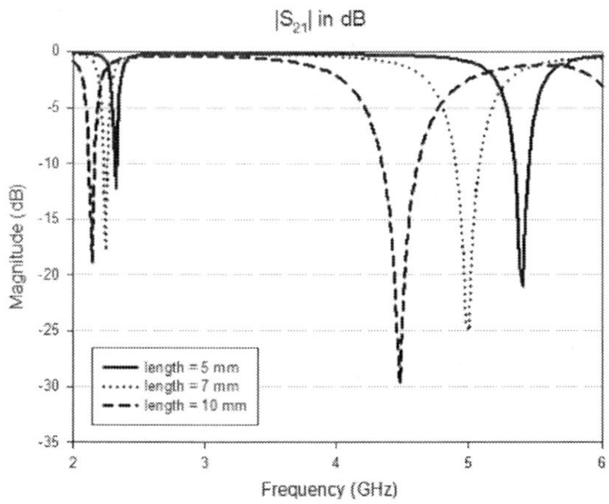

(a)

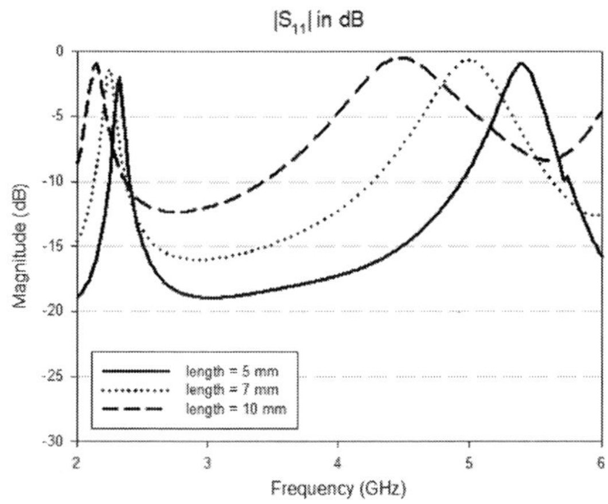

(b)

Fig. 4. DGS with different length (a) Insertion loss, S_{21} (b) Return loss, S_{11}

Analysing the simulated results in Figure 4, it is realized that if the length increased the center frequency decreased. Hence, the best DGS length for both resonators are 5 mm since it can deliver the nearest target resonant frequencies which at 2.32 GHz and 5.4 GHz. To improve the S-parameters and resonant frequencies, the proper DGS width is needed to be analysed.

Figure 5 shows the effects of the DGS width on S-parameters and resonant frequencies. The simulated DGS widths are 0.3 mm, 0.6 mm, 0.9 mm and the DGS length is permanently set at 5 mm.

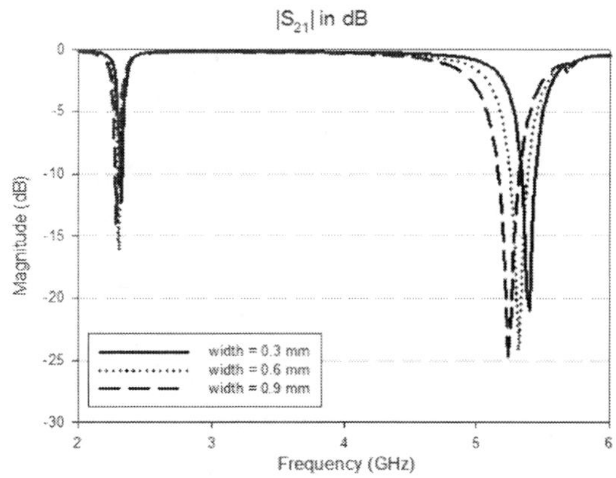

(a)

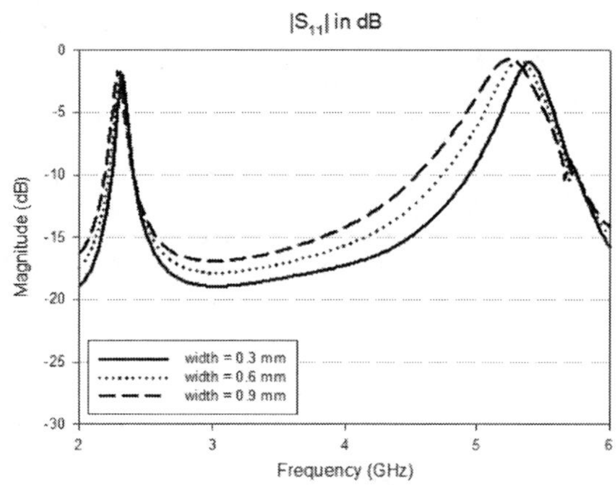

(b)

Fig. 5. DGS with different width (a) Insertion loss, S_{21} (b) Return loss, S_{11}

The simulated results are tabulated in the Table 2. As a conclusion, from the simulated results as shown in Figure 4 and 5, the best width and length of the DGS are 0.3 mm and 5 mm. At these DGS dimension values, it delivers 2.32 GHz (S_{21} = - 12.31 dB) and 5.4 GHz (S_{21} = -22.93 dB) that are nearest results to the resonant frequency at 2.4 GHz and 5.8 GHz.

Table 2. Result for varying dimension of length and width of DGS

		Resonance Frequency (GHz)		Insertion loss, S_{21}(dB)		Return loss, S_{11} (dB)	
		f_1 =2.4	f_2 =5.8	f_1 =2.4	f_2 =5.8	f_1 =2.4	f_2 =5.8
Length of DGS (mm)	5	2.32	5.40	-12.31	-22.93	-1.75	-0.78
	7	2.24	5.00	-17.69	-25.00	-1.31	-0.43
	10	2.14	4.48	-19.03	-29.94	-0.77	-0.34
Width of DGS (mm)	0.3	2.32	5.40	-12.31	-22.93	-1.75	-0.78
	0.6	2.30	5.30	-18.19	-23.00	-1.75	-0.69
	0.9	2.28	5.20	-19.12	-24.86	-1.57	-0.60

The comparison results between dual band bandstop microstrip filter with and without DGS are shown as in Figure 6. From the Table 3,it is shows that the resonant frequencies slightly changes when the design are added with slot DGS. It is changes from 2.4 GHz to 2.32 GHz and 5.8 GHz to 5.4 GHz.

It is due to the etched area in the ground plane that modify the characteristics of transmission line. It will produced the effective capacitance and inductance. The path of the ground current will increase when the etched area of DGS increase, hence increase the inductance. Thus, the capacitance will decreased since the distance between the ground and the signal line has increased.

As the increasing the inductance, the resonance frequency will decrease to a lower cut off frequency. With the result that, it indicates that the etched area in the ground plane provides for impact of the cut off frequency whether it is shift up or down. Therefeore, to obtain the target frequencies (at 2.4GHz and 5.8 GHz) both resonator will be optimized by varying the dimensions of the width and length.

However, the resonant frequency shifted, the S-parameters of insertion loss(S_{21}) and return loss(S_{11}) is improved to high performances. The slot DGS increase the slow wave factor that enhances the coupling between lines. As a results, it is shows that defected ground structure (DGS) can increase the rejection level of S-parameters performance.

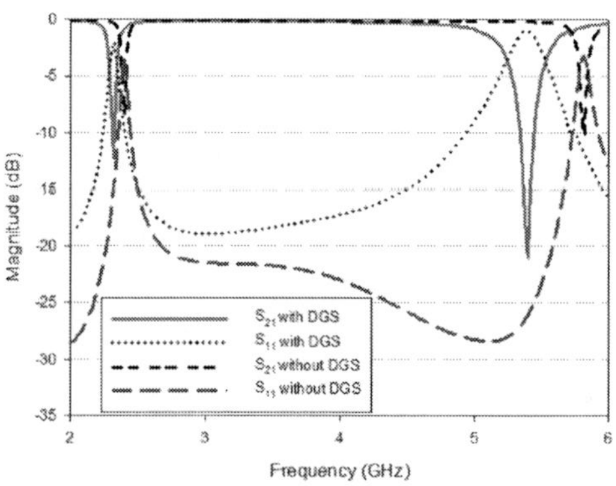

Fig. 6. Comparison between dual band bandstop filter with and without DGS

Table 3. Comparison results between dual band bandstop microstrip filter without and with DGS

	Without DGS	With DGS
Center Frequency	2.4 GHz 5.8 GHz	2.32 GHz 5.4 GHz
Insertion loss, S_{21}	-8.45 dB -10.42 dB	-12.31 dB -22.93 dB
Return loss, S_{11}	-3.23 dB -3.11 dB	-1.75 dB -0.78 dB

4 Conclusions

In this paper, dual band bandstop microstrip filter are designed by applied the rectangular shaped defected ground structure (DGS). The dimension of width and

DOI: 10.1051/epjconf/201716201063

length of DGS is analyzed by varying with different parameters. By varying the parameter, it is tuning the resonant frequency and the S-parameters. Moreover, the comparisons between dual band bandstop microstrip filter with and without DGS are shown in this paper.

References

1. D. Cui, Y. Liu, Y. Wu, S. Li, and C. Yu, Progress Electromagnetic Research, **121**, pp. 271–279, (2011).
2. Y. Luo and Q. Chu, *European Microwave Conference,* pp. 25–28, (2013).
3. M. K. Zahari, B. H. Ahmad, N. A. Shairi, and P. W. Wong, *IEEE Symposium Wireless Technology Applications ISWTA*, pp. 71–74, (2012).
4. W. Wang, M. Liao, Y. Wu, and Y. Liu, Electronics Let., **50**, 4 pp. 286–288, (2014).
5. H. Ning, J. Wang, Q. Xiong and L. Mao, *Progress in Electromagn. Res.,* **131**, pp. 259–274, (2012).
6. Y. Huang, S. Yang, L. Su, and C. C. Tzuang, *Proceedings of APMC,* pp. 235–237, (2012).
7. J. Qiu, F. Chen, and Q. Chu, *IEEE Int. Wireless Conf.*, pp. 9–11, (2013).
8. S. U. Rehman, A. F. Sheta, and M. Alkanhal, *IEEE SIECPC-Saudi International Electronics, Communications and Photonics Conference, Riyadh, Saudi Arabia,* **1**,(2011).
9. L. H. Weng, Y. C. Guo, X. W. Shi, and X. Q. Chen, *Progress in Electromagnetic Research B,* **7**, pp. 173–189, (2008).
10. A. Rahman, A. Boutejdar, A. K. Verma, G. Nadim, and A. S. Omar, Antennas and Propagation Society International Symposium, 2004. IEEE **1**, pp.998-1001, (2004).
11. B. G. Breed, High Frequency Electronics, November, pp. 50-54,(2008).
12. P. Singh, R. Tomar and P. Bhartia, "*International Journal of Advanced Research in Computer and Communication Engineering*, **2**, 10, pp. 3942–3948, (2013).
13. S. Saxena, S. Porwal, K. Soni, P. Chhawchharia, and S. K. Koul, IEEE International Conference. Microwaves, Communications, Antennas and Electronics Systems, COMCAS, (2009).
14. S. U. Rehman, A. F. a. Sheta, M. a. S. Alkanhal, and R. S. Aziz, *Saudi International Electronisc, Communications and Photonics Conference*, **1**, pp. 1–4, (2013).
15. N. C. Karmakar, S. Member, S. M. Roy, and I. Balbin, IEEE Transactions on Microwave Theory and Techniques, **54**, 5, pp. 2160–2168, (2006).

 DOI: 10.1051/epjconf/201716201064

Mathieu functions and its useful approximation for elliptical waveguides

Shamini Pillay [1,*], *Deepak* Kumar[2]

Faculty of Engineering Multimedia University Cyberjaya Malaysia

Abstract. The standard form of the Mathieu differential equation is $\frac{d^2y}{d\eta^2}+(a-2q\cos 2\eta)y=0$ where a and q are real parameters and $q>0$. In this paper we obtain closed formula for the generic term of expansions of modified Mathieu functions in terms of Bessel and modified Bessel functions in the following cases: (i) $\frac{Ce_1'(\xi_i,\gamma_1^2)}{Ce_1(\xi_i,\gamma_1^2)}$ (ii) $\frac{Fey_1'(\xi_i,\gamma_1^2)}{Fey_1(\xi_i,\gamma_1^2)}$ (iii) $\frac{Gey_1'(\xi_i,\gamma_1^2)}{Gey_1(\xi_i,\gamma_1^2)}$ (iv) $\frac{Ce_1'(\xi_i,-\gamma_2^2)}{Ce_1(\xi_i,-\gamma_2^2)}$ (v) $\frac{Se_1'(\xi_i,-\gamma_2^2)}{Se_1(\xi_i,-\gamma_2^2)}$. Let $\xi_0=\xi_i$, where i can take the values 1 and 2 corresponding to the first and the second boundary. These approximations also provide alternative methods for numerical evaluation of Mathieu functions.

1 Introduction

Mathieu functions were introduced in 1868 by E Mathieu[1] when he was investigating the vibrational modes of a stretched membrane having an elliptical boundary. Here are the solutions of the equation:

$$\frac{d^2y}{d\eta^2}+(a-2q\cos 2\eta)y=0 \qquad (1)$$

where a and q are real parameters and $q>0$. A related equation is

$$\frac{d^2y}{d\eta^2}-(a-2q\cosh 2\eta)y=0 \qquad (2)$$

The two solutions of the above equation are known as the modified Mathieu functions. For periodic solutions when q is given, equation (1) can be solved only for an infinite sequence of particular values of a and these are called the eigen values. A very detailed treatment of Mathieu functions and modified Mathieu functions can be found in McLachlan's [2]. A list of important formulae for the Mathieu functions is given in that book. Mathieu functions are written as $ce_\nu(\eta,q)$, $se_\nu(\eta,q)$ etc. and the modified Mathieu functions are generally written as $Ce_\nu(\eta,q)$, $Se_\nu(\eta,q)$ etc. Other solutions of the Mathieu equation are written as $fe(\eta,q)$ and $ge(\eta,q)$. Similarly other solutions of modified Mathieu equation are written as $Fe(\eta,q)$ and $Ge(\eta,q)$. To solve the problem of elliptic core, Mathieu differential equation (1) is used. The particular periodic solution of this differential equation is $ce_1(\eta,q)$. The notation *ce* is for "cosine elliptic". Thus the periodic solution of the first kind of the differential eq. (1) is cosine elliptic. Another periodic solution of the first kind of differential equation (1) is designated as $se_1(\eta,q)$ and is abbreviated as sine elliptic. It is important to note a solution of Mathieu's differential equation is periodic in η with period 2π. The behaviour of Mathieu functions has been described visually showing how Mathieu function can be applied in various physical systems[3].Solutions for many physical and astronomical problems further utilize these functions. The applications of other functions of mathematical physics such as Bessel have a large number of applications when compared to the Mathieu functions. A probable reason can be seen in the work of Whittaker [4].

2 Concept of Even and Odd Mathieu function

The scalar wave equation [4] for a field component $\psi(\xi,\eta)$ where $\psi(\xi,\eta)$ depends on z and the time is written as

$$\frac{\partial^2\psi}{\partial\xi^2}+\frac{\partial^2\psi}{\partial\eta^2}+\frac{q^2}{2}(\cosh 2\xi-\cos 2\eta)(n_j^2k^2-\beta^2)\psi=0$$
$$for(j=1,2)$$

Assuming separable solutions $\psi(\xi,\eta)=f(\xi)g(\eta)$ and a separation constant h gives

* Corresponding author: shamini.pillay@mmu.edu.my

$$\frac{d^2 f}{d\xi^2} - \left[h - \frac{q^2 \cosh 2\xi}{2} (n_j^2 k^2 - \beta^2) \right] f = 0 \tag{2}$$

and

$$\frac{d^2 g}{d\eta^2} + \left[h - \frac{q^2 \cos 2\eta}{2} (n_j^2 k^2 - \beta^2) \right] g = 0 \tag{3}$$

where $j = 1$ for $\xi \le \xi_0$ and $j = 2$ for $\xi \ge \xi_0$.

Equation (2) is the modified Mathieu differential equation and Equation (3) is the Mathieu differential equation. Solution of (2) which is divided in to even and odd functions [5] are given by

$$\psi(\xi,\eta) = \begin{cases} Ce_v(\xi,\gamma_1^2)ce_v(\eta,\gamma_1^2)(\xi \le \xi_0 even) \\ Se_v(\xi,\gamma_1^2)se_v(\eta,\gamma_1^2)(\xi \le \xi_0 odd) \end{cases}$$

Where

$$\gamma_1^2 = \frac{(k^2 n_1^2 - \beta^2)q^2}{4} = \frac{u^2 q^2}{4}$$

$$\gamma_2^2 = \frac{(\beta^2 - k^2 n_2^2)q^2}{4} = \frac{w^2 q^2}{4}$$

Due to the asymmetry of the elliptical waveguide we have two orientations for the field configuration. The axial magnetic field of even wave, i.e. of even function are represented by even and odd Mathieu functions. The axial electric field of even wave, i.e. of even function are also represented by even and odd Mathieu function. The axial magnetic and electric fields of odd wave, i.e. of odd functions are in similar way represented by odd and even Mathieu functions [4].

For Even wave (small eccentricity)

$$H_Z = \begin{cases} ACe_v(\xi,\gamma_1^2)\cos v\eta \to \xi \le \xi_0 \\ LFek_v(\xi,-\gamma_2^2)\cos v\eta \to \xi \ge \xi_0 \end{cases}$$

$$E_Z = \begin{cases} BSe_v(\xi,\gamma_1^2)\sin v\eta \to \xi \le \xi_0 \\ PGek_v(\xi,-\gamma_2^2)\sin v\eta \to \xi \ge \xi_0 \end{cases}$$

For Odd wave (small eccentricity)

$$H_Z = \begin{cases} ASe_v(\xi,\gamma_1^2)\sin v\eta, & \xi \le \xi_0 \\ LFek_v(\xi,-\gamma_2^2)\sin v\eta, & \xi \ge \xi_0 \end{cases}$$

$$E_Z = \begin{cases} BCe_v(\xi,\gamma_1^2)\cos v\eta, & \xi \le \xi_0 \\ PFek_v(\xi,-\gamma_2^2)\cos v\eta, & \xi \ge \xi_0 \end{cases}$$

3 Useful Approximation

The approximation presented in this paper is developed based on the approximation used in [6] Using the standard expressions for the Mathieu functions in terms of Bessel functions, the required ratios of Mathieu functions [4] are:

$$\frac{Ce_1'(\xi_0,\gamma_1^2)}{Ce_1(\xi_0,\gamma_1^2)} \approx u(1-e^2)^{1/2} \left[\frac{J_1'(u_e) + \left(\frac{u^2 e^2}{32}\right) J_3'(u_e)}{J_1(u_e) + \left(\frac{u^2 e^2}{32}\right) J_3(u_e)} \right]$$

$$\frac{Se_1'(\xi_0,\gamma_1^2)}{Se_1(\xi_0,\gamma_1^2)} \approx \frac{e^2}{(1-e^2)^{1/2}} + u(1-e^2)^{1/2} \left[\frac{J_1'(u_e) + \left(\frac{3u^2 e^2}{32}\right) J_3'(u_e)}{J_1(u_e) + \left(\frac{3u^2 e^2}{32}\right) J_3(u_e)} \right]$$

$$\frac{Fek_1'(\xi_0,-\gamma_2^2)}{Fek_1(\xi_0,-\gamma_2^2)} \approx w(1-e^2)^{1/2} \left[\frac{K_1'(w_e) + \left(\frac{w^2 e^2}{32}\right) K_3'(w_e)}{K_1(w_e) + \left(\frac{w^2 e^2}{32}\right) K_3(w_e)} \right]$$

$$\frac{Gek_1'(\xi_0,-\gamma_2^2)}{Gek_1(\xi_0,-\gamma_2^2)} \approx \frac{e^2}{(1-e^2)^{1/2}} + w(1-e^2)^{1/2} \left[\frac{K_1'(w_e) + \left(\frac{3w^2 e^2}{32}\right) K_3'(w_e)}{K_1(w_e) + \left(\frac{3w^2 e^2}{32}\right) K_3(w_e)} \right]$$

On the basis of the above ratios, we can do some modifications. Let $\xi_0 = \xi_i$, where i can take the values 1 and 2 corresponding to the first and the second boundary. Further it can be shown that u_e and w_e in the above formula is equal to ua_i and wa_i respectively. Where a_i is the semi-major axis of the ellipse corresponding to the first or to the second boundary. The following new ratios are then produced.

Thus for example,

$$\frac{Ce_1'(\xi_i,\gamma_1^2)}{Ce_1(\xi_i,\gamma_1^2)} \approx ua_i(1-e^2)^{1/2} \left[\frac{J_1'(ua_i) + \left(\frac{u^2 a_i e^2}{32}\right) J_3'(ua_i)}{J_1(ua_i) + \left(\frac{u^2 a_i e^2}{32}\right) J_3(ua_i)} \right] \tag{4}$$

There are some more important ratios which are given below:

 DOI: 10.1051/epjconf/201716201064

$$\frac{Fey_1'(\xi_i,\gamma_1^2)}{Fey_1(\xi_i,\gamma_1^2)} = ua_i(1-e^2)^{1/2}\left[\frac{Y_1'(ua_i)+\left(\frac{u^2a_ie^2}{32}\right)Y_3'(ua_i)}{Y_1(ua_i)+\left(\frac{u^2a_ie^2}{32}\right)Y_3(ua_i)}\right] \quad (5)$$

$$\frac{Gey_1'(\xi_i,\gamma_1^2)}{Gey_1(\xi_i,\gamma_1^2)} = \frac{e^2}{(1-e^2)^{1/2}} + ua_i(1-e^2)^{1/2}\left[\frac{Y_1'(ua_i)+\left(\frac{3u^2a_ie^2}{32}\right)Y_3'(ua_i)}{Y_1(ua_i)+\left(\frac{3u^2a_ie^2}{32}\right)Y_3(ua_i)}\right] \quad (6)$$

$$\frac{Ce_1'(\xi_i,-\gamma_2^2)}{Ce_1(\xi_i,-\gamma_2^2)} = wa_i(1-e^2)^{1/2}\left[\frac{I_1'(wa_i)+\left(\frac{w^2a_ie^2}{32}\right)I_3'(wa_i)}{I_1(wa_i)+\left(\frac{w^2a_ie^2}{32}\right)I_3(wa_i)}\right] \quad (7)$$

$$\frac{Se_1'(\xi_i,-\gamma_2^2)}{Se_1(\xi_i,-\gamma_2^2)} = \frac{e^2}{(1-e^2)^{1/2}} + wa_i(1-e^2)^{1/2}\left[\frac{I_1'(wa_i)+\left(\frac{3w^2a_ie^2}{32}\right)I_3'(wa_i)}{I_1(wa_i)+\left(\frac{3w^2a_ie^2}{32}\right)I_3(wa_i)}\right] \quad (8)$$

The approximation which is given in this section is useful for a two-layer elliptical waveguide in the core and cladding. These approximation was given in Adams [4] and utilized by Deepak [6] for an elliptical structure with helical windings. In the present work presented in this paper additional approximation is derived which is useful for a multilayer elliptical waveguide. Equations of multilayer elliptical waveguides are usually in terms of Mathieu and Modified Mathieu function. It is very difficult to tackle these equations due to its complex nature. Applying approximations (4)-(8) to the complex equations of Mathieu and Modified Mathieu functions, it will be converted into Bessel and Modified Bessel functions. These functions are less complicated in comparison to Mathieu and Modified Mathieu functions. Apart from complexity issues, numerical simulation will be simpler to perform as subroutines of Bessel and Modified Bessel functions are complete and readily available in software's such as Mathlab.

4 Conclusion

Equations (4-8) constitute the central new idea of this theoretical work. The authors want to point out that the Mathieu functions itself is a complex in nature and very difficult to tackle analytically. This complicated analysis brought to a point where we are getting the modified Mathieu functions in terms of Bessel and modified Bessel functions. This is a good initial step to convert modified Mathieu functions in terms of Bessel and modified Bessel functions. Such type of approximation is very useful for multilayer waveguides of elliptical cross section. An overall Mathieu function is difficult to apply in any condition because of the impossibility of analytically representing them in a simple and handy way.

References

1. E.Mathieu, J Math Pure Appl,**13** 137-203 (1868)
2. N W McLachlan, *Theory and Applications of Mathieu functions*, (1964)
3. J. C. Gutie´rrez-Vega, Am. J. Phys., **71**, 233-242 (2003)
4. E T Whittaker, Proc Edinburgh Math Soc,75-80 (1914)
5. M.J Adams, M. J., *An Introduction to Optical Waveguides*, 251–257 (1981)
6. K. Deepak, Ph.D Thesis I.I.T B.H.U Varanasi, *Propagation Characteristics of Helically cladded Elliptical step index Fiber* (1999)

 DOI: 10.1051/epjconf/201716201065

Laboratory demonstration of lightning strike pattern on different roof tops installed with Franklin Rods

Irshad Ullah[1,*], *MNR* Baharom[1], *H.*Ahmed[1], *HM.* Luqman[1], *Zainab* Zainal

[1]Department of Electrical Power Engineering, University Tun Hussein Onn, 86400, Batu Pahat Johor, Malaysia

Abstract. Protection against lightning is always a challenging job for the researcher. The consequences due to lightning on different building shapes needs a comprehensive knowledge in order to provide the information to the common man. This paper is mainly concern with lightning pattern when it strikes on the building with different shape. The work is based on the practical experimental work in high voltage laboratory. Different shapes of the scaled structures have been selected in order to investigate the equal distribution of lightning voltage. The equal distribution of lightning voltage will provide the maximum probability of lightning strike on air terminal of the selected shapes. Building shapes have a very important role in lightning protection. The shapes of the roof tops have different geometry and the Franklin rod installation is also varies with changing the shape of the roof top. According to the ambient weather condition of Malaysia high voltage impulse is applied on the lightning rod installed on different geometrical shape. The equal distribution of high voltage impulse is obtained as the geometry of the scaled structure is identical and the air gap for all the tested object is kept the same. This equal distribution of the lightning voltage also proves that the probability of lightning strike is on the corner and the edges of the building structure.

1 Introduction

This paper presents the pattern of lightning on different shapes of the roof top of the building. Five different shapes of scaled building have been selected. The work done in this paper is experimental work performed in high voltage laboratory. High voltage and low current is used under the ambient weather condition of Malaysia. The selected building shapes have been tested from the front and back view.

Lightning, for the old civilizations was a big mystery. People used to worship and thought that lightning is able to be praised. This thinking is still alive in the minds of many modern customs [1]. For the first time in 1746 Franklin and his colleagues started work on lightning. Till 1752 they did some important experiment on the physics and the behaviour of lightning. Their work done by them published in 1752 as a result of some letters, Franklin and his colleagues wrote [2]. The forces related with lightning are irregular. To protect buildings, human and electronic equipment is a big issue for electrical engineers [3]. Lightning is mainly divided into to four main types such as intra cloud, cloud to cloud, cloud to air and cloud to ground. Mainly researchers are concern with cloud to ground lightning [4]. During negative lightning the lightning flash is consist of one or many strokes. The stroke further consist of step leader and consequent leader which also include the return stroke [5].

2 Lightning protection system (LPS)

Franklin rod is the most important element of lightning protection system (LPS). It is installed on geometrical structure with different arrangement. During lightning stroke when the building is installed with the lightning rod it hit the rod and the rest of the building remains safe. The rod is connected with down conductor and earth wire. All these three collectively ground the current safely [6]

There are different international standard of lightning protection system. It can be divided into NFPA 780 and 781 which is an American standard. Similarly British standard BS 6652, IEC 61024, 62305-1-2. All These standard discuss the lightning rod installation, protection techniques, protection zone of lightning rod and design of protection system [7].

Lightning protection system can reduce the damages to the building. Lightning is very random phenomenon and it can hit the building directly or by induced effect indirectly. On the basis of this lightning protection system has two categories which is external protection

* Corresponding author: irshadullah95@yahoo.com

system. In this case lightning is diffused through franklin rod, grounding conductor and earth wire. In the same way for internal lightning protection a protection zone is produced which reduces the effect of electromagnetic field[8].

3 Lightning rods and building structures

As the technology is developing and the building protection becomes smarter therefor, it changes the traditional protection system of the buildings. Due to this it is most important to protect the building form electromagnetic effects [9]. When lightning strikes directly on the building it damages the building and also causes fire. To protect the building structure lightning rod is placed on a preferred place which can protect the building structure and the surrounding. In this, lightning rod is also connected with down conductor and earth wire [10]. The interception of lightning is based on the air terminals. The air terminal can attract the lightning and can make it ground from the surface of any building to make sure its safety. The position of the air terminal is of the most importance [9].

4 Lightning strike probability

The receiving ability of the lightning flash by the lightning rod is under discussion for a very long time. It has been proved that blunt rods are more effective in receiving of lightning flashes. Moreover the sharp rods produce corona around it which can reduce the probability of lightning receiving ability [11]. Taking in concentration a certain building, Monte Carlo techniques has proven that the lightning strike probability is mostly on the corners and the edges [12]. During lightning rods installation it has been investigated that the lightning rods, smaller in size on the corner has more probability of lightning attraction than the one which is bigger in size and installed on the center of a building [13].

5 Experimental procedures

For the selected shapes of the scaled building, high voltage impulse generator has been used. Five different shapes including rectangular, Gable, Circular, Inclined and hexagonal shapes are test. Every structure excluding circular shape is tested with front and back view. The front view is from the east to west direction when the impulse voltage is applied, while when the position of the lightning rods is reversed to the west-east direction the view of the test object becomes as back view .The Terco HV91500 impulse generator is basically a high voltage and low current generator. The average applied impulse voltage is 82k V. The material of the top electrode and lightning rod is metal, while the material of the scaled structure is insulator. The installed lightning rod is grounded to the earth and therefore the impulse voltage is grounded safely to the ground. Top electrode receives the impulse voltage which is connected to the measuring capacitor of HV91500.

Keeping in consideration the specifications of the high voltage impulse generator a single story 3m building is scaled down to 10 cm height according to [14],[15],[16]. A 3m single story building can be scale down as

$$\text{Measure Scale} = m_i = \frac{M_P}{M_M} = \frac{\text{Prototype measured}}{\text{Style measures}} \tag{1}$$

$$\text{Model measures} = M_M = \frac{M_P}{M_i} = \frac{\text{Prototype measured}}{\text{Measure Scale}} \tag{2}$$

Where M_P is the characteristic height in the real-world prototype and M_M is the corresponding height in the model (subscript M). The simple structures are scaled as 3 m = 300 cm~10 cm, hence, Eq. (1) gives,

$$m_i = \frac{300}{10} = 30 \tag{3}$$

The inverse of equation (3) shows the scale factor 1:30. Therefor from equation (2) we can calculate the height of a single story buildings as

$$\text{Model height} = \frac{M_P}{M_i} = \frac{300\,\text{cm}}{30} = 10\,\text{cm} \tag{4}$$

The scaled models are tested under the applied high voltage D.C and impulse. A top electrode is used to get the high voltage from the control disk. Top electrode works as cloud. The weather conditions including temperature, pressure and humidity are taken into account. The air gap between top electrode and the tip of the rod is kept same which is 3cm.Thirty flashes are applied to every shape. All the shapes are test in front and back view except the circular shape as it only the edges. Similarly the height of the lightning rod is kept same throughout the experiment.

The scaled models are tested under the applied high voltage D.C and impulse. A top electrode is used to get the high voltage from the control disk. Top electrode works as cloud. The weather conditions including temperature, pressure and humidity are taken into account. The air gap between top electrode and the tip of the rod is kept same which is 3cm.Thirty flashes are applied to every shape. All the shapes are test in front and back view except the circular shape as it only the edges. Similarly the height of the lightning rod is kept same throughout the experiment.

6 Experimental results

The results of the experimental work can be explained for every shaped tested under the high voltage condition in high voltage lab.

DOI: 10.1051/epjconf/201716201065

6.1 Experimental results for rectangular shape

For the rectangular shape the number of air terminals are six which are installed the edges and the corners. The applied voltage under the weather condition are given in Table 1.

Table 1. Lightning strike pattern on rectangular shape

Temperature, 28.30C⁰ **D.C Voltage ,** 86.50 kV	**Pressure,** 1.03 Pa **Impulse ,** 81.00k V		**Humidity** 71.20 % **Current** 1.6 A			
	Number of strikes on air terminal					
View	T1	T2	T3	T4	T5	T6
Front	7	7	3	Nil	6	6
Back	7	5	4	Nil	6	8

From Table 1 the number of lightning strikes for the front and back view are given. The air terminals on the front and back view are strike which shows that the distribution of lightning is equally distributed. While terminal 4 is missing as it is installed on the edges and almost at the middle while terminal receive a very few lightning flashes.

In the same way Figure 1 (a) and (b) show the front and back vies of the stricken air terminal

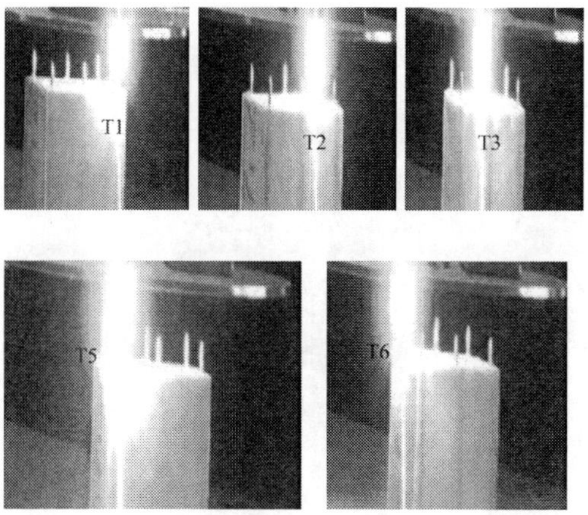

(a)

Fig.1. (a), (b) Lightning strike pattern on front and back view of rectangular shape

Figure 1 (a) and (b) clearly shows the strike distribution on the rectangular shape for both the front and back view

6.2 Lightning strike pattern on circular shape

For circular shape the nombres of lightning rods are 4 while it has been tested for only front view. Interestingly for the circular shape all the air terminals on the edges have been hit by lightning. The value of voltage and the ambient condition are given in table2.

Table 2. Lightning strike pattern on circular shape

Temperature, 28.30C⁰ **D.C Voltage ,** 86.50 kV		***Pressure,*** 1.03 Pa **Impulse ,** 81.00k V		***Humidity*** 71.20 % **Current** 1.6 A
Number of strikes on air terminal				
View	*T1*	*T2*	*T3*	*T4*
Front	*9*	*8*	*6*	*7*

Table 2 shows the number of lightning strikes on the circular shape with the applied voltage and the ambient weather condition

 DOI: 10.1051/epjconf/201716201065

Fig. 2. Lightning strikes on circular shape.

Figure 2 represents the lightning strike pattern on circular shape.

6.3 Lightning strike pattern on inclined shape

Inclined shape is provided with six air terminals. As the shape is inclined therefore only the first two terminals will have the same height and the rest will be different. The number of strikes on the terminal can be shown in Table 3.

Table 3. Lightning strike pattern on inclined shape

Temperature, 26.60C⁰	Pressure, 1.03 Pa	Humidity 70.80 %	
D.C Voltage, 86.70 kV	Impulse, 83.33k V	Current 1.6 A	
Number of strikes on air terminal			
View	*T1*	*T2*	*T3-T6*
Front	*16*	*15*	*Nil*
Back	*13*	*17*	*Nil*

Table 3 shows the number of lightning strike on inclined shape. The lightning pattern of inclined shape is shown in figure 3. Both the front and back vies are given.

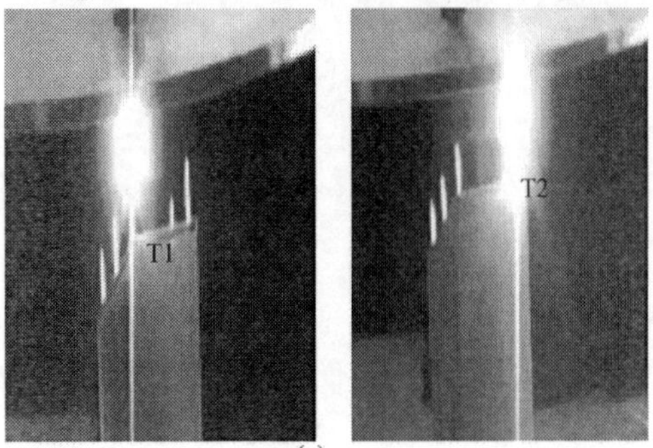

(a)

(b)

Fig.3. Lightning strike pattern on inclined shape

Figure 3 shows the lightning strike pattern for the front and back view of inclined shape. It is clear from the figure that only terminal 1 and 2 are struck in both in front and back views while the remaining are missing.

6.4 Lightning strike pattern for gable shape

The gable shape is tested with three air terminal installed on it. Same 30 flashes are given in both the front and back views. The ambient condition is given and the applied voltage is also shown. Table 4 shows the number of strikes on the gable shape and figure 4 shows the pattern of the lightning strikes on it.

Table 4. Lightning on gable shape

Temperature, 26.60C⁰	Pressure, 1.03 Pa	Humidity 70.80 %	
D.C Voltage, 86.70 kV	Impulse, 83.33k V	Current 1.6 A	
Number of strikes on air terminal			
View	T1	T2	T3
Front	14	3	13
Back	13	2	15

Table 4 shows the number of lightning strike on the different air terminal of the gable shape.

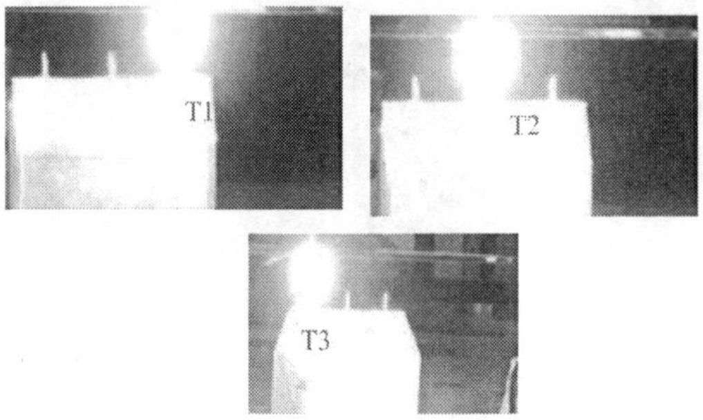

(a)

DOI: 10.1051/epjconf/201716201065

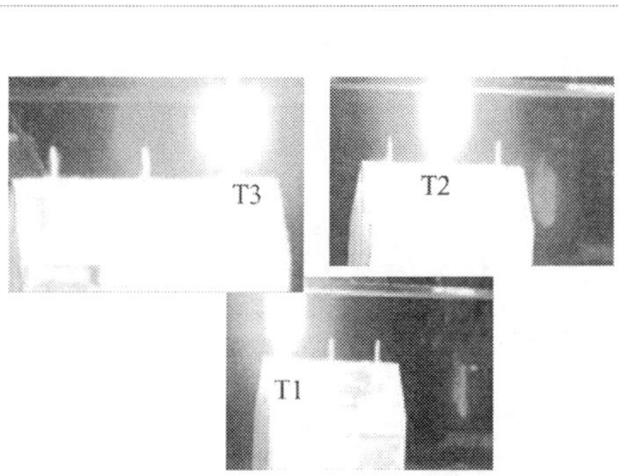

(b)

Fig. 4. Lightning pattern on the gable shape

From the figure given above the lightning pattern is represented on the gable shape. Terminal 2 has the very few lightning strikes on it as it is installed in the middle of the gable structure.

6.5 Lightning strike pattern on hexagonal shape

Hexagonal shape has five air terminal installed. Terminal 1 is in the middle and on the height as compare to other air terminal therefor the pattern will be the same for both the fort view and the back view.

Table 5 . Lightning pattern on hexagonal shape

Temperature,	Pressure,	Humidity
29.o0C^0	1.03 Pa	69.80 %
D.C Voltage ,	**Impulse ,**	**Current**
88.70 kV	83.90k V	1.7A
Number of strikes on air terminal		
View	T1	T2-T5
Front	30	Nil
Back	30	Nil

From the Table 5 the number of strike on each air terminal of the hexagonal shape is shown. All the strikes for the front and the back vies are on terminal 1 as it is on the more height and can attract the lightning easily.

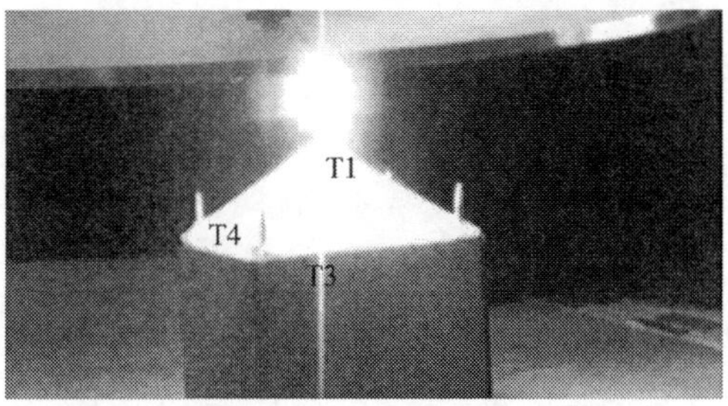

Fig.5. Lightning pattern for hexagonal shape

All 30 strikes are on terminal 1 for the front and back vies of the hexagonal shape. As terminal 1 has more height than all other therefore lightning will be attracted by terminal one firstly and easily.

7 Discussions

Lightning protection of building is a very important issue for the researcher of high voltage engineering. Building structures are on the high risk of the lightning strike. In this paper the equal distribution of lightning impulse has analysed in detail. Five different shapes have been investigated regarding lightning strikes distribution. For the stability of the proposed system two views of the building front and back are tested with 30 flash of impulse. The changing in the views means to make sure that experimental work performed is balance and if the position of the lightning rod is correct and geometry of the shape is scaled properly the system will perform even if the position is changed. All the experiments have been done on the STP condition of Malaysia. The shapes in this work is on the macro scale which means that for the impressive results every single element should be well balance and the geometry of everything should have the same geometry. In this work almost everything is properly balance and the results in the front view are same with the back view which shows that the voltage is equally distributed on each air terminal. The equal distribution of the lightning voltage can leads the consultant, government contractor and high voltage engineering researcher to develop a comprehensive protection system for the building structure.

8 Conclusion

From the overall experimental work of this paper it can be concluded that shape of the building having lightning rods installed has an important role in lightning protection and lightning attraction as well. The lightning pattern in this work is the direct lightning strike phenomenon. Every shape has different air terminal arrangement therefor the effect of lightning also will be different. For the same height of the building the installed lightning rods should also have the identical height. This will make sure the equal distribution of lightning even if it is in the different times of the year. The identical geometry of rods and the building element makes sure the proper protection against lightning.

This work is based on high voltage and low current applied to the macro geometrical structure. The height of the building and the air terminal can be varied and the effect of high voltage and high current effect could also be studied on different geometrical shapes. Similarly the shapes can also be more in number so the effect of high voltage and high current can be analysed. The effect of lightning distribution can also be investigated for the scaled models of complex building structures. Electric field simulation can also be performed as a result of the applied voltage to the structure.

DOI: 10.1051/epjconf/201716201065

The authors would like to thank the ORICC for supporting this research under the grnat contract using Vot. U563. Special thanks also to FKEE for giving the moral support to this research.

References

1. M.Szczerbinski, J. Electrostat., **48**, pp. 145–154, (2000).
2. E. P. Krider, *Proceedings of the International Commission on History of Meteorology*,**1**, pp. 1-13 (2004).
   D. W. Zipse, IEEE Trans. Ind. Appl., **30**, 5, (1994).
3. D. W. Zipse, IEEE Trans. Ind. Appl., **37**, 2, pp. 407–414, (2001).
4. Z. a. Baharudin, N. A. Ahmad, J. S. Mäkelä, M. Fernando, and V. Cooray, J. Atmos. Solar-Terrestrial Phys., **108**, pp. 61–67, (2014).
5. R. Hartono Zainal Abidin, Ibrahim, *Forum on Lighting Protection*, Petaling Jaya, Malaysia 2004, pp. 1–39, (2004).
6. M. Abu et al., *4th Student Conference on Research and Development (SCOReD)*, pp. 27–28, (2006).
7. G. Nourirad, M. Baojahmadi, and R. Ahmed, *IEEE 7th International Power Engineering and Optimization Conference (PEOCO2013)*, pp. 417–421, (2013).
8. P. N. Mikropoulos and T. E. Tsovilis, IEEE Transation Power Deliv., **24**, 2, pp. 863–873,(2009).
9. G. S.Ait Amar and Berger, Power Tech. (2005).
10. F. D'Alessandro, J. Electrostat., **65**, 2, pp. 113–121, Feb. (2007).
11. A. Srivastava and M. Mishra, J. Electrostat., **76**, pp. 201–207, (2015).
12. A. Kern, C. Schelthoff, and M. Mathieu, Atmos. Res., **117**, pp. 2–11, (2012).
13. V. Heller, "Scale effects in physical hydraulic engineering models," J. Hydraul. Res., vol. **49**, no. 3, pp. 293–306, Jun. 2011.
14. J. L. Bermudez, F. Rachidi, A. Chisholm, M. Rubinstein, M. Hussein, and S. Chang, *2003 IEEE Symposium on Electromagnetic Compatibility*, **2**, pp. 501–504, (2003).
15. E. A. Davidson and P. A. Lefebvre, Biogeochemistry, **22**, 2, pp. 107–131, (1993).

 DOI: 10.1051/epjconf/201716201066

ASIL determination for motorbike's Electronics Throttle Control System (ETCS) mulfunction

Fakhrul Zaman Rokhani[1], *Muhammad Taqiuddin* Abdul Rahman[1,3], *Noor Ain* Kamsani[2], *Roslina* Mohd Sidek[2], *M Iqbal* Saripan[1], *Khairulmizam* Samsudin[1], *Mohd Khair* Hassan[2]

[1]Department of Computer and Communication Engineering, Universiti Putra Malaysia, Malaysia
[2]Department of Electrical and Electronics Engineering, Universiti Putra Malaysia, Malaysia
[3]Department of Electronics and Computer Engineering Technology, Faculty of Engineering Technology, Universiti Teknikal Malaysia, Melaka

Abstract. Electronics Throttle Control System (ETCS) is the principal electronic unit in all fuel injection engine motorbike, augmenting the engine performance efficiency in comparison to the conventional carburetor based engine. ETCS is regarded as a safety-critical component, whereby ETCS malfunction can cause unintended acceleration or deceleration event, which can be hazardous to riders. In this study, Hazard Analysis and Risk Assessment, an ISO26262 functional safety standard analysis has been applied on motorbike's ETCS to determine the required automotive safety integrity level. Based on the analysis, the established automotive safety integrity level can help to derive technical and functional safety measures for ETCS development.

1 Introduction

Motorbike is one of the vehicles that contribute towards air pollution. Many emissions standards focus on regulating air pollutants released by vehicles are adopted around the world and varies between countries. For example, currently, European Union countries implemented EURO3 emissions standards for two-wheel vehicles to reduce air pollutant emission [1]. For instance, EURO3 standard outlines an average CO_2 emission of 2.3g/km from the motorbike and expected to be more stringent over the years.

In-line with this standard, fuel injection engine has been widely adopted for motorbike's engine that can reduce exhaust emissions and give greater control of the air/fuel ratio compared to a standard carburetor. There are two types of fuel injection system namely the Multi-point fuel injection (MPFI) and Direct fuel injection (DFI) system [2]. The advantage of DFI system is better in fuel control that leads to improved fuel consumption efficiency and lower emission level. However, DFI system requires rugged engine to sustain higher heat and pressure in comparison to MPFI engine, resulting in higher overall cost.

The operation of fuel injection engine is controlled by Engine Management System (EMS) that is made of several main sub-systems such as fuel injection control system, ignition timing control system, air induction control system and others [28][29]. The air induction control system, also known as Electronics Throttle Control system (ETCS) comprises of dedicated sensors, actuators and engine control unit (ECU) [3], purposely to filter, meter and measure the intake air flow into the engine. The operation of ETCS involve with real-time data, fast transient responses and precise control [5-6] in order to achieve optimal air-fuel mixtures in the combustion chambers, thereby maximizing performance and fuel economy, and minimizing emissions [4]. Thus, ETCS malfunction can cause large throttle plate opening that lead to unintended acceleration hazard, which may result in life-threatening injuries [7-12].

ISO 26262:2011 provides a standard approach to functional safety management of electrical and electronics (E/E) systems for road vehicles [26]. By definition, functional safety means the absence of unreasonable risk due to hazards caused by the malfunctioning of E/E systems. These functional safety features form an integral part of each automotive product development phase, ranging from the specification to design, implementation, integration, verification, validation, and production release. The current standard addresses passenger cars with a maximum gross vehicle mass up to 3500kg and the scope will be extended to cover other vehicle types such as motorbikes, trucks and buses in the second edition.

Hazard Analysis and Risk Assessment (HARA) forms the basis of the entire functional safety activity in ISO26262, where first it determines the safety goals and its automotive safety integrity level (ASIL) for the system under design. There are four ASIL levels including A, B, C and D with the latter signifies increasing in risks and hazardous level to riders. Once the ASIL level has been established, functional safety and technical safety concepts for the system are derived, followed by applying ASIL-oriented and safety-oriented analysis to benchmark the derived concepts against the target ASIL.

* Corresponding author: fzr@upm.edu.my

Previous work in ASIL determination has been studied for various car sub-systems including airbag [13], power steering [14], battery management system [15] and electric vehicle inverter [17]. Motorbike safety integrity level (MSIL) ISO19695:2015 has been studied for sub-systems such as anti-lock braking system (ABS), combined braking system and throttle-by-cable system [16]. While throttle-by-wire system in many aspects is similar to ETCS, [16] has not concluded the finding on its MSIL level and no establishment of ASIL level either.

In this study, HARA has been applied on motorbike's ETCS to determine the required ASIL level for two types of hazard namely, unintended acceleration and deceleration. Hazardous events that comprise of hazards and operational scenarios (i.e. riding situation) have been derived. Finally, the combination of three parameters that are controllability, severity, and exposure for each hazardous event are analyzed to determine the required level of Automotive Safety ASIL level.

The next section outlines as follows. In Section II, we discussed on the methodology to determine ASIL level of ETCS functional safety by using hazard analysis and risk assessment. In section III, the ETCS's Automotive Safety Integrity Level is verified by comparing with other automotive subsystems.

2 Hazard analysis and risk assessment (HARA) for motorbike's ETCS

Hazard Analysis and Risk Assessment as stated in the ISO26262:2011 starts with hazard and operation scenario identification, followed by safety goal identification and finally the ASIL determination. For motorbike's ETCS, hazards can be in the form of unintended acceleration and unintended deceleration.

Two safety goals associated with the selected hazards are then identified as the following:

1. The throttle plate does not open (i.e. lead to acceleration) unless user twist throttle handle.
2. The throttle plate does not close (i.e. lead to deceleration) unless user release throttle handle.

Next, ASIL can be determined by evaluating hazardous events (i.e. hazard combined with operational scenario) using the parameters Severity (S), probability of Exposure (E) and Controllability (C) as accordance with Table 1. Note that ASIL A and ASIL D classes represent the lowest and highest safety integrity level compliance respectively, while class QM (quality management) denotes no requirement to comply with ISO 26262.

The definition and classification of the each parameter is described as follows:

Severity (Sx) is the estimation of the extent of harm (i.e. physical injury or damage to the health of person) to one or more individuals that can occur in a potentially hazardous event. The severity can be classified to one of the severity classes S0, S1, S2 or S3 where the class descriptions are, no injuries, light and moderate injuries, severity and life-threatening injuries with survival probable and life-threatening injuries with survival uncertain and fatal injuries, respectively.

Exposure (Ex) is the state of being in an operational situation (i.e. scenario that can occur during a vehicle's life) that can be hazardous if coincident with the failure mode (i.e. manner in which an element or item) under analysis. The classes of probability of exposure regarding operational situations will be based on extremely unusual (E0), Very low probability (E1), Low probability (E2), Medium probability (E3) and High probability (E4).

Controllability (Cx) is the ability to avoid a specific harm or damage through the timely reactions of the persons involved, possibly with support from external measures. The C classes were evaluated from C0 (control in general) to C3 (difficult to control or uncontrollable) depending on the speed level and road condition.

Table 1. ASIL Determination.

Severity[a]	Exposure[b]	Controllability class[c]			
		C0	C1	C2	C3
S0	E1	QM	QM	QM	QM
	E2	QM	QM	QM	QM
	E3	QM	QM	QM	QM
	E4	QM	QM	QM	A
S1	E1	QM	QM	QM	QM
	E2	QM	QM	QM	QM
	E3	QM	QM	QM	A
	E4	QM	QM	A	B
S2	E1	QM	QM	QM	QM
	E2	QM	QM	QM	A
	E3	QM	QM	A	B
	E4	QM	A	B	C
S3	E1	QM	QM	QM	A
	E2	QM	QM	A	B
	E3	QM	A	B	C
	E4	A	B	C	D

[a] The levels S0 to S3 classify the severity of an accident. S0 denotes lowest and S3 highest severity

[b] The levels E0 to E4 classify the exposure. E0 denotes lowest and E4 highest exposure

[c] The levels C0 to C3 classify the controllability. C0 denotes best and C3 worst controllability

Table 2 shows the ASIL determination of malfunction of motorbike's ETCS. Note that, ASIL determination of each hazardous event and type of accident will be based on worse case scenario of the result reported in [21], [17] and [18]. Also, we assumed that the rider follows closely behind a car in all hazardous events. The results are divided based on unintended acceleration and deceleration. For unintended acceleration the result are as follows:

i. Riding on expressway give two different ASIL (i.e. C and D) because of the probability of exposure for a driving situation for cruising and overtaking is different.
ii. ASIL C is assigned for two different type of accident (i.e. rear-end and overtaking) of riding on country road due to similar probability of exposure.

iii. The ASIL for rear-end accident is higher than head-on collision for similar operational scenario (i.e. riding on country road) because of higher severity class.
iv. ASIL A is given for all type of accident for riding on town road because of low speed.

Meanwhile, the results of the deceleration hazards will be as follows:

i. Riding on expressway give two different ASIL (i.e. QM and B) because of the probability of exposure to the driving situation for overtaking is higher compared to cruising.
ii. Riding on the country road has similar ASIL A for both rear-end and head-on accident due to similar probability of exposure.
iii. The operational scenario of riding on country road for rear-end and head-on accident is different which is ASIL A and ASIL B, respectively due to different controllability.
iv. Riding on town road is not considered to comply with any ASIL due to lowest risk occurring.

Table 2. ASIL determination of malfunction of Motorbike's ETCS

Hazardous event		**Type of accident**	**ID**	**S [21]**	**E [17]**	**C [18]**	**ASIL**	**Justification**
Hazard	**Operational Scenario**							
Unintended acceleration	Riding on: 1. **Straight** road with large bends[1] 2. **Cruising** at speed (> 90km/h)[2]	Rear-end [7]	Hazard_1	S3	E3	C3*	C	S3: Life-threatening injuries (survival uncertain), fatal injuries E3: The probability of exposure that 'rider follow closely behind a car' in expressway during overtaking is low. [17] C3* : Uncontrollability because of high speed as compared to [18]
	Riding on: 1. **Straight** road with large bends 2. **Overtaking** at speed (> 90km/h)	Overtaking [8]	Hazard_2	S3	E4	C3*	D	E4: The probability of exposure that 'rider follow closely behind a car' in expressway during overtaking is high.
	Riding on: 1. **Curve** with medium bend road [3] 2. **Cruising** at speed (>60km/h ≤90km/h)[4]	Rear-end[7]	Hazard_3	S3	E4	C2	C	C2: It is normally controllable respectively due to lower speed.
		Head-on	Hazard_4	S2	E4	C2	B	S2: Severe and life-threatening injuries (survival probable)
	Riding on: 1. **Curve** & medium bend road 2. **Overtaking** at speed (>60km/h ≤90km/h)	Overtaking [8]	Hazard_5	S3	E4	C2	C	Similar as Hazard_3
	Riding on: 1. **Straight** and small bend road [5] 2. **Cruising** at speed (≤60km/h)[6]	Rear end[7] Crossing[9] Turn left[10] Turn right[11]	Hazard_6	S2	E4	C1	A	C1: The controllability during riding in town road is slightly more easier because of lower speed.
Unintended deceleration	Riding on: 1. **Straight** road with large bends[1] 2. **Cruising** at speed (> 90km/h)[2]	Rear end[12]	Hazard_7	S3	E3	C0	QM	C0: Controllable in general
	Riding on: 1. **Straight** road with large bends 2. **Overtaking** at speed (> 90km/h)	Rear end[12]	Hazard_8	S3	E4	C1	B	C1: The controllability during riding in town road is slightly more easier because of lower speed.
	Riding on: 1. **Curve** with medium bend road[3] 2. **Cruising** at speed (>60km/h ≤90km/h)[4]	Rear-end[12]	Hazard_9	S3	E4	C0	A	C0: Controllable in general
		Head-on	Hazard_10	S3	E4	C1	B	Similar as Hazard_8

	Riding on: 1. **Straight** road with large bends 2. **Overtaking** (>60km/h ≤90km/h)[4]	Overtaking[13]	Hazard_11	S3	E4	C0	A	Similar as Hazard_9
	Riding on: 1. **Straight** and small bend road [5] 2. **Cruising** at speed (<60km/h)[6]	Rear end[12]	Hazard_12	S2	E4	C0	QM	S0: Assume no injuries if riding speed below than 2km/h
		Head on						
		Crossing[9]						
		Turn left[10]						
		Turn right[14]						

[1]>70m circular radius road, [2](eg. Expressway), [3]70m circular radius road, [4](eg. country road), [5]20m circular radius road, [6]e.g town road), [7]Front-rear collision, [8]Side-side collision, [9]Front-front, Front-side, Side-front collision, [10]Front-side and side-side collision, [11]Front-front and front-side collision, [12]Rear-front collision, [13]Rear-front collision, [14]Side-front collision, In all cases the distance between motorbike with vehicle is short.

3 Verification of motorbike's ETCS automotive safety integrity level

The result of ETCS's ASIL D requirement for motorbike is compared to other ASIL requirement for the automotive electronics systems equipped in car and motorbike as shown in Table III. The lowest and highest reported ASIL among the systems is QM and D respectively. The ASIL QM was established for ETCS-Drive by cable due to the maximum possible acceleration caused by fuel injector system can easily be controlled by the brake [16]. In addition, the ISO26262 standard is only applicable to electrical and electronics in automotive system.

The ASIL for car's ETCS is different with motorbike's ETCS because of the differences in severity and controllability to/of the driver and rider. Car driver exposure to harm during accident generally less harmful compared to motorbike rider due to protection from the car frame. While rider controllability is generally less than driver due to two-wheels and four wheels vehicle respectively.

4 Conclusion

In this paper, it is mainly focused on the malfunction analysis of motorbike's ETCS-Drive by wire based on unintended acceleration and unintended deceleration during overtaking hazard. The unintended deceleration in motorbike causes less risk with maximum level of ASIL A comparing to unintended acceleration. The motorbike's ETCS (i.e. Drive by wire) ASIL D was compared to the car's ETCS (Drive by wire) ASIL B whereby there were differences in severity and controllability to/of the driver and rider. The ASIL for motorbike must be appropriately evaluated because the vehicle dynamics and riding maneuvers of motorcycles differ from those of passenger cars even for similar hazardous event and type of accidents. The established automotive safety level can guide designer to equip the right safety measures for ETCS hardware development.

Table 3. The Differences of ASIL determination between CAR and Motorbike

Automotive sub-system		**ASIL/MSIL**				
		QM	**A**	**B**	**C**	**D**
Car	Airbag [13]					□
	Electric power steering [14]					□
	Power window [24]		□			
	BMS [23]				□	
	EV inverter [17]			□		
	ETCS-Drive by wire [25]			□		
Motor-bike	Anti-lock braking system (ABS) [31]				□	
	ETCS-Drive by cable [16]	□				
	ETCS-Drive by wire **(Proposed)**					□

References

1. Greening, P., European vehicle emission legislation—present and future. Topics in Catalysis, **16(1)**, 5-13, (2001).
2. Karthikeyan, G., Ramajayam, M., & Pannirselvam, A., International Journal on Engineering and Advanced Technology (IJEAT) ISSN, 2249-8958.
3. Hedrick, J. R., Pursifull, R. D., & Yang, C. U.S. Patent No. 6, 874,470, (2005). .
4. Loh, R. N., Thanom, W., Pyko, J. S., & Lee, A., Engineering, **5(07)**, 587, (2013).
5. Rossi, C., Tilli, A., & Tonielli, A., IEEE Transactions on Control Systems Technology, 8(6), 993-1002, (2000).
6. Loh, R. N. K., Pornthanomwong, T., Pyko, J. S., Lee, A., & Karsiti, M. N., International Conference

on Intelligent and Advanced Systems, pp. 1029-1035, (2007).
7. Koopman, P., A case study of Toyota unintended acceleration and software safety, (2014).
8. Kane, S., Liberman, E., DiViesti, T., & Click, F., Toyota sudden unintended acceleration, Safety Research & Strategies, (2010).
9. Pandy, A., Bennett, M. A., & Chan, B. J., U.S. Patent Application No. 13/368,443, (2012).
10. Belt, R., An Electronic Cause for Sudden Unintended Acceleration. Center for Auto Safety, (2012).
11. National Research Council (US). Committee on Electronic Vehicle Controls, & Unintended Acceleration, The safety promise and challenge of automotive electronics: Insights from unintended acceleration (Vol. 308). Transportation Research Board, (2012).
12. Anderson, A. F., Case Study: NHTSA's Denial of Dr Raghavan's Petition to Investigate Sudden Acceleration in Toyota Vehicles Fitted With Electronic Throttles. IEEE Access, 4, 1417-1433, (2016).
13. Aljazzar, H., Fischer, M., Grunske, L., Kuntz, M., Leitner-Fischer, F., & Leue, S., IEEE Sixth International Conference on Quantitative Evaluation of Systems, (QEST'09), pp. 299-308, (2009).
14. Kwon, H., Itabashi-Campbell, R., & McLaughlin, K., IEEE International Systems Conference (SysCon), pp. 655-661, (2013).
15. Taylor, W., Krithivasan, G., & Nelson, J. J., IEEE Symposium on Product Compliance Engineering (ISPCE), pp. 1-6, (2012).
16. Hasegawa, M., & Kaneko, T., SAE International Journal of Passenger Cars-Electronic and Electrical Systems, 10(2016-32-0058), 95-101, (2016).
17. Wang, Y. C., Lee, C. S., Kuo, P. C., & Lin, Y. L., IEEE International Conference on Industrial Technology (ICIT), pp. 1287-1292, (2016)
18. Kawakoshi, M., Kobayashi, T., & Hasegawa, M., SAE Technical Paper, No. 2015-32-0746, (2015).
19. Kawakoshi, M., Kobayashi, T., & Hasegawa, M., SAE International Journal of Passenger Cars-Electronic and Electrical Systems, **10**, 2016-32-0059, (2016).
20. Arai, Y., Hasegawa, M., & Harigae, T., SAE International Journal of Engines, 9, 2015-32-0714, 397-404, (2015).
21. Arai, Y., Hasegawa, M., & Harigae, T., SAE Technical Paper, No. 2016-32-0057, (2016).
22. Hasegawa, M., & Kaneko, T., SAE International Journal of Passenger Cars-Electronic and Electrical Systems, **10**(2016-32-0058), 95-101, (2016).
23. Yang, W., Yanwen, L., Chunshu, L., & Xiyang, W., IEEE 2nd Asia-Pacific Conference on Intelligent Robot Systems (ACIRS), pp. 98-103, (2017).
24. Hamann, R., Kriso, S., Williams, K., Klarmann, J., & Sauler, SAE Technical Paper, No. 2011-01-1000, (2011).
25. Bergmiller, P., Springer Berlin Heidelberg on Automotive Systems Engineering, pp. 147-202, (2013).
26. ISO, I., International Standard ISO/FDIS, 26262, (2011).
27. Ong, H. C., Mahlia, T. M. I., & Masjuki, H. H. Renewable and Sustainable Energy Reviews, 15(8), 3516-3522, (2011).
28. Guzzella, L., & Onder, C., Springer Science & Business Media, (2009).
29. Ganesan, V., Internal combustion engines. McGraw Hill Education (India) Pvt Ltd, (2012).
30. JOHANSSON, D., & KARLSSON, P. Safety mechanisms for random ECU hardware failures in compliance with ISO 26262.

Speed and area analysis on hierarchy multiplier

Moon Cheng Wong, *Razaidi* Hussin

School of Microelectronic Engineering, Universiti Malaysia Perlis, Pauh Putra Campus, 02600 Arau, Perlis, Malaysia.

Abstract. This paper proposes designs of hierarchy multiplier by utilising different designs on 4:2 and 7:3 compressor and multiple compressors. The hierarchy multipliers is optimised in the term of speed or area of hierarchy multiplier by redesigning 4:2 compressor units and introducing a combination of 4:2 compressor and 7:3 compressor units in a Vedic multiplier block. All designs are simulated using Altera Quartus II software. The aim of this paper is to improve the performance in speed by moderately increasing the area without considering the power consumption. The proposed design is 4.5% to 8.3% faster and consumes -0.5% to 5.8% less area.

1 Introduction

Multiplication is a basic arithmetic operation in processor, computation and digital signal processing, image processing application and so on. High processing speed is needed for fast multiplication, and complexity of VLSI design increases every 18 months to satisfy Moore's law in VLSI advanced technology [2-4].

Partial production generation, partial product reduction and partial production addition are three basic stages of multiplication. First, the multiplicand and multiplier are multiplied to generate partial products. Then, 4:2 or 7:3 compressor units are used to reduce the delay of partial production [5-7]. Finally, the final addition is done by Binary-to-Excess-one converter (BEC). To optimise speed, Carry Save Adder tress is implemented by utilising 4:2 and 7:3 compressor. The detail will be discussed in the next section.

Hierarchy multiplier make use of fine-grained parallelism to reduce propagation delay of computer intensive operation. Four Vedic base multiplier in hierarchical are allowed to perform multiplication in parallel to speed up the hierarchical multiplier. The compressor blocks are implemented in the four Vedic blocks [1].

2 Hierarchy multiplier

A 16-bit hierarchy multiplier make use of four 8-bit base multiplier units using Vedic multiplication algorithm. Vedic multiplication is an ancient Indian Mathematics which is based on Urdhva Tiryakbhyam algorithm. The Vedic multiplier is time, area and power efficient due to the gate delay and space increase slowly when increasing the number of Vedic multiplier bit compared to other multipliers [8-10]. All the base multipliers consist of AND-gates, compressors, half adders and full adders. Furthermore, they also contain CSA adder, Carry Select Adder (CslA) and BEC with multiplexer (MUX) for final addition of partial product [1]. However this paper is only focusing on how is the implementation of different design of 4:2 compressors and 7:3 compressors, as well as the use of both compressors in an architecture in base multipliers to improve the performance of hierarchy multiplier.

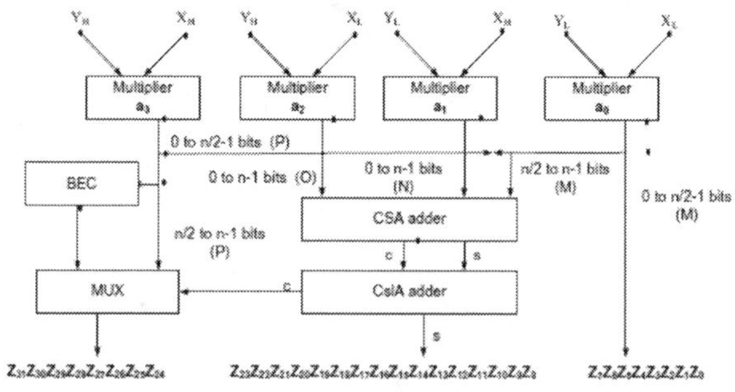

Fig. 1. Block diagram of hierarchy multiplier [1].

3 Base multiplier Architecture

A base multiplier is built from AND gates, compressors and adders. AND gates are used to generate partial product followed by compressors and adders, which are to reduce partial product. The base multiplier is illustrated in Figure 2.

Corresponding author: miowmc93@gmail.com

 DOI: 10.1051/epjconf/201716201067

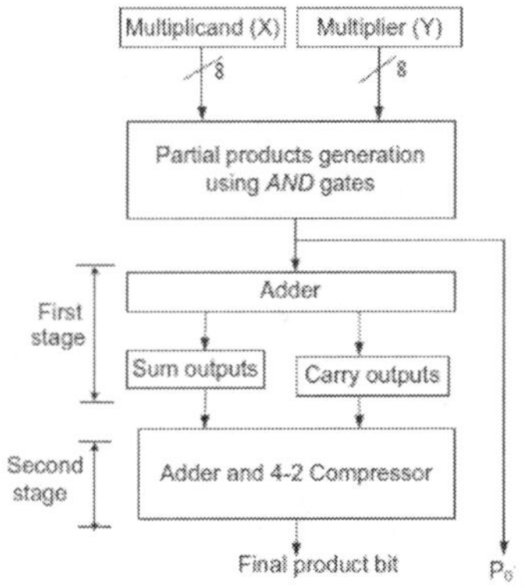

Fig. 2. Block diagram of a base multiplier [1].

3.1 A 4:2 Compressor Architecture

A standard 4:2 compressor is implemented by cascading two full adders unit as shown in Figure 3. A 4:2 compressor has four inputs (X1, X2, X3 and X4), carry input (Cin) and two outputs (Sum and Carry which along with Cout) [11-12]. The design in [1], design 1 and design 2 can be divided into two full adder blocks as shown in Figure 3.

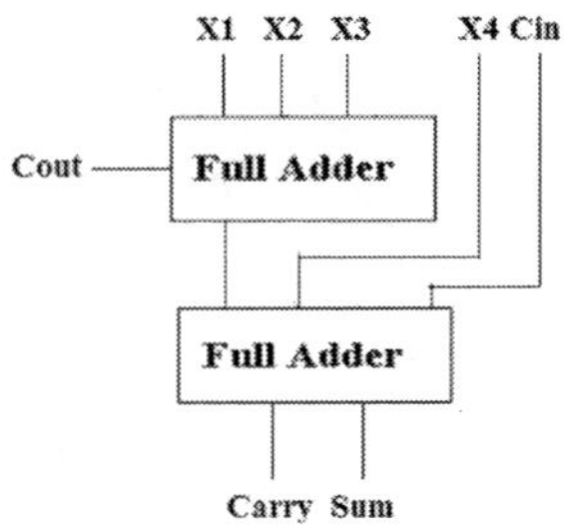

Fig. 3. Block diagram of 4:2 compressor.

3.1.1 Design in journal [1]

The 4:2 compressor in [1] utilises cascading of two full adder unit which both has two XOR gates and a MUX respectively [1]. The 4:2 compressor can be divided into two groups of full adders as shown in Figure 4.

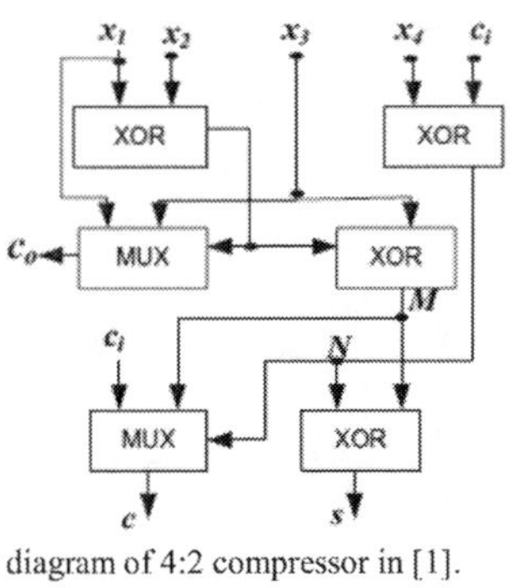

Fig. 4. Block diagram of 4:2 compressor in [1].

$$Sum = x_1 \oplus x_2 \oplus x_3 \oplus x_4 \oplus c_{in} \quad (1)$$

$$c_{out} = (\overline{x_1 \oplus x_2})x_1 + (x_1 \oplus x_2)x_3 \quad (2)$$

$$Carry = (\overline{x_4 \oplus c_{in}})c_{in} + (x_4 \oplus c_{in})(x_1 \oplus x_2 \oplus x_3) \quad (3)$$

3.1.2 Design 1

Cascading two Mux-Add units is similar as cascading two full adders in order to implement a 4:2 compressor. The mux-add in [13] suffers in number of transistors. The modified Mux-Add reduces transistor count. A XOR or XNOR consists of eight CMOS transistors respectively, both AND or OR use six CMOS transistors, MUX has four CMOS gates and NOT has two transistors. The total number of transistors in a full adder block is summarised in Table 1. The larger the design, the less the number of transistors will save cost.

Table 1: Comparison a MUX-Add unit between [13] and a Full adder unit in design 1.

Design	**Number of transistor (CMOS)**
[13]	36
Design 1	30

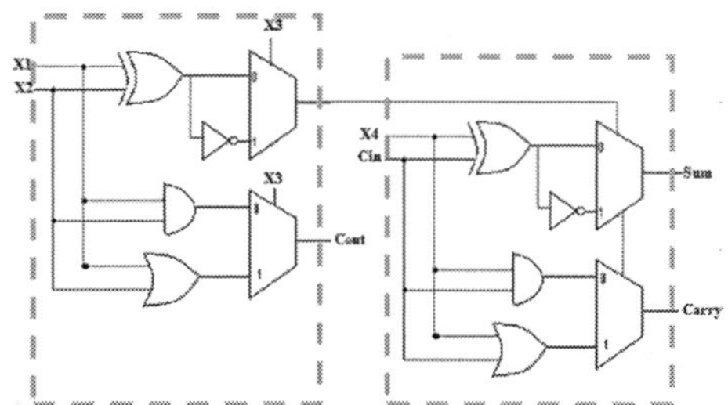

Fig. 5. Modified 4:2 compressor using two Mux-Add units.

Equations of design 1 is as below:

$$Sum = \overline{\left[(x_1 \oplus x_2)\overline{x_3} + \overline{(x_1 \oplus x_2)}x_3\right]}\,\overline{(x_4 \oplus c_{in})} + \left[(x_1 \oplus x_2)\overline{x_3} + \overline{(x_1 \oplus x_2)}x_3\right](x_4 \oplus c_{in}) \quad (4)$$

$$c_{out} = (x_1 x_2)\overline{x_3} + (x_1 + x_2)x_3 \quad (5)$$

$$Carry = \overline{\left[(x_1 \oplus x_2)\overline{x_3} + \overline{(x_1 \oplus x_2)}x_3\right]}(x_4 c_{in}) + \left[(x_1 \oplus x_2)\overline{x_3} + \overline{(x_1 \oplus x_2)}x_3\right](x_4 + c_{in}) \quad (6)$$

3.1.3 Design 2

Design 2 is designed with same number of CMOS transistors in journal 1 as shown in table2. This new 4:2 compressor utilises the sum part of Mux-Add (Figure 5) and carry part of the design in [1] (Figure 4) as shown in Figure 6.

Table 2: Comparison a full adder unit between [1] and a Full adder unit in design 2.

Design	Number of transistor (CMOS)
Journal 1	16
Design 2	16

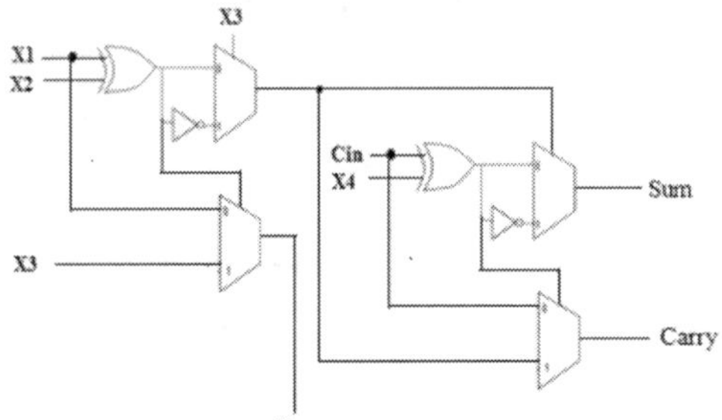

Fig. 6. A new 4:2 compressor design.

$$\text{Sum} = \text{equation (4)}$$

$$c_{out} = \text{equation (2)}$$

$$Carry = \overline{(x_4 \oplus c_{in})}c_{in} + (x_4 \oplus c_{in})\left[\overline{(x_1 \oplus x_2)}x_3 + (x_1 \oplus x_2)\overline{x_3}\right] \quad (7)$$

4 A New Base Multiplier Architecture

A new base multiplier architecture is almost similar with Figure 2 but it uses a combination of compressors which are 4:2 compressor and 7:3 compressor for optimization. The architecture is shown in Figure 7.

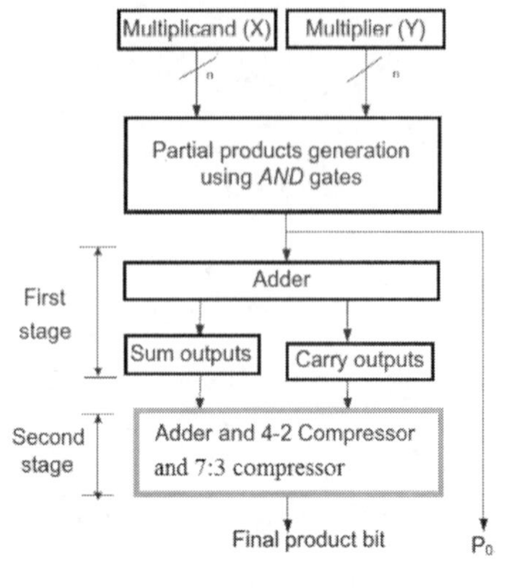

Fig. 7. Block diagram of a new base multiplier.

4.1 A 7:3 Compressor Architecture

A 7:3 compressor consists of four Full Adder (FA) blocks. The carry and the summation of inputs (X1-X6) are cascaded to different next stage full adder respectively. The FA which carries summation output from the first two FA will add up with X7 to produce sum output. The summation of carries from the first two full adders will produce output Carry and Cout. However the carry ($Carry_{i+1}$) is propagate to next two stage. The 7:3 compressor allows parallel addition when the sum and carry propagate to different next stage of FA. The 7:3 compressor is used in partial product reduction, so seven partial products which are generated from AND gate can be added up at the same time. A standard 7:3 compressor is shown in Figure 8. This section highlights the existence of 7:3 compressor for summation of seven partial products in parallel at the centre of compressor tree, which lead to faster multiplication process. This is due to hidden critical path of carry signal that is reduced by optimising the connection of compressor topology [14-16].

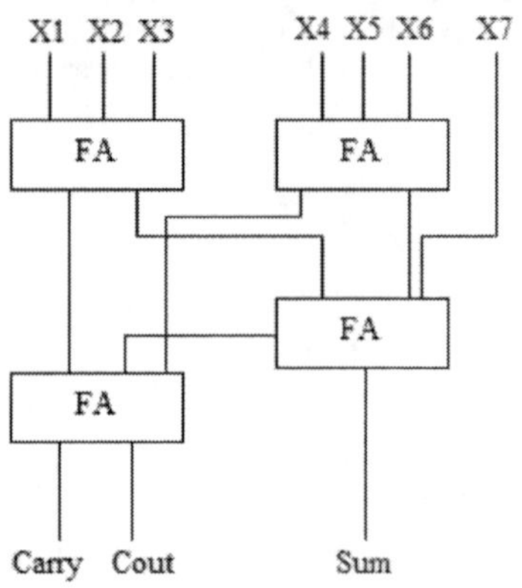

Fig. 8. A standard 7:3 compressor design [14-15].

4.1.1 Design 3

Design 3 is implemented using the design in journal [1] (Figure 4) with only a full adder unit design and duplicated by four times. The block diagram of design 3 is shown in Figure 9. Each full adder unit contains two XOR gates and a MUX.

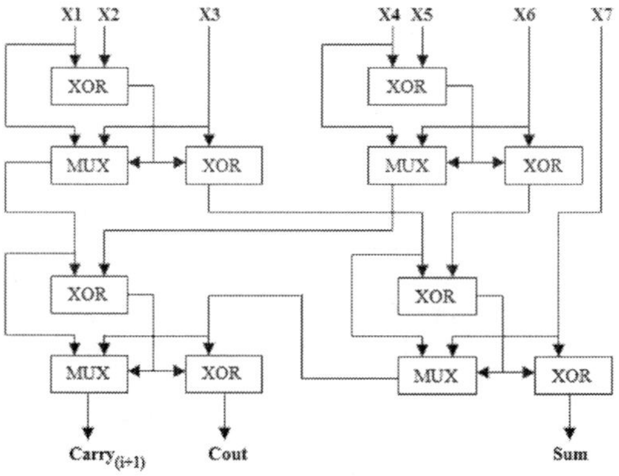

Fig. 9. A 7:3 compressor design by utilise a full adder design unit in [1].

4.1.2 Design 4

Design 4 utilises only a full adder block in design 2 and connect the other three full adder units as shown in Figure 10. Each full adder consists of a XOR gate, a NOT gate and two MUX.

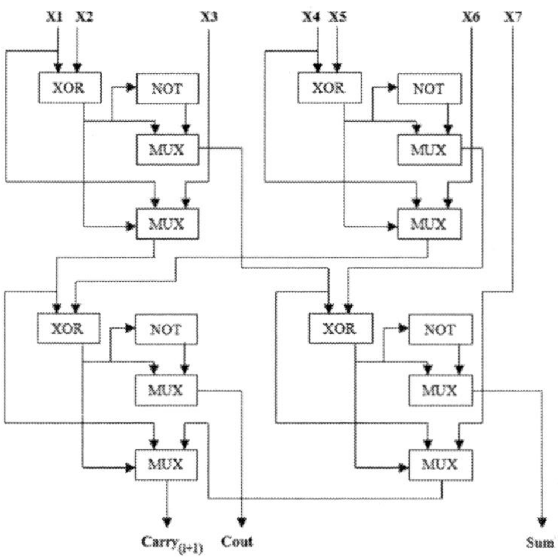

Fig. 10. A 7:3 compressor design by utilise a full adder design 2.

5 Simulation result and Analysis

All simulation is simulated in EP2C70F896C6 device from Cyclone II family in Altera Quartus II software by describing using Verilog Hardware Description Language. The functional verification of the 16-bit hierarchy multiplier is done by simulator tools in the software. The performance is analysed after each design is successfully debugged.

5.1 Functional Verification

The function of multiplier is verified before proceeding to performance analysis. The inputs are set from the lower to the higher range to ensure that the multiplier perform correct output and propagation of carry to every components is accurate as shown in Figure 11. The boundary condition is considered in this design since the use of four base multipliers in performing multiplication. The consideration of boundary condition is to verify that the input is propagating properly between base multiplier blocks. A set of inputs is used to verify the functionality of boundary issue as shown in Figure 12.

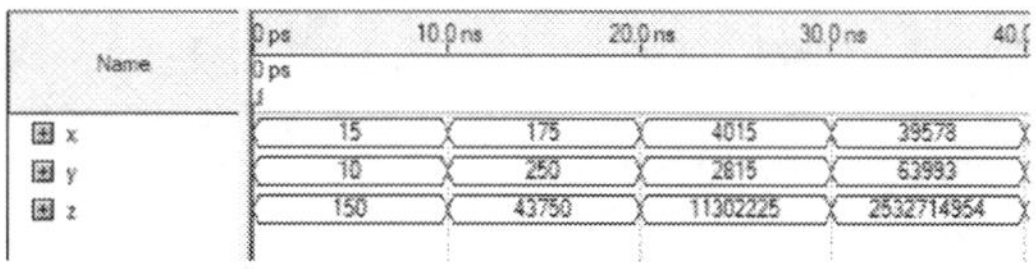

Fig. 11. Waveform from lower to higher range.

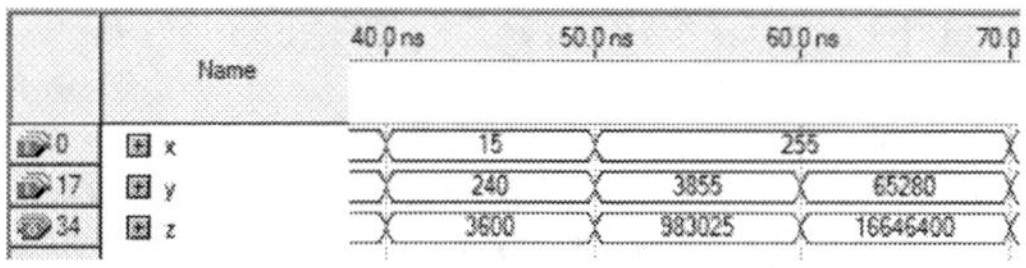

Fig. 12. Waveform for boundary condition.

5.2 Performance Analysis

The longest total propagation delay (tpd) and total logic elements or area is used to analyse the performance of hierarchy multiplier. The performance is summarized in Table 3. Since the aim of this paper is to optimise speed and area of the hierarchy multiplier, the percentage (%) improved is calculated using the formula below.

$$\%\,improved = \frac{Performance\ of\ journal\ [1] - performance\ of each\ design}{Peformance\ of\ journal\ [1]} \times 100\%$$

Table 3: Comparison the performance among designs.

Design	Longest delay tpd (ns)	% improved in tpd	Total logic elements	% improved
Journal [1]	32.758	-	636	-
Design 1	31.297	4.5	599	5.8
Design 2	31.102	5.1	619	2.7
Design 3	30.252	7.7	639	-0.5
Design 4	30.030	8.3	636	0.0

 DOI: 10.1051/epjconf/201716201067

From Table 3, both design 1 and design 2 are improved in terms of speed and area compared to journal [1] design. Design 1 reduce in 4.5% delays and consume less 5.8% area as well as design 2 improve 5.1% in terms of speed and 2.7% less area compare to journal [1]. Nevertheless, design 3 sacrifices 0.5% area and design 4 has no impact on area but both designs perform more significant effect in optimising the speed which are 7.7% and 8.3% respectively.

Design 1 is suggested for designers who suffer in limitation of size without increase in latency but designer who have plenty of space is more suitable to use design 4. Design 2 is more appropriate to have overall performance of hierarchy multiplier.

Conclusion

An efficient hierarchy multiplier is developed by redesigning 4:2 compressor and utilities multiple compressors which are 4:2 compressor and 7:3 compressor in partial production reduction. The use of design 1 to modify the 4:2 compressor architecture consumes less 5.8% area than previous design in journal [1]. The proposed designs improve the speed of hierarchy multiplier from 4.5% to 8.3%. The proposed multiple compressor has disadvantage in area consumption compare to design 1 and design 2, but it is faster compare to only using 4:2 compressor. Design 2 performs the best in terms of speed and area, with reduction of 5.1% and 2.7% respectively.

We appreciate and thanks Universiti Malaysia Perlis for providing financial support to produce this paper.

References

1. M. Shoba and R. Nakkeeran, Eng. Sci. Technol. an Int. J., **20**, no. 1, pp. 321–331 (2017)

2. P. J. Denning and T. G. Lewis, Communications of the ACM, **60**, no. 1, pp. 54–65 (2016)

3. S. Saravanan, I. Vennila, and S. Mohanram, Circuits and Systems, 7, pp. 630–642 (2016)

4. J. Wu et al., Appl. Comput. Intell. Soft Comput., **2013**, pp. 1–13 (2013)

5. P. D. Gopineedi, H. Thapliyal, M. B. Srinivas, and H. R. Arabnia, *Proceedings of the 2006 International Conference on Embedded Systems & Applications*, pp. 1–7 (2006)

6. K. Prasad and K. K. Parhi, Asilomar Conf. Signals, Syst. Comput., pp. 129–133 (2001)

7. R. Hussin, A. Y. M. Shakaff, N. Idris, Z. Sauli, R. C. Ismail, and A. Kamarudin, *2008 Int. Conf. Electron. Des. ICED*, (2008)

8. D. Kayal, P. Mostafa, A. Dandapat, and C. K. Sarkar, J. Signal Process. Syst., **76**, no. 1, pp. 1–9 (2014)

9. S. R. Vaidya, D. R. Dandekar, *Proceedings of the 2011 International Conference on Communication, Computing & Security*, pp. 383–386 (2011)

10. N. B. Reddy and K. Wasimakram, International Journal of Emerging Technology in Computer Science & Electronics (IJETCSE) **19**, no. 1 (2015)

11. T. U. Anusree and P. L. Bonifus, International Journal of Computer Trends and Technology (IJCTT) **36**, no. 4, pp. 213–218 (2016)

12. A. Pishvaie, G. Jaberipur, and A. Jahanian, Canadian Journal of Electrical and Computer Engineering, **36**, no. 3, pp. 6–8 (2013)

13. I. Banerjee, Int. Journal of Engineering Research and Applications, **3**, no. 6, pp. 75–78 (2013)

14. S. Mehrabi, K. Navi, and O. Hashemipour, *2013 IEEE 5th International Nanoelectronics Conference (INEC)*. pp. 322–325 (2013)

15. N. Ravi, T. Jayachandra Prasad, M. Umamahesh, and T. Subba Rao, *Proc. Int. Conf. "Recent Adv. Sp. Technol. Serv. Clim. Chang. - 2010", RSTS CC-2010*, **8**, no. 3, pp. 189–193 (2010)

16. V. Oklobdzija, D. Villeger, and S. Liu, IEEE Transactions on Computers, **45**, no. 3, pp. 294–306 (1996)

RTL validation methodology on high complexity wireless microcontroller using OVM technique for fast time to market

Nurul Zhafirah Muhammad[1], *A.* Harun[1,2], *N.A.M.A.* Hambali[1,2], *S. A. Z.* Murad[1], *S. N.* Mohyar[1], *M.N.* Isa[1], *AB* Jambek[1]

[1]School of Microelectronic, Universiti Malaysia Perlis, 02600 Arau Perlis
[2]Centre of Excellence for Advanced Sensor Technology (CEASTech)

Abstract. Increased demand in internet of thing (IOT) application based has inadvertently forced the move towards higher complexity of integrated circuit supporting SoC. Such spontaneous increased in complexity poses unequivocal complicated validation strategies. Hence, the complexity allows researchers to come out with various exceptional methodologies in order to overcome this problem. This in essence brings about the discovery of dynamic verification, formal verification and hybrid techniques. In reserve, it is very important to discover bugs at infancy of verification process in (SoC) in order to reduce time consuming and fast time to market for the system. Ergo, in this paper we are focusing on the methodology of verification that can be done at Register Transfer Level of SoC based on the AMBA bus design. On top of that, the discovery of others verification method called Open Verification Methodology (OVM) brings out an easier way in RTL validation methodology neither as the replacement for the traditional method yet as an effort for fast time to market for the system. Thus, the method called OVM is proposed in this paper as the verification method for larger design to avert the disclosure of the bottleneck in validation platform.

1 INTRODUCTION

1.1 The growth of integrated circuit

Over the last decade, rapid advancement in portable technology, wireless communication, audio, video and compact computers has drawn interest from electronic industry to venture themselves into the field of integrated circuit [1]. The growth in the integration density of microcontroller system followed closely the Moore's Law. In articles of [2], the author presented Moore's Law as a law that predicts that the number of transistors that can be placed on a single integrated circuit doubles about every two years.

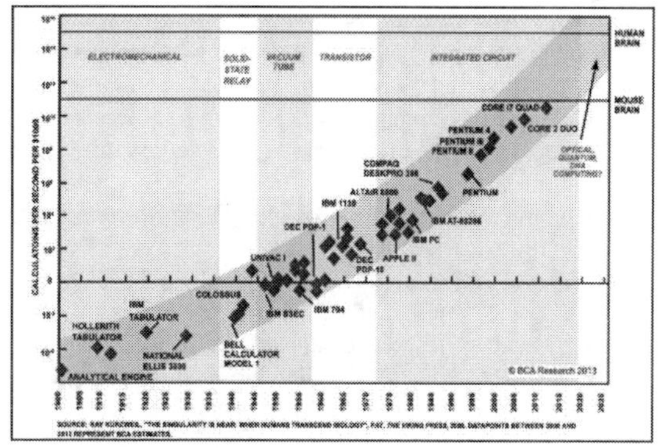

Fig. 1. Moore's Law [3]

Based on Figure 1, Moore's Law reported that the number of transistor on microchip doubled for every two years as well as the performance of the chip such as the transformation of the first home computer to more sophisticated machines. From that transformation also gave rise to high speed internet, smart phones, refrigerators and other devices that are becoming prevalent today [2-4]. However, this principle that powered the information technology since 1960s is near its end. Thus, in [4], Sutherland focused on the 'More than Moore's' strategy which starts with applications from smartphones and supercomputers to data center in

* Corresponding author: aziziharun@unimap.edu.my

DOI: 10.1051/epjconf/201716201068

the cloud and work downwards to see what chips are needed to support them such as power management and other silicon. This can be related to the introduction of sophistication on Internet-of-Things (IoT) which asserts that every object are connected to the internet and participating together in a system [5] as shown in the figure below.

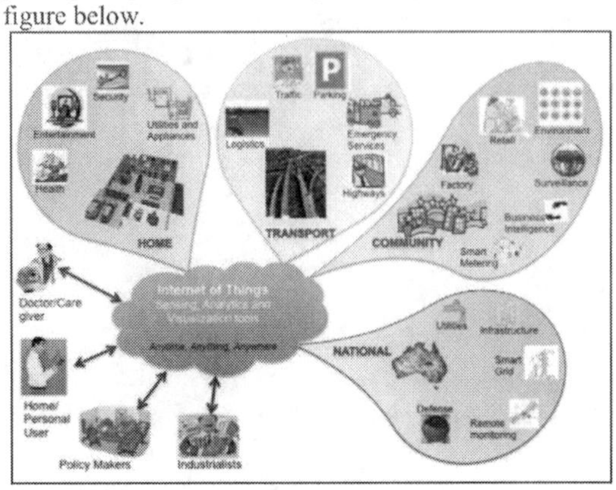

Fig. 2. Visualize of Internet of thing's application [6]

1.2 System-on-chip (SoC)

Hence, this has led to the advancements in the integrated circuit design in order to fulfill demand driven development by IoT solution, product requirement and potential intellectual property (IP) in the system. A well-known, the integrated circuit is a technology advancement that allows the integration of various parts to be put together into a chip such as digital and analog circuit, mix signal and others resulted in complexity of the chip and its design flow [7]. However, there is a limit to integrated circuit (IC) technology to allow only simple integrated transistor into a chip. Nevertheless, to overcome these limitation of integration, the large scale integration has been proposed such as system on chip (SoC) and wafer-scale integration (WSI) [7]. Even so, as the demand from clients to have technologies that allow the system to be able to integrate all the sophisticated electronic systems on a single chip tend the designers to design microcontroller based on system on chip (SoC), with the intension of reusing IP modules described in hardware description language (HDL) [8]. An example of these IP modules are microprocessors, digital signal processing, memory blocks, PLL and other peripherals as shown in Figure 2. Figure 3 shows that the SoC is designed based on AMBA bus design which comprised of ARM processors, ARM support modules, and memory [9]. Nowadays, the AMBA is one of the leading busing system used in high-performance SoC design that has been developed to become a family with different bus protocol such as ASB, APB, AHB, AXI and et al [9]. These IP then will be integrated together with the bus system to form the complex SoC. This is the starting point to start the validation process known also as a functional verification or validation. When choosing methodology to follow for such platform validation, it is imperative to understand the requirement and the restriction a program had on the development to ensure the successful execution. This paper reviews the various methods used in SoC validation in the industry and proposes a suitable RTL validation methodology for wireless microcontroller system.

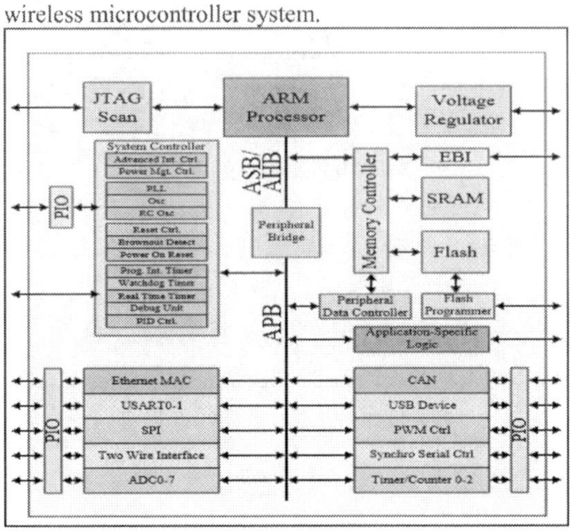

Fig. 3. Microcontroller based system on chip [10]

As presented in [11-13], the functional verification is divided into three methods: 1. Dynamic verification (simulation based), 2. Formal verification (static) and 3. Static Timing Analysis. However, in this paper the static timing analysis is not going too discussed in details because it is the method for full chip validation. This paper is organized as follows: An introduction is discussed in section 1, the methodology used is discussed in section 2 that divided by three subsection which is overview of the simulation will be presented in subsection 2.1, the formal verification methods are described in subsection 2.2 , while the recommended methodology by industry: hybrid technique is presented in subsection 2.3. Finally, section 3 summarizes the validation method available and propose the suitable validation methodology for wireless microcontroller at Register Transfer level.

2 Methodology

2.1. Simulation techniques

Dynamic verification is known as traditional approaches [14] of the existing method in industry and generally carried on at two levels [15]: module level and chip level. The module level is defined as logical unit which is represented by a set consisting of timing information and functions which are applicable to design verification of system which the main components are medium-scale integration (MSIs) and large-scale integration (LSIs) [16]. For each level, the proposed technique performed is by initiating the logic simulation technique for analyzing models to verify its function and performance by using simulators. Advertently, the authors in [13] presented the classification of simulators in three broad categories: HDL- based, schematic based, emulators and et al. However, in this paper focusing only on the HDL- based approaches to be applied for industrial environment. The HDL- based can be divided into two categories: event-driven simulator and cycle-based simulator [13]. Authors in [17-18] reported event-driven simulator operates by taking events, one at a time through the design until a steady state condition is achieved and any change in input stimulus is identified as an event. The design models include timing and functionality. Nevertheless, the event-driven simulator limited only for small design otherwise the speed of simulation becomes slow due to evaluation of output multiple times produced by this simulator [10] and [18-19]. Meanwhile, the approach by cycle-based simulator is introduced as the complementary of the limitation in event-driven simulator due to its ability for simulating very large designs. In articles of [17-18] and [20], the authors presented the cycle-based simulator which does not consider the time within clock cycle and only functions on synchronous logic circuit. There are some applications where this approach can be used [17], such as digital signal processing (DSP) chip, SoCs with processor and peripherals and others. Furthermore, the authors in [21] stated in his paper to use this approaches as one of simulators in performance evaluation and validation of microprocessor. This cycle-based simulator can be used also by combining it with other techniques: Decision diagram (DD) in order for achieving significant speed up of the simulation execution [22]. This DD techniques is used for representing the function of a design at RT and behavioral level. On the other hand, the author in [23] presented activity cycle diagrams (ADC) for modelling construction process using activities as the basic element into an equivalent executable Simulation Language for Alternative modelling (SLAM) [24] simulation model. Even so, the cycle based simulators also has a limitation for cannot detect glitches in the design and do not take the timing of design into consideration resulted the simulator to be performed using static timing analysis tool [19]. There are few of commercial tools provided by EDA to support the simulation for both techniques [13]: Verilog-XL (Verilog) Cadence, Sciriccor (VHDL) Synopsys, VCs (Verilog) Synopsys and others. Even though, the simulation is easy, flexible and economical, the simulation speed is extremely slow causing the simulation times became unreasonably long. Hence, the simulation using emulator is suggested to overcome the limitation [11].

Emulator is where the design is mapped into FPGA hardware for prototype simulation [13]. This stage is known as co-emulation: when hardware and software are simulated simultaneously [11]. The emulators are pre-packaged prototyping FPGA arrays that offer comparatively high debug visibility and run faster than simulator [25]. In [26], the authors presented four modes of co-simulation which can occur: RTL-FPGA co-simulation, SystemC-FPGA, Vector prototyping and In-circuit prototyping. However, there is still a continuous debate about FPGA prototyping versus emulation. Hence, in [27], the authors stated that FPGA prototyping addresses mostly validation of smaller design or a single IP while emulation is being used for large SoC designs. Nevertheless the authors in [28] reported that FPGA can be also uses as a basic for software development, hardware development, etc. as long as before the final silicon is available. In addition, FPGA prototyping is a method to prototype SoC on FPGAs for hardware verification and early software development [28].

2.2 Formal verification

The formal techniques show the correctness of a design by providing analytical proofs. This methodology is being introduced in validation methodology because not all the modules are well suited for simulations techniques. In [29], the authors presented the formal technique adopted by Brazil Semiconductor Technology Center (BSTC) which further divide into three additional techniques: model checking, theorem proving and equivalence checks.

Model checking is an important technique for verifying sequential. In this technique, the authors in [30] mentioned that the specification of a design is expressed in temporal logic and the implementation is designed as finite state machine. This technique has two main approaches that have been presented [30]: SAT – based and BDD-based. BDD-based [31-32] is a canonical form for Boolean expressions which is used to symbolically represent the set of states. This approach is capable of handling system with hundreds of state variables causing no space efficient variable ordering [31]. Fortunately, the SAT- based approaches is more recommended by researcher due to its potential to handle much larger design [31]. Same goes to the equivalence checking, there are approaches that have been presented such as equivalence checking using cuts and heap [32] and Boolean reasoning for equivalence checking [33]. The various verification techniques offer different advantages and features. **Table 1** shown the comparison between the verification techniques discussed above.

Table 1. Comparison of various verification techniques

	Event-based simulation	Cycle-based simulation	Emulation	Formal verification
Function	Yes	Yes	Yes	No
Abstraction level	Behavioral, RTL, Gate	RTL, Gate	RTL, Gate	RTL, Gate
Functional equivalence	Yes	Yes	Yes	Yes
Timing	Yes	No	No	No
Gate capacity	Low	Medium	Very high	High
Run time	<10 cycles	1K Cycles	1M Cycles	Medium
Cost	Low	Medium	High	Medium

2.3 Hybrid technique

In addition, the hybrid technique has been introduced to complement the imperfection of the simulation and formal verification techniques in order to perform full validation of the designs. In [34-35], the authors presented a convincing argument supporting hybridization to obtain better functional test coverage of large design. Same goes to [36] which also supported the hybrid techniques wherein the formal verification is used for test models to derive set of functional vector following simulation for design validation. This technique can be classified in four hybrid functional-verification methods [36]: 1. Combining formal and informal techniques 2. Combining two formal techniques 3.combining two informal techniques 4. Combining multiple verification techniques. Nevertheless, in [35], the authors presented hybrid incremental assertion based verification (HI-ABV) approaches for accurate verification of RTL implementations derived from TL to RTL wherein ABV also supported two methods; dynamic verification using simulation and formal using model checking. ABV in [37] approaches also allowed verification procedure and enable early bug detection. There is another approach that have been proposed by EDA and academic researcher for support of this hybrid methodology known as Transactor-based verification used to avoid the error from transaction level to RTL level [38].

On top of that, recently there is a new methods that have been introduced and already used in a few electronics companies even it might be unfamiliar for some but this is the suggested method by verification academy called Open Verification Methodology. The OVM is a methodology for functional verification using System Verilog that has been introduced to the world with the first version of OVM 1.0 in January 2008 followed by latest version is OVM 2.1.2 released in January 2011 [39], [40]. Advertently, the traditional RTL code and test bench written in traditional language (Verilog/VHDL) is converted to a suitable format to verify the functionality of the designs, includes the formal checking, logic simulation, transaction-based verification and coverage analysis which cause the unreasonable long time for verified it [19], [40]. Inversely, the OVM test bench are complete verification environment that composed of reusable verification components and used as part of overarching methodology of constrained random, coverage-driven verification and functional checking [19], [40]. Evidently, the superiority of the OVM specification reduce the gap of time for wireless microcontroller unit to verify the designs thus fast time to market. Moreover, the verification environment of this method can be developed by providing the System Verilog based supplementary class library and such environments are easy to build and maintained since all the components in OVM interact with each other via standard transaction level modeling [40]. In addition, all simulation-based verification unintentionally suffers from the issues that not run enough the test vectors to exhaustively test the whole design. Fortunately, the technique of constrained random stimulus able to overcome this issues as much as it is fully supported by SystemVerilog and OVM. As such, to improve validation lead time and overcome the limitations of traditional methods, we would like to propose using OVM method as the main RTL validation method for our wireless microcontroller.

3 Conclusion

This paper presented the existing of various RTL validation methodologies used in current IC design platform in electronic industries. RTL platform is being validated by following the design flow of SoC. The RTL is normally being validated by three methods: dynamic verification, formal verification and hybrid technique. However, the first two traditional methods seemed to be unsuitable for current electronic industries due to the complexity of the designs and the short lead time required for effective time to market. Hence, the more prevalent, well-established methodology used in verification is OVM. This technique is not only reducing the effort of verification but also more economical for whole system design. Due to these factors, we propose that the RTL validation effort for wireless microcontroller is performed using the OVM method.

References

1. G. Templeton, W.I.M'L (2015)
2. M. Miti, M. Stojč, and Z. Stamenkovi, A.O.S.B, *Digit. Syst. Appl.* **11**, 7, 1–17 (2007)
3. M.L (2017)
4. W. Ahmed, S. Brandstätter, and M. Huemer, V Open Source CPU Cores using Instruction Set Simulators in OVM Environments."
5. M. Cadence, OVMUG (2008)
6. P. Rashinkar, P. Paterson, and L. Singh, SACVMT (Kluwer academic)
7. S. Iman, SFV SVOVM (Hansen Brown) (2008)
8. A. H. C. Oliveira, R. Takiguti, L. Francisco, and P. J. De Andrade, D8-BMUSOACM
9. S. Sutherland and T. Fitzpatrick, UVMRA: APSUVM, *DVCon,* March, 1–28 (2015)
10. A. Biere, A. Cimatti, E. M. Clarke, M. Fujita, and Y. Zhu, SMCUSATP IBDDs (1999)
11. P. Chauhan, E. Clarke, and Y. Lu, SSI, **1**, 12 (2004)
12. W. Soto Encinas Jr and C. M. Augusto Dueñas CesarDuenas, FVi8bMACS
13. W.CommonsARMSBD
14. A. Dissertation, ASHVSS, August (2003)
15. M. Tokoro, M. Sato, M. Ishigami, E. Tamura, T. Ishimitsu, and H. Ohara, AMLSTFSCOLsi's Msi's, 418–427 (2006)
16. DVSoC
17. E. Staff, DLS:ECHB
18. A. Kupriyanov and F. Hannig, HSEDRTLCS, **3133**, 519–529 (2004)

19. R. Showcase et al. SMCBDDsSMCBDDs (1999)
20. L. Pierre and Z. Bel Hadj Amor, ARRVTSDF, *International Conference on Hardware/Software Codesign and System Synthesis, CODES+ISSS* (2013)
21. C. Y. Huang, Y. F. Yin, C. J. Hsu, T. B. Huang, and T. M. Chang, SHW/SWVV, *Proceedings of the Asia and South Pacific Design Automation Conference, ASP-DAC* (2011)
22. R. Ubar, A. Morawiec, and J. Raik, C-Bsdd, *Proc. - Design, Autom. Test Eur.* 454–458 (1999)
23. RTLDVHDL
24. ICDesign (2015)
25. M. Y. Vardi, FTSLV
26. IC
27. BDESCS
28. A. Gupta and S. Malik, TFVMUSC
29. N. Amla, X. Du, A. Kuehlmann, R. P. Kurshan, and K. L. McMillan, AASATMCTIE. *Lecture Notes in Computer Science (including subseries Lecture Notes in Artificial Intelligence and Lecture Notes in Bioinformatics)* (2005)
30. M.Cadence, EFPGAP-SDV
31. P.P.Chu, FPVHDLE (2008)
32. A. Kuehlmann and F. Krohm, ECCH
33. A. Kuehlmann, V. Paruthi, F. Krohm, and M. K. Ganai, RBRFECFPV, *IEEE Trans. Comput. Des. Integr. Circuits Syst* (2002)
34. N. Bombieri, F. Fummi, G. Pravadelli, and A. Fedeli, HIABVTLMDF *IEEE Des. Test Comput.* (2007)
35. N. Bombieri, F. Fummi, and G. Pravadelli, OE TBVRTLM AT RTL
36. J. Bhadra, M. S. Abadir, L.-C. Wang, and S. Ray, A SHTFV
37. J. J. O. Reilly, *Simulation* (1989)
38. P. Bose, PEVM, *ACM SIGMETRICS Perform. Eval. Rev.* (1999)
39. A. Kuehlmann, V. Paruthi, F. Krohm, and M. K. Ganai, RBRECFPV, *IEEE Trans. Comput. Des. Integr. Circuits System.* **21**, 12 (2002)
40. Doulos GSOVM

Implementation of floating gate MOSFET in inverter for threshold voltage tunability

F.A.S. Musa[1,*], *Nurulain* D.[1], *N.* Ahmad[1], *M.* Mohamad Isa[1] *and Muhammad M.* Ramli[1]

[1]School of Microelectronic Engineering, Universiti Malaysia Perlis, Kampus Pauh Putra, 02600 Arau, Perlis Malaysia

Abstract. This paper presents the ability of floating gate MOSFET (FGMOS) threshold voltage to be programmed or tuned which is exploited to improve the performance of electronic circuit design. This special characteristic owns by FGMOS is definitely contributes towards low voltage and low power circuit design. The comparison of threshold voltage between FGMOS and conventional NMOS is done in order to prove that FGMOS is able to produce a lower threshold voltage compared to conventional NMOS. In addition, in this paper, an implementation of FGMOS into inverter circuit is also done to show the programmability of FGMOS threshold voltage. The operations of the inverter circuits are verified using Sypnopsys simulation in 0.1μm CMOS technology with supply voltage of 1.8V.

1 Introduction

The vast progress of portable electronic devices has encouraged further studies on low voltage and low power analog circuit design. In portable devices in the field of biomedical, it is very crucial to have longer battery life because it needs to be implanted within the patients for a long term, and such requirement can be achieved with low power consumption. To achieve low voltage and low power circuit design, a distinct number of techniques have been approached by researchers, such as bulk-driven, self-cascode, current mirror, and floating gate MOSFET (FGMOS) techniques [1]–[14]. Moreover by scaling down the feature size of the device, the reduction of operating supply voltage is obvious whereas the speed of the device might be outflowed. Therefore, to achieve the optimum performance and low power digital circuits, a new and alternative circuit design is required to be implemented. Due to ability of FGMOS in providing multiple inputs structure, the number of transistors required in analog circuit design can be reduced, which results in circuit with less complexity. By implementing fewer number of transistors within the circuit, less current branches will be required resulting in lower power consumption. Low voltage supply circuit is able to be realized by implementation of FGMOS in which it offers the capability of tuning the effective threshold voltage [8],[6],[2],[15]. With the ability to tune and scale down the effective threshold voltage, the supply voltage required to operate the circuit is decreased [15]. Another advantage of FGMOS technique is that it is compatible in all CMOS circuit design. The paper is organized as follows: in Section 2, the basics of FGMOS are explained; the simulation of FGMOS circuit and the comparison between FGMOS and conventional NMOS is discussed in Section 3, while Section 4 present the implementation of FGMOS in inverter circuit and Section 5 conclude the paper.

2 Floating gate MOSFET (FGMOS)

The equivalent circuit and symbol of n-inputs FGMOS are shown in **Fig 1 (a)** and **(b)** respectively. FGMOS is an altered version of the standard MOSFET in which it is fabricated by isolating the gate of conventional MOSFET electrically, therefore resulting zero resistive connections to its gate or more specific known as floating gate. Meanwhile, a number of secondary gates are electrically isolated from the floating gate and deposited above it [15].

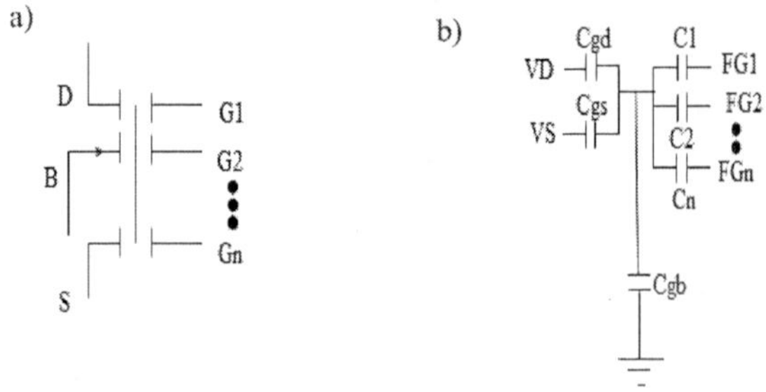

Fig. 1. a) Symbol of n-inputs FGMOS; b) Equivalent schematic circuit of n-inputs FGMOS.

* Corresponding author: F.A.S.M @amirasyakirahmusa@yahoo.com

By utilising the charge of conservation law on the floating gate, the voltage at floating (V_{FG}) can be expressed below in (1) [2], [6], [8], [15]:

$$V_{FG} = [C_1V_1 + C_2V_2 + ... + C_nV_n + C_{gd}V_d + C_{gs}V_s + C_{gb}V_b + Q_{fg}] / C_T \quad (1)$$

In which, $C_T = C_1 + C_2 + ... + C_n + C_{gd} + C_{gs} + C_{gb}$ and the parasitic capacitances of floating gate with drain, source and bulk are represented by C_{gd}, C_{gs} and C_{gb} respectively. The residual charge trapped at the floating gate during fabrication process is represented by Q_{fg} which can be eliminated by utilizing ultraviolet (UV) light, tunnel effect hot electron injection or forcing an initial condition with a switch [15],[16]. From the equation (1), it clearly show that FGMOS has no much differences compared to conventional MOSFET with a weighted sum of voltages applied onto its gate. The weights are produced by the ratios between each input capacitance and the total capacitance at the floating gate node. In consequences of the addition of the weights, in which is able to be realised without in need of using any additional circuitry, other than a number of capacitors. Therefore, the application of FGMOS can simplify certain circuit topologies and have a very positive significance on the circuit design in terms of area and power consumption. Furthermore, for the circuit designs that required to be operated by pushing the limits of the current technology, it is very crucial for the circuits to be designed in a simple manner [15].

Meanwhile the effective threshold voltage, $V_{th\ (eff)}$ is presented as below in (2) [2], [6], [8], [15]:

$$V_{th\ (eff)} = [V_T - k_nV_n] / k_1 \quad (2)$$

Where, $k_1 = C_1 / C_T$, therefore, it can be seen that the $V_{th,\ (eff)}$ is able to be controlled by a suitable selection of bias voltage, V_T and ratios of capacitance. Hence, making it possible to reduce the value of $V_{th(eff)}$ that contributes towards low voltage and programmable electronic devices [2].

3 Comparisons of FGMOS and NMOS

The ability of tuning and scaling down the threshold voltage enhances the gate delay time leading to a higher operation speed. Moreover, it is very crucial to reduce the threshold voltage when reducing the supply voltage.

In **Fig. 2**, a proposed test circuit of FGMOS with 2 inputs and a conventional NMOS is presented.

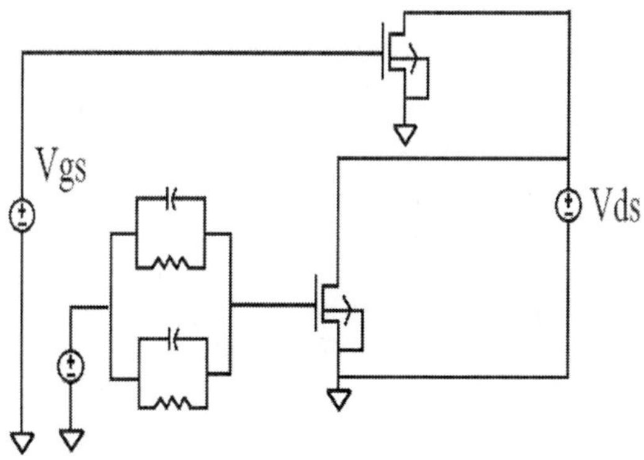

Fig. 2. Proposed test circuit of 2 inputs FGMOS and a conventional NMOS.

In the proposed test circuit, the transistors are using technology of 0.1μ with width of 20μ m and the supply voltage of 750mV. The output characteristics on comparative basis of drain current, I_d versus source-gate voltage, V_{gs} is simulated and shown in **Fig. 3.**

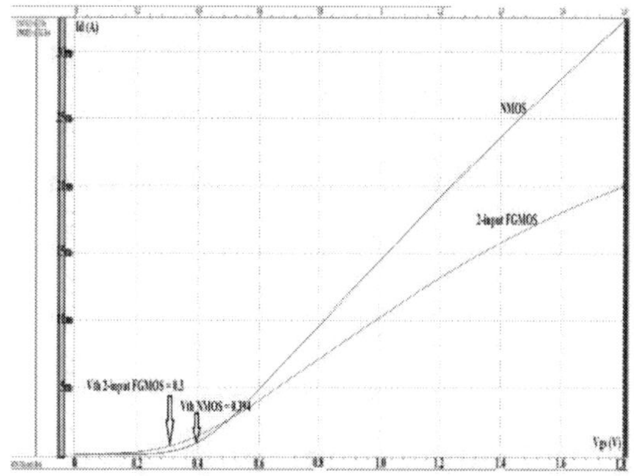

Fig. 3. Graph of drain current, I_d versus source-gate voltage, V_{gs}.

It can be clearly seen in **Fig. 2** that the threshold voltages, V_{th} produced by FGMOS is 0.3V which is lower than the one that is produced by the conventional NMOS, 0.397V. Hence, it is proven that FGMOS is able to produce lower V_{th} in which can be done by controlling it electrically through changes made at the voltages of the additional inputs and the tuning value of capacitors.

This unique advantage owned by FGMOS allowing the programming of the signal levels individually on each device as required by the needs of the specific circuit design. In addition, by referring to equation (2), and the results from the simulated graph of I_d versus V_{gs}, we are likely be able to scale down the effective threshold voltage, make it zero. Consequently, this makes the FGMOS devices most suitable for application of circuits involving high degree of tunability or programmability. Therefore, for instance, the FGMOS devices are able to compensate for huge variations in performance due to mismatch [15]. The simulated graph of drain current, I_d versus source-drain voltage, V_{ds} is presented in **Fig. 4**.

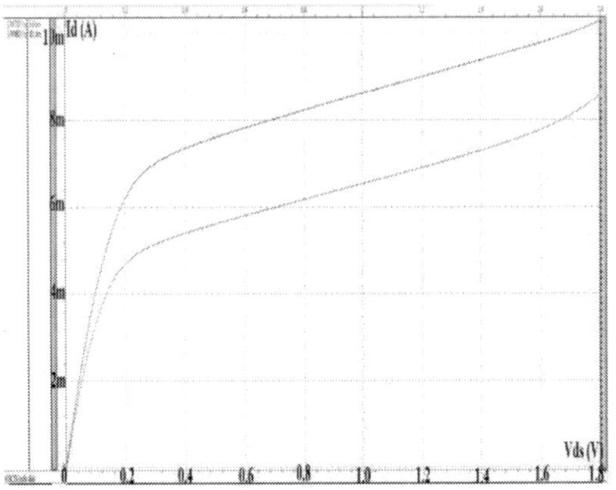

Fig. 4. Graph of drain current, I_d versus source-drain voltage, V_{ds}.

4 FGMOS based inverter.

In order to observe whether FGMOS are compatible with MOSFET circuit design, an FGMOS device is placed in an inverter circuit which is shown in **Fig. 5** and the resulted transient analysis is presented in **Fig. 6**.

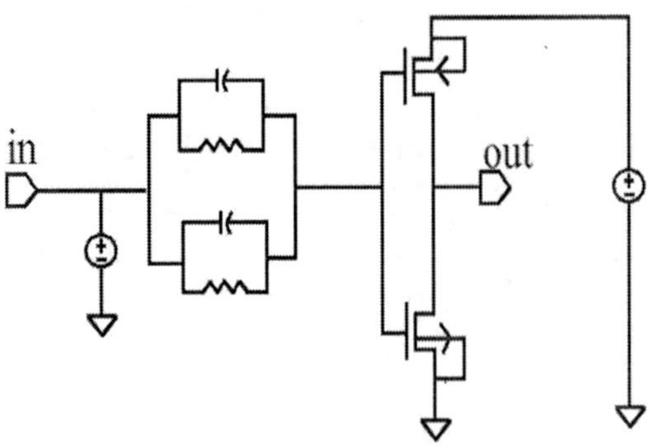

Fig. 5. The FGMOS implementations in an inverter circuit.

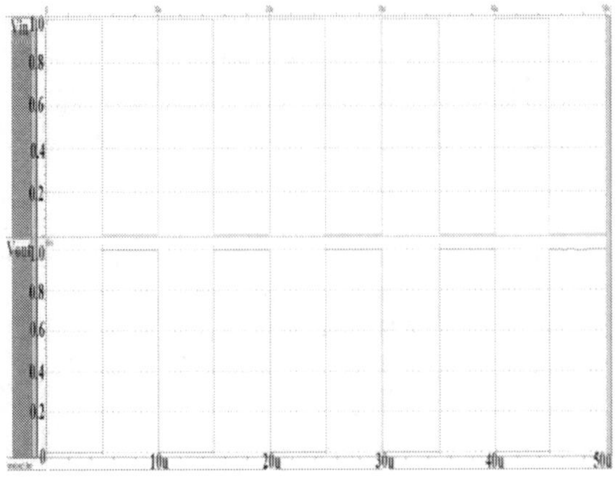

Fig. 6. The transient analysis of inverter FGMOS.

When both n-channel and p-channel MOSFETs are changed with 2 inputs n-channel and p-channel FGMOS with common floating gate, one of these transistors connected to a bias voltage and the other act as effective input, causing the threshold voltage to remain the same. In which, the width and length of the channel for both n-channel and p-channel are similar to the conventional MOSFET implementation [15]. The switching threshold, $U_{inverter}$ of a conventional CMOS inverter is shown in equation (3) [15].

$$U_{inverter} \approx [V_{DD} - |V_{Tp}| + \sqrt{(\beta_n / \beta_p)} \times V_{Tn}] / [1 + \sqrt{(\beta_n / \beta_p)}] \quad (3)$$

Where $\beta_n = \mu_n C_{ox} (W/L)_n$ *and* $\beta_p = \mu_p C_{ox} (W/L)_p$ are the transconductance parameters of n and p channel FGMOS FGMOS [16].

Fig. 7 shows another FGMOS inverter circuit followed by a MOSFET inverter [15]. In the buffer circuit, the switching threshold value is not significant because it serves as the value of voltage at which the floating gate is not the input anymore since the floating gate and the output share the operating point ($V_{FG} = V_{out}$). In consequence, it is more suitable to find the value of switching threshold at the effective input. The voltage is able to be obtained by combining the equation (1) and (3) resulting equation (4) [15].

$$V_{in(threshold)} = [(C_T / C_{in}) - ((C_{gdn} + C_{gdp}) / C_{in})] \times [(V_{dd} - |V_{Tp}| + \sqrt{(\beta_n / \beta_p)} \times V_{Tn}) / (1 + \sqrt{(\beta_n / \beta_p)})] - [(C_c / C_{in}) V_c] - [((C_{gsp} + C_{gbp}) / C_{in}) / V_{dd}] \quad (4)$$

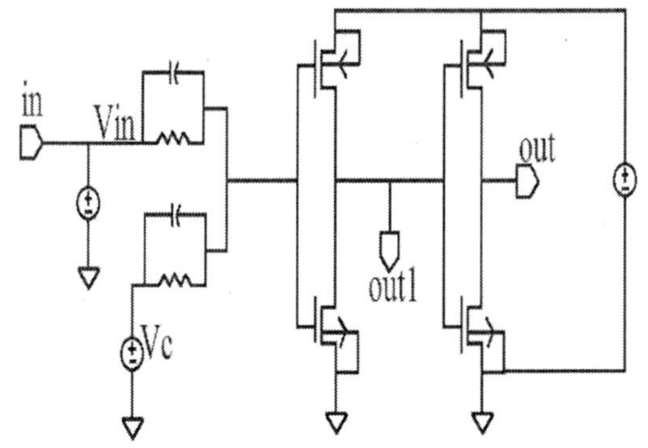

Fig. 7. FGMOS inverter with two inputs followed by a MOSFET.

In which, C_c and C_{in} are capacitances of FGMOS connected to V_c and V_{in} respectively. By altering the value of V_c, from the equation (4), it is shown that the switching threshold of FGMOS inverter is able to be programmed.

The objective of installing the second inverter is bifold, in which first it regenerates the digital levels and secondly, it helps to provide the comparisons of performances between FGMOS and MOS inverters

The results of this circuit is shown in **Fig. 8**. The output of the FGMOS inverter is represented by V_{out1} in which is able to be achieved by sweeping V_{in} between 0V to 2V and changing V_c in steps of 0.2V. Meanwhile, V_{out} is representing for the output of MOSFET inverter that connected in series with the FGMOS inverter and the aspect ratios are similar to the devices in FGMOS application. **Fig. 9** shows the transient analysis of the buffer circuit proving that it can operate and producing result compatible with the CMOS buffer.

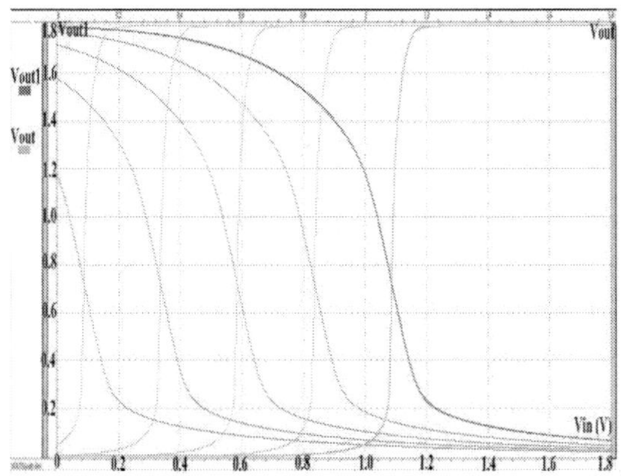

Fig. 8. Input-output characteristic of FGMOS inverter followed by MOSFET inverter circuit.

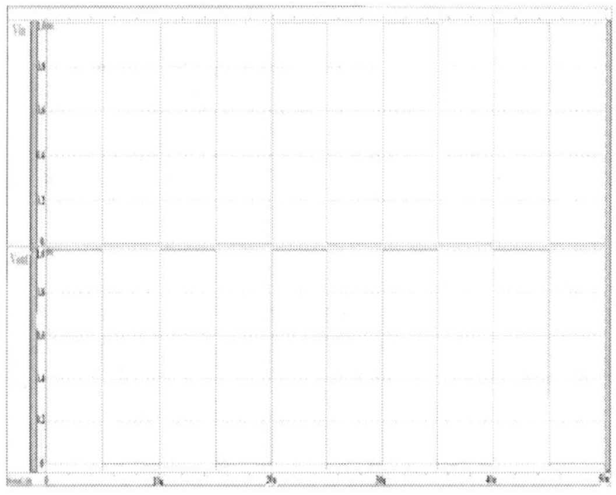

Fig. 9. Transient analysis of FGMOS buffer circuit.

From the simulated graph of input-output of FGMOS inverter in series with MOSFET inverter, it can be seen that by varying the V_c, the switching threshold is possible to be programmed and eventually can be reduced. Therefore, FGMOS implementation is definitely suitable as an alternative technique in order to achieve low voltage and low power circuit design [17].

5 Conclusions

In this paper, we have studied the special advantage of FGMOS, in which it owns the ability to be programmed through the tuning of the threshold voltage. Meanwhile, through the comparison done between FGMOS and NMOS shows that the threshold voltage produced by FGMOS lowers than the conventional NMOS. Furthermore, the tunability of threshold voltage is initiated by the programmability of parameter k. This contributes towards achieving low voltage and low power circuit design. In addition, by changing the bias voltage of FGMOS in the inverter circuit, the inverter voltage transfer characteristics is able to be appropriately altered leading to the ability of changing the switching threshold, making it to be zero or even change its sign.

This research is supported by the Fundamental Research Scheme (FRGS) with Grant number 9003-00421 and sponsored by the Malaysian Ministry of Higher Education (MoHE).

References

1. J. M. Algueta-miguel, C. A. De, and C. Blas, ISCAS, **1**, no. 3,. 277–280 (2015)
2. U. Bansal and M. Gupta, AEU - Int. J. Electron. Commun., **73**, 59–67 (2017)

3. B. Aggarwal, M. Gupta, and A. K. Gupta, Microelectronics J., **44**, no. 3, 225–235 (2013)
4. A. Kumar, Analog Integr. Circuits Signal Process., **80**, no. 2, 305–314 (2014)
5. A. J. Lopez-Martin, J. M. Algueta, C. Garcia-Alberdi, L. Acosta, R. G. Carvajal, and J. Ramirez-Angulo, Microelectronics J., **44**, no. 10, 920–929 (2013)
6. M. Gupta, R. Srivastava, and U. Singh, Microelectronics J., **46**, no. 2, 125–134 (2015)
7. M. B. K. Jamal, S. P. Chew, B. I. Khadijah, and S. B. M. Noormiza, Adv. Mater. Res., 433–440, 4189–4193 (2012)
8. M.Moghavvemi and S. S. J. A. A., 2012 Int. Conf. Informatics, Electron. Vis., 588–592 (2012)
9. M. Gupta and R. Pandey, *Microelectronics J.*, **42**, no. 6, 903–912 (2011)
10. J. R. E. Hizon and E. Rodriguez-Villegas, Proc. - IEEE Int. Symp. Circuits Syst., no. 1, 2093–2096 (2014)
11. J. Lu *et al.*, IEEE Journal of Solid-State Circuits, **50**, no. 1, 270–281 (2015)
12. J. Lu and J. Holleman, Proc. - IEEE Int. Symp. Circuits Syst., 1600–1603 (2013)
13. R. Srivastava, M. Gupta, and U. Singh, Analog Integr. Circuits Signal Process., **78**, no. 1, 245–252 (2014)
14. R. Srivastava, M. Gupta, and U. Singh, Int. J. Electron., **102**, no. 4, 563–581 (2015)
15. E. Rodriguez-Villegas, *Low Power and Low Voltage Circuit Design with the FGMOS Transistor*, **20**, London, United Kingdom: The Institution of Engineering and Technology, London, United Kingdom (2006)
16. R. Gupta, R. Gupta, and S. Sharma, ELELIJ, 5, no. 2, 37–49 (2016)
17. W.-L. G. Kiat-Seng Yeo, Samir S. Rofail, *CMOS/BiCMOS ULSI: Low Voltage, Low Power*, 1st ed. (Prentice Hall., 2001)

Design and analysis of microcontroller system using AMBA-Lite bus

Wang Hang Suan[1,*], *and Asral Bahari Jambek*[1]

[1]School of Microelectronic Engineering, Universiti Malaysia Perlis, Perlis, Malaysia

Abstract. Advanced Microcontroller Bus Architecture (AMBA) is one of the well-designed on chip communication system. It is designed for right first-time development with many processor and peripherals. In this paper, the different family of AMBA architecture such as AXI, APB, AHB are reviewed. In this work, the AMBA-Lite is used and implemented with a few peripherals and an ARM processor. The work is simulated using Synopsys and demonstrated on the Digilent Nexys4 DDR board and the software use to synthesis the design is Vivado 2016.2.

1 Introduction

The demand of microcontroller is increasing rapidly in this few decades. It can be found in many daily application such as washing machine, refrigerator, and automobile. The technology for miniaturising the system is now a major concern by the designer. The silicon technology has been developed from a few transistor to hundreds of million transistors since the invention of microprocessor. Today, the technology allows the integration of different part of component into a single chip which also known as System on Chip (SoC) [1].

One of the main advantage of SoC is that the intellectual property (IP) can be reuse in other design. The Advanced Microcontroller Bus Architecture (AMBA) developed by ARM Ltd. is also designed for this purpose. Therefore, AMBA architecture is most widely used in communication interface [2]. The AMBA-Lite is a simplified version of the AMBA bus family, but it can only support a single master. This cause some of the signals for multi-master system is removed like HGRANT, HBUSREQ.

Besides, the AMBA-Lite also support high-bandwidth operation. This include burst transfer, single clock edge operation, non-tristate implementation and so on [3]. The bus consists of one address decoder and one multiplexor to select the correct slave. The master starts giving the address and control signals. The write data then moves data from master to slave or vice versa. All slaves will be able to extend the data phase by making a request to master and send a response to the master.
In this work, an ARM soft processor and the AMBA-Lite are implemented on the Digilent Nexys4 DDR board. Section II discusses other research work on the different AMBA architecture. Next, section III show the methodology to implement a simple microcontroller using IP developed by ARM. In section IV, the result is shown on the board and on the monitor and the result is discussed and analysed.

2 Literature review

In paper [4], an extensible interface bus (AXI) was used in the design Input Output (IO) system. The AXI is similar to other Advanced Microcontroller Bus Architecture (AMBA) which are a high performance on chip communication buses. The slaves and master interface were developed with AXI on chip communication standard interface. The master was an Arithmetic Logic Unit (ALU) which can perform 22 operations, while the slave peripherals were SRAM, ROM, and FIFO. The AXI act as a medium between the master and the slaves, where the master will start to initiate the address. The round robin arbiter was used to select the slave where this type of arbiter depends upon the request from master. Fig. 1 shows the block diagram of the AMBA IO system.

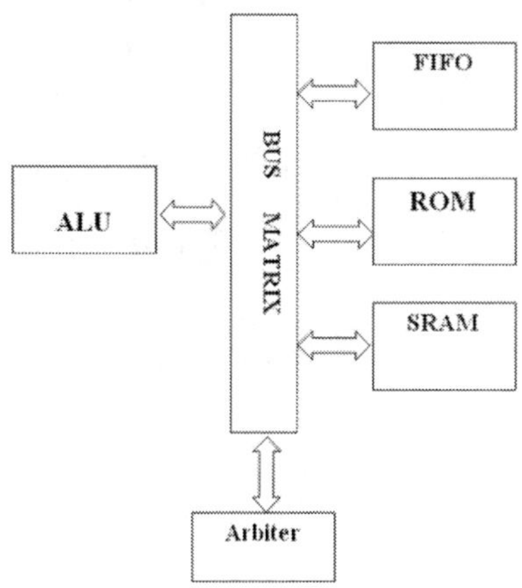

Fig. 1. AMBA IO block diagram. [4]

*wanghs92@gmail.com

DOI: 10.1051/epjconf/201716201070

In paper [5], a system had been designed with a processor that will send the address to the synthesized memory and read the data. First, a .bin file had been generated using Keil software. Then the file was implemented into a synthesized VHDL module. A Divider Clock Manager (DCM) was used to generate 10 MHz. The Core Generator tool was set to 4KB RAM and using the 32bit of data access which is shown in Fig 2. The result is shown by turning on or turning off the Light Emitted Diode (LED), and the maximum clock speed was 40MHz.

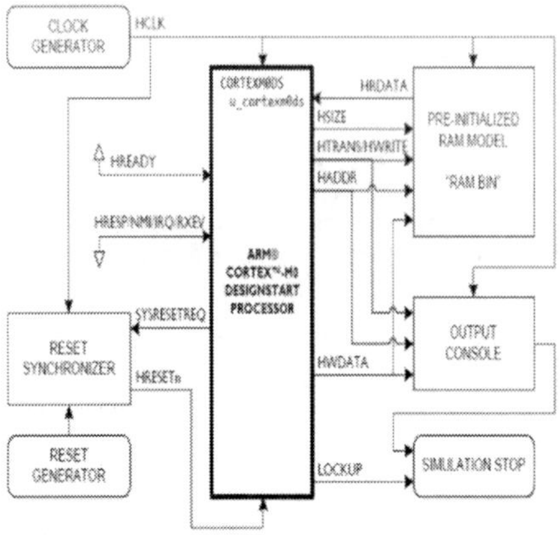

Fig. 2. Block diagram of the system. [5]

In paper [6], a design with an arbiter was used in the system. Fig 3 shows the AMBA AHB interconnect matrix for multichannel technique. The arbiter will decide which master can access the bus, then the master only can initiate read and write. This system used a combination of round robin and dynamic arbiter. The master will be chosen if any of the master request and will reduce one priority. The system is designed with a few master, an arbiter, slave and decoder. The result show that the output data was successfully obtained after the simulation.

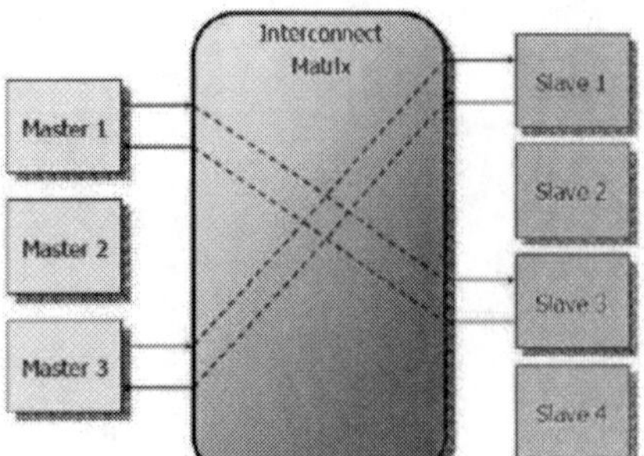

Fig. 3. Block diagram of interconnect matrix.[6]

In paper [7], an Advanced Peripheral Bridge (APB) bridge was used. The system was designed to minimise the clock skew by adding a 3 bits ripple down counter. The flip flop was clocked sequentially, so that the data output of source flip flop can only toggle the clock of sink flip flop. The bridge act as a single master for the connected peripherals which perform read and write transfers. The bridge consists of a state machine, a reset controller, and the ripple down counter sub-module. The undesired clock skew can be avoided by using this approach. The result shown that by using the ripple down counter, it can also reduce the power on chip and total clock domain.

In paper [8], the Advanced System Bus (ASB) was implemented. The operation start from the master contact with the bus. Then, the arbiter will determine its status before the master start transferring the data. The decoder will then select the slave and a response will be sent back to the master. The bridge acts as a slave to the master so it continues putting wait state on the master only when they are needed. The bridge consisted of a state machine and decoding logic to select signals. The state machine of the system is shown in Fig 4 which the decoding cycles start before the slave response. The system bus used was 50MHz with 32 bits of address and data buses. Besides, the system was assumed to connect with 4 peripherals.

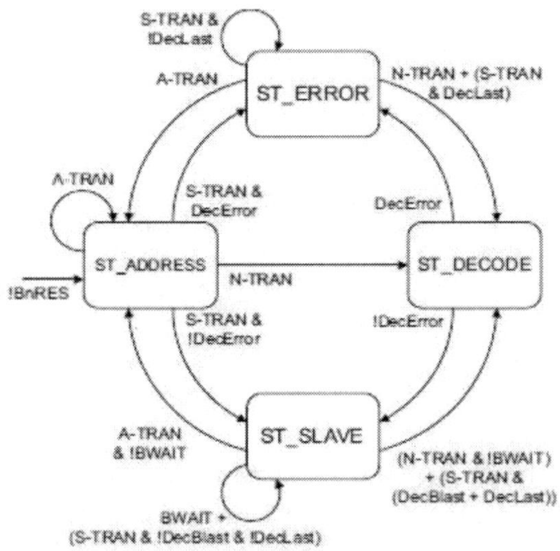

Fig. 4. State machine of the APB bridge.[8]

The summary of different bus architecture made by others is shown in table 1. The purpose of paper [7] and [8] is more to reduce power consumption while others are focusing on the functional of the system. The usage of arbiter depends on the application; it is used when more than one master is implemented. Besides, the bridge needs to avoid any clock skew or timing violation when designing the system.

DOI: 10.1051/epjconf/201716201070

Table 1. Summary of the methods by others paper.

	[4]	[5]	[6]	[7]	[8]
Master	ALU	Cortex M0 DSK	-	-	-
Bus Architectu re	AXI	AHB-Lite	AHB	APB	ASB
Slave	SRAM ROM FIFO	LED Memory	-	APB bridge	APB bridge
Arbiter	Round Robin arbiter	-	Round robin and dynamic	-	-
Total power	-	-	-	0.57mW	0.66m W

3 Methodology

In this section, the design of the top module and the individual module and how it work will be explained. The program of the microcontroller is stored inside the memory which is written in Verilog code. The size of the memory can be defined inside the code. In this system, 4KB of memory is sufficient to store the program. Since it is an internal memory, the program and the architecture is implemented together into the board. In this research, an ARM Cortex-M0 Design Start Processor, an Advanced High-performance bus (AHB) lite, an internal memory, an Universal Asynchronous Receiver Transmitter (UART) peripheral, a Video Graphics Array (VGA) peripheral, a timer peripheral, a General Purpose Input Output (GPIO) peripheral, a Light Emitting Diode (LED) peripheral, and a 7-segment peripheral are used. All these module are written in the form of Verilog code and can be reusable in other design. Fig 5 shows the system block diagram with each has the interface to interact with the AMBA-Lite. The most important aspect to integration is the address mapping. The address need to be correct so that the bus is able decode it to the right IP. Besides, inside the application program, the programmer need to know where is the IP address in order call the respective IP.

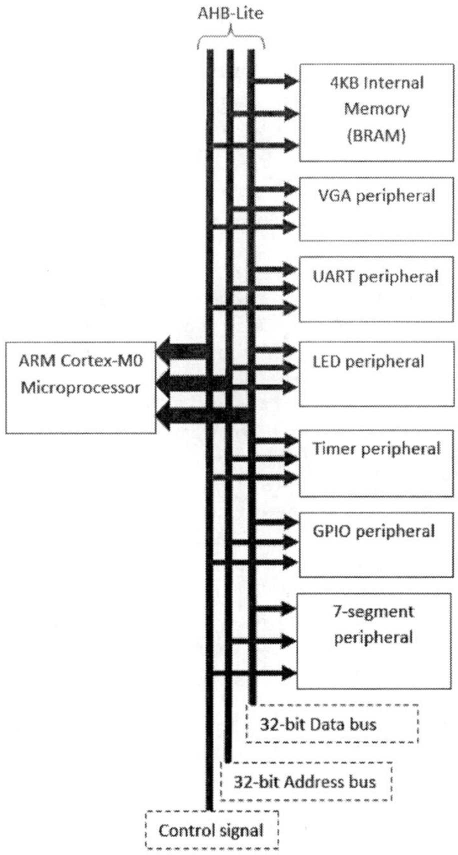

Fig. 5. The peripheral used in the system.

First, the program which the instruction to be executed by the processor is compiled from assembly code into machine code using Keil μVision before targeting into the board as mention above. The LED peripheral is a simple block to interact with the 8 LED on the board for other application purposes. In this paper, the LED is connected to GPIO, when the GPIO receive the signal from switch, it will turn on the LED. The GPIO has a data direction register to indicate either it is input or output signal.

On the other hand, the VGA is a more complex module which consists of a few sub-module inside. One of the sub-module is VGA interface, it is used to generate synchronization signals to the VGA port and transfer the colour to the pin on the board. Besides, it has image buffer and text console, which are used to store the colour information in the image region and displaying text respectively. The text console is implemented on the hardware logic so that it can save the valuable on-chip memory. The VGA connector is coming along the function to convert the digital output to analogue using resistor divider. The hierarchy of the VGA module is as depicted in Fig 6.

DOI: 10.1051/epjconf/201716201070

```
AHBVGASYS.v
|--- vga_console.v
|      L--- font_rom.v
|--- vga_image.v
|      L--- dual_port_ram_sync.v
L--- vga_sync.v
       L--- counter.v
```

Fig. 6. Hierarchy of the VGA module. [9]

Next, the timer module is a software timer to act like a clock to count the event and the result is displayed on the 7-segment display and it is programmed to reset when it reaches zero. It is based on the hardware timer but when it compared to a hardware timer, it has a lower level of precision. In this module, it has a 32-bit counter, which start from 'FFFF'.

The module has a programmable register to control the frequency of the counter and the load value. The timer prescale value can be set as 1 bit, 16 bits or 256 bits. Besides, there are two modes available for interrupt to occur, which are free running mode and periodic mode. These types of interrupt enable different program can be run on the same time and without modifying the timer architecture.

In this paper, the application will not use any interrupt, therefore, the four digit of 7-segment display will be used to show the result. In order to use the 7-segment display, a decoder is needed. It is designef based on the four registers to display the 32-bit counter with a 1 kHz looping frequency. The peripheral is set for common anode display since the 7-segment display of Digilent Nexys4 DDR board is an anode type.

Besides, this architecture also integrate the UART module which can receive or send signal through serial port. The UART also consists of some sub-module, such as module to generate a fixed transmission baud rate, a First In First Out (FIFO) register to buffer the data to be sent or received. The UART module is designed with a start bit and a stop bit to indicate finishing transfer.

In this paper, the baud rate is set to be 19200 bps, and an open source software called Tera Term is used to communicate between the laptop and the board. The Table 2 shows the memory map of each peripheral in this design. The peripherals module used are all available on the ARM university program website [9].

Table 2. Memory map design.

Module	Base address	End address
BRAM	0x0000_0000	0x4FFF_FFFF
VGA	0x5000_0000	0x50FF_FFFF
UART	0x5100_0000	0x51FF_FFFF
Timer	0x5200_0000	0x52FF_FFFF
GPIO	0x5300_0000	0x53FF_FFFF
7-seg	0x5400_0000	0x54FF_FFFF
LED	0x5500_0000	0x55FF_FFFF

4 Result and discussion

In this paper, the simulation was carried out using Synopsys. The result shown is similar as intended and carried out in different part based on IP. Based on Fig 7, an input act as a switch is added to test the GPIO IP. The LED then has the same output based on the input (sw). Fig 8 shows the UART IP sending out a signal from the system. The word "TEST" is stored in the memory and the application program is transmitted out from RsTx. It can be interpreted using the ASICII character. For example, from the ASICII, the "T" is 1010100 as shown in figure, with a zero start bit and a one stop bit.

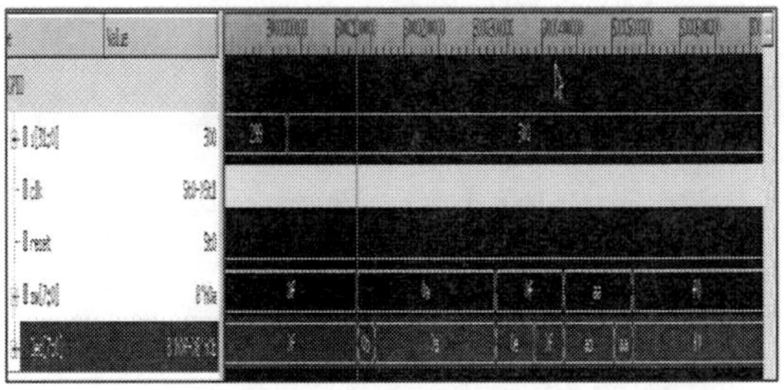

Fig. 7. The simulation on GPIO and LED IP.

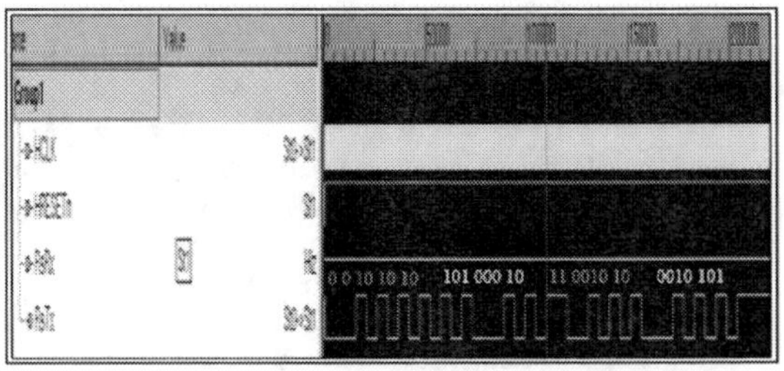

Fig. 8. The content of memory is send out on UART IP.

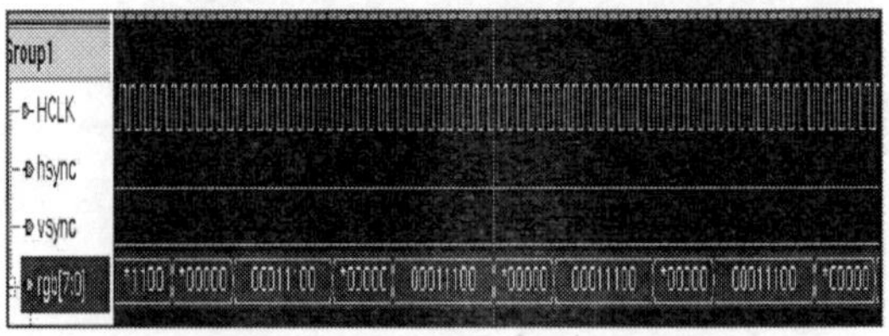

Fig. 9. The content of memory is send out on RGB IP.

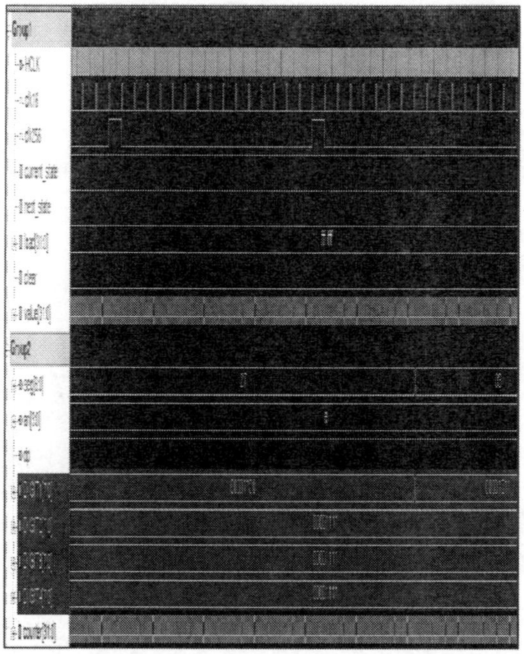

Fig. 10. The timer IP is count down and shown on 7segdec IP.

Fig 9 shows the content stored in memory is displayed on RGB and the colour is set as green which is 00011100. The word "HELLO" is displayed which is also demonstrated in FPGA.

Fig 10 shows the timer is counting down and the 7 segment decoder decoding the time. Four digits are displayed and for the first digit is 00001100 and the next state is 00001011. The figure for the timer is used to debug and to check for the state is running from one state to another. The timer has a prescalar that can be set the clock for an extra 16 count or 256 count which can be seen as clk16 and clk256.

Meanwhile, the Digilent Nexys4 DDR FPGA board is used to run the ARM IP architecture using Vivado 2016.2. The UART communicates with the laptop using Tera Term with a baud rate 19200bps. Fig 7 shows the word 'HELLO' is stored in the memory is printed out on the screen. In this figure, the result shown has two IP used. The word 'HELLO' which previously stored in the hex file is shown on the first line of another monitor using VGA IP. Another IP used is the UART IP and the result is transferred from the board to the laptop screen via UART terminal.

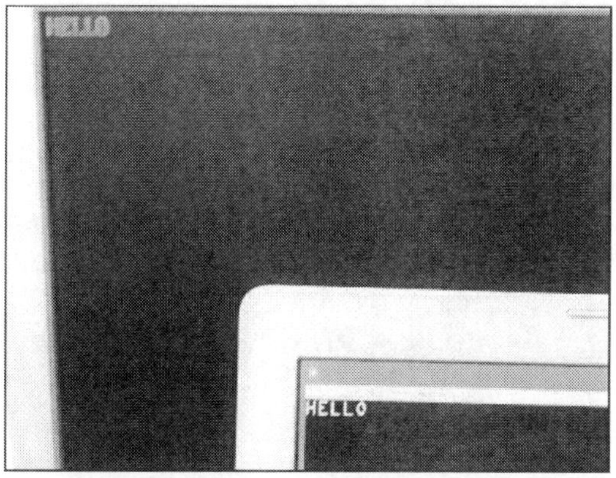

Fig. 11. The content of memory is printed out on screen.

In Fig 8, the result shows that the timer is decreasing in the 7-segment display from FFFF to 0000, with a 16-bit prescaler. The prescaler can be set using the program to determine the speed of the count down. There is no interrupt function in this case, but the program can be added to this IP that is developed by ARM [10]. Besides, the figure also show that the LED is turned on according to the respective switch below by using GPIO module.

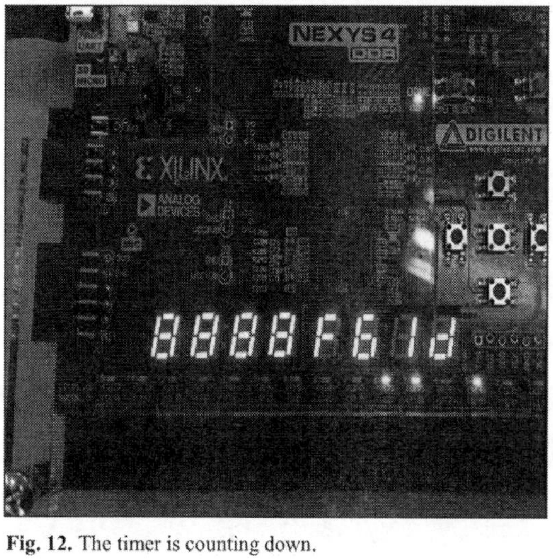

Fig. 12. The timer is counting down.

4 Conclusion

In this paper, an AMBA-Lite with a few peripherals was implemented on to the FPGA board and the result showed was similar as expected. The result is verified using Synopsys and implemented on the FPGA Digilent Nexys4 DDR board. The ARM Cortex-M0 Design Start was used as the processor to run the program inside the memory where the application was compiled using ARM tool chain.

This research was supported by TalentCorp Malaysia.

DOI: 10.1051/epjconf/201716201070

References

1. K. S. Kumar, P. Deepthi, IJERA **3**, 1005-1010 (2013)
2. A. Laurentiu, B. Angel, WCE **I**, 459-464 (2015)
3. "AMBA 3 AHB-Lite Protocol," **1.0 ed**: ARM Ltd. (2006)
4. K Swetha, G. Ramakrishna, IJEERT **2**, 51-59 (2014)
5. M. J. Oviedo, P. A. Ferreyra, VII DF 87-93 (2011)
6. S. Divekar, A. Tiwari, ICCSP, 1854-1858 (2014)
7. K. Rawat, K. Sahni, S. Pandey, J. Rawat, and S. Tripathi, IC4, 1-4 (2015)
8. K. Rawat, K. Sahni, and S. Pandey, SPIN, 927-930 (2015)
9. "DesignStart - ARM," **1.0 ed**: ARM Ltd.
10. "Cortex-M0 Technical Reference Manual Copyright," ed: ARM Ltd., (2009)

DOI: 10.1051/epjconf/201716201071

On-chip temperature-based digital signal processing for customized wireless microcontroller

Siti Farhah Razanah Faezal [1,*], *Mohd Nazrin* Md Isa[1], *Azizi* Harun[1], *Shaiful Nizam* Mohyar[1], and *Asral Bahari* Jambek[1]

[1]School of Microelectronic Engineering, Universiti Malaysia Perlis, 02600 Malaysia.

Abstract. Increases in die size and power density inside system-on-chip (SoC) design have brought thermal issue inside the system. Uneven heat-up and increasing in temperature offset on-chip has become a major factor that can limits the system performance. This paper presents the design and simulation of a temperature-based digital signal processing for modern system-on-chip design using the Verilog HDL. This design yields continuous monitoring of temperature and reacts to specified conditions. The simulation of the system has been done on Altera Quartus Software v. 14. With system above, microcontroller can achieve nominal power dissipation and operation is within the temperature range due to the incorporate of an interrupt-based system.

1 Introduction

The measurement of temperature is very critical especially in portable devices with densely packed circuits which can dissipate power in the form of heat. The continuous increase of the operating frequency increases the density of dissipated power [1]. As power consumption, P is directly proportional to the operational frequency, f by the following equation:

$$\boldsymbol{P} = c.\ V^2.f + Ps \qquad (1)$$

where c is scaling constant with the dimension of capacitance (F). V is an input voltage and Ps is the static power dissipation which is the power at a zero clock frequency [2]. Rise in heat density creating vast difficulties which lead to serious thermal problem even though in low power applications.

Thermal considerations should become one of the important part of the electrical design process. To increase system reliability, the temperature of the operating circuits should be continuously watched during the operation of system [3]. Local overheating in one spot of a high-density circuit as in Fig.1, such as high-speed mixed-signal circuits, can cause a whole system to crash due to clock synchronization problems, parameter mismatches or other coefficient changes due to the uneven heat-up on a single chip [3].

Heat sinks and fans, known as passive heat dissipation mechanisms, are widely used in system design. Advanced circuit designs recently have included active mechanisms that detect and handle an over-heating even [4]. The improvement in cooling technology are quite slow. In the mobile system and external skin hot spots, the implemented cooling fan can be a major source of acoustic noise [5].

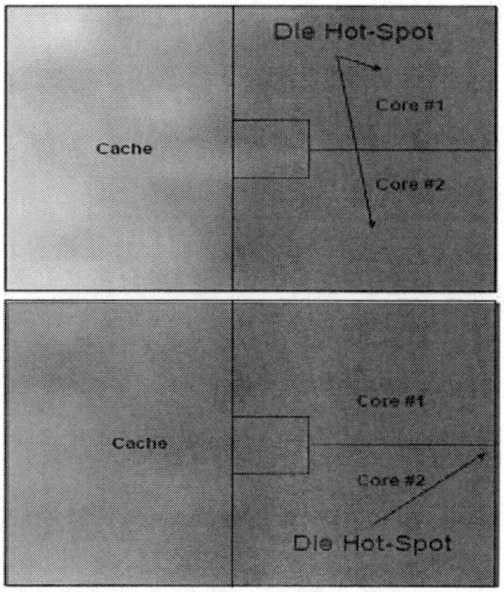

Fig. 1. Die hotspot of high density circuit [5].

Thus, the needs for efficient thermal management system are very important as the increasing demand for compute density. Several systems have been proposed, for example the Advanced Configuration and Power

* Corresponding author: farhafaezal@gmail.com

Interface (ACPI), thermal management in personal computer systems [6]. ACPI is an industry standard infrastructure that facilitate thermal management of computer platforms. Thermal management is done by the use of active cooling devices, such as fans, or passive cooling actions. However, the ACPI standard is quite limited, as it simply supports extra control to turn on or off a cooling mechanism and shift the alert level, which is feedback to the system [7]. Meanwhile, later, Dynamic Frequency Voltage Scaling (DVS) is proposed [8]. These mechanisms were implemented in CPUs such as the Intel® Centrino® Processor [9]. Thus, the important feedback required for power and thermal management systems is temperature measurement and monitoring.

Increases in die size and power density inside system-on-chip (SoC) design have brought thermal issue inside the system. Uneven heat-up and increasing in temperature offset on-chip has become a major factor that can limits the system performance [10]. In order to meet this challenge, we proposed a system that is temperature-aware system that is able to monitor the temperature distributions on-chip without any knowledge about the material properties or the chip structure.

In this project, we are developing a temperature-based digital signal processing IP circuit (TDSP). Moreover, this research covers Verilog HDL and design of a temperature-based digital signal processing module for an on-chip temperature monitoring system using Altera Quartus software. This circuit is optimized in architecture and circuit implementation to fit system-on-chip designs. Therefore, an interrupt-based system will be implemented.

2 Proposed Architecture

2.1. Circuit Blocks

The block diagram of the temperature-based digital signal processing circuit is shown in Fig. 2. The function of every block is describe in Section 2.2.

2.2 Function

Function of each block is describe in details in this section;

• ***Input Controller***: This unit used to control the input data received and providing an input interface synchronously.

• ***Average Module***: This unit contains five registers to store the data received from ADC and provided average data output.

• ***Voltage to Temperature Module***: This unit contains a look-up table to convert the input ADC which is in voltage value into the corresponding temperature value in Celsius (°C).

• ***Parameterizable Watchdog Unit***: This unit contains two monitoring circuits: the threshold monitor for each temperature sensor, and the offset temperature monitor. Both circuit providing sufficient speed to compare temperature provided in the system.

• ***Interrupt Generator***: This unit provides data outputs that are read by the system CPU, like temperature value, offset temperature value, and interrupt types.

• ***Output Converter***: This unit convert the output into [31:0] bit form (including temperature threshold, interrupt).

• ***Controller***: This unit control the interaction between the system and CPU.

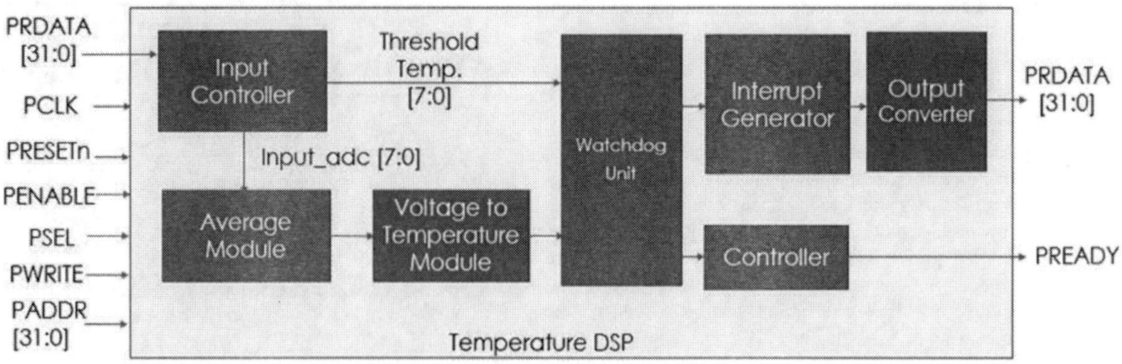

Fig. 2. Temperature-based DSP circuit block diagram

DOI: 10.1051/epjconf/201716201071

3 Flowchart of The System

Consider the flowchart in Fig.3 below is indicating the process flow for temperature-based digital signal processing.

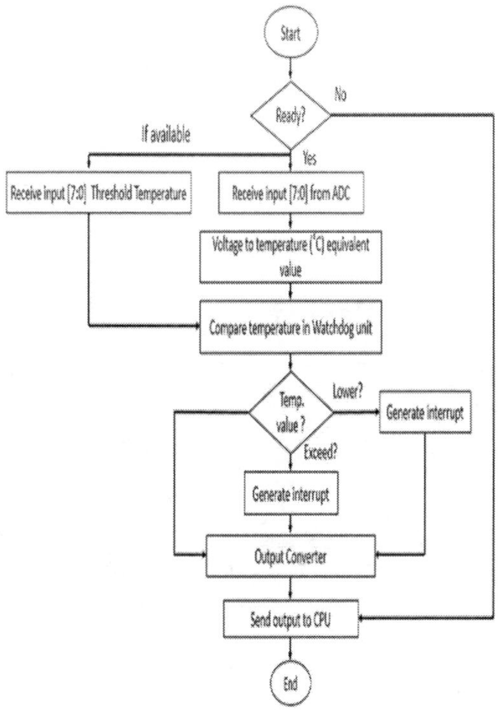

Fig. 3. Flowchart of the TDSP system

When the system is ready, the input controller would receive an input from ADC in 8-bit binary form of voltage value. After received in total of 5 different input, average module will automatically calculate the average value of the inputs and generate the average output.

Then, simultaneously, voltage to temperature module would process the input received and generated an output equivalent to temperature in degree Celsius (°C). Parameterizable watchdog unit would monitor the threshold temperature (if available) as in flowchart Fig. 4 below and offset temperature and send signals to the interrupt and output generator.

Interrupt is produced when the temperature is very lower (-25°C - 0°C) and also when the temperature is very high (61°C - 80°C). Output converter will then convert all information (including temperature value, offset temperature value, and interrupt types) into output [31:0] bit form.

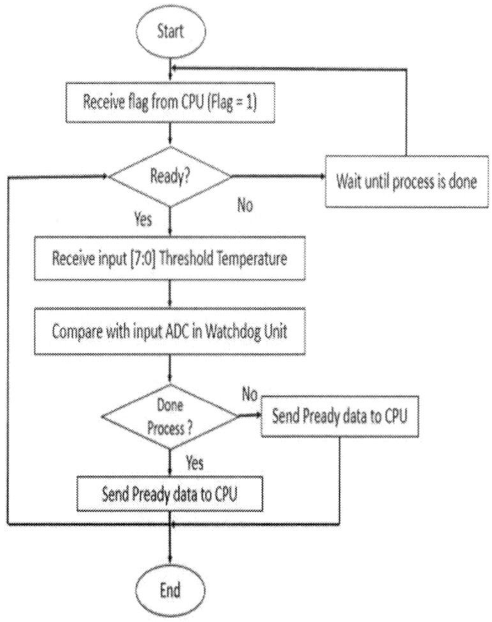

Fig. 4. Flowchart of the TDSP system when threshold temperature is available.

3 Architecture on APB Bus

The temperature sensor ranges from -25°C to 80°C and convert it into equivalent 1.38mV/°C. This analog voltage gets converted into equivalent 8-bit digital voltage through ADC. These 8-bit inputs are applied to TDSP where TDSP will monitor the threshold and offset temperature inside the system and implement the interrupt-based system if the condition needed. The programming in TDSP is done by Verilog HDL language and that builds TDSP system module. The block diagram for architecture on APB:

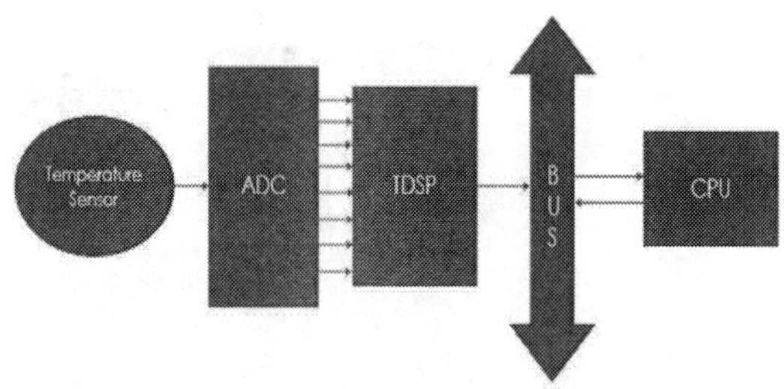

Fig. 5. Architecture on APB

DOI: 10.1051/epjconf/201716201071

5 Simulation Result

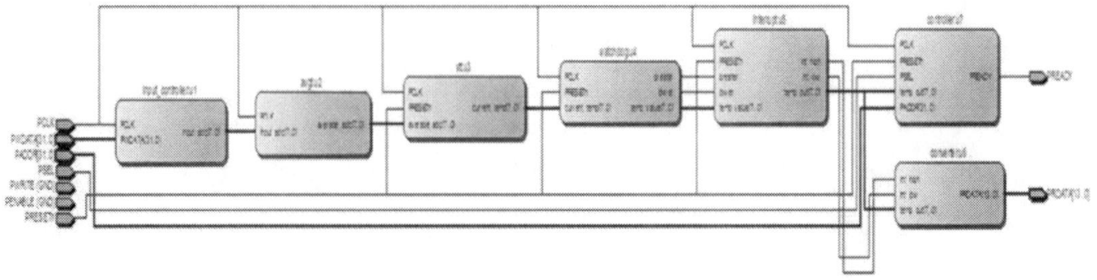

Fig. 6. RTL Level Synthesis of TDSP System

The ALTERA QUARTUS software generates the output waveform corresponding to input change of ADC or threshold temperature on simulation. This simulation is executed by random values chosen in inputs. RTL level synthesis of simulation is generated by QUARTUS software after code is compiled. The RTL level synthesis is the lowest level representation of circuitry from QUARTUS software. The RTL generated for TDSP is shown in Fig. 6 above.

6 FPGA Implementation

TDSP system is implemented using DE2-70 Altera Board to check the flexibility and functionality of the system. DE2-70 is used due to its capability to be used in wide range of experiments development of the FPGA-based systems, as well as design of hardware controllers. Altera DE2 board is connected the PC using USB-blaster provided with DE2 package. Fig.7 & Fig. 8 below shows the result on FPGA Board.

Fig.7. Result when temperature exceeding threshold temperature

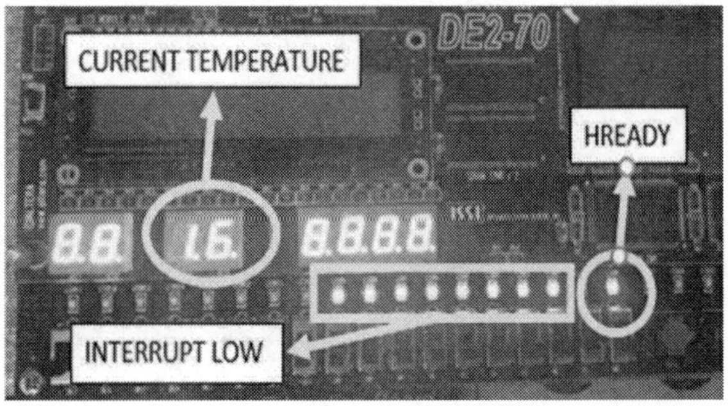

Fig.8. Result when temperature lower than threshold temperature

7 Conclusion and Future Work

A temperature-based digital signal processing is successfully implemented using FPGA and its performance is verified and tested for temperature in different level. FPGA based TDSP features the current temperature and interrupt signals. Simulation have been done using Verilog HDL code for the purpose of simulation in ModelSim. Such type of implementation can be used with different applications, such as temperature control and monitoring system. Thus, this design provides an intricate control and optimal thermal management on system-on-chip (SoC) devices, upon which a complete thermal management system for modern computer designs can be implemented with an advantage of minimum chip area required. This design can be further extended in terms of area and power at layout and logic level by using Synopsys software.

 DOI: 10.1051/epjconf/201716201071

This work is supported in part by Talentcorp Malaysia grant also collaborated with Intel Malaysia and CEDEC USM for 3D (Demand, Driven, Develop) program. I would also like to thank to my supervisor, Dr. Mohd Nazrin Bin Mohd Isa, and my 3D teammates, Azwan, Zhafirah and Wang Hang Suan for being so helpful throughout this project. And lastly, special thanks goes to my family members for their outstanding supports.

References

1. M. Szermer, Z. Kulesza, M. Janicki, and A. Napieralski, **3**, pp. 529–532 (2008)
2. N.H.E Weste, D.M. Harris, *CMOS VLSI Design A Circuits and System Perspective, 4th ed.* (Addison Wesley, 2011)
3. V. Szekely, M. Rencz, and B. Courtois, *Proc. 13th IEEE SEMITHERM Symposium*, pp. 210–217 (1997)
4. J. Draper, J. Block, J. Koller, and C. Steele, *Proc. Lecture Notes in Computer Science 1388 (IPPS/SPDP'98 Workshops Proceedings)*, pp. 900–901 (1998)
5. E. Rotem and M. Platform, *Temperature measurement in the Intel ® Core TM Duo Processor*, pp. 1–5.
6. Compaq, Intel, Microsoft, Phoenix, and Toshiba, *Advanced configuration and power interface specification* (2000)
7. J. Steele, *Proc. Wescon 98*, pp. 169–182 (1998)
8. D. Brooks, M. Martonosi, *Proceedings of the HPCA-07* (2001)
9. http://download.intel.com/design/mobile/datashts/30218908.pdf
10. H. Chiueh, J. Draper, L. Luh, and J. Choma Jr., *Proc. 2nd Internationl Workshop on Design of Mixed-Mode Integrated Circuits and Applications*, pp. 109–113 (1998)

 DOI: 10.1051/epjconf/201716201072

Architectural design proposal for real time clock for wireless microcontroller unit

Muhammad Nor Azwan Mohd Alias[1, *], and *Shaiful Nizam* Mohyar[1]

[1]School of Microelectronic Engineering, University Malaysia Perlis, 02600 Malaysia

Abstract. In this project, we are developing an Intellectual properties (IP) which is a dedicated real-time clock (RTC) system for a wireless microcontroller. This IP is developed using Verilog Hardware Description Language (Verilog HDL) and being simulated using Quartus II and Synopsys software. This RTC will be used in microcontroller system to provide precise time and date which can be used for various applications. It plays a very important role in the real-time systems like digital clock, attendance system, digital camera and more.

1 Introduction

A real-time clock (RTC) is a clock system that provide real time value used with system that deals with time. For example, it is included in a computer motherboard. It is an electronic device in the form of an Integrated Chip (IC) available in various packaging options. Memory on the RTC stores system description or setup values including current time values stored by the real-time clock. The time values are for the year, month, date, hours, minutes, and seconds. When the computer is turned on, the Basic Input-Output Operating System (BIOS) that is stored in the computer's read-only memory (ROM) reads the current time from the memory in the RTC [1]. Using an RTC, you can keep track of long timelines, even if you reprogram your microcontroller or disconnect it from USB or a power plug [2]. In this project, we are developing a dedicated Real Time Clock (RTC) for Intel Customized Microcontroller unit. This research will be covered only the logic part of RTC.

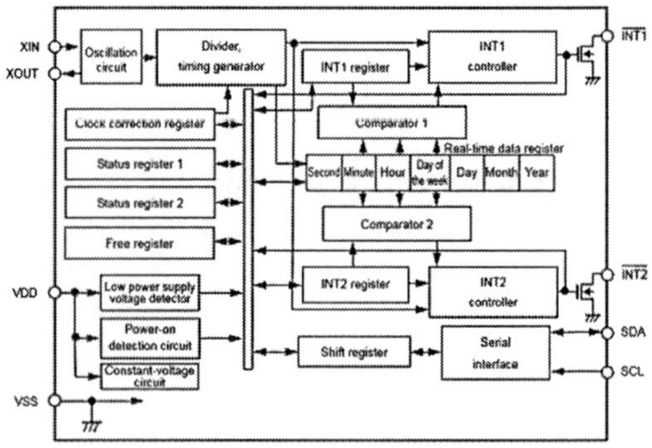

Fig. 1. RTC Block Diagram from SII-IC Semiconductor(S-35390A) [3].

1.1 Design comparison of conventional RTC

From the literature review, there are many design of RTC nowadays. Every RTC has their own features according to their specification to be use on any application. Even they have different features, they also have a similar part to make the RTC function correctly such as power supply, oscillation circuit, communication interface and many more. Fig. 1-4 shows a different RTC design from different company such as SII-IC Semiconductor, Dallas Semiconductor, Maxim Integrated, and NXP Semiconductor. The comparison in term of bus interface, block unit is tabulated in Table 1.

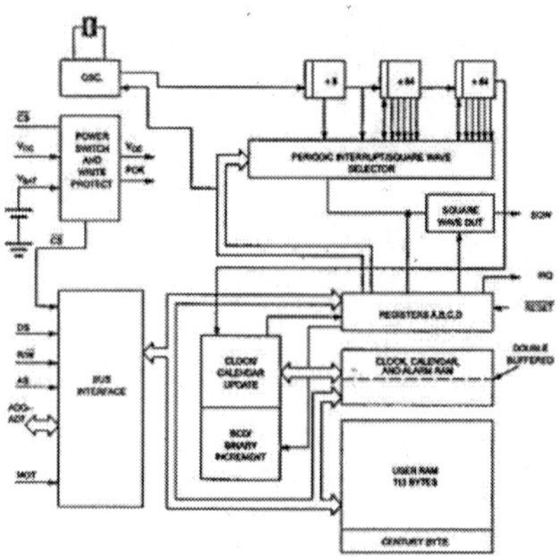

Fig. 2. RTC Block Diagram from Dallas Semiconductor (DS12C887) [4]

* Corresponding author: azwan_mjg93@yahoo.com

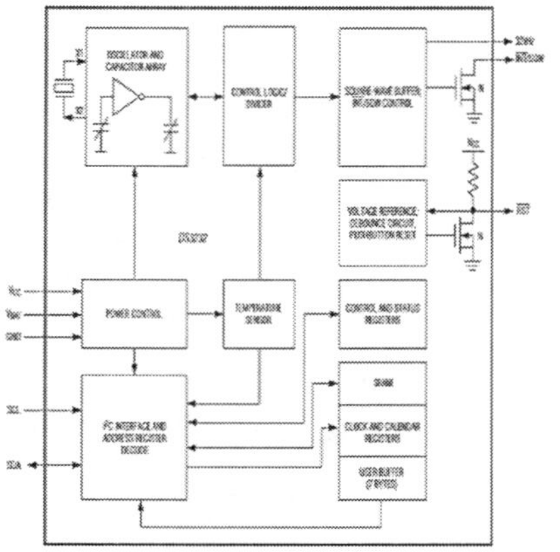

Fig. 3. RTC Block Diagram from Maxim Integrated (DS3232) [5]

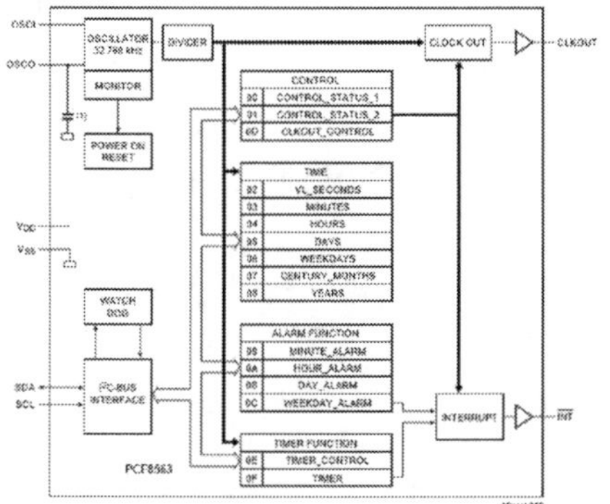

Fig. 4. RTC Block Diagram from NXP Semiconductor (PCF8563) [6]

Table 1 shows the comparison of different RTC from different company. As we can see from Fig. 1-4 and Table 1 some of them using I2C bus interface and some using SPI interface. This interface has their own advantage depending on their application. As for their oscillation frequency, all the RTC use the 32.768kHz crystal oscillator due to the stability of the crystal. This frequency is the most stable to be used in the RTC. The most important part on this comparison is their block unit. Even their block and design are different, but on every RTC, they must have the basic block that is oscillation circuit, bus interface, power management, register, frequency divider and counter. For this research, the part covered is on the logic part of the RTC that is bus interface, register, frequency divider and counter. The power management and oscillation circuit will be done by another researcher.

1.2 RTC counter increment flow

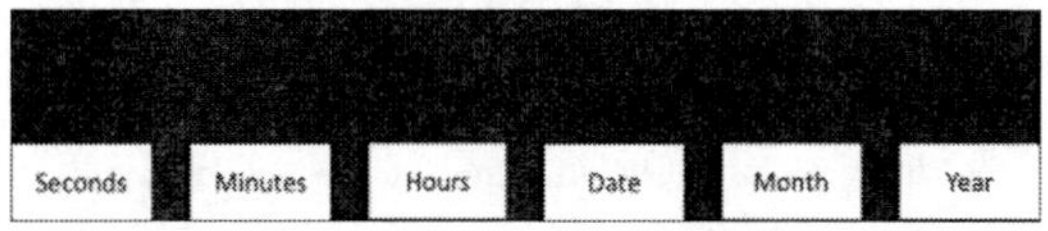

Fig. 5. Counter increment flow [7]

Fig. 5 show the increment of the counter on the RTC. Every 1Hz counter, it will represent 1 second of a real time. This increment will continue up until years. The seconds counter will count to 60 second then the minutes will increment by one. After the second reach exactly 60 second, the second counter will reset to 0. Then minutes counter will continue increase for 60 minutes. Then the hours will increment. After 24 hours, the date counter will increase. This increment flow will continue exactly as a real time until it reaches 1 years. After that it will

Table 1. PRDATA and PWDATA bit assignment

Company	SII-IC Semiconductor[3]	Dallas Semiconductor[4]	Maxim Integrated[5]	NXP Semiconductor[6]
Bus interface	I2C	SPI	I2C	I2C
Crystal oscillator frequency	32.768kHz	32.768kHz	32.768kHz	32.768kHz
Operating voltage	1.3 V- 5.5 V	5.5V-5.5 V	2.3-5.5 V	1V – 5.5V
Current consumption	0.25 μA typ. (VDD = 3.0 V, Ta = +25°C)	4mA	200 μA at 3.3V	0.25 μAat VDD = 3.0 V and Tamb = 25°C
Operating temperature	Operating ambient temperature : -40°C to +85°C	0°C to 70°C	Commercial: 0°C to +70°C Industrial: -40°C to +85°C	Ambient temperature : -40°C to +85°C Storage temperature : -65°C to +150°C
Block unit	• Oscillation circuit • Divider timing generator • Register • Power control • Comparator • Real time data register • Serial interface	• Bus interface • Power management • Register • User RAM • Clock Calendar Update • BCD Binary increment • Oscillation circuit • Square wave output	• Oscillation circuit • Power control • Bus interface • Control and status register • SRAM • Clock and calendar register • User buffer • Temperature sensor • Voltage reference • Push button reset	• Oscillation circuit • Control unit • Alarm function • Time register • Timer function • Bus interface • Watchdog unit

continue for the next year. The date will be depending on month and year as the month has different amount of days in every month and for leap years that will decide by the amount of days in February. This are the basic counter increment flow used to design this research RTC.

1.3 Advance Peripheral Bus

As for this RTC is dedicated to wireless microcontroller, this RTC needs to follow the bus protocol to be interface with the microcontroller. The microcontroller will use the Advanced Microcontroller Bus Architecture (AMBA) bus protocol. To easily interface the IP with microcontroller, the RTC need to match with Advanced Peripheral Bus (APB). APB is part of the AMBA hierarchy of buses. It is optimized to reduce the complexity of interfacing and for minimalize power consumption. The AMBA APB should be used to interface to any peripherals which not require high performance of bus interface and low bandwidth. The latest revision of the APB ensures that all signal transitions are only related to the rising edge of the clock. This improvement means the APB peripherals can be integrated easily into any design flow, with the following advantages [8, 9]:

• High-frequency operation
• Easy integration with cycle based simulators.
•It simpler to interface with new Advanced High-performance Bus (AHB)

Since this IP will be integrated with the microcontroller that using AMBA Bus Architecture, the RTC design then will follow the APB Specification as show in Fig. 6

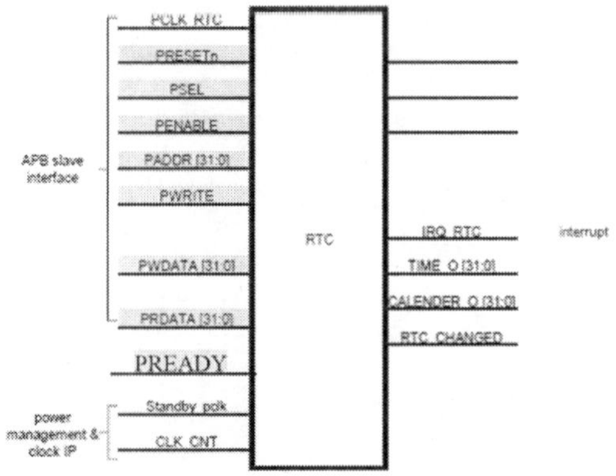

Fig. 6. APB Standard for RTC [8,9]

The highlighted input/output pin is the important pin for the APB specification. The non-highlighted is not important and can be ignore if the RTC does not have that input/output pin. That pin also can be implement for additional features [8, 9].

2 Propose RTC block diagram

Fig. 7 show the propose block diagram with bus interface which follow the APB Specification. The block diagram consists of bus interface, frequency divider, function controller, RTC, alarm and out data module. The RTC module is the module that will do all the counting and store the real-time value. The function of every module in the block diagram is explain in Table 2

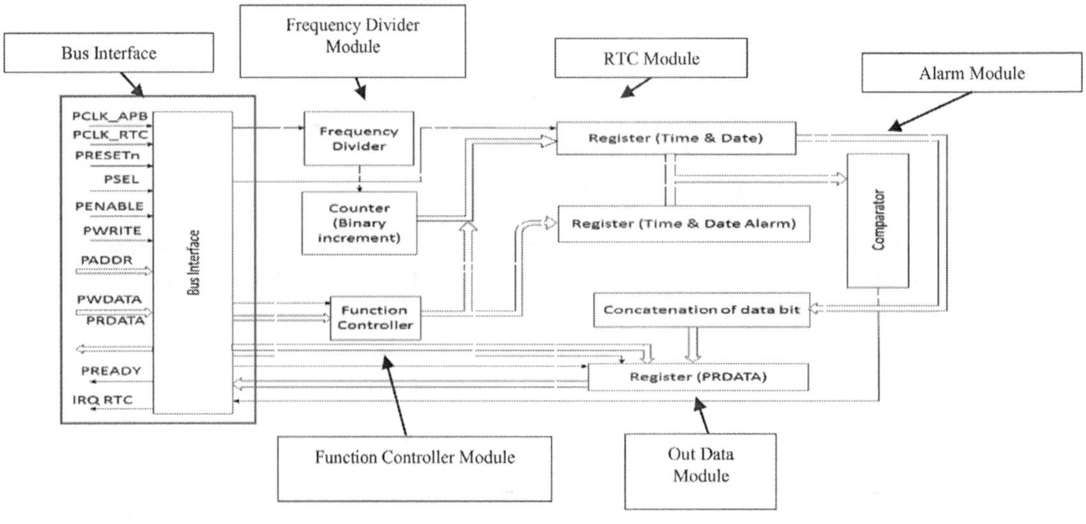

Fig. 7. Propose block diagram of the RTC

DOI: 10.1051/epjconf/201716201072

Table 2. Function of every module on the propose RTC

UNIT	FUNCTION
Frequency Divider	Generate 1Hz frequency from 32.768kHz
Function Controller	To control function (e.g. Manual set time and date, set alarm)
RTC	Generate time and date from counter and 1Hz frequency generated by frequency divider
Out data	Store output data (32 bit) to be read (time and date value)
Alarm	To compare the alarm set with the real-time clock

3 Full operation flows

Full flow of the RTC operation is as shows in Fig. 8. Firstly, the RTC will check for read and write data event. If there is data writing for manual set the time and date, the register inside the RTC module then will be overwritten by a new input value and will continue counting starting from the new value. If there is no writing data event or only the set alarm event occurs, the RTC will start counting from its initial value as stated in the flowchart. If reset is toggle, the RTC will goes into its initial condition that is 00:00:00 (hh:mm:ss) & 01/01/2017 (dd/mm/year). If not the current time and date value will be store into outdata module register. This value will be update and store for every 1 second. To read the data from the RTC, all the condition need to be meet which is the address is call, PSEL and PENABLE is toggle, then the date and time value will be concatenated and transferred to PRDATA (output pin) in form of 32 data bit. For the alarm module, if the alarm is set, the alarm value will be compared with the real-time value. When the value is matched, an interrupt signal will be sent out to CPU as alarm signal.

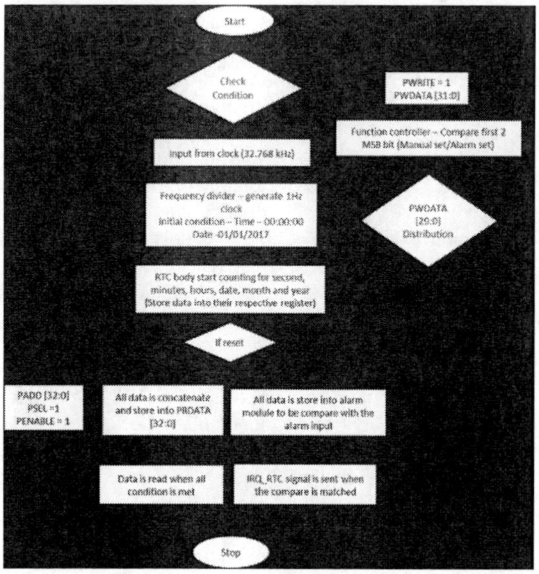

Fig. 8. Flowchart of RTC operation

4 Result (Simulation and FPGA testing)

The developed RTC is simulated in Quartus II software. The functionality of the RTC is being test out from waveform simulation and testing on FPGA Board (Altera DE2-70). The RTL view of the Top Module RTC is as in Fig. 9.

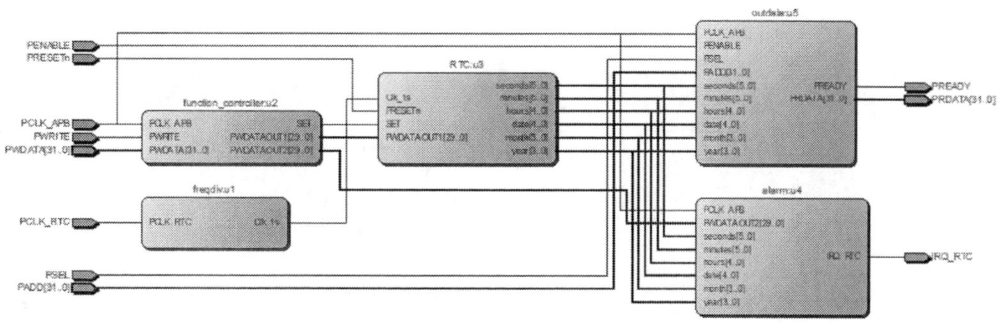

Fig. 9. RTL view of Top Module RTC

Table 3. PRDATA and PWDATA bit assignment

Data	Bits																															
	31	30	29	28	27	26	25	24	23	22	21	20	19	18	17	16	15	14	13	12	11	10	9	8	7	6	5	4	3	2	1	0
PWDATA	Function bit		Hours					Minutes						Seconds						Date					Month				Years			
PRDATA	Unused		Hours					Minutes						Seconds						Date					Month				Years			

The Top Module of RTC consist of 5 sub-module that is frequency divider, function controller, RTC, alarm and outdata module. Their functionality is as stated on Table 2 before. 3 of the sub module will be running by APB clock that is function controller, alarm and outdata module. APB clock resolution is 62.5ns which is 16MHz. For frequency divider and RTC, it will be running by RTC clock source that is 32.768 kHz with 30.52μs resolution. The features that this RTC had is write data features (to manual set time & date and to set alarm). The data bit description for write (PWDATA) and read (PRDATA) is as shown in Table 3. The first two MSB bit of the PWDATA is used as a function bit that is to manual set the time & date and to set the alarm. For now, it only had two functions to be controlled. The function bit description is as shown in Table 4. For 00 and 11 bit function, this is reserved for future improvement features if need to be added.

Table 4. Function bit description

Bits	Function
00	
01	Set Time & Date (Manual Set)
10	Set Time & Date (Alarm)
11	

Lastly, the full RTC module then has been tested using FPGA Board (Altera DE2-70) with adding test module (red box in Fig.10) as a dummy input and the result is as shown in Fig. 11 (a)-(f). The dummy input description is as shown in Table 5. The LCD module is added to show the result on the LCD display of the FPGA Board.

Table 5. Sample input data used for PWDATA

PWDATA							
	Func. bit	Hours	Minutes	Second	Date	Month	Year

Fig. 11(a-f) shows the LCD display result with their condition. Box below the LCD display show their input data.

PRESETn	PENABLE	PSEL	PADD	PWRITE	PWDATA
0	0	0	0	0	Z*

Fig. 11(a). Condition: Reset (Initial condition)

Fig. 11(a) shows the reset condition of the RTC module. This RTC is active low on the reset. So, when the PRESETn is 0, the RTC will reset all the date and time value to its initial condition that is 00:00:00 (time= hh:mm:ss) and 01/01/2017 (date=dd/mm/yyyy)

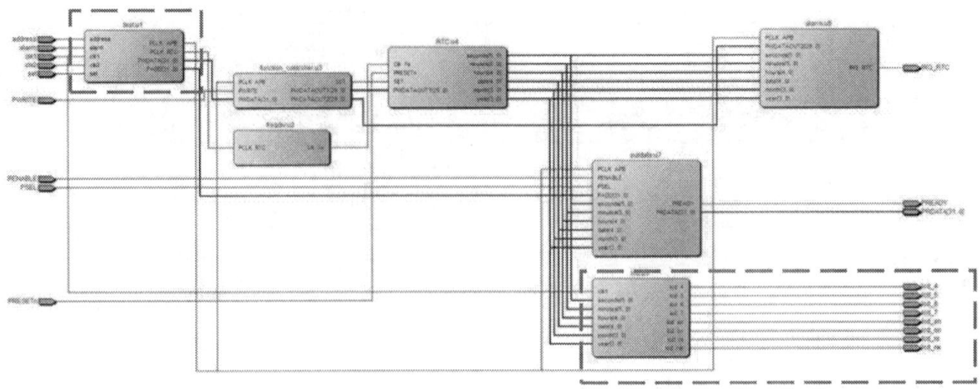

Fig. 10. RTL view of Top Module RTC with added test module

DOI: 10.1051/epjconf/201716201072

For Fig. 11(b), the condition shows the RTC running. Since the PRESETn is toggle to 1, the RTC will start counting from its initial value. The figure shows the RTC already counting for 9 second.

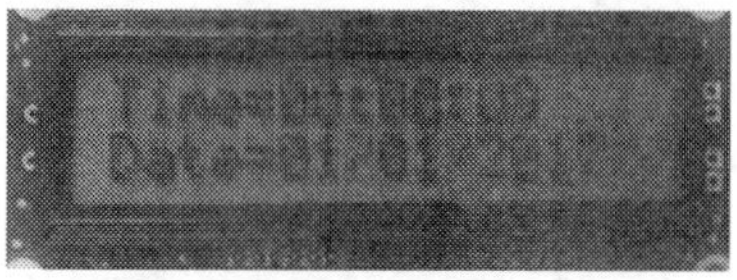

PRESETn	PENABLE	PSEL	PADD	PWRITE	PWDATA
1	0	0	0	0	Z*

Fig. 11(b). Condition: Real Time Clock Running

When the PWRITE=1 and PWDATA sent a 32bit data, The RTC is on the writing event. Its either manual set or alarm is set. Fig. 11(c) shows the manual set time and date since the 32bit data sent is using 01 function bit to manual set as in Table 5(sample data input). When the PWDATA use the 10 function bit, it will set the alarm in the alarm module as shown in Fig. 11(d).

PRESETn	PENABLE	PSEL	PADD	PWRITE	PWDATA
1	0	0	0	1	(1)*

Fig. 11(c). Condition: Manual Set (Date and Time)

PRESETn	PENABLE	PSEL	PADD	PWRITE	PWDATA
1	0	0	0	1	(2)*

Fig. 11(d). Condition: Set alarm

Fig. 11(e) shows the interrupt signal is sent out. The interrupt signal is sent out because the alarm set (in Fig.11(d)) is already match the real time. So, the alarm is on.

PRESETn	PENABLE	PSEL	PADD	PWRITE	PWDATA
1	0	0	0	0	Z*

Fig. 11(e). Condition: Alarm triggered (Interrupt)

To read the output data, The PSEL,PENABLE and PADD need to fill the requirement as stated in the operation flow. When the requirement is meet, the PREADY signal will be sent out and the PRDATA can be read is form of 32bit. The Fig. 11(f) shows that condition.

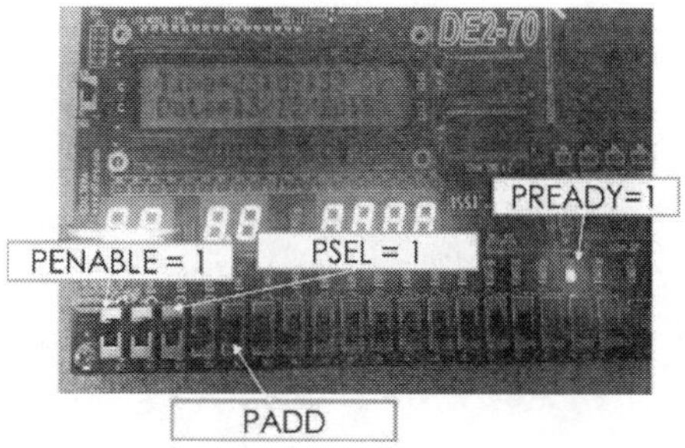

PRESETn	PENABLE	PSEL	PADD	PWRITE	PWDATA
1	1	1	address	0	Z*

Fig. 11(f). Condition: IP is ready to be read (PRDATA can be read).

5 Conclusion

A logic part of real time clock is successfully developed and implemented on FPGA board with accurate time counter. This work will be improved in the future and then be continue to the layout level. After that, it can be integrated into the wireless microcontroller with the other IP to be fabricated. All the reference design and idea is referred from various application note and technical papers of conventional RTC [3-6,10-12].

This project is sponsored by Talent Corp. Malaysia in collaboration with Intel Malaysia and CEDEC USM. First of all, I would like to thank my supervisor, Dr. Shaiful Nizam Bin Mohyar for all his guidance and support throughout this project. I would like to thank my fellow friends, Farhah, Fira and Wang Hang Suan for outstanding support and helpful comment.

References

1. http://whatis.techtarget.com/definition/real-time-clock-RTC. (Accessed: 27- Jul- 2016)
2. https://learn.adafruit.com/ds1307-real-time-clock-breakout-board-kit/what-is-an-rtc. (Accessed: 27- Jul- 2016)
3. SII-IC Semiconductor, S-35390A, rev. 4.2, Datasheet (2016)
4. Dallas Semiconductor, "Real Time Clock," DS12C887

DOI: 10.1051/epjconf/201716201072

5. Maxim Integrated," Extremely Accurate I2C RTC with Integrated Crystal and SRAM," DS3232, Rev 5 (2010)
6. NXP semiconductor, "Real-time clock/calendar," PCF8563, Rev.11 (2016)
7. https://www.maximintegrated.com/en/app-notes/index.mvp/id/5791. (Accessed: 30- Apr-2017)
8. *AMBA specification 2.0*, ed: ARM Ltd. (1999)
9. "Cortex-M System Design Kit Technical Reference Manual," ed: ARM Ltd., (2011)
10. https://www.maximintegrated.com/en/products/digital/real-time-clocks.html. (Accessed: 27- Jul- 2016)
11. Dallas Semiconductor," *64 X 8 Serial Real Time Clock*", DS1307, Datasheet
12. EPSON Semiconductor, "*Real Time Clock Module*," ETM49E-01, Datasheet (2016)

 DOI: 10.1051/epjconf/201716201073

Study on data compression algorithm and its implementation in portable electronic device for Internet of Things applications

Nor Asilah Khairi[1,*], and *Asral* Bahari Jambek[1]

[1]School of Microelectronic Engineering, Universiti Malaysia Perlis, Pauh Putra Campus, 02600 Arau, Perlis, Malaysia

Abstract. An Internet of Things (IoT) device is usually powered by a small battery, which does not last long. As a result, saving energy in IoT devices has become an important issue when it comes to this subject. Since power consumption is the primary cause of radio communication, some researchers have proposed several compression algorithms with the purpose of overcoming this particular problem. Several data compression algorithms from previous reference papers are discussed in this paper. The description of the compression algorithm in the reference papers was collected and summarized in a table form. From the analysis, MAS compression algorithm was selected as a project prototype due to its high potential for meeting the project requirements. Besides that, it also produced better performance regarding energy-saving, better memory usage, and data transmission efficiency. This method is also suitable to be implemented in WSN. MAS compression algorithm will be prototyped and applied in portable electronic devices for Internet of Things applications.

1 Introduction

Internet of Things (IoT) is known as a network of physical objects containing embedded technology that communicates and senses or cooperates with internal states or an external environment via wireless and wired connection [1]. The authors in [2] indicated that the IoT could be defined as the next generation of internet evolution, where it can connect everyday objects such as refrigerators, air conditioners, and automotive devices, through embedded wireless devices that allow interaction with each other and provide new services. The existence of IoT creates a seamless intelligence that improves human life quality.

Cisco predicted that 50 billion new connections would be part of the IoT by 2020 [3]. These billions of connected devices will produce a significant amount of data to be stored. The current methods to store data have proven to be insufficient for the demands of the IoT. Therefore, a few papers discussed the possibilities of embedding a compression algorithm into IoT devices or sensor nodes. According to the authors [4], compression is the perfect solution that allows the reduction in both storage requirements and input/output operations. Therefore, it is very useful during the transmission of data via a network.

Wireless Sensor Network (WSN) is popular regarding monitoring the different types of physical environments. However, sensor nodes were designed with limited battery power. Hence, it limits the energy to function and necessitates a low power to save energy and in turn extend the lifetime of the sensor node [4]. Furthermore, radio transmission is a major contribution to energy consumption in sensor nodes. An efficient data compression approach is highly recommended to overcome this problem. Therefore, several past studies were selected to investigate the concepts of data compression algorithm.

This paper is organized as follows: Section 2 reviews previous compression algorithms; the comparison between different types of data compression is discussed in Section 3; Section 4 focuses on the project proposed of MAS compression algorithm; and finally, conclusions are drawn in Section 5.

2 Latest trend of data compression algorithm

Data compression is a popular issue among researchers due to its ability to be used in nearly every functional area. According to various researches, data compression is divided into two categories: lossless and lossy compressions. The Autoencoders (AE) compression algorithm in [5] is known as a lossy compression. The improvement of the IoT device's lifetime such as energy constrained, optimization of the internal memory space, and efficient data transmission via wireless connection was brought into focus in this experiment, while biomedical signals such as ECG, photo-plethysmography, and respiratory traces were used as the dataset. The performance of the AE compression algorithm was then compared with Discrete Cosine Transforms (DCT), Wavelet transforms, Principal Component Analysis (PCA), and linear approximations. Based on the results produced in the paper [5], the

* Corresponding author: asilah.khairi91@gmail.com

energy consumption required for the online compression of the signal is quite small when compared with a scheme using linear approximations. According to researchers, this project also posed a weakness. It requires an offline training phase, which means it has to be executed only once for each class of signals.

Vecchio, Giaffreda, and Marcelloni [6] proposed to endow the lossless compression algorithm (LEC) for wireless sensor networks using two simple adaption schemes that relied on the new concept of the appropriately rotated prefix-free tables. This project focused on improving the power consumption of IoT devices during transmission and the reception of data. According to the researchers, the primary idea of the LEC algorithm relied on dividing the alphabets or numbers into a group in which the sizes will increase rapidly. Hence, the LEC codeword was a hybrid of unary and binary codes, where the unary code (a variable-length code) specified the group and the binary code (a fixed-length code) represented the index within the group that is almost similar to Golomb and Elias's coding. In this project, the researchers introduced two more different adaption blocks known as GA-LEC and FA-LEC. Air temperature, surface temperature, relative humidity, and solar radiation were used as the dataset. Both GA-LEC and FA-LEC compressions produced a high percentage of efficiency.

Domingo-Prieto et al. [7] focused on the relationship between the amounts of messages each one sends that affects the lifetime of the battery in mote. Huffman coding is applied as data compression in this project. The dataset was extracted from the IoT device, and the performance of the Huffman compression algorithm was later compared against address clustering. Based on the results, the motes that send more messages have improved their lifetime compared to the motes that send fewer messages. According to researchers, the implementation of Huffman coding does not seem to be a trivial task in a real deployment, while the Address Clustering is straightforward.

In paper [8], Hsu et al. proposed the development of the WSN system that collects vibration acceleration from bridges in real time and sends the data via the internet. This project not only focused on remote monitoring, but also in achieving the lowest production cost, shortest installing time, and convenient relocation. The Huffman coding was used as the data compression in this project, whereas the vibration frequency of bridge was used as a dataset. The performance of the Huffman compression algorithm was then compared against three other compression algorithms, which included LZ77, Run Length Encoding (RLE), and Rice coding. Based on the results, the number of WSN end device node grew 300% more compared to the similar sampling rate that did not have an algorithm implemented on the WSN network. According to researchers, the wireless transmission payload was reduced to about 60% and the node number of the implemented increased about three times.

According to Assi et al. [9], the MAS compression algorithm is one of the first algorithms to compress floating-point data specifically. In this project, the researchers focused on exploiting the disproportionality in energy consumption between data transmission and processing. MAS compression algorithm stands for Minimalist, Adaptive, and Streaming. Carbon dioxide, water levels, radioactivity, temperature, humidity, and low sea pressure were used as the dataset in this experiment. The researchers conducted six different types of experiments: compression ratio, computation time and energy, transmission energy, total consumed energy, energy-savings, and memory requirements. The results showed that MAS produced higher compression ratio than the S-LZW and K-RLE compression algorithms. Hence, using MAS algorithm can achieve better results compared to other algorithms. Besides, MAS consumed the least amount of transmitted energy among the other algorithms. Despite K-RLE consumption of the least computation energy, it still consumed the most amount of total energy, as it sent significant amounts of data over the network. Meanwhile, MAS consumed the least energy, thus proving that MAS is the strongest candidate for compression in WSN. It also managed to save more energy than S-LZW and K-RLE. MAS consumed the largest amount of flash memory because it had a longer programming code, with two representations for integers and real numbers. However, it did not cause any problem, since the flash memory was reserved for program code and not for random access. Moreover, MAS consumes only 44 bytes regarding RAM usage. According to researchers, the results of the experiment showed that MAS produced energy-saving on an average of 54%, defeating K-RLE and S-LZW while maintaining the highest compression ratios.

Marcelloni and Vecchio [10] proposed a straightforward and efficient data compression algorithm that is precisely suited to be used on nodes of WSN, where energy, memory, and computational resources were limited. They focused on developing a simple compression algorithm that reduced memory and computational resources of the WSN node. A simple lossless compression algorithm was introduced as data compression in this project, and it used a combination of the Huffman and JPEG algorithms. The researchers used temperature and humidity as the dataset. Based on the results, the proposed compression algorithm achieved higher compression ratio than the S-LZW. Last but not least, it proved that this algorithm could outperform gzip and bzip2.

3 Comparison between different types of compression algorithms

Every single data of compression algorithms is built differently. There are different types of compression algorithms in this world, some of which can support image files, while some can only support text files. Some data compressions cannot preserve the originality of the data after decoding, for example, the Autoencoder algorithm [5]. In Section 3, the data compression algorithms that were previously discussed in Section 2 are analyzed and displayed in a table for comparison.

Table 1 shows the summarization of the comparison between the different types of compression algorithms.

Table 1. Comparison of data compression algorithms.

Description	[5]	[6]	[7]	[8]	[9]	[10]
Year	2015	2014	2014	2014	2013	2008
Focus	Improves lifetime of IoT devices (energy constraint, optimization on internal memory space, and efficient transmission of data).	Improves power consump tion-n of IoT devices	Focuses on the relationship between battery lifetime in mote and number of messages each one sends	Remote monitoring, low cost, short installation time, and convenient relocation	Exploit the disproportionality in energy consumption between data transmission and processing	Simple compression algorithm suited to reduce memory and computational resources of WSN node
Types of Data Compression	Autoencoder (AE)-Lossy Compression	GAS-LEC, FAS-LEC	Huffman coding	Huffman coding	Minimal, Adaptive, and Streaming compression algorithm (MAS)	Simple lossless compression algorithm - Combination of Huffman, and JPEG coding
Comparison of Other Data Compression	Discrete Cosine Transforms (DCT), DCT-LW, Discrete Wavelet Transform (DWT), DWT-LW, LTC (based on linear approximation), Principal Component Analysis (PCA)	Lossless Compression Algorithm (LEC)	Address Clustering (non-data compression)	LZ77, Run Length Encoding (RLE), Rice	S-LZW, K-RLE	S-LZW, gzip, bzip2
Dataset	Biomedical signal (ECG, photo-plethysmographic, respiratory traces)	Air, surface, temperature, solar radiation, relative humidity	Extracted from the IoT device deployment	Vibration frequency of bridge	Carbon dioxide, water levels, radioactivity, temperature, humidity, sea low-level pressure	Temperature, humidity
Advantage	Small reconstruction, high compression efficiency, low energy consumption	Improves the compression efficiency of LEC	Has better overall improvement because it used variable-size packets and also compressed data	Wireless transmission payload can be reduced by 60%, node number of the implemented network can be increased by three times	Better regarding compression ratio, compression speed, memory requirements, and energy-saving	Higher compression ratio, com
Disadvantage	Requires an offline training phase (has to be executed only once for each class of signals)	Not stated	Does not seem a trivial task in real deployment	Not stated	Not stated	Not stated

Based on the summarization in Table 1, project [9] has the most potential and a more suitable project to be used as a prototype of this experiment. The similarities of the project [9] with past studies can be seen in the advantage, method, and focus. Project [9] produced better compression ratio, which is also similar to the project [10]. Regarding its method, MAS project concentrated on the implementation of the data compression algorithm in WSN. This approach is similar to projects [5], [6], [7], [8], and [10]. The last similarity of the project [9] was that it focused on improving the performance of the data compression. Project [9] was focused on reducing the usage of power consumption during data transmission, which was similar to projects [5], [6], and [7].

Paper [9] was selected as the reference paper not only for its similarities but also for its uniqueness. One of it is the data compression algorithm. This algorithm is one of the first algorithms to compress, floating-point data specifically. Besides that, it supported not only the floating-point data but also integer number data. Another factor that made this paper unique is its use of measurement other than the compression ratio, which included compression speed, memory requirement, energy-saving, and energy cost. The measurements of the project [9] are not similar to the other projects. Therefore, it produces various results that will ease the observation process.

Based on the analysis, paper [9] was chosen as the reference project due to its high similarity concept to the one proposed. However, this project required certain adjustments to meet the IoT concept. The recommendation of the work was the implementation of the compression algorithm in the end devices and the end user devices. The structure of the MAS algorithm required minor adjustments to allow the MAS compression to be implemented inside the end devices and MAS decompression inside the end user devices. Different types of collected data sets were suggested in this project with the intention of producing various kinds of results. Furthermore, it was also done to test the compatibility of MAS algorithm with the devices.

DOI: 10.1051/epjconf/201716201073

4 Proposed MAS compression algorithm in IoT Device

Once the description of the data compressions was compared with each other in Section 3, the suitable compression algorithm was selected to be used as a reference and prototype for this project. On that regard, the MAS compression algorithm was chosen as a prototype project due to its potential to meet the project requirement.

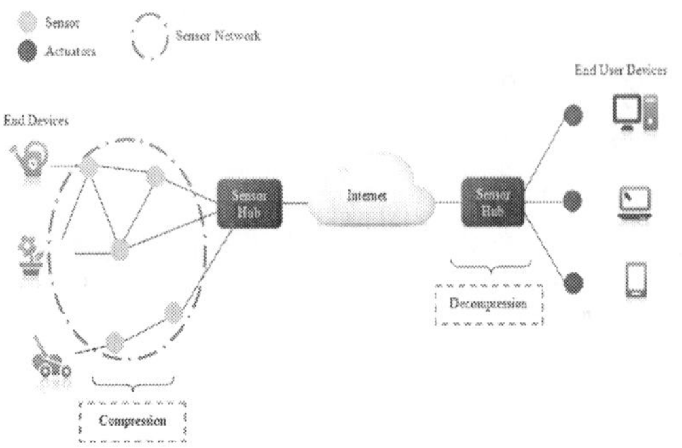

Fig. 1. MAS data compression framework for IoT device.

The proposed data compression framework for IoT device is illustrated in Figure 1, which consisted of the compression and decompression processes in the sensor network and the sensor hub. The figure shows that the watering can, flowering plant, and lawnmower are considered as the end devices. However, other relevant items can also be used as the end devices.

According to the authors in [1], the sensor network is the first phase where all the data are generated. The sensors collect the data from the end devices and compress the data with the intention of reducing power consumption and memory space while increasing the speed of the data transmission. After the compression process completed, the data are sent to the sensor hub via a wireless sensor network. Later, this modem redirects the data via the internet using basic protocols.

The second sensor hub receives data from the first sensor hub through the internet. The modem then decompresses the data and triggers the actuators on the other side. The intention of the data decompression is to convert the data into its original form. Any smart gadgets such as smartphones, tablets, laptops, and desktop computers are classified as the end user devices. After the decompression process is completed, the data is displayed on the end user devices.

5 Conclusion

IoT devices are always powered by a small battery, and this battery does not usually last long. As a result, saving energy in IoT devices has become the main focus of this paper. Radio communication is known for its high power consumption. Therefore, several researchers have proposed data compression algorithms with the sole purpose of overcoming this particular problem. Several compression algorithms in previous reference papers were analyzed and discussed in this paper. The descriptions of previous compression algorithms were collected and tabulated. Based on the analysis, the MAS compression algorithm was selected as the project prototype due to its high potential in meeting the requirements such as high energy-saving, low memory usage, and efficient data transmission. Besides, this method is suitable to be implemented in WSN. This proposed algorithm will be prototyped and applied in portable electronic devices for Internet of Things applications.

References

1. P. Srivastava and N. Garg, *2015 International Conference on Computing, Communication & Automation (ICCCA)*, pp. 720-723 (2015)
2. Z. Wang, R. Bie, and M. Zhou, *2012 IEEE 11th International Conference on Privacy in Computing and Communications (TrustCom)*, pp. 1799-1804 (2012)
3. S.E. Collier, *2015 IEEE Rural Electric Power Conference (REPC)*, pp. 65-68 (2015)
4. S. Sheikh and H. Dakhore, International Journal of Computer Science and Information Technologies (IJCSIT), **6**, 1, pp. 818-821 (2015)
5. D.D. Testa and M. Rossi, IEEE Signal Processing Letters, **22**, 12, pp. 2304-2308 (2015)
6. M. Vecchio, R. Giaffreda and F. Marcelloni, IEEE Transactions on Wireless Communications, **13**, 2, pp. 1088-1100 (2014)
7. M. Domingo-Prieto, B. Martinez, M. Monton, I. Vilajosana-Guillen, X. Vilajosana-Guillen and J. Arnedo-Moreno, *2014 Eighth International Conference on Complex, Intelligent and Software Intensive Systems (CISIS)*, pp. 170-176 (2014)
8. C-H. Hsu, C-T. Lin, H-P. Tserng and J-Y. Han, *2014 IEEE World Forum on Internet of Things (WF-IoT)*, pp. 548-552 (2014)
9. M.E. Assi, A. Ghaddar, S. Tawbi and G. Fadi, International Journal of Ad hoc, Sensor & Ubiquitous Computing (IJASUC), **4**, 5, pp. 13-28 (2013)
10. F. Marcelloni and M. Vecchio, IEEE Communications Letters, **12**, 6, pp. 411-413 (2008)

EPJ Web of Conferences **162**, 01074 (2017) DOI: 10.1051/epjconf/201716201074
InCAPE2017

Comparison of edge detection techniques for M7 subtype Leukemic cell in terms of noise filters and threshold value

Afifah Salmi Abdul Salam [1,*], *Mohd. Nazrin* Md. Isa [1], *Muhammad Imran* Ahmad[1] and *Rizalafande* Che Ismail[1]

[1]School of Microelectronic Engineering, University Malaysia Perlis, Pauh Putra Campus, 02600, Arau, Perlis, Malaysia
[2]School of Computer and Communication Engineering, University Malaysia Perlis, Pauh Putra Campus, 02600, Arau, Perlis, Malaysia

Abstract. This paper will focus on the study and identifying various threshold values for two commonly used edge detection techniques, which are Sobel and Canny Edge detection. The idea is to determine which values are apt in giving accurate results in identifying a particular leukemic cell. In addition, evaluating suitability of edge detectors are also essential as feature extraction of the cell depends greatly on image segmentation (edge detection). Firstly, an image of M7 subtype of Acute Myelocytic Leukemia (AML) is chosen due to its diagnosing which were found lacking. Next, for an enhancement in image quality, noise filters are applied. Hence, by comparing images with no filter, median and average filter, useful information can be acquired. Each threshold value is fixed with value 0, 0.25 and 0.5. From the investigation found, without any filter, Canny with a threshold value of 0.5 yields the best result.

1 Introduction

Leukemia cells in humans are growing rapidly each year. In 2016, statistics show that 24,400 people are expected to die from this disease [1] Leukemia is one of the deathly diseases amongst humans and it is known as a bone marrow disorder. This disease involves abnormalities of white blood cells proliferation that disables the cells to function normally [2].This disease is grouped by how quickly the illness develops. It can be either acute or chronic [3].

In addition, Leukemias are grouped by affected blood cell type (lymphocytes or myelocytes). This disease can be categorized into four main types which includes acute lymphocytic leukemia (*ALL*), chronic lymphocytic leukemia (*CLL*), acute myelocytic leukemia (*AML*) and chronic myelocytic leukemia (*CML*) [4]. Each leukemia type has its particular properties or shape to differentiate them from others. Acute Leukemia grows briskly which can invade the body within a few weeks or months. Chronic Leukemia on the other hand, takes time growing but progressively worsens over the years.

Acute myelocytic leukemia (Fig.1) is a heterogenous group of clonal disorders which is usually not detected until it has spread into other organs. AML is classified by a system known as French-American British (FAB) classification which is categorized into eight subtypes, namely M0 until M7[5].

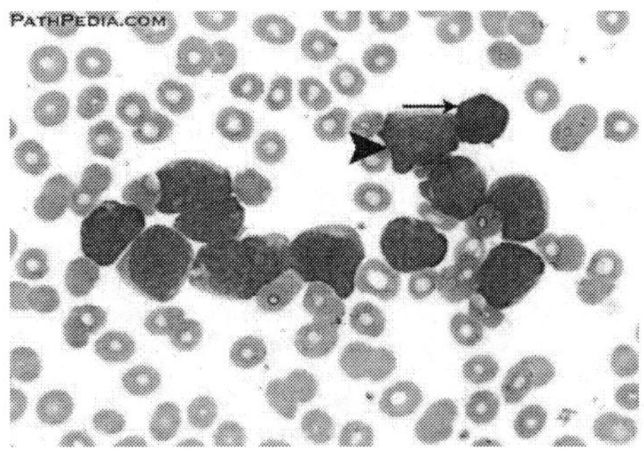

Fig.1. Acute Myelocytic leukemia (*AML*)

Table 1. French-American-British Classification [6]

FAB SUBTYPE	NAME
M0	Undifferentiated acute myeloblastic leukemia
M1	Acute myeloblastic leukemia with minimal maturation
M2	Acute myeloblastic leukemia with maturation
M3	Acute promyelocytic leukemia (APL)
M4	Acute myelomonocytic leukemia
M4 eos	Acute myelomonocytic leukemia with eosinophilia
M5	Acute monocytic leukemia
M6	Acute erythroid leukemia
M7	Acute megakaryoblastic leukemia

* Corresponding author: afifahsalmi.79@gmail.com

Table 1 shows the subtypes of *AML* classification. Each subtypes varies in properties like the size and number of leukemia cells. The usage of edge detection becomes a major part in detecting these changes of cell shape. M7 subtypes have been distinguished over the years by the uses of cytochemical and morphological criteria which were found lacking [7]. Moreover it has been found that 3-10% of primary childhood *AML* and children may consists varieties of symptoms such as low-grade fever, diarrhea and easy bruising[8]. Studies have been conducted that this type of leukemia is usually abnormally abundant in children with Down Syndrome cases.

Image segmentation (Preprocessing) is one of the crucial steps in an automatic leukocyte recognition system[9]. This is because the last two steps after segmentation depends greatly on the result of segmentation. Edge detection is part of image segmentation. Edge detection contains significant information. It reduces the image size and filters out information that are less convenient [10]. Edges typically occur on the boundary between two different region in an image [11]. Frequently, this detection is the first step in recovering information from images. Detecting edges greatly rely on the noise, intensity, brightness and blur. Thus, by working with different edges of the same image, differences can be observed and the selected suitable algorithm will be chosen to further to the next stage of image processing (feature extraction). Edge detection can come with a threshold value which functions to detect edges. Typically the lower the threshold value, more edges can be detected [12]. However, it is not necessarily that many edges found can make an image looks great. Due to it's advantages, edge detection continues to be an active research area.

Thus, this paper will study different edge detection techniques (Sobel and Canny) which then will be compared with one another to select which is the best techniques to be applied on detecting edges on an M7 subtype leukemic cell (Figure 2). The system should able to recognize the cell pattern and shape. A threshold value of 0, 0.25 and 0.5 will be chosen. The threshold value will determine how many edges are available to be detected in an image. Filters (median and averaging) which is for noise removal will also be studied to find which is the most suitable filter to minimise/remove noise filters. The sofware to implement this project will be MATLAB 2014a.

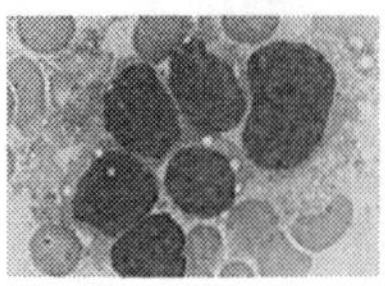

Fig.2. M7 subtype of AML cell

2 Proposed methodology

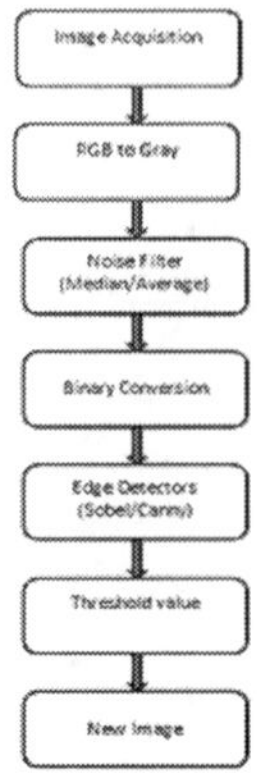

Fig.3. Flowchart of the whole system

Figure 3 shows the overall flowchart of the edge detection system, starting from image acquisition until new image is achieved. Since the highlight of this paper is about noise filter, edge detection and threshold value, thus the next subsequent section will explain briefly about these three things.

3 Noise Filter

Filtering is removing some frequencies in order to suppress interfering signals and reduce background noise [13]. In other words, filtering is a device or process that removes signal of unwanted component or feature. Filters can come in high pass or low pass depending on the amount of noise needs to be filtered. This paper involves noise removing using low pass filter of averaging filter and median filter.

Average filter or mean filter functions to smooth the image into non-noise image [14]. The smoothing filters are used for image blurring and noise reduction in spatial domain.

Median filter is a nonlinear image processing operation used to remove the impulsive noise from images [15]. They are most preferred against impulsive noise due to their robustness and denoising power.

4 Edge detection

Edge detection comprises of algorithm varieties [11]. Each type has pros and cons whereby the usage depends on the image requirement. Two of popular used edge detection are Sobel and Canny. Finding the correct algorithm is essential as it brings up the correct information desired by a researcher. Hence, a good edge detection will help to strengthen the result for diagnosing early detection of leukemic cell.

Table 2. Advantages and Disadvantages of Sobel and Canny [12]

Edge Detection	Advantages	Disadvantages
SOBEL	-Simplicity - Detection of edges and their orientations	- Sensitivity to noise - Inaccurate
CANNY	- Improving signal to noise ratio - Better detection especially in noise detection - Good for finding error rate	- Complex computation - False zero crossing -Time consuming

Table 2 shows the pros and cons of each edge detection. Aforementioned that edge detection varies in algorithm, thus the table shows how important it is to choose a correct detection that fit best to the application.

4.1. Sobel edge detector

The function of Sobel edge detector is for detecting vertical and horizontal edges in an image known as Gx and Gy [16]. It also combines information into a single matrix. The operator consists of 3x3 convolution kernels by which one kernel is simply rotated by 90° by the other as shown in Figure 2 [11].

-1	0	+1
-2	0	+2
-1	0	+1

Gx

+1	+2	+1
0	0	0
-1	-2	-1

Gy

Fig.4. Sobel Masks

The kernels can be joined together to find the absolute magnitude and the orientation of the gradient as shown in Equation (1).

$$|G| = \sqrt{Gx^2 + Gy^2} \quad (1)$$

4.2 Canny Edge detector

Canny edge detection is also known as the optimal edge detector [16], which satisfies all of the performance criteria. Canny comes in several steps as follow [11, 12], [16-18]:

1. Gaussian filtering for noise removal

$$G\sigma = \frac{1}{\sqrt{2\pi\sigma^2}} exp[-\frac{m^2+n^2}{2\sigma^2}] \quad (2)$$

2. Finding the intensity and gradient of image whereby it uses four filters to find out the vertical, horizontal and diagonal edges to find the blurred image.

$$M(m,n) = \sqrt{g^2m(m,n) + g^2n(m,n)} \quad (3)$$

and

$$\theta = tan^{-1}\left[\frac{gn(m,n)}{gm(m,n)}\right] \quad (4)$$

3. Non maximum suppression for thinning of the edge.
4. Double threshold to get rid of spurious responses from bothering factors such as noise and colour variation.
5. Track edge by hysteresis to achieve accurate result by getting rid of weak edges caused by latter reasons. To track down the edge connection, Binary Large Object analysis is applied by looking at weak edge pixel and its 8-connected neighbourhood pixels.

5 Threshold

The aim of threshold is to determine which pixels fall into each category [19]. A straightforward thresholding method uses a single value of intensity and relying on it, every pixel can belong to one of two categories:

a. Low intensity of pixel = pixel set to black
b. High intensity of pixel = pixel set to white

In Matlab, BW = edge(I,'edge',thresh) is one of the easiest method to detect edge by using threshold. The thresh specifies the sensitivity threshold for the particular edge method [13]. If the thresh is not specify or empty, edge chooses the value automatically. In this paper a threshold value of 0, 0.25 and 0.5 are used to compare which value gives out the best effect of M7 subtypes of *AML*.

6 Experiments

This section explains the experiment that has been conducted to find which threshold values (0, 0.25 and 0.5) and noise filters that gives the best result for either Sobel edge detector or Canny edge detector in finding the M7 subtype *AML* cell. The image is obtained from *clinicalflow* website. The experiment starts with the test of these two operators with no filters then followed by testing with averaging noise filters and median filters. The difference between each images are tabulated in Table 3 and Table 4 of this section.

6.1 Sobel operator with no filter

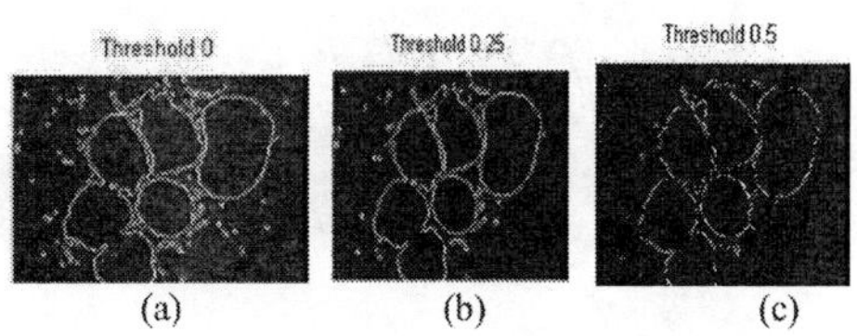

Fig.5. (a) Threshold 0 of sobel, (b) Threshold 0.25 of sobel and (c)Threshold 0.5 of sobel

6.2 Canny operator with no filter

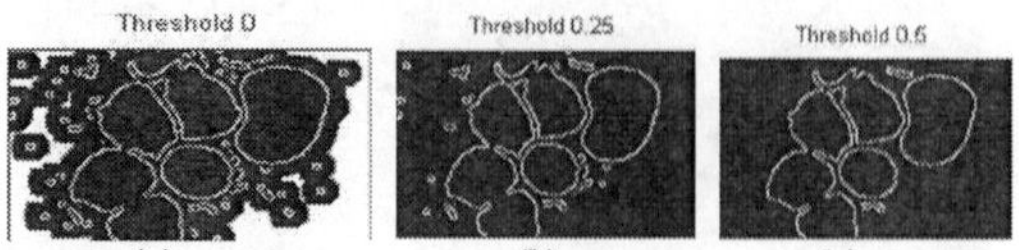

(a) (b) (c)

Fig.6. (a) Threshold 0 of canny, (b) Threshold 0.25 of canny, and (c) Threshold 0.5 of canny

6.3 Sobel operator with averaging filter

(a) (b) (c)

Fig.7. (a) Threshold 0 of sobel averaging filter, (b) Threshold 0.25 of sobel averaging filter and (c) Threshold 0.5 of sobel averaging filter

6.4 Canny operator with averaging filter

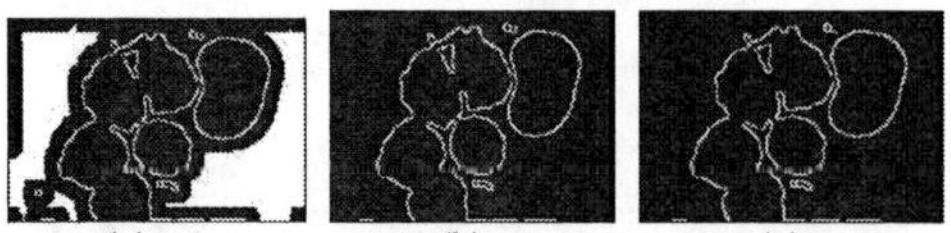

(a) (b) (c)

Fig.8. (a)Threshold 0 of canny averaging filter, (b) Threshold 0.25 of canny averaging filter and (c) Threshold 0.5 of canny averaging filter

6.5 Sobel operator with median filter

(a) (b) (c)

Fig.9. (a)Threshold 0 of sobel median filter, (b) Threshold 0.25 of sobel median filter and (c) Threshold 0.5 of sobel median filter

6.6 Canny operator with median filter

(a) (b) (c)

Fig.10. (a) Threshold 0 of canny median filter, (b) Threshold 0.25 of canny median filter and (c) Threshold 0.5 of canny median filter

Figure 5 to 10 shows the results of one image of M7 that has been tested with different values of threshold, noise filters and edge detection (Sobel and Canny). Table 3 and 4 shows the overall result of M7 subtypes *AML* image with different edge detection techniques, thresholding and filter. From Table 3 that shows the result of Sobel edge detection, it proves that noise will decrease as threshold increase. However, the edge lines will gradually disappear as threshold increase. For Table 4, for non-filtering cases of Canny using threshold 0.5, the cell looks much clearer with only a few noise left and two cells overlap but are able to identify that they are different because the overlapping case does not make much difference. Overall, with average filter when there is no threshold, the image will produce a border but gradually fades as threshold increase. Other than that, Sobel edge's line fades as threshold increase but for Canny, the edge still maintains with an increase sharpness of line. This is due to Canny edge detection having to do double threshold and finding intensity and gradient of image.

Table 3. Overall result for Sobel edge Operator

Filter	Thresholding		
	0	0.25	0.5
No Filter	- Lots of background noise - Two cells overlap	- Background noise less	- Noise are less - Lines are not clear
Averaging	- Hardly background noise but there is a border	- Some noise are missing	- Lines are not clear -Border gradually fades away
Median	-Cells overlap - Hardly background noise	- Not much different as threshold 0 -Some lines are shortened than other	- Lines gradually fades

Table 4. Overall result for Canny edge operator

Filter	Thresholding		
	0	0.25	0.5
No Filter	- Lots of background noise - Two cells overlap	- Background noise less	- Noise are less - Lines are not clear
Averaging	- Hardly background noise but there is a border	- Some noise are missing	- Lines are not clear -Border gradually fades away
Median	-Cells overlap - Hardly background noise	- Not much different as threshold 0 -Some lines are shortened than other	- Lines gradually fades

7 Conclusion

Briefly, this paper is about finding suitable edge detection techniques (Sobel and Canny) for M7 subtype *AML* to further on the next step of image processing after segmentation step, which is feature extraction. The image is tested with different threshold values (0, 0.25, and 0.5) and filters (averaging filter and median filter) to give out the best result. Based on the obtained results, without any filter and a threshold of 0.5, canny edge detection is suitable to be picked since noise decreases greatly compare to Sobel. With the factor of threshold as mentioned in [12], 0.5 value is apt in producing clear edge lines because the value is not too high and too low. Also aforementioned with its ability to give better detection in terms of noise detection, Canny also filters noise earlier in one of it steps hence it does not require additional filter to remove the noise. Albeit that there are still one or two background noises left, Canny managed to produce a clearer edge line compare to Sobel. For future reference it is good to try on different low pass filter so that the background noise will completely be removed and solve the problem of overlapping cell cases.

References

1. https://www.lls.org/http%3A/llsorg.prod.a cquia-sites.com/facts-and-statistics/facts- and-statistics-overview/facts-and-statistics. [Accessed: 14-Mar-2017].
2. S. Mohapatra, D. Patra, and S. Satpathy, Neural. Comput. Appl., **24**, no. 7–8, pp. 1887–1904, (2014)
3. http://www.medicinenet.com/leukemia/article.htm. [Accessed: 14-Mar-2017].
4. http://www.empowher.com/leukemias/content/type s-leukemia. [Accessed: 14-Mar-2017].
5. H. Mahmood *et al.*, J. Hematop., **7**, no. 1, pp. 3–8, (2014)
6. https://www.cancer.org/cancer/acute- myeloid-leukemia/detection-diagnosis-s taging/how-classified.html. [Accessed: 15-Mar-2017].
7. a Carroll *et al.*, Blood, **78**, no. 3, pp. 748–752, (1991)
8. P. D. Mueller and W. S. Korey, Iran. J. Paediatr. Haematol. Oncol., **4**, no. 4, pp. 7–10, (1998)
9. G. Gu and D. Cui, Comput. Informatics, **30**, no. 1, pp. 189–199, (2011)
10. Y. Zheng, J. Rao, and L. Wu, *2010 5th Int. Conf. Comput. Sci. Educ. (ICCSE)*, pp. 471–473, (2010)
11. V. Sucharita, S. Jyothi, and D. M. Mamatha, Int. J. of Comp. Applications, **78**, no. 6, pp. 1–5, (2013)
12. G. T. Shrivakshan and C. Chandrasekar, Int. J. Comput. Sci. Issues, **9**, no. 5, pp. 269–276, (2012)
13. B. Poornima, Y. Ramadevi, and T. Sridevi, Int. J. of Eng. Science and Technology (IJEST), **3**, no. 4, pp. 400–403, (2011)
14. C. M. Tsai and Z. M. Yeh, *Proc. - 2012 Int. Symp. Comput. Consum. Control. IS3C 2012*, no. **1**, pp. 153–156, (2012)
15. L. A. Aranda, P. Reviriego, and J. A. Maestro, IEEE Trans. Nucl. Sci., **X**, no. X, pp. 1–1, (2017)
16 V. G. Narendra and K. S. Hareesh, Int. J. of Agricultural and Biological Eng. **4**, no. 2, (2011)
17. K. Ogawa, Y. Ito, and K. Nakano, *2010 First Int. Conf. Netw. Comput.*, pp. 279–280, (2010)
18. T. Nadu and T. Nadu, *2nd International Conference on Electronics and Communication Systems (ICECS 2015)*, pp. 282–285, (2015)
19. P. Anghelescu, V. G. Iliescu, C. Mara and M. B. Gavriloaia, *2016 8th International Conference on Electronics, Computers and Artificial Intelligence (ECAI)*, (2016)

EPJ Web of Conferences 162, 01075 (2017) DOI: 10.1051/epjconf/201716201075
InCAPE2017

FPGA-based protein sequence alignment : A review

Mohd. Nazrin Md. Isa [1*], *Ku Noor Dhaniah* Ku Muhsen[1], *Dayana* Saiful Nurdin[1], *Muhammad Imran* Ahmad[2], *Sohiful Anuar* Zainol Murad[1], *Shaiful Nizam* Mohyar[1], *Azizi* Harun[1] and *Razaidi* Hussin[1]

[1]School of Microelectronic Engineering, Universiti Malaysia Perlis, Pauh Putra Campus, Perlis, Malaysia

[2]School of Computer and Communication Engineering, Universiti Malaysia Perlis, Pauh Putra Campus, Perlis, Malaysia

Abstract. Sequence alignment have been optimized using several techniques in order to accelerate the computation time to obtain the optimal score by implementing DP-based algorithm into hardware such as FPGA-based platform. During hardware implementation, there will be performance challenges such as the frequent memory access and highly data dependent in computation process. Therefore, investigation in processing element (PE) configuration where involves more on memory access in load or access the data (substitution matrix, query sequence character) and the PE configuration time will be the main focus in this paper. There are various approaches to enhance the PE configuration performance that have been done in previous works such as by using serial configuration chain and parallel configuration chain i.e. the configuration data will be loaded into each PEs sequentially and simultaneously respectively. Some researchers have proven that the performance using parallel configuration chain has optimized both the configuration time and area.

1. Background

Protein sequence alignment searches for sequence homology between query (unknown sequence) and subject sequence. For the case of pairwise sequence alignment, a query sequence of length *n* with sequence characters of $(x_1x_2...x_n)$ and a subject sequence characters of $(y_1y_2...y_m)$ of length *m*, is The Smith-Waterman algorithm shown in (1), using dynamic programming (DP) to solve the sequence alignment problems in finding the best score as shown in Fig. 1.

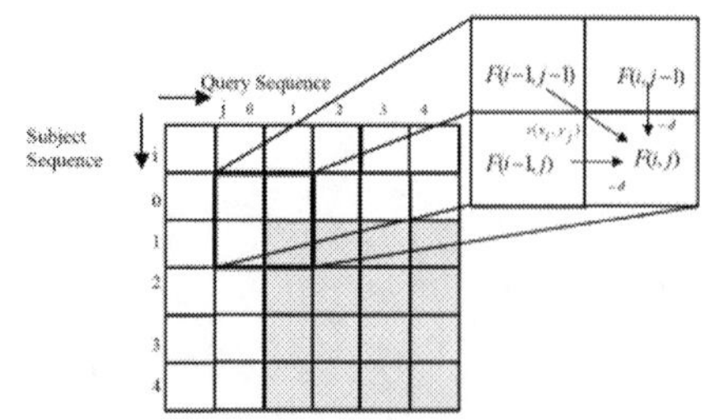

Fig. 1. The DP illustrate how SW algorithm obtain the final output which is the best score of *F(i,j)* and form a matrix F of scores [17].

Based on (1), *F(i,j)* is the similarity score of node at the *(i,j)* position. The best score is the maximum value among the four possible values based on (1).

$$F(i,j) = \max\begin{cases} F(i-1,j-1)+s(x_i,y_j) \\ F(i-1,j)-d \\ F(i,j-1)-d \\ 0 \end{cases} \quad (1)$$

Here $s(x_i,y_j)$ represents the similarity score by residue comparison of *x* and *y* sequence which can be obtained from the substitution matrix.

* Corresponding author: nazrin@unimap.edu.my

A. Protein Sequence Alignment

In protein sequence alignment, substitution matrix score has been substitute into the equation of DP algorithm for computation process in finding the optimal alignment between two sequences. Substitution matrix is a two dimensional matrix which shows the relationship between each protein residues since the penalty score of substituting a character with another character is different for each residue due to the biology meaning. There are several examples of substitution matrices for protein residue such as the BLOck SUbstitution Matrix (BLOSUM) e.g. BLOSUM50 and BLOSUM62 and Point Accepted Mutation (PAM) e.g. PAM160. The BLOSUM50 substitution matrix is illustrated as in Fig. 2. It has been widely used in aligning biological sequence.

	A	C	D	E	F	G	H	I	K	L	M	N	P	Q	R	S	T	V	W	Y
A	5	-1	-2	-1	-3	0	-2	-1	-1	-2	-1	-1	-1	-1	-2	1	0	0	-3	-2
C	-1	13	-4	-3	-2	-3	-3	-2	-3	-2	-2	-2	-4	-3	-4	-1	-1	-1	-5	-3
D	-2	-4	8	2	-5	-1	-1	-4	-1	-4	-4	2	-1	0	-2	0	-1	-4	-5	-3
E	-1	-3	2	6	-3	-3	0	-4	1	-3	-2	0	-1	2	0	-1	-1	-3	-3	-2
F	-3	-2	-5	-3	8	-4	-1	0	-4	1	0	-4	-4	-4	-3	-3	-2	-1	1	4
G	0	-3	-1	-3	-4	8	-2	-4	-2	-4	-3	0	-2	-2	-3	0	-2	-4	-3	-3
H	-2	-3	-1	0	-1	-2	10	-4	0	-3	-1	1	-2	1	0	-1	-2	-4	-3	2
I	-1	-2	-4	-4	0	-4	-4	5	-3	2	2	-3	-3	-3	-4	-3	-1	4	-3	-1
K	-1	-3	-1	1	-4	-2	0	-3	6	-3	-2	0	-1	2	3	0	-1	-3	-3	-2
L	-2	-2	-4	-3	1	-4	-3	2	-3	5	3	-4	-4	-2	-3	-3	-1	1	-2	-1
M	-1	-2	-4	-2	0	-3	-1	2	-2	3	7	-2	-3	0	-2	-2	-1	1	-1	0
N	-1	-2	2	0	-4	0	1	-3	0	-4	-2	7	-2	0	-1	1	0	-3	-4	-2
P	-1	-4	-1	-1	-4	-2	-2	-3	-1	-4	-3	-2	10	-1	-3	-1	-1	-3	-4	-3
Q	-1	-3	0	2	-4	-2	1	-3	2	-2	0	0	-1	7	1	0	-1	-3	-1	-1
R	-2	-4	-2	0	-3	-3	0	-4	3	-3	-2	-1	-3	1	7	-1	-1	-3	-3	-1
S	1	-1	0	-1	-3	0	-1	-3	0	-3	-2	1	-1	0	-1	5	2	-2	-4	-2
T	0	-1	-1	-1	-2	-2	-2	-1	-1	-1	-1	0	-1	-1	-1	2	5	0	-3	-2
V	0	-1	-4	-3	-1	-4	-4	4	-3	1	1	-3	-3	-3	-3	-2	0	5	-3	-1
W	-3	-5	-5	-3	1	-3	-3	-3	-3	-2	-1	-4	-4	-1	-3	-4	-3	-3	15	2
Y	-2	-3	-3	-2	4	-3	2	-1	-2	-1	0	-2	-3	-1	-1	-2	-2	-1	2	8

Fig. 2. BLOSUM50 substitution matrix [12].

Commonly, alignment matrix computation involves two stages; the configuration stage and the computation stage. Flow chart in Fig. 3 describes the process flow of filling in an alignment matrix for protein pairwise sequence alignment. The configuration stage involves with access or load data into the PE. One query residue xi will be preloaded into each PE and then the substitution matrix's column corresponding to the query residue will be loaded in the form of a Look-Up Table (LUT). This initial stage relates mostly with memory accesses. The next stage dealing with arithmetic instructions which is the computation of best match score. There are several challenges on filling in an alignment matrix which consists of frequent memory access and high data dependency during PE configuration. The next section discussed on related works that have been done in order to optimized both configuration time and PE area complexities.

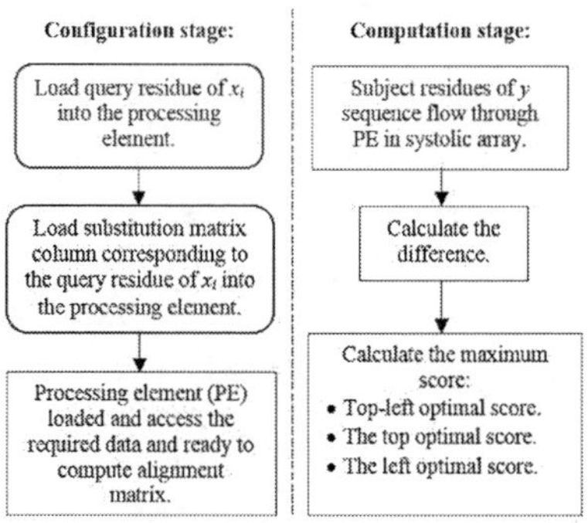

Fig. 3. The flow that shows steps involved in calculating the optimal alignment score by using DP algorithm. The curved edge rectangle represents process involving memory access [23].

2 A Survey of Hardware Implementation and Several Approach of PE Configuration

Reference [10] was among the first in 1990s that used FPGA to accelerate DP algorithm with systolic array. It was reported that they proposed the implementation on SPLASH reconfigurable array which is among the earliest off-the-shelf FPGA-based sequence-edit distance accelerators. In early 2000s, FPGAs has become more favored platform especially for acceleration hardware since the size are become larger and the performance is also almost proportional to the size of the FPGA. Moreover, FPGAs is capable of reprogrammable and consequently leads to implementable of DP algorithms on FPGAs. However, the FPGAs suffering with not have enough hardware resources especially memory bandwidth which is constrain on the memory storage for PE configuration. Thus, configuring the memory access need to be enhanced to achieve faster performance in homology search especially during the initial PE configuration stage.

In 2004, another method that have been considered in FPGA implementation is the modification of the basic DP algorithm into more simplify algorithm as presented by Dydel and Bala [11] and Gok and Yilmaz [12]. Dydel and Bala claimed that large scale of protein sequence alignment can be done with some modification to the SW algorithm with used the only one row version of an alignment matrix named 'one row c' as reference for the next computation during pipelined operation in alignment matrix computation process. This leads to performance speed-up especially when dealing with long sequence length. Commonly systolic array is used as accelerator but this work used different parallelism scheme with 88 copies of PE in one chip. Thus, it can execute the algorithm by processing one fixed query sequence against many different subject sequence in parallel. It leads to better PEs utilization since number of PEs utilized is independent to the sequence length unlike in systolic array. It used five BlockRAMs for each algorithm copy during alignment matrix computation and a full substitution matrix scores were uploaded in one of it. Thus, is used the dedicated limited memory in FPGA which become a bottleneck in hardware implementation using FPGA.

Reference [13] and [14] claimed that they improved the performance of aligning a query sequence with longer length against a database sequence by the used of folded systolic array approach. The alignment matrix computation processed sequentially by multiple passes over the same size of systolic array. The multiple passes concept partitioning the recursive algorithm into smaller alignment due to long length sequence thus the computation can be done without having difficulties in limited hardware resources. This architecture need first-in-first-out (FIFO) to hold the subject sequence and temporarily store the results of each pass during computation stage before fed back into the array for next pass usage as shown in Fig. 4. During configuration stage, each PE is preload with one query sequence and its corresponding substitution matrix column in LUT form. However, this technique suffers with increasing of configuration time and number of LUTs since the alignment process is performed in multiple passes over the linear array i.e. the PE requires different column of substitution matrix for each pass computation.

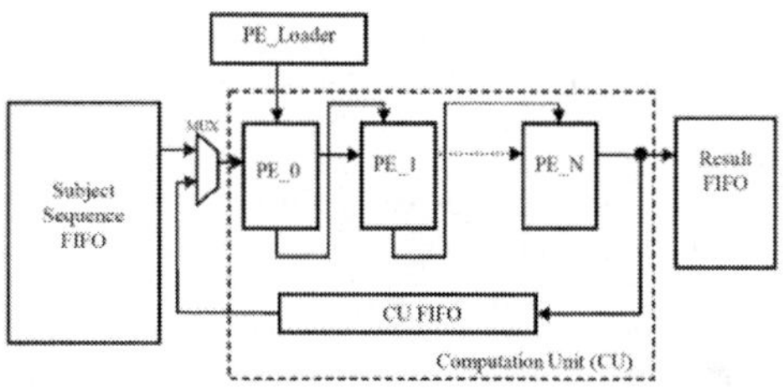

Fig. 4. The PE loader loads configuring data for each PE with different substitution matrix column corresponding to the query residue, through the serial configuration chain during configuration state [14].

* Corresponding author: nazrin@unimap.edu.my

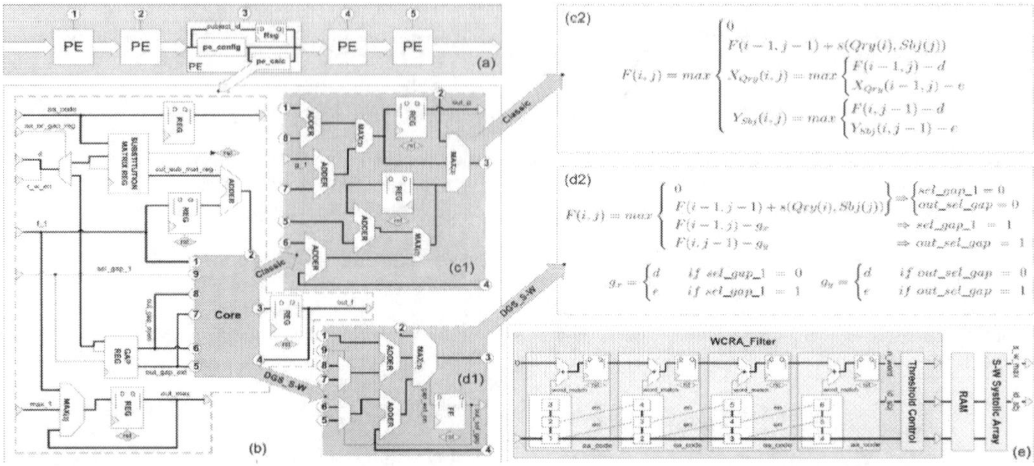

Fig. 5. (a) SW systolic array. (b) PE architecture of SW. (c1) Classic SW core implementation. (c2) Classic SW algorithm. (d1) DGS_SW core (optimized). (d2) DGS_SW SW algorithm. (e) Complete connection between WCRA_Filter architecture and SW systolic array [15].

In 2012, Urgese et al. [15] proposed enhanced hardware acceleration by the used of filtering effect and a flexible gap selector which also claimed to be the first to implement. The hardware consists of a Dynamic Gap Selector Smith-Waterman (DGS_S-W) which reduced the computation time compared to the classical solution of SW algorithm, and also a Word Counter Reduced Alphabet Filter (WCRA_Filter) which be able to remove the insignificant residue hence reduced the amino acids residue for repetitive computation of SW in the huge reference database as illustrated in Fig. 5. It stored a column of substitution matrix corresponded to the query sequence in a register in each PE sequentially thus, the PE configuration time is increasing with the number of PEs. Even though the filter managed to reduce the sequence length, it faces the same challenge as sequence alignment using heuristic approach endure. The result showed that it has increasing error which proportionally to the query sequence length due to low sensitivity issue of alphabet filtering approach.

In 2011, Isa et al. [14] proposed the PE design with multiple number of configuration element (CE) holding the desired column of substitution matrix. This architecture allows alignment matrix computation using folded systolic array which is suitable in computation for longer length of query sequence. Thus, number of CEs (*n*CEs) in each PE depends on the total number of passes (n-pass) or folding factor as shown in Fig. 6. During configuration stage, the CEs is preloaded with the substitution matrix column related to its query residues held by each PEs as shown in Fig.6 by the dotted line. It will be loaded in the form of LUTs. Thus, the alignment computation process can be progressed without the need of reconfiguring the PE with new LUT for the next fold of computation process since the PE loader can update the PE with coefficient needed for each passes in simultaneously. This PE stored the substitution matrix columns using the abundant CLB logic slices, therefore the dependency of the design with the limited memory resources was reduced.

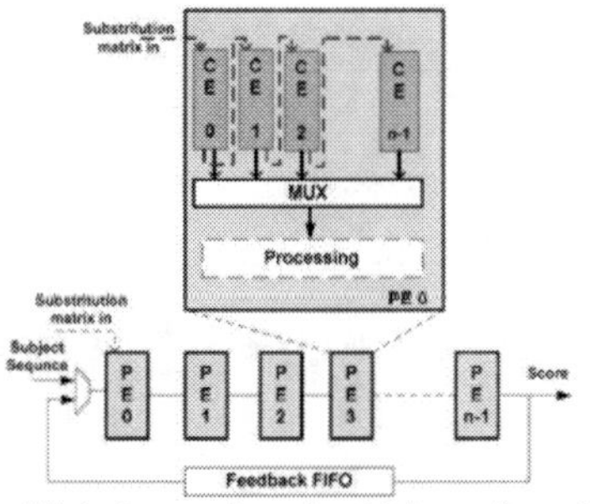

Fig. 6. The PE design consists of multiple configuration elements (*nCEs*) [14].

Besides, it also allows PE scalability due to the *n*CEs required in each PE but it suffers with *n*CE area overheads per PE. However, the CEs is being updated in serially, thus it increases the PE configuration time since it required n-pass in initial configuration stage due to the long sequence length. Consequently, in order to achieve better PE configuration design, the new PE architecture with fixed number of CEs are proposed in 2014 as reported in [16]. It is called the overlap computation and configuration (OCC). This architecture had optimized the logic resources by having only two CEs instead of having nCEs in each PEs. Each PE is used for supplying the substitution matrix for alignment matrix computation by one after another.

Regarding to have folded or n-pass computation with only two CEs, this approach used a well-organized scheduling strategy based on double buffering technique which efficiently organized the CEs to operate one after another through a controller. Both CE works with overlapping operation where one CE will be preloaded with the coefficients needed for the next computation pass while another CE is being used for the current pass of computation. Both of these processes occurred at the same time. The overlapping operation have proven that this architecture have used parallel configuration chain where it updates all the PEs with the desired column of

* Corresponding author: nazrin@unimap.edu.my

substitution matrix in simultaneously during the current subsequence pass computation occurs. However, the CEs is being updated in serially during PE configuration thus the configuration time is still independent to the number of CE.

In 2015, a new approach of PE parallel loader have been proposed by the used of parallel configuration chain technique. The parallel loader comprises of several circular buffers (CBs) using shift registers (SRL32). The total number of columns, n_{col} and rows, n_{row} of the substitution matrix represents the total number of shift registers within each CBs and the total number of CB respectively as shown in Fig. 7. The n_{col} and n_{row} is parameterizable for this loader. The substitution matrix scores are loaded column by column into the circular buffers. After the configuration data completely loaded into the buffers, it keeps on shifting within the buffers to avoid data loses. When the probability scores are ready as an output by the circular buffers, then it is all set to be transmitted to the PEs through a large data bus. The data bus connected to all circular buffers output and broadcast the requested substitution matrix column corresponding to the query sequence resides in each PEs. The loader can be connected to each PEs in parallel and allows the configuration elements to be transferred simultaneously into all the PEs. Hence, the configuration time for this parallel loader is independent to the number of PEs. In PE architecture with folded systolic array which have multiple passes for long length query sequence alignment, this parallel loader can supply the substitution matrix columns required for one fold simultaneously for each PEs without any delay. The configuration elements can be selected by a multiplexer during PE configuration in order to select the required substitution matrix columns corresponding to the query residues within the PEs for a particular fold. Thus, all the PEs can be updated concurrently for each pass of computation during PE configuration stage. Consequently, the parallel loader can reduce the configuration time from $t_{config.}$ x (nCE x nPE) to $t_{config.}$/(nCE x nPE) due to it is independent to both the number of CE and PE. Moreover, this loader used the FPGA's LUT instead of using the limited dedicated memory in FPGA. Hence, the parallel loader managed to avoid the bottleneck of using FPGA.

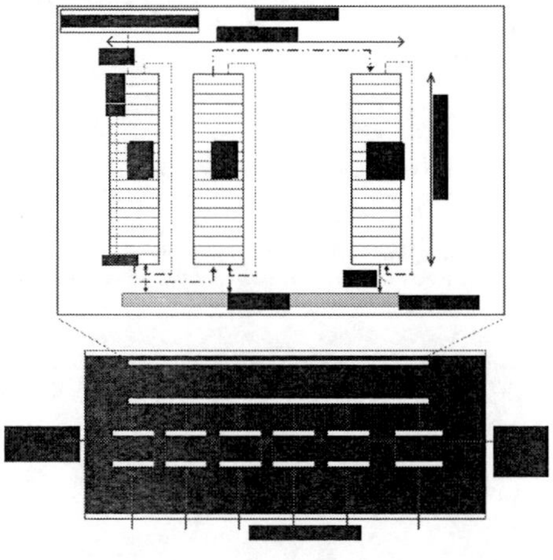

Fig. 7. The PE parallel loader with circular buffers [23].

3 Results and Discussion

This section discusses the FPGA-based implementation performance with varies methods used of PE loader which had been reviewed in previous section. The results from proposed PE architecture which used between serial configuration chain [13]-[15] and parallel configuration chain [16], [23] is summarized in Table 1. Some researchers include the total number of slices utilized for PE configuration in the performance result and some of them only consider the total slice utilization for the core architecture of the PE. Slices is actually the elementary programmable logic block in Xilinx FPGAs. Thus, an estimation results in terms of configuration area has been made for [13] and [15] due to lack of information on the initial configuration performance. A conclusion has been made that most of designs have the same number of slice utilization with ~77% of the total number slices utilized per PE just for the PE configuration elements. Hence, it gives the estimation slices utilized on the PE configuration for [13] is ~65 slices while [15] utilized about ~140 slices. However, the new approach of PE parallel loader from [23] has the least slice utilization.

In terms of the clock frequency or speed performance of the PE configuration, the results from prior works which used the parallel configuration chain technique shown to be faster than the works that used serial configuration chain. Typical core system operating between 100 to 200 MHz. Thus, both [16] and [23] shows that they achieved by ~4 times faster.

Table 1. Performance Comparison Against various FPGA implementation

Ref.	Device	Total PEs	Slices Utilized		Clock Freq. (MHz)	
			/PE	*Config.*	*Core/ System*	*Config.*
[13]	XC2VP 100-6	135	85	~65	40	4.098 kHz
[14]	XC5V LX110	195	88	68	200	-
[15]	XC5V LX330 T-3	285	-	~140	167	-
[16]	XC5V LX110	140	118	87	209.6	396.10
[23]	XC5V LX50T	-	-	52	-	480.25

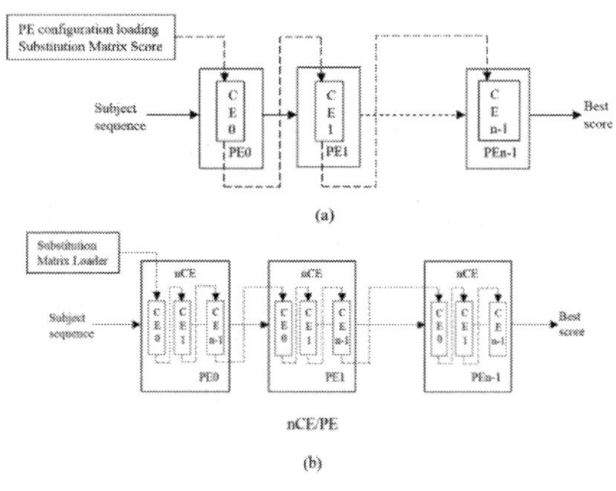

* Corresponding author: nazrin@unimap.edu.my

*** Formatting Issue - Best Available Paper/Graphic ***

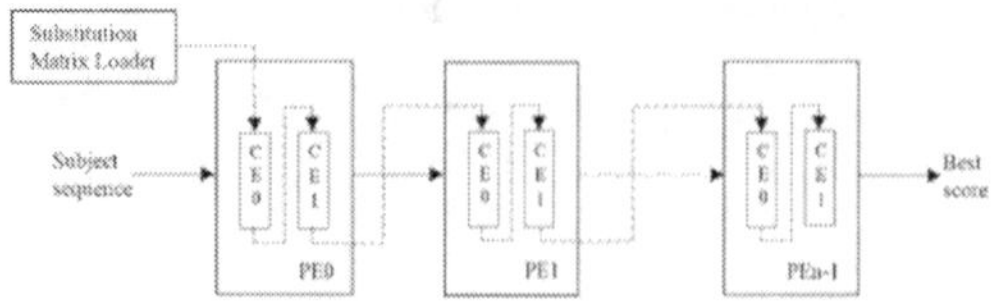

2 Fixed CEs/PE

(c)

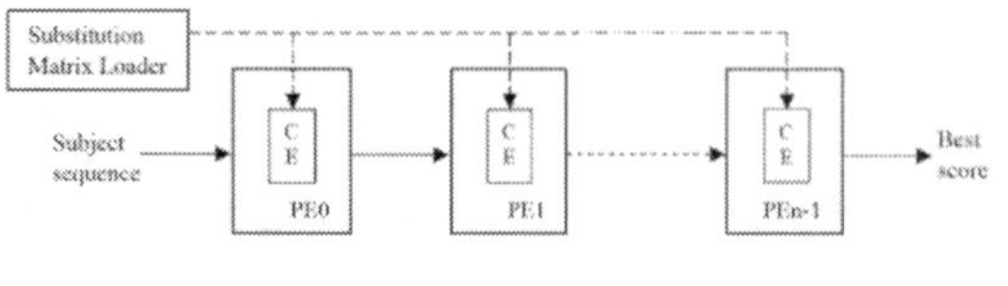

(d)

Fig. 8. Various approach of substitution matrix loader; by serial configuration chain approach (a) with one CE per PE [15], (b) with multiple CEs per PE [14], and by parallel configuration chain approach (c) with two fixed CEs per PE [16], and (d) single CE per PE whereby all the PE connected in parallel with the loader [23].

Based on PE configuration architecture for the substitution matrix storage during the configuration stage as illustrated in Fig.8, some researcher proposed to store only a column of substitution matrix in a register such as in [15] as shown in Fig.8 (a). Nevertheless, other PE architecture literally that had reported stored the entire substitution matrix score in the restricted memory resources to enable rapid access of the scores as in [13]. On the other hand, work by Isa et al. in [14] and [16] used PE loader by storing a column of substitution matrix in each CE. Though it updates the PEs with required substitution matrix column even for each passes in folded systolic array concurrently, the CEs still updates by serially as shown in Fig.8 (b) and (c). Consequently, it yet takes higher PE configuration time. Hence, another approach of PE loader used parallel configuration chain technique as in [23] is an efficient way to update each PE with a desired column of substitution matrix simultaneously as shown in Fig.8 (d).

4 Conclusion

This paper reviewed PE configuration methods for various pairwise protein sequence alignment architectures. There are two widely used techniques for PE configuration; serial and parallel configurations. Results have shown that, parallel configuration chain technique achieved better configuration speed i.e. ~4 times faster than the PE systolic array core frequency. However, not all PE configuration with parallel configuration technique be able to optimize in terms of area due to design optimization and tradeoff. Nevertheless, the new approach of PE parallel loader has proven that it can be optimized in terms of both area and speed complexities. It managed to perform at 480.25 MHz of clock frequency due to its ability to update all the PEs with the configuration elements simultaneously. Moreover, it also achieved least slice utilization of only 52 slices, which is almost ~2 to 3 times smaller than reported work in literature.

* Corresponding author: nazrin@unimap.edu.my

This work was supported by the FRGS under Grant 900300480. Modelling of an Efficient DNA/Protein Sequencer using Hardware Description Language (HDL) for Biomedical Applications.

References

1. D. Lavenier and M. Giraud, Reconfigurable Computing, pp. 157-182 (2005)
2. T. F. Smith and M. S.Waterman, J. Molec. Biol., **147**, no. 1, pp. 195–197 (1981)
3. S. B. Needleman and C. D. Wunsch, Journal of Molecular Biology, **48**, pp. 443–453 (1970)
4. J. Xianyang, et al., IEEE Transactions on Circuits and Systems II, **54**, no. 12, pp. 1077-1081 (2007)
5. K. Dohi, K. Benkrid, C. Ling, T. Hamada, and Y. Shibata, *21st IEEE International Conference on Application-specific Systems Architectures and Processors (ASAP)*, pp. 29-36 (2010)
6. Y. Munekawa, F. Ino, and K. Hagihara, *8th IEEE International Conference on BioInformatics and BioEngineering*, pp. 1-6, (2008)
7. X. Feng, H. Jin, R. Zheng, L. Zhu, and W. Dai, International Journal of Parallel Programming, pp. 1-22 (2013)
8. W. Liu, B. Schmidt, G. Voss, A. Schroder, and W. Muller-Wittig, *20th IEEE International Parallel and Distributed Processing Symposium*, pp. 8, April (2006)
9. L. Hasan, M. Kentie, and Z. Al-Ars, *IEEE Annual International Conference, Engineering in Medicine and Biology Society, EMBC*, pp. 2442-2446, August (2011)
10. D. T. Hoang, *International Workshop on Field Programmable Logic and Applications*, Vienna, Austria (1992)
11. S. Dydel and P. Bala, *Field Programmable Logic and Application*, pp. 23-32 (2004)
12. M. Gök and C. Yılmaz, *Computer and Information Sciences–ISCIS* pp. 277-285 (2006)
13. K. Benkrid, Y. Liu, and A. S. Benkrid, IEEE Transactions on Very Large Scale Integration (VLSI) Systems, **17**, pp. 561-570 (2009)
14. M. N. Isa, K. Benkrid, T. Clayton, C. Ling, and A. T. Erdogan, *NASA/ESA Conference on Adaptive Hardware and Systems (AHS)*, pp. 344-351 (2011)
15. G. Urgese, M. Graziano, M. Vacca, M. Awais, S. Frache and M. Zamboni, *19th IEEE International Conference, in Electronics, Circuits and Systems (ICECS)*, pp. 145-148, December (2012)
16. M. N. Isa, M. I. Ahmad, S. A. Z. Murad, R. C. Ismail & K. Benkrid, *Region 10 Symposium, 2014 IEEE*, pp. 39-44, April (2014)
17. H. T. Kung, *International conference on Management of data (SIGMOD '80), New York, USA* (1980)
18. R. J. Lipton and D. Lopresti, *Proc. of the Chapel Hill Conference on VLSI,H Fuchs, Ed. Rockville*, pp. 363-376 (1985)
19. J. M. Marmolejo-Tejada, V. Trujillo-Olaya, C. P. Renteria-Mejia, and J. Velasco-Medina, *2014 IEEE 5th Latin American Symposium on Circuits and Systems, Santiago*, pp. 1-4, February (2014)
20. L. Hasan and Z. Al-Ars, *Annual International Conference of the IEEE, Engineering in Medicine and Biology Society, EMBC 2009*. pp. 3845–3848 (2009)
21. Z. Nawaz, M. Nadeem, H. van Someren, and K. Bertels, *International Conference on Field-Programmable Technology (FPT), 2010*, pp. 454–459 (2010)
A. Pulka and A. Milik, *Proceedings of the 18th International Conference, Mixed Design of Integrated Circuits and Systems (MIXDES) 2011*, pp. 283–288 (2011)
22. K.N.D. Muhsen, "*Design and analysis of an efficient repository system for protein coefficients in systolic array-based architecture by using Xilinx Virtex-5 FPGA*," M.S. thesis, Dept. Microelectronic Eng., Univ. Malaysia Perlis, Perlis, Malaysia, (2015)

Monostatic ultra-wideband GPR antenna for through wall detection

Jawad Ali[1, 2,*], *Noorsaliza* Abdullah[1], *Roshayati* Yahya[1], and *Taimoor* Naeem[2]

[1]Department of Communication Engineering, Faculty of Electrical and Electronic Engineering, Universiti Tun Hussein Onn Malaysia (UTHM), Johor, Malaysia
[2]Electrical Engineering Department, COMSATS Institute of Information Technology (CIIT), Lahore, Pakistan

Abstract. The aim of this paper is to present a monostatic arc-shaped ultra-wideband (UWB) printed monopole antenna system with 3-16 GHz frequency bandwidth suitable for through-wall detection. Ground penetrating radar (GPR) technique is used for detection with the gain of 6.2 dB achieved for the proposed antenna using defected ground structure (DGS) method. To serve the purpose, a simulation experiment of through-wall detection model is constructed which consists of a monostatic antenna act as transmitter and receiver, concrete wall and human skin model. The time domain reflection of obtained result is then analysed for target detection.

1 Introduction

Ultra-wideband (UWB) is a wireless technology, which is mainly referred to a system with very larger operating bandwidth, usually this bandwidth is considered to be greater than 500 MHz. Where, after 2002 the free localization of UWB spectrum for industrial and scientific purposes, opened new gates for the researchers to develop different technologies for the use of UWB [1]. UWB covers the frequency band ranges from 3.1-10.6 GHz with 7.5 GHz bandwidth, there are several advantages of this very large bandwidth as it consumes less power, have low implementation cost, with high data rate and time-resolution, also have the capability of obstacle penetration [2]. Such advantages widen the number of prominent applications in the field of communications such as wireless local area network (WLAN), wireless body area network (WBAN), WiMAX, wireless personal area network (WPAN), also in the field of radar like ground penetrating radars (GPR), imaging and positioning [3].

UWB technology, from sometime has been in use for many GPR applications [4]. GPR have a number of applications in the field of engineering for military and civilian applications like the detection of landmines, through-wall imaging, in remote sensing techniques such as non-destructive testing of concrete and detection of trapped people under debris or in the opaque environment [5-6]. For the implementation of UWB technology specifically in the applications of GPR different types of antenna like a horn, bow tie, tapered slot, patch, spiral, and dipole antennas have been designed previously. These type of antennas are considered for their high gain, design complexity, directivity and frequency bandwidth which are the basic requirements in antenna design for GPR applications [7-9]. In this paper, an arc-shaped printed monopole UWB antenna with defected ground structure (DGS) is designed that have the ability to operate from 3-16 GHz bandwidth with the gain of around 6 dB and GPR model for through-wall detection prototype is simulated using CST-MWS, the reflection of the target is then computed and demonstrated using MATLAB.

The remainder of the paper is organized as follows: in Section 2 antenna design and complete simulation model for through-wall imaging by using GPR technology is presented, the discussion about the analysis of obtained results is in Section 3. Finally, Section 4 summarizes the complete discussion in the form of conclusion.

2 Design and modeling

This section is further categorized into two sub-sections. Section 2.1 is about antenna design for ultra-wideband (UWB) ground penetrating radar (GPR) use in detection, while Section 2.2 discusses the simulation model designed using CST-MWS for experimental purpose.

2.1 Proposed antenna design

For the implementation of ultra-wideband (UWB) antenna ground penetrating radar (GPR) to address the through wall penetration, an arc shape printed monopole antenna with defected ground structure (DGS) is introduced in this paper. The arc shape design of the antenna is achieved by cutting a circular slot towards the minor axis of elliptical shape patch with microstrip line feed. The proposed UWB arc antenna design is shown in Fig. 1, the dimensions of the substrate, microstrip line shows that designed antenna also have a compact shape.

* Corresponding author: engrjawad.ali@outlook.com

The main challenges in designing UWB compact shape antenna is the bandwidth matching at low frequencies [10] of the band from 3.1-10.6 GHz, to achieve the matching over complete band a modified half ground is introduced in the design. But the gain obtained as a result was very low, which was again another challenge because for GPR applications the gain should be nominal. For the gain to be around 6 dB DGS [11] technique is used on the ground and as a result of this technique, slots are introduced at the different locations on the ground as shown in Fig. 1. The substrate material used in this design is FR-4 with $\varepsilon r = 4.2$. Table 1 shows the dimensions along with the substrate material, ground, and the major and minor axis of the elliptical arc.

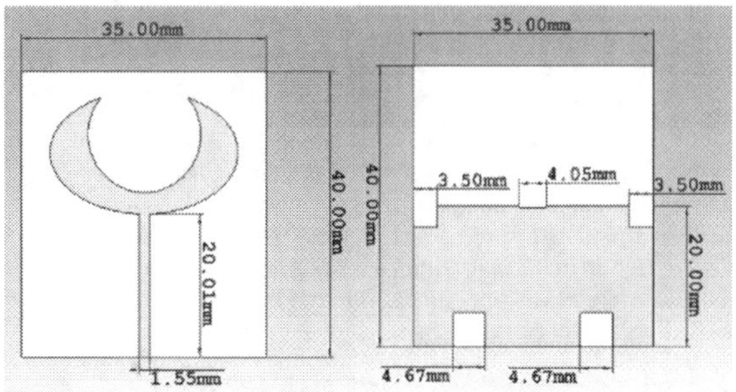

Fig. 1. Arc Antenna Design

Table 1. Design Parameters of Arc Antenna

Parameter Name	Material	Dimension (mm)
Major arc axis	Copper	27
Minor arc axis	Copper	17.28
Substrate (LxW)	FR-4	40x35
Ground (LxW)	Copper	20x17.5

2.2 Simulation model

The antenna discussed in Section 2.1 is specifically designed using CST-MWS for through-wall detection application using ground penetrating radar (GPR) technique in order to enhance the detection capability of compact antenna design. The proposed through-wall detection experiment is designed and simulated in CST-MWS environment. Fig. 2 shows the prototype of complete designed model where monostatic antenna used for transmission and reception is placed at some distance from 10 cm thick and 1 year old concrete wall, and human skin targeted model is on the other side of the wall. The distance from antenna to the wall is measured as D_Wall whereas between wall and target it is computed as D_Target. The dielectric constant of wall throughout the ultra-wideband (UWB) is $\varepsilon_r = 5.46$, while for skin it varies from higher to lower frequencies with the value $\varepsilon_r = 26–36$. The reflection based results of the complete model will be discussed in Section 3.

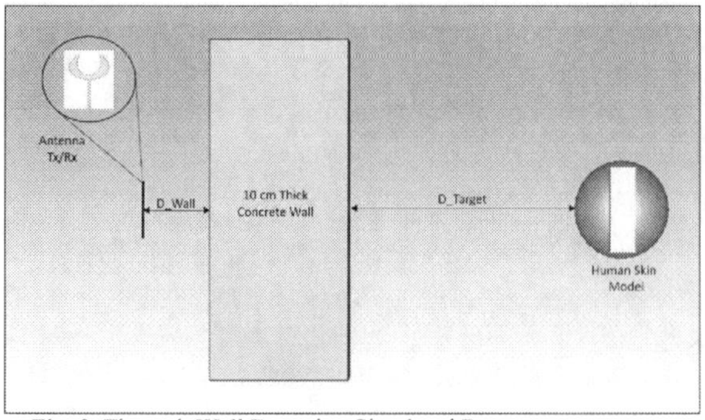

Fig. 2. Through Wall Detection Simulated Prototype

3 Results and discussions

The proposed antenna is simulated using CST-MWS environment and the obtained results are shown in Fig. 3. The return loss (S_{11}) in Fig. 3a shows the result is significant as it covers the complete ultra-wideband (UWB), even the simulated band is from 3-16 GHz which is more than the actual band from 3.1-10.6 GHz. The voltage standing wave ratio (VSWR) of the proposed design is also less than 2 as shown in Fig. 3b, this VSWR value shows that the designed antenna is radiating from 3-16 GHz without any interference. Fig. 3c demonstrates the information about the gain of the proposed antenna is 6.21 dB, this value of gain is achieved by introducing slots in the ground plane using defected ground structure (DGS) method. The obtained gain is suitable for through-wall detection as its radiation is also directional.

After the successful simulation of proposed antenna design, a complete monostatic ground penetrating radar (GPR) experiment is performed for detection purpose as discussed in Section 2.2. The obtained results from experimentation are further analyzed in the time domain using inverse Fourier transformation technique to study the difference between the reflections from antenna, wall, and target.

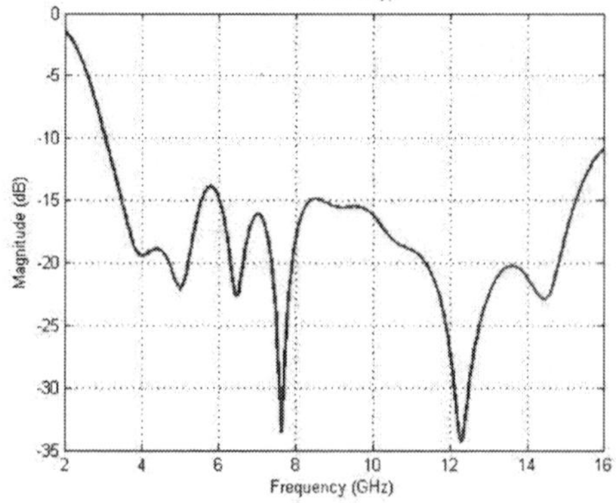

(a)

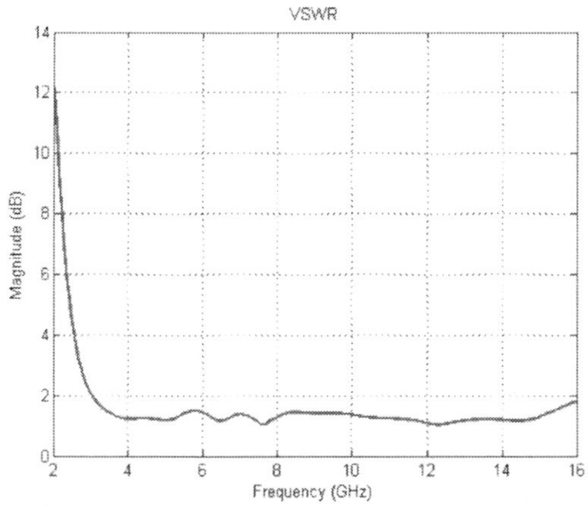

(b)

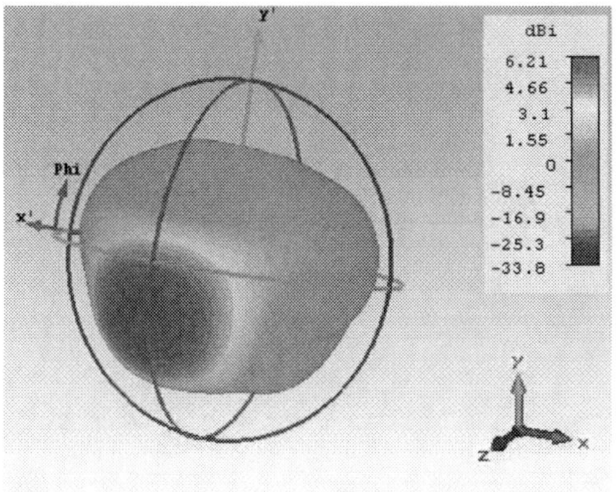

(c)

Fig. 3. Simulated Results of Arc Antenna (a) S-Parameter Result from 3-16 GHz (b) VSWR < 2 for proposed antenna (c) 3D Simulated Antenna Gain

Fig. 4 shows the time reflection results of a monostatic antenna system using GPR technique, it mainly consists of two times interval resolutions the first with dashed line is only from antenna and wall while the second with the continuous line is the reflection from antenna, wall, and target. From the figure, it is observed that the reflection from the wall is received at 0.33ns, while the reflection before 0.33ns is considered to be the antenna coupling. The second time interval shows a peak at 2.8ns, this peak is the obvious reflection from the targeted object behind the wall. From different time intervals (t), the depth resolution (d) of the target and the wall can also be determined using [12]:

$$d = v\frac{t}{2} \tag{1}$$

$$v = \frac{c}{\sqrt{\varepsilon_r}} \tag{2}$$

Where (v) is the propagation velocity in terms of speed of light (c) and dielectric constant (ε_r).

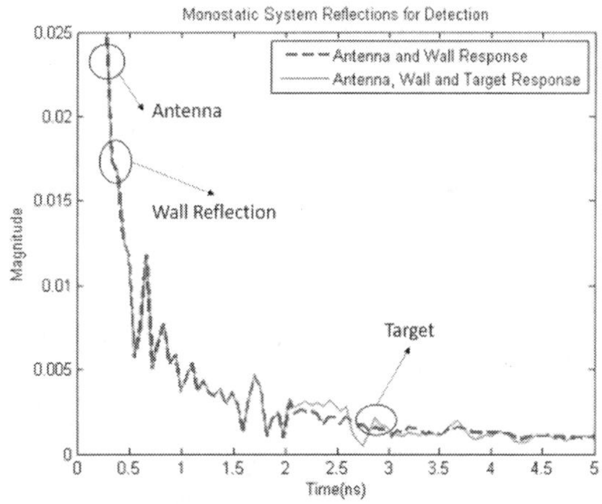

Fig. 4. Antenna, Wall and Target Reflections

It has also been observed that there is an insignificant difference between antenna and wall reflections, the reason of this insignificance is aliasing that occurs mainly due to the inadequate sampling rate [13].

4 Conclusions

A complete monostatic ultra-wideband (UWB) model for through-wall detection is discussed during the paper. The proposed arc shape antenna having defected ground structure (DGS) with the gain of 6.2 dB and 13 GHz bandwidth is designed using FR-4. This antenna is then modeled and simulated with wall and human skin structure as a targeted object for detection purpose. Later the results are analyzed to see the time domain reflections obtained from simulation. This demonstration hence makes the UWB monostatic antenna one of the potential candidates for detection applications.

The authors would like to thanks ORICC Universiti Tun Hussein Onn Malaysia (UTHM) for supporting this work.

References

1. M. M. Islam, M. Samsuzzaman, M. R. I. Faruque, Micro. and Opt. Tech. Lett. **57**, 1470–1475 (2015) .
2. D. Valderas, X. Chen, C. Ling, J. I. Sancho, D. Puente, *Ultrawideband antennas: Design and applications*, Imperial College Press, London, (2010)
3. F. B. Zarrabi, Z. Mansouri, N.P. Gandji, H. Kuhestani, AEU-Int. J of Electro. and Comm. **70**, 64-69 (2016)
4. A. Elboushi, A. R. Sebak, T. Denidni, ACES, **28**, 411-418 (2013)
5. C. Warren, A. Giannopoulos, IEEE J of Sel. Top. in Appl. Earth Obs. and Rem. Sens. **9**, 29-36 (2016).

6. C. Ozdemir, B. Yilmaz, S. I. Keceli, H. Lezki, O. Sutcuoglu, 15th International Radar Symposium (IRS), Gdansk, (2014).
7. A. Ahmed, Y. Zhang, D. Burns, D. Huston, T. Xia, IEEE Geosci. and Rem. Sens. Lett. **13**, 92-96, (2016).
8. E. A. Etellisi, M. A. Elmansouri, D. S. Filipovic, IEEE Trans. on Ant. and Prop. **64**, 6-15, (2016).
9. E. A. Etellisi, M. A. Elmansouri, D. S. Filipovic, IEEE International Symposium on Antennas and Propagation (APSURSI), Fajardo,(2016).
10. S. Koziel, A. Bekasiewicz, IEEE Ant. and Wire. Prop. Lett. **15**, 750-753, (2016).
11. R.A. Pandhare, P.L. Zade, M.P. Abegaonkar, Engg. Sci. and Tech., an Int. J. **19**, 1360-1367, (2016).
12. D. J. Daniels, *Ground Penetrating Radar* (The Institution of Engineering and Technology, 2004)
13. X. Zeng, A. Fhager, M. Persson, P. Linner and H. Zirath, IEEE Trans. on Ant. and Prop. **59**, 4279-4285, (2011).

Effect of Thermoelectric Cooling (TEC) module and the water flow heatsink on Photovoltaic (PV) panel performance

A.R. Amelia[1,*], *MA* Jusoh[1], and *Ida Shamira* Idris[1]

[1]Centre of Excellence for Renewable Energy (CERE), School of Electrical System Engineering, University Malaysia Perlis, Malaysia

Abstract. Photovoltaic (PV) panel suffers in low conversion efficiency of the output performance affected by the elevated operating temperature of the PV panel. It is important to keep the PV panel to operate at low temperature. To address this issue, this paper proposes the cooling system using thermoelectric cooling (TEC) and water block heatsink for enhancing the PV panel output performance. These both types cooling system were designed located on the back side of the PV panel to cool down the operating temperature of the PV panel. To evaluate the function for the existing cooling systems, the experiment was subsequently performed for PV panel without and with different design of the cooling system in outdoor weather conditions. By comparing the experimental results, it is concluded that by the hybrid cooling system which combining TEC module and the water block heatsink could improve the output performance of the PV panel. By the reduction temperature of the PV panel by 16.04 %, the average output power of the PV panel has been boosted up from 8.59 W to 9.03 W. In short, the output power of the PV panel was enhanced by the reduction of the operating temperature of the PV panel.

1 Introduction

Nowadays, the rapid incorporation of renewable energy technologies within the global energy supply infrastructure growing demand, due to the surging oil prices and increasing environmental awareness. Especially, solar energy with clean energy source, renewable and environmental-friendly is received attention. Solar energy is artificially extracted in the forms of electricity and heat [1]. However, the actual performance of solar energy is still below its actual efficiency. The weakness is that the unemployed infrared energy is converted into waste heat, leads to the rising operating temperature of the PV panel. Increasing of the operating temperature causes the efficiency of PV panel decreases [2]. Therefore, application cooling system on a PV panel is required to enhance the efficiency.

Various methods can be employed to achieve cooling action for PV panels. Water cooling is a favorable for cooling PV panel, mostly integrated using the water spray [3]. D. Kim et al. employed the water circulation over the surface PV panel in improving its outputpower. There is about 11.6 % increasing in maximum powers when attached cooling system on a PV panel. Moreover, the extra power can be generated when the existing water flow removed the particles that interfere with solar radiation from the surface cell [4]. T. Chinamhora et al. cools the back and front PV panel by using the water coolant then compared to the PV panel without cooling system. The measured results indicated the electrical efficiency increases around 10-22 % when using water-coolant during summer day, unfortunately decreased to 3-20 % on cloudy days [5]. Y.M.Irwan et al. investigates the effect of water cooling using DC pump under indoor testing. Under different level of solar radiation, the water cooling was observed to be performed rather than PV panel without cooling system when about 9-22 % increasing in output power. The increasing in output power caused by the reduction of PV panel temperature [6].

Recently, the hybrid system involves the combination of TE and PV technology also attracted as an alternative method for enhancing conversion efficiency by cooling the PV panels. A standard TE module utilizes the Seebeck effect composed of several tens to hundreds of pairs of thermocouples connected together electrically in series and thermally in parallel [7]. TE module can operate as a heat pump or providing heating or cooling. Research on PV panel with TE modules, such as Kane and Verma [8] proved that the hybrid system by combining the PV panel thermoelectric can be operated at PV panel temperature of 53 °C. From this review paper, the results show that the PV panel can operate at elevated operating temperature without loss of the output power. Besides that, the efficiency of the PV panel also had been improved when PV panel was cooling at the temperature of 10 °C. Park et al. investigated the effect of 2-wire hybrid configuration on conventional crystalline silicon of PV cells. The experimental results indicate the importance of accurate internal resistance between the PV cells and TE devices, contributes to output power increase ranging from 30 % to 250 % [9]. Recently, Alomair et al. [10] investigated the PV/TE air-conditioning system for remote area application through a theoretical and experimental. They stated that by using PV/TE air-conditioning system to cool space is possible.

* Corresponding author: amelia_razak87@yahoo.com

 DOI: 10.1051/epjconf/201716201077

It is because the desired temperature may be achievable if the size and capacity of the TEC are well-designed.

In this present work, the TEC and water cooling system have been proposed in order to improve the output performance of the PV panel. These both devices act as the cooling system to reduce the PV panel temperature by extracting the excesses heat. Different design of cooling system also has been compared with the same type of the PV panel. Section 2 explains materials used in the experimental study. This section also presents the methodology of the experiment. Section 3 analyses and discussed briefly on the results produced. Meanwhile, the last section shows the conclusion of this proposed study.

2 Methodology

2.1. Material preparation

2.1.1 PV panel

For the purpose of the present work, a polycrystalline silicon PV panel with the rating power of 10 W has been used. With the surface area of 0.105 m^2, this PV panel able to generate the open-circuit voltage boost to 18.72 V, meanwhile 0.59 A for short-circuit current.

2.1.2 TEC module

In this study, the TEC module was used as an alternative method to reduce the operating temperature of the PV panel. In Table 1, the specification data for the use of the TEC module are listed. As showed in this table, the typical TEC module is made of Bismuth Telluride (Bi2Te3) based alloy which known as the most commercial module in TEC market [11]. Through Table 1, it shows that the selected TEC module which was constructed by 127 numbers of thermocouples can generate 6 A of maximum current and 15.2 V for the maximum voltage. Besides that, this TEC module is operated at 12 V of a direct current (DC).

Table 1. The specification of TE module.

Specifications	Data
Model	TEC1-12706
Material	Bismuth Telluride (Bi_2Te_3)
No. of thermocouples	127
Dimension module	40 x 40 x 4.3 mm
Maximum current	6 A
Maximum voltage	15.2 V
Operational voltage	12 V (DC)

2.1.3 Water block heatsink

Water is the best coolant naturally available with the high heat capacity when compared to the air cooling system [12]. Therefore, the water block heatsink is used to cool down the operating temperature of the PV panel. Besides that, this present work also used the water block heatsink as the coolant to cool down the cool side of the TEC module. Water is flowing through the channels such shown in Figure 1. The water from inlet will flow across the heatsink block, the structure design of the heatsink affects the water circulation process.

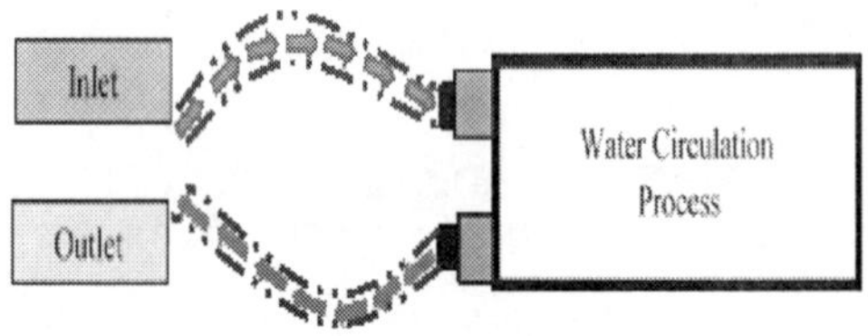

Fig. 1. Flow of water in water block heatsink.

2.2. System description and experimental setup

In order to evaluate the effect of cooling systems for heat extraction from the PV panel, four different systems were investigated. Schematic diagrams and experimental setup for four different systems are shown such in Figure 2. Figure 2 (a) is represented for PV panel without a cooling system. Meanwhile, Figure 2 (b) shows the diagram of PV panel with TEC modules. Figure 2 (c) consists of a PV panel cooled by the only water block heatsink. Lastly, Figure 2 (d) represents for the PV panel using a hybrid system which combining the TEC module and water block heatsink as the cooling system. In analyzing all different cases, the experiment was conducted in a sunny day from 11.30 a.m. to2.15 p.m. under outdoor condition.

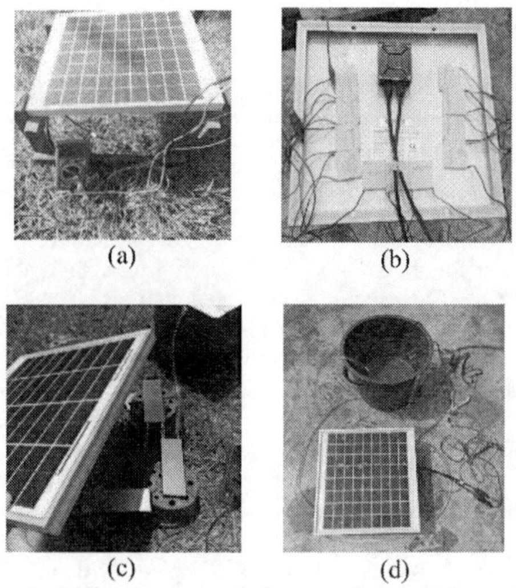
(a) (b) (c) (d)

Fig. 2. Different cases of the experiment under outdoor condition; (a) Without cooling, (b) With TEC module, (c) With water block heatsink, (d) With TEC and water block heatsink.

3 Experiment results

In analyzing the distribution of temperature, all recorded hot surface and cold surface temperature were plotted. The hot surface TE module and operating temperature of the PV panel are supposed to be equal since thermocouple was attached on the backside of the PV panel. The operating temperature of hot surface curves, obtained following the analysis of experimental results, is presented in Figure 3. Observing this figure, the variation in temperature were produced caused by the different design of the cooling system integrated on the PV panel. In more details, the highest average of the hot surface temperature is 56.03 °C for PV panel without a cooling system. On the other hand, when hybrid cooling system (TEC with water block heatsink) is taken into consideration, a significant temperature drop was observed compared to other cooling system. There was about the reduction by 16.04 % when this cooling system was integrated on a PV panel. The average of the hot surface temperature for this case is 47.04 °C.

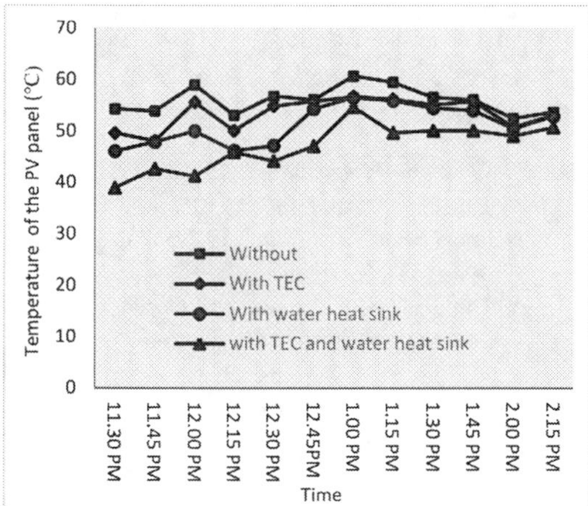

Fig. 3. Operating temperature of the PV panels.

Figure 4 shows the open-circuit voltage produced from each case of experiment. Based on this figure, the lowest average of open-circuit voltage is produced by a PV panel without a cooling system. The average of open-circuit for this case was determined approximately 19.43 V. In theoretical, the output voltage of the PV panel reduced with the decrease in band gap. The decrease in the band gap is caused by the increase in photon generation, which is influenced by the rising of PV panel temperature [13]. Inversely, the highest average of open-circuit voltage is represented by the PV panel with hybrid cooling system (PV panel with TEC and water heatsink). The open-circuit voltage was increased by 2 % when integrated hybrid cooling system on PV panel. This is due to the reduction temperature of the PV panel, enhances the output performance from PV panel [14].

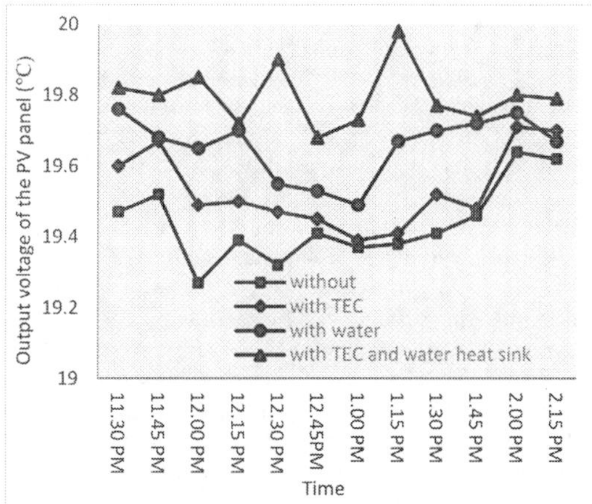

Fig. 4. Open-circuit voltage of the PV panels.

Figure 5 describes the calculated output power by using Equation 1 [15]. The output power of each case was calculated based on distribution operating temperature of the PV panel. By analyzing this figure, the lowest average of the output power is 8.59 W for PV panel without a cooling system. Conversely, by applying the TEC module as a cooling system, the output power of the PV panel increased by 1.5 %, followed by PV panel with water cooling with 2.5 %. Meanwhile, the highest average of the output power is represented by a PV panel with attaching TEC module and water as the cooling system. With this hybrid combination, the output power of the PV panel could be increased approximately5.12 %. This is because of the reduction of the operating temperature of the PV panel. This fact is supported by [16] which states that the output power of the PV panel was increased when its operating temperature was decreased.

$$P_{max_{corrected}} = P_{max_{stc}} \times \{1 + [[\frac{\gamma}{100}] \times (T_{panel} - T_{stc})]\} \quad (1)$$

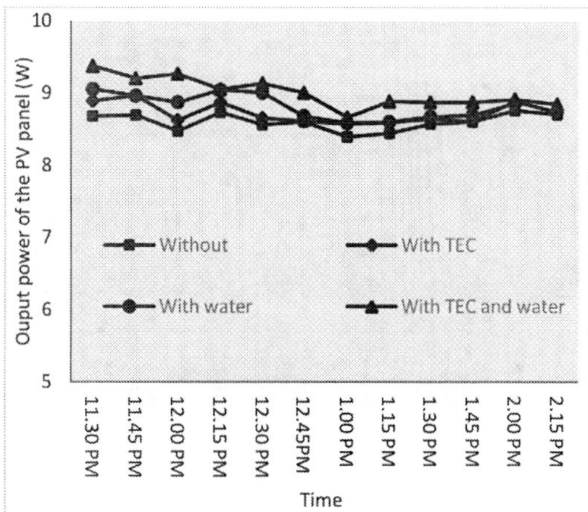

Fig. 5. Output power of the PV panels.

Figure 6 and Figure 7 present the temperature difference between hot and cold side surface of TEC module. Figure 6 is represents for the PV panel with only TEC module, meanwhile Figure 8 shows the results for PV panel with hybrid system (TE module and water heatsink). By referring to Figure 6, the observation result indicated that there is small differences between hot and cold side surface temperature. The average of cold side temperature was determined by 49.13 °C. Therefore, the average of difference temperature between hot and cold surface was approximately 2.2 °C. Conversely, the difference temperature of TEC module for PV panel with hybrid system such shown in Figure 7 is quite larger with 4.8 °C. The average of cold side temperature of TEC module for this cases was approximately 42.24 °C. The performance TEC could be enhanced with the lower temperature on cold side surface of TEC module. It is because of a high amount of output voltage of TEC can be generated affected by the large temperature difference between the hot side and cold side surface [17].

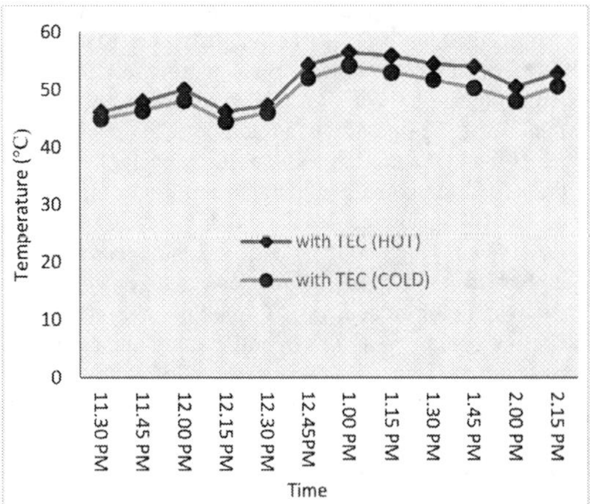

Fig. 6. Temperature difference of TEC module for the case of PV panel with TEC module.

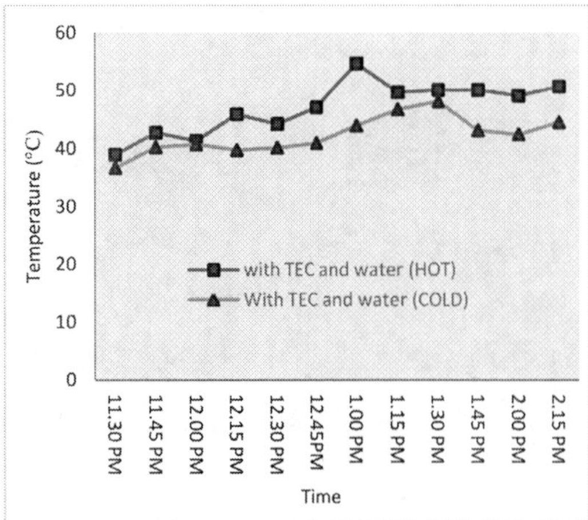

Fig. 7. Temperature difference of TE module for the case of PV panel with TEC module and water heatsink.

4 Conclusion

In short, this paper studies the effect of cooling system for PV panel output performance. Different cooling system were integrated on the PV panels to compare their functionality. For the purpose study, the output performance of the PV panel without cooling system was compared with the PV panels which used water block heatsink or TEC modules as the cooling system. The experimental results briefly shows that the elevated operating temperature of the PV panel gives the negative impact on PV panel output performance. From the results, PV panel without cooling system has the lower output power when compared to the PV panel with cooling systems. By developing the hybrid (TE modules and water) cooling system, the output power of the PV panel has increased by 5.12 %. This is because of the reduction in operating temperature of the PV panel from 56.03 °C to 47.04 °C. Therefore, it shows that the existing cooling system makes possible to make PV panel operate at maximum output performance by decreasing the operating temperature of the PV panel.

The authors thank the Fundamental Research Grant Scheme (FRGS) for providing the fund to complete this study.

References

1. G.K. Singh, *Energy*, **53**, 1-13, (2013)
2. S. Chander, A. Purohit, A. Sharma, Arvind, S.P. Nehra, M.S. Dhaka, *Energy Reports*, **1**, 104-109, (2015)
3. Y. Cui and Q. Zhu, *Power and Energy Engineering Conference (APPEEC)*, (2012)
4. Dong-Jun. K, D.H. Kim, S. Bhattarai and J. Oh, *Journal of Solar Energy Engineering*, **133**, (2011)
5. T. Chinamhora, G. Cheng, Y. Tham and W. Irshad, *Proceeding of International Conference on Energy and Sustainability*, (2013)
6. Y.M.Irwan , W.Z. Leow , M. Irwanto, Fareq.M, A.R. Amelia, N. Gomesh and I.Safwati, *Energy Procedia*, **79**, 604-611, (2015)
7. U. Pau and P. Adour, *Energy Conversion and Management*, **140**, 167-181, (2017)
8. A.N. Kane and V. Verma, *International Journal of Renewable Energy Research*, **3**, 320-324, (2013)
9. K. Park, S. Shin, A. S. Tazebay, H. Um, J. Jung, A. Jee, M. Oh, S. Park, B. Yoo, C. Yu and J. Lee, *Scientific Reports*, **3**, (2013)
10. M. Alomair, Y. Alomair, S. Mahmud and H.A. Abdullah, *Journal of Thermal Science and Engineering Applications*,**7**, (2015)
11. P. Huen and W.A. Daoud, *Renewable and Sustainable Energy Reviews*,**72**, 1295-1302, (2017)
12. K. Moradi, M. Ali Ebadian and C. Lin, *International Journal of Heat and Mass Transfer*, **64**, 483-500, (2013)
13. S. Kumar and D.T. Kumar, *International Journal of Innovative Research in Electrical, Electronics,*

DOI: 10.1051/epjconf/201716201077

Instrumentation and Control Engineering, **2**,2168–2172, (2014)

14. A.A. Hachicha, C. Ghenai and A.K. Hamid, *International Journal of Environmental, Chemical, Ecological, Geological and Geophysical Engineering*, **9**, 999–1002, (2015)
15. T. Ishii, K. Otani, A. Itagaki and K.Utsunomiya, *Japanese Journal of Applied Physics*, **51**, (2012)
16. A.R. Amelia, Y.M. Irwan, M. Irwanto, W.Z. Leow, N. Gomesh, I. Safwati and M.A.M. Anuar, *International Journal of Electrical and Computer Engineering*, **6**, 526-534, (2016)
17. D. Vadhel, S. Modhavadiya and Prof. J. Zala, *International Research Journal of Engineering and Technology*,**4**, 1904-1908, (2017).

Substrate integrated waveguide (SIW) 3 dB coupler for K-Band applications

Nurehansafwanah Khalid[1,*], *Siti Zuraidah* Ibrahim[1], *Fwen Hoon* Wee[1] and *Farah Shazuani* Mahmud[1]

[1]Advanced Communication Engineering Centre (ACE) School of Computer and Communication Engineering, Universiti Malaysia Perlis, Arau, Perlis, Malaysia

Abstract. This paper presented a designed coupler by using Rogers RO4003C with thickness (h) 0.508 mm and relative permittivity (ε_r) 3.55. The four port network coupler operates in K-band (18-27 GHz) and design by using substrate integrated waveguide (SIW) method. The reflection coefficient and isolation coefficient of propose Substrate Integrated Waveguide (SIW) coupler is below than -10 dB. Meanwhile the coupler requirements are phase shift 90° between coupled port and output. SIW are high performance broadband interconnects with excellent immunity to electromagnetic interference and suitable for use in microwave and communication electronics, as well as increase bandwidth systems. The designs of coupler are investigated using CST Microwave Studio simulation tool. This proposed couplers are varied from parameters that cover the frequency range (21 -24 GHz) and better performance of scattering (S-parameter).

1 Introduction

At this moment in time, rectangular waveguide components are used in communication and microwave system, radar with outstanding features such as increase power capability and small insertion loss [1]. However, it is in bulky size that strict and limits the accuracy and precision in manufacturing. Unfortunately, Printed Circuit Board (PCB) is sensitive toward about radiation by others active and passive devices in the same dielectric substrate [2] [3].

Substrate Integrated Waveguide (SIW) is a techniques can achieve a little loss in the highest frequency that can come up with very good frequency performance on low dielectric permittivity substrate materials compared to coplanar waveguide (CPW) and planar circuits in microstrip [4] [5]. The development of Substrate Integrated Waveguide (SIW) methods build-up from double rows of conducting via (slots) filled in a dielectric substrate to connect two plates (copper planes) of substrate [6][7]. The advantages of Substrate Integrated Waveguide (SIW) technology are to miniaturize size and minimization losses [3]. The technology is widely applied to the implementation of Microwave field [8]. Moreover, one of emerging topics in communication application is the Internet of Things (IoTs).

Coupler is a passive device which couples with generating the transmission power with properties requirements such as operational bandwidth, frequencies, and size [9]. Moreover, coupler may have 3-port or 4-port components which denoted as input, through (transmitted), coupled and isolated [10]. The coupling of mechanism divided into two which depends on the interspacing between them, and the coupled line.

In this paper, a Substrate Integrated Waveguide (SIW) coupler for microwave and communication are proposed. The operating frequency for the coupler investigated from 18 GHz to 27 GHz. Simulate performance results are show in this research paper.

2 Design Descriptions

2.1 Configuration of Substrate Integrated Waveguide (SIW) Coupler

Basic structure of coupler is shown in Fig. 1 below with parameters.

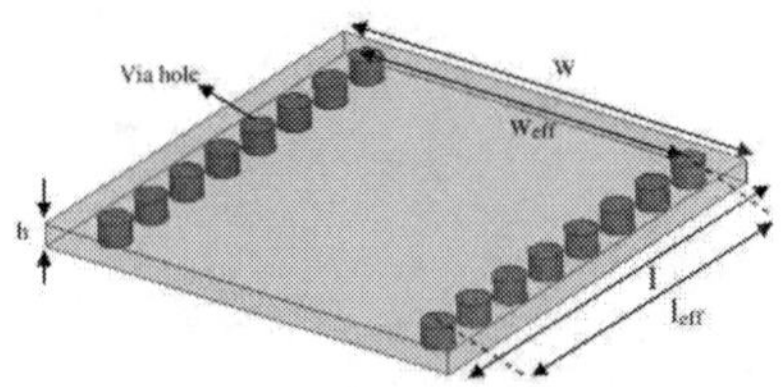

Fig. 1. Structure of Substrate Integrated Waveguide (SIW) [9].

As depicted in Fig. 1, the structure Substrate Integrated Waveguide (SIW) consists of two rows of circular metallic via holes which immersed in dielectric substrate. Moreover, there is also effective width (W_{eff}) and effective length (L_{eff}) of Substrate Integrated

* Corresponding author: nurehansafwanah@yahoo.my

DOI: 10.1051/epjconf/201716201078

Waveguide (SIW) cavity can be calculated from Equation (1) and (2) [8] :

$$W_{eff} = w - \frac{d^2}{0.95p} \tag{1}$$

$$L_{eff} = l - \frac{d^2}{0.95p} \tag{2}$$

Where: w and l are the width and length of Substrate Integrated Waveguide (SIW) cavity.

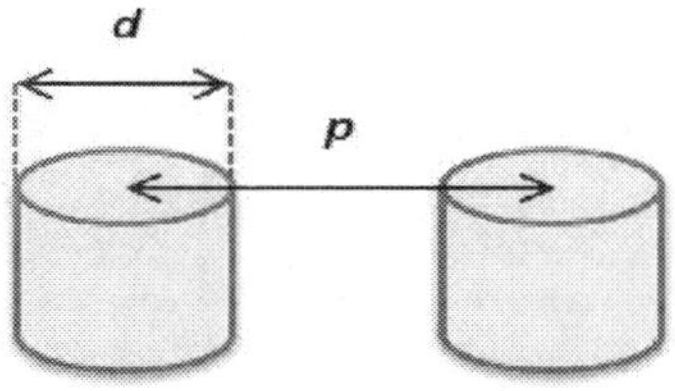

Fig. 2. The Structure of Circular Metallic Via

From Fig. 2, to minimize the leakage between metallic via holes, the pitch distance between circular metallic via (p) needs to be reduced and the diameter of circular metallic via holes (d) needs to be considered. The practical of d/p ratio value needs to follow $0.5<d/p<0.8$ are specified. To determine the diameter (d) and pitch (p) expression (3), (4) and (5) is used [6]:

$$d < \frac{\lambda_g}{5} \tag{3}$$

$$p \leq 2d \tag{4}$$

$$\lambda_g = \frac{2\pi}{\sqrt{\frac{\varepsilon_r (2\pi f)^2}{c^2} - (\frac{\pi}{w})^2}} \tag{5}$$

Where f is the frequency and c is the speed of light.

3 Design Techniques

The proposed design of coupler is designed on Roger 4003 substrate materials where copper layers on both planes. The size of coupler is 25 mm x 15 mm which is considered as compact size. The design of Substrate Integrated Waveguide (SIW) coupler is shown in Fig. 3.

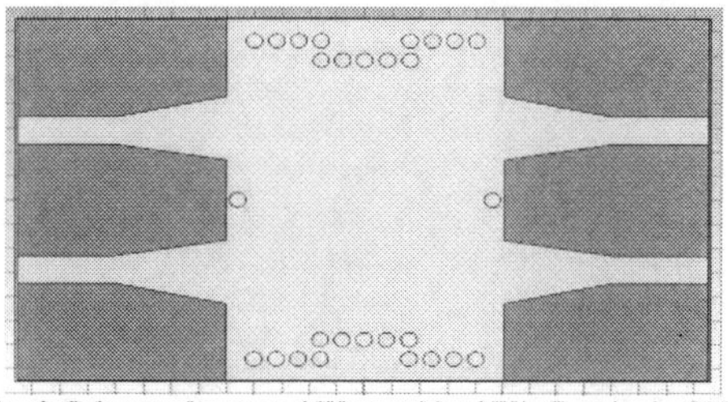

Fig. 4. Substrate Integrated Waveguide (SIW) Coupler in CST

Fig. 4 illustrates the proposed coupler using Substrate Integrated Waveguide (SIW) techniques. In this section, there are 13 metallic via holes on both side wall of substrate and 2 metallic via holes at the centre. Coupler using this technique needs to build transition tapered proposed to match impedance between Substrate Integrated Waveguide (SIW) and 50 Ohm microstrip. Thus, it increases the insertion loss in proposed design.

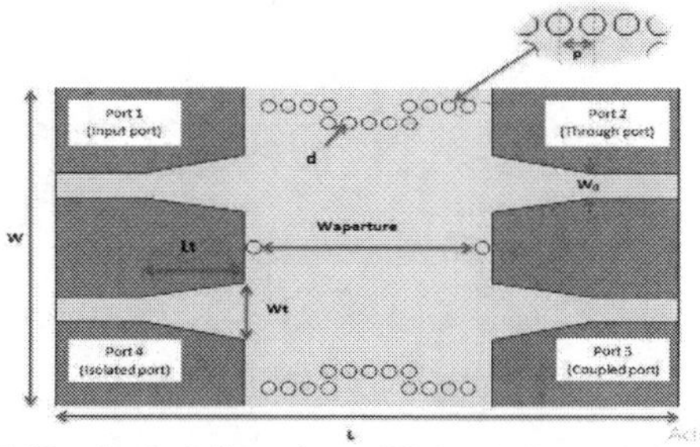

Fig. 5. The physical dimensions of Substrate Integrated Waveguide (SIW) Coupler

Table 1. Dimensions of Coupler

Parameters	Descriptions	Dimensions (mm)
L	The length of substrate	25.00
W	The width of substrate	15.00
d	Diameter of circular via	0.40,0.50,0.60,0.85
p	Pitch of circular via	0.80
Wt	Width of microstrip taper	2.63
Lt	Length of microstrip taper	4.00
$W_{aperture}$	Width of aperture	9.20
W_{siw}	Width of siw	6.60
W_o	Width of microstrip line	1.15

4 Analysis of Previous Research

As depicted in Fig. 6 the paper presented a design and fabrication of X-band Substrate Integrated Waveguide (SIW) directional coupler. From above table, the overall size of design is 54 mm x 48 mm at operating frequency 8 to 10.5 GHz. The results show the flat coupling only at narrow bandwidth with 7.97%. For measurement result, there are high amplitude imbalances [7].

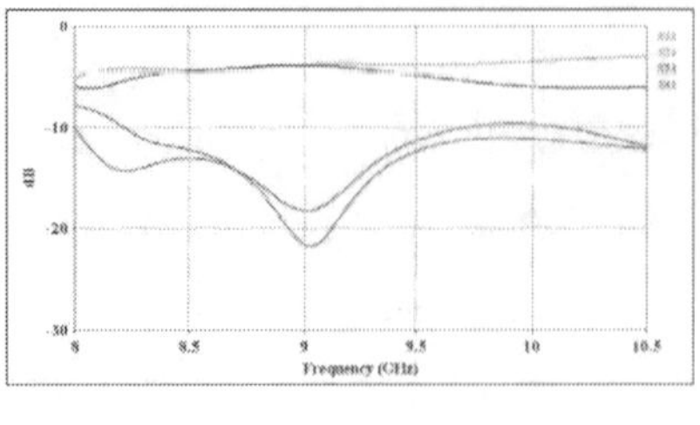

Fig. 6. Simulated Results S-Parameter

The paper that reviewed in fig. 7 is design of broadband planar Substrate Integrated Waveguide (SIW) Transvar coupler. The size of coupler is 80 mm x 60 mm. From the table, the insertion loss is higher below than -5dB although flat coupling is achieved. Then, the simulated and measured result are quiet similar in insertion loss performance [10].

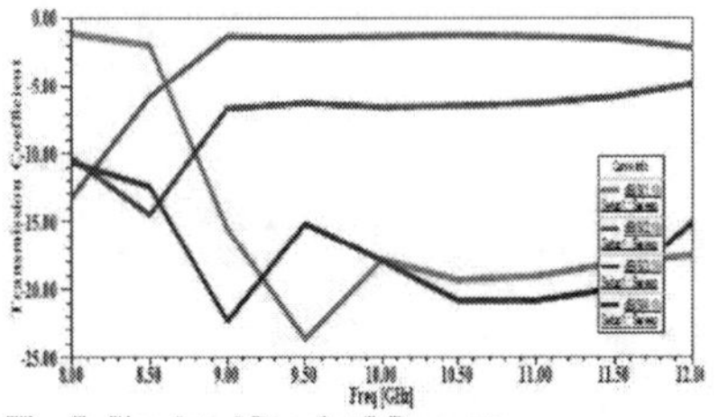

Fig. 7. Simulated Results S-Parameter

5 Results and Discussions

5.1 Parametric Study for Substrate Integrated Waveguide (SIW) coupler

This parameter such as Waperture, Metallic via holes and pitch between holes with rectangular waveguide structure is studied to start the initial design of proposed Substrate Integrated Waveguide (SIW) coupler. This section was discussed about the modification of coupler based on the diameter of circular metallic via (*d*). The variable diameter (*d*) is varied from 0.4 mm, 0.5 mm, 0.6 mm and 0.85 mm.

5.1.1 The Simulated Reflection Coefficient (S_{11})

The results illustrated with Fig. 8, indicate that the return loss (S_{11}) better than 10 dB. The simulation results show the investigated in variable diameter which are 0.4 mm, 0.5 mm, 0.6 mm and 0.85 mm have a good performance on return loss (S_{11}). It is less than -30 dB and the better return loss (S_{11}) with -66.676 dB but there are shifting in frequency is occurred to diameter 0.85 mm. All the proposed coupler is operating in K-Band (18-27 GHz). Therefore, the diameter of circular metallic via is affecting the return loss of waveguide.

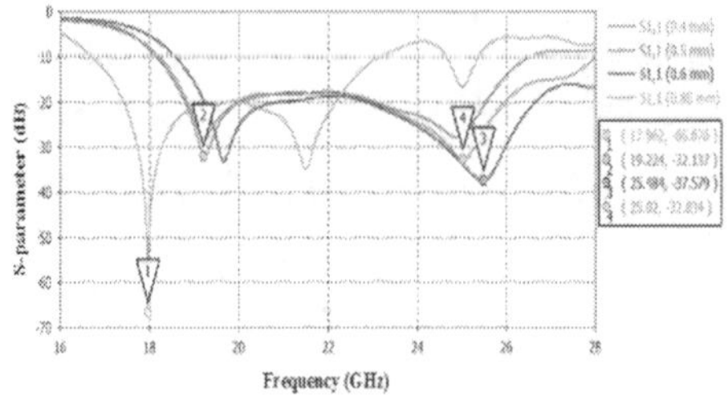

Fig. 8. Simulated result of Reflection Coefficient (S_{11})

5.1.2 The Simulated Isolation Coefficient (S_{41})

As depicted in Fig. 9, the isolation coefficient (S_{41}) presented a good performance are lower than -28 dB for all optimization diameter of circular metallic via (*d*). Moreover, as can be seen In Fig. 9 below, the better isolation response is at -44.46 dB compared to the other variable diameters (*d*). The simulations results highlight, the coupling between port 1 (input) and port 3 (coupled) became weak, S_{11} and S_{41} are increases.

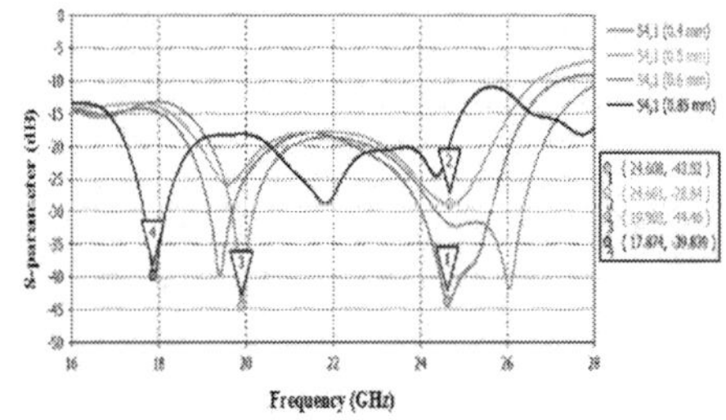

Fig. 9. Simulated result of Isolation Coefficient (S_{41})

5.1.3 The Simulated Phase Difference (S_{31}-S_{21})

From Fig. 10 shows a coupler with d=0.4 mm shows the S_{21} and S_{31} are close to -4.005 dB (22 GHz) and to -3.3242 dB (21.99 GHz). The phase difference in fig. 11 is nearly 90° from 19.0 GHz to 26.0 GHz.

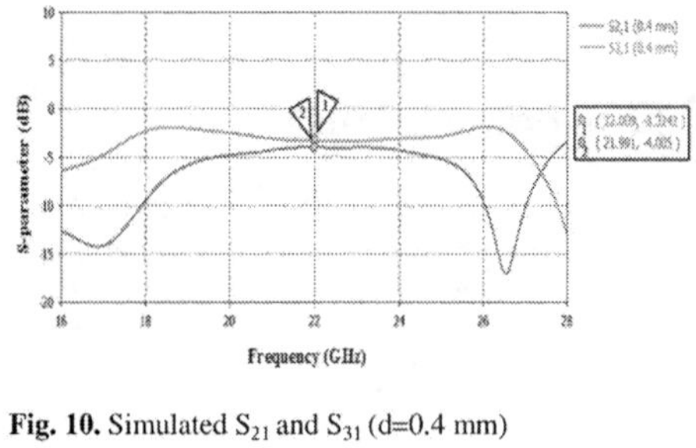

Fig. 10. Simulated S_{21} and S_{31} (d=0.4 mm)

 DOI: 10.1051/epjconf/201716201078

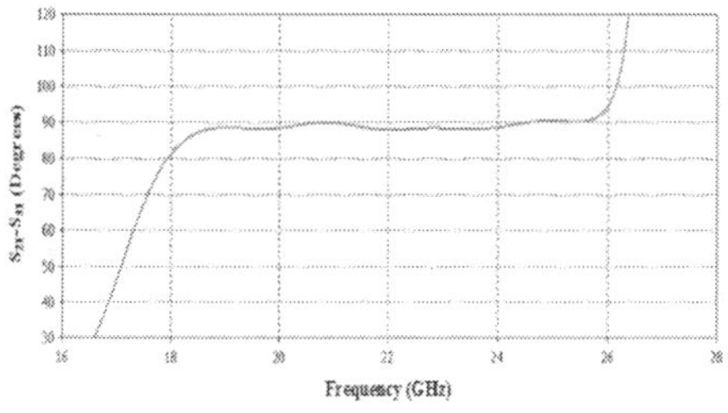

Fig. 11. Simulated phase of coupled S_{21} and S_{31} (d=0.4 mm)

Simulation parameters in Fig. 12 are shown the value S_{21} and S_{31} are closely to -3.5025 dB and -3.6317 dB at 22.0 GHz. The phase difference is close to 90° from 18.0 GHz to 26.0 GHz are presented in Fig. 13 respectively.

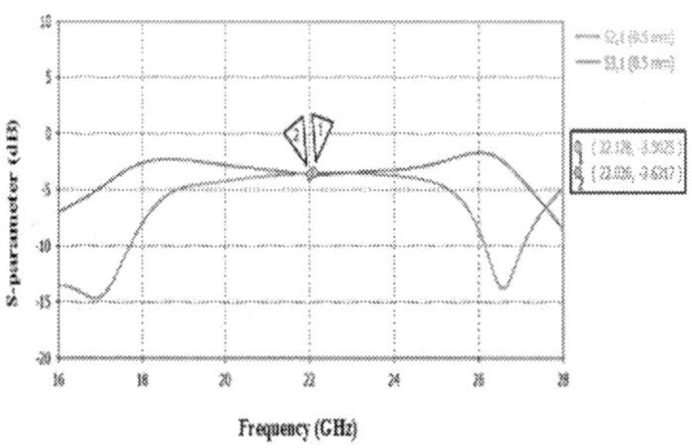

Fig. 12. Simulated S_{21} and S_{31} (d=0.5 mm)

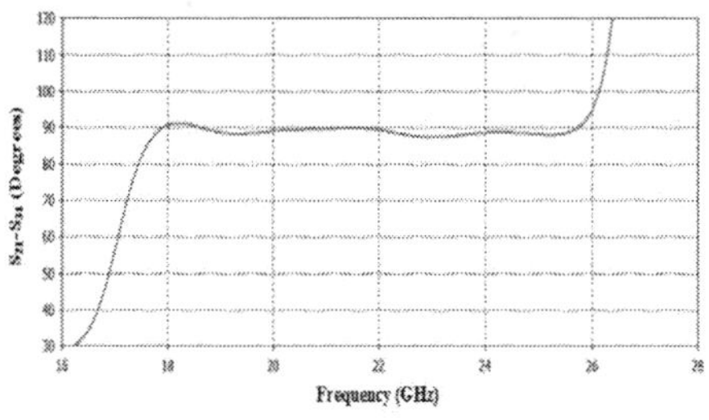

Fig. 13. Simulated phase of coupled S_{21} and S_{31} (d=0.5 mm)

The phase difference in Fig. 14 and Fig. 15 illustrates that approximately 90° which is are the simulated S_{21} and S_{31} shown close to -3.7907 dB and -3.341 dB at 0.6 mm. The frequency covered in 19.0 GHz to 26.0 GHz.

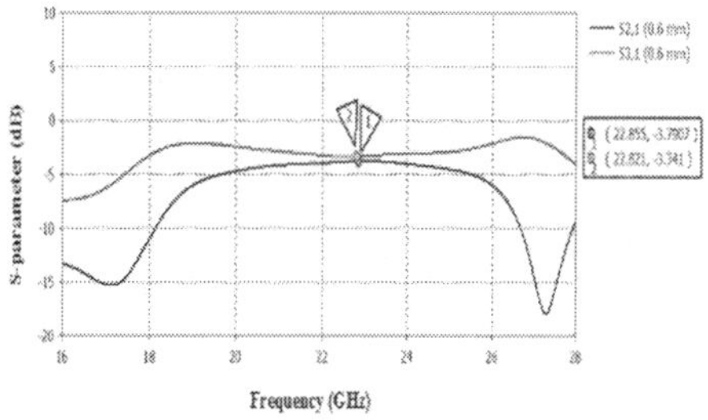

Fig. 14. Simulated S_{21} and S_{31} (d=0.6 mm)

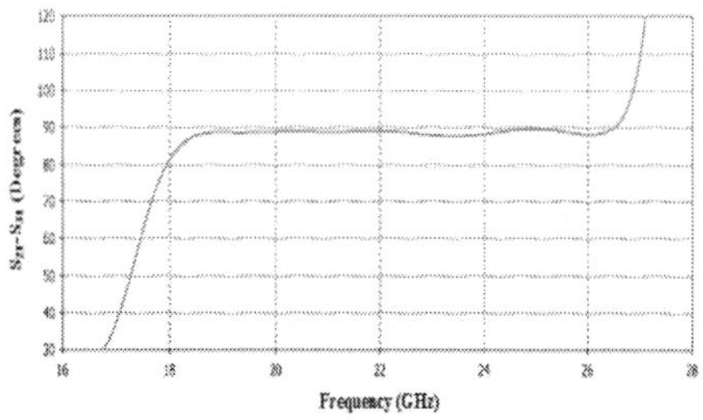

Fig. 15. Simulated phase of coupled S_{21} and S_{31} (d=0.6 mm)

In Fig. 16, their S_{21} and S_{31} at d=0.85 mm are near to -4.5932 dB and -3.1409 dB. The phase difference (S_{21} – S_{31}) at diameter 0.85 mm is varying from 3° and nearly 90° from 16.0 GHz to 22.0 GHz in Fig. 17.

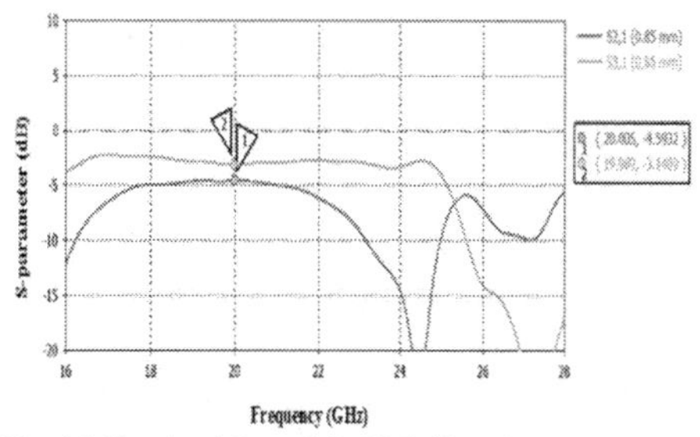

Fig. 16. Simulated S_{21} and S_{31} (d=0.85 mm)

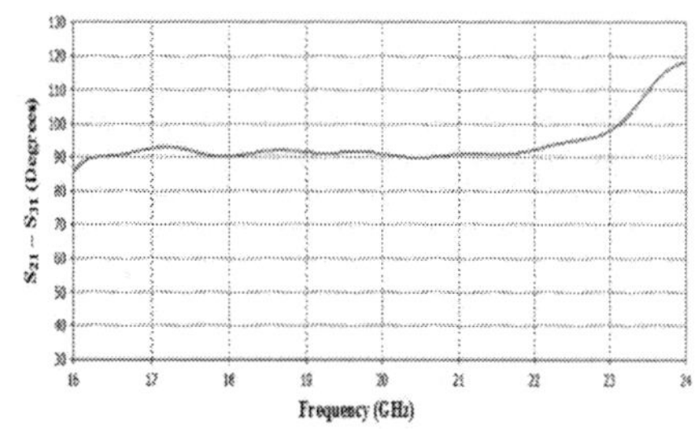

Fig. 17. Simulated phase of coupled S_{21} and S_{31} (d=0.85 mm)

5.2 Final Results for Substrate Integrated Waveguide (SIW) coupler

From Fig. 18 shows the bandwidth is 11.93% with flat coupling between S_{21} and S_{31}. The Reflection coefficient and isolation performance are below -10 dB. Fig. 19 shows the phase difference is 90 degree at operating frequency 17.8 to 26 GHz. The dimension of coupler is 25 mm x 15 mm.

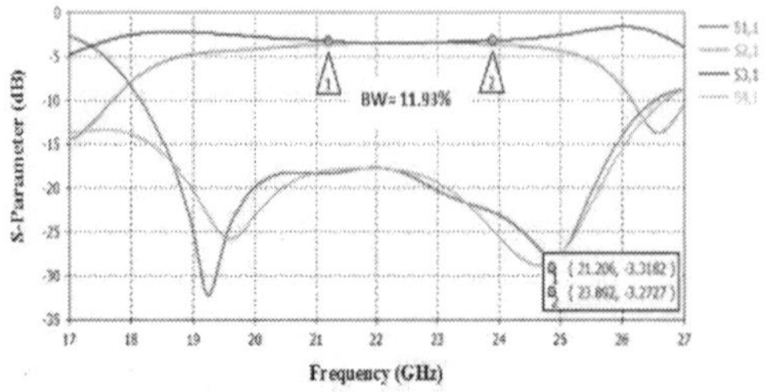

Fig. 18. Simulated Results S-Parameter (d=0.5 mm)

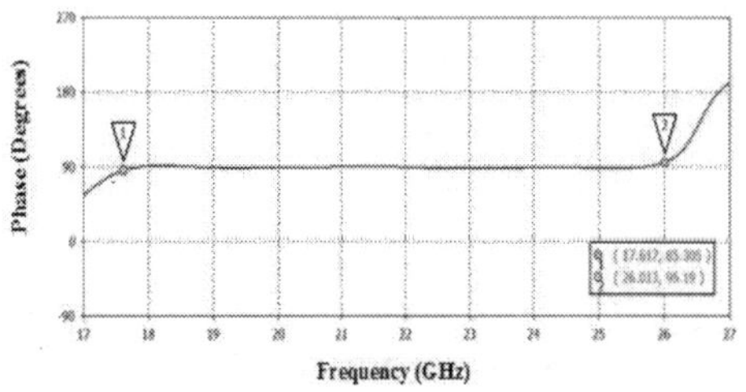

Fig. 19. Phase Difference (d=0.5 mm)

6 Conclusions

The variable diameters of circular metallic via are designed using CST software and the simulation performance are analysed. The design coupler is simulated using Computer Simulation Tools (CST) software. The parameter studies are designing on substrate material Rogers RO4003C, in the K-band has performed. The coupler with variable diameter of circular metallic via obtained a phase difference in 90°. By using Substrate Integrated Waveguide (SIW) techniques the reduced in size and improved bandwidth (11.93%) of coupler is produced. These types of coupler are suitable and favourable choice for microwave applications.

The author would like to thank the Ministry of Higher Education of Malaysia (MOHE) for the financial support for this research work in terms of FRGS grant of 9003-00555.

References

1. Z. C. Hao, W. Hong, J. X. Chen, H. X. Zhou, K. Wu, **153**, 5, (2006).
2. A. Nasri, H. Zairi, A. Gharsallah, Am. J. Appl. Sci., **13**, 3, pp. 276–280, (2016).
3. D. Feature, *Fabricate C-Band with Rectangular SIW, Microwave and RF*, May, (2016).
4. B. Rahali, M. Feham, *IJCSI, Int. J. of Comp. Sci.* **4**, 1, pp. 1–11,(2014).
5. M. Bozzi, A. Georgiadis, K. Wu, IET Microwaves, Antennas Propag., **5**, 8, pp. 909–920,(2011).
6. A. Kumar, G. Saini, S. Singh, A. Kumar, G. Saini, S. Singh, Cogent Eng., **48**,1, pp. 1–12, (2016).
7. B. H. Ahmad, S. S. Sabri, A. R. Othman, *IJET Int. J. Eng. and Tech.*, **5**, 2, pp. 1905–1911, (2013).
8. D. Pozar, *Microwave Engineering Fourth Edition*. (2005).
9. Z. Kordiboroujeni, J. Bornemann, *IEEE Trans. Microw. Theory Tech.*, **62**,12, pp. 2983–2989, (2014).
10. R. Srivastava, S. Mukherjee, A. Biswas, *IEEE Antennas Propag. Soc. AP-S Int. Symp.*, **2015**–Octob, pp. 1402–1403, (2015).

EPJ Web of Conferences 162, 01079 (2017) DOI: 10.1051/epjconf/201716201079
InCAPE2017

Investigation of the performances of PZT vs rare earth ($BaLaTiO_3$) vibration based energy harvester

Nehemiah Pak[1], *Hasnizah* Aris[1] *and Bibi Nadia* Taib[1]

[1]School of Microelectronic Engineering, University Malaysia Perlis (UniMAP), Kampus Alam, Pauh Putra, 02600 Arau, Perlis, Malaysia.

Abstract. This study proposes the investigation of two piezoelectric material namely PZT and Lanthanum Doped Barium Titanate ($BaLaTiO_3$) performance as a vibration based energy harvester. The piezoelectric material when applied mechanical stress or strain produces electricity through the piezoelectric effect. The vibration energy would exude mechanical energy and thus apply mechanical force on the energy harvester. The energy harvester would be designed and simulated using the piezoelectric material individually. The studied outputs are divided to frequency response, the load dependence, and the acceleration dependence whereby measurement are observed and taken at maximum power output. The simulation is done using the cantilevers design which employs d_{31} type of constants. Three different simulations to study the dependence of output power on the resonant frequency response, load and acceleration have found that material that exhibit highest power generation was the $BaLaTiO_3$.

1 Introduction

The piezoelectric energy harvester is one of the most promising devices to generate electric energy from renewable energy sources. In particular, it is of interest in wireless and mobile devices as an alternative energy supply since the feature size of the devices becomes smaller and power consumption increases with diversified functions. In order to extract the maximum power from the environment, the resonant frequency of such device has to match the dominant frequency in the surroundings and this research delves into the investigation of the performance of two piezoelectric based energy harvesters. Vibrations based energy harvesting usually employs an electromagnetic, electrostatic or piezoelectric transduction mechanism [1]. For the design of a structure to convert mechanical vibration to electricity, the cantilever is one of the basic structures since displacement can be easily generated in a transverse direction upon vibration. A piezoelectric transduction device normally consists of an elastic cantilever supporting a piezoelectric layer with metal electrodes on either side. The electrical output energy attains a peak value if the vibration frequency of the environment matches the resonant frequency of the cantilever, and dies out dramatically when it deviates from the resonant frequency of the device [2]. The advantage of choosing a cantilever structure is that the cantilever has the lowest stiffness for a given size, and therefore it is easier to design a low frequency system that can generate highest average strain for a given load, the d_{33} mode PEH has the advantage in obtaining higher voltage while the d_{31} mode can be superior in large current generation. In the case of output power determined by the product of voltage and current, better performance was reported from d_{31} mode PEH than the d_{33} mode device due to low capacitance in the d_{33} mode device

2 Design and material characteristic

The relative permittivity of the material especially for the Lanthanum doped barium Titanate is one of the defining features that sets it apart from the likes of Barium Titanate due to fact that doping of the rare earth (Lanthanum) causes a change in the grain size of the Barium Titanate due to the smaller lanthanum ion stabilizing the cubic form of the crystal after the ion substitution between the rare earth and Barium Titanate. With the large number of holes means a large surface area with dangling bonds of the ceramic particles that tend to absorb species and water. These bonds absorb electrical energy [3]. This reaction can be described as in equation 1 [4]:

$$Ba^{2+}Ti^{4+}O_3^{2-} + xLa \rightarrow (Ba_{1-x}^{2+}La_x^{3+})Ti_{1-x}^{4+}(Ti^{4+}e^{1-})_x O_3^{2-} \quad [1]$$

Fig. 1 shows the design architecture used to simulate power generation out of the piezoelectric material.

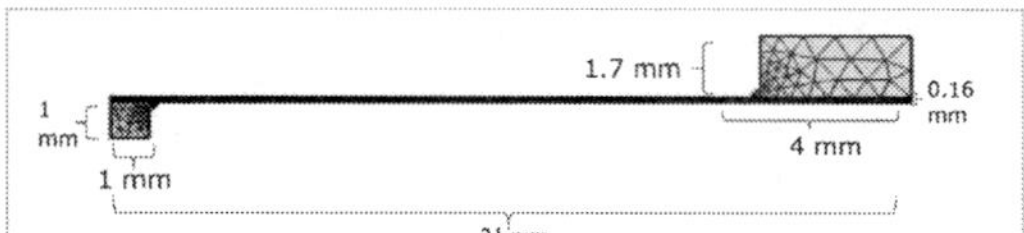

Fig. 1. The dimensions of meshed cantilever

It is noted that Lanthanum doped barium Titanate possess twice the relative permittivity than that of PZT and the following Table 1 shows the material properties of a PZT material.

* Corresponding author: nehe_92@yahoo.com

DOI: 10.1051/epjconf/201716201079

Table 1. Material properties of PZT

Properties		Notation	Value
✓	Density	Rho	7750 [kg/m^3]
✓	Elasticity matrix	cE	{1.20346e+011}[Pa]
✓	Coupling matrix	cES	{0[C/m^2],
✓	Relative permitttivity	epsilonrS	{919.1,919.1,826.6}
	Relative permitttivity	epsilonr	{919.1,919.1,826.6}
	Compliance Matrix	sE	{1.64e-011[1/Pa]}

The calculation for piezoelectric constant, d_{31} and power generation, P_{31} are based on the equations 2-6 where ω, C_{p31}, d_{31}, g_{in}, k, and ζ are the resonance frequency, capacitance of piezoelectric layer, piezoelectric constant, vibration input, electromechanical coupling coefficient, and damping ratio, respectively.

$$P_{31} = \frac{1}{\omega^2} \cdot \frac{R_o C_{p31}^2 \left(\frac{E_p d_{31} t_p b^*}{\varepsilon_0 \varepsilon_r} g_{in}\right)^2}{(4\zeta^2 + k^4)(R_o C_{p31}\omega)^2 + 4\zeta k^2 (R_o C_{p31}\omega) + 4\zeta^2} \quad (2)$$

$$R_o = \frac{1}{\omega C_{p31}} \frac{2\zeta}{\sqrt{4\zeta^2 + k^4}} \quad (3)$$

$$b^* = \frac{3b}{l_c} \frac{(2l_c + l_{pm} - l_e)}{\left(2l_c - \frac{3}{2} l_{pm}\right)} \quad (4)$$

$$b = t_s + \frac{t_p}{2} - n_s \quad (5)$$

$$n_s = \frac{\left(\frac{E_s}{E_p}\right)\frac{t_s^2}{2} + t_p\left[t_s + \frac{t_p}{2}\right]}{\left(\frac{E_s}{E_p}\right)t_s + t_p} \quad (6)$$

4 Results and discussion

The results that were obtained can be analysed by the three different type of output based on the simulation namely frequency response (Voltage and Power), Load dependence and lastly Acceleration dependence. The resonant frequency of the material is determined based on the frequency response towards voltage and power. The highest power is obtained by the cantilever when it resonated at the resonant frequency. As shown in Fig. 2 and Fig. 3, the resonant frequencies of the cantilever are 75.5 Hz and 102.5 Hz for PZT5A and $BaLaTiO_3$ respectively. Once the resonant frequency is determined, that frequency would be sampled for the following two studies load dependence and lastly acceleration dependence. The reason behind this is because at resonant frequency the material would vibrate at its maximum and thus this translates to the maximum output and can be proven based on the simulation output of the frequency response.

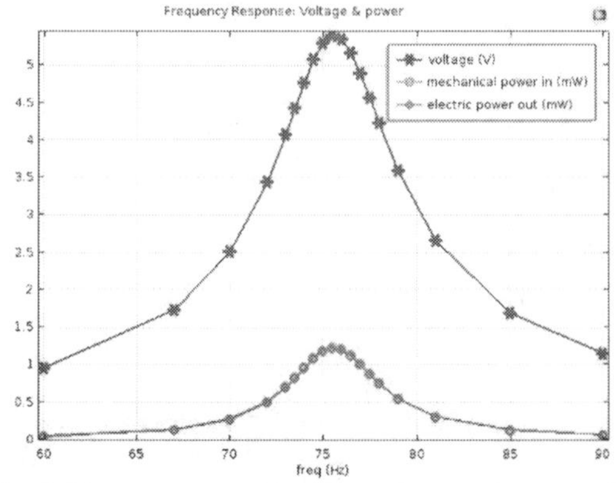

Fig. 2. Frequency responses for voltage and power of PZT at 75.5 Hz

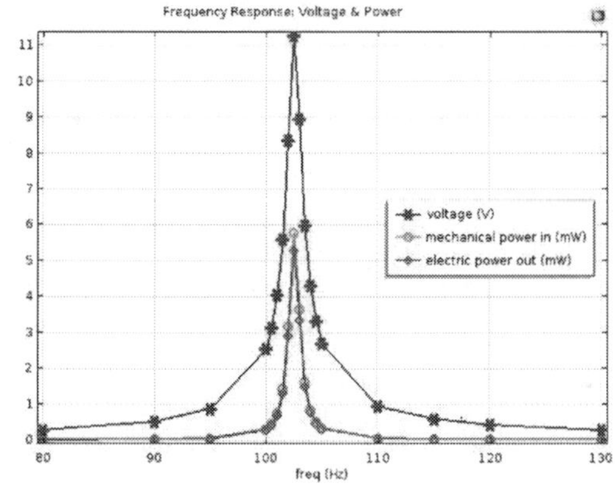

Fig. 3. Frequency responses for voltage and power of BaLaTiO3 at 102.5Hz

Table 2 has summarized the results of power when both materials were simulated with variety load values. It was found that the maximum power is obtained when the applied load is 5623.4 Ω for PZT5A and 1000 Ω for $BaLaTiO_3$. As for the acceleration dependence, 1.2 mW of output power is obtained from PZT5A and 5.02 mW from the $BaLaTiO_3$ when 1g of acceleration is applied onto the structure.

Table 2. Summary of simulation results

PZT5A	Specification	$BaLaTiO_3$
1.228 mW at 75.5 Hz	Frequency Response : Voltage & Power	5.395 mW at 102.5 Hz
1.4436 mW at 5623.4 Ω	Load Dependence : Voltage & Power	17.15 mW at 1000 Ω
1.2 mW	Acceleration Dependence : 1g	5.02 mW

DOI: 10.1051/epjconf/201716201079

5 Conclusion

In conclusion the piezoelectric cantilever that doped with rare earth material i.e. $BaLaTiO_3$ generates for a higher power than the PZT. Application of Lanthanum doped Barium Titanate as energy harvester would be of positive impact to the environment and consumer in comparison to PZT as the future material for vibration based energy harvester utilising a greener piezoelectric would seem very possible in the near future. The performance of the energy harvester can be made more stable with the inclusion of polyvinyl material due to its robust properties thus enabling longer shelf life of the energy harvester since piezo-ceramics can undergo fatigue crack when submitted to high frequency cyclic loading [5]. Such example is that a complete energy harvester system installed within the car's engine that produces vibration energy as a side effect of the engine combusting within is a great source to harvest the wasted energy.

References

1. M. M. Vijatovi, J. D. Bobi, T. Ramo, J. Banys, and B. D. Stojanovi, Materials Characterization, vol. 2, (2011).
2. K. Prashanthi, N. Miriyala, R. D. Gaikwad, and W. Moussa, Nano Energy, vol. **2**, no. 5, pp. 923–932, (2013).
3. J. B. 2 and S. G. 1 M.S. Dash 1*, ICSD'07, pp. 281–284, (2007).
4. H. Liu, C. Quan, C. Jui, T. Kobayashi, and C. Lee, Phys. Procedia, vol. **19**, pp. 129–133, (2011).
5. J. Nunes-pereira, V. Sencadas, V. Correia, V. F. Cardoso, W. Han, and J. G. Rocha, Compos. Part B, vol. **72**, pp. 130–136, 2015.

 DOI: 10.1051/epjconf/201716201080

Stress-strain relationship of PDMS micropillar for force measurement application

Shazlina Johari[1,*] and *L. Y.* Shyan

[1]School of Microelectronic Engineering, Universiti Malaysia Perlis, Pauh Putra Kampus, 02610 Arau, Perlis, Malaysia

Abstract. There is an increasing interest to use polydimethylsiloxane (PDMS) based materials as bio-transducers for force measurements in the order of micro to nano Newton. The accuracy of these devices relies on appropriate material characterization of PDMS and modelling to convert the micropillar deformations into the corresponding forces. Previously, we have reported on fabricated PDMS micropillar that acts as a cylindrical cantilever and was experimentally used to measure the force of the nematode *C. elegans*. In this research, similar PDMS micropillars are designed and simulated using ANSYS software. The simulation involves investigating two main factors that is expected to affect the force measurement performance; pillar height and diameter. Results show that the deformation increases when pillar height is increased and the deformation is inversely proportional to the pillar diameter. The maximum deformation obtained is 713 um with pillar diameter of 20 um and pillar height of 100 um. Results of stress and strain show similar pattern, where their values decreases as pillar diameter and height is increased. The simulated results are also compared with the calculated displacement. The trend for both calculated and simulated values are similar with 13% average difference.

1 Introduction

C. elegans is a non-parasitic worm that has been extensively utilized for genetic model purposes to explore the correlation concerning genes and movement at the neuronal level. The reason for this is predominantly due to its transparency, having a nervous system so simple with merely 302 neurons, and a completely sequenced genome. The understanding of the nervous system, which coordinates the movements/actions and transmits signal between different body parts can potentially help to discover basic mechanism behind the actions of organism. In order to study *C. elegans* propulsion, research has been conducted to measure the worm's movement force which is induced by the contraction of their body wall muscle. The relationship between the nerves and the muscles in charge for the force generation in the neuromuscular system during the worm's motion can be obtained. Information regarding the locomotion forces can also be beneficial for the progress of therapy for muscle disorders, neurodegenerative [1] and humans genetic disease such as muscular dystrophy [2-4]. As the worm is small with the length of approximately 1 mm and the estimated width of 100μm, the existing sensors used in millimeter and larger scales are not capable of sensing the forces at the micro-Newton level.

Polydimethylsiloxane (PDMS) is a silicon based material consisting of a base part and a curing agent. The viscous polymer liquid changes into a solid elastic structure when both components are mixed and cured at a certain temperature. The use of PDMS pillar as force sensing mechanism have been applied in preceding works, as summarized in Table 1. The force obtained is based on the dimension of pillar used.

Table 1. PDMS micropillars as force sensor

Target	Pillar Dimension	Force applied
Mouse oocyte [5] (D: 75 um)	H: 45 um D: 12 um	50 nN - 550 nN
Cardiac myocyte [6] (H: 100-150 um D: 15-20 um)	H: 8 um D: 2 um	47.9 nN - 175 nN
D. melanogaster larvae [7] (H: 1000-1300 um D: 450-550 um)	H: 153 um D: 50 um	~ 32.61 ± 8.68 μN

Previously, the application of PDMS micropillars as a force sensor to measure *C. elegans* locomotion has been demonstrated [8], where the measurement of the worm locomotion forces were conducted experimentally. The microfabricated device consists of several parallel rows of vertical cantilever-like micropillars supported by the channel base. Each pillar functions as an independent force-measuring unit dedicated to one individual *C. elegans*. The worm locomotion force was measured experimentally with force resolution of 3.13 μN for worm body width of 100 μm. Since the nematode body sizes ranges according to their developmental stages [9],

* Corresponding author: shazlinajohari@unimap.edu.my

it is essential to predict the worm locomotion forces based on their body width.

In this work, PDMS micropillars with various dimension are designed and simulated. From the simulation conducted, the force measurement of worm locomotion could be predicted from the displacement obtained by applying reasonable range of input forces. The distributed force was represented by an equivalent point load on the pillar. This load results in an equivalent deflection of the pillar tip [10]. Deflection of a cantilever beam is attributed to both bending and shear incurred by the equivalent force load. When the aspect ratio (i.e. height-to-diameter ratio) of a cantilever beam is greater than 5, deflection due to the shear can be safely ignored as it contributes less than 5% to the total deflection [11]. However, as the aspect ratio of the micropillar is smaller than 5, hence both bending and shear must be considered in the force-deflection mechanics model. The equivalent displacement of the micropillar is calculated using Equation (1),

$$\Delta = \left\{ \left(\frac{l^3}{3EI} + \frac{d^2(1+\gamma)l}{4EI} \right) + \frac{l^2}{2EI}(h-l) \right\} f \qquad (1)$$

where f is the force applied, Δ is the displacement, I is the area moment of inertia, γ is the Poisson's ratio for PDMS, l, d and h are pillar length, diameter and height respectively, and E is the PDMS Young's Modulus.

2 Design and Simulation

The design of the micropillar is based on a cylindrical cantilever as shown in Figure 1. The pillar height, h and diameter, d is varied and their effect on the displacement, stress and strain was studied. The material properties used during the simulation are indicated in Table 1. Based on previous research, [8], the force exerted by the *C. elegans* is in the range from 5 - 70 µN. Thus, the forces used in ANSYS simulation are selected to be 5, 10, 20, 40, 60 and 80 µN. The deformation of the micropillar is observed and recorded. The force applied is set as a nodal or point force being exerted from the side of the micropillar, similar to their natural locomotion, where the worm moves by pushing their body and exerting force against their surroundings.

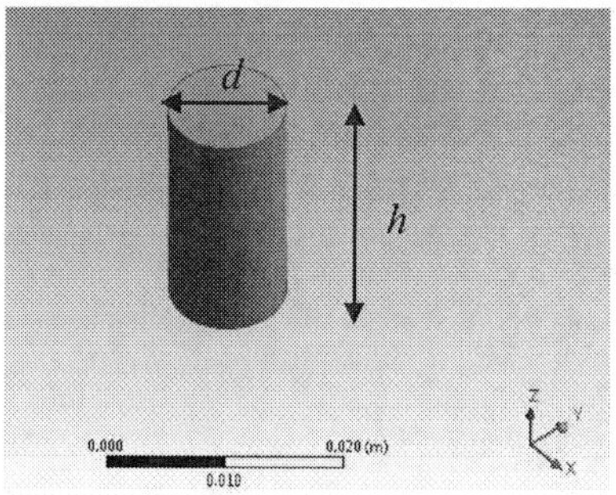

Fig. 1. Micropillar design with diameter, *d* and height, *h*

Table 2. PDMS Material Properties

Parameter	Value
Young's Modulus	1.47 MPa
Poisson's Ratio	0.48
Bulk Modulus	12.25 MPa
Shear Modulus	0.497 MPa

3 Results and Discussion

There are three output obtained from simulation, namely the deformation of the micropillar at a certain force input, stress distributed along the pillar height and the strain produced on the pillar during deformation.

(a)

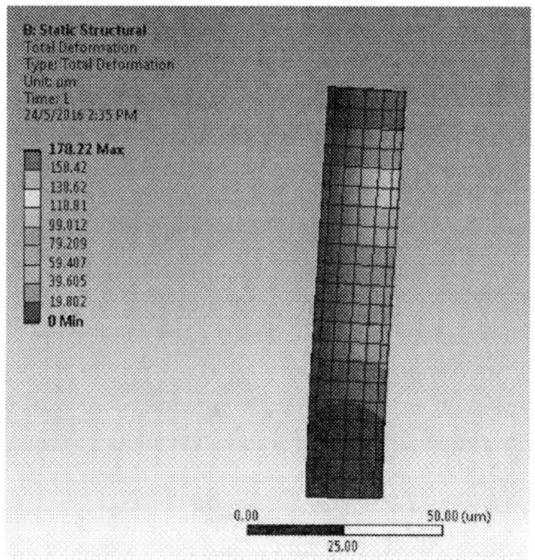

(b)

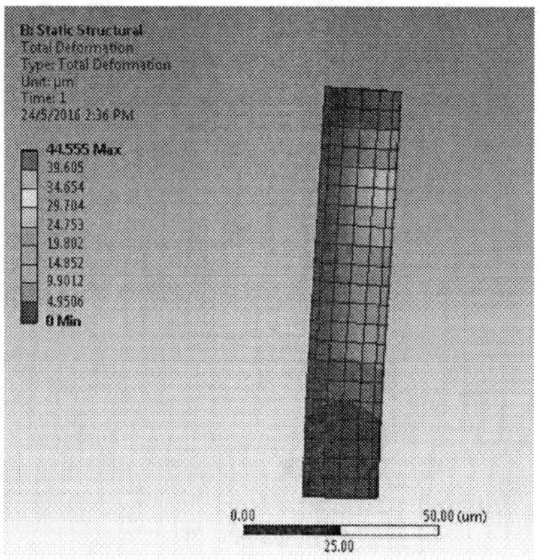

(c)

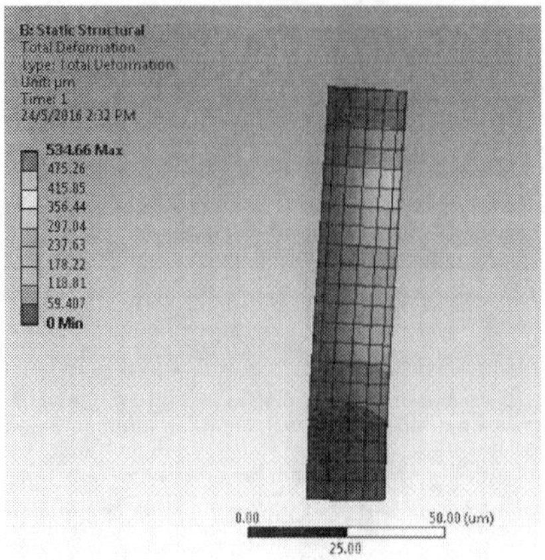

Fig. 2. Total deformation generated by ANSYS at diameter 20 µm when applying different forces. (a) 5µN (b) 20µN (c) 60µN

When point load are applied to the microcantilever, it will cause different stress levels generated on the microcantilever, thus, resulting in the displacement of the microcantilever. Fig. 2 shows the deformation results for pillar diameter of 20 µm and pillar height of 100 µm with applied force of 5, 20 and 60 µN. The highest displacement can be seen at the free end of the microcantilever while the maximum stress point is located at the fixed end of the microcantilever. It can be observed that the deformation occurs at the same place which is at the edge of the free end of the pillar. However, the maximum deformation of each force applied is different, which are 44.555 µm, 178.22 µm, and 534.66 µm for force of 5 µN, 20 and 60 µN respectively. The blue colour indicates the smallest deformation, since the part is the fixed support, where the fixed position is held in position. The changes of colour indicated different level of deformation of a micropillar. Each colour indicates different bending level of the micropillar.

During the simulation, the pillar diameter are varied while the pillar height is fixed to 100 um in order to see the effects of the micropillar dimension on the deformation, as shown in Fig. 3. For each pillar diameter, the simulated displacement show similar increasing trend as the force is increasingly applied. When pillar diameter is increased from 40 um to 60 um, the displacement roughly decrease 74%. When pillar diameter is bigger, the pillar aspect ratio will decrease, hence will result in the decrease of the pillar displacement. For pillar diameter of 40 um, the highest displacement is recorded at 49 um with applied force of 80 uN. The minimum recorded displacement is 125 nm when the pillar diameter is set to 120 um subjected to 100 um pillar height.

Next, the pillar diameter is fixed at 50 um while its height were varied from 80 um to 200 um. Fig. 4 depicts the variation in the micropillar displacement with respect to varying input force ranging from minimum of 5uN to maximum of 80 uN. The maximum deformation obtained at the applied input force of 80 µN is 18.201 µm when pillar height is 80 µm. The displacement value

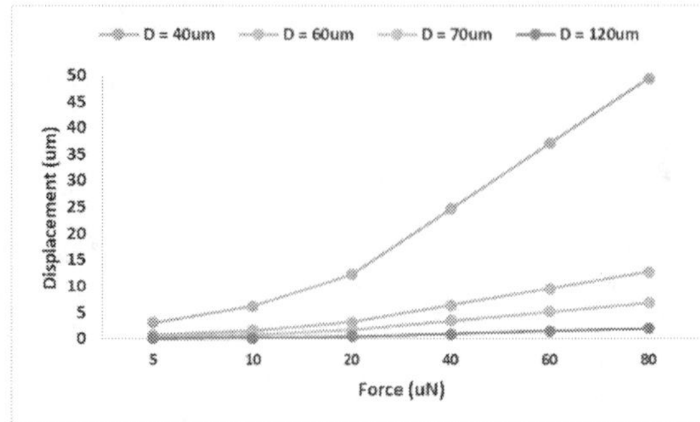

Fig. 3. Graph of displacement versus force for varying pillar diameter

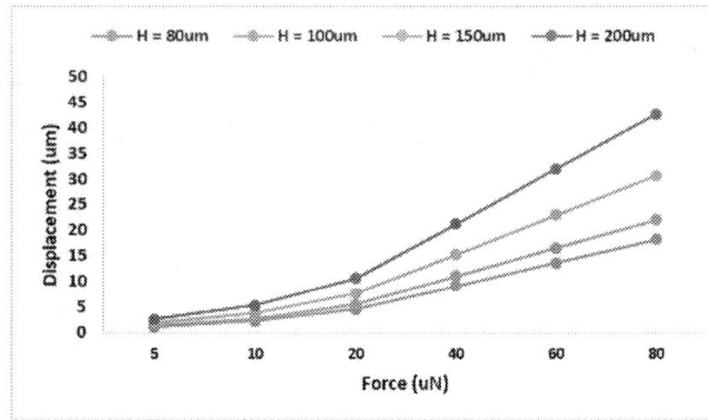

Fig. 4. Graph of displacement versus force for varying pillar height

increases 17.44% to 22.05 µm when pillar height is increased by 20 um. Then, this value increase 28.06% to 30.65 µm for pillar height of 150 µm and continue to increase by 28.03% to 42.6 µm at pillar height of 200 µm. The deviation in the displacement obtained is relatively smaller when compared to the displacement results from changes made in pillar diameter. A change of 100 µm (from 200 µm to 300 µm) only showed a deformation different of 13.033 µm (different of deformation at height 200 µm and 300 µm), which is increasing by 23.43%. The range of deflection obtained for varying pillar height and pillar diameter is similar, which is from 100 nm to 40 um. This shows that the aspect ratio critically influence the pillar bending performance.

We also measured the stress-strain relationship of the PDMS micropillar based on varying pillar height and diameter. At pillar diameter of 20 µm and fixed pillar height of 100 µm, the simulated stresses are 0.31 MPa, 0.62 MPa, 1.24 MPa, 2.47 MPa, 3.71 MPa and 4.94 MPa when force applied is 5 to 80 uN. Since stress is related to the force applied against a certain surface area, the stress value is increasing when the applied force is also increased. The effect of the pillar diameter on the stress distribution shows that as the pillar diameter increases, stress required to bend the pillar also increases. The changes in height showed opposite result as when the diameter is varied, where the stress distributed decreased when pillar height is increased.

 DOI: 10.1051/epjconf/201716201080

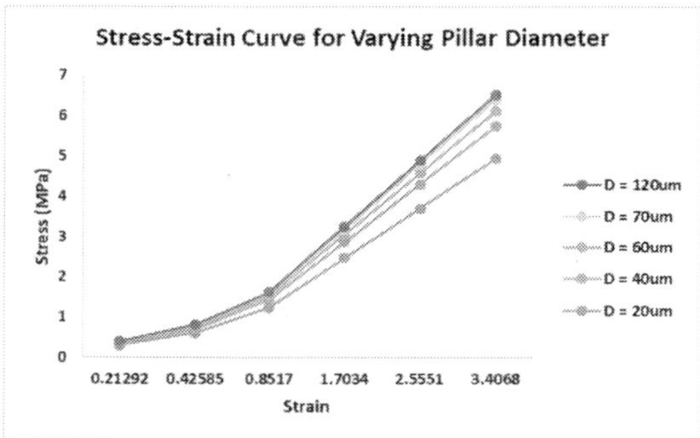

Fig. 5. Stress-strain curve of micropillar force for varying pillar diameter

The pillar with 50 μm diameter and 80 μm height shows a large stress deviation, which is 0.055 MPa, 0.11 MPa, 0.22 MPa, 0.44 MPa, 0.66 MPa and 0.87 MPa for applied force of 5 μN, 10 μN, 20 μN, 40 μN, 60 μN and 80 μN respectively. These simulated stresses are much different then 100 μm height, increasing height by 20 um only showed 0.029 MPa, 0.057 MPa, 0.113 MPa, 0.23 MPa, 0.34 MPa and 0.46 MPa for applied force of 5 μN, 10 μN, 20 μN, 40 μN, 60 μN and 80 μN correspondingly. In essence, materials experience strain when they are subject to stress. The relationship between stress and strain is different for different materials, and can be depicted by plotting stress against strain. The stress-strain curve that we obtained is comparable with existing work that reported on the stress-strain relationship of PDMS [12]. It is also important to note that the stress-strain curve is highly influenced by the PDMS material properties, hence resulting in a non-linear curve. Since this work is purely simulation, the material properties have been predefined prior to simulation, as shown in Table1.

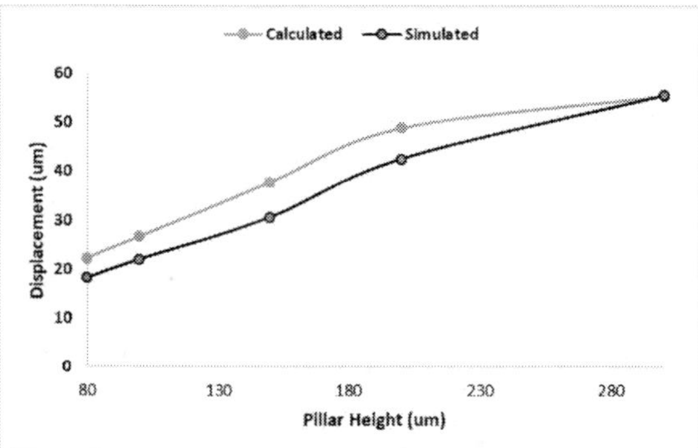

Fig. 6. Calculated and simulated value of pillar displacement for varying pillar height

The simulated results are also compared with the calculated displacement using Equation 1. The simulated and calculated value of displacement for constant input force of 80 μN are plotted in Fig. 6. For example, for pillar height of 80 μm, the calculated displacement is 22.2 μm compared to 18.2 μm simulated value. As the pillar height is increased, the trend for both calculated and simulated values are similar with 13% average difference. When pillar height is 300 μm, both calculated and simulated displacement value is 55 μm. This shows that the conducted simulation of this work is in agreement with the theoretical calculated displacement.

4 Conclusion

The high elastic property of PDMS allows this material to be widely used as force measurement apparatus of biological organism. We have design and investigated the bending behaviour of micropillar with varying height and diameter by observing its displacement. We found that pillar aspect ratio influence the output displacement the most, where for pillar diameter of 20 μm and pillar height of 100 μm, the maximum displacement is obtained. In terms of stress-strained curve, the result is comparable with preceding works reported. The simulated results are also compared with the calculated displacement using bending equation and both simulated and calculated value depict similar increasing trends.

We are grateful for the funding provided by FRGS 9003:00405 and the authors would like to thank School of Microelectronic (SoME), Universiti Malaysia Perlis for their technical assistance and supports.

References

1. M. Dimitriadi, J. N. Sleigh, A. Walker, H. C. Chang, A. Sen, G. Kalloo, J. Harris, T. Barsby, M. B. Walsh, J. S. Satterlee, C. Li, D. Van Vactor, S. Artavanis-Tsakonas, and A. C. Hart, PLoS Genetic, **6**, p. e1001172 (2010)
2. C. I. Bargmann, Science, vol. 282, pp. 2028-33 (1998)
3. J. E. Mendel, H. C. Korswagen, K. S. Liu, Y. M. Hajdu-Cronin, M. I. Simon, R. H. Plasterk, and P. W. Sternberg, Science, **267**, pp. 1652-5 (1995)
4. L. S. Nelson, M. L. Rosoff, and C. Li, Science, vol. 281, pp. 1686-90 (1998)
5. X. Liu, J. Shi, Z. Zong, R. Fernandes, R. F. Casper, A. Jurisicova, K.T Wang and Y. Sun, *15th International Conference on Miniaturized Systems for Chemistry and Life Sciences* (2011)
6. Y. Zhao and X. Zhang, Sensors and Actuators A 125 398–404 (2006)
7. S.M. Kharel, V. Venkataraman1 and S.P. Koushika, *18th International Conference on Miniaturized Systems for Chemistry and Life Sciences* (2014)
8. S. Johari, V. Nock, M. M. Alkaisi, W Wang, Lab Chip, **13** (9), pp. 1699-1707 (2013)
9. S. E. Hulme, S. S. Shevkoplyas, A. P. McGuigan, J. Apfeld, W. Fontana, and G. M. Whitesides, Lab Chip, **10**, pp. 589-97 (2010)

DOI: 10.1051/epjconf/201716201080

10. A. C. Ugural and S. K. Fenster, Advanced Strength and Applied Elasticity (2003)
11. A. Ghanbari, V. Nock, S. Johari, R. J Blaikie, X. Q. Chen, W. Wang, Journal of Micromechanics and Microengineering 22(9): 095009 (2012)
12. K. Khanafer, A. Duprey, M. Schlicht and R. Berguer, Biomed Microdevices 11:503–508 (2009)

Effects of pH and concentration on the capability of E. coli and S. epidermidis with bentonite clay as biosorbent for the removal of Copper, Nickel and Lead from polluted water

Delia B. Senoro[1,2*], *Josel B.* Godezano[1], *Meng-Wei* Wan[3] ,*Lemmuel L.* Tayo[1,4], *Zaliman* Sauli[5] and *Hasnizah* Aris[5]

[1] Environmental Engineering Graduate Program, Mapua University, 658 Muralla St., Intramuros, Manila, Philippines 1002
[2]Office of International Linkages for Research and Development, 2/F Research Building, Mapua University, 658 Muralla St., Intramuros, Manila Philippines 1002
[3] Department of Natural Resources Management, College of Sustainable Environment, Chia Nan University of Pharmacy and Science, Tainan, Taiwan
[4]Biological Engineering Program, Mapua University, 658 Muralla St., Intramuros, Manila, Philippines 1002
[5]School of Microelectronic Engineering, University Malaysia Perlis (UniMAP), Kampus Alam, Pauh Putra, 02600 Arau, Perlis, Malaysia.

Abstract. This paper discusses the effects of pH and concentration on the capability of E. coli ATCC29522 and S. epidermidis RP62A biofilm with bentonite in removing divalent copper, nickel and lead from wastewater. Batch adsorption study at laboratory scale was utilized to evaluate the potential use of bacterial biomass (E. coli ATCC29522 and S. epidermidis RP62A) aided with geosynthetic clay (bentonite) for the removal of Cu^{2+}, $Ni^{2+}$and Pb^{2+}. Results revealed that removal of Cu^{2+}, $Ni^{2+}$and Pb^{2+} by both types of organisms supported with bentonite were high in the first 4 hours of the experiment. This illustrates that the binding site on that particular time was abundant. Hence, the removal rate was evident at high concentration depicting the line adsorption equilibrium. It also revealed that S. epidermidis RP62A supported with bentonite had the highest affinity to Copper and Lead with Qm = 277.7 mg/g and 5.0075 mg/g, respectively. While E. coli ATCC 29522 had the highest affinity to Nickel (Qm= 58.82 mg/g). Hence, the sorption of Cu^{2+}, $Ni^{2+}$and Pb^{2+} onto E. coli ATCC29522 and S. epidermidis RP62A biofilm supported with bentonite clay occurred through monolayer chemisorption on the homogeneous surface of E. coli ATCC29522 and S. epidermidis RP62A biofilm with bentonite clay. Batch kinetics studies revealed that the sorption of Cu^{2+}, $Ni^{2+}$and Pb^{2+} onto E. coli ATCC29522 and S. epidermidis RP62A biofilm supported with bentonite clay was well described by a pseudo-second-order equation model of type 1 (R2 = 0.9999), which implies that chemisorption is the rate limiting step.

1 Introduction

Heavy metals pollution has become a global issue of great concern due to their higher toxicities, higher bioaccumulation in human body, food chain, nature of non-biodegradability and most likely carcinogenicities to human. Lead, mercury, chromium, arsenic, cadmium, zinc, copper and nickel are the most common contaminants found in contaminated surface water and groundwater as well as industrial wastewater.

The occurrence of these heavy metals in water causes great threats to humans and other living organisms [1]. According to Vijayaraghavan and Yun [2], there are many different sources of water pollutants and they were categorized into two: direct and indirect contaminants. Direct contaminants include those that were discharge by industries, refineries, mining vessels and waste treatment plants; whereas indirect contaminants are those that enter the water bodies which came from soil/ground water system and from the atmosphere via water rain. Heavy metals are regarded as toxic material due to their non-biodegradable characteristics and are very dangerous to living organism even in low concentration [3]. Thus , attention in removing these heavy metals requires effective removal methods.

Vast methods based on physicochemical such as extraction, ion exchange, chemical precipitation and membrane separation processes has been demonstrated for emoval of heavy metals from aqueous solutions [3,4]. However, the mentioned methods do have some limitation when the concentration of metal in waste water is under 100 ppm. Hence, it is necessary to develop a cost effective and efficient separation method.Biosorption is a passive and metabolically-independent process such as the passive uptake of metals by microbial biomass that can be performed either by dead biomass, fragment of cells and tissues or live cells [5].

On other hand, removals of heavy metal using clay materials are also arising in remediation environment.

* Corresponding author: dbsenoro@mapua.edu.ph

Example of the clay use in remediation process is bentonite. Natural bentonites are argillaceous materials that have been widely used to remove toxic metal ions, dyes, chlorophenols and drugs [6]. Due to their high specific surface area, high cation exchange capacity (CEC), their chemical and physical stability, they can be effectively employed as adsorbents for many wastewater pollutants [7]. Combination of biological microorganism and natural clay such as bentonite will form a new remediation system that will efficiently improve its capacity in removal of heavy metals.

The main goal of this study focuses on investigating the potential removal of Cu^{2+}, Ni^{2+} and $Pb^{2+}$from Simulated Wastewater using Biofilms of Escherichia coli ATCC 25922 (E.coli) and Staphylococcus epidermidis RP62A (S. epidermidis) supported with bentonite in a single-component system, with focus on evaluating the effects of pH and concentration on the removal.

Significantly, this study attempt to augment results that are helpful in removing heavy metals from polluted water specially contaminants coming from industrial facilities. Furthermore, this is an avenue for future researchers to thoroughly develop and ameliorate biosorption technology for the removal of heavy metals. This study will be limited to the analysis of heavy metals concerned, which are mentioned above with the concentration of 10, 50, 100 and 200 ppm (mg/L), pH ranging from 4-8 with constant temperature of 37°C. Agitation rate will be considered at constant speed of 150 rpm and bentonite as a support.

2 Methodology

This study was conducted at Innovation Green Technology Laboratory of Chia Nan University of Pharmacy and Science, Taiwan in collaboration with Sustainable Development Research of Mapua Institute of Technology, Philippines.

2.1 Materials

Bentonite was obtain from Choneye Pure Chemicals (Taipei, Taiwan). The bacterium Escherichia coli ATC25922 and Staphylococcus epidermidis RP62A was obtained at Chia Nan University of Pharmacy and Sciences, Tainan Taiwan. Sodium hydroxide (NaOH) and Hydrochloric acid (HCl) were used to vary the pH of the system. NaOH (0.1 M), was prepared by dissolving 2.0g of NaOH crystals in 500 mL deionized water. Hydrochloric acid (0.1 M) was prepared by diluting 4.229 mL concentrated HCl (36.5%) with 500 mL deionized water.Glassware utilized wereeroded in 10% nitric acid and rinsed with deionized water to avoid intrusion of foreign metals.

2.2 Metal Solutions Preparation

Stock of heavy metals concentrations were 10 mg/L, 50 mg/L, 100 mg/L and 200 mg/L which was prepared by dissolving Ni $(NO_3)_2$. $6H_2O$ (Ferak, Berlin), Cu $(NO_3)_2$. $2.5H_2O$ (JT Baker, USA), Pb $(NO_3)_2$. $6H_2O$ (Darmstadt, Germany) in deionized (DI) water (18. 2 Ω, ELGA Purelab Ultra).

2.3 Agar Medium Preparation

An agar (Bacto Agar, BD, USA) amounting to 1.50 g and 3.70 g of BHI (Bacto Brain Heart Infusion, BD, USA) with 7.70 g/L calf brain extract, 9.80 g/L beef heart extract, 10.0 g/L of proteose peptone, 2.0 g/L dextrose, 5.0 g/L of Sodium Chloride, and 2.5 Disodium Phosphate were dissolved in 100 ml DI water. The medium was autoclave (Speedy Autoclave, TOMIN) at 121 °C for 45 min, cool to room temperature. Prepared medium was used for bacterial growth.

2.4 BHI Broth Medium Preparation

BHI (Bacto Brain Heart Infusion, BD, USA) with 7.70 g/L calf brain extract, 9.80 g/L beef heart extract, 10.0 g/L of proteose peptone, 2.0 g/L dextrose, 5.0 g/L of Sodium Chloride, and 2.50 Disodium Phosphate amounting to 3.70 g was dissolved in 100 ml DI water, was then adjusted to a pH of 4.6. The medium was sterilized in an autoclave (Speedy Autoclave, TOMIN) at 121.0 °C for 45 min, cool to room temperature. Prepared medium was used to enhance bacterial growth.

2.5 Bacterial Cultivation

Streak plating was used to culture bacteria from bacterial tube stored at -20.0°C to the prepared agar medium by drawing a three-zoned area in the plate with an sterilized loop in a laminar flow and was incubated for 16-18 hours at 37.0°C. After incubation overnight, single colony was selected from the agar plate and transferred in test tubes with 3.0 mL BHI broth medium for its cultivation.

2.6 Enhance Bacterial Suspension

Bacterial growth was enhanced by diluting 1.0 ml of incubated bacterial suspension (from 3.0 ml BHI broth + single colony) with 20.0 ml BHI broth (1:20 ratio). Bacterial suspension was incubated for 24 hours at 37.0°C. So that the number of bacteria will be within a given range, bacterial suspension absorbance was read using UV-Vis Spectrophotometer (2800 Series, UNICO, Taiwan) at a wavelength of 600 nm.

2.7 Minimum Inhibitory Concentration

Minimum inhibitory concentration (MIC) was conducted at the Department of Medical Laboratory Science and Biotechnology, Kaohsiung Medical University, Kaohsiung Taiwan.Bacteria were cultured overnight at 37 °C in Lactose Broth agar and bacterial concentration was adjusted to 106 CFU/mL in Mueller- Hinton broth. The 100 μL bacterial suspensions were added to 96- well micro titer plates and incubated with 100 μL various concentrations of metal solutions (4000- 62.5 ppm). After incubation at 37 °C for 20 hours, the absorbance was read with a microplate reader at a wavelength of 600 nm.

2.8 Biosorbent Formation with Bentonite as Support

Agitation 0.50 g of bentonite clay with 7.50 mL of bacterial suspension and 75 mL of 10, 50, 100 and 200 mg/L heavy metal solutions in 250 mL Erlenmeyer flasks.The Erlenmeyer flask was kept at constant temperature of 37 ^{0}C for 1 day with moderate agitating at 150 rpm and 4 days without agitation in orbital shaker incubator (721 SR HIPOINT, Taiwan).

2.9 Particle Separation Process

After every contact time, the bacterial solution with heavy metal was centrifuge (KUBOTA 2420, Japan) at 3000 rpm for 5 minutes and filtered by 0.20 μmm micro pore syringe filter (Whatman, USA) to separate aqueous phase from solid phase for Physical and Chemical analysis.

2.10 Metal Ion Analysis using ICP- OES

To quantitatively determine the metal ion concentration, inductively coupled plasma optical emission spectrometry (ICP-OES) (DV2000 Series Perkin Elmer, USA) was used. Standard solutions for the instrument's calibration curve were prepared. Samples were diluted to 10 mg/L of ICP standards [8].

2.11 Biosorption Optimization

The effect of pH on metal ions uptake was studied by performing equilibrium sorption experiments at pH range of 4 to 8. Adjustments in the solution pH were done by adding 0.1 M NaOH or 0.1 M HCl while metal ion concentration of 100 mg/L, temperature of 37 ^{0}C, amount of biosorbent and contact time of 24 hours were kept constant.

2.12 Batch Biosorption Studies

Study was conducted by agitating sample solution with 0.5 g of bentonite, 7.5 mL of bacterial suspension and 75 mL of 10, 50, 100 and 200 mg/L heavy metal solutions in an orbital shaker incubator (721 SR HIPOINT, Taiwan) at 150 rpm with contact time of 4, 12 and 24 hours and optimum pH of 5.0. After attaining equilibrium, biosorbent was centrifuge and filtered using 3000 rpm at 5 minute and 0.20 μmm micro pore syringe filter respectively. The aqueous-phase concentration of metal is determined using ICP- OES.

3 Result and Discussion

3.1 Minimum inhibitory concentration (MIC)

Results of the MIC diagnostic showed that *E.coli* ATC 25922 and *S. epidermidis* RP62A can survive and resist in lead solution with concentration greater than 2000 mg/L. On the other hand, *E.coli* ATC 25922 is resistant to Copper and Nickel solution at 250 mg/L and 125 mg /L respectively. However, *S. epidermidis* RP62A was resistant to copper and nickel up to 125 mg/L and 62.5 mg/L, respectively.

3.2 Effect of pH

The effect of pH on Cu^{2+}, Ni^{2+} and Pb^{2+} uptake were investigated by performing equilibrium sorption experiments by varying the solution pH from 4 to 8.The effect of pH in the solution (Figure 1) at initial concentration of 100 mg/L and temperature of 37 oC showed that E. coli ATCC 29522 removed 99.94% at pH 5 while S. epidermidis RP62A removed Cu^{2+} by 100% at pH 5.The increase in sorption was attributed to fewer hydrogen ions in solution that may compete with the copper for adsorption to negatively-charged surfaces [9].Ni^{2+} provide an slight increase of biosorption from pH 5 to 6 as showed in Figure 2. Sari [10], reported a maximum biosorption for nickel at pH 5–6 with low Ni^{2+} adsorption at lower pH values.Both E. coli ATCC29522 and S. epidermidis RP62A supported with bentonite clay in Figure 3 showed a sharp increase of Pb^{2+} removal as pH rose from 4.0-5.0 due to absorbent binding sites were not yet occupied. At this pH, both E. coli ATCC29522 and S. epidermidis RP62A removed 99.21% -99.89 % and 99.92 -100.0% respectively.In the present study, the pH = 5.0 had an average percent removal of 99.71% and 99.81% in both E. coli ATCC29522 and S. epidermidis RP62A respectively as shown in Figure 1 to 3. This pH was chosen for subsequent studies of Cu^{2+}, Ni^{2+} and Pb^{2+} removal.

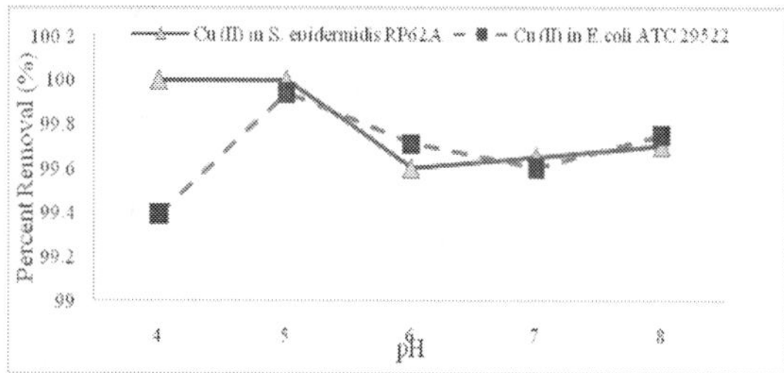

Fig. 1. Effect of pH on the adsorption efficiency of Cu^{2+} ions using *E. coli* ATCC 29522 and *S. epidermidis* RP62A biofilm with bentonite clay

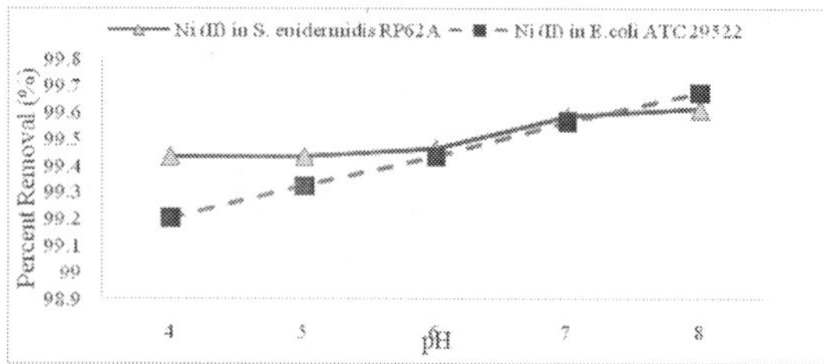

Fig. 2. Effect of pH on the adsorption efficiency of Ni^{2+} ions using *E. coli* ATCC 29522 and *S. epidermidis* RP62A biofilm with bentonite clay

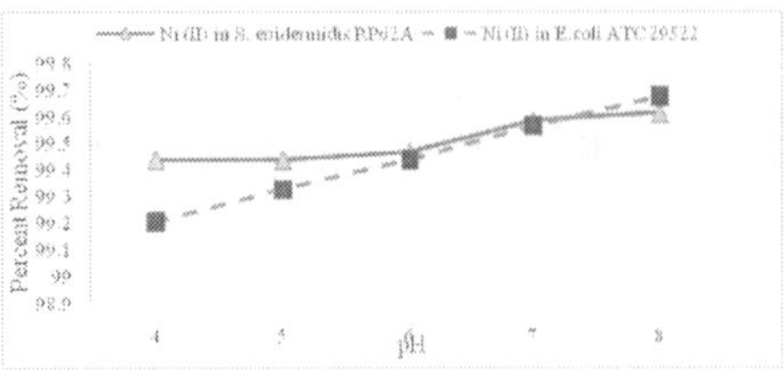

Fig. 3. Effect of pH on the adsorption efficiency of Pb^{2+} ions using *E. coli* ATCC 29522 and *S. epidermidis* RP62A biofilm with bentonite clay

3.3 Batch Biosorption Studies

Batch experiments were carried out at different initial metal concentrations of 10, 50, 100 and 200 mg/L with contact time at 4, 12 and 24 hours at pH 5 with constant temperature of 37 °C. Uptake rates on copper (Fig. 3 & 4), nickel (Fig. 5 & 6), and lead (Fig. 7 & 8) were recorded sharp increase of biosorption at the first four hours. This was attributed to the abundance of available adsorption sites. At low metal concentrations, adsorption sites are available and could easily be occupied since the ratio between the number of metal moles in solution and available surface area is low. Therefore, adsorption is independent of the initial concentration. However, the adsorption rate was expected to gradually decrease with an increase in process time leading to the adsorption equilibrium [11]. The equilibrium time is defined as the time required to attain this state of equilibrium and the adsorbent capacity is determined based on the concentration of the copper remaining in the solution under the stated operating conditions [12].

Nevertheless, the removal efficiency is dependent on the initial concentration. At higher concentration, the number of ions competing for the available binding sites on the biomass surface increases thus reduces the number of binding sites. The sorption rate gradually declined with time until equilibrium was achieved starting from 4 hours up to 24 hours. The intensification in removal efficiency of Ni^{2+} ions is due to the augmentation surface area resulting from the increase in adsorbent mass [13]. The initial rapid phase in Fig. 7 & 8 was due to the availability of an initial large number of vacant sites which later became difficult due to the repulsive forces between Pb^{2+} adsorbed on the microsphere surface and Pb^{2+} in bulk solution [14] and slowed down as the sites are gradually filled up. Therefore, the kinetics will be more dependent on where Pb^{2+} are transported from the liquid phase to the sorption sites [15].

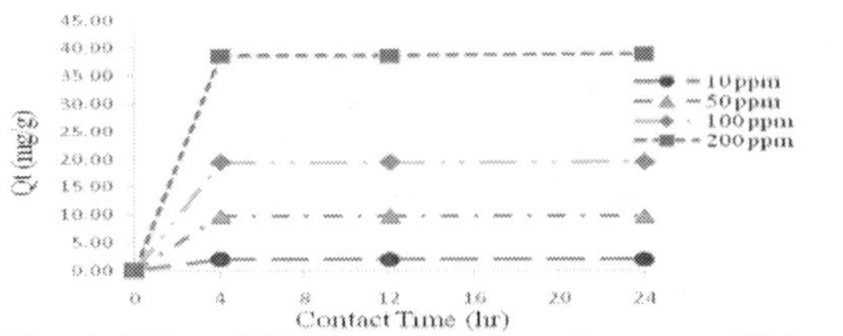

Fig. 4. Effect of initial metal concentration on equilibrium biosorption capacity of Cu^{2+} using *E. coli* ATCC 29522 biofilm supported by bentonite clay

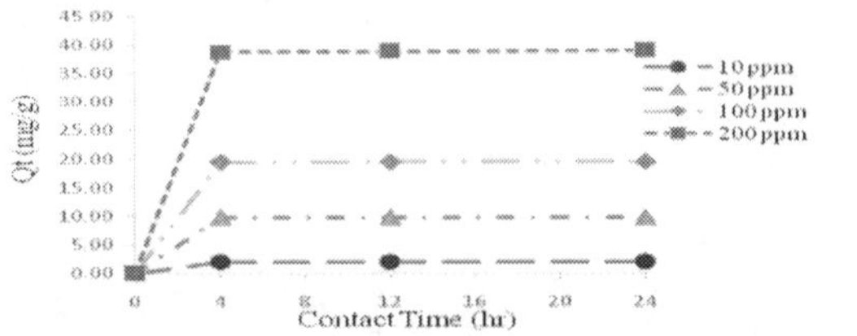

Fig. 5. Effect of initial metal concentration on equilibrium biosorption capacity of Cu^{2+} using *S. epidermidis* RP62A biofilm supported by bentonite clay

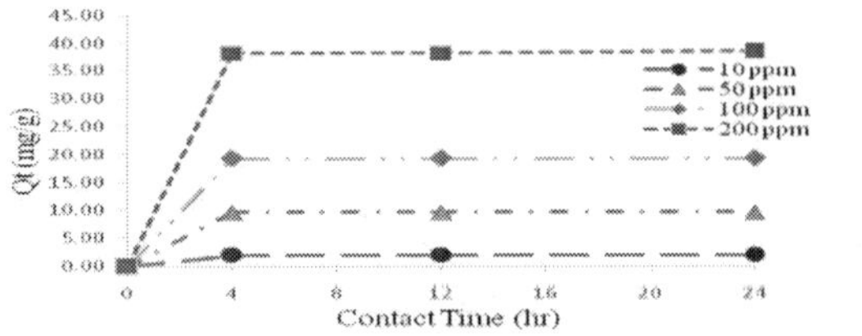

Fig. 6. Effect of initial metal concentration on equilibrium biosorption capacity of Ni^{2+} using *E. coli* ATCC 29522 biofilm supported by bentonite clay

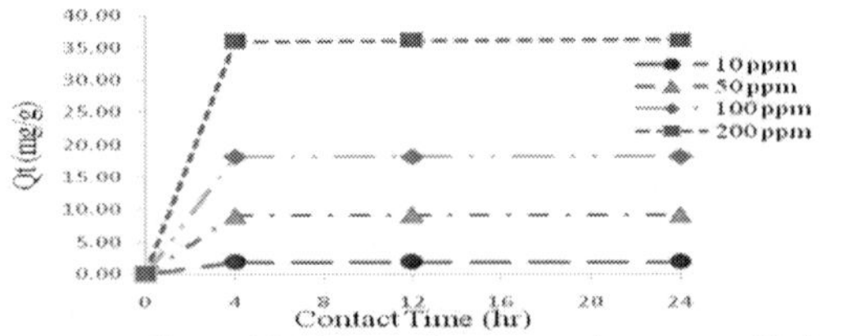

Fig. 7. Effect of initial metal concentration on equilibrium biosorption capacity of Ni^{2+} using *S. epidermidis* RP62A biofilm supported by bentonite clay

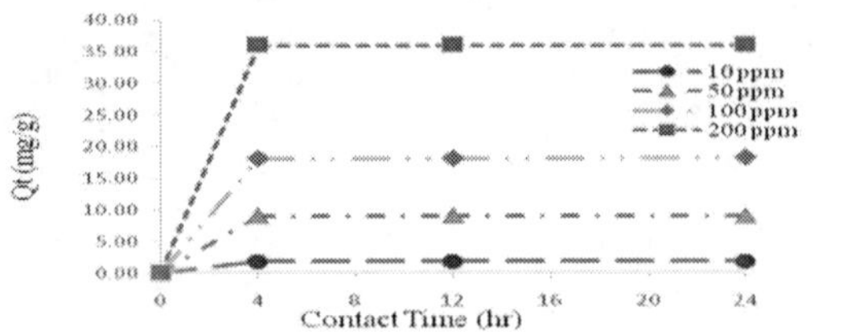

Fig. 8. Effect of initial metal concentration on equilibrium biosorption capacity of Pb^{2+}using *E. coli* ATCC 29522 biofilm supported by bentonite clay

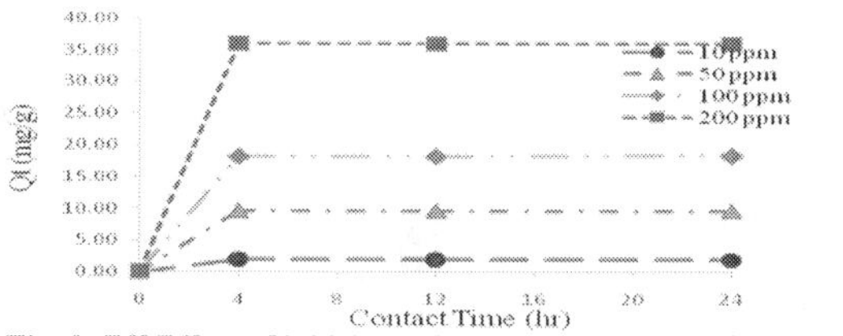

Fig. 9. Eff Effect of initial metal concentration on equilibrium biosorption capacity of Pb^{2+} using *S. epidermidis* RP62A biofilm supported by bentonite clay

Further, the kinetic parameters, and correlation coefficients are listed in Table 1. The correlation coefficient (R^2) for the pseudo-second order model is 0.9999 [13]. The calculated *Qe* value is found to be much closer to the experimental *Qe* value [13]. These results confirm that the adsorption kinetics of Cu^{2+}, Ni^{2+}, and Pb^{2+} ions onto the *E. coli* ATCC 29522 and *S. epidermidis* RP62A supported with bentonite is mainly governed by pseudo-second order equation and hence the rate-limiting of the biofilm is chemisorption

Table 1. Kinetic Data for the Biosorption of Cu^{2+}, Ni^{2+} and Pb^{2+} ions

Biosorbent	Heavy Metal	Pseudo- First Model			Pseudo Second Model		
		Qe (mg/g)	K1 (min-1)	R2	Qe (mg/g)	K2 (min-1)	R2
E. coli ATCC 29522	Cu	4.276	0.167	0.5766	19.418	194.6252	0.9999
	Ni	4.364	0.156	0.5714	18.149	5.47E+13	0.9999
	Pb	5.090	0.248	0.7969	19.268	9.676514	0.9999
S. epidermidis RP62A	Cu	3.930	0.193	0.6222	19.455	32.36659	0.9999
	Ni	3.320	0.310	0.7896	18.149	43.23335	0.9999
	Pb	4.459	0.316	0.8403	19.305	22.80649	0.9999

4 Conclusion

The results revealed that removal of Cu $^{2+}$, Ni $^{2+}$and Pb $^{2+}$ by *E. coli* ATCC 29522 and *S. epidermidis* RP62A biofilm with bentonite were 99.94% and 100%, 99.32% and 99.43%, 99.89 and 100% respectively, at pH 5.0 with the temperature of 37°C and at concentrations of 10, 50, 100, 200 ppm. The data from the minimum inhibitory concentration (MIC) showed that *E.coli* ATCC 25922 and *S. epidermidis* RP62A can live and resist in lead solution with concentration higher than 2000 mg/L. Additionally, *E.coli* ATCC 25922 is resistant to Cu $^{2+}$ and Ni $^{2+}$ at 250 mg/L and 125 mg/L, respectively, as opposed to *S. epidermidis* RP62A which was at 125 mg/L and 62.5 mg/L, respectively. The uptake performances of both organisms supported with bentonite were high in the first 4 hours of the experiment. It was associated with the high availability of adsorption sites that were spontaneously available and easily occupied. Hence, the removal rate was very evident at high concentration. Further, it could be concluded that *E. coli* ATCC 29522 and *S. epidermidis* RP62A biofilm with bentonite is a potential sorbent for the removal of Cu $^{2+}$, Ni $^{2+}$and Pb $^{2+}$ from polluted water.

References

1. He, J., Chen, J. P. *Bioresour. Technol.* ***160*** 67. (2014)
2. Vijayaraghavan, K.; Yun, Y.-S. *Biotechnol. Adv.* ***26*** **(3),** 266, (2008).
3. Farooq, U., Kozinski, J. A., Khan, M. A., Athar, M. *Bioresour. Technol.* ***101*** **(14),**5043 (2010).
4. Fomina, M., Gadd, G. M. *Bioresour. Technol.* ***16,*** *03*,(2014).
5. Volesky, B., Holan, Z. R. *Biotechnol. Prog.* ***11*** **(3),** 235, (1995).
6. Ahluwalia, S. S., Goyal, D. *Bioresour. Technol.* ***98*** **(12),** 2243, (2007).
7. Polat, H., Erdogan, D. *J. Hazard. Mater.* ***148*** **(1),** 267, (2007).
8. Quintelas, C., Rocha, Z., Silva, B., Fonseca, B., Figueiredo, H., Tavares, T. *Chem. Eng. J.* ***149*** **(1),** 319, (2009).
9. Chang, N.-B., Houmann, C., Lin, K.-S., Wanielista, M. *Chemosphere* ***144,*** 1280, (2016).
10. Sarı, A., Tuzen, M., Uluözlü, Ö. D., Soylak, M. *Biochem. Eng. J.* ***37*** **(2),** 151, (2007).
11. Horsfall, M., Ogban, F., Akporhonor, E. E. *Chem. Biodivers.* ***3*** **(2),** 161, (2006).
12. Banerjee, K.; Ramesh, S. T., Gandhimathi, R., Nidheesh, P. V, Bharathi, K. S., *Iran. J. Energy Env.* ***3*** **(2),** 143, (2012).
13. Sudha, R., Srinivasan, K., Premkumar, P. *Ecotoxicol. Environ. Saf.* ***117,*** 115, (2015).
14. Ge, Y., Qin, L., Li, Z., *Mater. Des.* ***95,*** 141, (2016).
15. Shahat, A., Awual, M. R., Khaleque, M. A., Alam, M. Z., Naushad, M., Chowdhury, A. M. S. *Chem. Eng. J.*, *273*, 286, (2015).

Biosorption of Pb^{2+} and Cu^{2+} in aqueous solutions using agricultural wastes

Aileen D. Nieva [1,*], *Bonifacio* T. Doma Jr [1], *Huan-Ping* Chao[1] and *Lai* Siang Leng[2]

[1] School of Chemical Engineering and Chemistry, Mapúa University 658 Muralla St., Intramuros, Manila 1002, Philippines
[2]School of Microelectronic Engineering, University Malaysia Perlis (UniMAP), Kampus Alam, Pauh Putra, 02600 Arau, Perlis, Malaysia.

Abstract. This study aimed to determine and compare the adsorptive capacity of Pb^{2+} and Cu^{2+} in simulated wastewater onto three agricultural wastes The adsorption capacities of Pb^{2+} onto the agricultural wastes can be arranged as *Litchi chinensis* (4.30 mg of sorbate per g of sorbent (mg g^{-1}), 85.68% adsorption) > *Bambusa vulgaris* (3.83 mg g^{-1}, 76.19% adsorption) > *Annona squamosa* (2.70 mg g^{-1}, 53.66% adsorption) while the adsorption capacities of Cu^{2+} onto the same agricultural wastes can be arranged in the order: *Bambusa vulgaris* (3.86 mg g^{-1}, 77.17% adsorption) > *Annona squamosal* (3.58 mg g^{-1}, 71.58% adsorption) > *Litchi chinensis* (3.42 mg g^{-1}, 68.32% adsorption). The biosorbents had relatively higher adsorptive capacities with Cu^{2+} as compared to that of Pb^{2+} except for *Litchi chinensis*. Although the results show lower adsorptive capacity as compared to a number of treated agricultural wastes showing 80% up to almost 100% adsorption of Pb^{2+} and Cu^{2+}, the results show that *Annona squamosa*, *Bamubusa vulgaris*, and *Litchi chinensis* are potential biosorbents and promote sustainable treatment process.

1 Introduction

Industrial effluents from different industries contain various contaminants such as heavy metals i.e. Pb^{2+} and Cu^{2+} come from semiconductor industries, paint and battery productions, manufacture of ballast keel of sailboats and polyvinyl chloride (PVC) and coal-based power generation and mining industries. Due to the significant adverse impacts on humans and to the environment of these contaminants, there is a need for more stringent effluent regulatory standards which the different industries need to comply by employing sustainable treatment methodology. Adsorption has been found to be a sustainable treatment as compared to conventional treatments in terms of cost, minimal sludge produced, higher efficiency, no additional nutrient requirements and possible regeneration and recovery of biosorbents and the heavy metal contaminants. Among the adsorbents, agricultural wastes are currently being utilized due to cost and effectivity while helping in minimizing the solid wastes generation problem. A number of agricultural wastes showing the high % removal ranging from 80% up to almost 100% by adsorption were conducted with treated agricultural wastes with chemicals like HCl and HNO_3 [1]; with NaOH [2]; or using other treatments like burned until the wastes turned activated carbon [3,4,5]. This study aimed to determine and compare the adsorptive capacity of Pb^{2+} and Cu^{2+} in simulated wastewater onto three agricultural wastes, namely, peels of *Annona squamosa* (custard apple) and *Litchi chinensis* (lychees), and shoot of *Bambusa vulgaris* (bamboo) without any utilization of pre-treatments of the biosorbents to promote sustainability of the process.

2 Methodology

2.1 Preparation of Biosorbents

Peels of *Annona squamosa* and *Litchi* chinensis, and shoot of *Bambusa vulgaris* were obtained individually in Chungli, Taiwan, washed several times with distilled water to ensure that all adhering dirt was removed, air-dried under room temperature until water was evaporated and oven-dried at 50°C for 24 to 48 hours. Dried materials were ground and sieved to obtain a particle size of 30 mesh screen. The properties of the biosorbents were analyzed using the following laboratory analyses: (1) surface properties using scanning electron microscopy (SEM, Hitachi S-300N), (2) particle size distribution by a laser diffraction particle size analyzer (LS™ 13 320, Beckman, USA), (3) Fourier transform infrared (FTIR, Perkin Elmer Model 1600) spectroscopy, (4) zeta potential analyzer (ZEN3600, MALVERN Nano-ZS) to determine the surface charges and (5) an elemental analyzer (HORIBA 7021H) to determine the composition in terms of the major elements, (6) Cationic Exchange Capacity (CEC) (Bio-Rex 70) analyzer.

* Corresponding author: adnieva@mapua.edu.ph

2.2 Preparation of Sorbate Solutions.

Stock solutions of $Pb(NO_3)_2$ and $Cu(NO_3)_2$ were prepared to have concentration of 200 mg L^{-1} stock solution by dissolving the solutes in 1000 ml of double distilled water. Appropriate dilutions of 50 ml volumes containing initial concentrations (Co) of 2, 4, 6, 8, 10, 12, 18, and 20 mg L^{-1} were made from the stock solution.

2.3 Equilibrium Studiess.

All batch adsorption experiments were conducted in 50-mL plastic bottles with stoppers shaken in the dark on an orbital shaking incubator at 180 rpm until equilibrium was reached. The temperature was controlled at room temperature and the equilibrium time of 24 h was chosen. Adsorption measurement was determined by batch experiments of 0.2 g adsorbent with 50 ml of aqueous sorbate solutions of known concentration in a series of 50 ml conical plastic containers at pH adjusted to 4.0, 5.0, and 6.0 ± 0.1 using 0.1N HCl or 0.1N NaOH aqueous solution. The samples were withdrawn from the shaker and the solutions were separated from adsorbent by settling for few minutes followed by filtration using 0.45μm Whatmann filter. At time $t = 0$ and at equilibrium, sorbate concentrations in the supernatant solution were estimated using atomic absorption spectrophotometer at suitable maximum wavelength to represent the amount of contaminants adsorbed on the biosorbents and concentration was computed from the calibration curve. For each series of measurements, the adsorption calibration curve was constructed composed of a blank and three or more standards from Merck (Germany). Accuracy and precision of the sorbate measurement were confirmed using external standard reference material for trace elements in water.

2.4 Isotherm Models

The equilibrium concentration results for the adsorption system were plotted against adsorption uptake in mol kg^{-1}. The amount of sorbate adsorbed onto the adsorbent biomass at time t, q_E (mg/g), is calculated by the following mass balance relationship.

$$q_e = (C_0 - C_e)\frac{V}{W} \qquad (1)$$

where C_0 and C_e were the initial and equilibrium liquid phase concentrations of sorbate, respectively (mg L^{-1}), V the volume of the solution (L), and W the weight of the biosorbent used (g). The equilibrium data obtained were fit into different isotherms i.e. Langmuir, Freundlich and Temkin linearized isotherm sorption models for comparison to predict which among the models would have the best fit and would give the maximum adsorption capacity among the three biosorbents.

3 Results and Discussion

3.1 Preparation of Biosorbents.

The preparation method of drying ensured that the cellulose, hemicelluloses, xylose and lignin present on the surface of the agricultural wastes were protonated by drying at 50°C for 24 to 48 hours to produce basic components of functional groups i.e. acetamido groups, carbonyl, phenolic, structural polysaccharides, amide, amino, sulphydryl carboxyl groups alcohols and esters [6] that may serve as the binding sites to the contaminant with strong affinity to the contaminant with positive charges in aqueous solution. The preparation did not involve any chemical nor higher energy requirement to turn the wastes into char like previous studies.

3.2 Characterization of Biosorbents.

The surface properties of dried and powdered peels of *Annona squamosa* and *Litchi chinensis*, and the shoots of *Bambusa vulgaris* were determined using scanning electron microscopy (SEM, Hitachi S-300N) as shown in Figure 1.

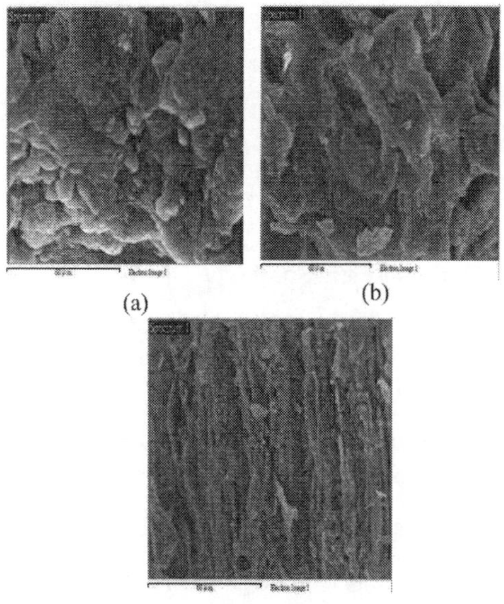

Fig. 1. SEM of the samples (6) biosorbents (a) Annona squamosal, (b) Litchi chinensis, (c) Banbusa vulgaris.

Sample of *Annona squamosa* has a rough multilayer surface with the compact surfaces lumped together which may provide accessible large surface areas during adsorption. Sample *Litchi chinensis* has rough multi-layer surface formed circular with small amount of pores on the surface which also indicates good adsorbing capacity. Shoot of *Bambusa vulgaris* has compact flat surface with some fissures in its structures mainly on the cross section of the fiber that can initiate contact with the sorbate. The Fourier transform infrared (FTIR) spectroscopy analyses of the biosorbents are shown in Figure 2 indicating the available functional groups on the surface of each biosorbent that provide binding sites and strong affinity to the ions of manganese in aqueous solution of Pb^{2+} and Cu^{2+}.

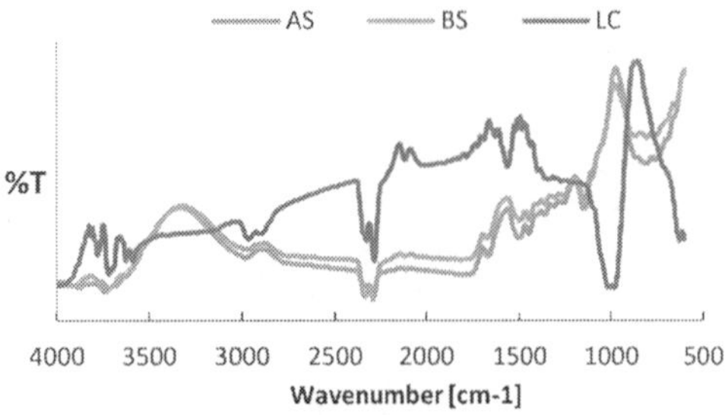

Fig. 2. Fourier transform infrared (FTIR) spectroscopy analyses of the biosorbents: *Annona squamosa* (AS), *Bambusa vulgaris* (BS), and *Litchi chinensis* (LC).

The FTIR spectroscopy analyses indicate that *Annona squamosa* has C–CH_3 alkane stretching with wavelength of 2318.02 cm^{-1}, C=C aromatic with wavelength 1535.06 cm^{-1}, C–H with wavelength 1186.01 cm^{-1}, R_2C=CHR with wavelength 863.95 cm^{-1} while *Bambusa vulgaris* has the following results: 3781.72 cm^{-1} indicating presence of –OH stretching, 3596.59 cm^{-1} for –OH, 2980.45 cm^{-1} for C–H stretching, 2356.59- 2307.41 cm^{-1} for C–H, 1600.00 cm^{-1} for –COO, and 1022.09 cm^{-1} for carboxyl and hydroxyl C–OH. Common functional groups were observed present in both *Annona squamosal* and *Bambusa vulgaris* samples. *Litchi chinensis* shows results having 3726.76 cm^{-1} for –OH, 2900 cm^{-1} for C–H, 1536.99 cm^{-1} for –COO, 1185.04 cm^{-1} for carboxyl and hydroxyl C–OH, and 853.34 cm^{-1} for C–H.

3.3 Equilibrium Studies

The preparation method of drying ensured that the cellulose, Figure 3 shows comparison of the equilibrium studies of Pb^{2+} with the three sorbents and Cu^{2+} with the three sorbents. At the highest initial sorbate concentration of Pb^{2+} 20 mg L^{-1}, the adsorption capacities of the three sorbents can be arranged as *Litchi chinensis* (4.30 mg of sorbate per g of sorbent (mg g^{-1}), 85.68% adsorption) > *Bambusa vulgaris* (3.83 mg g^{-1}, 76.19% adsorption) > *Annona squamosal* (2.70 mg g^{-1}, 53.66% adsorption). As the initial sorbate concentrations were increased, the amount adsorbed also increased. With regards to Cu^{2+} 20 mg L^{-1}, the adsorption capacities of the three sorbents can be arranged as *Bambusa vulgaris* (3.86 mg g^{-1}, 77.17% adsorption) > *Annona squamosa* (3.58 mg g^{-1}, 71.58% adsorption) > *Litchi chinensis* (3.42 mg g^{-1}, 68.32% adsorption). This can be attributed to the available binding sites on the surface of *Litchi chinensis* and *Bambusa vulgaris* due to the presence of counterions as shown in the elemental analysis and selected properties of the metals which are summarized in Table 1.

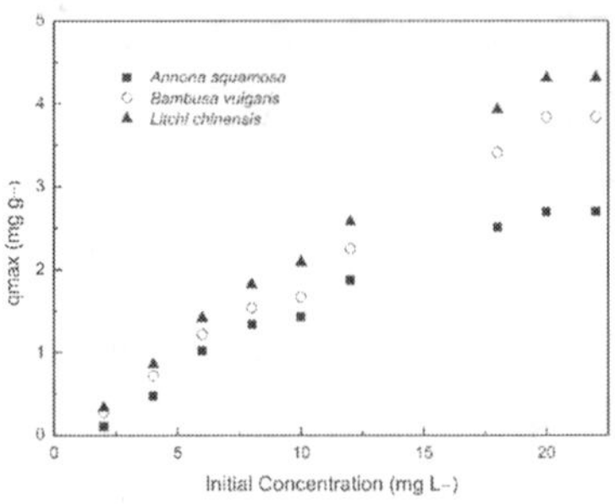

(a)

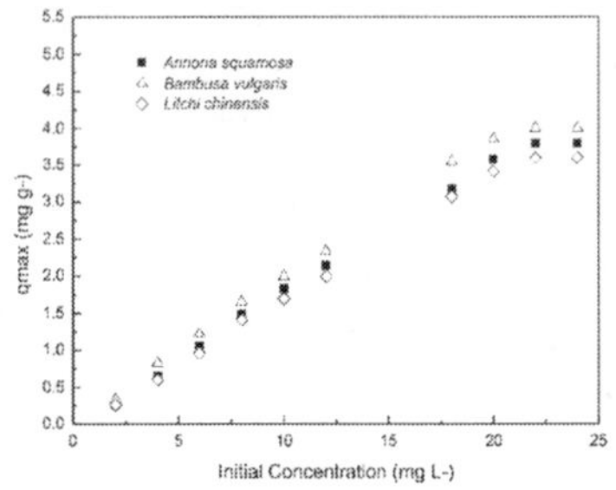

(b)

Fig. 3. Results of batch adsorption equilibrium studies of Pb^{2+} (a) and Cu^{2+} (b) onto the three biosorbents.

Table 1. Selected Properties of Metals in the Sorbates and the Cations Present in Each Biosorbent

Metal/Cation	Ionic Radius (Å)	Hydrated Ionic Radius (Å)	Pauling's Electronegativity
Pb^{2+}	1.20	2.61	2.33
Cu^{2+}	0.72	2.95	2.00
K^{+}	1.49	3.31	0.82
Zr^{2+}	0.80	5.50	1.33
Mg^{2+}	0.73	6.00	1.31

Based from the elemental analysis of the biosorbents, all biosorbents contain K^{+} which has higher ionic radius than that of Pb^{2+} which may allow possible cationc exchange [7]. *Annona squamosa* has the lowest percentage composition of K^{+} (2.54%). Although *Bambusa vulgaris* has the highest percentage composition of K^{+} (6.81%), it also has 6.34% of Zr^{2+} that may serve as a competing ion during the reaction. Mg^{2+} in *Litchi chinensis* is only 0.53%, thus it is negligible to cause this competition. According to Arshadi, the smaller the ion's hydration than that of the counterion, the greater the adsorption is and the easier the ion exchange occurs [7]. Pb^{2+} and Cu^{2+} have relatively smaller ionic radius and smaller hydrated ionic radius as compared to all the counterions present on the surface of all the three biosorbents allowing possible cationic exchange. The difference in adsorptive capacities can be explained by the amount of counterion present on the

surface of the biosorbents. Although *Litchi chinensis* has higher percentage composition of counterions on its surface as compared to that of *Annona squamosa*, the presence of another counterion, playing as a possible competing site, decreased its adsorptive capacity.
Since the electronegativity of both Pb^{2+} and Cu^{2+} is higher than the counterions on the biosorbent surfaces thus Pb^{2+} was adsorbed readily by the biosorbents [8]. Cationic Exchange Capacity (CEC) in meq per 100g of the biosorbents also explains the probability of having an ion exchange of the sorbate ion with the counterions on the surface of the biosorbents. *Litchi chinensis* has a CEC of 38.3 meq per 100g, *Bambusa vulgaris* has a CEC of 16.9 meq per 100g and *Annona squamosa* has a CEC of 22.6 meq per 100g.

3.4 Isotherm Study

The batch equilibrium adsorption data show that for Pb^{2+} adsorbed onto the three different biosorbents, the best fit observed was in the linearized form of Langmuir isotherm.. The isotherm fitting results are summarized in Table 2. The q_{max} refers also to the amount of sorbate adsorbed per unit mass of the adsorbent to complete monolayer coverage and b_L represents the energy of adsorption.

Table 2. Isotherm Fitting for Sorbate Adsorbed onto the three biosorbents.

Langmuir Fitting for Pb^{2+} Adsorbed onto the three biosorbents.					
Biosorbent	Slope	Intercept	RSS	q_{max} mg g^{-1}	b_L L mg^{-1}
Annona squamosa	1.563	0.703	0.999	2.695	2.222
Bambusa vulgaris	1.207	0.193	0.964	3.827	6.258
Litchi chinensis	1.612	-0.178	0.970	4.304	-9.064
Langmuir Fitting for Cu^{2+} Adsorbed onto the three biosorbents.					
Biosorbent	Slope	Intercept	RSS	q_{max} mg g^{-1}	b_L L mg^{-1}
Annona squamosa	23.397	0.440	0.981	2.272	0.019
Bambusa vulgaris	11.789	0.510	0.941	1.668	0.051
Litchi chinensis	34.910	0.273	0.903	3.664	0.008

Annona squamosa and *Bambusa vulgaris* have relatively higher adsorptive capacities with Cu^{2+} as compared to that of Pb^{2+}. The ionic radius of Cu^{2+} is 0.72 Å while that of Pb^{2+} is 1.20Å as compared to 1.49 Å of K^+. Theoretically, the smaller the ionic radius, the greater is the adsorption. However, the valence electron also plays a role in the adsorption process. The greater the valence electron, the greater is the adsorption [7]. The valence electron of copper is at $3d^9$ while that of lead is at $6p^2$. Although the valence electron of copper is greater than that of lead, its valence electrons are not as accessible as those of Pb^{2+}. This can explain the better binding capacity of Pb^{2+} over Cu^{2+} onto the surface sites of *Annona squamosa* and *Bambusa vulgaris*. However, if the adsorption of Pb^{2+} and Cu^{2+} onto *Litchi chinensis*, Pb^{2+} has higher percentage of adsorption since Pb^{2+} has a shorter hydrated radius (2.61 Å) compared to that of Cu^{2+} (2.95 Å). When the ion's hydration is smaller than the counterion, ion exchange is easier and adsorption is greater [7]. This allows Pb^{2+} easier exchange with counterions present onto *Litchi chinensis* which have relatively higher hydrated radii than that of Pb^{2+}. Hydrated radius is the size of the water molecule in comparison to normally stabilizing cations that causes the physical expansion or swelling of the biosorbent which further causes reduction in permeability. In certain situations, the expansion of biosorbent can also cause other embedded materials to be discharged from the pore walls that may allow additional bridging or blocking [8]

4 Conclusion

At 20 mg L^{-1} of Pb^{2+}, the adsorption capacities of the three sorbents can be arranged as *Litchi chinensis* (4.30 mg of sorbate per g of sorbent (mg g^{-1}), 85.68% adsorption) > *Bambusa vulgaris* (3.83 mg g^{-1}, 76.19% adsorption) > *Annona squamosa* (2.70 mg g^{-1}, 53.66% adsorption). As the initial sorbate concentrations were increased, the amount adsorbed also increased. With regards to 20 mg L^{-1} of Cu^{2+}, the adsorption capacities of the three sorbents can be arranged as *Bambusa vulgaris* (3.86 mg g^{-1}, 77.17% adsorption) > *Annona squamosal* (3.58 mg g^{-1}, 71.58% adsorption) > *Litchi chinensis* (3.42 mg g^{-1}, 68.32% adsorption). The biosorbents had relatively higher adsorptive capacities with Cu^{2+} as compared to that of Pb^{2+} except for *Litchi chinensis*. Although the results show lower adsorptive capacity as compared to a number of treated agricultural wastes showing 80% up to almost 100% adsorption of Pb^{2+} and Cu^{2+}, the results show that *Annona squamosa*, *Bamubusa vulgaris*, and *Litchi chinensis* are potential biosorbents and promote sustainable treatment process.

References

1. J.C.P. Vaghetti, E. C. Lima, B. Royer, B. M. da Cunha, N. F. Cardoso, J. L. Brasil, S. L. P. Dias, Journal of Hazardous Materials, 162, pp. 270–280, (2009).
2. A. E. Ofomaja, E. B. Naidoo, S. J. Modise, Journal of Environmental Management, 91 (8), pp. 1674-1685, (2010).
3. L.V.A. Gurgel, P. de Freitas, G. Rossimiriam, F. Laurent, Carbohydrate Polymers, 74, pp. 922–929, (2008).
4. P. L. Homagai, K. N. Ghimire, K. Inoue, Bioresource Technology, 101, pp2067–2069, (2010).
5. M. R. Lasheen, N. S. Ammar, H. S. Ibrahim, Solid State Sciences, 14, pp. 202-210, (2012).

DOI: 10.1051/epjconf/201716201082

6. D. Choudhurya, J. K. Sahub, G. D. Sharmaa, Industrial Crops and Products, 33, pp. 211–216, (2011).
7. M. Arshadi, M. J. Amiri, S. Mousavi, Water Resources and Industry, 6, pp. 1-17, (2014).
8. K. Jiang, T. Sun, L. Sun, H. Li, Journal of Environmental Sciences, 18 (6), pp. 1221-1225, (2006).

DOI: 10.1051/epjconf/201716201083

Determination of diffusion coefficients of various livestock antibiotics in water at infinite dilution

Allan N. Soriano [1,*], *Kristoni* G. Adamos[1], *Pauline* B. Bonifacio[1] *Adonis* P. Adornado[1], *Vergel* C. Bungay [1] and *Rajendaran* Vairavan[2]

[1]School of Chemical Engineering and Chemistry, Mapúa University 658 Muralla St., Intramuros, Manila 1002, Philippines
[2]School of Microelectronic Engineering, University Malaysia Perlis (UniMAP), Kampus Alam, Pauh Putra, 02600 Arau, Perlis, Malaysia.

Abstract. The fate of antibiotics entering the environment raised concerns on the possible effect of antimicrobial resistance bacteria. Prediction of the fate and transport of these particles are needed to be determined, significantly the diffusion coefficient of antibiotic in water at infinite dilution. A systematic determination of diffusion coefficient of antibiotic in water at infinite dilution of five different kinds of livestock antibiotics namely: Amtyl, Ciprotyl, Doxylak Forte, Trisullak, and Vetracin Gold in the 293.15 to 313.15 K temperature range are reported through the use of the method involving the electrolytic conductivity measurements. A continuous stirred tank reactor is utilized to measure the electrolytic conductivities of the considered systems. These conductivities are correlated by using the Nernst-Haskell equation to determine the infinite dilution diffusion coefficient. Determined diffusion coefficients are based on the assumption that in dilute solution, these antibiotics behave as strong electrolyte from which H^+ cation dissociate from the antibiotic's anion.

1 Introduction

The fate of antibiotics entering the environment has many concerns regarding on its possible effects on the disturbance of the natural bacterial ecosystem. The presence of antibiotics in municipal sewage and aquatic environment are able to spread to the microbial communities. This will then result to the development of antibiotic-resistant bacteria, which can lead to a serious threat to human health. Not only does antibiotic resistance threaten the effective prevention and treatment of bacterial infectious disease but it could also make the medical procedures very dangerous to perform. However, limited information, which gives significant data for the detection and remediation of antibiotic in aquatic environment, is available to evaluate the transport properties of antibiotics. Therefore, it is important to know the fate and transport of antibiotics released in the environment to be able to properly evaluate the risks, the effects and the potential impacts of these products [1]. Fate and transport of particles is an important matter, which is concerned with the ground water, wastewater, atmosphere and soil transport of different hazardous contaminants and Pharmaceutically Active Compounds (PHAC) that describe their physical form and behavior in different media. Its primary concern is to determine the transport speed and combined effects of the chemicals in their discharge environments [2]. In order to predict the fate and transport in the environment, physicochemical properties of different substances are taken into account. There are several parameters considered such as molecular weight, aqueous solubility, chemical concentration, diffusion coefficient, Henry's law constant, boiling point and vapor pressure [3]. The study mainly focuses on the determination of diffusion coefficient of antibiotics, which can be helpful for future studies in detecting and removing antibiotics in wastewater. Diffusion coefficient indicates how individual molecules and ions move through a medium under a concentration gradient at a certain pressure and temperature. It often controls the source of a contaminant (antibiotic) released to the environment and flow usually becomes its main transport process. The diffusion coefficient can also be utilized in many remediation processes such as adsorption or chemical filtering as well as a rate-determining step for extraction and distillation [4]. There are several methods in determining the diffusion coefficient of antibiotics in water such as diaphragm cell, interferometry, Taylor's dispersion analysis (TDA), electric conductivity, group contribution, and liquid diffusion coefficient apparatus. The most convenient way is by using electric conductivity, because it is the simplest, less expensive, rapid and clean when it comes to detection of emerging environmental pollutants. Electrolytic conductivity measures the transport of ionic substances with equivalent conductance over a homogenous solution between two electrodes [5].
In the present work, the diffusion coefficient in water at infinite dilution of the common livestock antibiotics

* Corresponding author: ansoriano@mapua.edu.ph

 DOI: 10.1051/epjconf/201716201083

which is rampantly used in the Philippines was indirectly determined using electrolytic conductivity data. Nernst-Haskell equation was then used to estimate the diffusion coefficient in water at infinite dilution of the considered system.

2 Methodology

The conceptual framework shown in Figure 1 shows the methodology in determining the diffusion coefficient of antibiotics in water through measurements of electrolytic conductivity using the Continuous Stirred Tank Reactor (CSTR) and correlation using the Nernst-Haskell equation. In the succeeding section, only the brief discussions of the conductivity measurements and diffusion coefficient determination was presented.

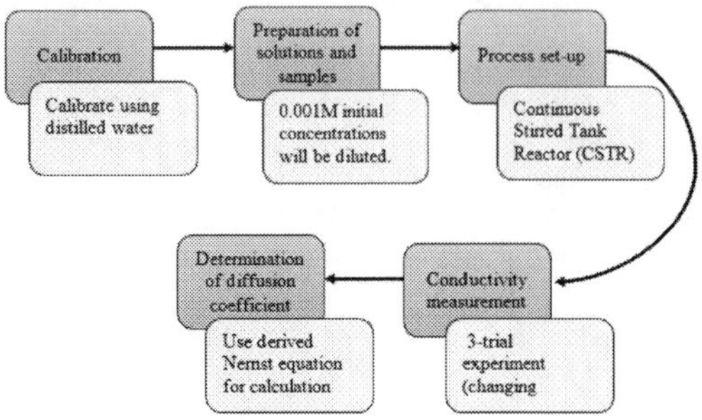

Fig. 1. Conceptual framework of the methodology applied in this work.

2.1 Electrolytic Conductivity Measurements

The over-all process flow diagram using the Armfield CEX Chemical Reactors Teaching Equipment, which summarizes the experimental procedure considered was presented in Figure 2. In this study only the tank reactor and hot water circulator was utilized. A one liter of distilled water was poured into the tank reactor and a small volume of aqueous solution of antibiotic, from the prepared mother solution, was introduced into the system to step up the concentration gradient. The conductivity data was collected as the temperature of the reactor increases from 293.15 to 313.15 K that was controlled by the hot water circulator. The recorded data was sent through the computer and the data logger where the data output port has to be connected, or simply through the direct reading from the conductivity meter in the console. Measurements were done using the desired temperatures and concentrations in three independent replicate runs for each of the five antibiotics namely: Amtyl, Ciprotyl, Doxylak Forte, Trisullak, and Vetracin Gold. The calculated volume of a known concentrated aqueous antibiotic solution was poured into the reactor to increase the concentration from 0.0002 M to 0.0014 M. The heater installed in the equipment was operated using a demineralized water and must reach the water level of 30 mm from the top of the vessel. The temperature adjustment was done and controlled through the PID controller software installed in the computer.

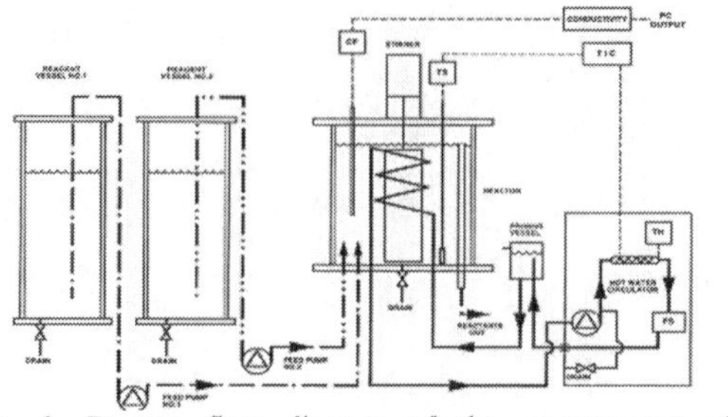

Fig. 2. Process flow diagram of the measurement of electrolytic conductivity

2.2 Diffusion Coefficient Determination

Using the Nernst-Haskell equation, as in Eqn. (1), the diffusion coefficient of the considered systems in water at infinite dilution was evaluated [7].

$$D_{AB}^{0} = \frac{RT}{F^2}\frac{|z_+| + |z_-|}{|z_+ z_-|}\frac{\Lambda_+^0 \Lambda_-^0}{\Lambda_+^0 + \Lambda_-^0} \quad (1)$$

where D_{AB}^{0} is denoted as the diffusion coefficient of the solute, which is the antibiotic (A) in water (B) at infinite dilution in m^2/s, R is the universal gas constant in J/mol-K, T is the absolute temperature in K and F is the Faraday's constant. The z_+ and z_- designate the charge number of the cation and anion, respectively, and Λ_+^0 and Λ_-^0 are the molar conductivities of the cation and anion in Sm^2/mol, respectively at infinite dilution.

The molar conductivity was determined using the measured electrolytic conductivity, κK, in S/m, which depends primarily on the type of electrolyte and the molar concentration of the solution in $kmol/m^3$. Equation (2) shows the given formula for the evaluation of molar conductivity, Λ, in $S \cdot m^2/mol$ [8]

$$\Lambda = \left(\frac{\kappa}{C \cdot 1000}\right) \quad (2)$$

The molar conductivities of the cation and anion at infinite dilution are determined using the Kohlrausch's law of the Independent Migration of Ions which states that the molar conductivity of an electrolyte at infinite dilution can be expressed in terms of the sum of contribution from the cation and anion of antibiotic in dilute solution as presented in Eqn. (3) [8]

$$\Lambda^0 = z_+\Lambda_+^0 + z_-\Lambda_-^0 \quad (3)$$

The molar conductivity at infinite dilution, Λ^0, is determined using the modified form of the Robinson and Stokes equation and is presented as Eqn. (4) [9].

$$\Lambda = \Lambda^0 + a_1 C^{1/2} + a_2 C \quad (4)$$

3 Results and Discussion

3.1 Electrolytic Conductivity Data

The measured electrolytic conductivity data for different concentrations (0.0002 to 0.0014 M) of the considered antibiotics at various temperature (293.15 to 313.15 K) are presented in Table 1.

Table 1. Measured electrolytic conductivities, κ, of antibiotics at different concentrations and temperatures.

Amtyl (amoxicillin + tylosin): $[H^+][AMT^-]$*					
C / (kmol·m^{-3})	0.0002	0.0005	0.0008	0.0011	0.0014
T/K	κ/ mS·cm^{-1}				
293.15	0.193319	0.246322	0.294432	0.339281	0.385761
298.15	0.215135	0.273031	0.32848	0.379853	0.424701
303.15	0.233790	0.300656	0.357736	0.417263	0.475159
308.15	0.254891	0.326649	0.394330	0.457934	0.512568
313.15	0.275078	0.354175	0.429194	0.497691	0.567003
Ciprotyl (ciprofloxacin): $[H^+][CIP^-]$					
C / (kmol·m^{-3})	0.0002	0.0005	0.0008	0.0011	0.0014
T/K	κ/ mS·cm^{-1}				
293.15	0.180272	0.221859	0.258553	0.30014	0.336019
298.15	0.200457	0.244491	0.284447	0.335819	0.376591
303.15	0.217482	0.268854	0.316149	0.366706	0.411555
308.15	0.236952	0.289955	0.346220	0.397592	0.452226
313.15	0.257953	0.313403	0.377007	0.434087	0.494429
Doxylak Forte (doxycyline tiamulin hydrogen fumarate): $[H^+][DOX^-]$					
C / (kmol·m^{-3})	0.0002	0.0005	0.0008	0.0011	0.0014
T/K	κ/ mS·cm^{-1}				
293.15	0.215335	0.287909	0.345805	0.399623	0.446918
298.15	0.237152	0.316249	0.379853	0.438564	0.494013
303.15	0.257438	0.340612	0.411555	0.476789	0.530608
308.15	0.276908	0.369052	0.446518	0.512568	0.579434
313.15	0.298725	0.394946	0.479751	0.555586	0.619190
Trisullak (contramazine): $[H^+][TRI^-]$					
C / (kmol·m^{-3})	0.0002	0.0005	0.0008	0.0011	0.0014
T/K	κ/ mS·cm^{-1}				
293.15	0.177010	0.211695	0.247788	0.283811	0.315634
298.15	0.196380	0.236844	0.273754	0.315484	0.354574
303.15	0.213405	0.257130	0.300563	0.344811	0.387907
308.15	0.232875	0.281565	0.329890	0.375769	0.424501
313.15	0.251430	0.304197	0.359118	0.410776	0.462627
Vetracin Gold (doxycycline tiamulin vitamin A+B12): $[H^+][VET^-]$					
C / (kmol·m^{-3})	0.0002	0.0005	0.0008	0.0011	0.0014
T/K	κ/ mS·cm^{-1}				
293.15	0.204735	0.272416	0.359667	0.450996	0.535801
298.15	0.225736	0.29994	0.396977	0.493198	0.603281
303.15	0.244391	0.32675	0.436018	0.530608	0.655369
308.15	0.264677	0.352743	0.467719	0.581880	0.704195
313.15	0.285678	0.380268	0.508291	0.622452	0.760260

**Assumed ions comprising the antibiotics.*

The conductivity of all solution changes as the temperature changes. Based from the graph, the relationship between the two parameters is as the temperature increases the electrolytic conductivity increases as well. Therefore, this trend line indicates a linear relationship between the two parameters which can be seen in all systems of antibiotics tested in this study. This relationship can be associated with the interionic attractions of the given solution. An increase in temperature can cause the number of ions to increase because of dissociation of molecules which is a factor that affects the electrolytic conductivity of a solution [6]. Therefore, as the conductivity is affected by this factors, then an increase in a system's temperature will lead to an increase in the electrolytic conductivity. Figure 3 shows the calculated values of molar conductivity of the studied systems as function of concentration and temperature. The behavior of the molar conductivity as a function of concentration can be observed to be decreasing exponentially at lower concentration and seems to becoming linear as the concentration increases. As the concentration of ions increases, the interionic attractions also increases. These interactions affect the conductivity of an electrolytic solutions because these interionic attractions can reduce the capability of ions in an electrolytic solution to conduct the electric current between the electrodes [6]. Higher concentrations will hinder the motion of ions which results to a decrease in the conductance. At lower concentration or at more dilute solutions, the ions are far apart which decreases the interionic attractions and increase the dissociation of ions and at the same time increases the molar conductivity. Ions become unaffected by each other's motion which causes them to migrate independently as was proven by Kohlrausch's principle.

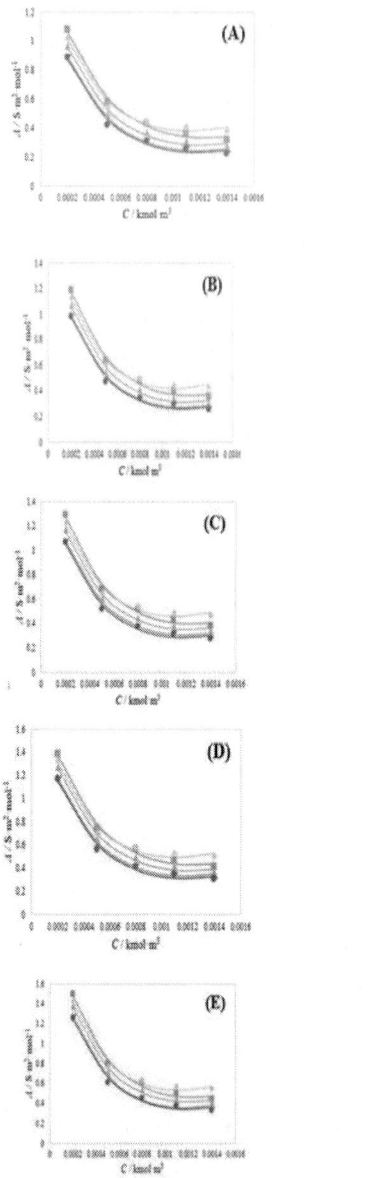

Fig. 3. Molar conductivity of the studied systems as function of concentration and temperature (A. T = 293.15 K; B. T = 298.15 K; C. T = 303.15 K; D. T = 308.15 K; E. T = 313.15 K): (●, Trisullak; -, Vetracin Gold;▲, Amtyl; ◆, Doxylak Forte; ■, Ciprotyl), and lines are calculated values using Eqn. (4).

3.2 Determined Diffusion Coefficient at Infinite Dilution

There are few studies which focus on the diffusion of antibiotics in water. In this study, the conductivities of antibiotic solutions were determined at an infinite dilution state. It was assumed and estimated that the antibiotics have strong electrolyte ions in which H^+ cations dissociate from the CIP^{1-}, TRI^{1-}, AMT^{1-}, VET^{1-}, and DOX^{1-} anions. The diffusion coefficient at infinite dilution of the considered systems was estimated using the Nernst-Haskell equation given in Eqn. (1). The Nernst-Haskell equation is used for diffusion of electrolytes which dissociate into ions of unequal valence similar to the system. The infinite dilution diffusion coefficient calculated using the Nernst-Haskell equation is given in Table 2 and is representatively shown in the plot against the absolute temperatures as shown in Figure 4, where it was found that an increase in temperature causes an increase in diffusion coefficient due to the increasing mobility of ions.

Table 2. Determined diffusion coefficient at infinite dilution, D^0_{AB}, of the studied antibiotics in water.

	$10^7\ D^o_{AB}$ / m^2·s (Nernst-Haskell eqn.)				
T/K	Amtyl	Ciprotyl	Doxylak Forte	Trisullak	Vetracin Gold
293.15	0.084332	0.084269	0.084422	0.084261	0.084373
298.15	0.091660	0.091598	0.091751	0.091585	0.091699
303.15	0.099377	0.099316	0.099469	0.099299	0.099414
308.15	0.107490	0.107431	0.107582	0.107408	0.107525
313.15	0.116004	0.115947	0.116095	0.115918	0.116037

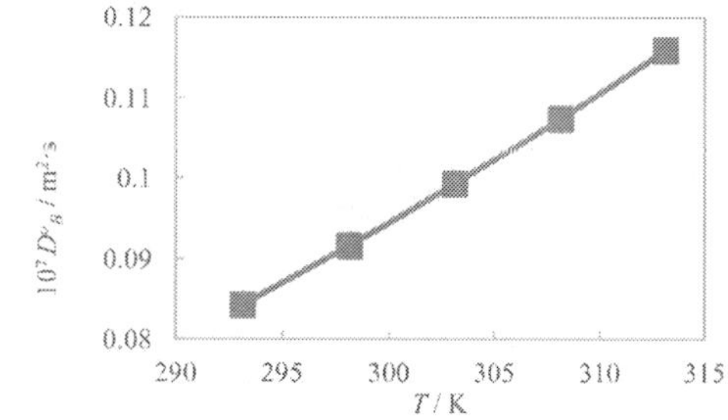

Fig. 4. Infinite dilution diffusion coefficient of Trisullak in water as function of temperature

4 Conclusion

This study mainly focuses on the estimation of diffusion coefficient of antibiotics in water at infinite dilution through the use of electrolytic conductivity method. The measurable data obtained from the experiment is correlated using the models and correlations available. The Nernst-Haskell equation was utilized for the determination of infinite dilution diffusion coefficient. The study was conducted at different temperatures to further investigate the different conditions of antibiotic when present in the aquatic environment

References

1. P. Segura, A. Garcia-Ac, A. Lajeunesse, D. Ghosh, C. Gagnon, and S. Sébastien, J. Environ. Monitoring, vol. 9, pp. 307-313, (2007).
2. J. G. Sczechowski, ENVE 436: Fate and Transport. Retrieved 1, February 2016, Available:

http://ceenve3.civeng.calpoly.edu/cota/enve436/fate.html, (1997).

3. Bester-Rogac, M., Boncina, M., Apelblat, Y., and Apelblat, A., Journal of Physical Chemistry, **111**. 11957-11967, (2007). .
4. Soriano, A.N., Entrolizo, R.V., Era, C.M.A., San Diego, J.S.H., Li, M.-H., PIChE Journal, **15**, 54-63, (2014).
5. L. van der Wielen, M. Zomerdijk, J. Houwers, K. Luyben, Chem. Eng. J., vol. 66, pp. 111-121, (1997).
6. Apelblat, A. and Bester-Rogac, M., Journal of Molecular Liquids, 421, (2015).
7. F. Fiegenbaum, G. Peres, M. O. de Souza, E. M. A. Martini, and R. F. de Souza, J. Mol. Liq., vol. 215, pp. 302–307, (2016).
8. S. Thawarkar, N. D. Khupse, and A. Kumar, Phys. Chem. Chem. Phys., vol. 17, no. 1, pp. 475–482, (2015).
9. J. T. Edward, J. chem. Educ, vol. 47, no. 4, p. 261, (1970).

Effects of air temperature and velocity on the drying kinetics and product particle size of starch from arrowroot (*Maranta arundinacae*)

Alvin R. Caparanga [1,*], *Rachael Anne L.* Reyes[2], *Reiner L.* Rivas[1] , *Flordeliza C.* De Vera [1], *Vithyacharan* Retnasamy[2] *and Hasnizah* Aris[2]

[1]School of Chemical Engineering and Chemistry , Mapúa University 658 Muralla St., Intramuros, Manila 1002, Philippines
[2]School of Microelectronic Engineering, University Malaysia Perlis (UniMAP), Kampus Alam, Pauh Putra, 02600 Arau, Perlis, Malaysia.

Abstract. This study utilized the 3^k factorial design with k as the two varying factors namely, temperature and air velocity. The effects of temperature and air velocity on the drying rate curves and on the average particle diameter of the arrowroot starch were investigated. Extracted arrowroot starch samples were dried based on the designed parameters until constant weight was obtained. The resulting initial moisture content of the arrowroot starch was 49.4%. Higher temperatures correspond to higher drying rates and faster drying time while air velocity effects were approximately negligible or had little effect. Drying rate is a function of temperature and time. The constant rate period was not observed for the drying rate of arrowroot starch. The drying curves were fitted against five mathematical models: Lewis, Page, Henderson and Pabis, Logarithmic and Midili. The Midili Model was the best fit for the experimental data since it yielded the highest R^2 and the lowest RSME values for all runs. Scanning electron microscopy (SEM) was used for qualitative analysis and for determination of average particle diameter of the starch granules. The starch granules average particle diameter had a range of 12.06 – 24.60 µm. The use of ANOVA proved that particle diameters for each run varied significantly with each other. And, the Taguchi Design proved that high temperatures yield lower average particle diameter, while high air velocities yield higher average particle diameter.

1 Introduction

Drying is the oldest preservation technique and one of the most common methods used in removing moisture content. It is used in various industries since it is cost – effective and the most practical means of preservation [7]. The drying process employed affects the quality of a dried product. Drying methods used for root crops and starches include solar drying, drum drying, spray drying, freeze drying, microwave drying and hot air drying [5].

Maranta arundinacae L. (Arrowroot) is a root crop coming from the genus *maranta* which is a large, perennial herb found in rainforest habitats. The crop is cultivated for starch, commonly called as arrowroot, obtained from the rhizomes or rootstock of the plant. Arrowroot starch is widely considered as a nutritious starch and can be easily digested [3]. Some of its characteristics include, being not weakened by acidic ingredients, has a more neutral taste, and it is not affected by freezing [2]. When full grown, its composition in 100 parts include 26% starch, 6% woody fiber, 1.5% albumin, 1% gummy extract, volatile oil and salts, and 65.5% water [8]. Extensive extraction and drying is needed to extract pure arrowroot starch and remove most of its water content.

Different studies regarding the drying kinetics of other root crops and other starches have been made. However limited studies about kinetic drying of arrowroot have been reported. Therefore, the study aims to determine the effects of temperature and varying air condition rates on the drying characteristics and particle diameter of arrowroot using a tray dryer. This research will fit a suitable model that represents its drying behaviour and evaluate its initial moisture content and drying rate curves.

Conducting this study will help improve the understanding of the drying phenomenon involve in the drying of arrowroot as well as contribute to existing studies regarding the root crop. Determination of the drying characteristics of arrowroot will aid in predicting and improving its shelf life, which is mostly important for novelty foods that are easily spoiled. The results of this study can be used by other researchers in designing a better drying process of arrowroot through the optimization of present techniques and technologies available in the industry.

The drying characteristics of *Maranta arundinacae L.* is the only variety of arrowroot that will be used in this study. Drying procedures will be conducted using the 3^k factorial design with k=2 for the two factors to be

* Corresponding author: arcaparanga@mapua.edu.ph

 DOI: 10.1051/epjconf/201716201084

varied namely, temperatures at 30, 48 & 65°C, and air velocity at 0.4, 0.5 & 0.6 m/s at constant relative humidity. The starch property to be fitted in the models proposed will be the moisture ratio plotted against the drying time. The Lewis, Page, Henderson and Pabis, Midilli and Logarithmic Models will be the only models used for curve fitting on the drying characteristics of the starch. The flavour and nutritional aspects of the arrowroot will not be addressed in this study. Also, it should be noted that the economic aspect will not be considered as well.

2 Methodology

2.1 Preparation of Arrowroot

The arrowroot was first verified by the Botany Department of the National Museum of the Philippines so as to authenticate its species. The arrowroot sample was bought from the local municipality of Cuyo, Palawan and was stored at room temperature not exceeding 30 °C. The arrowroots were then washed with tap water and then cleaned of the paper-like scale. After trimming, the arrowroots were washed again and then drained.

2.2 Extraction of the liquid

The arrowroot was cut into small pieces and grounded using a blender. The milky liquid obtained was passed through a coarse cloth or hair sieve and the pure starch, which is insoluble, was allowed to settle at the bottom. Decantation was used in order to separate the liquid from the insoluble powder.

2.3 Initial moisture content determination

The initial moisture content, X_o, of the insoluble powder was determined by placing freshly decanted and dried samples of about 3.5g to 5g in a crucible inside an oven dryer at 105 °C then weighed every 40 minutes until constant. X_o was calculated using equation 1, where W_i is the initial weight of the sample and W_f is the final weight of the sample. Moisture analysis was done in triplicate.

$$X_o = \frac{W_i - W_f}{W_i} x100 \quad (1)$$

2.4 Drying equipment and drying procedure

The equipment used was a computer controlled tray dryer (UOP8MKII) located at the pilot plant of the school. Decanted starch extracted from samples weighing approximately 2kg± each, was spread evenly in a lightweight coated aluminum tray then placed in a tray dryer subjected to moisture analysis at constant relative humidity. The drying procedure followed the 3^k factorial design, with k being the two factors being varied namely, temperature and air velocity. Table 1 shows the factorial design used for the experiment, along with the temperature and air velocity values used. The drying progress was followed by weighing the samples, which was done electronically by the equipment since it has a built in integrated electronic weight measurement. The drying process was set to stop once the measured weight of the sample by the equipment becomes constant.

Run no.	Temperature (° C)	Air velocity (m/s)
1	30	0.4
2	30	0.5
3	30	0.6
4	48	0.4
5	48	0.5
6	48	0.6
7	65	0.4
8	65	0.5
9	65	0.6

Table 1. 3^k factorial design for arrowroot starch drying

2.5 Drying rate curves

The equilibrium moisture content, X*, at constant relative humidity was set as the constant weight generated by the dryer. The free moisture content at a certain time will be calculated using equation 2,

$$X = \frac{W - BDS}{BDS} \quad (2)$$

where X is the free moisture content at a certain time, W is the mass of the wet sample and BDS is the mass of the calculated bone dry sample using the initial moisture content, X_o. This free moisture content was then plotted against time to obtain the equation that was used to get the drying rate. The drying rate was obtained using equation 3.

$$R = -\left(\frac{W}{BDS}\right)\left(\frac{dX}{dt}\right) \quad (3)$$

where R is the drying rate in g H_2O/g BDS-min and dx/dt as the slope of the tangent line drawn on the X vs. t curve. Moisture content was also plotted against the drying rate to determine the constant rate of drying, R_c and the critical moisture content, X_c.

2.6 Drying Models

With the data obtained from the experiment and the determination of initial moisture content, the moisture ratio per minute of drying was determined then fitted using 5 models, namely: Lewis, Page, Henderson and

Pabis, Midili and Logarithmic. These mathematical models along with their corresponding equations are shown in Table 2. Regression analysis was performed using the drying models and the experimental data. The best fit model was chosen based on the highest value of R^2 and lowest RSME values.

Table 2. Thin layer drying models to be used in the experiment [11]

Model Name	Equation
Lewis	$M_R = \exp(-kt)$
Page	$M_R = \exp(-kt^n)$
Henderson and Pabis	$M_R = a \exp(-kt)$
Midili	$M_R = a \exp(-kt^n) + bt$
Logarithmic	$M_R = a \exp(-kt) + c$

2.7 Scanning Electron Microscope (SEM) and Statistical Analysis Data

Particle diameter of the dried starch granules were measured using a scanning electron microscope (JEOL JSM-5310, Japan). Micrographs were generated for each run along with measured particle diameters. Statistical analysis of data was applied to the obtained particle diameters to determine the significance of the varying conditions for each run. All values will be reported as mean ± standard deviation of 3 trials. One-way analysis of variance (ANOVA) using Minitab was conducted to determine if there are significant differences between means and design of experiment (DOE) using Taguchi Design was applied to determine the significance of the varying conditions to the particle diameter of the sample.

3 Results and Discussion

3.1 Initial Moisture Content

From Figure 1, the % moisture removed from the fresh sample becomes constant at 49.4%. The initial moisture content of the fresh sample can then be assumed to be equal to this value. In a previous study of arrowroot starch, it was found that the initial moisture content is equal to 40.4% [1]. The higher value obtained for the initial moisture content is due to the arrowroot sample used in this study was not fully grown which corresponds to a higher water content of the crop.

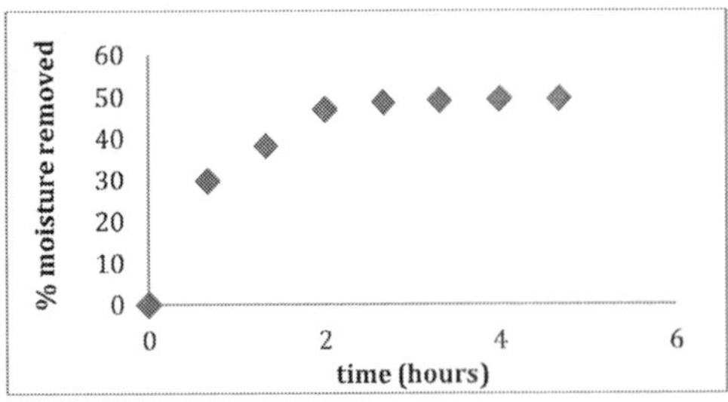

Fig. 1. Percent moisture removed vs. time

The final moisture content of the dried samples was also obtained to check whether they fall within 10%, since microbial damage, non – enzymatic browning and enzymatic reactions occur at moisture contents above 10% for dried food materials [10]. Table 3 shows the final moisture content obtained for each run of the dried samples.

Table 3. Final moisture content of dried arrowroot starch samples

Run no.	Final moisture content (%)
1	9.9167
2	9.2889
3	8.7016
4	9.9443
5	7.6416
6	9.1049
7	4.5600
8	4.1736
9	5.7636

3.2 Drying Rate Curves

Figures 2 and Figure 3 shows the plot of free moisture content versus time for runs 2 and 5 respectively. In reference with these figures, it is evident that, at a certain time, the free moisture suddenly rises then falls again. This trend can be seen in other runs as well and is due to the uneven distribution of the starch in the tray during drying [1]. Comparing Figures 2 and 3, it can be inferred that different drying temperatures operating at the same air velocity corresponds to different drying times. Runs having higher temperatures result in lower drying time compared to runs having lower temperatures. Thus, an increase in drying temperature increases the amount of moisture removed at a certain time.

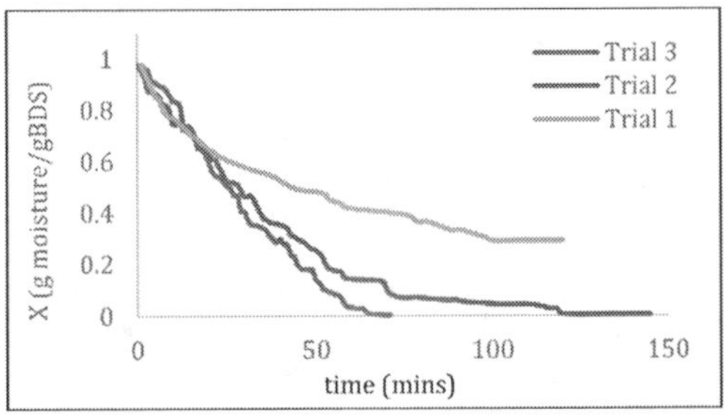

Fig. 2. Free moisture content, X vs. time for run 2

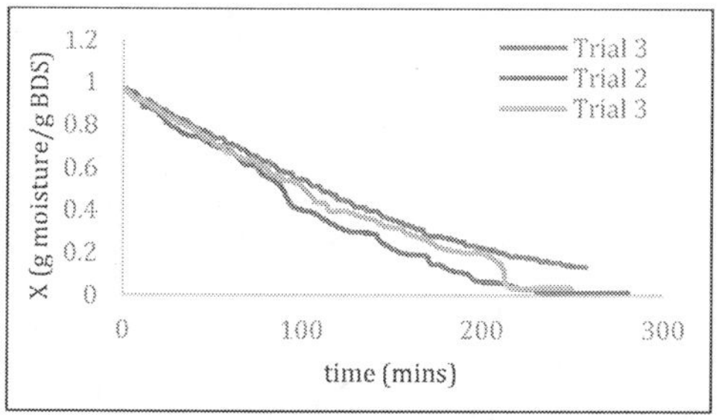

Fig. 3. Free moisture content, X vs. time for run 5

Shown in Figures 4 and 5 and are the drying rate curves plotted against moisture content of runs 2 and 5. From these figures, it is observed that the drying rate of the sample has no constant rate period. It only has a continuous falling rate period until it reaches its equilibrium moisture content. Starchy foods have no true constant drying rate period, however the concept of constant rate is still used as an approximation [4].

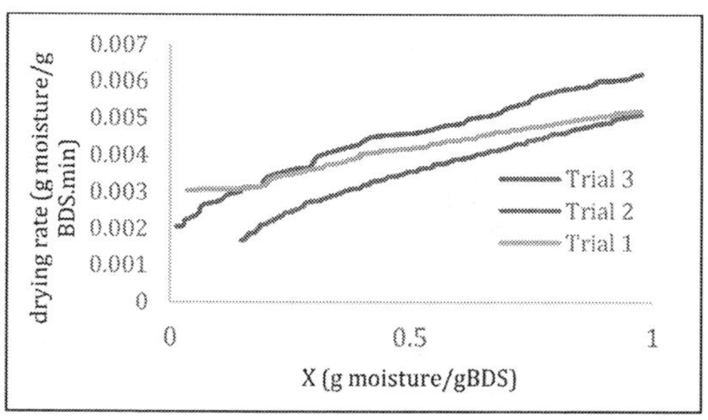

Fig. 4. Drying rate vs. moisture content (wet basis) of run 2

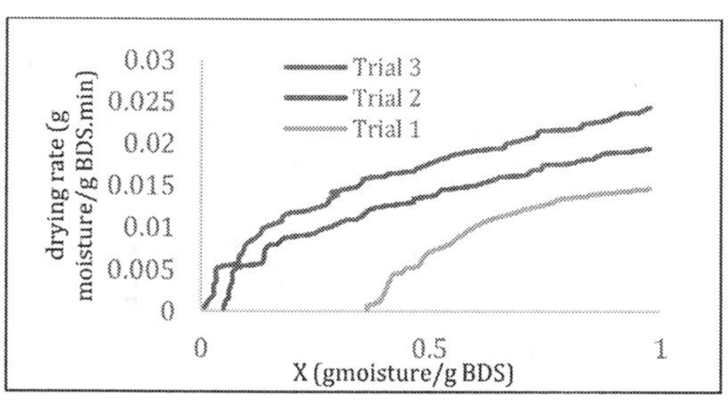

Fig. 5. Drying rate vs. moisture content (wet basis) of run 5

In Table 4, the drying rates per run with varying conditions, is shown. From here, it can be observed that as the temperature increases, the drying rate also increases It is also observed that air velocity has no or little effect in the value of the drying rate as the value is approximately constant with different values of air velocities. Since air velocity has little effects in rate of moisture removal, based from the moisture content vs. time curve, it would imply that its effects on the value of the drying rate would be negligible. This then implies that the drying rate is a strong function of temperature and time.

Table 4. Comparison of drying rates of different runs

Temperature (°C)	Air velocity (m/s)	Run no.	Drying rate (g moisture/g BDS.min)
30	0.4	1	0.0055 ± 0.0005
	0.5	2	0.0055 ± 0.0006
	0.6	3	0.0053 ± 0.0020
48	0.4	4	0.0179 ± 0.0011
	0.5	5	0.0195 ± 0.0048
	0.6	6	0.0145 ± 0.0010
65	0.4	7	0.0252 ± 0.0081
	0.5	8	0.0204 ± 0.0023
	0.6	9	0.0227 ± 0.0056

3.3 Mathematical Modelling of Arrowroot

The data on moisture ratio versus drying time obtained is plotted and fitted on the five mathematical models chosen. Figure 6 shows the plotted models for run 5. The best fit model is selected based from R^2 and RSME values. Based on the computed values, it was observed that the Midili Model gave the best fit on the experimental data since it yielded the highest R^2 value and lowest RSME value. This indicates that the temperature has the greatest effect in the drying rate since the coefficients of the Midili Model are dependent on the drying air temperature [6].

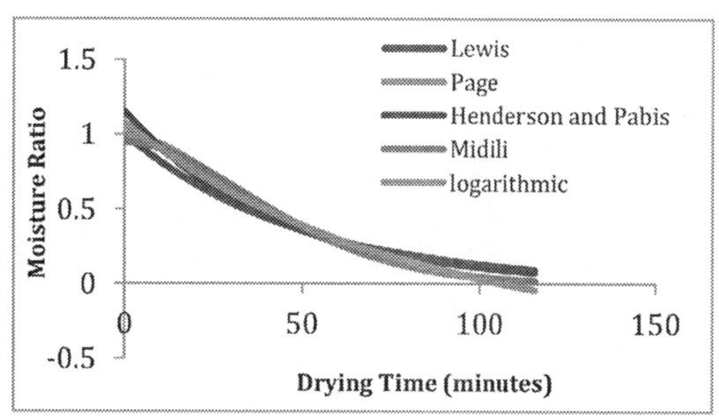

Fig. 6. Moisture Ratio versus Time for Run 5 with the temperature at 48°C.

3.4 Scanning Electron Microscopy and Statistical Analysis

Scanning electron microscopy was used to determine the range of the average particle diameter of the arrowroot granules along with the effect of temperature and air velocity on the particle diameter. Figures 7 show the micrographs from the scanning electron microscopy. From these figures, it can be seen that the arrowroot starch granules are oval in shape and show an off –

 DOI: 10.1051/epjconf/201716201084

center hilum similar to that of potato, canna and sago starches [9]. The average particle diameter for each run was determined by selecting random samples of granules from the micrograph and measuring the individual particle size using the scale of figures on the left side if need be. The obtained particle diameter for each run is within the general range of 10 – 50 µm for arrowroot starch granules [9]

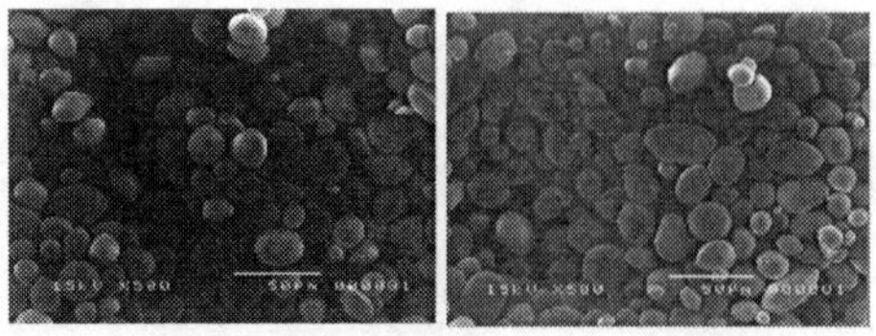

(a) (b)

Fig. 7. Micrograph for (a) Run 2 and (b) Run 5

Using one-way analysis of variance or ANOVA, the P-value that was determined was 0.027 which is lower than the α-value of 0.05. This implies that the average particle diameter from runs 1 to 9 varies significantly from each other because of the temperature and velocity settings. The Taguchi Design was used to determine the specific effect of temperature and air velocity on the particle diameter. Figure 12 shows the plot generated for Taguchi Design which shows that the particle diameter increases at lower temperature and decreases at higher temperature. In terms of velocity, higher velocity means an increase in average particle diameter and a lower velocity means a lower average particle diameter. It can also be inferred that the temperature has more effect on the average particle diameter than the air velocity.

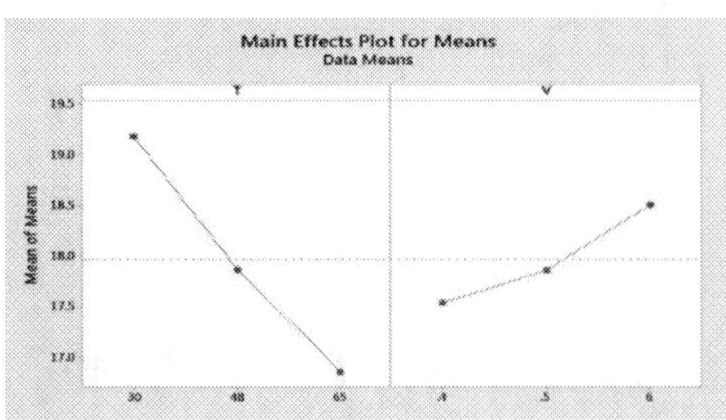

Fig. 8. Effect of Temperature and Velocity on particle diameter

4 Conclusion

Modelling and determination of the drying characteristics of the arrowroot starch (*Maranta arundinacae*) was determined and evaluated using a factorial design of 3^k. The initial moisture content was determined to have a value of 49.4% due to the arrowroot samples used which were not fully grown. At higher temperatures, the drying time for the samples were shorter compared to those operating at lower temperatures. Air velocity has little or negligible effect to the drying time and drying rate. The drying rate of arrowroot starch is therefore a function of temperature and time, and the drying curves of arrowroot starch has no constant rate period and only has steady falling rate period. The Midili Model was found to be the best fit model for drying of arrowroot starch since it yielded the highest R^2 values and lowest RSME values for all runs. Using scanning electron microscopy, the average particle diameters of arrowroot starch was determined to have a range of 12.06 – 24.60 µm and have an oval shape with an off – center hilum. Using one – way analysis of variance, it was found that the average particle diameters for each run, varies significantly with each other. Lastly, with the aid of the Taguchi Design, it was determined that higher temperatures correspond to lower average particle diameter, while higher velocity corresponds to higher average particle diameters.

References

1. Alcaraz, J.M., Capati, A.M. & Alipio, J.E.C.S, (2011).
2. Jyothi, A. N., Sheriff, J. T., & Sajeev, M. S., Journal of food science, 74(2), (2009).
3. Charles, A. L., Cato, K., Huang, Y., Ciou, J., Chang, J., & Lin, H. (2015), Food Hydrocolloids, 53, 187-191
4. Earle, R. L., Unit Operations in Food Processing, New Zealand Institute of Food Science and Technology, Inc., (1983).
5. Guine, R. P. F., Pinho, S. & Barroca, M. J., Food and Bioproducts Processing, 89, 422 – 428, (2011).
6. Idlimam, A., Kane, C. S. E. & Kouhila, M., Revue des Energies Renouvelables, 10(2), 191 – 203, (2007).
7. Rahman, S. N. F. S. A., Wahid, R., & Rahman, N. A., Procedia – Social and Behavorial Sciences, 195, 2731-2741, (2015).
8. Ukpabi, U. J., Ukenye, E., & Olojede, A. O., Journal of Food Technology, 7(4), 135-138, (2009).
9. Snyder, E. M., Industrial Microscopy of Starches, p. 661 in Starch: Chemistry and Technology, 2nd ed., Academic Press, New York, N.Y., (1984).
10. Wilhelm, L.R., Suter, D.A. & Brusewitz, G.H., American Society of Agricultural Engineers. 259 – 284, (2005).
11. Alibas, I., Journal of Biological and Environmental Sciences, 6(17), (2012).

Kinetics, mass transport characteristics, and structural changes during air-drying of purple yam *(Dioscorea Alata L.)* at different process conditions

Flordeliza C. De Vera [1,*], *Leif Anthony B.* Comaling [1], *Iya Ray Alyanna M.* Lao[1], *Alvin R.* Caparanga[1] and *Zaliman* Sauli[2]

[1]School of Chemical Engineering and Chemistry, Mapúa University 658 Muralla St., Intramuros, Manila 1002, Philippines
[2]School of Microelectronic Engineering, University Malaysia Perlis (UniMAP), Kampus Alam, Pauh Putra, 02600 Arau, Perlis, Malaysia.

Abstract. This experiment was designed to follow the 2^k factorial design to study the effects of the three drying parameters on the drying characteristics and effective moisture diffusivity and to fit each run performed on the best thin-layer drying kinetics model. Raw purple yam samples were pre-treated and undergone the designed drying procedures at which the weight of the samples were recorded every minute until such time that the sample weights become constant. Scanning Electron Microscopy (SEM) is utilized for qualitative analysis of the dried samples. The number of pores per unit area and the overall aesthetics of the surface of the dried samples were compared also using SEM. Considering the qualitative analysis conducted on the samples from the images of SEM, dried samples from run 2 has the most desirable conditions such as high temperature and low air velocity for drying because the samples from this run have large pore diameters with minimal cell breakages.

1 Introduction

Drying refers to the removal of relatively trace amounts of liquid (generally water) from a material. The use of heat to expel liquids recognizes drying from mechanical dewatering techniques like filtration, decantation or sedimentation, and centrifugation, in which no adjustment in stage from liquid to vapor occurs [5]. In the food industry, drying is being used as a preservation technique. Since microorganisms in food require water for growth and reproduction; and enzymes, which cause chemical alterations in food and biological materials, cannot function in the absence of water, removal of water results in reduced rate of spoilage and prevention of chemical changes in food. For microorganisms to be inactive, water content of the food must be reduced below 10 wt % while it is desirable to reduce the water content below 5 wt % for nutrient and flavor preservation [3]. Drying in food industry is usually the final processing step before packaging because drying promotes increased shelf-life of food and preservation of food characteristics. It is used to dry meats, garlics and onions, edible mushrooms, fruits, vegetables, and other starchy foods.

In this study, purple yam was dried to serve as a potential source of starch. Starch is commonly produced by isolating it from the food source through drying. Starch is the term applied to a white, powdery or granular, odorless and tasteless complex carbohydrate $(C_6H_{10}O_5)x$, which is commonly derived from the seeds of cereal plants, and in bulbs, roots and tubers. Starch consists of 20-30% amylose and 70-80% amylopectin. It can be utilized as a significant raw material in the food, pharmaceutical, cosmetic, chemical and oil industries, and is commonly extracted from the source material in aqueous medium. This material is usually packaged and supplied in granular or powdery form making the drying process a fundamental unit operation in starch production. The extraction of starch commonly involves dry milling, wet milling or the combination of both. Some of the commonly used raw materials as the natural source of starch are yams. Yams are from the genus Dioscorea of the family Dioscoreaceae; which has around 600 species, one of which is the Dioscorea. alata L. [1]. D. alata L., generally known as purple yam, is a vital tuber crop and purple pigmented elite cultivar has recently become popular due to its associated health benefits. It provides protein three times more superior than cassava and sweet potato [2]. Purple yam begun in the Asian tropics and is a boss commodity in some tropical countries (Fang et al., 211). In a previous study, it was reported purple yam's moisture content of 82.91±.41% (wet basis). To gain control of the high perishability because of high moisture content, purple yams are typically subjected to drying. It is necessary to extend the duration of storage of purple yam for its supply throughout the entire time of its off-season without losing healthful abilities [6].

There are literatures about drying of different food products but until now, characterization of drying and

* Corresponding author: fcdevera@mapua.edu.ph

 DOI: 10.1051/epjconf/201716201085

drying kinetics of purple yam specifically using tray dryer, is not yet completely established for there is still limited information on the said topic. A study claimed that dehydration of fruits, vegetables, and crops deal challenging problems involving complex drying conditions that involve many interconnected and opposing phenomenon associated with material's complex nature of drying. Another problem specifically upon drying of purple yam is the determination of the desired temperature that should be utilized during the process. If the temperature is too high, the food matrix ruptures which results to reduction of the sizes of pores. Large pores sizes are desirable for good rehydration of the dried product [4].

The major objective of this study is to determine the drying kinetics of purple yam. The specific objectives include 1) to study the effects of different process conditions on the drying behavior of purple yam slices; 2) to determine the most efficient process condition for purple yam through scanning electron microscopy; 3) to generate drying curves for purple yam; and.

This study will help establish drying protocol for purple yams. The protocol will include the appropriate drying air temperature with the corresponding air velocity and thickness as well as the proper pretreatments for the purple yam upon using tray dryers. Additionally, this study will lead to the equation that fits the drying phenomena of purple yam. The experimental data that will be obtained from this study may be of good considerations upon the design of future dryers.

The drying process will be conducted at temperatures 50°C and 60 °C with air velocities 1 and 1.3 m/s and sample thickness of 1 and 2 mm. The tray dryer that will be used is the model UOP8MkII computer controlled tray dryer of Armfield. Characterization of the samples will be done through scanning electron microscopy.

2 Material and methods

2.1 Preparation of purple yam

The purple yam samples (*Discorea alata L.*) were purchased from Balintawak Market and were stored at a room temperature not exceeding 30°C prior to usage. A sample was brought to the Bureau of Plant Industry for species authentication. Yams were washed and manually sliced using mandolin slicer. Uniform disk shaped was maintained for all samples through a round shape cookie cutter having a 55 mm. In this way disk-shaped samples having uniform area and thickness were prepared. The slices were steam blanched for 7 min at 98±2°C as adapted from Hsu [7]. The purpose of blanching is to deactivate enzymes responsible for deterioration of samples. Subsequently, the samples were then soaked in sodium sulfite solution at 45°C for 10 minutes to avoid discoloration and browning. The soaked yam slices were drained and laid on paper towels to remove any adhering dirt and excess water and weighed on an analytical balance.

2.2 Initial moisture content (IMC) determination

The Initial moisture content was determined in triplicate by drying known weight of samples prepared at a temperature of 105°C until constant weight is achieved. Samples were then cooled in a desiccator for about 5 to 10 minutes. The difference between the initial weight and final weight of the samples is considered as the *IMC*.

$$IMC_{wb} = \frac{W_i - W_f}{W_i} x100 \quad (1)$$

2.3 Tray dryer

The tray dryer that was used is the model UOP8MkII computer controlled tray dryer of Armfield. This small-scale bench top tray drier dries solids by passing a stream of hot air over trays of wet material. The tray dryers specifications are as follows: i) a compact desktop tray dryer for laboratory use, ii) computer control of temperature and air flow rate, with sophisticated data logging and analysis software, iii) capacity up to 2.1 kg of wet material, iv) flow rates, 0.4-3.0 m/s over trays, v) temperature up to 80 °C at 0.4 m/s (less at higher flows), vi) integrated electronic weight measurement to determine drying rate, vii) electronic measurement of temperature and humidity before and after the drying trays, viii) electronic measurement of air flow, and ix) Stainless steel construction. The dimensions of the tray dryer are 0.325x1.160x 0.345 m in height, length and depth.

2.4 Drying

The sliced samples were spread evenly across the customized stainless wire mesh trays. Trays were placed inside the tray dryer at different drying conditions. Sixteen pieces of samples per tray and three rays per trial were utilized during the drying process. Table 1 shows the design matrix at different varying parameters such as temperature, air velocity, and thickness. Table 1 displays the drying conditions to be used in relation with the symbols used in the 2^3 factorial designs.

Table 1. Drying Conditions to be Designated in the 24 Factorial Designs.

Drying Conditions	+	-
Temperature (°C)	60	50
Air Velocity (m/s)	1.3	1.0
Thickness (mm)	3.02 ± 0.024	2.00 ± 0.042

2.5 Experimental design

Three drying parameters, namely drying air temperature, drying air velocity and slice thickness, were considered during the drying process. High and low values for the said parameters were utilized and so an experimental design of 2^3 factorial was followed. Thus, there are 8 treatment combinations or 8 runs and there were three trials for each combination. Table 2 presents the design matrix was used during the drying process.

Table 2. Design Matrix with factors Temperature, Velocity and Thickness.

Run	Temperature	Air Velocity	Thickness	Label
1	-	-	-	(1)
2	+	-	-	A
3	-	+	-	B
4	+	+	-	AB
5	-	-	+	C
6	+	-	+	AC
7	-	+	+	BC
8	+	+	+	ABC

2.6 Drying-time curve and drying-rate curve

The drying time curves were generated by the computer which controls the dryer. Drying time curves are plots of weight of the sample as the ordinate with elapsed time as the abscissa. The drying-rate curves were generated based on the slopes of the drying-time curve, which were placed in the vertical axis and the corresponding moisture content in the horizontal axis. The equations used for the drying curves determination were obtained [3]. The moisture content is being defined as:

$$M_t = \frac{W_i - W_f}{W_i} x100 \quad (2)$$

Where M_t is the moisture loss at t time, W_i is the initial weight of the sample and W_f is the weight of the dried sample.

2.7 Determination of drying time curves and drying rate curves

The drying time curves were determined for each run by plotting the Moisture Ratio (MR) vs. time (min) data. The drying rate curves were determined for each run by plotting the Drying Rate vs. Free Moisture content data.

2.8 Drying shrinkage and pore Size

The shrinkage of the dried samples were determined by comparing the thickness of the purple yam slices, before and after drying using a Vernier caliper [8]. The value percentage shrinkage must be minimal to consider the drying procedure to be successful and also represents a good rehydration quality of the dried product.

The pore sizes of dried samples were obtained through scanning electron microscope (JEOL JSM-5310, Japan). Samples were preserved in ziplock plastic bags prior to testing. The samples were coated with gold by electron beam evaporation in a fine coater (EOL KFC-1200, Japan) and were then placed in the SEM. Micrographs were be generated for each sample.

2.9 Statistical analysis of data

All results were reported as mean of 3 trials and standard deviation. One-way analysis of variance was conducted to determine if there are significant differences between means. Post hoc analysis was be performed to study further significantly different means and minitab software was used for all statistical analysis.

3 Results and discussions

3.1 Initial moisture content

The plot for the percent decrease in weight of fresh samples vs. time was presented in Figure 1. Based on the plot, the percentage decrease in weight of the fresh samples were 82.24% and 80.18% for thick fresh samples and thin fresh samples respectively. The average percentage decrease in weight for the fresh samples is 81.21%. It is assumed that the decrease in weight of the fresh samples is mainly composed of evaporated water from the fresh samples due to the exposure of the fresh samples in the high drying temperature catered by the oven drier. Thus, the average initial moisture content of the fresh samples is 81.21%. Based on a previous study, the initial moisture content for *D. alata* is 82.91±0.41% [6].

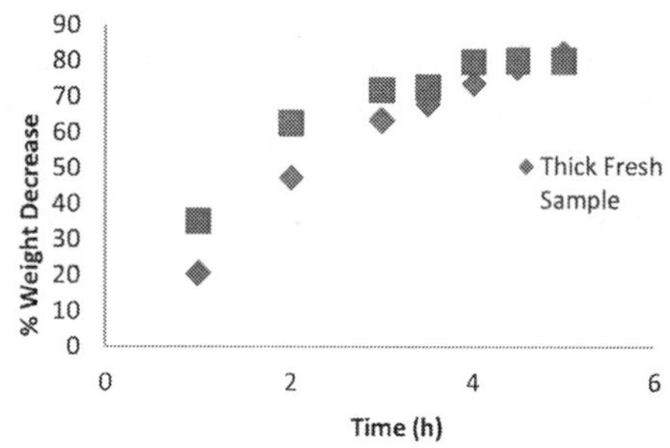

Fig. 1. Percentage Weight Decrease of Fresh Samples Vs. Time (h)

3.2 Scanning electron microscopy (SEM)

In this study, scanning electron microscopy is used to conduct a qualitative analysis to determine the most efficient condition among the generated conditions from the 2^3 factorial design. Figures 2 to 3 show the micrographs generated for thin dried purple yam samples.

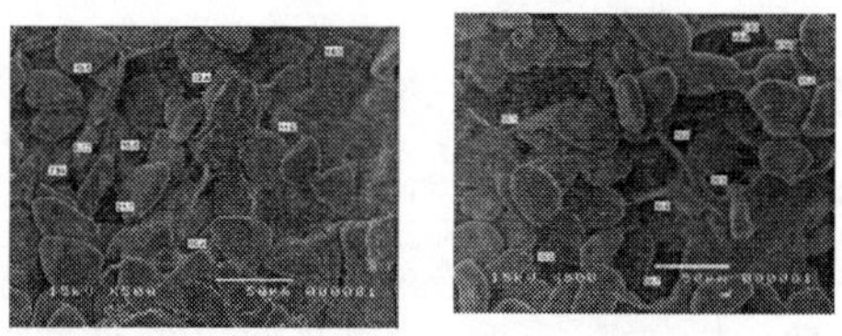

Fig. 2. Micrographs of Dried Purple Yam for Run 1 and Run 2

 DOI: 10.1051/epjconf/201716201085

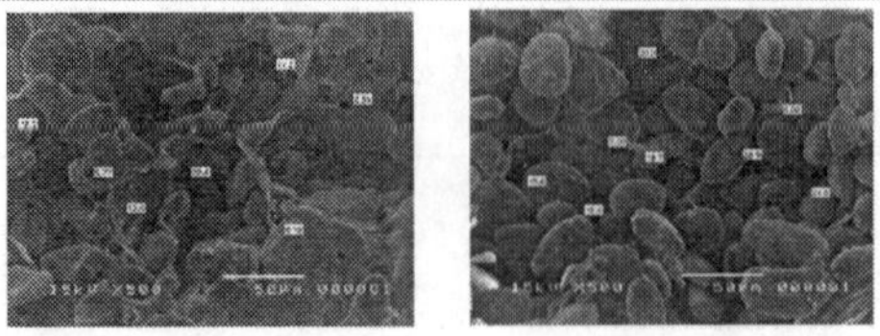

Fig. 3. Micrographs of Dried Purple Yam for Run 3 and Run 4

Based on the micrographs presented for thin dried purple yam samples, desirable qualities of dried purple yam are obtained from Run 2. High temperature and low air velocity were able to present large pores sizes which are being desired for good rehydration of the dried product. It can also be observed through that condition that the sample was able to maintain its food matrix which means that minimal cell breakages occurred or rupturing of the food structure have been avoided. Figures 4 to 5 present the micrographs generated for thick dried purple yam samples.

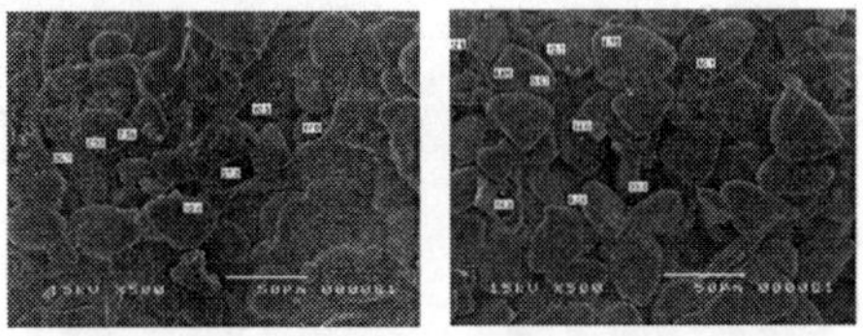

Fig. 4. Micrographs of Dried Purple Yam for Run 5 and Run 6

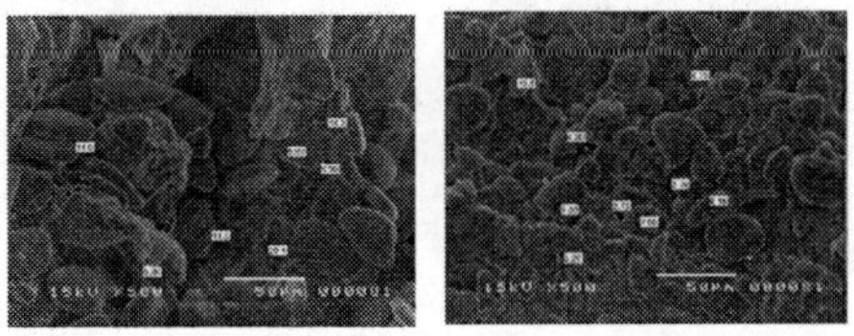

Fig. 5. Micrographs of Dried Purple Yam for Run 7 and Run 8

Same observations as to that of thin dried samples were observed for thick dried samples. High temperature and low air velocity under Run 6 give the desirable qualities of thick dried purple yam samples. Large pores sizes as well as intact cell structure were maintained. According to [9], low air velocity contributed to a lower pressure in the food product tissues resulting to a nearly intact cellular structure despite of the high temperature applied to the samples. Both for thin and thick slices, the desired qualities of dried product were observed for high temperature and low air velocity however as to comparing Run 2 with Run 6, smoother surfaces were observed under Run 2 and so Run 2 has the better drying condition.

3.3 Drying characteristics

In analyzing the pore size distribution in each sample, one-way analysis of variance using Minitab 17 was performed to determine if there is a significant difference between the means of the runs which are of different drying conditions. The null hypothesis which states that all means are equal was rejected after having a P-value of 0.044 which is smaller than the α-value equal to 0.05. This means that varying drying conditions would give significant effect on the pore size distribution of the dried samples. As presented in Figure 6, runs with high temperature give higher pore sizes than runs with lower temperature. Increasing the air velocity from the chosen low to high value has almost negligible effect. Runs with low thickness give higher pore sizes than those with high thickness

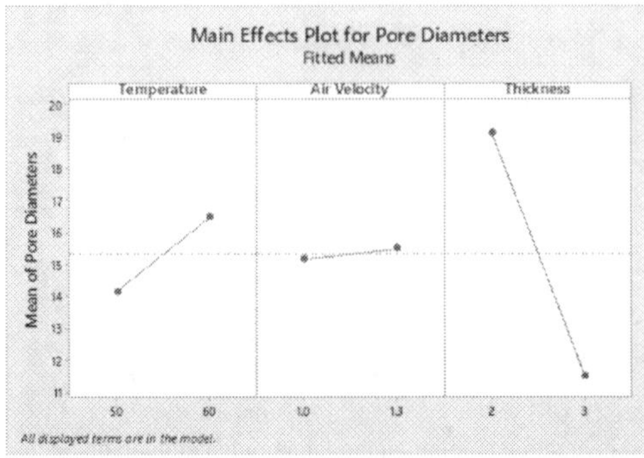

Fig. 6. Effect of Temperature, Air velocity and Thickness on Pore Diameters

As can be seen from Figure 7, thickness and the interaction of temperature and thickness are of equal magnitude therefore are of same effect. Both having the highest magnitude which and so both greatly affects the pore size distribution of the dried samples. The chart shows that thickness affects the effective moisture diffusion coefficient most as compared to temperature and air velocity as well as the combinations of the said factors. It also shows that air velocity has the least effect among the given factors. And so to have high diffusion coefficient, thickness is potentially the most important factor to consider while air velocity is the least one

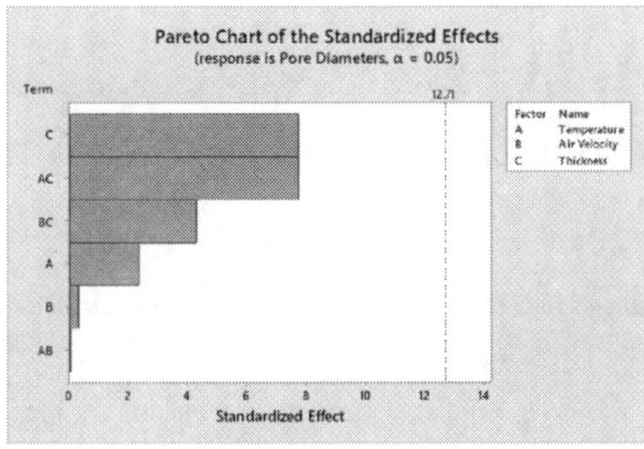

Fig. 7. Pareto Chart on Standardized Effects of Temperature, Air velocity and Thickness on Pore Size Distribution of Dried Samples

3.4 Drying curves

Figure 8 displays the change in the moisture content of the dried samples as the drying time progressed. It can be seen that each run decreased their moisture content during drying which corresponds to the decrease in weight of the samples. It can be observed that runs having higher drying temperature and air velocity have much lower drying time compared to runs of lower drying temperature and air velocity. Based on an analysis of factorial design using Minitab 17, the thickness of the sample has the highest effect on the drying time as compared to the other varying parameters. On the other hand, air velocity has the next major effect on drying time while the temperature has the least effect on the drying time.

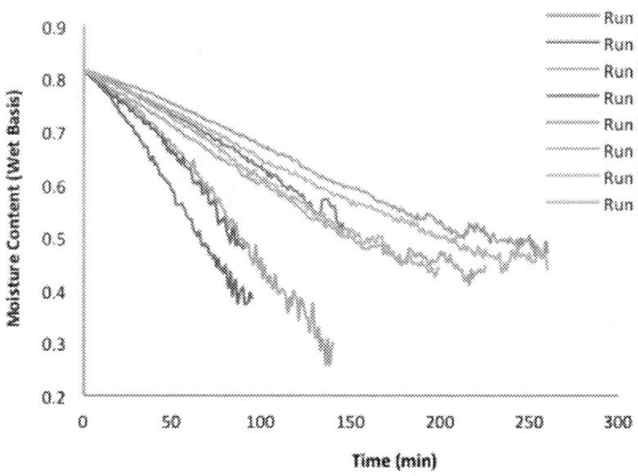

Fig. 8. Moisture Content of Samples in Wet Basis vs. Drying Time (min)

4 Conclusion

In this study, the drying characteristics of purple yam *(Dioscorea Alata L.)* were assessed by subjecting the purple yam samples into different drying conditions at which eight runs were generated using 2^3 factorial design. The three varying drying parameters considered were temperature,, air velocity and thickness. After drying, the tabulated values generated were treated to obtain the drying curves for each run. Based on the generated drying curves, it was determined that the drying rate decreases as the moisture content of the sample decreases. Considering the individual effects of each drying parameters on the drying process itself, it was found that the thickness of the samples directly affect the drying time. Thicker samples took much longer to achieve their corresponding equilibrium moisture content. The drying air temperature is responsible in determining the equilibrium moisture content that can be achieved on each sample regardless of the thickness. It was found that samples dried on higher temperature achieved much lower equilibrium moisture content as compared with the thinner samples. The last drying parameter, air velocity directly affects the drying time to achieve the equilibrium moisture. Samples dried at higher air velocity approached the equilibrium moisture content much faster than samples dried at lower air velocity. The three varying drying parameters also affected the shrinkage of the dried samples and the effective moisture diffusivity. It was found that the thicker samples have the highest shrinkage of around 21.69% as compared to the 14.42% shrinkage of thinner samples. It was also found that the effective moisture diffusion coefficient is higher at higher drying temperature. it was found that run 2, high temperature-low air velocity, yielded the most desirable quality of the dried samples. Samples from run 2 have large pore diameters but with minimal cell breakages, thus allowing optimal rehydration

References

1. Akanbi, C. T., Gureje, P. O., & Adeyemi, I. A., Journal of Food Engineering, 28, 45–48, (1996).
2. Bourret-Cortadellas D., Ethnobotanic Study of the Dioscorea, Yams New Caledonia. Doctorate 3rd cycle. Faculty of science, Paris, France. p. 135, (1973).
3. Geankoplis, C. J., Transport processes & separation process principles (includes unit operations). 4th edition. Saffron House, London, (2014).
4. Kiranoudis, C. T., Maroulis, Z.S., & Morinos-Kouris, D., International Journal of Heat and Mass Transfer, 38, (1995).
5. Marshall, William R., Jr., Drying. InAccessScience. McGraw-Hill Education, (2014).
6. Shajeela, P.S., Mohan, V. R., Jesudas Louis, L., Tresina Soris, P., Nutritional and Antinutritional Evaluation of Wild Yam (Dioscorea spp.). 723-730, (2011).
7. Hsu, C., Food Chemistry, 83(1), 85-92, (2003).
8. Mohsenin, N. N., Physical properties of plant and animal materials. Gordon and Breach, (1986).
9. Xiao, H. W., Yao, X. D., Lin, H., Yang, W. X., Meng, J. S., & Gao, Z. J., Journal of Food Process Engineering, 35(3), 370-390, (2012).
10. Ertekin and Yaldiz, Journal of food engineering, Volume **63**, 349-359, (2003).
11. Condorı, M., Echazu, R., & Saravia, L. Renewable Energy, 22(4), 447-460, (2001).

DOI: 10.1051/epjconf/201716201086

Determination of drying kinetics and sorption isotherm of black pepper (Piper Nigrum)

Flordeliza C. De Vera [1,*], *Vanessa Bernadette B.* Atienza [1], *Jomicah B.* Capili [1] and *Zaliman* Sauli [2]

[1] School of Chemical Engineering and Chemistry, Mapúa University , 658 Muralla St., Intramuros, Manila 1002, Philippines
[2]School of Microelectronic Engineering, University Malaysia Perlis (UniMAP), Kampus Alam, Pauh Putra, 02600 Arau, Perlis, Malaysia.

Abstract. In the present study of food products, determination of the drying characteristics of black pepper using an oven is not yet completely established. This study aimed to determine the drying kinetics and sorption isotherm of black pepper using a convective oven at 30°C, 40°C and 50°C. The data gathered in this study were used to fit in selected mathematical models for drying kinetics and sorption isotherm. Among these models, the Midilli model ($MR=0.5338exp(0.7273t^{-0.0551})+-0.0005t$ for 30°C, $MR=0.5814exp(0.6293t^{-0.0764})+ -0.0008t$ for 40°C and $MR=0.3187exp(1.1777t^{-0.0466})+ -0.0011t$ for 50°C) was the best fit to explain the moisture transfer in black pepper, while the GAB Model ($m/0.1302=((0.1906)(\ 0.7811)a_w)/(1-(0.7811)a_w)[1-(0.7811)a_w+(0.1906)(\ 0.7811)a_w])$) was for the equilibrium moisture content and water activity relationship. After evaluating the data, the drying characteristics of black pepper at 40°C yielded better results than 30°C and 50°C. XLSTAT and ANOVA Add-in of Microsoft Excel was the software used to compute for the necessary values in the assessment of the mathematical models for this study.

1 Introduction

Drying of food material occurs when water vapor is removed from its surface into the surrounding space, resulting in a relatively dried form of the material. The drying process as a whole has to involve the generation of vapor (transforming from the liquid phase inside the moist material to a gaseous phase) on and within the moist material being dried, and the transport of the water phases outwards of the material. Sufficient reduction of moisture content inhibits the growth of microorganisms [1]. The extent of this phenomenon can be quantified by its water activity

For centuries, people have been preserving foods to extend shelf life and protect the characteristics of food. Drying is an essential process in the post – harvest operation for spices [2]. There have been a lot of conducted researches on drying of spices such as paprika (*Capsicum annum*) [3], rosella flower (*Hibiscus sabdariffa*) and chili (*Capsicum annuum*) [4] because of its purpose on food application. Today, it is considered to be the king among the spices [5]. It is widely used to enhance the flavor of the foods.

There are literatures about conducted research on drying of different food products but nothing is published about drying of the fruit of black pepper in the Philippines, specifically drying using an oven drier. The Department of Agriculture (2012) has published an article but only about the varieties of black pepper, which are fitting to their respective source of origin such as Quezon, Batangas, Laguna, Zamboanga, Davao, or Basilan black pepper.

The process used on black peppercorn is still by open sun drying, which is conventional but inconvenient. This drying technique is slow, uncontrolled, exposed to environmental contamination and requires hard labor [6]. Another difficulty encountered is the drying time, since there is no proper evaluation of the moisture released with respect to time.

With so much difficulty in the traditional open sun drying, this study established the drying kinetics of black peppercorn variety from Indang, Cavite. The resulting data from the drying kinetics were used to determine the drying rate curve. The sorption isotherm was established and the models that will best fit the data were identified. Determining the sorption isotherm and drying kinetics of black peppercorn helps in the prediction of its shelf life. Other researchers can also use the resulting data and drying curves to design a better manufacturing process of black peppercorn through optimization of the present drying techniques and technologies. Thus, this can result to better quality black peppercorn and increase the efficiency of the drying equipment.

This study deals only with the drying characteristics of black peppercorn variety from Indang, Cavite. Drying will be conducted at temperatures 30, 40 and 50 degrees Celsius. Drying rate, moisture content and drying time will be determined. The models that will be used for the drying kinetics are thin layer drying models of Lewis, Page, Henderson and Pabis, Two Term Exponential, and

* Corresponding author: fcdevera@mapua.edu.ph

Midilli. The determination of sorption isotherm of black peppercorn will be done by gravimetric method. Results of the sorption isotherm will be analyzed using GAB and BET models. This study will not cover the effect of drying in the physical qualities such as color and texture of black peppercorn.

2 Material and methods

2.1 Sample and equipment preparation

The peppercorn, which grows in cluster, from Cavite State University was harvested and placed in clean plastic bags. The batch was washed with water to remove dirt then be laid on a paper towel to remove excess water. Using the analytical balance (d=±0.0001), triplicate in each proposed weight was prepared and served as trial 1, 2 and 3.

2.2 Determination of initial moisture content (IMC or MC_o)

The IMC or MC_o was determined by drying triplicate of 30g, 40g and 50g samples at a temperature of 100℃. Drying continued until the weight of the samples becomes constant.

2.3 Drying mechanism

In this study, Memmert UNB 400 oven was used. The oven has a mechanism of natural convection thus allows the air to circulate around the sample and distribute heat evenly. The set temperatures were 30℃, 40℃ and 50℃. In each temperature, a triplicate of 30g sample was placed in a single layer on aluminum pans. Weighing was done with intervals of two minutes for the first ten minutes, and five minutes until the desired moisture content of 7.505% is achieved.

2.4 Mathematical modeling of drying curves

In general, knowledge of the mass of the samples (m) as a function of drying time (t) allows characterizing the drying kinetics. Table 1 shows the drying curves model.

Table 1. Thin layer drying curve models.

Name	Model Equation
Lewis	$MR = \exp(-kt)$
Page	$MR = \exp(-kt^n)$
Henderson & Pabis	$MR = a\exp(-kt)$
Two term exponential	$MR = aexp(-k_o t) + bexp(-k_1 t)$
Midilli	$MR = aexp(-kt^n) + bt$

2.5 Moisture sorption isotherm

The sorption isotherms were obtained by gravimetric method. In this method, the sample is weighed regularly at fixed pressure and temperature. Peppercorns were dried to the desired moisture content using 40°C – considering that its drying rate is typical compared to 30°C and 50°C. The dried sample was placed in desiccators where the inside environment is controlled. Each desiccator contained a triplicate of the dried peppercorns placed in open crucibles. This method also involved the use of saturated salt solutions to maintain constant relative humidity [6].

A range of water activity was set from 0.113 to 0.806 which necessitated the use of the following salts: LiCl, $MgNO_3$, KI, NaCl and $(NH4)_2SO_4$ [7]. The salt solution volume was ¼ of the total volume of the total space for the desiccant. Salt solutions tend to loss or gain moisture when subjected to a constant temperature.

2.6 Determination of equilibrium moisture content (EMC or MC_e)

EMC is associated with constant weighing. The weight of each sample for every salt was monitored until it became consistent.

2.7 Mathematical modeling of sorption isotherm

Data obtained from the gravimetric method of getting the sorption isotherm were fitted on the models found in Table 2.

Table 1. Models for Sorption Isotherm

Name	Model Equation
BET	$\frac{m}{M_o} = \frac{Ca_w}{(1-a_w)[1+(C-1)a_w]}$
GAB	$\frac{m}{M_o} = \frac{CKa_w}{(1-Ka_w)[1-Ka_w+CKa_w]}$

2.7 Statistical method of analysis

For the selection of drying model, drying curves were fitted to five thin layer-drying models. The suitability of the models was evaluated in terms of statistical errors – coefficient of determination (R^2), mean sum of squares errors (MSE) and root mean square error (RMSE).

3 Results and discussion

3.1 Drying characteristics of black pepper drying-time curve

The drying-time curve of black pepper shows that drying time increases as temperature decreases. Heat greatly affects the drying time of black pepper but it also modifies the characteristic of black pepper, which must be considered. As shown in Figure 1, 30℃, 40℃ and 50℃ took 2168 minutes (36 hours), 1167 minutes (19 hours) and 801 minutes (13 hours), respectively, to dry

 DOI: 10.1051/epjconf/201716201086

the black pepper up to the desired moisture content of 7.505%, which is approximately 12g of the final product.

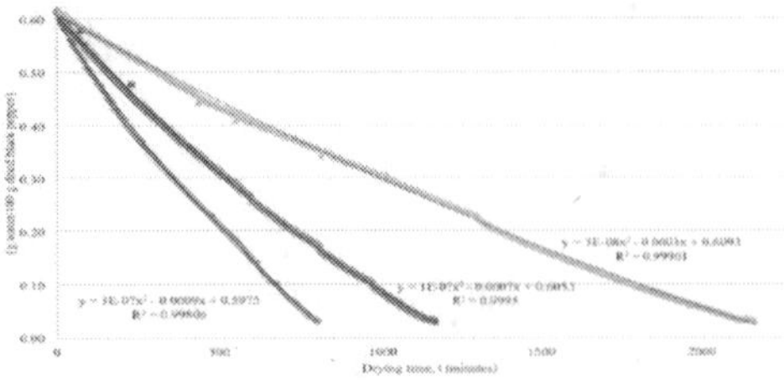

Fig. 1. Drying-time curve at 30°C (Green), 40°C (Red) and 50°C (Blue)

3.2 Drying-rate curve

Based on observation from Figure 2, there was no constant rate period thus the rate continued decreasing all throughout the drying process. Similar results were obtained for cocoa [8] and coffee [9], only falling rate period was observed both food products. For temperatures 30℃ and 40℃, it was noticeable that the rate was very fast at first then it slowly decreased as it proceeded. Some irregularities on drying rate were due to weighing of samples. Results were presented as the average of three trials per temperature.

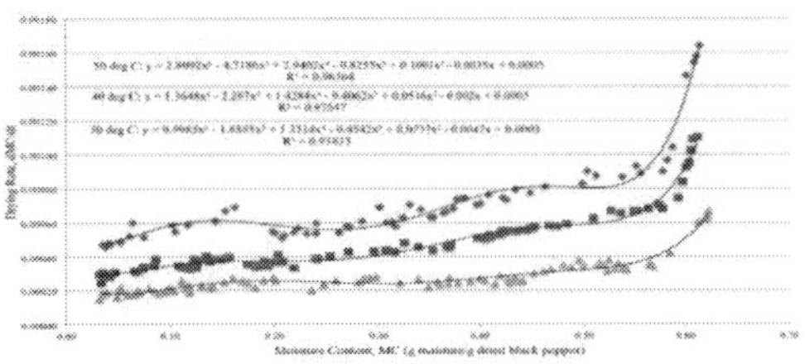

Fig. 2. Drying-rate curve at 30°C (Green), 40°C (Red) and 50°C (Blue)

3.3 Mathematical modeling for drying kinetics

Model parameters were obtained using XLStat Add –in of MS Excel by non – linear regression. These model parameters were listed in Table 3. Table 4 shows the statistical analysis for the modeling of the moisture ratio and drying time at different temperatures for black pepper.

Table 3: Comparison of the drying models and its corresponding modeling constants at different temperatures for black pepper

Model	30℃	40℃	50℃
Lewis	k=-0.0005	k=-0.0009	k=-0.0013
Page	k=3.83E-05 n=1.4737	k=0.0001 n=1.4375	k=0.0007 n=1.2697
Henderson and Pabis	a=5.52E07 k=8.9003	a=1.0735 k=0.0022	a=1.0072 k=0.0034
Two Term Exponential	a=0.0005 k=2.0436	a=0.0006 k=3.5200	a=0.0036 k=0.9317
Midilli	a=0.5338 k=-0.7273 n=-0.0551 b=-0.0005	a=0.5814 k=-0.6293 n=-0.0764 b=-0.0008	a=0.3187 k=-1.1777 n=-0.0466 b=-0.0011

Table 4: Statistical analysis for the modeling of the moisture ratio and drying time at different temperatures for black pepper

Model	Temperature ℃	R^2	RMSE	MSE
Lewis	30	0.9940	0.0245	0.0006
	40	0.9925	0.0275	0.0008
	50	0.9916	0.0287	0.0008
Page	30	0.9734	0.0543	0.0029
	40	0.9755	0.0523	0.0027
	50	0.9532	0.0722	0.0052
Henderson and Pabis	30	0.0083	0.5046	0.2546
	40	0.9505	0.0752	0.0057
	50	0.9407	0.0821	0.0067
Two Term Exponential	30	0.9558	0.0846	0.0072
	40	0.9584	0.0795	0.0063
	50	0.9427	0.0810	0.0066
Midilli	30	0.9976	0.0156	0.0002
	40	0.9973	0.0166	0.0003
	50	0.9984	0.0126	0.0002

Of all the models tested, the Midilli model gave the highest value of R^2 and the lowest value of RMSE and MSE for all temperatures used in the experiment, which indicates that Midilli model is suitable in describing the drying characteristics of black pepper. Same as for apple and red bell pepper [10] and for soybean [11], these foods were best described by Midilli model. From the result of ANOVA, there is no significant difference between the R^2 values. Thus, the models are temperature-independent.

3.4 Moisture sorption curve

The curve given on Figure 3 illustrates the relationship of the two parameters, which aids in the understanding of the water sorption properties of food. Based on Figure 3, it is assumed that black pepper can reach the maximum equilibrium moisture content of 0.0769, which was the average of the results of MCe of the four salts used. Conclusions on studies canola [12] and *Capsicum annuum* [13] state that increasing water activity leads in upward trend in value of the equilibrium moisture content. In this study, there's a sudden decrease in equilibrium moisture content at water activity value of 0.679 from 0.0987 at water activity of 0.514. This was due to exposure to moisture during the weighing of samples. Data were presented as the average for the three trials.

DOI: 10.1051/epjconf/201716201086

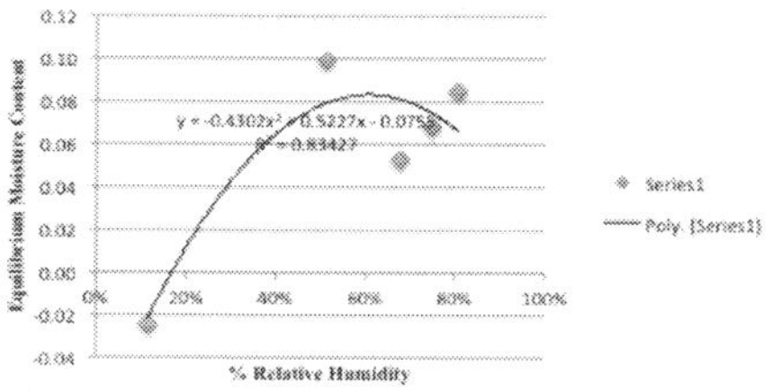

Fig. 3. EMC vs RH

3.5 Mathematical modeling for sorption isotherm

In this study, two models were used in predicting the sorption isotherms of black pepper, Bet and GAB model were used since these two are the most commonly used in sorption isotherms of foods [7]. Based on the statistical analysis in Table 5, GAB model has an R^2 value of 0.9725 which is higher than BET. This means that GAB model is considered to be the most suitable model for describing the relation between equilibrium moisture content and water activity of black pepper. GAB was also the best model to describe the sorption isotherm of *Capsicuum annuum* [13]. Table 6 shows the statistical analysis for the modeling of the moisture ratio and drying time at 40 degrees Celsius for black pepper.

Table 5. Comparison of the sorption models and its corresponding modeling parameters at 40 degrees Celsius

Model	Model Parameters
GAB	Mo=0.1302,C=0.1906, K=0.7811
BET	Mo=0.0199, C=7.1097

Table 6: Statistical analysis for the modeling of the moisture ratio and drying time at 40 degrees Celsius for black pepper

Model	R^2	RMSE	MSE
GAB	0.9725	0.0273	0.0007
BET	0.4067	0.0519	0.0027

4 Conclusion

Based on the drying kinetics of black pepper (*piper nigrum*) by oven drying, it was observed that the data on moisture content decreases with time and drying rate increases with increasing temperature. Also, drying time decreases as temperature increases since the heat greatly affects the evaporation of moisture to the surrounding. Moisture ratio – time relationship at temperatures 30°C, 40°C and 50°C was described using five thin – layer model but Midilli model gave the highest R^2 value of 0.9976, 0.9973 and 0.9984 for 30°C, 40°C and 50°C, respectively. Midilli model also generated the lowest value of RMSE and MSE. The result indicated that Midilli model could be used to explain the moisture transfer in black pepper. Sorption curves give valuable information about the equilibrium moisture content of black pepper at different relative humidity. In this study, GAB and BET models were applied to fit the data. Results confirmed that GAB model best describes the relationship of EMC – Aw of black pepper with an R^2 value of 0.9725. These data were important in storage and preservation of black pepper.

References

1. Heldman, D.R., Food Preservation Process Design, Academic Press, (2011).
2. Janjai, S. and P. Tung, Renewable Energy, Volume **30** (14), 2085-2095, (2005).
3. Ramesh, M. N., et al, Journal of Food Engineering, Volume **49** (1), 63-72, (2001).
4. Janjai, S., et al, Energy, Volume **33** (1), 91-103, (2008).
5. Jose, K. P., et al, Pertanika J. Trop. Agri. Sci., Volume **25** (1), 39-45, (2002).
6. Raji, A. O., & Ojediran, J. O., Food and bioproducts processing, **89**(3), 178-184, (2011).
7. Chen, H. H., Huang, T. C., Tsai, C. H., & Mujumdar, A. S., Drying Technology, **26**(4), 503-507, (2008).
8. Chinenye N.M., et al., Chilean Journal of Agricultrural Research, Volume **70**(4), 633-639, (2010).
9. Correa, P.C., et al, Drying Characteristics and Kinetics of Coffee Berry. Revista Brasileira de Produtos Agroindustriais, Campina Grande, v.8, n.1, p.1-10, (2006).
10. Schössler, K., Jäger, H., & Knorr, D., Journal of Food Engineering, **108** (1), 103-110, (2012).
11. Rafiee, S., Keyhani, A., Sharifi, M., Jafari, A., Mobli, H., & Tabatabaeefar, A., J. Agr. Sci. Tech, **11**, 289-300, (2009)..
12. Zare D. et al., Iran Agricultural Research, Vol. **30**, No.1, 2, (2011) & Vol. **31**, No.1, (2012).
13. Akin, A., et al., GIDA **34** (4): 205-211, (2009).

AUTHOR INDEX

AUTHOR INDEX

AUTHOR INDEX